Modern and Ancient Fluvial Systems

EDITED BY J. D. COLLINSON
AND J. LEWIN

SPECIAL PUBLICATION NUMBER 6 OF THE
INTERNATIONAL ASSOCIATION OF SEDIMENTOLOGISTS
PUBLISHED BY BLACKWELL SCIENTIFIC PUBLICATIONS
OXFORD LONDON EDINBURGH
BOSTON MELBOURNE

First published 1983

British Library
Cataloguing in Publication Data
    Modern and ancient fluvial systems
      – (Special publication/International Association of
      Sedimentologists, ISSN 0141–3600; 6)
      1. Watersheds
      2. Sediments (Geology)
      3. Rivers
      I. Collinson, J. D.
      II. Lewin, J.
      ISBN 0–632–00997–7

Printed and bound in Great Britain at the
University Press, Cambridge

Contents

Facies models

Economic aspects

Spec. Publs int. Ass. Sediment. (1983) **6**, 1–2

Modern and ancient fluvial systems: an introduction

J. D. COLLINSON*† *and* J. LEWIN‡

† *Department of Geology, University of Keele, Keele, Staffordshire ST5 5BG and*
‡ *Department of Geography, University College of Wales, Aberystwyth, Dyfed SY23 3DB*

Study of the transport and deposition of sediment by rivers has a long history, originally deriving from basic engineering and agricultural necessity. The success or failure of navigation and irrigation projects, some of great antiquity, depended upon an empirical understanding of sediment behaviour. Rivers are also manifestly important agents in the development of the landscape and there is an old and important tradition, dating back well into the last century, of studying their behaviour and geomorphological products. Similarly, the appreciation of the more detailed processes of sediment movement and bedform behaviour dates back to Sorby (1859) and Gilbert (1914).

In so far as sedimentology is a relatively young branch of geology, the more detailed understanding of fluvial processes and products within the geological record is a fairly recent development. Certainly one may argue for fluvial sedimentology having its origins in nineteenth-century discussions of 'diluvium' but it is perhaps better to regard it as having its roots in the late 1950s and early 1960s when a more rigorous application of our knowledge of processes and present-day environments was established.

It has been clear for a long time that the three strands contributing to fluvial sedimentology—engineering, geomorphology and geological sedimentology—share much common ground and should be able to offer much to one another. For the most part, however, any cross-fertilization appears to have taken place through the literature or through private contacts. Few forums for meeting and discussion between members of the three branches existed prior

to the first International Conference on Fluvial Sedimentology, held in Calgary in 1977. The success and value of this meeting was apparent to all those lucky enough to be present and the subsequent publication (Miall, 1978) reflects much of the achievement of the meeting.

The second conference, held at Keele, England, in September 1981, attempted to continue this progress and to encourage further the participation of engineers and applied sedimentologists. Whilst this latter aim was perhaps not so well achieved as the organizers had hoped, many old contacts were re-established and new ones formed. The meeting itself attracted some 220 participants and about 130 papers were presented either as talks or as posters. Abstracts for all the papers were published in a separate volume.

The present book comprises 44 papers which present a selection of those submitted as full manuscripts. It is not, therefore, in any sense a full record of the meeting. The papers included do however reflect reasonably well the overall balance of the meeting. They are grouped into broad categories behind review papers written by the invited chairmen of some of the conference sessions.

It is not our intention here to comment on all the papers included in the volume, but it may be useful to try to identify any shifts of interest or emphasis which have taken place in the last few years, particularly since the Calgary meeting.

In the field of hydrodynamics and bedforms, the continued application of a knowledge of turbulence seems to be allowing a more fundamental understanding of bedform development. The response of bedforms to unsteady flow continues to receive both theoretical and, most appropriately, controlled experimental study.

* Present address: Britoil, 150 St Vincent Street, Glasgow G2 5LJ.

0141-3600/83/0106-0001 $02.00

In geomorphology, a growing number of field studies of contemporary fluvial systems has been concerned with the analysis of transported and deposited sediments, and especially with identifying the sources of such sediments within catchments, and the routeing of identified or 'tagged' material through channel reaches. Sustained field monitoring of meandering, 'wandering' and other channel pattern changes has been very usefully extended by the analysis of air photos and historical maps. Amongst other things, these demonstrate or suggest interactions between large mobile bedforms or sedimentation zones and planform changes, and such features as channel recession *away* from arcuate cutbanks leading to sedimentation in what may be termed counterpoint locations. Late Quaternary sediments are also being examined with renewed vigour given the possibility of more sophisticated palaeohydraulic and facies interpretations. However, there are parallel changes now taking place in our understanding of the variable effects of contemporary high-magnitude events, of the complexity of within-reach channel flow/sediment transport relationships, and of the notable importance of inherited controls on present channel and sediment characteristics. Thus the apparent precision of some palaeohydraulic reconstructions may eventually prove too facile. The careful qualifications and assumptions made by contributors to this volume in this field must be borne in mind.

In geological sedimentology, several major themes which date back to the time of the Calgary meeting continue to develop. The idea of 'alluvial architecture', meaning the large-scale spatial organization of channel sand bodies within their associated non-channel sediments, seems to be moving towards a consideration of different types of alluvial system. This is a development of greater realism than the pioneering studies which dealt rather mechanistically with meanderbelt sandstones. Another significant development seems to be the recognition of the importance of non-migrating, commonly anastomosed, channels.

Point bars, which have served for so long as the basis for facies models of many fluvial sandstones, came under close scrutiny at Calgary. In particular, their great variability became apparent. This trend

continues with descriptions of both coarse and fine-grained types being presented here. However, the products of supposed low-sinuosity streams seem to be attracting most attention. Perhaps this is a natural reaction to the long predominance of point bars in sedimentological thinking. In almost all the papers which document channel-fill organization we appear to be dealing with individual case histories. The temptation to present more general synthetic models for the deposits of particular river types seems to be being resisted.

The accumulation of well-documented examples is a necessary prelude to any future attempts at formulating more general facies models. Perhaps, however, the very complexity and variability of fluvial sedimentation will make general facies models so unrealistic as to be often worthless. It seems likely that, for the near future at least, a wide experience of many case histories will be a better basis for interpretation of new examples than will knowledge of a restricted number of highly distilled (*sensu* Walker, 1979) models. The papers on facies organization presented here can only help in this widening of experience.

The fact that this volume has been published in little over a year of the date of the conference is a testament to the remarkable cooperation which we as editors have received from the authors. They have kept to our deadlines and have helped greatly by carrying out revisions very thoroughly. We are also greatly indebted to the many referees who helped us and to Jane Lewin and Tina Mousley who have acted as editorial assistants. The conference was sponsored by a variety of organizations, and these are listed at the front of the volume.

REFERENCES

GILBERT, G.K. (1914) The transportation of debris by running water. *Prof. Pap. U.S. geol. Surv.* **86**, 263 pp.
MIALL, A.D. (Ed.) (1978) Fluvial sedimentology. *Mem. Can. Soc. Petrol. Geol., Calgary*, **5**, 859 pp.
SORBY, H.C. (1859) On the structures produced by the currents present during the deposition of stratified rocks. *Geologist*, **2**, 137–147.
WALKER, R.G. (1979) Facies and facies models: a general introduction. In: *Facies Models* (Ed. by R.G. Walker). *Reprint Series*, **1**, 1–7. Geoscience Canada.

Hydrodynamics and bedforms

Spec. Publs int. Ass. Sediment. (1983) **6**, 5–18

On the interactions between turbulent flow, sediment transport and bedform mechanics in channelized flows

M. R. LEEDER

Department of Earth Sciences, University of Leeds, Leeds LS2 9JT, Yorkshire, U.K.

ABSTRACT

The interrelationships between turbulent flow, sediment transport and bedform theory are critically reviewed. Progress in these fundamental fields of physical sedimentology is hampered by lack of evidence and theory on: (1) the precise nature of the link between inner and outer zone flow structures at high Reynolds numbers, (2) the effect of grain roughness and bedforms upon inner and outer zone structures, (3) the effects of transported grains upon turbulence generation and dissipation, (4) the possible anisotropy of turbulence and its role in sediment suspension. A number of proposals are made concerning these problems and some suggestions for future research are presented.

INTRODUCTION—THE 'TRINITY' OF FLOW, TRANSPORT AND BEDFORM

The erosional and depositional processes of channelized flows depend upon the interaction between turbulent flow, sediment transport and bedforms. As shown in Fig. 1, a continuous feedback system is at work. In such a system it is clearly dangerous to try to investigate one portion in isolation, yet such 'blinkered' approaches predominate in the literature. A good example arises in the theory of bedform mechanics, involving the application of linear stability theory to ripple and dune development (e.g. Richards, 1980). Although very sophisticated in mathematical development, such approaches ignore the basic experimental fact that flow separation dominates the development and equilibrium of such bedforms and clearly cannot be dismissed as a second-order effect.

The complexity of the relationships sketched in Fig. 1 is enormous and there is no doubt that much work remains to be done. This paper seeks to review critically the simplified theory of the flow–transport–bedform 'trinity' and to propose approaches towards a unified bedform theory (largely descriptive) based

0141-3600/83/0106-0005 $02.00

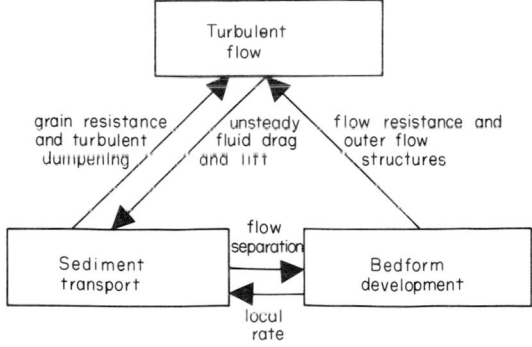

Fig. 1. The 'trinity' of flow, sediment transport and bedform as a feedback loop.

upon a simple interpretation of interrelationships within the trinity.

STRUCTURE OF TURBULENT FLOW

Dual structure of turbulent flow

Observations concerning the structure of turbulent boundary layers and the relevance of this structure to alluvial sediment transport and bedform development were drawn together by Jackson (1976) in an influential paper. The core of his approach was the so-called burst–sweep cycle (Fig. 2) with:

(1) an inner layer ($y^+ < 40$; where $y^+ = yu_*/v$, with y = height above bed, u_* = shear velocity, v = kinematic viscosity) of spanwise streaks of low- and high-speed fluid (in the viscous sublayer) subject to periodic disruption and lift-up, passing upwards into

(2) an outer layer ($y^+ > 40$) with large burst 'break-up' vortices and accelerated, downward-moving sweep fluid (Kline *et al.*, 1967; Nychas, Hershey & Brodkey, 1973; Offen & Kline, 1975). Events of the burst cycle are responsible for most of the turbulence and thus for a high proportion of the Reynolds stresses in the turbulent boundary layer (Grass, 1971; Kim, Kline & Reynolds, 1971; Nychas *et al.*, 1973).

In the inner zone the spanwise, low-speed streak spacing (using two 'wall' variables) is

$$\lambda u_*/v \equiv \lambda^+ \simeq 100, \qquad (1)$$

where λ is the mean spanwise spacing of low-speed wall streaks, u_* is the shear velocity and v the kinematic viscosity.

Data over a wide range of Reynolds numbers assembled by Rao, Narasimha & Narayanan (1971), and many later authors, show that, even in the inner layer, the mean burst periodicity scales with outer layer flow variables. Thus, for any point in a flow,

$$u_\infty T/h \equiv T^+ \simeq 5, \qquad (2)$$

where T is the mean periodicity of bursts at a point in the flow (Eulerian observer) u_∞ is the free stream flow velocity and h is the flow depth.

As pointed out by Rao *et al.*, equation (2) means that the energy balance of turbulent boundary layers can only be understood if a coupling mechanism exists between the inner and outer layers. Indeed, it must also imply that coherent structures exist in the outer flow, not the complex pattern of uplifted streaks caused by the bursting mechanism revealed by the instantaneous photographs taken of hydrogen bubble strain markers by Kline *et al.* (1967) and Grass (1971).

Coherent outer layer structures

Early flow visualization studies using smoke photographed in the Eulerian frame showed large coherent structures (Fielder & Head, 1966) composed of many smaller-scale eddies. These were later confirmed by velocity studies and by further flow visualization at Reynolds numbers of up to 3×10^6 (Dimotakis & Brown, 1976).

Studies of single-particle tracers in low Re flows (< 1000) by Nychas *et al.* (1973) and Praturi & Brodkey (1978) revealed that burst events were not directly dependent upon wall conditions, but upon the passages overhead of transverse vortices of Helmholtz type at heights of $100 < y^+ > 300$, thus directly confirming Rao *et al.*'s hypothesis of a link between outer and inner processes. At these low Reynolds numbers, bulges periodically form at the upper junction of the turbulent boundary layer, causing high-speed fluid to descend towards the wall. According to this model: (1) events in the outer region initiate a wall response, (2) wall streamwise vortices occur, (3) bursts extend no further than $y^+ = 100$ on average (rarely, up to 300) and move at angles of 45°–90° to the wall, (4) a regular outer layer structure is produced by the deep indentations of the boundary layer by high-speed irrotational fluid. Considering the average picture, Praturi & Brodkey note that it fits a horseshoe vortex structure although their own method of visualization could not confirm this (see below).

Large-scale outer motions were perhaps best visualized by Falco (1977) by the smoke process (see also Head & Bandyopadhyay, 1981; Bradshaw, 1981, fig. 4) and by simultaneous hot-wire velocity measurements. Coherent motions of large-scale type showed superimposed 'typical' eddies whose dimensions were comparable to sublayer streaks and thus scaled on inner variables. The large-scale motions in the outer layer are not burst-like features themselves, but according to Falco (1977) and Brown & Thomas (1977) they produce a slowly varying component in the wall shear stress. The high-frequency, large-amplitude fluctuation that occurs near the maximum in the slowly varying wall shear is associated with bursting. According to Falco, only about one-half of the large-scale motions show this feature. Since, on average, the wavelength of the motions is about $2 \cdot 5 h$, this leads to an *effective* large-scale motion wavelength of about $5 h$. The reader will note that this is the streamwise spacing predicted by equation (2).

The large-scale eddies visualized by Falco (1977) and inferred by Brown & Thomas (1977) clearly show a 20° interface (dipping upstream) between high- and low-speed fluid. Photographs by Head & Bandyopadhyay (1981) confirm this, but these authors state that at higher Reynolds numbers (> 4000) such large, organized structures are more rarely seen. According to Falco the parasitic 'typical' eddies are not produced from the sublayer. But these scale, like streak spacings, on inner variables and, importantly, produce large Reynolds stress peaks. The very recent

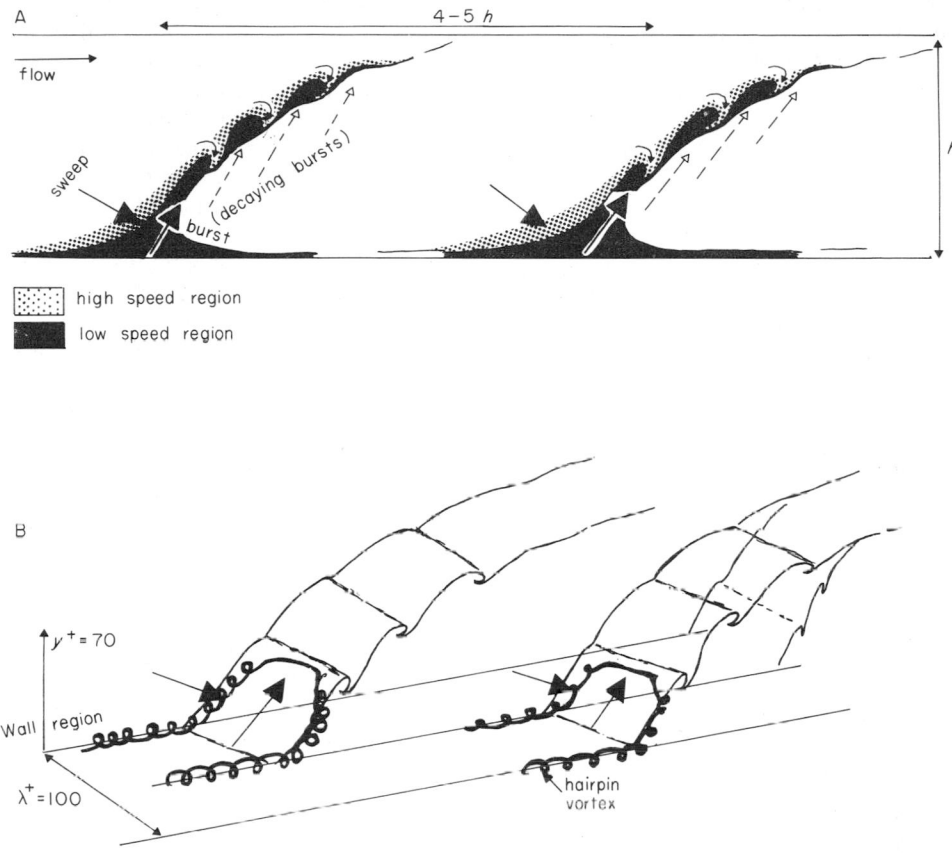

Fig. 2. Sketches to show (A) the two-dimensional, and (B) the three-dimensional structure of turbulent boundary layers. After Nakagawa & Nezu (1981), with the visualization studies of Head & Bandyopadhyay (1981) also taken into account.

studies of Head & Bandyopadhyay (1981) make it quite clear that these 'typical' eddies (also the transverse vortices of Praturi & Brodkey, 1978) are the tips of various types of burst-produced hairpin vortices lifted up at angles of about 45° into the outer layer (Fig. 2).

Relevance of the link between inner and outer zones

We have seen above that recent flow visualization studies reveal links between small-scale wall structures (microturbulence) and large-scale outer structures (macroturbulence) in turbulent flows. Such links are important in sediment transport studies, particularly regarding the role of bursts in suspension and bedform development (Jackson, 1976; Nakagawa, Nezu & Tominaga, 1981). Mathes (1947) and Znamenskaya (1964) observed periodic ejections of macroturbulence from dune troughs. Grass (1971) observed disturban-

ces on the free water surface during his experiments over rough boundaries and inferred that these resulted from burst events traversing the entire flow. Jackson (1976) assembled data from natural flows over dune bedforms, which suggested enhanced burst development from dune leesides where there exists an adverse pressure gradient to encourage bursting.

Nakagawa *et al.* (1981) suggest that bursting motions occurring in the wall region will be advected outward from the wall and downstream to decay in the outer region (see Figs 2 and 3). The structure of macroturbulent large-scale structures is such that at a given point in time new bursts will be occurring, on average, every 5 water depths in the flow direction (Fig. 2) and that the Eulerian burst periodicity (at a point) will be given by $5h/u_\infty$. The feedback effect between flow and bedform is clearly of great importance. Nakagawa *et al.* (1981) point out that macroturbulence (of surface boil variety) cannot be observed over hydraulically smooth beds, but is common over rough and wavy (dune) beds. Such observations indicate that boils are caused by particular bed features (Jackson, 1976) that are neither developed nor artificially induced in any reported fluid dynamics experiments save those of Znamenskaya (1964).

Effects of stationary bed roughness on turbulence

The effects of bed roughness elements on the pattern of macroturbulence are unknown (but see Townsend, 1976, p. 142 for one prediction) since no relevant experiment with the smoke visualization technique has yet been reported.

Grass (1971) has provided the most comprehensive data regarding the influence of bed roughness upon turbulence intensity, fluctuating components of velocity, Reynolds stresses and patterns of turbulence in the xy-plane. He found that burst and sweep events existed regardless of the surface roughness but that the streaky structure of a smooth bed was disrupted by roughness elements. Grass's turbulence intensity data scale directly with u_* independently of boundary roughness for $y/h > 0.2$, where $y =$ flow depth, $h =$ height from bed (his fig. 5), implying that, beyond a certain boundary distance, the turbulence intensity becomes dependent solely on boundary distance and boundary shear stress but independent of the conditions producing the shear stress. By way of contrast, close to the bed, the velocity measurements from smooth and rough beds become divisible into two groups, the longitudinal intensity decreasing with increasing roughness whilst the vertical intensity increases. Burst events over the rough bed rose vertically to affect the surface flow as boils. Nakagawa & Nezu (1977) point out that, contrary to the case of a smooth bed, the intensity of ejection decreases towards a rough wall to become less intense than sweeps in the vicinity of the rough wall (see also Raupach, 1981).

GRAIN MOVEMENT AND TRANSPORT MODES

Single grain movement and the burst/sweep cycle

Two outstanding experimental contributions, both involving multi-exposure photography, have shed much light on this subject.

Abbott & Francis (1977), experimenting with solitary grains in water flows over rough beds,

presented a mass of information chiefly concerning saltation trajectories and their kinematic properties. It should be noted that the suspensive mode referred to by Abbott & Francis is not the development of full suspension but that of incipient suspension (*sensu* Leeder, 1979), where vertically upward fluid impulses exerted on the particle are insufficient to reverse its downward momentum (Abbott & Francis, 1977, p. 230).

Sumer & Deigaard (1979, 1981), following earlier studies by Sumer & Ogoz (1978), traced the paths of suspended particles in water streams over smooth and

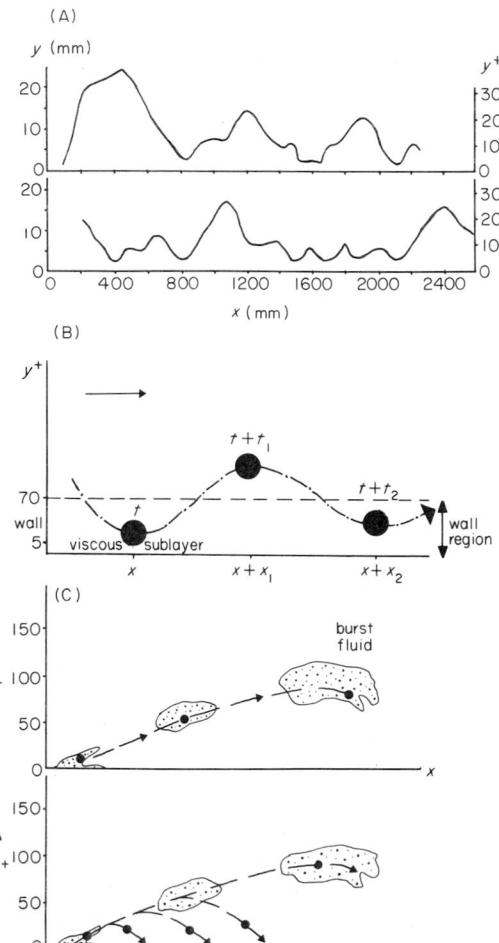

Fig. 3. Burst/sweep interactions and sediment suspension. (A) Typical paths of suspended particles from photographic data. $X =$ streamwise distance. (B) A complete cycle of particle trajectory; $t =$ time, $t_1 =$ burst duration, $t_2 =$ burst periodicity. (C) Sketches to show bursting fluid with suspended and saltant grains. Note the advective nature of suspension (after Sumer & Deigaard, 1979).

rough beds using a stereo-photogrammetric system coupled with a stroboscope. Close agreement was found between the kinematics of particle motions and the burst/sweep process (Fig. 3). Grain-rise from near the bottom occurred within the lifted fluid of a burst event (a point first discovered by Grass, 1974). The apogee of the grain path coincides with burst dissipation, followed by grain fall encouraged by high-speed fluid sweeps which penetrate to the bottom, spread out sideways and deliver the particle into an adjacent low-speed wall streak; whereupon the cycle begins again. Sumer & Deigaard (1979, 1981) note that for heavier grains, the grain ejected from the near-bottom region falls out of the lifting fluid *before* burst break-up (Fig. 3). The authors are here obviously describing a perfect transition from saltation through incipient suspension to the true, fully developed suspension that most of their work discusses. At the limit, the suspension wavelength should clearly approximate to the burst wavelength/repeat distance. With decreasing transport stage the suspension wavelength should tend towards a saltant wavelength much smaller than the burst wavelength. Unfortunately, Sumer & Deigaard's (1979) analysis of the limiting stage of suspension is based upon a faulty definition of suspension (*op. cit.*, p. 90) which simply states that a suspended particle is one that has been 'lifted up from near or from the bottom'. This definition equally applies to saltation and so their quantitative analysis based upon instantaneous pressure gradients is giving us a 'limiting' stage for saltation jumping'.

Multi-grain movement and transport modes

The application of the results of single-grain studies to natural sediment transport systems is greatly hindered by a lack of data concerning the influence of grain upon grain. At a certain transport stage both saltant and suspensive trajectories will become modified by grain 'collisions'. A dispersion of moving grains will now be present. There should be little effect of collisions upon mean forward grain speeds (see Francis, 1973) but a significant effect upon both grain path and turbulent structure close to the bed (see later section). Application of simple kinetic theory to saltant grains sheds light on this effect and enables a *rough* estimate of the critical transport stage for collision effects to be arrived at (Leeder, 1979). The concept of a moving concentrated granular dispersion close to the bed at high transport stages is a major development of Bagnold (1956, 1966), with key

applications to both sediment transport and bedform theory (e.g. Allen & Leeder, 1980).

Velocity of forward grain movement

Much of Abbot & Francis's work, together with that of Meland & Norrman (1966) and Luque (1974), was designed to give estimates of the mean forward grain speed, U, as a function of transport stage. The grain velocity assumes great importance in many bedload sediment transport theories (e.g. Bagnold, 1973; Engelund & Fredsøe, 1976). According to Bagnold (1973),

$$U = \bar{u} - 5 \cdot 75 u_* \log(0 \cdot 37 h / n_\mathrm{D}) - V_\mathrm{g}, \qquad (3)$$

where $\bar{u}$ is the mean flow velocity, u_* is the shear velocity, h is the flow depth, n_D is the distance from the bed, in grain diameters, of the centre of fluid thrust on the grain and V_g is the grain fall velocity. This equation is derived from simple first principles by assuming that a lag between grain and fluid velocity exists, being equal to the grain fall velocity. The postulate is confirmed by a consideration of Sumer & Deigaard's results (1979, fig. 1.15). Bridge (1981) has shown that equation (3) is equivalent to the expression

$$U = k(u_* - u_{*\mathrm{c}}) \qquad (4)$$

where $u_{*\mathrm{c}}$ is the critical shear velocity for grain motion. Equation (4) fits experimental data well when $k \approx 10$.

DISPERSION MECHANICS AND THE EFFECTS OF MOVING GRAINS UPON TURBULENCE

Dispersion mechanics

Following Bagnold (1956, 1966) we may divided the resistance to applied fluid shear stress in a water flow transporting bedload over a plane bed into solid and liquid components. Thus

$$\tau_{yx} = T + \tau_\mathrm{c}, \qquad (5)$$

where τ_{yx} is the applied shear stress (tractive), T is the solid transmitted resisting shear stress and τ_c the fluid resisting shear stress, assumed by Bagnold to be equal to the critical fluid stress for grain motion.

When T is regarded as the *directly* transmitted solid resisting stress then it may be calculated from data on sediment transport rates, being the downslope immersed weight component due to bedload grains times the coefficient of dynamic friction. Bagnold's

2

postulate was tested by Leeder (1977) and found to be in error, T being a much smaller portion of the resisting stress than postulated in equation (5). Clearly, some additional resisting stress is present to balance the left-hand side of equation (5).

I now believe that T must also include effects due to added fluid masses (see Hamilton & Lindell, 1971; Hamilton & Courtney, 1977) associated with every grain in the dispersion, and also to fluid friction generated by spinning grain collisions. Both of these effects will give rise to additional resisting stresses, though their magnitude is not easy to calculate.

We may now write equation (5) in the notional form

$$\tau_{yx} = T_s + T_f + \tau_c, \qquad (6)$$

where T_s is now the *direct* solid stress and T_f is the indirect solid stress transmitted via added fluid mass and grain spin friction. My recalculated values for these stress components are given in Fig. 4.

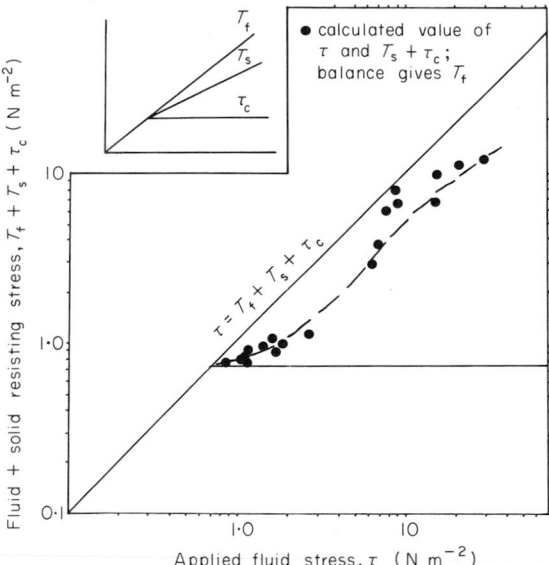

Fig. 4. Stress balance graph for bedload transport calculated from the revised data of Williams (1970) and Leeder (1977). Data points give the calculated values of τ_{yx} and $T_s + \tau_c$; the value of $\tau_{yx} - (T_s + \tau_c)$ gives the residual value ascribed to T_f. T_s calculated as in Leeder (1977), with $\tan \alpha = 0.6$.

Dispersions, viscosity and turbulent

The effects of appreciable ($\sim 10\%$ by volume) concentrations of solid grains upon fluid turbulence are little known, yet are of the greatest significance in theoretical sediment transport studies.

The effects of grains upon pure fluid dynamic viscosity, η, are better known, being reviewed by Roscoe (1952). He derives the expression

$$\eta_r = (1 - 1.35c)^{-2.5}, \qquad (7)$$

where η_r is the relative viscosity in the presence of grains and c is the grain concentration by volume ratio.

Thus Newton's law for viscous flow becomes

$$\tau_{yx} = (1 - 1.35c)^{-2.5}\, \eta\, du/dy \qquad (8)$$

for the flow of suspensions.

Bagnold (1954) found from experiment that, for the 'viscous' regime of suspension, shear

$$\tau_{yx} = (1 + \lambda)(1 + \tfrac{1}{2}\lambda)\, \eta\, du/dy \qquad (9)$$

where λ is linear concentration. In equations (8) and (9), τ_{yx} is a mixed stress due to the effects of fluid viscosity as modified by the presence of grains.

Evaluation of equations (8) and (9) shows that there seems to be a 2- to 3-fold discrepancy between the Roscoe and Bagnold formulae for viscous shear. The reasons for this are not clear.

In the inertial regions of shear the extra apparent viscosity is vastly increased and cannot be due, in the main part, to simple concentration effects (Bagnold, 1954; Davidson, Harrison & Guedes de Cavalho, 1977) but must be due instead to the effects of interparticle collisions.

The dampening effect of grains upon turbulence in flows has been predicted by Bagnold (1955, 1956, p. 242). Later studies in a variety of applied fields have generally confirmed Bagnold's postulate, although the results are still considered controversial (see Radin, Zakin & Patterson, 1975). Sproull (1961) found that dust particles in air reduced the resistance coefficient due to turbulence dampening. Saffman (1965) pointed out that any heavy particle has a much larger inertia than the equivalent volume of fluid and will not therefore participate as readily in the turbulent fluctuations. The relative motion of the particles will dissipate energy because of the drag between particle and fluid, and extract energy from the turbulent fluctuations. Saffman explains the drag reduction as due to a reduction in turbulence intensity which decreases the Reynolds stresses, and thus the force required to maintain a given flow rate will be reduced. Rossetti & Pfeffer (1972) point out that Saffman's analysis will be most important close to the wall where there is a large slip velocity between fluid and particle. Lumley has recently written extensively on two phase flows (Lumley, 1976, 1977), but with most relevance to those flows with very low grain concentrations. In his notes on shear flows he confirms that grains should suppress turbulence. Significantly, he notes in his 1977 paper (p. 64) that the addition of grains to a flow affects only the dissipative scales of turbulence, suppressing the dissipative eddies, and increasing the scale of dissipation.

SEDIMENT TRANSPORT THEORY

The most significant advance in recent years has been Bagnold's progressive honing of the ideas concerning sediment transport theory expressed in his long 1956 paper. The power approach has been refined (Bagnold, 1966), particularly the bedload efficiency concepts used (Bagnold, 1973) and the role of water depth upon transport rates has been identified with the introduction of a universal empirical relation (Bagnold, 1977, 1980) for bedload.

Bedload theory

The final Bagnold (1973) equation is, in its simplest form,

$$i_b = k\tau_{yx}\bar{U}/\tan \alpha \qquad (10)$$

where i_b is the bedload transport rate, τ_{yx} is the applied bed fluid shear stress, $\bar{U}$ is the mean forward grain speed (equation 3), and $\tan \alpha$ is the dynamic friction coefficient. k is a constant which gives that proportion of τ_{yx} transferred to the bed via the saltating bedload grains. k is given by Bagnold as $(u_* - u_{*c})/u_*$, being the simplest function which ranges from zero at threshold towards 1 at high transport stage.

Bridge (1981) has shown how close equation (10) is to the Engelund & Fredsøe (1976) bedload equation, which may be given as

$$i_b = (\tau_{yx} - \tau_c)\bar{U}/\tan \alpha, \qquad (11)$$

where $\bar{U}$ is given by equation (4). Neither equation (10) nor equation (11) has ever been tested with high-quality flume data for the strict conditions under which they are applicable (bedload transport alone, plane bed states). Using the Williams (1970) coarse sand data for lower and upper plane beds and assuming a value of $\tan \alpha$ of 0·6, I have calculated i_b over 5 orders of magnitude (Fig. 5). Note how i_b is overestimated by both equations, the Engelund & Fredsøe/Bridge equation being more faulty. Assuming that $\tan \alpha$ is about right, which is by no means certain, I suspect that the rather arbitrary k factor in Bagnold's

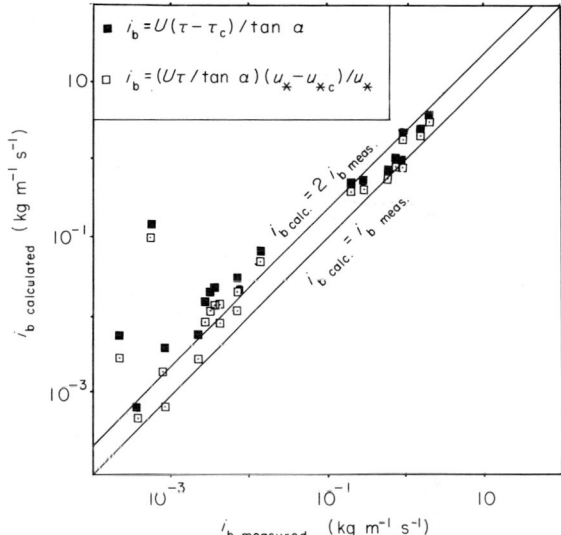

Fig. 5. Test of the Bagnold and Engelund-Fredsøe bedload transport theories against the experimental lower- and upper-stage plane bed data of Williams (1970). See discussion in text.

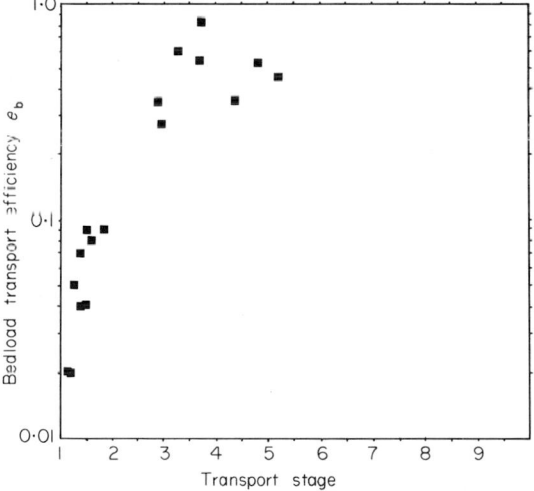

Fig. 6. The variation of bedload transport efficiency with transport stage as calculated from the lower- and upper-stage plane bed data of Williams (1970).

equation and the omission of such a function in equation (11) is responsible. Incidentally, equation (10) is perfectly successful for a value of $k = (u_* - u_{*c})/2u_*$. The relationship of the bedload efficiency factor, e_b, to transport stage for these plane bed runs is shown in Fig. 6. These values are of importance in calculating suspended and total load in the Bagnold methodology.

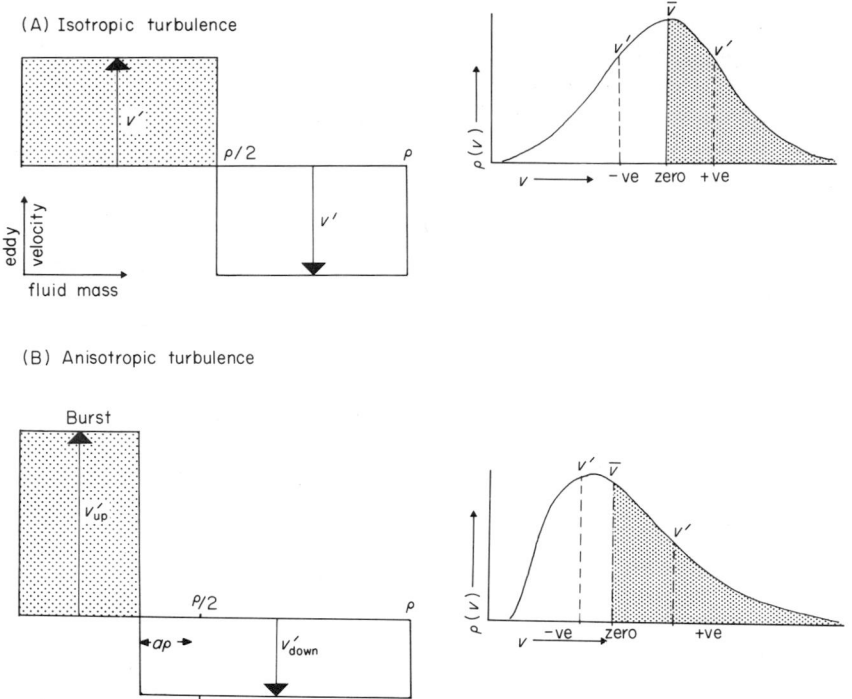

Fig. 7. Sketches to illustrate Bagnold's hypothesis of anisotropic turbulence. In isotropic turbulence, the rms fluctuation v' is assumed symmetrical about $\bar{v}$. Considering unit volume of fluid, the upward and downward masses ($\rho/2$; where ρ = fluid density) and amounts of fluid momentum are equal. In anisotropic turbulence the fluctuations, v'_{up} and v'_{down} are unequal because of a skewed distribution. There is thus a greater net absolute upward fluctuation, than downward net fluctuation. In order to conserve momentum the fast upward-moving fluid must be of smaller mass than the slow downward-moving fluid. There thus exists an asymmetry, a, of the fluid masses involved in turbulence. Adopting Bagnold's ideas to turbulent structural models (see Fig. 2) we must assign the fast-moving upward masses to burst motions and the slow-moving downward masses to sweep motions. Although momentum is conserved, there exists a momentum flux (residual stress τ_{yy}) away from the bed whose magnitude is controlled by a.

Suspension theory

Ignoring the very large amount of work done on suspension transport from the kinematic viewpoint of diffusional theory, no advance has been made concerning theories for the *rate* of suspended transport since Bagnold's innovative approach in his 1966 paper. This approach was dynamic, based upon the concept of anisotropic turbulence (see also Irmay, 1960) such that a residual upward normal stress τ_{yy} supports the suspended load. Bagnold postulated that τ_{yy} arises because of an asymmetry in shear turbulence such that upward turbulent excursions were fast-moving minor masses whereas downward excursions were slow-moving major masses of fluid (Fig. 7). Although the total normal momentum must be zero, the normal momentum *fluxes*, per unit area of a shear

plane, are unequal. Thus a residual momentum flux, τ_{yy}, is directed into the flow, which is balanced by an equivalent excess of mean static pressure at the boundary.

Making certain assumptions about the value of a, the asymmetry of the fluid masses involved in turbulence (Fig. 7), Bagnold was able to estimate the value of τ_{yy} as about $0.4\tau_{yx}$ and to arrive at a value for the magnitude of the efficiency of suspended sediment transport.

Surprisingly, Bagnold's approach to suspended sediment transport has attracted very little attention, although Jackson (1976) made the suggestion that bursts might produce the vertical anisotropy in turbulence needed to suspend sediment. This statement slightly obscures the distinction between the mechanics of suspension, already seen to be due to bursts (first discovered by Grass, 1974) in a previous section, and the dynamics of maintaining a net suspended load against gravity.

The crux of Bagnold's theory concerns turbulent anisotropy. Such anisotropy may only be identified from turbulence data that distinguishes upward and downward components. This effectively rules out the McQuivey (1973) hot-wire data. The only practicable way of obtaining such data is from hydrogen-bubble visualization studies. To my knowledge Grass (1967) is the only researcher who has ever presented 'up' and 'down' turbulent measurements which are of sufficient number and of high enough quality to test Bagnold's theory. I am grateful to Dr Grass for his permission to calculate the necessary parameters from his unpublished thesis data of turbulent fluctuations over perfectly smooth boundaries. The results of these preliminary calculations are shown in Fig. 8, where it may be seen that a very clear turbulent anisotropy certainly exists in the region $y^+ < 40$. However, this anisotropy clearly reflects the existence of positive values of $\bar{v}$ due to time-mean secondary circulations in the experimental flow. Thus the data cannot be used to test Bagnold's theory and the status of the theory remains unclear.

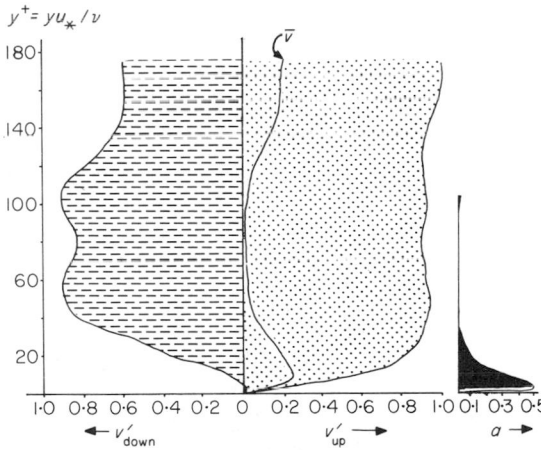

Fig. 8. The Grass (1967) data on smooth-bed vertical turbulence plotted to show the position and magnitude of the asymmetry, a, which approaches 0·5 very close to the smooth bed, at $y^+ = 3·6$. Note a second problematic maximum in a at about $y^+ = 150$. Water surface at $y^+ = 424$. Maximum τ_{yy} calculated at $y^+ = 16$, where $a = 0·21$, in complete agreement with Bagnold's theoretical prediction.

BEDFORM MECHANICS

What should a bedform theory explain?

Bedform theory is in its infancy. Strong growth requires a focus of attention away from the mathematical appeals of linear stability theory (applicable only to antidune forms in supercritical flows) towards a recognition of the basic physics of subcritical bedforms. Such an approach needs to recognize: (1) the interactions of the 'flow–sediment transport–bedform' trinity outlined previously; (2) the real pattern of turbulent, separated flow and macroturbulence over certain forms.

A bedform theory ought to seek to explain bedform magnitude, shape, migration rate and limits to stability. In the following account I simply wish to go through each bedform state (ripples, dunes, upper and lower plane beds, bars) and make some remarks concerning these matters. The approach lends itself to a broad dynamic classification of bedforms but cannot yet be said to be true theory since many of the ideas presented remain to be quantified.

Current ripples

These forms are present only on hydraulically smooth boundaries. They grow to equilibrium from burst/sweep-produced defects (Williams & Kemp, 1971) by the action of flow separation and reattachment (Raudkivi, 1963; Allen, 1968, 1969). The stable ripple assemblage is relatively insensitive to variations in water depth (Pratt, 1973) and, to a lesser extent, bed shear (Allen, 1968), there being a weak tendency for mean ripple height and a stronger tendency for wavelength to increase with τ_{yx}. Ripple forward speed shows a strong positive correlation with τ_{yx} (Allen, 1973). There is a tendency for the rms deviation of both height, wavelength and velocity to increase with increased τ_{yx} (Allen, 1969, 1973; Jain & Kennedy, 1971).

Ripple magnitude can only be understood in terms of the dimensions and effectiveness of flow separation in the ripple lee (see also Yalin, 1977, p. 263). Ripple height presumably reflects the maximum scouring depth possible by turbulent stresses at flow reattachment over the negative step. Ripple wavelength must be controlled by the length of the separation bubble (roughly equal to $\frac{1}{3}$ ripple wavelength; Mercer & Haque, 1973) and the downstream transport distance of grains eroded at reattachment and subsequently redeposited.

As flow strength increases and the rms deviations of height, wavelength and ripple velocity increase there will exist an increasing tendency for 'cannibalization' processes to occur, with large, fast ripples overtaking and assimilating smaller, slower forms. From such 'rogue' ripples (perhaps these are bars or flattened dunes) it is possible that 'megabursts' in ripple lees are produced, so increasing scour and encouraging the appearance of dune bedforms.

As grain size increases, ripples occupy an increasingly small portion of the τ/d phase plane. Ultimately, as grains disrupt the viscous sublayer at $d \approx 0.65$ mm, increasing turbulent mixing begins to inhibit enlargement of defects because of ineffective flow reattachment (Leeder, 1980). The coarse sand bed remains flat as a stable, lower-phase plane bed.

There is no evidence that current ripples interact with the outer flow region, or that pre-existing outer flow structures have any effect on the morphology of the bedform. The morphology and magnitude of the bedforms seem to be entirely controlled by *flow separation dynamics*.

Dunes

In contrast to current ripples, dunes show positive correlations of both height (Allen, 1968) and wavelength (Jackson, 1976) with water depth, strongly suggesting that the form is adjusted to flow outer zone processes (Yalin, 1977; Jackson, 1976). Note that the form of these positive correlations is considerably more complex than generally noted (see Znamenskaya, 1964; Korchokha, 1968). Following the work of Mathes (1947) on 'kolks', Znamenskaya undertook flow visualization studies during her study of dune morphology. She found that dunes (her type II dunes) developed periodic ejections of macroturbulence from their troughs that rose through the flow to the water surface as 'boils'. This was confirmed by Korchokha (1968), Coleman (1969) and Jackson (1976). In his extensive discussion of macroturbulent boils Jackson proves that their periodicity corresponds to that predicted by equation (2).

Yalin (1977) is the only author who has considered macroturbulence in terms of the theory of bedform origins. He develops the idea that a discontinuity at the bed disturbs the structure of turbulence at repeat distances, L, of 4–7 flow depths. Let us develop these ideas a little more. Current ripples provide many discontinuities at the bed, yet macroturbulence is apparently ineffective in controlling bedform characteristics (NB this does not mean that there are no outer

flow structures present). The bed region is dominated by 'micro'-flow separation dynamics until some effect triggers the development of effective macroturbulence. The trigger may be the build-up of a critical excess pressure in the lee of certain higher individual bedforms, causing an unfavourable $(dp/dx > 0)$ pressure gradient sufficient to encourage major leeside bursting (see Kline *et al.*, 1967; Jackson, 1976). The real trigger for this change is probably the growth of occasional 'rogue' ripples (?bars or flattened dunes) due to enhanced cannibalization processes at relatively high values of τ_{yx} (see above).

We must also consider exactly *how* macroturbulent bursts might cause the ripple → dune catastrophe. The bursts must cause major erosion in the 'rogue' ripple troughs, thus increasing their effective height, influencing the leeside separation and enhancing the bursting process even more. Following the work of Sumer & Deigaard (1979) discussed earlier, we may be sure that a large proportion of this mass of eroded, suspended sand will travel downstream a *maximum* of about 5 flow depths before deposition occurs. Continued bursting, erosion and deposition will eventually cause the bedforms to adjust to a relationship between bedform wavelength, height and flow depth. A stable dune bed will gradually develop in which macroseparation as well as burst macroturbulence plays an important role in determining dune morphology and magnitude.

A most interesting problem arises. Are macroturbulent leeside bursts produced quite independently of the outer flow, or does the existing outer flow structure cause the periodic appearance of macroturbulent bursts? Future flow visualization studies should set out to answer this question, as well as to indicate how bursts interact with the active dune leeside separation eddy.

With increasing applied fluid shear, dunes become progressively flatter until a stable flat bed, the upper phase plane bed, results. For reasons discussed by Southard (1971) there is considerable overlap between the two bed states in the τ_{yx}–d phase plane. The disappearance of dunes at high values of τ_{yx} is attributed by Engelund & Fredsøe (1974) to the development of appreciable suspension.

Upper-stage plane beds

In a recent contribution, Allen & Leeder (1980) examined the stability limits to this bedform and found strong support for the Bagnold (1966) 'universal' plane-bed instability criterion in flows decelerating through the upper-stage plane bed τ_{yx}–d phase plane. This criterion is

$$\tau_{\text{crit}}/(\sigma-\rho)gD = \theta_{\text{crit}} = C\tan\alpha, \qquad (12)$$

where τ_{crit} is the critical shear stress for instability, σ and ρ are solid and fluid densities, D is grain diameter, θ_{c} is the dimensionless critical shear, C is the fractional volume concentration in the static bed and $\tan\alpha$ is the dynamic friction coefficient.

A reinterpretation of the mechanism of upper plane-bed instability in decelerating flows was based upon the extent to which high grain concentration in plane bed flows decreases turbulence production over potential bed defects, thereby preventing effective erosion at flow reattachment and subsequent redeposition, thus prohibiting current ripple or dune propagation and growth (*op. cit.*, fig. 4). Upper plane bed flows are thus dominated by dispersion mechanics (see previous section) at grain concentrations of about 10%. The hypothesis remains untested.

An alternative view would be that the frequency of sweeps on the upper plane bed is so high that individual small defects are literally destroyed by impacting sweeps before leeside reattachment eddies can induce an effective erosive response from the bed. This contrasts with the situation on a fine, initially planed bed at threshold when flow-produced defect mounds will be unaffected by other sweeps for the time interval in excess of that required to cause erosion at reattachment. Further experimental and theoretical studies on this problem are clearly needed.

Allen & Leeder (1980, p. 214) also showed that current ripples succeed upper-stage plane beds in decelerating flows if the boundary is smooth, and that dunes succeed the plane bed on a transitional to rough boundary (*op. cit.*, fig. 3). This result implies little effect of high bedload and suspended load concentration on the structure of the viscous sublayer.

Lower-stage plane beds

As briefly noted above, current ripples do not form at close-to-threshold shears on a transitional or rough bed (Sundborg, 1956; Raudkivi, 1963; Williams & Kemp, 1971; Jackson, 1976; Yalin, 1977). Instead, an initially planed bed remains stable as grains roll and saltate over it. Leeder (1980), in an as yet untested hypothesis, proposed that this lower-stage plane bed remains stable and does not ripple because flow separation over natural chance defects (or artificial defects) is inhibited by the enhanced vertical mixing of boundary layer fluid over transitional to rough boundaries. This vertical mixing neutralizes the increased static pressure over bed defects and prevents erosion by the weakly reattaching leeside flow so that defects are unable to amplify and propagate downstream.

Bars and 'intermediate flattened dunes'

In his careful experiments, Costello (1974) observed that straight-crested bar bedforms appeared on the lower plane bed as the fluid shear was increased. These long, low bedforms with irregular wavelengths seem to arise from chance bed defects produced from greater than average grain mounds caused by grain hindrance and sheltering effects (induced by grain rolling at low transport stage over rough beds) or where a discontinuity in transport is produced by some wall or bed irregularity. They, and their equivalents in gravel-grade sediments, appear to be generally due to grain–grain hindrance effects. They are, in effect, a form of kinematic wave analogous to traffic queues (Langbein & Leopold, 1968). Their transition to dunes must imply that the larger bars trigger off macroturbulent bursts and effective leeside macroseparation eddies at a critical value of applied shear.

Costello also found that bar-like forms succeed current ripples in sands down to diameters of 0·51 mm. These bars were also recognized by Pratt, (1971, 1973) with a 0·49 mm sand; he called them 'intermediate flattened dunes'. The origin of bar-like forms that succeed current ripples remains problematical. They are clearly a stable bedform yet it is not clear whether they should be regarded as a separate form from true 'dunes' (see Costello & Southard, 1981). They clearly represent macroforms which do not greatly influence the whole outer flow. Perhaps, as briefly noted above, they represent extreme cannibalization effects of large current ripples on small. There is obviously much scope for further work here.

Preliminary genetic bedform classification

Much theoretical and experimental work remains to be done before bedform theory can be said to have advanced beyond the realm of informed speculation. At this early stage it may be helpful to propose a preliminary classification of forms (Fig. 9) in terms of their genesis (see Jackson, 1975, for a more wide-ranging but essentially flow-structure-dominated classification).

(1) *Separation*-dominated forms (current ripples)
Only present on smooth boundaries. Bedform

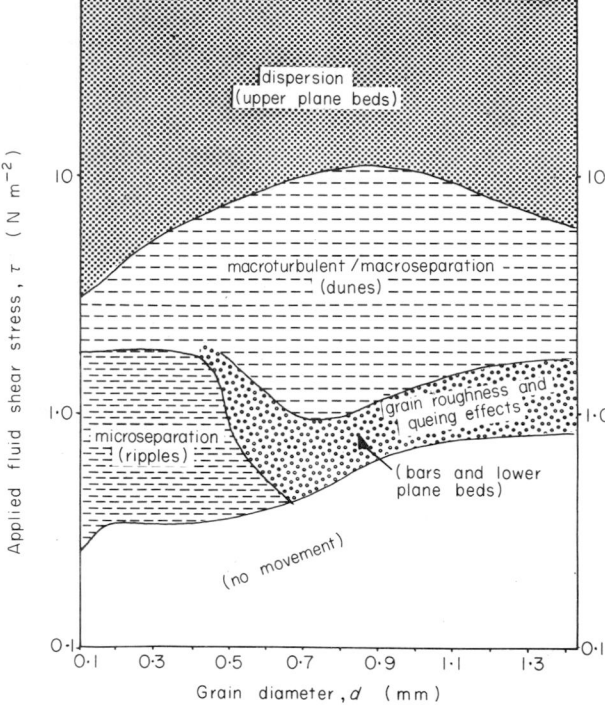

Fig. 9. Bedform phase diagram in τ_{yx}–d space (boundaries after Leeder, 1980) to show how bedforms may be classified according to the physical processes which cause their size limits and stability (see discussion in text).

growth, magnitude and morphology due to flow-separation effects.

(2) *Separation/macroturbulence*-dominated forms (dunes)

Bedform growth, magnitude and morphology due to the combined effects of leeside macroturbulent bursts and flow separation.

(3) *Bed/grain roughness*-dominated forms (lower flat beds/bars)

Bedform growth, magnitude and morphology due to the somewhat random effects of grain–grain hindrance and grain/fluid drag effects on transitional to rough boundaries.

(4) *Grain dispersion*-dominated forms (upper flat beds)

Bedform morphology due to the inhibiting effects of high bed and suspended load concentrations on flow separation and/or the erosive action of very frequent fluid sweep phases on small, chance defects.

ACKNOWLEDGMENTS

I thank John Bridge, Tony Grass and Henry Pantin for their critical reactions to this paper. Useful seeds of scepticism were sown by Gerry Middleton and John Southard in conversations at the Keele Conference on Fluvial Sedimentology, where this paper formed the basis of a 'keynote' address.

REFERENCES

ABBOTT, J.E. & FRANCIS, J.R.D. (1977) Saltation and suspension trajectories of solid grains in water streams. *Phil. Trans. R. Soc.* A, **284**, 225–254.

ALLEN, J.R.L. (1968) *Current Ripples*. North-Holland, Amsterdam. 433 pp.

ALLEN, J.R.L. (1969) On the geometry of current ripples in relation to stability of fluid flow. *Geog. Annlr.* **51 A**, 61–96.

ALLEN, J.R.L. (1973) Features of cross-stratified units due to random and other changes in bed forms. *Sedimentology*, **20**, 189–202.

ALLEN, J.R.L. & LEEDER, M.R. (1980) Criteria for the instability of upper-stage plane beds. *Sedimentology*, **27**, 209–218.

BAGNOLD, R.A. (1954) Experiments on a gravity-free dispersion of large solid spheres in a Newtonian fluid under shear. *Proc. R. Soc.* A, **225**, 49–63.

BAGNOLD, R.A. (1955) Some flume experiments on grains but little denser than the transporting fluid, and their implications. *Proc. Inst. civ. Engrs* **4**, 174–205.

BAGNOLD, R.A. (1956) The flow of cohesionless grains in fluid. *Phil. Trans. R. Soc.* A, **249**, 235–297.

BAGNOLD, R.A. (1966) An approach to the sediment transport problem from general physics. *Prof. Pap. U.S. geol. Surv.* **422–I**. 37 pp.

BAGNOLD, R.A. (1973) The nature of saltation and of 'bed-load' transport in water. *Proc. R. Soc.* A, **332**, 473–504.

BAGNOLD, R.A. (1977) Bed load transport by natural rivers. *Wat. Resour. Res.* **13**, 303–312.

BAGNOLD, R.A. (1980) An empirical correlation of bedload transport rates in flumes and natural rivers. *Proc. R. Soc.* A, **372**, 453–473.

BRADSHAW, P. (1981) Turbulence. *Sci. Prog.* **67**, 185–204.

BRIDGE, J.S. (1981) A discussion of Bagnold's (1956) bedload transport theory in relation to recent developments in bedload modelling. *Earth Surf. Proc.* **6**, 187–190.

BROWN, G.L. & THOMAS, A.S.W. (1977) Large structure in a turbulent boundary layer. *Phys. Fluids*, **20**, S243–S252.

COLEMAN, J.M. (1969) Brahmaputra River: channel processes and sedimentation. *Sedim. Geol.* **3**, 129–239.

COSTELLO, W.R. (1974) *Development of Bed Configuration in Coarse Sands*. Rep. 74.1, Earth and Planetary Science Department, Massachusetts Institute of Technology. 120 pp.

COSTELLO, W.R. & SOUTHARD, J.B. (1981) Flume experiments on lower-flow-regime bed forms in coarse sand. *J. sedim. Petrol.* **51**, 849–864.

DAVIDSON, J.F., HARRISON, D. & GUEDES DE CAVALHO,

J.R.F. (1977) On the liquid-like behaviour of fluidised beds. *Ann. Rev. Fluid Mech.* **9**, 55–86.

DIMOTAKIS, P.E. & BROWN, G.L. (1976) The mixing layer at high Reynolds number: large-structure dynamics and entrainment. *J. Fluid Mech.* **78**, 535–560.

ENGELUND, F. & FREDSØE, J. (1974) Transition from dunes to plane bed in alluvial channels. *Inst. Hydrodyn. Hydraul. Engng, Techn. Univ. Denmark, Ser. Pap.* **4**, 46 pp.

ENGELUND, F. & FREDSØE, J. (1976) A sediment transport model for straight alluvial channels. *Nordik Hydrol.* **7**, 293–306.

FALCO, R.E. (1977) Coherent motions in the outer region of turbulent boundary layers. *Phys. Fluids*, **20**, S124–S132.

FIELDER, H. & HEAD, M.R. (1966) Intermittency measurements in the turbulent boundary layer. *J. Fluid Mech.* **25**, 719–735.

FRANCIS, J.R.D. (1973) Experiments on the motion of solitary grains along the bed of a water-stream. *Proc. R. Soc. A*, **332**, 443–471.

GRASS, A.J. (1967) *Boundary layer turbulence in open channel flows*. Unpublished Ph.D. Thesis. University of London.

GRASS, A.J. (1971) Structural features of turbulent flow over smooth and rough boundaries. *J. Fluid Mech.* **50**, 233–255.

GRASS, A.J. (1974) Transport of fine sand on a flat bed: turbulence and suspension mechanics. *Proc. Euromec. 48, Tech. Univ. Denmark*, pp. 33–34.

HAMILTON, W.S. & COURTNEY, G.L. (1977) Added mass of sphere starting upward near floor. *J. Engng Mech. Div. Proc. Am. Soc. civ. Engrs* **103**, EM1, 79–97.

HAMILTON, W.S. & LINDELL, J.E. (1971) Fluid force analysis and accelerating sphere tests. *J. Hydraul. Div. Proc. Am. Soc. civ. Engrs* **97**, HY6, 804–817.

HEAD, M.R. & BANDYOPADHYAY (1981) New aspects of turbulent boundary layer structure. *J. Fluid Mech.* **107**, 297–338.

IRMAY, S. (1960) Accelerations and mean trajectories in turbulent channel flow. *J. Basic Eng. Trans. Am. Soc. mech. Engrs* **82**, 961–971.

JACKSON, R.G. (1975) Hierarchical attributes and a unifying model of bedforms composed of cohesionless material and produced by shearing flow. *Bull. geol. Soc. Am.* **86**, 1523–1533.

JACKSON, R.G. (1976) Sedimentological and fluid dynamic implications of the turbulent bursting phenomenon in geophysical flows. *J. Fluid Mech.* **77**, 531–560.

JAIN, S.C. & KENNEDY, J.F. (1971) The growth of sand waves. *Proc. int. Symp. Stochastic Hydraul.* pp. 449–471. Pittsburgh.

KIM, H.T., KLINE, S.J. & REYNOLDS, W.C. (1971) The production of turbulence near a smooth wall in a turbulent boundary layer. *J. Fluid Mech.* **50**, 133–160.

KLINE, S.J., REYNOLDS, W.C., SCHRAUB, F.A. & RUNDSTADLER, P.W. (1967) The structure of turbulent boundary layers. *J. Fluid Mech.* **30**, 741–773.

KORCHOKHA, Yu.M. (1968) Investigation of the dune movement of sediments on the Polomet River. *Soviet Hydrol.* **6**, 541–549.

LANGBEIN, W.B. & LEOPOLD, L.B. (1968) River channel bars and dunes–theory of kinematic waves. *Prof. Pap. U.S. geol. Surv.* **422–L**, 16 pp.

LEEDER, M.R. (1977) Bedload stresses and Bagnold's bedform theory for water flows. *Earth Surf. Proc.* **2**, 3–12.

LEEDER, M.R. (1979) 'Bedload' dynamics: grain–grain interactions in water flows. *Earth Surf. Proc.* **4**, 229–240.

LEEDER, M.R. (1980) On the stability of lower stage plane beds and the absence of current ripples in coarse sands. *J. geol. Soc. London*, **137**, 423–429.

LUMLEY, J.L. (1976) Two-phase and non-Newtonian flows. In: *Topics in Applied Physics* (Ed. by P. Bradshaw), **12**, chapter 7. Springer-Verlag, Berlin.

LUMLEY, J.L. (1977) Drag reduction in two phases and polymer flows. *Phys. Fluids*, **20**, S64–S71.

LUQUE, R.F. (1974) Erosion and transport of bed sediment (*Diss.*), *Krips Repro B.V.-Meppel*.

MATHES, G.H. (1947) Macroturbulence in natural stream flow. *Trans. Am. geophys. Un.* **28**, 255–262.

MCQUIVEY, R.S. (1973) Summary of turbulence data from rivers, convergence channels and laboratory flumes. *Prof. Pap. U.S. geol. Surv.* **892–B**, 66 pp.

MELAND, N. & NORRMAN, J.O. (1966) Transport velocities of single particles in bed load motion. *Geogr. Annlr.* **48A**, 165–182.

MERCER, A.G. & HAQUE, M.I. (1973) Ripple profiles modeled mathematically. *J. Hydraul. Div. Proc. Am. Soc. civ. Engrs* **99**, HY3, 441–459.

NAKAGAWA, H.R. & NEZU, I. (1977) Prediction of the contribution to the Reynolds stress from bursting events in open-channel flows. *J. Fluid Mech.* **80**, 99–128.

NAKAGAWA, H. & NEZU, I. (1981) Structure of space-time correlations of bursts phenomenon in an open channel flow. *J. Fluid Mech.* **104**, 1–43.

NAKAGAWA, H., NEZU, I. & TOMINAGA, A. (1981) Spanwise streaky structure and macroturbulence in open-channel flows. *Mem. Fac. Engng Kyoto Univ.* **43**, 34–66.

NYCHAS, S.G., HERSHEY, H.C. & BRODKEY, R.S. (1973) A visual study of turbulent shear flow. *J. Fluid Mech.* **61**, 513–540.

OFFEN, G.R. & KLINE, S.J. (1975) A proposed model of the bursting process in turbulent boundary layers. *J. Fluid Mech.* **70**, 209–228.

PRATT, C.J. (1971) *An experimental investigation into the flow of water and the movement of bed material in alluvial channels*. Unpublished Ph.D. Thesis. University of Southampton.

PRATT, C.J. (1973) Bagnold approach and bed-form development. *J. Hydraul Div. Proc. Am. Soc. civ. Engrs* **99**, HY1, 121–137.

PRATURI, A.K. & BRODKEY, R.S. (1978) A stereoscopic visual study of coherent structures in turbulent shear flow. *J. Fluid Mech.* **89**, 251–272.

RADIN, I., ZAKIN, J.L. & PATTERSON, G.K. (1975) Drag reduction in solid–fluid systems. *A. I. Ch. E. Jl* **21**, 258–371.

RAO, K.N., NARASIMHA, R. & NARAYANAN, M.A.B. (1971) The bursting phenomenon in a turbulent boundary layer. *J. Fluid Mech.* **48**, 339–352.

RAUDKIVI, A.J. (1963) A study of sediment ripple formation. *J Hydraul. Div. Am. Soc. civ. Engrs* **89**, 15–33.

RAUPACH, M.R. (1981) Conditional statistics of Reynolds stress in rough-wall and smooth-wall turbulent boundary layers. *J. Fluid Mech.* **108**, 363–382.

RICHARDS, K.J. (1980) The formation of ripples and dunes on an erodible bed. *J. Fluid Mech.* **99**, 597–618.

ROSCOE, R. (1952) The viscosity of suspensions of rigid spheres. *Br. J. appl. Phys.* **3**, 267–269.

ROSSETTI, S.J. & PFEFFER, R. (1972) Drag reduction in dilute flowing suspensions. *A. I. Ch. E. Jl* **18**, 31–38.

SAFFMAN, P.G. (1965) On the stability of laminar flow of a dusty gas. *J. Fluid Mech.* **22**, 120–128.

SOUTHARD, J.B. (1971) Representation of bed configurations in depth–velocity–size diagrams. *J. sedim. Petrol.* **41**, 903–915.

SPROULL, W.T. (1961) Viscosity of dusty gases. *Nature*, **190**, 976.

SUMER, B.M. & DEIGAARD (1979) Experimental investigation of motion of suspended heavy particles and the bursting process. *Inst. Hydrodyn. Hydraul. Engng, Techn. Univ. Denmark, Series Publ.* **23**.

SUMER, B.M. & DEIGAARD (1981) Particle motions near the bottom in turbulent flow in an open channel. Part 2. *J. Fluid Mech.* **109**, 311–338.

SUMER, B.M. & OGOZ, B. (1978) Particle motions near the bottom in turbulent flow in an open channel. *J. Fluid Mech.* **86**, 109–128.

SUNDBORG, Å. (1956) The River Klarälven: a study of fluvial processes. *Geogr. Annlr.* **38**, 125–316.

TOWNSEND, A.A. (1976) *The Structure of Turbulent Shear Flow*. Cambridge University Press. 429 pp.

WILLIAMS, G.P. (1970) Flume width and water depth effects in sediment-transport experiments. *Prof. Pap. U.S. geol. Surv.* **562-H**, 37 pp.

WILLIAMS, P.B. & KEMP, P.H. (1971) Initiation of ripples on flat sediment beds. *J. Hydraul. Div. Proc. Am. Soc. civ. Engrs* **97**, 505–522.

YALIN, M.S. (1977) *Mechanics of Sediment Transport*. 2nd ed. Pergamon Press, Oxford. 298 pp.

ZNAMENSKAYA, N.S. (1964) Experimental study of the dune movement of sediment. *Soviet Hydrol. Selected Paps* 1963, No. 3, 253–275.

Spec. Publs int. Ass. Sediment. (1983) **6**, 19–33

River bedforms: progress and problems

J. R. L. ALLEN

Department of Geology, The University, Reading RG6 2AB, U.K.

ABSTRACT

Despite great recent advances in our understanding of river bedforms and related phenomena, much remains to be done. Our knowledge of bedform existence fields under steady-state equilibrium conditions needs improvement by more work at extreme grain sizes and sediment transport stages, each of which calls for the use of deeper currents. Bedforms in natural environments are shaped by varying flows. The influence of flow non-uniformity and unsteadiness will not be understood until we have many more well-documented case histories and parallel attacks from theory and experiment. An understanding of the factors that determine the size and shape of bedforms that have grown to quasi-equilibrium requires a shift of emphasis in mathematical studies away from the rather restricted conditions that seem to govern the initial instability of granular beds. Experimental studies still have much to contribute, since we have little empirical knowledge of sediment entrainment and transport conditions at sloping beds, and many aspects of turbulent fluid flow over wavy surfaces have yet to be completely defined.

INTRODUCTION

Sediment transport in rivers is manifested by several kinds of bedform, each of which has a characteristic shape and internal structure. From these structures, sedimentologists can already infer the directions of ancient river currents, and now wish to draw conclusions about current speed, depth and strength and their spatio-temporal variation. To fluvial geomorphologists, it is perhaps the channelized river itself that is of most interest, as a major feature of riverine landscapes and as an agent of floodplain construction and modification. Bedforms to the river engineer are sources of boundary roughness and hindrances to navigation and commerce, in addition to being objectionable and unpredictable manifestations of sediment movement.

The attack upon bedforms and channel accumulations led by sedimentologists, fluvial geomorphologists and river engineers, supported by applied mathematicians, has afforded over the past 25 years many substantial advances in knowledge and understanding.

A host of issues remain, however, of which three are of especial concern, namely:

(1) our incomplete empirical grasp of the general hydraulic conditions defining the existence of bedforms under steady-state equilibrium conditions, and particularly our ignorance of high transport stages and extreme grain sizes;

(2) our limited theoretical understanding of the stability of cohesionless beds, and especially our gross ignorance of the mechanisms maintaining bedforms grown to quasi-equilibrium;

(3) our meagre understanding of bedforms in unsteady and/or non-uniform flows (i.e. the conditions prevailing in all natural environments).

The first issue is perhaps not so much a problem as a matter of the will to extend already established techniques, particularly towards the use of deeper flows. A solution to the second problem demands a radical change of direction, away from the analysis of the initial instability of granular beds, and towards the mathematically more difficult area of bedforms which have completed a path of wavelength-selection and amplitude-growth. The third issue seems to me

0141-3600/83/0106-0019 $02.00

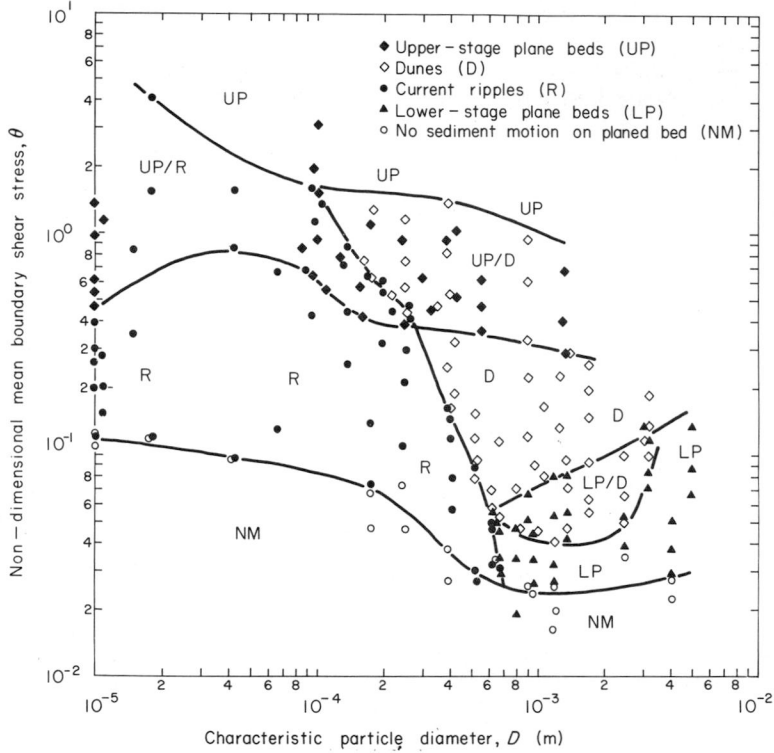

Fig. 1. Bedform existence fields in the non-dimensional mean boundary shear stress–grain size plane, based on 601 steady-state equilibrium laboratory experiments using quartz sands (data sources fully listed in Allen, 1982a, with additional points from Mantz, 1980). The shear stress is corrected for wall effects, and the grain size adjusted to plain water at 25°C. A selection of data points only is shown.

perhaps the most important of all, demanding attacks on theoretical and experimental fronts, as well as in the field. A new attitude to the field study of bedforms is necessary, in which bedform history and the time-scales of hydraulic changes are assigned their rightful importance.

EMPIRICAL KNOWLEDGE OF BEDFORM EXISTENCE FIELDS

Current knowledge

Interpretations of fluvial sedimentary structures, and the prediction of channel roughness, rest on an ability to define the general hydraulic conditions appropriate to each kind of fluvial bedform, especially current ripples, dunes, and plane beds with sediment motion. The problem has been attacked in two ways, from theory, as discussed in a later section, and from controlled, graded experiments under steady-state, equilibrium conditions. Sorby (1859, 1908), Owens (1908), and Gilbert (1914) began the work of empirical

definition, with Sheldon (1928, 1929) early making use of Gilbert's findings, but the main researches are less than 25 years old, and are particularly associated with Colorado State University (Guy, Simons & Richardson, 1966). The Colorado results, combined with the data of other investigators, have afforded many graphs purporting to define bedform existence fields. In one popular broad class (Simons, Richardson & Nordin, 1965; Allen, 1968 ; Allen & Leeder, 1980; Leeder, 1980), related to Shields's (1936) widely quoted entrainment curve, bedforms are shown in a plot of flow force (boundary shear stress, non-dimensional boundary shear stress), or specific stream power against grain size or grain Reynolds number. The other popular broad class is a three-variable plot involving mean flow velocity, flow depth and grain size (Simons, Richardson & Albertson, 1961; Southard, 1971; Costello & Southard, 1981).

Figure 1 summarizes current empirical knowledge in terms of a graph of the first kind. It is based on 601 steady-state equilibrium experiments made by 20 investigators using 39 grades of quartz sediment ranging in diameter between 1×10^{-5} and 4.94×10^{-3} m, the data sources appearing in Allen (1982a) but with additional runs by Mantz (1980). The non-dimensional mean boundary shear stress $\theta = \tau/(\sigma - \rho)gD$, in which τ is the wall-corrected shear stress, σ and ρ the sediment and fluid densities respectively, g the acceleration due to gravity, and D the grain diameter adjusted to plain water at 25 °C. The lowest plotted threshold defines the field of no sediment motion on an artificially planed bed. An important threshold rising steeply across the graph divides current ripples in the finer sediments from lower-stage plane beds and dunes developed in the coarser grades. The lower part of this bound, together with the no-motion threshold, was extensively explored by Southard & Boguchwal (1973) and by Costello & Southard (1981). The triple point between upper-stage plane beds, dunes and current ripples depends critically on work by Hill, Srinivasan & Unny (1969). In Allen & Leeder's (1980) graph, the entire bound is defined by a unique grain Reynolds number, corresponding to a grain diameter comparable to the thickness of the viscous sublayer. The controls on bedforms are therefore expressed partly through grain roughness, as has long been suspected (e.g. Sundborg, 1956). Substantial overlaps appear in Fig. 1, between current ripples and upper-stage plane beds, and between dunes and both upper-stage and lower-stage plane beds.

Extreme grain sizes and transport stages

Inspection of Fig. 1 reveals a general lack of information on bedforms at relatively high sediment-transport stages and at grain sizes outside the sand range.

Future work at high transport stages will inevitably necessitate studies in large water tunnels and in substantially deeper flumes than have traditionally been used, in order to avoid the free-surface effects so soon encountered with conventional apparatus. It seems important that we should further elucidate the bedforms arising close to the dune–plane bed (upper stage) transition (see Saunderson & Lockett, 1983), as well as establish more widely than hitherto the conditions under which ripples and dunes definitely cannot exist.

The limitations of grain size coverage stem partly from practical difficulties and partly from a preoccupation with sand. However, the importance of cohesionless silt should not be overlooked, especially in the shallow-marine environment where such

materials can be abundant, and we should also remember that many rivers have gravel or gravel–sand beds. Our ignorance of silts remains a particularly glaring gap, despite the pioneering achievements of Kalinske & Hsia (1945), Vanoni & Brooks (1957), Rees (1966) and Mantz (1978, 1980). Suitable experimental materials are difficult to obtain, however, and are subject to rather unpredictable surface chemical effects absent from sands. Mantz (1978, 1980) ensured the cohesionless behaviour of his sediments by carefully adjusting the electrochemistry of the flume water. Jopling & Forbes's (1979) quantitative results on inspection seem anomalous, perhaps because of the partly cohesive behaviour of the sediment, for which there is some evidence. A gap just as serious is our ignorance of bedforms in granules and gravels. Extrapolation of thresholds in Fig. 1 suggests that dunes may not form in gravels coarser than about 0.01 m, in apparent contradiction to the field evidence from, say, breakout floods (Baker, 1973). More experimental effort should be directed towards these coarser sediments, which also will demand larger and deeper flumes, as well as to silts under stringent chemical as well as physical control.

Field overlaps

It will be noted from Fig. 1 that current ripples and dunes range far upward into the field of upper-stage plane beds, and that dunes also overlap downward with plane beds of lower stage. The first-mentioned and more important overlap is probably attributable to the fact that the boundary shear stress appearing in the ordinate of the graph is influenced, so far as current ripples and dunes are concerned, by both grain roughness and form roughness (Einstein & Barbarossa, 1952), whereas the sediment transport conditions reflect primarily the grain-related part of the stress. The form roughness depends on the size, shape and spacing of the bed waves, as well as on the flow, and cannot yet be easily predicted (Alam, Cheyer & Kennedy, 1966; Engelund, 1966; Vanoni & Hwang, 1967). Because of its unpredictable magnitude, form roughness has never been successfully eliminated from the data underpinning bedform existence fields (e.g. Fig. 1). Generally applicable methods for the accurate and reliable prediction of form roughness remain to be devised. However, while the dual contribution to the total stress in the case of ripples and dunes seems satisfactorily to explain the first-mentioned overlap, the cause of the second remains at present elusive.

Velocity–depth–grain size plots (Southard, 1971)

reveal much less overlap between fields, but suffer from other weaknesses. They are difficult to use, being three-dimensional, and cannot yet be confidently applied to other than very shallow flows, because of the small depths ($\lesssim 0.3$ m) permitted by conventional laboratory flumes.

Dunes a divisible class?

In Fig. 1 only one class of large-scale transverse bedform is depicted, namely, dunes. It is widely believed, however, that either two substantially distinct kinds of dune are recognizable, or that dune-like bedforms fall into at least two hydro-dynamically, as well as morphologically, distinct categories. Structures of the first kind—relatively flat, long-crested and two-dimensional in appearance—are called 'simple sand waves' (Klein, 1970), 'diminished dunes' (Smith, 1971), 'bars' (Costello, 1974; Leeder, 1980), 'sand waves' (Boothroyd & Hubbard, 1974; Dalrymple, Knight & Middleton, 1975; Cant, 1978), 'transverse bars' (Jackson, 1976), and 'type 1 mega-ripples' (Dalrymple, Knight & Lambiase, 1978). Here also belong certain of the 'transitional' bedforms of Guy *et al.* (1966). The second kind of structure is markedly three-dimensional, with a relatively steep vertical profile and a strongly curved and in some instances short crest. It has attracted such names as 'megaripples' (Boothroyd & Hubbard, 1974; Dal-rymple *et al.*, 1975), 'dunes' or 'sand waves' (Jackson, 1976; Cant, 1978), and 'type 2 megaripples' (Dal-rymple *et al.*, 1978).

Is the claimed distinction valid, even at the morphological level? The evidence seems to me unconvincing, particularly as most of the claims originate from studies of natural environments. In these settings, where currents are unsteady and non-uniform, bedforms created at a number of different periods in the history of the flow, and therefore under a wide range of flow conditions, may coexist within a single limited area. Perceived differences of size and/or shape need not then imply fundamental hydrodynamic differences. Steady-state equilibrium experiments furnish evidence of the wide morphological variation possible within a single bedform class.

Harms (1969) distinguished between what he called 'low energy' and 'high energy' forms of current ripple, the former being rather uniform and two-dimensional in character, but the latter strongly three-dimensional and commonly very variable in shape and arrangement. However, these categories are

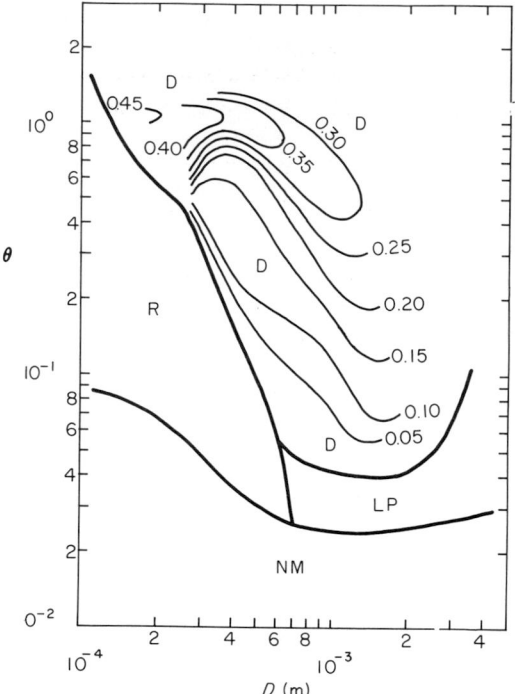

Fig. 2. Curves of equal inverse relative roughness (group mean dune height/mean flow depth) for equilibrium steady-state dunes produced experimentally using quartz sands. Data of Shinohara & Tsubaki (1959), Stein (1965), Guy *et al.* (1966), and Williams (1970). Field symbols as in Fig. 1.

merely end-member specifications with an arbitrary boundary. Allen (1969) and Banks & Collinson (1975) showed, from graded series of steady-state equilibrium experiments, that current ripples change continuously from the first to the second of Harms's types in response to increasing flow strength and decreasing relative roughness. The larger transverse bedforms could respond in a similar manner to changing flow conditions, in view of their similarity to current ripples in shape and movement processes.

Experiments on large bedforms in straight channels seem to support this view (Allen, 1982a). The work of Shinohara & Tsubaki (1959), Stein (1965) and Williams (1970) and, particularly, the detailed accounts by Guy *et al.* (1966) and by Costello & Southard (1981), point to the following sequence of dune forms (Fig. 2). The dunes which appear at the lowest boundary shear stresses are relatively flat, long-crested, and of small relative roughness. As the stress is raised, the features gradually increase to a maximum of relative roughness and three-dimension-ality, the crests becoming strongly curved and of considerable variability in height. Any further increase in stress results in dunes of declining dimensionality and relative roughness. As can be seen from the graph, an increase in grain size severely limits the maximum relative roughness and three-dimensionality attainable by dunes. Fredsøe (1975) and Yalin & Karahan (1979) find that in parallel the steepness of dunes passed through a maximum with increasing shear stress. These changes occur smoothly and continuously, giving no hint that more than one class of bedform is present in the single field assigned to dunes in Fig. 1. The more three-dimensional kind of dune therefore seems to arise only in fine-medium sands, and when conditions are well removed from the thresholds to the dune field.

BEDFORM LAG

Ideal and real sedimentary systems

The experimental bedforms yielding Fig. 1 were steady-state equilibrium features, that is, the statistical characteristics of each population remained constant in time and in deterministic adjustment to a single bed-material and a steady discharge. However, river currents change in direction and magnitude with space and with time, and these changes occur frequently and commonly rapidly. How do bedforms then respond? Since bedforms can change in size and/or shape only in consequence of sediment erosion, transport and deposition—all proceeding at a finite rate—they should show a delayed response to a change of flow. The delay or lag may be measured in terms of distance travelled and/or time.

The degree of lag will increase with both bedform and river size. Dunes and bars, amongst the largest of river bedforms, should lag the flow most, for their removal or substantial modification demands a comparably large total sediment transport. The amount of lag will also increase with river size. The mean velocity U, depth d and width w of rivers increase downstream with increasing discharge Q at respectively $Q^{0.1}$, $Q^{0.4}$ and $Q^{0.5}$ (Leopold, 1953; Leopold & Maddock, 1953). Now on a simple view, the bedload transport rate varies as U^3 and the streamwise cross-sectional area of a dune or bar roughly as d^2 or w^2. As rivers, irrespective of size, experience essentially the same seasonal controls on discharge, bars and dunes will respond to discharge changes relatively less in big than small streams, because bedform size increases more steeply with discharge than the sediment transport rate.

Bedform lag is particularly relevant to fluvial sedimentology and river engineering. It may be expected to distort the uniform preservation into the stratigraphic record of the evidence of sedimentary events, and weakens the sedimentologist's ability to give unique quantitative interpretations. Bedforms contribute significantly to channel roughness and therefore strongly influence stage-discharge relationships. Lag makes these relationships non-unique, so complicating the economical design of training works, take-off schemes and dredging programmes.

Case histories

Bedform lag was first demonstrated experimentally by Simons & Richardson (1962). They observed double-valued depth–discharge relationships on admitting time-dependent discharges to a sand-bedded flume, and in particular noted how the bedforms responded with a delay to the changing flow. A concept of lag was more implicit than explicit however, and it was some years before experiments were resumed (Bayazit, 1969; Jensen, 1973; Gee, 1975; Wijbenga & Klaasen, 1983).

Lag has been demonstrated from several rivers or their tidally influenced lower reaches. NEDECO (1959) reported dune lag from the Niger, the larger forms being related to recessive flows. Allen, Deresseguier & Klingebiel (1969) found that the height of intertidal dunes in the Gironde lagged the tidal range. A similar effect related to river stage was described by Stückrath (1969) from the Paraná. Further observations confirming dune lag were given by Pretious & Blench (1951) from the Fraser (see Allen, 1974), by Peters (1971) from the Congo (see Allen, 1982a), by Nasner (1974) from the Weser (see Allen, 1976a), by Jackson (1976) from the Wabash, and by Levy, Kjerfve & Getzen (1980) from the Congaree. These studies seem to confirm the suggestion that bedforms lag least in the smaller rivers.

Claims of bedform lag are convincing only when a rigorously defined bedform population has been surveyed quantitatively and relevant flow measurements made at a density in time appropriate to the time-scale of flow change. Nasner's (1974) data from four reaches of the Weser are unquestionably the most satisfactory so far published but are not without limitations. His dune population was defined one-dimensionally as those forms occurring on a fixed sounding line, and an interval of several weeks was permitted between surveys, the flow being characterized by the freshwater discharge. Nasner gave his

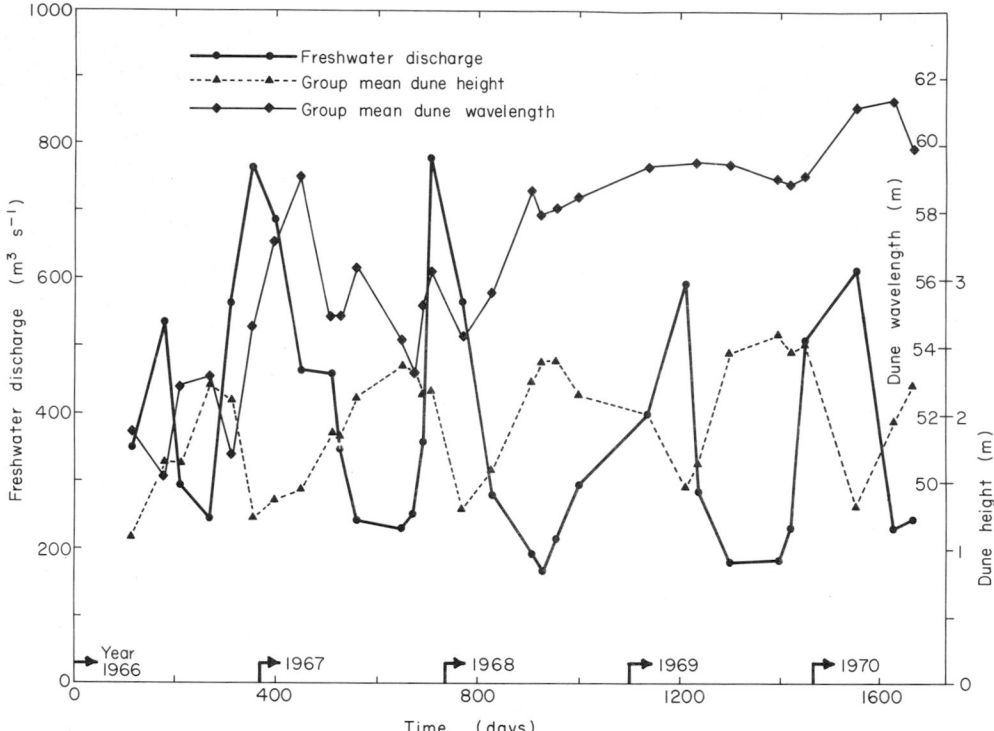

Fig. 3. Time-series for freshwater discharge, group mean dune height, and group mean dune wavelength in Reach 1 of the River Weser, between early 1966 and late 1970 (data of Nasner, 1974).

results in the form of time-series, such as appear in Fig. 3, representing Reach 1, 20 km downstream from Bremen. Evidently mean dune height lags the discharge more than mean dune wavelength. More instructive, however, are plots such as phase diagrams (Stückrath, 1969; Allen, 1974, 1976a), from which explicit time is eliminated. These emphasize the double-valued nature of the bedform-discharge relationships, and make it quite clear that dune characteristics and discharge cannot be specified uniquely the one from the other, without knowing how the system is changing.

Mechanisms of lag

Some analogies can be drawn (Allen, 1974, 1976b) between the behaviour of bedform and of biological populations (Maynard Smith, 1968; Solomon, 1969; Williamson, 1972), to which dynamical systems theory is so comprehensively applicable (Rosen, 1970; May, 1973, 1976).

Experience teaches that populations of bedforms are affected by a creation–destruction process driven by sediment transport, under the influence of which

new individual forms arise to replace older features (Allen, 1968, 1974; Jain & Kennedy, 1974). Therefore one way in which a population of (say) dunes may respond to changed flow conditions is through a change in its composition. Given ultimately deterministic controls on bedform characteristics, newly created individual forms must be better adjusted in shape and size to the flow conditions that prevailed when they were created than the older dunes accompanying them or destroyed to make way for them. The rate of population change by this mechanism is therefore inversely related to the dune persistence, expressible in two ways through

$$E = \frac{1}{L} \int^{T} V(t) . dt \qquad (1)$$

in which E is the non-dimensional distance travelled by a dune (excursion) during its lifespan T (interval between creation and destruction), L a characteristic dune wavelength, and $V(t)$ the dune celerity as a function of time t. The magnitude of the excursion may vary with the size and general vigour of the flow system, implicit in L and $V(t)$, the latter increasing with the sediment transport rate but varying inversely with dune height. Little is known of dune excursions. Wijbenga & Klaasen's (1983) laboratory experiments suggest a characteristic dune excursion of the order of one wavelength. Field studies point to dune excursions in the order of several to many wavelengths. Systematic research is now required to establish the magnitude of and the controls on excursions.

Although none but the simplest organisms have any power to vary their essential physical character in response to environmental change, it does seem likely that bedforms can to some extent respond as individuals to changing flow conditions. An individual dune or bar is probably most changeable in terms of its height H, for its wavelength and breadth are largely fixed by the behaviour of its neighbours within the statistical tessellation that the population represents when viewed in plan. Hence we can tentatively suggest for the population as a whole that

$$\partial \overline{H}/\partial t = f_1(k, \ \partial \overline{H}_\infty/\partial t, \ J/\overline{H}\gamma, \ R/\gamma), \qquad (2)$$

in which $\overline{H}$ is the mean height of the bedforms actually in the population, $\overline{H}_\infty$ the mean height if the changing flow was a series of steady-state equilibrium flows, J the dry-mass bedload transport rate, γ the sediment dry bulk density, and R the overall sediment deposition/erosion rate. The quantity k is the coefficient of height change (or changeability), a function of the kind of bedform and, possibly, of the flow system. The quantity $\partial \overline{H}_\infty/\partial t$ is the 'pressure' on the population to respond through individual change to the changing flow. The third term denotes the necessity for sediment transport and is directly proportional to the bedform celerity. The rate of change will grow with increasing transport rate, which measures the activity of the system, but fall with increasing bedform height, expressing the scale of the bedforms and hence their 'inertia'. The overall sediment deposition/erosion rate appearing as the fourth term may also be expected to influence the rate of change. Wijbenga & Klaasen's (1983) experiments suggests that individual change could be an important mechanism of population adjustment. However, as with Gee's (1975) work, the contribution to the total change from individual changeability was not separated from that arising from other mechanisms. Comprehensive field and laboratory investigations are again required.

Models for lag

A simple numerical model for dunes lagging in strongly varying periodic flows was developed by Allen (1976c, d, e, 1978a, b), who assumed that each of the above mechanisms was effective. Several empirical constants were introduced and simplifying assumptions made, and reliance was placed on the simpler of the empirical rules describing how dune dimensions vary with flow properties. Wijbenga & Klaasen's (1983) work suggests that the changeability may not be constant, and it is certainly undesirable to use a friction coefficient independent of discharge and bedform geometry. Yalin's (1964) early rules for dune geometry are oversimplified, though Stein's (1965) formula for height, used in one series of experiments (Allen, 1978b), is more realistic. In the model, dunes in a set of lineages are iterated away from arbitrary starting dimensions to yield loop-shaped phase diagrams (stable limit cycles) connecting mean wavelength and height with the periodic discharge. These loops (e.g. Allen, 1976c, 1978b) closely resemble corresponding plots of field data, and suggest that dune height can lag discharge much more than wavelength.

The model was designed to output the behaviour of individual dunes as well as that of the population as a whole, and thus affords insights into the origins of population behaviour and structure, defined by such properties as dune creation and destruction rates, dune survivorship, and the instantaneous frequency distribution of dune age, lifespan, height and wavelength. As a partial illustration, Fig. 4(B) shows from a typical experiment involving a population constant at a total of 36 dunes, how the wavelength distribution changes in response to discharge over one whole cycle of flow (Fig. 4A). The population is bimodal over much of the low-water period and the early part of the rise, as a consequence of the creation during this period of comparatively large numbers of short-wavelength dunes of short lifespan. The later rise, and particularly the recession, see the creation of dunes of intermediate and large wavelength, the distribution consequently changing to unimodal. Some of these forms have lifespans approaching the flow cycle in length. The particular dunes created during any one of the 72 steps of the flow cycle constitute a cohort, to use the language of the ecologist, and Fig. 4(C) shows the fate of all of these in the form of a survivorship graph. As is to be expected, Fig. 4 shows that the larger rates of dune creation are associated with: (1) the higher flow stages, when the rate of sediment transport is relatively large, and (2) the later part of the low-water period, the time when the large-wavelength dunes in this particular experiment reach the end of their lives.

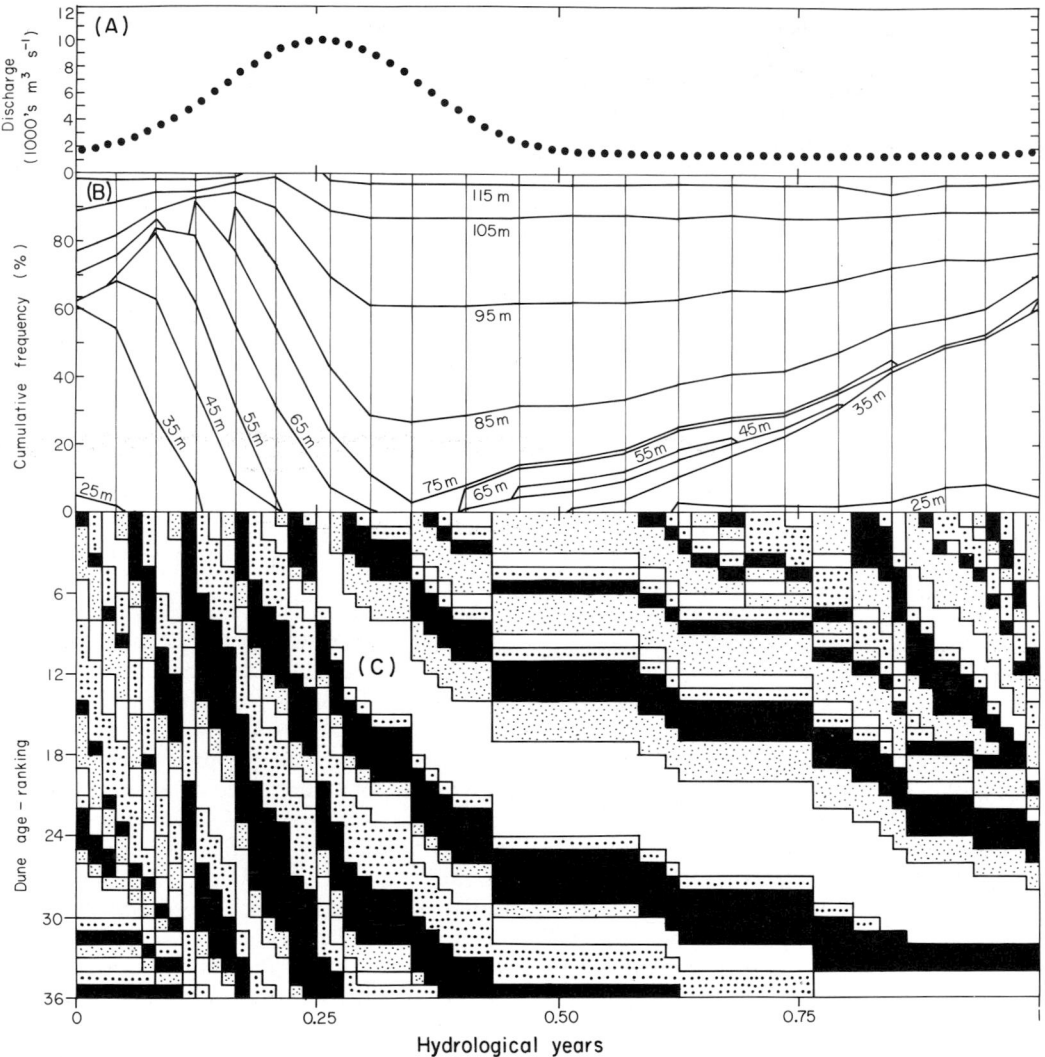

Fig. 4. Selected results from a numerical experiment on dune time-lag in a unidirectional flow (see Allen, 1976d). The behaviour of a population totalling 36 dunes at any instant is examined over a hydrological year discretized into 72 discharge events. The dunes have an assigned mean excursion of 18 wavelengths, equivalent to an actual mean lifespan of 15·7 flow increments. (A) Discharge pattern. (B) Variation in wavelength composition shown by cumulative frequencies sampled at intervals of approximately five discharge increments. (C) Survivorship amongst the dunes as shown by changes in age-ranking, increment by increment. The dunes created during any one flow increment constitute a cohort, a repeating cycle of four ornaments being sufficient to render the cohorts distinct from each other. The dunes which compose each cohort enter the population with age-rankings between 1 and n, where n is the total number of dunes in the cohort generated at that particular flow increment. The vertical height of the block beginning at the increment on the upper horizontal axis of the graph is proportional to the corhort size, n the block shifting downward and to the right as the cohort ages and dune desctruction reduces its size. The dune creation rate is proportional to the slope of the lines of blocks at the upper horizontal axis of the graph.

As the limit cycle generated in the numerical experiment was not exactly stable, and different individual flow cycles were analysed, there are small but insignificant discrepancies between (B) and (C) and between the start and finish of (C).

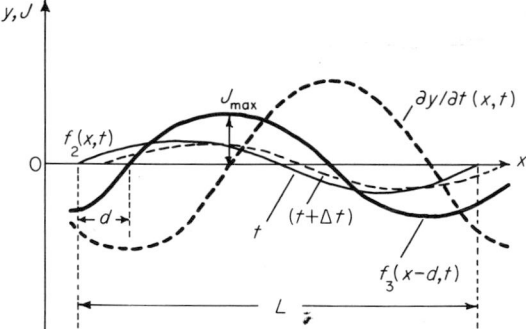

Fig. 5. Definition diagram for sediment transport and transfer at a wavy bed.

Fredsøe (1979, 1981) tackled dune lag analytically, developing a model for a stepwise change in discharge which was then extended through linearized equations to the case of weakly periodically varying flows. This scheme lacks the empirical features and assumptions of Allen's model, but supposes that the only mechanism of change is through the response of individual features. Fredsøe's model yields similar phase diagrams to those calculated by Allen and afforded by field data. Wijbenga & Klaasen (1983) find discrepancies between their experimental results and Fredsøe's model, and his restriction of the model to the one mechanism of lag seems unrealistic in view of the time-scales for change actually calculated. The degree of lag is probably overestimated by this model based on a single cause, but the approach, if it can be extended to include compositional change as a mechanism, is very promising.

STABILITY OF A GRANULAR BED BENEATH A SEDIMENT-LADEN STREAM

Mechanism of instability

Current ripples, dunes and antidunes manifest a bed waviness dominated by transverse elements. These bedforms imply the instability of a plane, cohesionless, granular surface over which there is sediment transport. Since Kennedy (1963) introduced the idea, it has generally been accepted that the instability mechanism involves the creation of a resultant streamwise lag-distance between small perturbations to the bed and the sediment transport over the bed. The resultant lag-distance itself is compounded algebraically from: (1) delays between a change of bed elevation and the response of the flow, and (2) delays

between the local flow properties and the local rate of sediment transport, in suspension and/or as bedload. The relative magnitude of the lag distance determines which kind of bedform will arise.

Figure 5 shows a steady, free-surface flow carrying sediment over a transversely wavy bed. The bed is described by

$$y = f_2(x, t) \tag{3}$$

and the sediment movement by

$$J = f_3(x - d, t) \tag{4}$$

in which y is the bed elevation relative to the undisturbed level, x the downstream distance, t the time, d the resultant lag distance, and J is now the deviation of the dry-mass bedload transport rate from the spatial average. The change in bed elevation consequent on the streamwise variation in transport rate is given by the steady-flow continuity equation

$$\frac{\partial y}{\partial t} = -\frac{1}{\gamma}\frac{\partial J}{\partial x}, \tag{5}$$

where γ is the sediment dry bulk density as before. When equations 3 and 4 are simple-harmonic, the deposition rate $\partial y/\partial t$ lags the transport rate by a fixed $L/4$, where L is the wavelength of the bed undulations (Fig. 5).

How does the resultant lag-distance d affect bed-wave development? Compare in Fig. 5 the bed at times t and $(t + \Delta t)$. For the particular resultant lag-distance chosen, putting J_{max} in the second quadrant on the bed wave as graphed in Fig. 5, the waves advance downstream at the celerity $c = dx/dt > 0$ while decreasing in amplitude a ($da/dt < 0$), until the bed eventually becomes flat. The chosen lag is therefore damping and excludes the permanent existence of a wavy bed. Table 1 lists the consequences of choosing other quadrants for J_{max}. Bed waves corresponding to current ripples and dunes arise only when J_{max} occurs in the first quadrant, and only when J_{max} lies in the fourth quadrant are growing antidunes possible.

Initial instability

Many workers exploited the above mechanism in mathematical studies of the initial instability of a slightly perturbed, planed, granular bed beneath a unidirectional sediment-laden stream. Attacks were made using models based on potential flow (Kennedy, 1963, 1964, 1969; Reynolds, 1965; Tsuchiya & Ishizaki, 1967; Falcon, 1969; Engelund & Fredsøe, 1970; Hayashi, 1970; Gradowczyk, 1971; Shirasuna,

Table 1. Effect of lag distance on stability of a wavy bed

Bed-wave quadrant including J_{max}	Bed-wave property	Interpretation
1st	$da/dt > 0,\ c > 0$	Amplifying current ripples or dunes
2nd	$da/dt < 0,\ c > 0$	Damping current ripples or dunes
3rd	$da/dt < 0,\ c < 0$	Damping antidunes
4th	$da/dt > 0,\ c < 0$	Amplifying antidunes

1973; Parker, 1975; Nakagawa & Tsujimoto, 1980, Eltayeb & Hassan, 1981), rotational flow models (Tsubaki & Saito, 1967; Engelund, 1970; Smith, 1970; Fredsøe, 1974a, b; Richards, 1980), and hydraulic models (Reynolds, 1965; Engelund & Hansen, 1966; Raudkivi, 1966; Gradowczyk, 1968). Excepting Eltayeb & Hassan (1981), who explored non-linear effects, the analyses were conducted on the basis of linearized equations. Most workers introduced an arbitrary lag-distance, and in only a few cases was an implicit lag obtained (Falcon, 1969; Engelund, 1970; Parker, 1975; Richards, 1980). The bed perturbations were assumed to be sinusoids of a single wavelength. In reality—and no real granular bed is mathematically plane—the defects arise from random variations in grain size and packing over the surface and/or through the dynamic activities of the current itself (e.g. viscous sublayer streak-bursting, grain-queuing).

These studies have been remarkably successful at the qualitative level but are less successful when compared in detail against observational data and when accurate predictions of the scale of quasi-equilibrium bedforms are required. These limitations derive from: (1) the simplified modelling of the flow and sediment transport, and (2) the restriction to initial instability, when non-linear effects can be discounted. The latter is arguably much less important to the sedimentologist or river engineer than the question of which factors determine the ultimate scale and shape of bedforms which have grown up to a state of quasi-equilibrium.

Stability of quasi-equilibrium bedforms

The considerations underlying Table 1 reveal that a bedform can be of a constant size and shape only if the resultant lag places J_{max} at particular positions, in the case of ripples and dunes at their actual or effective crests ($d = 0$), and in the case of antidunes in the troughs ($d = L/2$). Now the resultant lag-distance can be either zero or $L/2$ only when its component

parts, some denoting amplification and others damping, exactly cancel as regards the strength of their effect. Stated formally, we require

$$(da/dt)_1 + (da/dt)_2 + \ldots (da/dt)_n = 0, \qquad (6)$$

where the subscripts refer to the individual sources of lag. A possible maximum of six contributions—from bed shape, pressure gradients, bed-material entrainment, bedload-transport lag, suspension-transport lag, and the sediment-transport work-rate—are identifiable (Table 2), of which only two or three have ever figured in published analyses of initial instability. This is not of course a criticism but merely an indication of the difficulties of the problem and of where future research could, with profit, be directed. Note that Table 2 separates the effects due to the several possible components of the resultant lag, whereas Table 1 shows only the influence of the resultant lag no matter how compounded.

A contribution from *bed shape* was early appreciated (Benjamin, 1959; Kendall, 1970; Hsu & Kennedy, 1971; Richards & Taylor, 1981). In the flow over a sinusoidal bed, the mean boundary shear stress is a maximum just upstream of the crest, that is, assuming no other contributions to the lag-distance, J_{max} occurs on either the first (current ripples, dunes) or third (antidunes) quadrant of the bed wave as graphed in Fig. 5. If this factor operated alone, current ripples or dunes would therefore be amplified but antidunes would become damped (Table 2).

The contribution from streamwise *pressure gradients* reflects the influence of the alternate contractions and expansions in the flow over the bedform, and therefore, the changing combined local steepness of the bed and water surface. Compression implies a streamwise pressure decrease and therefore an enhancement of the boundary shear stress and sediment transport rate relative to the corresponding uniform flow. Conversely, expansion implies an adverse pressure gradient and a relative decrease of the stress and transport rate. Hence if the pressure-

Table 2. Influence of six factors inducing lag on the stability of ripples and dunes (unitalicized interpretations) and antidunes (italicized interpretations)

| | Bed-wave quadrant containing J_{max} | | | |
	1st	2nd	3rd	4th
Bed shape	Amplifying, downstream-moving		*Damping, downstream-moving*	
Pressure gradient	Amplifying, downstream-moving	*Damping, downstream-moving*	*Damping, downstream-moving*	Amplifying upstream-moving
Entrainment		Damping, downstream-moving *Damping, downstream-moving*	Damping, upstream-moving *Damping, upstream-moving*	
Sediment-transport work-rate		Damping, downstream-moving *Damping, downstream-moving*	Damping, upstream-moving *Damping, upstream-moving*	
Bedload-transport lag		Damping, downstream-moving		Amplifying upstream-moving
Suspension-transport lag	Amplifying, downstream-moving *Amplifying downstream-moving*	Damping, downstream-moving *Damping, downstream-moving*	Damping, upstream-moving *Damping, upstream-moving*	Amplifying, upstream-moving *Amplifying, upstream-moving*

gradient component acted alone, the effect on current ripples and dunes is likely to be amplifying, with J_{max} lying in the first or fourth quadrants, while for antidunes it is damping, with J_{max} restricted to the second or third quadrants (Table 2).

Bed-material *entrainment* contributes to the resultant lag through the finite steepness of real bedforms. Bagnold (1956) and Allen (1982b) amongst others pointed out that bed-material can be entrained in downslope flow at a smaller mean boundary shear stress (or mean flow velocity for constant bed friction) than in upslope flow. This is because in downslope flow the fluid drag and the tangential component of the particle weight act in the same sense, whereas beneath an upslope current the weight component opposes the drag. Hence if only this factor were operative, J_{max} would occur in the second or third quadrants, all bedforms being damped (Table 2).

The fourth source of lag is the variation of the *sediment-transport work-rate* with local bed steepness. Bagnold (1956) pointed out that a higher rate of sediment transport was possible in downslope than upslope flow, because the load did not have to be lifted up a gravity slope as well as maintained above the bed. Hence when this component acts alone, J_{max} lies in either the second or the third quadrants on the bed wave as graphed in Fig. 5, and the effect is damping for all bedforms.

The contribution from *bedload-transport lag* is the spatial delay between the mean boundary shear stress and the bedload transport rate. It will be proportional to grain size (Einstein, 1950) and to the mean distance travelled by bedload particles (one to several saltations may be involved) between episodes of lodgement on the bed. Since the length of an aqueous saltation is in the order of 10–100 times the grain diameter (Tsuchiya, 1969; Tsuchiya, Watado & Aoyama, 1969; Tsuchiya & Aoyama, 1970; Gordon, Carmichael & Isackson, 1972; Abbott & Francis, 1977), increasing with the boundary shear stress, the lag-distance should be small compared to both flow depth and bedform wavelength. Nakagawa & Tsujimoto (1980) calculated bedload-transport lags equivalent to J_{max} in the second quadrant (damping) on a rippled or duned bed, and in the fourth quadrant (amplifying) on a surface with antidune bedforms (Table 2).

The *suspension-transport lag* is the spatial delay between the mean boundary shear stress and the rate of suspension transport, and will be comparable with $h(W, U) . U/W$, where $h(W, U)$ is the height above the bed of the centre of gravity of the suspended load, W the mean falling velocity of the suspended material, and U the overall mean flow velocity. Now suspension transport is negligible unless the friction velocity $(f/8)^{\frac{1}{2}}U$, where f is the Darcy-Weisbach coefficient, is at least closely comparable with W (e.g. Bagnold, 1966; Middleton, 1977). Hence a significant influence from the suspension-transport lag makes the lag-distance at least of the same order as the flow depth.

Accordingly, no particular quadrant is singled out in Table 2, although with this component acting alone the likeliest sites for J_{max} are upstream quadrants on dune beds and downstream ones on antidune bedforms.

The six sources of lag are of three kinds. The first and second in Table 2 depend on how the flow is constrained by the forms of the bed and free surface. The third and fourth record the influence of bed steepness on sediment entrainment and transport rate, while the fifth and sixth relate to transport mode. A given effect may damp one kind of bedform but amplify another.

What combinations of lag effects seem likely to govern quasi-equilibrium river bedforms? For current ripples and dunes, the damping influence of entrainment, bedload transport, and sediment-transport work-rate must ultimately balance the amplifying effect of bed shape and pressure gradients, the uncertain contribution from the possibly unimportant suspension-transport lag being set aside. As regards antidune bedforms, the amplifying effect of the two transport modes must ultimately be neutralized by the damping supplied by the bed shape, pressure gradient, entrainment, and work-rate effects.

Can the scale of quasi-equilibrium bedforms be successfully predicted from the six effects? So far no attempt has been made for river bedforms, but there have been modest successes with tidal sand waves. Plausible dynamically stable forms result from balancing the pressure gradient against the entrainment and work-rate effects (Allen, 1982c, d). Statically stable sand waves apparently are due to the reshaping of dynamically stable ones in response to increasing water depth; a good match for static forms comes from applying just the entrainment effect (Allen, 1982b). Analysis on these lines has the advantage over a purely empirical approach—for example, the correlation of dune height with flow depth (e.g. Allen, 1968)—that a precise physical control is revealed. On the other hand, many of the above lag effects have, as yet, meagre foundations. For example, only Lysne (1969) has experimented on entrainment from sloping surfaces, and no investigator has tested Bagnold's (1956) views on transport over sloping beds. These questions demand more attention.

CONCLUDING DISCUSSION

Justifiable satisfaction can be drawn from the significant advances made in our understanding of river bedforms over the past 25 years. Some long-established modes of attack are yet to yield all their fruit, however, and several exciting new prospects are newly emerged.

It is undeniably important to secure a sound empirical knowledge of the general hydraulics of river bedforms under steady-state equilibrium conditions, before penetrating too deeply into the issues raised by bedforms in varying flows, their universal context in the real world. The core of that knowledge has been secured but we lack data for extreme grain sizes (silt and gravel), high sediment-transport stages, and dimensional scales which overlap significantly with prototype conditions. The contribution from equilibrium experiments is not yet complete, and the move in future will be towards apparatus permitting greater depths than seen hitherto.

Bedforms in varying flows provide an exciting challenge to field workers as well as to experimentalists and theoreticians. It seems vital that we quickly obtain many more well-documented case histories of bedform lag and its influence on the stratigraphic record, and at the same time extend our knowledge of lag mechanisms through quantitative experimental and theoretical studies. The behaviour of biological populations—a topic backed by a long history of quantitative empirical and theoretical research—has already provided useful analogies. Our late arrival at an interest in dynamical systems theory will not delay our advance, provided we have the imagination and tolerance to exploit what is helpful amongst the concepts and observations of other disciplines.

A fruitful approach to river bedforms has been to regard them as expressing states of stability or instability in the flow of a unidirectional sediment-driving fluid over a cohesionless granular bed. The initial instability of such a (plane) granular bed has been widely attributed to a spatial lag between perturbations of bed shape and the flow properties which control sediment transport. A strikingly different mechanism related to stationary recirculating mass-transport currents governs the initiation of ripple marks due to wind waves and, perhaps, sand waves shaped by oscillatory tidal currents (Allen, 1980). It is important to draw this contrast, because there is a tendency, particularly amongst field workers, to lump tidal and river bedforms together in empirical treatments of dubious validity. Perhaps of greater importance than the problem of initial instability, however, are the factors that govern the size and steepness of quasi-equilibrium bedforms substantially at the end of a path of height increase and wavelength change. At least six effects can be distinguished which may in combination control the neutral stability of matured river bedforms, but their quantitative significance remains largely unknown. This problem is a challenge to the ingenuity of applied mathematicians interested in problems of loose boundaries. It is also a challenge to experimentalists, for we are grossly ignorant of, for example, sediment entrainment and transport at sloping beds.

The fact that the solutions to the problems identified above will depend on skills of many kinds is the prime justification for our interdisciplinary symposium.

REFERENCES

ABBOTT, J.E. & FRANCIS, J.R.D. (1977) Saltation and suspension trajectories of solid grains in a water stream. *Phil. Trans. R. Soc.* A, **284**, 225–254.

ALAM, A.M.Z., CHEYER, T.F. & KENNEDY, J.F. (1966) Friction factors for flow in sand bed channels. *Res. Rep. Hydrodyn. Lab. Dep. civ. Engng, M.I.T.* **78**, 98 pp.

ALLEN, G.P., DERESSEGUIER, A. & KLINGEBIEL, A. (1969) Évolution des structures sédimentaires sur un banc sableux d'estuaire en fonction de l'amplitude des marées. *C. r. hebd. Séanc. Acad. Sci., Paris D* **269**, 2167–2169.

ALLEN, J.R.L. (1968) *Current Ripples*. North-Holland, Amsterdam. 433 pp.

ALLEN, J.R.L. (1969) On the geometry of current ripples in relation to stability of water flow. *Geogr. Annlr* A **51**, 61–96.

ALLEN, J.R.L. (1974) Reaction, relaxation and lag in natural sedimentary system: general principles, examples and lessons. *Earth Sci. Rev.* **10**, 263–342.

ALLEN, J.R.L. (1976a) Time-lag of dunes in unsteady flows: an analysis of Nasner's data from the R. Weser, German. *Sedim. Geol.* **15**, 309–321.

ALLEN, J.R.L. (1976b) Bedforms and unsteady processes: some concepts of classification and response illustrated by common one-way types. *Earth Surf. Proc.* **1**, 361–374.

ALLEN, J.R.L. (1976c) Computational models for dune time-lag: general ideas, difficulties and early results. *Sedim. Geol.* **15**, 1–53.

ALLEN, J.R.L. (1976d) Computational models for dune time-lag: population structures and the effects of discharge pattern and coefficient of change. *Sedim. Geol.* **16**, 99–130.

ALLEN, J.R.L. (1976e) Computational models for dune time-lag: an alternative boundary condition. *Sedim. Geol.* **16**, 255–279.

ALLEN, J.R.L. (1978a) Polymodal dune assemblages: an interpretation in terms of dune creation–destruction in periodic flows. *Sedim. Geol.* **20**, 17–28.

ALLEN, J.R.L. (1978b) Computational models for dune time-lag: calculations using Stein's rule for dune height. *Sedim. Geol.* **20**, 165–216.

ALLEN, J.R.L. (1980) Large transverse bedforms and the character of boundary-layers in shallow-water environments. *Sedimentology*, **27**, 317–323.

ALLEN, J.R.L. (1982a) *Sedimentary Structures*, vol. 1. Elsevier, Amsterdam. 593 pp.

ALLEN, J.R.L. (1982b) Simple models for the shape and symmetry of tidal sand waves: (1) statically-stable equilibrium forms. *Mar. Geol.* **48**, 31–49.

ALLEN, J.R.L. (1982c) Simple models for the shape and symmetry of tidal sand waves: (2) dynamically-stable symmetrical equilibrium forms. *Mar. Geol.* **48**, 51–73.

ALLEN, J.R.L. (1982d) Simple models for the shape and symmetry of tidal sand waves: (3) dynamically-stable asymmetrical forms without flow separation. *Mar. Geol.* (in press).

ALLEN, J.R.L. & LEEDER, M.R. (1980) Criteria for the instability of upper-stage plane beds. *Sedimentology*, **27**, 209–217.

BAGNOLD, R.A. (1956) The flow of cohesionless grains in fluids. *Phil. Trans. R. Soc.* A, **249**, 235–297.

BAGNOLD, R.A. (1966) An approach to the sediment transport problem from general physics. *Prof. Pap. U.S. geol. Surv.* **422-I**, 37 pp.

BAKER, V.R. (1973) Palaeohydrology and sedimentology of Lake Missoula flooding in eastern Washington. *Spec. Pap. geol. Soc. Am.* **144**, 79 pp.

BANKS, N.L. & COLLINSON, J.D. (1975) The size and shape of small-scale current ripples: an experimental study using medium sand. *Sedimentology*, **22**, 583–599.

BAYAZIT, M. (1969) Resistance to reversing flows over moveable beds. *J. Hydraul. Div. Am. Soc. civ. Engrs* **95**, HY 4, 1109–1127.

BENJAMIN, T.B. (1959) Shearing flow over a wavy boundary. *J. Fluid Mech.* **6**, 161–205.

BOOTHROYD, J.C. & HUBBARD, D.K. (1974) Bed form development and distribution pattern, Parker and Essex Estuaries, Massachusetts. *Misc. Pap. coastal Engng Res. Center U.S.* **1-74**, 39 pp.

CANT, D.J. (1978) Bedforms and bar types in the South Saskatchewan River. *J. sedim. Petrol.* **48**, 1321–1330.

COSTELLO, W.R. (1974) Development of bed configurations in coarse sands. *Dept Earth planet. Sci., MIT Rept No.* **74-1**, 120 pp.

COSTELLO, W.R. & SOUTHARD, J.B. (1981). Flume experiments on lower-flow-regime bed forms in coarse sands. *J. sedim. Petrol.* **51**, 849–864.

DALRYMPLE, R.W., KNIGHT, R.J. & LAMBIASE, J.J. (1978) Bedforms and their hydraulic stability relationships in a tidal environment, Bay of Fundy, Canada. *Nature*, **275**, 100–104.

DALRYMPLE, R.W., KNIGHT, R.J. & MIDDLETON, G.V. (1975) Intertidal sand bars in Cobequid Bay (Bay of Fundy). In: *Estuarine Research*, vol. 2 (Ed. by L.E. Cronin), pp. 295–307. Academic Press, New York.

ELTAYEB, I.A. & HASSAN, M.H.A. (1981) On the non-linear evolution of sand dunes. *Geophys. J. R. astr. Soc.* **65**, 31–45.

EINSTEIN, H.A. (1950) The bed-load function for sediment transportation in open channel flow. *Tech. Bull. U.S. Dep. Agric.* **1026**, 70 pp.

EINSTEIN, H.A. & BARBAROSSA, N.L. (1952) River channel roughness. *Trans. Am. Soc. civ. Engrs* **117**, 1121–1146.

ENGELUND, F. (1966) Hydraulic resistance of alluvial streams. *J. Hydraul. Div. Am. Soc. civ. Engrs* **92**, HY 2, 315–326.

ENGELUND, F. (1970) Instability of erodible beds. *J. Fluid Mech.* **42**, 225–244.

ENGELUND, F. & FREDSØE, J. (1970) Three-dimensional stability analysis of open channel flow over erodible bed. *Danish Centre appl. Math. Mech. Rept No.* **6**, 18 pp.

ENGELUND, F. & HANSEN, E. (1966) Investigation of flow in alluvial streams. *Acta polytech scand.* **35**, 100 pp.

FALCON, M. (1969) Theoretical description of free surface two-phase flow over a wavy bed. *Boll. lab. hidraul. Univ. Central Venezuela, Caracas*, **2**, 87–102.

FREDSØE, J. (1974a) The formation of sediment waves in closed channels. *Tech. Univ. Denmark, Inst. Hydro. Hydraul. Engng, Progr. Rep.* **32**, 29–36.

FREDSØE, J. (1974b) On the development of dunes in erodible channels. *J. Fluid Mech.* **64**, 1–16.

FREDSØE, J. (1975) The friction and height–length relation in flow over a dune-covered bed. *Tech. Univ. Denmark Inst. Hydro. Hydraul. Engng, Progr. Rep.* **37**, 31–36.

FREDSØE, J. (1979) Unsteady flow in straight alluvial channels: modification of individual dunes. *J. Fluid Mech.* **9**, 497–512.

FREDSØE, J. (1981) Unsteady flow in straight alluvial channels, Part 2. Transition from dunes to plane beds. *J. Fluid Mech.* **102**, 431–453.

GEE, D.M. (1975) Bed form response to nonsteady flows. *J. Hydraul. Div. Am. Soc. civ. Engrs* **101**, HY 3, 437–449.

GILBERT, G.K. (1914) The transportation of debris by running water. *Prof. Pap. U.S. geol. Surv.* **86**, 263 pp.

GORDON, R., CARMICHAEL, J.B. & ISACKSON, F.J. (1972) Saltation of plastic balls in a 'one-dimensional' flume. *Wat. Resour. Res.* **8**, 444–459.

GRADOWCZYK, M.H. (1968) Wave propagation and boundary instability in erodible-bed channels. *J. Fluid Mech.* **33**, 93–112.

GRADOWCZYK, M.H. (1971) Interfacial instability between fluids and granular beds. In: *Instability of Continuous Systems* (Ed. by H. Liepholz), pp. 143–150. Springer-Verlag, Berlin.

GUY, H.P., SIMONS, D.B. & RICHARDSON, E.V. (1966) Summary of alluvial channel data from flume experiments, 1956–61.

HARMS, J.C. (1969) Hydraulic significance of some sand ripples. *Bull. geol. Soc. Am.* **80**, 363–396.

HAYASHI, T. (1970) Formation of dunes and antidunes in open channels. *J. Hydraul. Div. Am. Soc. civ. Engrs* **96**, HY 2, 357–366.

HILL, H.M., SRINIVASAN, V.S. & UNNY, T.E. (1969) Instability of flat bed in alluvial channels. *J. Hydraul. Div. Am. Soc. civ. Engrs* **95**, HY 5, 1545–1558.

HSU, S.T. & KENNEDY, J.F. (1971) Turbulent flow in wavy pipes. *J. Fluid Mech.* **47**, 481–502.

JACKSON, R. G. (1976). Large-scale ripples of the lower Wabash River. *Sedimentology*, **23**, 593–623.

JAIN, S.C. & KENNEDY, J.F. (1974) The spectral evolution of sedimentary bed forms. *J. Fluid Mech.* **63**, 301–314.

JENSEN, P.D. (1973) Dune formation under non-steady conditions. *Proc. 15th int. Ass. Hydraul. Res. Congr.* **1**, 173–179.

JOPLING, A.V. & FORBES, D.L. (1979) Flume study of silt transportation and deposition. *Geogr. Annlr* A **61**, 67–85.

KALINSKE, A.A. & HSIA, C.H. (1945) Study of transportation of fine sediments by flowing water. *Bull. St. Univ. Iowa, Engng* **29**, 30 pp.

KENDALL, J.M. (1970) The turbulent boundary layer over a wall with progressive surface waves. *J. Fluid Mech.* **41**, 259–281.

KENNEDY, J.F. (1963) The mechanics of dunes and antidunes in erodible-bed channels. *J. Fluid Mech.* **16**, 521–544.

KENNEDY, J.F. (1964) The formation of sediment ripples in closed rectangular conduits and in the desert. *J. geophys. Res.* **69**, 1517–1524.

KENNEDY, J.F. (1969) The formation of sediment ripples, dunes and antidunes. *Ann. Rev. Fluid Mech.* **1**, 147–168.

KLEIN, G. DE V. (1970) Depositional and dispersal dynamics of intertidal sand bars. *J. sedim. Petrol.* **40**, 1095–1127.

LEEDER, M.R. (1980) On the stability of lower stage-plane beds and the absence of current ripples in coarse sands. *J. geol. Soc. London*, **137**, 423–429.

LEOPOLD, L.B. (1953) Downstream changes of velocity in rivers. *Am. J. Sci.* **251**, 606–624.

LEOPOLD, L.B. & MADDOCK, T. (1953) The hydraulic geometry of stream channels and some physiographic implications. *Prof. Pap. U.S. geol. Surv.* **252**, 511–537.

LEVEY, R.A., KJERFVE, B. & GETZEN, R.T. (1980) Comparison of bed form variance spectra within a meander bend during flood and average discharge. *J. sedim. Petrol.* **50**, 149–155.

LYSNE, D.K. (1969) Movement of sand in tunnels. *J. Hydraul. Div. Am. Soc. civ. Engrs* **95**, HY 6, 1835–1846.

MANTZ, P.A. (1978) Bedforms produced by fine, cohesionless, granular and flaky sediments under subcritical water flows. *Sedimentology*, **25**, 83–103.

MANTZ, P.A. (1980) Laboratory flume experiments on the transport of cohesionless silica silts by water streams. *Proc. Instn civ. Engrs* **69**, 2, 977–994.

MAY, R.M. (1973) *Stability and Complexity in Model Ecosystems*. Princeton University Press. 235 pp.

MAY, R.M. (1976) Simple mathematical models with very complicated dynamics. *Nature*, **261**, 459–467.

MAYNARD SMITH, J. (1968) *Mathematical Ideas in Biology*. Cambridge University Press. 152 pp.

MIDDLETON, G.V. (1977) Hydraulic interpretation of sand size distributions. *J. Geol.* **84**, 405–426.

NAKAGAWA, H. & TSUJIMOTO, T. (1980) Sand bed instability due to bed load motion. *J. Hydraul. Div. Am. Soc. civ. Engrs* **106**, HY 12, 2029–2051.

NASNER, H. (1974) Über das Verhalten von Transportkörpern im Tiedegebiet. *Mitt. Franzius-Inst.* **40**, 1–149.

NEDECO (Netherlands Engineering Consultants) (1959) *River Studies and Recommendations on Improvement of Niger and Benue*. North-Holland, Amsterdam, 1000 pp.

OWENS, J.S. (1908) Experiments on the transporting power of sea currents. *Geogrl J.* **31**, 415–425.

PARKER, W.R. (1975) Sediment inertia as a cause of river antidunes. *J. Hydraul. Div. Am. Soc. civ. Engrs* **101** HY 2, 211–221.

PETERS, J.J. (1971) *La Dynamique de la sédimentation de la région divagante du bief maritime du fleuve Congo*. Laboratoire de Recherches Hydrauliques à Borgerhout, Belgium. 120 pp.

PRETIOUS, E.S. & BLENCH, T. (1951) *Final Report on Special Observations in Lower Fraser River at Ladner Reach during 1950 Freshet*. Report National Research Council of Canada, Fraser River Model, Vancouver, Canada. 12 pp.

RAUDKIVI, A.J. (1966) Bed forms in alluvial channels. *J. Fluid Mech.* **26**, 507–514.

REES, A.I. (1966) Some flume experiments with a fine silt. *Sedimentology*, **6**, 209–240.

REYNOLDS, A.J. (1965) Waves on the erodible bed of an open channel. *J. Fluid Mech.* **22**, 113–133.

RICHARDS, K.J. (1980) The formation of ripples and dunes on an erodible bed. *J. Fluid Mech.* **99**, 597–618.

RICHARDS, K.J. & TAYLOR, P.A. (1981) A numerical model of flow over sand waves in water of finite depth. *Geophys. J. R. astr. Soc.* **65**, 103–128.

ROSEN, R. (1970) *Dynamical System Theory in Biology.* Wiley, New York. 302 pp.

SAUNDERSON, H.C. & LOCKETT, F.P.J. (1983) Flume experiments on bedforms and structures at the dune-plane bed transition. In: *Modern and Ancient Fluvial Systems* (Ed. by J. D. Collinson and J. Lewin). *Spec. Publs int. Ass. Sediment,* **6**, 49–58. Blackwell Scientific Publications. Oxford.

SHELDON, P.G. (1928) Some sedimentation conditions in middle Portage rocks. *Am. J. Sci.* **15**, (5), 243–252.

SHELDON, P.G. (1928) On the derivation of the Portage sandstones of central New York. *Am. J. Sci.* **17**, (5), 525–533.

SHIELDS, A. (1936) Anwendung der Ähnlichkeitsmechanik unter der Turbulenzforschung auf die Geschiebewegung. *Mitt. Preuss. Versanst. Wesserb. Schiffb.* **26**, 26 pp.

SHINOHARA, K. & TSUBAKI, T. (1959) On the characteristics of sand waves formed upon the beds of the open channels. *Rep. Res. Inst. appl. Mech. Kyushu Univ.* **7** (25), 15–45.

SHIRASUNA, T. (1973) Formation of sand waves. *Proc. 15th Congr. int. Ass. Hydraul. Res.* **1**, 107–114.

SIMONS, D.B. & RICHARDSON, E.V. (1962) The effect of bed roughness on depth-discharge relations in alluvial channels. *Wat-Supply Irrig. Pap., Wash.* **1498-E**, 26 pp.

SIMONS, D.B., RICHARDSON, E.V. & ALBERTSON, M.C. (1961). Flume experiments using medium sand. *Wat.-Supply Irrig. Pap., Wash.* **1498-A**, 76 pp.

SIMONS, D.B., RICHARDSON, E.V. & NORDIN, C.F. (1965) Sedimentary structures generated by flow in alluvial channels. In: *Primary Sedimentary Structures and their Hydrodynamic Interpretation* (Ed. by G.V. Middleton), *Spec. Publs Soc. econ. Paleont. Miner.,* Tulsa, **12**, 34–52.

SMITH, J.D. (1970) Stability of sand beds subjected to a shear flow of low Froude number. *J. geophys. Res.* **75**, 5928–5940.

SMITH, N.D. (1971) Transverse bars and braiding in the Lower Platte River, Nebraska. *Bull. geol. Soc. Am.* **82**, 3407–3420.

SOLOMON, M.E. (1969) *Population Dynamics.* Arnold, London. 60 pp.

SORBY, H.C. (1859) On the structures produced by the currents present during the deposition of stratified rocks. *Geologist,* **2**, 137–147.

SORBY, H.C. (1908) On the application of quantitative methods to the study of the structure and history of rocks. *Q. Jl geol. Soc. Lond.* **64**, 171–233.

SOUTHARD, J.B. (1971) Representation of bed configurations in depth–velocity–size diagrams. *J. sedim. Petrol.* **41**, 903–915.

SOUTHARD, J.B. & BOGUCHWAL, L.A. (1973) Flume experiments on the transition from ripples to lower flat bed with increasing sand size. *J. sedim. Petrol.* **43**, 1114–1121.

STEIN, R.A. (1965) Laboratory study of total load and apparent bed load. *J. geophys. Res.* **70**, 1831–1842.

STÜCKRATH, T. (1969) Die Bewegung von Grossrippeln an der Sohle des Rio Paraná. *Mitt. Franzius-Inst.* **32**, 267–293.

SUNDBORG, Å. (1956) The River Klarälven: a study of fluvial processes. *Geogr. Annlr* **38**, 127–316.

TSUBAKI, T. & SAITO, T. (1967) Regime criteria for sand waves in erodible-bed channels. *Ann. Fac. Engng Kyushu Univ.* **40**, 741–748.

TSUHIYA, A. & ISHIZAKI, K. (1967) The mechanics of dune formation in erodible-bed channels. *Proc. 12th int. Ass. Hydraul. Res. Congr.* **1**, 479–486.

TSUCHIYA, Y. (1969) On the mechanics of saltation of a spherical sand particle in a turbulent stream. *Proc. 13th int. Ass. Hydraul. Res. Congr.* **2**, 191–198.

TSUCHUIYA, Y. & AOYAMA, T. (1970) On the mechanism of saltation of a sand particle in a turbulent stream (2). On a theory of successive saltations. *Ann. Disas. Prev. Res. Inst. Kyoto Univ.* **13 B**, 199–216.

TSUCHIYA, Y., WATADO, K. & AOYAMA, T. (1969) On the mechanism of saltation of a sand particle in a turbulent stream (1). *Ann. Disas. Prev. Res. Inst. Kyoto Univ.* **12 B**, 475–490.

VANONI, V.A. & BROOKS, N.H. (1957) Laboratory studies of the roughness and suspended load of alluvial streams. *U.S. Army Corps Engineers, Missouri River Division, Sediment Series,* **11**, 121 pp.

VANONI, V.A. & HWANG, L.S. (1967) Relation between bedforms and friction in streams. *J. Hydraul. Div. Am. Soc. civ. Engrs* **93** HY 3, 121–144.

WIJBENGA, J.H.A. & KLAASEN, G.J. (1983) Changes in bedform dimensions under unsteady flow conditions in a straight flume. In: *Modern and Ancient Fluvial Systems* (Ed. by J.D. Collinson and J. Lewin). *Spec. Publs int. Ass. Sediment.* **6**, 35–48. Blackwell Scientific Publications, Oxford.

WILLIAMS, G.P. (1970) Flume width and water depth effects in some sediment-transport experiments. *Prof. Pap. U.S. geol. Surv.* **562-H**, 37 pp.

WILLIAMSON, M. (1972) *The Analysis of Biological Populations.* Arnold, London. 180 pp.

YALIN, M.S. (1964) Geometrical properties of sand waves. *J. Hydraul. Div. Am. Soc. civ. Engrs* **90**, HY 5, 105–119.

YALIN, M.S. & KARAHAN, E. (1979) Steepness of sedimentary dunes. *J. Hydraul. Div. Am. Soc. civ. Engrs* **105**, HY 4, 381–392.

Spec. Publs int. Ass. Sediment. (1983) **6**, 35–48

Changes in bedform dimensions under unsteady flow conditions in a straight flume

J. H. A. WIJBENGA* *and* G. J. KLAASSEN†

* *Project Engineer and* † *Project Advisor, Delft Hydraulics Laboratory, Rivers and Navigation Branch, P.O. Box* 152, *Emmeloord, The Netherlands*

ABSTRACT

Flume tests are being carried out at the Delft Hydraulics Laboratory to study the changes in bedform dimensions and resistance to flow for unsteady flow conditions. The tests are carried out in a straight flume with uniform bed material ($D_m = 0.77$ mm). Results are presented for a sudden increase or decrease of the discharge. A comparison is made with the experiments by Gee (1973) and the theoretical work by Allen (1976a and subsequent articles) and Fredsøe (1979). It is concluded that a linear first-order differential equation with constant coefficients for the change in bedform dimensions does not fit the measured change in bedform dimensions. The coefficient of adaptation appears to be higher for decreasing discharges than for increasing discharges. The dune excursion as used in Allen's (1976a) computational model is in the order of 1–2, slightly increasing for increasing water depth. Furthermore, it is concluded that Fredsøe's (1979) method does not correctly simulate the phenomena observed during the tests in the flume with a large change in discharge. To predict the resistance to flow of the minor bed of the Rhine branches in the Netherlands during extreme flood conditions an adaptation of both the computational method of Allen (1976a, b, c, 1978) and Fredsøe (1979) is needed.

INTRODUCTION

Until some decades ago, the height of the main levées along the large rivers in the Netherlands was based on the highest recorded stage. Nowadays, the design height of these levées is determined by means of a frequency analysis of a series of historical flood discharges. A discharge with a probability in excess of only 8% in 100 years has recently been adapted as the design basis.

To determine the water levels along the main rivers during the passage of this decisive flood, (two-dimensional) computations are carried out by the Dutch Water Control and Public Works Department (Jansen *et al.*, 1979; Ogink, 1981). The computed stages are governed by the resistance to flow of both the flood plain and the minor bed. The roughness of the flood plain is made up by the roughness of grasslands, (fruit) trees, hedges and fences. The resistance to flow of the alluvial minor bed consists of

the grain roughness, bedform roughness and additional roughness due to structures such as bridge piers and groynes. It has been demonstrated (Vreugdenhil & Wijbenga, 1982) that the lateral diffusion of momentum also plays an important role. For a correct reproduction of the flow field all roughness components have to be estimated correctly.

The determination of the roughness of the minor bed poses the largest problem. Several methods have been proposed (Einstein & Barbarossa, 1952; Engelund & Hansen, 1967; White, Paris & Bettess, 1980). From White *et al.* (1980) it may be concluded that even under steady-flow conditions the prediction of the water depth by these methods has only a limited accuracy.

Under unsteady-flow conditions the predicted value may differ even more from the actual level, due to the time-lag with which the dimensions of the bedforms change (Allen, 1976a; Nasner, 1978). A fairly accurate prediction of the resistance to flow for varying flow (as

0141-3600/83/0106-0035 $02.00

occurs during the passage of a flood wave) can therefore only be obtained if this time-lag is explicitly taken into account.

Unfortunately the knowledge of the changes in bedform dimensions during varying flow conditions is very limited. To deepen the insight into these phenomena flume tests are being carried out at the Delft Hydraulics Laboratory. In these tests the changes in both bedform dimensions and resistance to flow are studied. During the experiments the conditions are such that only dunes with some superimposed ripples are present in the sand flume.

This present paper reports on these investigations which are carried out, within the framework of a basic research programme, on the sediment transport and resistance to flow of rivers, in close co-operation between the Delft Hydraulics Laboratory and the Dutch Water Control and Public Works Department. A review of related experimental and theoretical work is given. Furthermore a description of the tests and some preliminary results are presented. The results enable a verification of some existing theories. The continuation of the flume tests and the related field measurements are briefly indicated.

REVIEW OF EARLIER STUDIES

Theoretical studies

Changes in bedform dimensions can be determined by means of mathematical models if the water movement over a dune-covered river bed and the local sediment transport along dunes, for varying flow conditions, can be determined accurately enough. The description of the water movement should explicitly take into account the occurrence of recirculation downstream of the dune crest and the extra turbulence generated in the shear layer between the main flow and this separation zone. Promising results on this subject have recently been reported by Klaassen (1978) and Sündermann, Vollmers & Puls (1980), but at present (1981) no operational models for detailed simulation of changes in dune dimensions are available that can be used for varying flow.

A less detailed approach has been presented by Fredsøe (1979, 1981). For the form roughness especially, the dune height is of importance (Engelund, 1978). Fredsøe concentrated on the prediction of the dune height changes. He derived a relation for the initial change in bed elevation at the top of the dune based on the local change in flow conditions, a change

in sediment transport, the continuity equation for sediment and the total derivative for the bed elevation. The derived expression reads as follows:

$$\frac{dH}{dt} = \frac{\sqrt{\Delta g D_{50}^3}}{(1-\epsilon)L} \left\{ 1 - \frac{(\Theta' F_2)}{(\Theta' F)_1} \right\} \phi_2, \qquad (1)$$

in which

$$F = \frac{1}{\phi} \frac{d\phi}{d\Theta'}, \qquad (2)$$

where g = acceleration of gravity (m sec^{-2}), H = dune height (m), D_{50} = characteristic grain diameter (m), ϵ = porosity ($-$), L = dune length (m), Θ = local value of the dimensionless shear stress ($-$) acting as a skin friction on the surface of the dune defined as $\Theta' = h'i/\Delta D_{50}$, h = water depth (m), i = energy slope ($-$), $d\Theta'$ = change in dimensionless shear stress due to the change in flow ($-$), ϕ = dimensionless sediment transport ($-$) defined as $\phi = s/\sqrt{\Delta g D_{50}^3}$, s = sediment transport per unit width (m^2 sec^{-1}), Δ = relative density of sediment under water ($-$) defined as $\Delta = (\rho_s - \rho)/\rho$, ρ = specific density of the water (kg m^{-3}), ρ_s = specific density of the sediment (kg m^{-3}), t = time (sec).

The following remarks should be made regarding Fredsøe's method.

(1) No expression is presented for the change in dune length.

(2) The transport relation is the one proposed by Engelund & Fredsøe (1976). Fredsøe claims that other transport formulae may also be used. However, from equations (1) and (2) it can be concluded that no changes in dune height would occur if the dimensionless sediment transport is defined as a power function of the dimensionless shear stress.

A totally different approach to the description of changes in bedform dimensions with varying flow conditions has been followed by Allen (1976a, and subsequent publications). This computational model assumes the creation/destruction of dunes to be a stochastic process and is furthermore based on the inability of dunes to respond perfectly to changes in flow conditions. For the flow conditions in the computational model it is assumed that they can be considered as quasi-steady, with a Darcy–Weisbach coefficient which is constant in place and time.

After a dune has travelled a certain assigned excursion it is destroyed and a new one, adjusted to the instantaneous flow conditions, is created. At the moment of creation the bedform dimensions correspond to the instantaneous flow conditions as if steady flow exists. The dimensions of the bedform at the moment of creation are random, with mean dimensions which are a function of the flow conditions and standard deviations to be specified. During the life of the dunes only the dune height is (partially) adapted to changes in flow conditions, in accordance with a linear first-order differential equation:

$$\frac{dH(t)}{dt} = \frac{Ac_b}{H(t)}(H_\infty - H(t)), \qquad (3)$$

where A = coefficient of change $(-)$, $H(t)$, H_∞ = dune height at time t and for $T \to \infty$, respectively, for the flow conditions considered (m), c_b = bedform celerity (m sec^{-1}).

With respect to Allen's computational model the following remarks have to be made.

(1) Several constants, such as the dune excursion C and the coefficient of change A, have been introduced. The influence of some of them on the time-lag of the dunes has been investigated in Allen's subsequent publications, but their values in flume tests or prototype have not been given.

(2) A number of relations have to be introduced in the computational model, viz. a method to predict the bedform dimensions under steady flow conditions (Yalin, 1964, or Stein, 1965) and the bedform celerity as a function of flow velocity. The results of the computational model are highly dependent on the applicability of the formulae used.

(3) If Allen's method is to be applied to predict the resistance to flow of the minor bed of rivers during the passage of flood waves, the assumption of a constant Darcy–Weisbach coefficient should be dropped. Instead this coefficient should be related to the bedform dimensions. According to a recent verification carried out at the Delft Hydraulics Laboratory, reasonably reliable results are obtained with the relationship suggested by Vanoni & Hwang (1967) and Engelund (1978).

Experimental studies

Simons, Richardson and Haushild (1962) carried out flume tests with a flood-wave type of discharge variation. Only qualitative conclusions are presented with respect to the occurrence of loop-type depth-averaged relations. Jensen (1969) measured bedform dimensions during triangular discharge variations in a scale model of the River Oubangue (Africa). He observed a time-lag in the bedform dimensions. The shorter the period of the discharge variation, the longer the time-lag that was recorded. Jensen's results are too limited to allow any generalization.

Gee (1973) carried out flume tests to study the response of bed roughness to changes in discharge. Given a rough bed composed of dunes the discharge was suddenly increased so as to obtain a flat bed in the following equilibrium situation (high Froude Numbers) and vice versa. During the transition between the two equilibrium conditions, the bedform dimensions were not recorded.

From the limited amount of prototype data available only those for the Rivers Weser and Niger and the Rio Paraña are mentioned, which were analysed by Nasner (1978). Measurements of the dune height changes during the hydrograph yielded contradictory results. For tidal conditions in the River Weser a decreasing dune height was measured with increasing discharge, while for the unidirectional conditions in the River Niger the dunes were smaller during the rising stage than during the falling stage, which implies an increase in dune height with increasing discharge.

Starting from an initially flat bed, Bishop (1977) carried out flume tests to study the growth of dunes. The developing time of dunes was determined by measuring the dune length as a function of time. Only very limited information on the initial growth in dune height was presented. For each bed material used (sand, $D = 1·1$ mm; sand, $D = 0·54$ mm and bakelite, $D = 1·0$ mm) a different growth curve was found.

It may be concluded from the earlier theoretical and experimental studies reviewed above that it is not yet possible to predict the changes in dune dimensions (length and height) by means of a mathematical model of the detailed flow over a dune-covered river bed. Promising methods have been proposed by Fredsøe (1979, 1981) and Allen (1976a, b, c, 1978), but no sufficiently detailed and accurate measurements of bedform changes are available that have been carried out under controlled conditions so as to enable testing of these methods.

FLUME EXPERIMENTS

To deepen the insight into the phenomenon of increase and decrease of dune dimensions under varying flow conditions, the Delft Hydraulics Laboratory is carrying out a series of experiments in a flume with a movable bed. The experimental facility, the testing procedure, the tests carried out and some results will be described below.

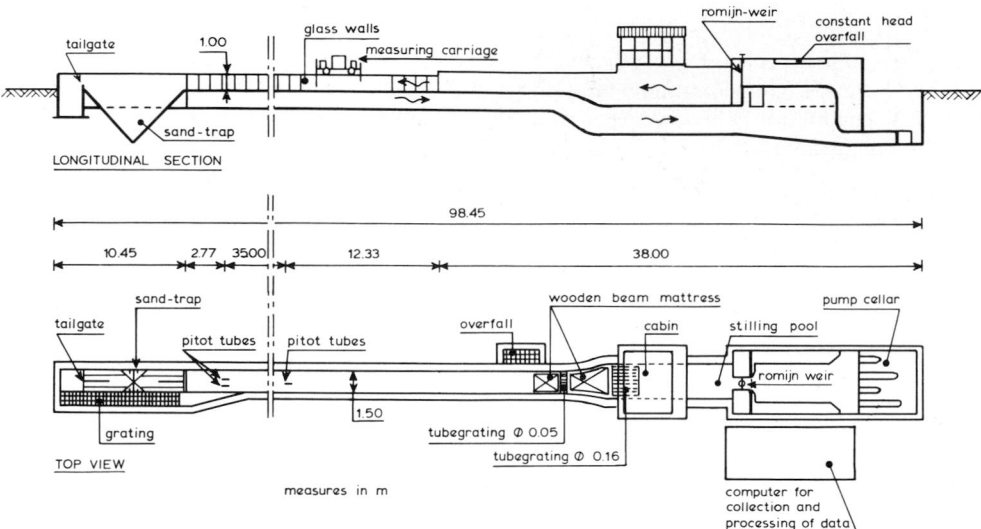

Fig. 1. Longitudinal section and top view of sand flume at the Delft Hydraulics Laboratory.

Experimental facility

The tests are carried out in a sand flume built especially for fundamental research into sediment transport and resistance to flow (Klaassen, 1978). The sand flume has been constructed in reinforced concrete, with a measuring section consisting of a steel frame with glass windows. The main dimensions of the flume are (see also Figs 1 and 2): overall length 98 m, length of section with glass windows 50 m, measuring length for bed level records 30 m, width of the measuring section 0·30 and 1·50 m, maximum water depth without sediment 1·00 m.

Various control and measuring devices have been installed in and around the flume. The discharge is regulated by means of a Romijn weir. The water discharge may be varied from 0·02 to 0·80 m³ sec⁻¹. In the return flume, located beneath the sand flume, a Rehbock measuring device has been installed to enable a constant control on the discharge. In the return flume a heating and cooling system has been installed to maintain a constant water temperature. Sediment can be supplied by means of a hydrocyclone to a maximum of 800 kg h⁻¹ submerged weight ($\simeq 0.8$ m³ h⁻¹). The amount of sediment which is supplied upstream and caught downstream of the flume is weighed under water by means of a hydrocyclone.

For measuring the energy slope, two sets of pitot tubes have been installed. One of these sets was installed recently for a slope-control system with which the energy slope can be kept constant through continuous adjustment of the tail gate by a feedback system. When the flume is operated with the slope control system, the water depth and the sediment transport become dependent variables. In the other case, the tail gate is set to a certain position and the sediment transport is imposed, the water depth and the energy slope thus becoming the dependent variables (Van Rijn & Klaassen, 1981).

A minicomputer has been installed for the acquisition and partial processing of data. Recently a microprocessor has been added. The minicomputer together with the microprocessor enable automatic acquisition of data and automatic imposition of changing boundary conditions (such as the discharge) as a function of time. Over the glass window section rails have been mounted for an instrument carriage. On the carriage three profile indicators and a water level indicator are installed. Thus three longitudinal bed level profiles are measured, usually one in the middle of the flume and two at $\frac{1}{6}$ of the flume width from the walls. Initially the recorded data collected by the minicomputer are stored in a disc memory, then a number of simple calculations are made to check the operation of the instruments and the progress of the test. Finally the data stored on disc are transferred to magnetic tape to be used for more complicated calculations at a later stage.

Fig. 2. Sand flume in Delft Hydraulics Laboratory.

Testing procedure

For the first series of tests a sudden increase or decrease of the discharge in the flume was selected. The energy slope was kept constant in order to simulate the conditions in the field as accurately as possible. For the discharge variation a sudden increase or decrease was preferred for the following reasons: (1) to enable the verification of the method of Fredsøe (1979, 1981); (2) to check whether the adaptation of the dunes develops according to a first-order system (equation 3); (3) Gee (1973) and Bishop (1977) also carried out their measurements for a discontinuous change in discharge.

The tests were carried out as follows. First the discharge and sediment supply were kept constant until equilibrium conditions (defined via sediment supplied upstream = measured sediment transport downstream) were reached. Once this equilibrium had been reached, the water depth, energy slope, bed slope, bed level and bedform celerity were measured. Then the discharge and sediment supply were suddenly changed in such a way that the energy slope during the

following equilibrium corresponded with the initial energy slope. The tail gate was set to a new position to avoid excessive backwater effects. This procedure had to be accepted because the slope control method had not yet been installed at the time the tests described in this paper were carried out.

The phase between the two equilibrium stages is called the transition. During the transition and during

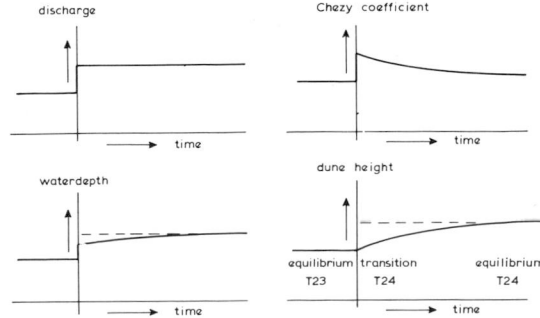

Fig. 3. Expected changes in Chézy coefficient, water depth and dune height after sudden increase in discharge (energy slope constant).

the following equilibrium the flow and bedform characteristics as mentioned above were measured. In Fig. 3 the expected changes in the resistance to flow (expressed in terms of the Chézy coefficient), water depth and dune height during the transition period after a sudden increase in discharge are indicated. For a sudden decrease in discharge the opposite behaviour may be expected.

Performed tests

The tests described in the present paper were all carried out for a flume width of 1·50 m. The water temperature was maintained at 18 °C. The energy slope was kept constant at a value of about $1·6 \times 10^{-3}$. The characteristic grain sizes of the applied uniform bed material are indicated hereafter: $D_{10} = 0·70$ mm, $D_{35} = 0·75$ mm, $D_{50} = 0·78$ mm, $D_{65} = 0·80$ mm, $D_{90} = 0·84$ mm, $D_m = 0·77$ mm.

A number of transitions were studied. A schematic overview is given in Fig. 4. A transition is preceded by the equilibrium stage of the previous test and followed by the equilibrium stage of the test under consideration. An example is given in Fig. 3 (sequence

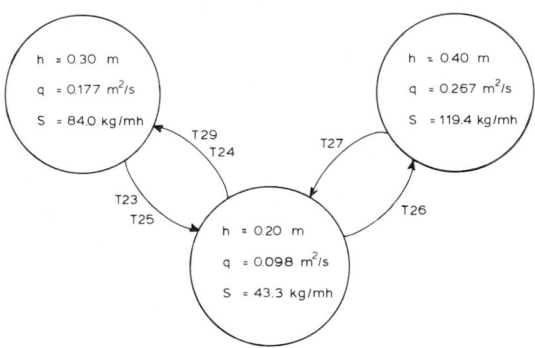

Fig. 4. Diagram of tests with sudden changes of discharge with constant slope of $1·6 \times 10^{-3}$ (sediment transport in kg $m^{-1} h^{-1}$ submerged weight).

equilibrium T23—transition T24—equilibrium T24). For administrative reasons the designation T28 was not used. A summary of the performed tests is presented in Table 1.

Measurements and data processing

The hydraulic and bedform characteristics were measured during the initial equilibrium phase, during the transition and during the equilibrium phase following the transition. Each measurement consisted of the following observations: the energy slope, by means of one pair of pitot tubes located 35 m apart, the discharge, the bed level in three longitudinal profiles at every centimetre of the 30 m long measuring section, the water level in the middle profile at every centimetre of the measuring section.

During the equilibrium phases about 20 of such measurements were carried out, at intervals sufficiently large to guarantee statistical independence. During the transitions the time interval between two measurements was about 6 min, which is the minimum that can be realized with the present data collection system.

After the operation of the flume had been checked, the data were stored on magnetic tape and processed on a larger computer system. This processing included the determination of the water-surface slope and the slopes of the recorded bed levels by means of a least-squares method. Bedform dimensions were determined on the basis of the zero-level crossings, and bed roughness was determined, applying a wall correction as proposed by Einstein (see Wijbenga, 1978). Details of the data processing are given in Bogirski (1977).

The obtained data were used not only to make a comparison with Allen (1976a) and Fredsøe (1979) but also to check whether the initial change in dune height can be considered as a linear first-order system, with constant coefficient, which can be written as:

$$\frac{dH(t)}{dt} = A_H(H_\infty - H(t)) \qquad (4)$$

Table 1. Transitions and equilibrium phases preceding and following the transitions

Transition	Equilibrium phase preceding transition	Equilibrium phase following transition
T23	T22	T23
T24	T23	T24
T25	T24	T25
T26	T25	T26
T27	T26	T27
T29	T27	T29

where A_H = coefficient of adaptation for the dune height (sec^{-1}).

Note that there is a relation between this coefficient of adaptation for the dune height and the coefficient of change according to Allen ($A_H = Ac_b/H$). To be able to determine the coefficient of adaptation A_H for the dune height immediately after the change of discharge, it is necessary to know the dune height during equilibrium conditions and the initial change in dune height. The equilibrium dune height as a function of the water depth is determined from a regression analysis of a limited number (15) of the flume tests (Fig. 5). This figure is only applicable for the specific conditions in the sand flume during the described tests.

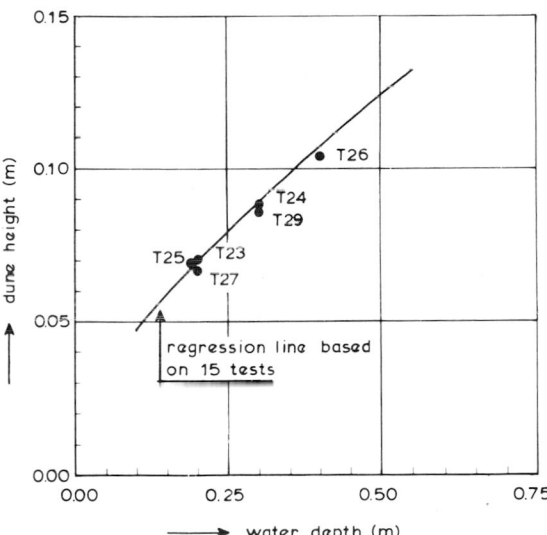

Fig. 5. Relation between dune height and water depth for steady flow conditions in sand flume ($i = 1.6 \times 10^{-3}$, $D_m = 0.77$ mm, uniform material).

Results

The most important results are presented in Table 2. The differences in observed dune celerities for tests with the same boundary conditions are probably caused by three-dimensional phenomena in the sand flume (Van Rijn & Klaassen, 1981). The complete results will be published in a forthcoming report.

During every transition a series of longitudinal bed-level records were obtained. These records were plotted against the time. An example, relating to

transition T27 (water depth decreases from 0.40 to 0.20 m), is presented in Fig. 6.

After the data had been processed, for each transition the water depth, Chézy coefficient, dune height and dune length were plotted versus time. The results obtained for a transition in water depth from 0.20 m to 0.40 m and vice versa are presented in Fig. 7. The initial change in dune height was measured from the plots of dune height versus time. The results are presented in Table 3, in which also the computed value of A_H (equation 4) is tabulated.

COMPARISON WITH EARLIER STUDIES

Comparison with Gee (1973)

For tests with a sudden change in discharge, Gee (1973) concluded that the transformation of a flat bed into a dune-covered bed required a lower total sediment transport than the reversed transition. This higher efficiency implies that the coefficient of adaptation is higher for decreasing discharge than for increasing discharge. Similar results were obtained in the present flume tests, as can be seen from Table 3 and Fig. 8, where the measured coefficient of adaptation for the dune height is plotted versus the ultimate change in dune height. During the tests the change in dune height takes place at a higher rate for decreasing discharges (and water depths) than for increasing discharges (and water depths). Furthermore it can be concluded from Fig. 8 that the change in dune height cannot be approximated by a linear first-order differential equation with a constant coefficient as assumed in equation (4) (see the variation of A_H in Fig. 7).

Comparison with Allen (1976a)

In his computational model Allen (1976a) applied two parameters, namely the dune excursion (length over which a dune travels before it loses its identity) and the coefficient of change A (equation 3). The value of these two parameters can be derived from the results of the present tests.

To characterize the life-span of dunes during equilibrium conditions a cross-correlation technique was applied for a set of two measurements with an increasing time lapse. The maximum correlation coefficient was plotted versus the time lapse. The intersection of the tangent for $t = 0$ with the horizontal axis is considered as a measure for the

4

J. H. A. Wijbenga and G. J. Klaassen

Table 2. Main results of measurements during equilibrium phases

Test number	Number of measurements (–)	Discharge per unit width q (m² sec⁻¹)	Sediment transport per unit width (kg m⁻¹ h⁻¹)	Energy slope i (–)		Water depth h (m)		Chezy coefficient C (m$^{\frac{1}{2}}$ s⁻¹)		Dune height H (m)		Dune length L (m)		Average bedform celerity c_b (mh⁻¹)
				i	σ_i	$\bar{h}$	σ_h	$\bar{C}$	σ_C	$\bar{H}$	σ_H	$\bar{L}$	σ_L	
T22	17	0·177	86·9	1·64	0·13	0·302	0·003	27·5	1·0	0·077	0·013	1·20	0·24	2·43
T23	18	0·097	43·3	1·60	0·09	0·200	0·001	28·1	0·9	0·071	0·010	1·38	0·35	1·38
T24	20	0·177	83·5	1·56	0·12	0·301	0·003	28·5	1·3	0·087	0·013	1·45	0·29	2·38
T25	20	0·098	43·5	1·65	0·10	0·200	0·002	27·7	0·9	0·071	0·008	1·39	0·33	1·44
T26	20	0·267	119·4	1·53	0·12	0·405	0·005	28·1	1·0	0·104	0·013	1·59	0·26	2·88
T27	23	0·098	43·6	1·61	0·11	0·201	0·002	27·9	1·2	0·069	0·009	1·25	0·26	1·93
T29	13	0·177	82·1	1·68	0·06	0·301	0·002	27·4	0·7	0·086	0·009	1·41	0·18	2·46

Note 1: $\bar{x}$ = mean value of x; σ_x = standard deviation of x.
Note 2: sediment transport has been measured under water.

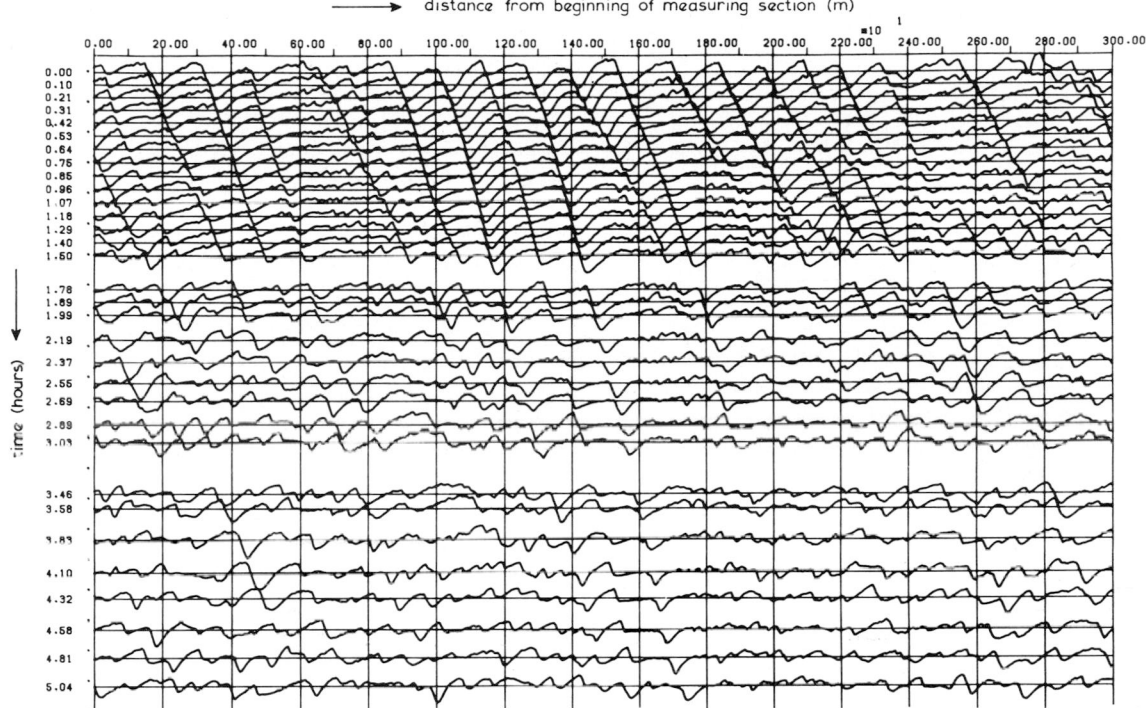

Fig. 6. Bed-level records of middle profile versus time for transition T27 (water depth decreases from 0·40 to 0·20 m, flume width 1·50 m, energy slope $1·6 \times 10^{-3}$, $D_{\mathrm{m}} = 0·77$ mm, uniform material).

life-span of dunes (see Fig. 9). The dune excursion C was calculated from:

$$C = \frac{c_{\mathrm{D}} \cdot T}{L} \qquad (5)$$

where T = life-span (sec).

The computed values for the dune excursion are presented in Table 4. For the present flume tests (see Fig. 10) the dune excursion is in the order of 1 to 2 (mean value of dune excursion for all tests equals 1·31 with a standard deviation in the single value of 0·26). A similar value for the dune excursion is found when the plots of bed-level records versus time (Fig. 6) are inspected visually. The increase of dune excursion is mainly caused by the increase in bedform celerity.

The coefficient of change, A, appears to vary with the ultimate change in dune height. This can be concluded from Table 4, in which the computed values of A are listed. Apparently the assumption of a constant value for A in equation (3) does not hold. It should be remarked that the computed value of A includes both the change in dune height according to equation (3) and the change in dune height due to the

selective process of creation and destruction of bedforms during the transition. It may be assumed, however, that during the initial phase of the transition the latter plays only a minor role. In Fig. 6 no significant creation or destruction can be seen during the first 30 minutes. Similar results are found in the bed-level records during the other transitions.

Comparison with Fredsøe (1979)

The results of the flume tests were also used to make a comparison between the measured value of the coefficient of adaptation A_{H} and the value calculated on the basis of the initial change in dune height (equation 1) as proposed by Fredsøe (1979). The results for two transport fomulas (Meyer-Peter & Müller, 1948; Engelund & Fredsøe, 1976) are plotted in Fig. 8.

Apparently there are differences between the coefficient of adaptation for a decrease in discharge (with the same final water depth) and an increase in discharge (starting from the same initial water depth). During the latter a higher coefficient of adaptation is

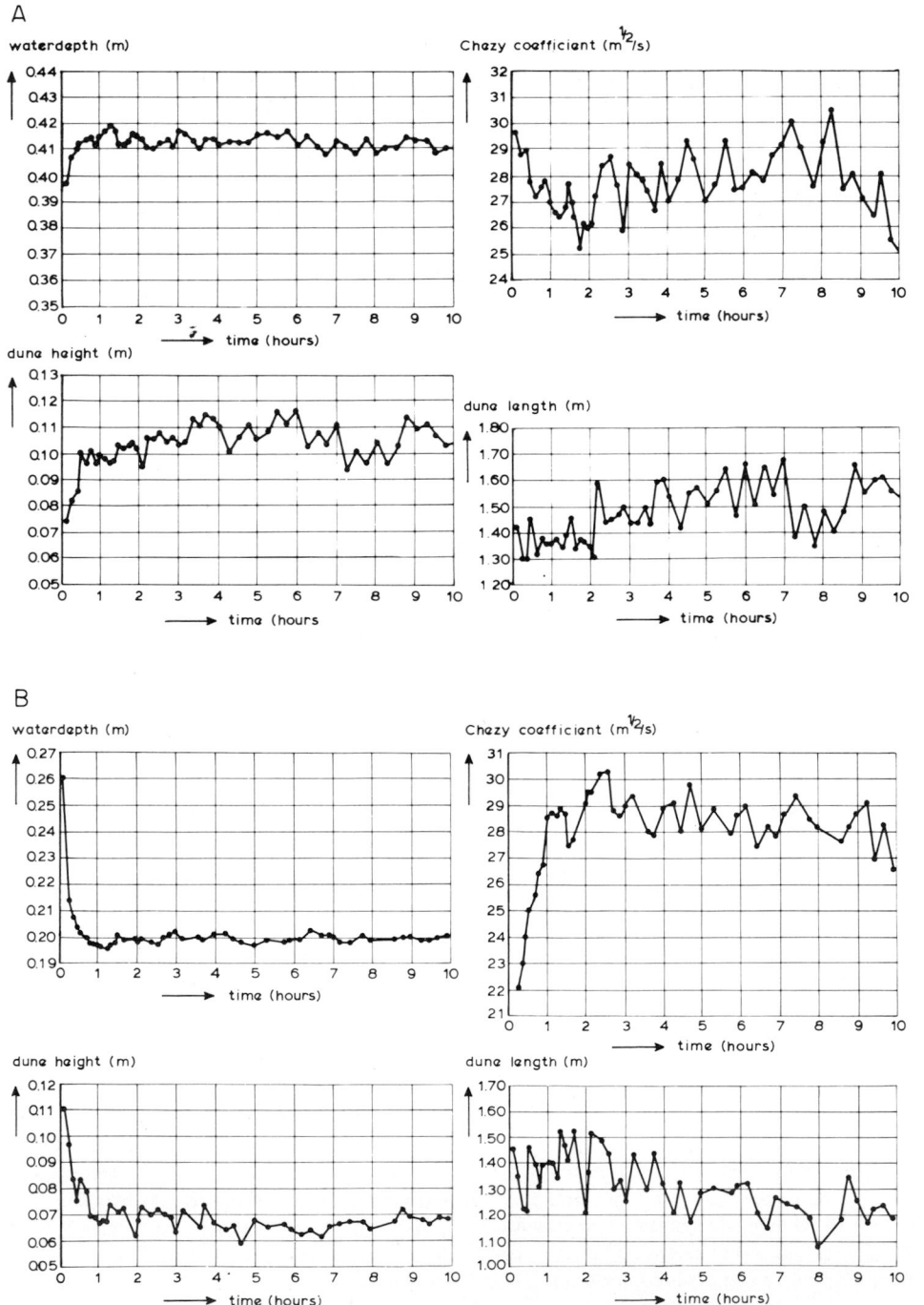

Fig. 7. Changes in water depth. Chézy coefficient, length and height of dunes for transitions T26 (A) and T27 (B) (flume width 1·50 m, energy slope $1·6 \times 10^{-3}$, $D_m = 0·77$ mm, uniform material).

Table 3. Main results at start of transition

Test number	Water-depth h (m)	Equilibrium dune height at $t = 0$ H_0 (m)	Equilibrium dune height at $t \to \infty$ H_∞ (m)	Initial change in dune height $\dfrac{dH}{dt}$ (10^{-6} m sec^{-1})	Coefficient of adaptation for dune height A_H (10^{-3} sec^{-1})
T23	0·30–0·20	0·085	0·067	−3·2	0·18
T24	0·20–0·30	0·067	0·085	3·2	0·18
T25	0·30–0·20	0·085	0·067	−1·5	0·09
T26	0·20–0·40	0·067	0·100	15·3	0·46
T27	0·40–0·20	0·100	0·067	−26·7	0·81
T29	0·20–0·30	0·067	0·085	3·0	0·17

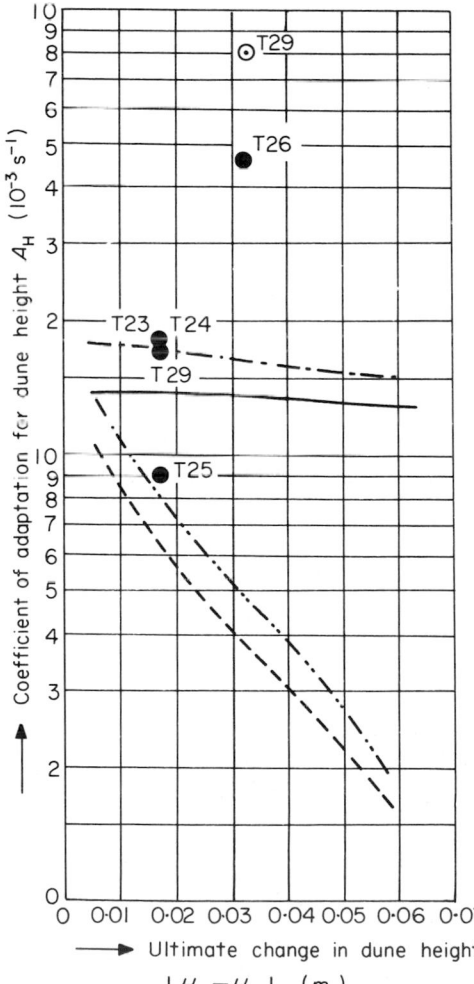

found. During the flume tests, however, lower values were measured for an increase in discharge and higher values during a decrease in discharge (see Fig. 8).

Furthermore, it may be concluded that for the present flume tests the coefficient of adaptation determined with Fredsøe's relation using the Meyer-Peter & Müller formula yields satisfactory results as long as the change in discharge (and thus the ultimate change in dune height) is not too large.

ADDITIONAL REMARKS

The following remarks should be made regarding the described flume tests and the whole study into the resistance to flow of the minor bed during the passage of flood waves.

(1) A similar set of tests with an increase or decrease in discharge was carried out for a flume width of 0·5 m. This was done because it was expected that three-dimensional phenomena in the flume (Van Rijn & Klaassen, 1981) might have influenced the results of the described tests. The results have not been processed completely, but it would appear that the results are approximately similar.

(2) Recently a number of tests have been started

Fig. 8. Measured coefficient of adaptation compared with prediction by method of Fredsøe (1979) with two different transport formulae (energy slope $1·6 \times 10^{-3}$, flume width 1·50 m, $D_{\rm m} = 0·77$ mm, uniform material). ●, measurement; initial water depth 0·20 m. ○, measurement; final water depth 0·20 m. ——, computed with Engelund & Fredsøe (1976); initial water depth 0·20 m. ––––, computed with Engelund & Fredsøe (1976); final water depth 0·20 m. –·–·–, computed with Meyer-Peter & Müller (1948); initial water depth 0·20 m. –··–··–, computed with Meyer-Peter & Müller (1948); final water depth 0·20 m.

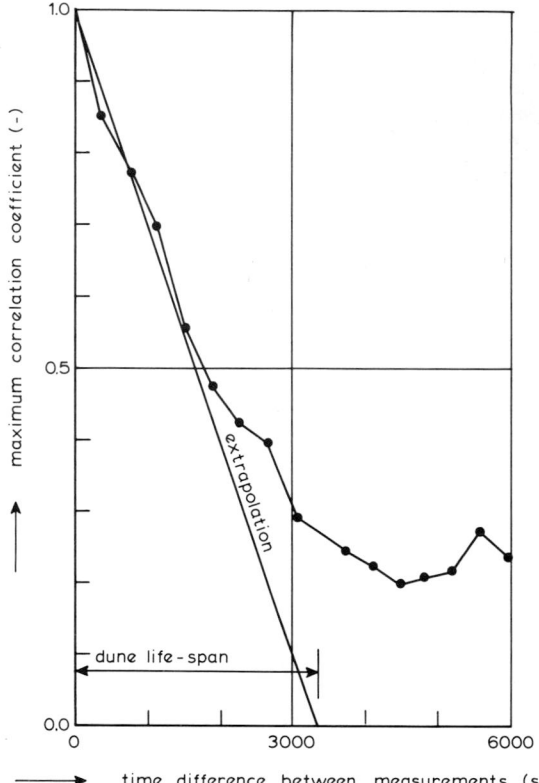

Fig. 9. Determination of the life-span of dunes.

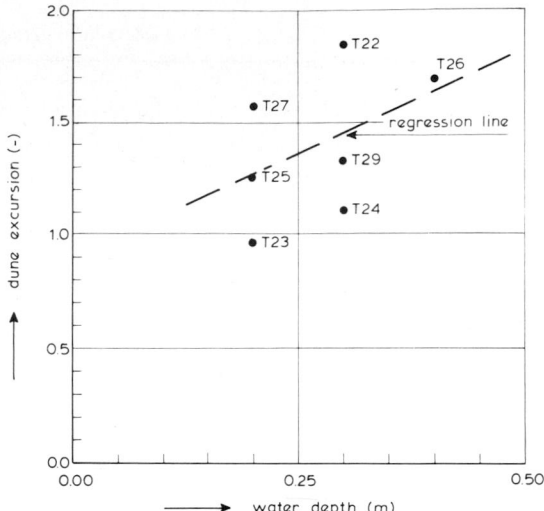

Fig. 10. Measured dune excursion as defined by Allen (1976a) for the sand-flume data.

with an imposed discharge variation similar to a real flood wave in a river. Also during these tests (with a constant slope and a water depth variation between 0·15 and about 0·50 m) the resistance to flow and the bedform dimensions are measured as a function of time.

(3) Apart from these laboratory tests, extensive field measurements of the resistance to flow and the change in bedform dimensions have also been carried out since 1979, and are continuing (Havinga & Van Urk, 1980). These measurements are made in two of the Rhine branches in the Netherlands. The results of these field measurements will be used for checking the applicability of the results of the present tests for prototype conditions.

(4) To be able to apply the methods of Allen (1976a, b, c) or Fredsøe (1979, 1981) it is necessary to have a prediction method for H_∞. Methods have been proposed by Yalin (1977) and Allen (1978), but verification of these methods, especially for field conditions, is required. For this reason also, field measurements are carried out in the various Rhine branches.

Table 4. Dune excursion and coefficient of change for sand-flume data

Test number	Water-depth h (m)	Measured dune length L (m)	Bedform celerity c_b (m hr⁻¹)	Life-span of dunes T (hr)	Dune excursion according to Allen (1976) (−)	Change in dune height $(H_\infty - H_0)$ (m)	Coefficient of change A (m)
T22	0·302	1·20	2·43	0·91	1·84	—	—
T23	0·200	1·38	1·38	0·97	0·97	0·018	0·040
T24	0·301	1·45	2·38	0·67	1·10	0·018	0·018
T25	0·200	1·30	1·44	1·22	1·26	0·018	0·019
T26	0·405	1·59	2·88	0·93	1·68	0·033	0·039
T27	0·201	1·25	1·93	1·02	1·57	0·033	0·151
T29	0·301	1·41	2·46	0·70	1·33	0·018	0·017

CONCLUSIONS

The following conclusions can be drawn from the results of the flume tests described in this paper and the comparison with earlier investigations.

(1) The change in bedform dimensions does not follow a first-order curve as assumed in equation (4).

(2) The coefficient of adaptation for the dune height is higher for decreasing discharges than for increasing discharge, if the ultimate change in water depth is the same (see Fig. 8).

(3) For the flume tests considered the dune excursion C from Allen's (1976a) computational model is in the order of 1 to 2, and increases slightly for increasing water depth due to an increase in bedform celerity. The coefficient of change A determined with the flume tests is not constant.

(4) The coefficient of change determined with the relation proposed by Fredsøe (1979) depends on the transport formula used. Besides, the difference in coefficient of adaptation for the dune height is slightly higher for increasing discharges than for decreasing discharges, which is just opposite to the measured results (see Fig. 8).

(5) Fredsøe's (1979) method does not adequately describe the observed increase and decrease of bedform dimensions during the described tests, while a number of adaptations should be made to the computational model of Allen (1976) before it can be applied for the simulation of the time-dependent behaviour of dune dimensions.

ACKNOWLEDGMENTS

This investigation has been commissioned by the Dutch Water Control and Public Works Department, which is gratefully acknowledged for permission to publish part of the results. The measurements in the sand flume and the data processing have been carried out under the guidance of Mr B. Bakker. Regular discussions on the progress of the investigation have been held with Mr H. N. C. Breusers and Professor M. de Vries as special advisers.

REFERENCES

ALLEN, J.R.L. (1976a) Computational models for dune time-lag: general ideals, difficulties and early results. *Sedim. Geol.* **15**, 1–53.

ALLEN, J.R.L. (1976b). Computational models for dune time-lag: population structures and the effects of discharge pattern and coefficient of change. *Sedim. Geol.* **16**, 99–130.

ALLEN, J.R.L. (1976c) Computational models for dune time-lag: an alternative boundary condition. *Sedim. Geol.* **16**, 255–279.

ALLEN, J.R.L. (1978) Computational models for dune time-lag: calculations using Stein's rule for dune height. *Sedim. Geol.* **20**, 165–216.

BISHOP, C.T. (1977) *On the time-growth of dunes.* Unpublished M.Sc. Thesis. Queen's University, Kingston, Ontario. 112 pp.

BOGIRSKI, H. (1977) Contribution to the analysis of sand waves, *Rep. Delft Hydraul. Lab,* **R 657-V**, 1–31.

EINSTEIN, H.A. & BARBAROSSA, N.L. (1952) River channel roughness, *Trans. Am. Soc. civ. Engrs* **117**, 1121–1146.

ENGELUND, F. (1978) Hydraulic resistance for flow over dunes, *Progr. Rep. Tech. Univ. Denmark, ISVA,* 44, 19–20.

ENGELUND, F. & FREDSØE, J. (1976) A sediment transport model for straight alluvial channels. *Nordic Hydrol.* 7, 293.

ENGELUND, F. & HANSEN, E. (1967) *A monograph on sediment transport in alluvial streams,* pp. 1–59. Teknisk Forlag, Copenhagen.

FREDSØE, J. (1979) Unsteady flow in straight alluvial streams: modification of individual dunes. *J. Fluid Mech.* **91**, 497–512.

FREDSØE, J. (1981) Unsteady flow in straight alluvial streams, part 2, transition from dunes to plane bed. *J. Fluid Mech.* **102**, 431–453.

GEE, D.M. (1973) Sediment transport in non-steady flow. *Univ. Calif. Rep.* 4EC 22–23, Berkeley.

HAVINGA, H. & VAN URK, A. (1980) Solving river problems in The Netherlands. *Proc. 13th IAHR Symp. River Engineering and its Interaction with Hydrological and Hydraulic Research,* Belgrade.

JANSEN, P. PH., BENDEGOM, L. VAN, BERG, J. VAN DEN, VRIES, M. DE & ZANEN, A. (Eds) (1974) *Principles of River Engineering.* Pitman, London, 509 pp.

JENSEN, P. (1969) Effets des régimes variables sur le modelage des dunes, EDF. *Bull. Étud Rech.* (Série A), **4**, 5–14.

KLAASSEN, G.J. (1978) Sediment transport and hydraulic roughness in relation to bedforms. *Proc. Conf. 'Advances in Sediment Transport',* p. 33, Jablonna, Poland (also DHL-publ. no. 213).

MEYER-PETER, E. & MÜLLER, R. (1948) Formulas for bed-load transport, *Proc. 3rd meeting IAHR,* pp. 39–64.

NASNER, H. (1978) Time-lag of dunes for unsteady flow conditions. *Proc. 16th Coastal Engineering Conf.* pp. 1801–1817. Hamburg.

OGINK, H. (1981) *Evaluation of Flood Wave Computations* (in Dutch). Dutch Water Control and Public Works Department, District South-East. 287 pp.

RIJN, L. VAN & KLAASSEN, G.J. (1981) Experience with straight flumes for movable bed experiments. *Proc. IAHR Workshop 'Particle Motion and Sediment Transport',* Rapperswihl, Switzerland.

SIMONS, D.B., RICHARDSON, E.V. & HAUSHILD, W.L. (1962) Depth-discharge relations in alluvial streams. *J. Hydraul. Div. Am. Soc. civ. Engrs* **88**, 5, 57–72.

STEIN, R.A. (1965) Laboratory study of total load and apparent bed load. *J. geophys. Res.* **70**, 1831–1842.

SÜNDERMANN, J., VOLLMERS, H.J. & PULS, W. (1980) A numerical model for dune dynamics. *Proc. 17th int. Conf. Coastal Engineering,* pp. 271–272. Sydney, Australia.

VANONI, H.A. & HWANG, L.S. (1967) Relation between bedforms and friction in streams. *J. Hydraul. Div. Am. Soc. civ. Engrs* **93**, HY3, 121–144.

VREUGDENHIL, C.B. & WIJBENGA, J.H.A. (1982) Computation of flow patterns in rivers. *J. Hydraul. Div. Am. Soc. civ. Engrs* in press.

WHITE, W.R., PARIS, E. & BETTESS, R. (1980) The frictional characteristics of alluvial streams: a new approach. *Proc. Inst. civ. Engrs* **69**, 737–750.

WIJBENGA, J.H.A. (1978) Flume tests with flat bed without sediment transport (in Dutch). *Rep. Delft Hydraul. Lab.* **R657–VIII**, 1–18.

YALIN, M. S. (1964) Geometrical properties of sand waves, *J. Hydraul. Div. Am. Soc. civ. Engrs* **90**, HY 5, 105–119.

YALIN, M.S. (1977) *Mechanics of Sediment Transport.* 2nd ed. Pergamon Press, Oxford. 298 pp.

Spec. Publs int. Ass. Sediment. (1983) **6**, 49–58

Flume experiments on bedforms and structures at the dune–plane bed transition

HOUSTON C. SAUNDERSON *and* FRANCIS P. J. LOCKETT

Department of Geography, Wilfrid Laurier University, Waterloo, Ontario N2L 3C5, Canada

ABSTRACT

A tiltable, recirculating flume, 18 m long and 76 cm wide, was used to investigate bedforms and structures near the transition between dunes and a plane bed for a moderately sorted coarse sand. At Froude numbers ranging from about 0·4 to 1·0, three dune types developed: (1) asymmetrical (triangular) dunes, (2) convex (symmetrical) dunes and (3) humpback dunes. Asymmetrical dunes had gentle, long stoss sides and steep, short lee sides, and contained cross-stratification with a maximum dip of about 30–35°. Flow separation and avalanching were strongly developed to the lee of these dunes. Convex dunes formed when the bed was thin and had longitudinal profiles that were convex-upwards, with stoss and lee sides of equal steepness. Internal cross-beds were likewise convex and formed from draping of sediment over the lee sides rather than from avalanching. Humpback dunes were the most distinctive bedforms in that on each dune profile the point of maximum elevation was offset from the top of the foreset (avalanche) slope. Immediately downstream from this maximum point, low-angle topset bedding merged uninterruptedly into steep foreset beds and these into bottomsets, producing sigmoidal bedding inside each dune. Although foreset slopes were much shorter in humpback dunes than in asymmetrical dunes, their steepness remained about 30–35° right up to the change to a plane bed at a Froude number of about 1·1. For one other run a plane bed also formed, but at a Froude number of about 1·7, a rather high value for a plane bed just beyond the dune bed phase. This second plane bed may be that which occurs at Froude numbers larger than those for in-phase waves.

INTRODUCTION

During the last twenty years or so a great deal has been written on bedform mechanics and the internal structures of bedforms in alluvial channels. Ripples and their internal cross-laminae, dunes and their cross-stratification, plane beds and parallel laminae, antidunes and backset laminae are all, by now, well-documented bedforms and structures from both experimental and field investigations. Reports on the transitional bedforms between these well-known types are less abundant, a deficiency which is evident from inspection of the boundaries on stability diagrams (Southard, 1971).

The transition from ripples to lower flat (plane) bed has been investigated for coarse sand (Southard & Boguchwal, 1973) and from ripples to dunes (Banks & Collinson, 1975) for medium sand. Mathematical treatment of the dune–plane bed transition has been done by Engelund & Fredsøe (1974), who tested their model against the experimental data of Guy, Simons & Richardson (1966). Jopling & Forbes (1979) made some experimental observations on the ripple to upper plane bed transition in a coarse silt. Allen & Leeder (1980) analysed stability criteria for the transition from an upper plane bed to dunes and ripples for decelerating flows. Still deficient though are detailed observations of bedform shape and associated stratification in flumes and natural streams for the dune–plane bed transition.

This report deals with the results of flume experiments conducted to investigate the nature of transition from lower-flow-regime dunes to upper-flow-regime plane beds.

0141-3600/83/0106-0049 $02.00

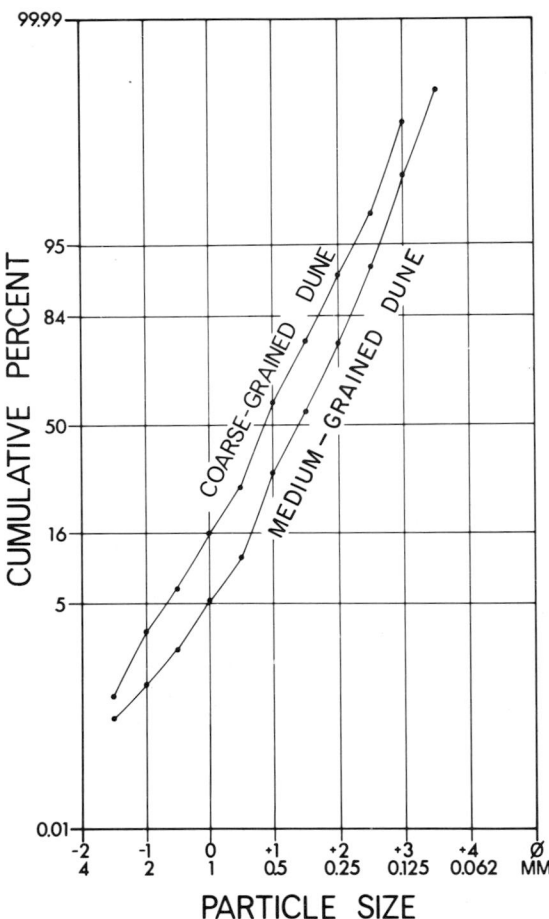

Fig. 1. Cumulative frequency–size distributions for sediments used in the experiments. Although the sediment was mixed thoroughly before being introduced to the return pipe and flume channel, sorting of sizes into coarse- and medium-grained dunes occurred along the channel. Median diameter is about 0·5 mm and 50% of the combined distribution contains coarse sand, very coarse sand and granule gravel.

EXPERIMENTAL PROCEDURE

The experiments were conducted in the flume housed in the Terrain Sciences Division of the Geological Survey of Canada. The flume is 18 m long and 76 cm wide; it is tiltable and can recirculate up to 4 mm diameter sediment (see McDonald, 1972 for more complete information on the apparatus). Moderately sorted coarse sand was used in the experiments (Fig. 1) and was introduced from the storage bins to the central end tank using conveyor belts and then pumped via the return pipe to the

headbox and into the channel. Although the flume is equipped with two pumps, only one was used to generate the flows. Fluid depth and discharge were controlled by opening and closing a valve operating a bypass from the return pipe back to the end tank and by adjusting the tailgate. Depth was measured by point gauge, point velocities by pitot tube and water temperature by thermometer. The water–sediment mixture was recirculated for 5–6 h during each run to ensure development of bedforms along most of the flume length (barring entrance and exit effects). Surface velocities were measured by timing the travel of surface floats over a known distance and were converted to a mean velocity for each run by using the approximation 0·8 times the average surface velocity, following the practice of Fahnestock (1963, p. A30) and Williams (1967, p. B5) for shallow flows. Bedform height, wavelength and steepness of stoss and lee sides were measured through the sidewalls of the flume and observations were made on stratification, degree of sorting, thickness of laminae and number of laminae deposited per unit of time. Between runs the bed was flattened artificially.

RESULTS OF EXPERIMENTS

Hydraulic conditions

The main objective was to delimit the range of hydraulic conditions over which dunes became transformed or 'washed out' (Simons, Richardson & Nordin, 1965; Simons & Richardson, 1966; among others). As is true for most other flumes, the prime limitation was the small range of attainable flow depths, which were never greater than 45 cm and were mostly between 20 and 40 cm. Table 1 is a summary of the hydraulic conditions that existed during the experiments.

Hydraulic radius decreased generally from the first to last runs as a consequence of slight decreases in water depth. This decrease, combined with a general increase in mean velocities from about 50 to 160 cm sec^{-1}, resulted in a corresponding increase in Froude number from about 0·4 to 1·7. Changes in Reynolds number resulted from velocity changes rather than temperature, which ranged only from 23° to 25 °C; the magnitude of these changes was not considered to be of physical significance as an important variable when explaining variations in bedforms, and so the Reynolds number was rounded off to the nearest order of magnitude which, for all runs, was $1·0 \times 10^5$. Subtle changes in bedform

Table 1. Summary of experimental measurements and hydraulic calculations*

Run	Surface velocity (cm sec⁻¹)	Mean velocity (cm sec⁻¹)	Hydraulic radius R (cm)	Temper- ature (°C)	Kinematic viscosity (stokes)	Reynolds number, Re	Bedform	Froude number, F
1	65·6	52·48	19·34	23	0·009325	$\simeq 10^5$	h/a	0·38
2	64·3	51·44	18·58	24	0·009111	$\simeq 10^5$	h/a	0·38
3	79·2	63·36	18·84	24	0·009111	$\simeq 10^5$	h/a	0·47
4	85·7	68·56	17·93	24	0·009111	$\simeq 10^5$	h/a	0·52
5	100·0	80·00	16·59	24	0·009111	$\simeq 10^5$	h/a	0·63
6	93·8	75·04	17·42	24	0·009111	$\simeq 10^5$	h/a	0·57
7	121·0	96·80	15·19	24	0·009111	$\simeq 10^5$	h/a	0·79
8	130·4	104·32	15·25	24	0·009111	$\simeq 10^5$	h/a	0·85
9	142·9	115·32	15·64	24	0·009111	$\simeq 10^5$	h/a	0·93
10	150·0	120·0	14·48	25	0·008904	$\simeq 10^5$	h	1·01
11	157·9	127·32	14·25	25	0·008904	$\simeq 10^5$	h/p	1·08
12	200·0	160·00	8·53	24	0·009111	$\simeq 10^5$	p	1·75

* No physical significance is attached to the decimal places in velocity calculations. They have been maintained just to reduce rounding errors in the calculation of the Froude number.

a = asymmetrical dune. h = humpback (whaleback) dune. p = plane (flat) bed.

Fig. 2. Three morphological dune types formed:

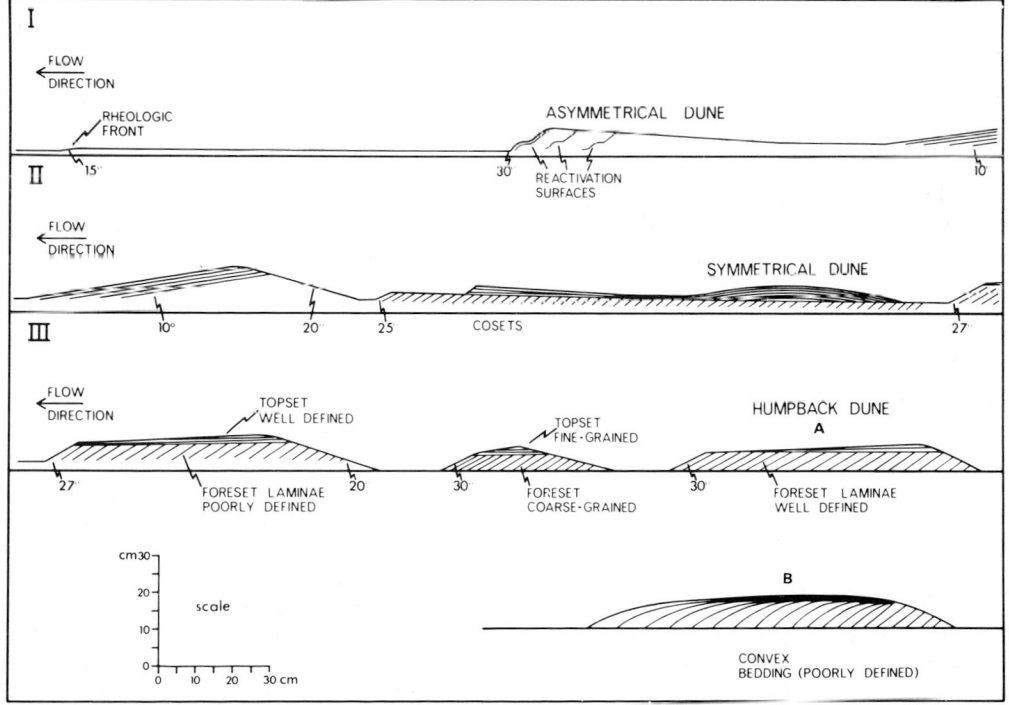

Fig. 2. Three morphological dune types formed: (I) asymmetrical, (II) symmetrical and (III) humpback dunes. Panels I, II and III represent a continuous series of dunes formed in one experimental run. Reactivation surfaces were cut by the reverse circulation, which changed in size to the lee of asymmetrical dunes. Symmetrical dunes contained convex bedding and humpback dunes topset and foreset bedding. Note that the maximum elevation on humpback dunes does not coincide with the tops of foreset slopes, and that one dune type may change to another (Panel III A, B). Run 10.

Fig. 3. (A) Asymmetrical dune resulting from transformation of humpback dune. Sigmoidal bedding inside the dune formed during humpback migration and continued during aggradation at the foreset during the change to the asymmetrical type. Note the bottomset deposition containing small, regressive ripples. (B) Asymmetrical dune showing long, gentle stoss side which is erosional and transportational, and shorter, steeper lee side maintained by flow separation, reverse circulation and undermining. Cross-beds formed by avalanching (predominantly) and settling of suspended load through the zone of flow separation.

characteristics and stratification were related more to variations in Froude number.

Bedform dimensions and mechanics

Three distinct shapes of dunes formed during the experiments and were designated: (a) asymmetrical dunes, (b) convex, symmetrical dunes and (c) humpback dunes. The asymmetrical dunes typically had a short, steep lee side which formed an avalanche slope and a gentler, longer stoss side over which erosion and bedload transport took place (Fig. 2). Flow separation occurred at the crest of the dune, and the zone of reverse circulation or backflow maintained

the steepness of the avalanche slope by undermining it. Occasionally though, the thickness of the zone of reverse circulation decreased and only the upper half or so of the foreset slope became eroded. The crest of the dune became planed off and a new crest formed slightly upstream of the first. Flow separation and avalanching of bedload at the new crest produced a small set of cross-laminae superimposed on the large-scale cross-laminae of the dune separated by a small erosion surface. This surface is similar to reactivation surfaces reported from the field investigation of modern streams (Collinson, 1970) but, unlike larger streams, it is related to changes in the size of the separation zone to the lee of the dune crest rather than

Fig. 3. (B) For legend see facing page.

to erosion and renewed deposition as a consequence of changes in fluid discharge. The observed mechanism was closer to that deduced by Jones & McCabe (1980, see their fig. 5). The dune then grew in height, the separation zone became thicker and a single set of cross-laminae formed by avalanching along the lengthened foreset slope.

Convex, or symmetrical, dunes (Fig. 2), were so designated because of their convex-upward profile as seen through the sidewalls of the flume. Upstream and downstream sides of this type were of equal steepness (symmetrical) and met at the highest point on the dune, about halfway along its length. Flow separation was absent and consequently no undermining of the lee side took place. Convex cross-laminae were deposited to the lee of the crest, but not by avalanching; bedload was simply carried on over the crest by inertia and draped over the lee side. This mode of deposition was very similar to that observed by one of us

(Saunderson, 1981) to the lee sides of antidunes in medium sand.

Humpback dunes differed from both asymmetrical and convex (symmetrical) dunes in that the point of maximum elevation was closer to the upstream limit of the dune than to the downstream avalanche face (Fig. 2). Moreover, the avalanche face itself was not a permanent feature of humpbacks, occasionally disappearing altogether or disappearing and reforming. Erosion of the stoss side occurred up to the point of maximum height, but immediately downstream from this point deposition produced long, low-angle to almost horizontal lamination which extended right along to, and over, the avalanche slope of the dune. The avalanche slope was much shorter than that of asymmetrical dunes but, like these dunes, was maintained at a steep angle by flow separation. The most obvious difference was that in humpback dunes the top of the avalanche slope was not the highest point of the dune profile; these two points were offset from one another. Suspended load settled through the zone of flow separation to form bottomset laminae, as is common for dunes and laboratory deltas (Jopling, 1965a; Allen, 1965, 1968a, b) and the backflow was often strong enough to generate regressive ripples in the bottomset (Fig. 3A).

In Runs 1–6 both asymmetrical and humpback dunes were abundant and so it was possible to measure the lengths and steepness of stoss and lee sides of a number of dunes in each run as well as wavelengths and dune heights. Average stoss-side lengths ranged from about 77 cm in Run 1 to 110 cm in Run 5, and average lee-side lengths from about 11 to 18 cm. Dune heights varied from an average of about 7 to 11 cm and average wavelengths from about 90 to 130 cm. Steepness of foreset slopes ranged from 30° to 37° and internal cross-stratification from 28° to 40°, but mostly between 30° and 35°. Dune migration rate was slowest at about 2 cm min^{-1} in Run 1 and fastest in Run 5 at about 10 cm min^{-1}. At higher Froude

numbers (Runs 7–11) bed morphology was more intermixed and complex, so it was not possible to obtain meaningful average figures for bedform dimensions. All three dune types, asymmetrical, convex and humpback, were often present in a single run and became transformed from one type to another of the three (Fig. 2). Convex dunes sometimes appeared at lower Froude numbers when only a small amount of sediment was present on the flume bottom, but they changed into asymmetrical dunes as the mound thickened (Fig. 4, I). Bedding also changed concomitantly from convex to more steeply dipping. After asymmetrical dunes formed, flow separation and reattachment became distinguishable, and near the point of reattachment (which oscillated rather than remaining stationary) bursting threw sediment temporarily into suspension. This suspended load then landed a short distance downstream on the stoss side of the same dune, thereby thickening the stoss side.

At the downstream end of the dune there is a maximum height to which the crest can aggrade

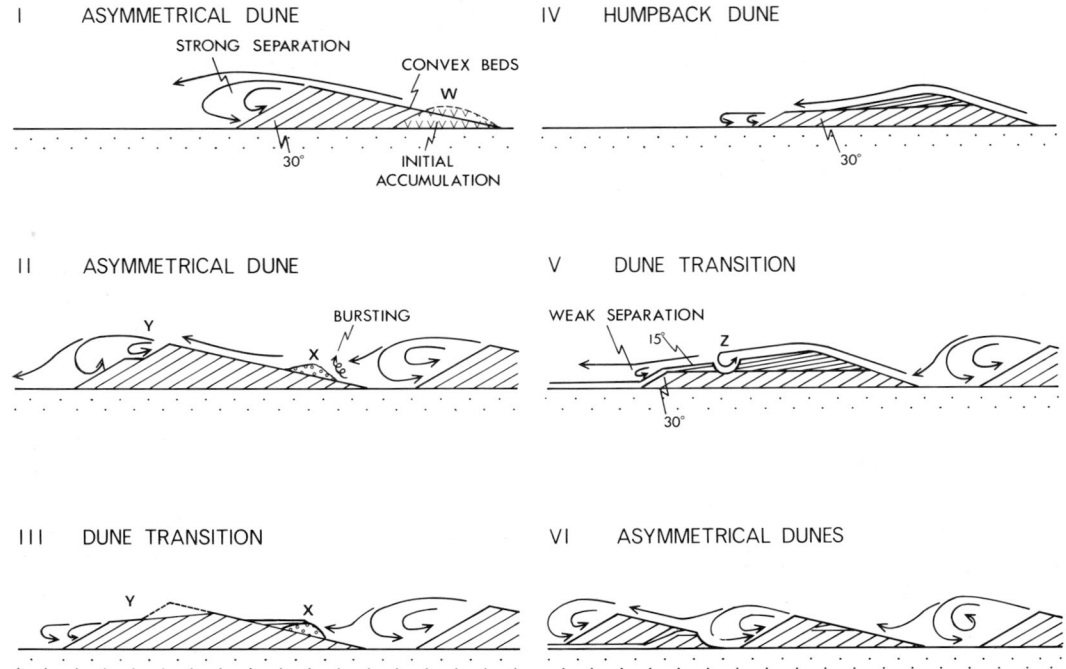

Fig. 4. Schematic diagram to show transitions asymmetrical dunes → humpback dune → asymmetrical dunes. A mound of sand forms a symmetrical, convex feature (I, W) with convex bedding which then changes to steeper bedding as the sediment thickens and progrades. After asymmetrical dunes established (II) bursting occurred near the reattachment point and sediment thrown into suspension was deposited on the stoss side of a dune (X). Note storage of sediment near X forming topsets and erosion of upper foresets near Y to produce a humpback dune (IV). Topsets prograde forming sigmoidal bedding (V). Flow separation occurs (Z) and intensifies, dissecting the humpback into two separate (asymmetrical) dunes (VI).

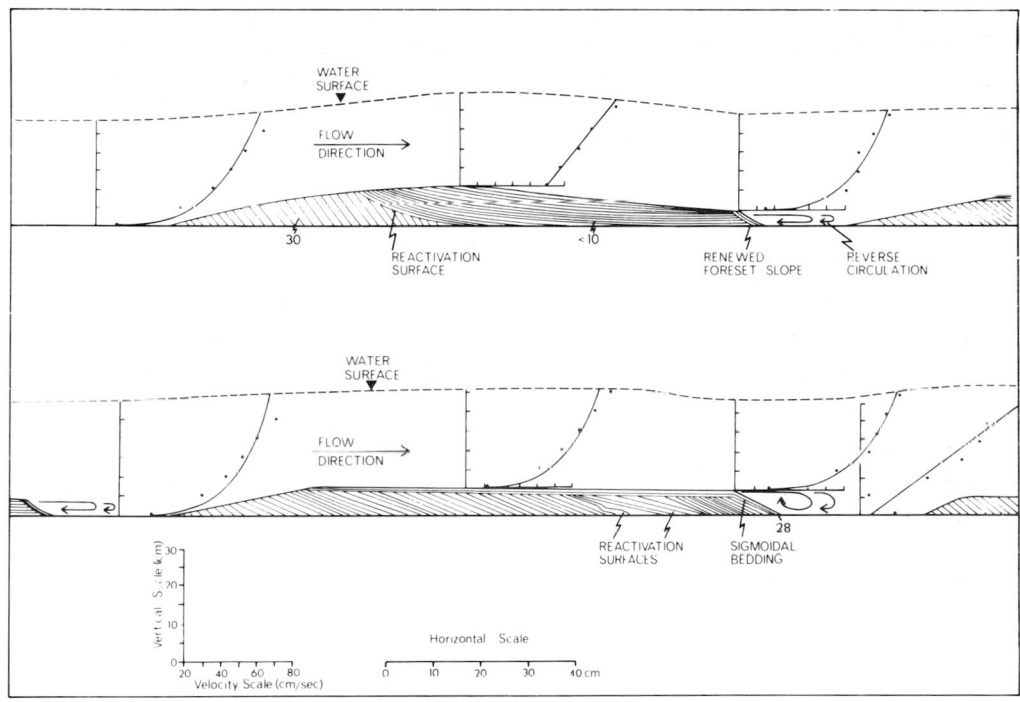

Fig. 5. Velocity profiles and phase relationships for humpback dunes. The upper half of the diagram shows water surface wave and dune almost in phase. Each point on velocity profiles respresents one point velocity measured by pitot tube. Note the considerable shear (a velocity of $\simeq 70$ cm sec^{-1}) right at the bed at the locus of maximum elevation on the humpback profile (upper half of diagram) and the flow separation to the lee of each dune. In the lower half of the diagram, topset and foreset laminae are separated by an erosional surface, except at the front of the dune where they are continuous and grade into bottomsets (see Fig. 3) to form sigmoidal bedding. Run 10.

because the length of the lee slope is determined by the amount of sediment available for avalanching at the crest. If this amount is insufficient to sustain the whole avalanche slope, the lower foreset progrades and the upper foreset either collapses or erodes (Fig. 4, III). In this way the downstream end of the dune flattens and the upstream end thickens, resulting in a transformation from asymmetrical to humpback dunes. Velocity profiles indicated that bed shear was greatest near the highest point on a humpback dune, in one case with a near-bed velocity of about 70 cm sec^{-1} (Fig. 5). However, a short distance downstream from the highest point, flow separation developed, probably at a location where the adverse pressure gradient reached a critical point, and this separation eddy grew in size and intensity until the humpback dune became eroded into two separate dunes, now asymmetrical (Fig. 4, VI).

At higher Froude numbers the asymmetrical dunes atrophied to low-amplitude features and rheological fronts (Moss, Walker & Hutka, 1980, pp. 52–53) became more abundant on a bed which was almost plane. At lower Froude numbers the water surface was out of phase with the bed undulations, whereas at higher values they were either in phase or almost so (Fig. 5).

Sediment size sorting occurred along the channel, but not simply from coarse to medium in the downstream direction. Instead, sorting was made manifest by the presence of medium-grained and coarser-grained dunes (Fig. 1). The mechanical details of how this sorting came about were not obtainable from observation, but it is certain that sorting was not an artefact of how the sediment was introduced, because all size grades were thoroughly churned up by using the by-pass system near the end tank before being pumped along the return pipe to the headbox and into the open channel of the flume.

Sigmoidal bedding formed as humpback dunes migrated along the channel (Fig. 5). Erosion occurred on the upstream sides of these dunes and sediment was carried up to the point of maximum height at the top of the stoss side. Low-angle ($< 15°$) laminae were deposited as 'topsets' between the highest point and the top of the foreset slope where the bedload accumulated and then avalanched along the foreset to form steeply dipping cross-beds. These cross-beds were deposited at the same time as bottomsets in each dune trough, but because suspended load was not plentiful, topsets were absent, the cross-beds of avalanche origin instead forming an angular contact with the bottomsets. Sigmoidal bedding was thus the result of the fusion of topset deposition (between the point of maximum height and the top of the foreset slope), foreset deposition by avalanching along the foreset slope and bottomset deposition in the bottom of the trough to the lee of the foreset. Although these three types of laminae were continuous, the topset laminae were much longer than either foreset or bottomset laminae, emphasizing a tendency for the dune top to lengthen and the foreset to shorten.

A plane bed formed at two quite distinct Froude numbers, at about 1·1 (Run 11) and about 1·7 (Run 12, Table 1). At $F \simeq 1·1$ the plane bed was clearly associated with humpback dunes and is interpreted as being the upper-stage plane bed found commonly at Froude numbers higher than dunes. The plane bed at $F \simeq 1·7$ is probably not the same, but is the plane bed observed at Froude numbers greater than for in-phase waves (Williams, 1970).

Velocity profiles over dunes

A number of velocity profiles were constructed from pitot tube measurements at specific locations on asymmetrical and humpback dune profiles. These included trough, stoss-side, 'hump' and crest locations. The shapes of profiles for asymmetrical dunes were much like those shown by Bridge & Jarvis (1977) and showed that bed shear was at a minimum in dune troughs and at a maximum near dune crests. This distribution was not quite the same over humpback dunes. Least bed shear was again at trough locations, but maximum bed shear occurred over the highest points, the humps, and then decreased slightly between the hump and the top of the avalanche slope (Fig. 5). Maximum bed shear at the highest point was emphasized during experiments by an intense sheet transport of bedload over the surface of the hump. In some dunes, when the hump became flattened velocity profiles had more of a uniform shape over the tops of dunes or dune platforms (Fig. 5), showing a uniformity of fairly high bed shear such as that found over a plane bed.

DISCUSSION OF RESULTS

Simons & Richardson (1966, p. J11) described the bed configuration in the transition between lower and upper flow regimes as being variable, depending largely on antecedent conditions. If a bed were initially plane, then a decrease in stream power would result in a transitional configuration with many of the properties of a plane bed, whereas an increase in power from an initial dune configuration would produce a transitional configuration with many of the characteristics of dunes. No doubt the same reasoning explains why the transitional forms discussed in this report have many of the structural attributes of dunes, because stream power was increased from the dune into the plane bed states (Table 1). Simons & Richardson (1966) also described dunes as becoming smaller in amplitude and longer in wavelength in the transition. Although longer and lower features appeared on the bed in the present runs (Runs 9 and 10), measurements of wavelength and height did not, in general, show a progressive change as the Froude number increased. Instead there was a modal grouping of wavelengths and heights with considerable variance from Runs 1 to 6 and, when a plane bed developed at a higher Froude number (Run 11), the transition from dunes to plane bed was quite abrupt. Rather than progressively less steep foresets and cross-beds, bedding steepness remained mostly in the 30–35° range right up to the transition, when a plane bed and rheological fronts became prominent. Nor did the transition consist solely of a mixture of dune and plane bed features; the convex dunes and their convex laminae are very similar in origin to the lee-side laminae of antidunes and standing waves.

The rounding of dunes to form convex dunes may be analogous to rounding of ripples observed by Jopling & Forbes (1979) for the ripple to plane-bed transition in a coarse silt. They also observed that bed and water surface were sometimes in phase at the transition. Perhaps similar morphological changes occur in the transition between an undulatory and a plane bed, regardless of whether the undulatory forms are dunes or ripples. Humpback bed undulations of much larger size than those reported here have been investigated in the Pitt River (Ashley, 1978), and dunes with convex-upward stoss sides have been found in the return pipes of the same flume as that used in the present study (McDonald & Vincent, 1972).

The in-phase relationship between some of the humpback dunes and water-surface waves (Fig. 5; Run 10, Table 1), the presence of steeply dipping foresets, and the development of long plane beds on the tops of dunes exemplify the complexity that may occur at the transition from dunes to a plane bed and at the transition from subcritical to supercritical flow (Runs 9–11, Froude number close to 1·0). This complexity results from mixing of bed phases, which is a normal expectation near bed-phase boundaries, particularly for non-equilibrium and quasi-equilibrium flows. Experiments of the present report are best considered quasi-equilibrium rather than truly equilibrium conditions, and the inherent, albeit small, unsteadiness of flow is probably related to the short recirculation time (5–6 hr) during each run.

Mixing of dune and plane-bed phases is deducible for slightly unsteady, non-uniform flows from inspection of Allen & Leeder (1980, fig. 3) whose stability criteria separating ripple, dune and plane-bed fields intersect, thus making it possible for more than one bedform to coexist as hydraulic and sediment transport conditions fluctuate slightly about some average, equilibrium intersection. Bridge (1981) considered the backs of dunes, when devoid of ripples, to be dynamically similar to upper-regime plane beds. The long, nearly horizontal topset portions of humpback dunes are likewise considered to be plane beds in the present report. Morphologically each forms a crestal platform, like that shown diagrammatically by Allen (1968a, p. 61) though unexplained by him at the time.

Sigmoidal bedding was the most distinctive structure found in the humpback dunes. This type of stratification has been reported from experiments on laboratory deltas (Jopling, 1965b, p. 789), hurricane washover deposits (Morton, 1978, fig. 7D), ancient point-bar deposits (Nami & Leeder, 1978) and from linguoid ripples in modern flood deposits of ephemeral streams (Williams, 1971, p. 21). Much earlier mention (Davis, 1890, fig. 2) may be found of the structure, though not with specific interpretation as a dune or sandwave structure. The relative lengths of the topset, foreset and bottomset segments of sigmoidal bedding probably reflect the proximity of conditions to the plane-bed state. In other words, the foreset atrophies whereas the topset lengthens as the plane-bed state is approached. As the foreset shortens, flow separation diminishes as an effective sorting mechanism and bedload is simply draped over the foreset slope, thereby rounding it off.

CONCLUSIONS

As the transition to a plane bed is approached from a dune configuration, mixing takes place of attributes from the dune stability field and from the plane-bed field. Dune attributes consist of cross-beds formed predominantly by avalanching and maintained at fairly high dips ($\simeq 30$–$35°$) by flow separation right up to the transition from dunes to a plane bed. Attributes from the plane-bed field consist of low-angle to horizontal (parallel) laminae superimposed as topsets on the dune form.

Overlapping of bed phases other than dunes and plane beds may also occur at the transition. The convex, symmetrical dunes with their convex, lee-side laminae are probably dynamically similar to in-phase waves (standing waves and antidunes), and indeed the in-phase nature of water surface waves and some humpback dunes is on the edge of the in-phase wave stability field.

The most distinctive structure of the experiments was the sigmoidal bedding found in the humpback dunes. The relative lengths of the topset, foreset and bottomset portions of this type of bedding are direct results of the proximity of the dune–plane bed transition. Specifically, the topset component is relatively short and the foreset component long at the low-velocity end of the transition, whereas the foresets and bottomsets eventually disappear and the topset component lengthens and becomes dominant at the high-velocity end where a plane bed finally becomes the stable form.

ACKNOWLEDGMENTS

The authors are most grateful to Drs B. G. Craig and T. J. Day for permission to use the flume facility at the Terrain Sciences Division, Geological Survey of Canada. Financial assistance was provided by the Natural Sciences and Engineering Research Council of Canada under Operating Grant No. A4281 to Saunderson. The research was done as part of an M.A. dissertation being written by Lockett under Saunderson's supervision.

We thank Drs J. S. Bridge and G. V. Middleton for valuable comments made in their critical reviews of the manuscript.

REFERENCES

ALLEN, J.R.L. (1965) Sedimentation to the lee of small underwater sand waves—an experimental study. *J. Geol.* **73**, 95–116.

ALLEN, J.R.L. (1968a) *Current Ripples*. North-Holland, Amsterdam. 433 pp.

ALLEN, J.R.L. (1968b) The diffusion of grains in the lee of ripples, dunes, and sand deltas. *J. sedim. Petrol.* **38**, 621–633.

ALLEN, J.R.L. & LEEDER, M.R. (1980) Criteria for the instability of upper-stage plane beds. *Sedimentology*, **27**, 209–217.

ASHLEY, G.M. (1978) Bedforms in the Pitt River, British Columbia. In: *Fluvial Sedimentology* (Ed. by A.D. Miall). *Mem. Can. Soc. Petrol. Geol., Calgary.* **5**, 89–104.

BANKS, N.L. & COLLINSON, J.D. (1975) The size and shape of small-scale current ripples: an experimental study using medium sand. *Sedimentology*, **22**, 583–599.

BRIDGE, J. S. (1981) Bed shear stress over subaqueous dunes, and the transition to upper-stage plane beds. *Sedimentology*, **28**, 33–36.

BRIDGE, J.S. & JARVIS, J. (1977) Velocity profiles and bed shear stress over various bed configurations in a river bend. *Earth Surf. Processes*, **2**, 281–294.

COLLINSON, J.D. (1970) Bedforms of the Tana River, Norway, *Geogr. Annlr* **52A**, 31–56.

DAVIS, W.M. (1890) Structure and origin of glacial sand plains. *Bull. geol. Soc. Am.* **1**, 195–202.

ENGELUND, F. & FREDSØE, J. (1974) Transition from dunes to plane bed in alluvial channels. *Tech. Univ. Denmark, Inst. Hydrody. Hydraul. Engng, ser. Pap.* **4**, 56 pp.

FAHNESTOCK, R.K. (1963) Morphology and hydrology of a glacial stream—White River, Mount Rainier, Washington, *Prof. Pap. U.S. geol. Surv.* **422–A**.

GUY, H.P., SIMONS, D.B. & RICHARDSON, E.V. (1966) Summary of alluvial channel data from flume experiments 1956–61. *Prof. Pap. U.S. geol. Surv.* **462–I**.

JONES, C.M. & McCABE, P.J. (1980) Erosion surfaces within giant fluvial cross-beds of the Carboniferous in northern England. *J. sedim. Petrol.* **50**, 613–620.

JOPLING, A.V. (1965a) Laboratory study of the distribution of grain size in cross-bedded deposits. In: *Primary Sedimentary Structures and their Hydrodynamic Interpretation* (Ed. by G.V. Middleton). *Spec. Publs Soc. econ. Paleont. Miner., Tulsa*, **12**, 52–65.

JOPLING, A.V. (1965b) Hydraulic factors controlling the shape of laminae in laboratory deltas. *J. sedim. Petrol.* **35**, 777–791.

JOPLING, A.V. & FORBES, D.L. (1979) Flume study of silt transportation and deposition. *Geogr. Annlr* **61A**, 67–85.

McDONALD, B.C. (1972) The Geological Survey of Canada sedimentation flume. *Pap. geol. Surv. Can.* **71–46**.

McDONALD, B.C. & VINCENT, J.S. (1972) Fluvial sedimentary structures formed experimentally in a pipe, and their implications for interpretation of subglacial sedimentary environments. *Pap. geol. Surv. Can.* **72-27**.

MORTON, R.A. (1978) Large-scale rhomboid bed forms and sedimentary structures associated with hurricane wash-over. *Sedimentology*, **25**, 183–204.

MOSS, A.J., WALKER, P.H. & HUTKA, J. (1980) Movement of loose, sandy detritus by shallow water flows: an experimental study. *Sedim. Geol.* **25**, 43–66.

NAMI, M. & LEEDER, M.R. (1978) Changing channel morphology and magnitude in the Scalby Formation (M. Jurassic) of Yorkshire, England. In: *Fluvial Sedimentology* (Ed. by A.D. Miall). *Mem. Can. Soc. Petrol. Geol., Calgary* **5**, 431–440.

SAUNDERSON, H.C. (1981) Tunnel transformation of anti-dunes. *Phys. Geog.* **2**, 125–145.

SIMONS, D.B. & RICHARDSON, E.V. (1966) Resistance to flow in alluvial channels. *Prof. Pap. U.S. geol. Surv.* **422-J**.

SIMONS, D.B., RICHARDSON, E.V. & NORDIN, C.F. (1965) Sedimentary structures formed by flow in alluvial channels. In: *Primary Sedimentary Structures and their Hydrodynamic Interpretation* (Ed. by G.V. Middleton). *Spec. Publs Soc. econ. Paleont. Miner., Tulsa*, **12**, 34–52.

SOUTHARD, J.B. (1971) Representation of bed configurations in depth–velocity–size diagrams. *J. sedim. Petrol.* **41**, 903–915.

SOUTHARD, J.B. & BOGUCHWAL, L.A. (1973) Flume experiments on the transition from ripples to lower flat bed with increasing grain size. *J. sedim. Petrol.* **42**, 1114–1121.

WILLIAMS, G.E. (1971) Flood deposits of the sand-bed ephemeral streams of central Australia. *Sedimentology*, **17**, 1–40.

WILLIAMS, G.P. (1967) Flume experiments on the transport of a coarse sand. *Prof. Pap. U.S. geol. Surv.* **562-B**.

WILLIAMS, G.P. (1970) Flume width and water depth effects in sediment-transport experiments. *Prof. Pap. U.S. geol. Surv.* **562-H**.

Present-day channel processes

Spec. Publs int. Ass. Sediment (1983) **6**, 61–83

River channel changes: retrospect and prospect

EDWARD J. HICKIN

Department of Geography, Simon Fraser University, Burnaby, British Columbia V5A 1S6, Canada

ABSTRACT

This paper reviews some of the major developments in the study of channel changes during this century and offers comment on the direction of future research. Topics addressed include the relation of channel equilibrium to time-scale, the process relations of the formal theory of equilibrium channel-morphology, regime theory and hydraulic geometry, complex channel response, human-induced channel adjustments, and channel changes of the Holocene and Pleistocene. It is argued that our perspective on the fluvial system would benefit greatly from research focused on channel changes during the last several hundred years rather than for shorter or longer time-scales.

INTRODUCTION

Although the manner in which rivers change the form and pattern of their channels in response to environmental change has been a recurring theme in river studies, it has recently enjoyed considerably increased attention from earth scientists. Perhaps the most significant recent evidence of this interest is the appearance of several collected works and reviews of studies of channel changes (for example, see Gregory, 1977b, 1979, Rhodes & Williams, 1979 and Park, 1981), and the fact that a session was devoted to the topic at the Second International Conference on Fluvial Sediments at Keele.

The study of river channel changes, in the broadest sense of the term, is no less than the study of equilibrium channel behaviour and the nature of excursions from those equilibrium conditions. As such it includes almost all that we know about the fluid mechanics and morphology of alluvial channels. But in a more narrow sense of the term it is the collection of empirical and theoretical studies concerned with the adjustment of channel cross-sectional size, form and pattern to shifts in environmental conditions, particularly those that promote changes in discharge and in sediment loads. In a still narrower sense, channel

0141-3600/83/0106-0061 $02.00

changes may be regarded as those that have been induced by the activities of human beings.

CHANNEL EQUILIBRIUM AND TIME-SCALES

Central to any discussion of river changes, however, is the concept of equilibrium, and as Schumm & Lichty reminded physical geologists in 1965, geomorphic equilibrium only has meaning in the context of particular time-scales. The full significance of this fact is only now achieving general recognition (see Cullingford, Davidson & Lewin, 1980). Organizing knowledge by time-scale may be useful and desirable for many purposes but it also constrains the way in which we think about the world. Time-scale selection largely determines the questions that we can ask. In the context of channel changes it has emerged as a serious limitation to our understanding of the ways of rivers.

Changes in river form and pattern have typically been examined at three arbitrary time-scales; for the sake of discussion we might call them geologic, geomorphic and engineering time-scales.

Geologic time, measured in terms of perhaps a million plus years, might also be termed cyclic time; it is the time-scale in which the completion of an erosion cycle is a significant event. The governing conditions for landscape change at the geologic time-scale include major tectonic events such as crustal plate displacement and related processes of regional warping and mountain building. These factors, in turn, influence climate, weathering, and general denudation rates. It is the time-scale that many of us will associate with the regional physiographic models of classical geomorphologists such as William Morris Davis, Lester King, and Walther Penck. It is the domain of the physical geologist or regional physiographer who relies on the rock record, on global radiometric dating, on earth magnetic-field reversals, on surveys of planation surfaces and of deep weathering, and on the methods of structural and historical geology, to piece together the evolution of the planet's surface.

The geomorphic time-scale is that of much of modern fluvial geomorphology, ranging from a hundred to a few hundreds of thousands of years. For example, the now classic palaeoclimatic reconstruction by G. H. Dury (see his review in 1977a) is based largely on Holocene and late Pleistocene river underfitness. Similarly, later attempts at refinement of this method by Schumm (see his review in 1977) are based largely on studies of contemporary river behaviour on the American Great Plains and on several tens of thousands of years of history of the Murrumbidgee River in the Riverine Plains of eastern Australia.

Fluvial geomorphologists generally have assumed that the governing conditions for landscape change at the geologic time-scale are sensibly constant and therefore largely irrelevant to discussions of river changes over geomorphic time. They have been far more concerned with the fluvial consequences of the profound environmental fluctuations associated with Pleistocene glaciation. Investigations of the history of river changes evidenced by valley fills, terraces, deranged drainage, channel-capacity contraction, and so on, have relied on extensive field observation employing the descriptive tools of the stratigrapher and soil scientist. Much of the interpretation has relied on the use of techniques such as the mechanical analysis of sediments, C^{14} dating, pollen and tree ring analysis, and more recently on identifying actual and laboratory analogues through studies of contemporary river form and process. It is the latter development that has seen research interests of geomorphologists, sedimentologists and engineers converge in recent decades.

Fluvial geomorphologists have also relied on the historical data contained in personal journals and in time-lapse photography and maps, although these are largely restricted to changes during the last century. There is something of an information gap between the domain of this type of record and that of reliable C^{14} dating.

Engineering time, at the high-frequency end of the time-scale continuum, is the domain of the basic science of the river engineer: fluid mechanics. Engineers traditionally have developed solutions to river problems on the assumption that the governing conditions at the geomorphic time-scale are constant. This is the basis of the grade, regime, or steady-state approach to river behaviour. Given various assumptions, fluid mechanics models provide precise quantitative information on the fluid flow-field and form of an open channel (for example, see the 1957 Rozovskii analysis of bend flow). Such models are usually developed for the time period necessary for the flow to pass through the channel reach in question. It is assumed that equilibrium channel morphology is quickly established and it follows from the steady-state assumption that long-term channel form is obtained by simple extension of a stationary series with respect to time. This assumption might be valid for short periods of time (say a couple of years or even a few decades) but it becomes less defensible as the time period in question is lengthened. Unfortunately, river engineers are commonly required to design structures with a life of the order of 100 years or more and in such cases the assumption is tenuous indeed.

Because problems of river behaviour have been perceived differently by engineers on the one hand, and by earth scientists on the other, until recently there has been little interaction among these two groups. But in the last few decades this situation has changed markedly for at least two reasons. First, earth scientists have looked to fluid mechanics and engineering research to provide explanations and integration of their largely qualitative observations. This is particularly evident in sedimentology and in fluvial geomorphology. Secondly, engineers have looked to the earth sciences to specify the non-stationarities in their steady-state models that all too often prevent accurate forecasts of long-term river behaviour. Engineers' concerns with long-term river adjustment have been legislated partly by governments responding to the new environmental awareness of the public. Engineers have been required to give answers to a broad range of environmental questions, many of which are well beyond their traditional area of expertise. These social forces have no doubt quickened the pace of scientific collaboration.

Although this collaboration of river engineers and

earth scientists has clearly been beneficial to both groups and to science in general, it has at the same time been a disappointing one in some respects. Engineers have gained only general insights into design problems as they relate to processes operating at geomorphic time-scales, and earth scientists have tended to become embroiled in the problems of the engineering time-scale without really finding these process studies very useful for explaining significant channel changes at the geomorphic time-scale. The fact is that, after a century of relatively intense effort to understand the ways of rivers, we remain unable to predict an accurate channel response to a simple change in flow regime at any time-scale. Some of the underlying problems are brought into sharper focus if we briefly review what we actually know about rivers in equilibrium at engineering time-scales. These are the facts and ideas that collectively might be termed the theory of equilibrium channel morphology.

THE THEORY OF EQUILIBRIUM CHANNEL MORPHOLOGY

Introduction

In its simplest and most general form it states that, if left undisturbed, rivers will establish some stable combination of morphological elements for a given discharge of water and sediment. Any disturbance to this equilibrium channel morphology will set in motion processes that will return the channel to its stable form and pattern. Any change in the character of fluid discharge will set in motion changes to establish a new state of channel equilibrium in a manner dependent on the degrees of freedom of channel adjustment and on the nature of the adjustment processes.

A river conveying a given discharge possesses the potential of at least eight significant degrees of freedom for change: width, depth, sediment calibre, sediment discharge, velocity, slope, boundary roughness and planform. It follows that it is necessary to specify as many processes linking these variables to characterize this fluvial system fully. The search to identify these process equations has constituted a major effort by river scientists during this century.

Three particular processes are thought to provide obvious candidate equations: flow continuity, flow resistance and sediment transport. The exact nature of the corresponding process equations is the subject of much debate, however, and the general character of the remaining five processes is even less well understood.

Perhaps one of the best-known attempts at closure of a somewhat simpler set is that of Langbein (1964) and Langbein & Leopold (1964). They postulated two further 'processes' (actually probable conditions) to the three listed above: a tendency for a river to expend equal power per unit bed area and another tendency towards equal power expenditure per unit channel length (minimum work rate). Although these two conditions allow closure and yield a supposed average hydraulic geometry, they are, nevertheless, probability statements rather than process equations and a unique solution simply is not possible. Indeed, the two conditions are mutually exclusive.

Others have attempted similar solutions (for example, see Yang, 1976 and Yang, Song & Woldenberg, 1981, for a minimum unit-power approach and Davies & Sutherland, 1980 and Kirkby, 1977, who respectively argue for maximization of flow resistance and for efficiency of sediment transport) while others have concluded that the fluvial system is intrinsically indeterminant (see Maddock, 1970).

Another view, perhaps most recently stated by Hey (1978), is that the lack of any physical justification for the minimization procedure (the arguments generally are rather anthropomorphic), coupled with its inability to provide unique solutions, simply suggests that additional process equations need to be defined to obtain a determinant solution. Hey (1978) suggests that the remaining equations should specify processes of bed deformation, bank erosion, and of meandering. Few would argue about the importance of these processes, but of course it is one thing to state the nature of the problem and altogether another to solve it.

It is useful to our developing a sense of the state of the fluvial art to consider some of the better-defined process equations.

The process equations

Flow continuity is a basic physical relationship which can be expressed as

$$Q = w\bar{d}\bar{v} \qquad (1)$$

where Q is discharge, $\bar{v}$ is mean flow velocity, and w and $\bar{d}$ are respectively the width and mean depth of the channel. This fundamental definitional equality, however, is deceptively simple. Although the continuity equation finds relatively direct application in laboratory experiments with controlled flows, highly unsteady natural flows are not conveniently reduced to a single summary discharge. The width and mean depth of a river channel are clearly determined by all discharges greater than that corresponding to threshold sediment transport, integrated over a time period limited by the rate of lateral or vertical channel movement. But most river scientists recognize that not all discharges in this domain are of equal effectiveness in forming a river channel; a far more limited range is thought to be dominant.

Although the dominant discharge concept has become firmly entrenched in fluvial geomorphology and hydraulic engineering, it remains a controversial idea (see Williams, 1978). Many researchers agree with Dury (1961) that, not only is the concept of a single dominant discharge valid, but also it can be equated to bankfull discharge (see Leopold, Wolman & Miller, 1964, and the more recent discussion of Ackers & Charlton, 1970). Furthermore, Dury and others embraced the notion that bankfull discharge corresponds with the most probable flood on the annual series (Dury, 1961; Dury, Hails & Robbie, 1963). Many researchers would argue, however, for several dominant discharges (Woodyer, 1968) or for a limited range of dominant discharge (Carlston, 1965). More recent work has developed the idea that the frequency of bankfull discharge and dominant discharge may differ and that both are dependent on the type of river regime (see Benson & Thomas, 1966, Harvey, 1969, Pickup & Warner, 1976 and Pickup & Rieger, 1979).

Difficulties of applying the dominant discharge concept will be compounded if it is shown that different components of channel morphology (say mean depth and meander wavelength) scale with different parts of the discharge distribution. One thing clearly is certain: the basic problem of specifying dominant discharge in rivers remains unresolved.

Flow resistance, the second of the three process equations, constitutes a still more elusive relationship although a commonly adopted empirical formula is

$$1/\sqrt{ff} = v/V_* = C_1 (\bar{d}/k_s)^{C_2}, \tag{2}$$

in which $V_* = \sqrt{g\bar{d}s}$, the shear velocity, and ff is the Darcy Weisbach resistance coefficient. If the Strickler relation $n = 0 \cdot 034 D^{\frac{1}{6}}$ for the resistance effect of grain size, D, applies then $k_s = D$, $C_1 = 8 \cdot 4$, and $C_2 = 1/6$, equation (2) yielding the Manning equation.

Although equation (2) has enjoyed successful application in channels with regular fixed boundaries, it becomes increasingly difficult to apply when allowances are made for the varied character of natural alluvial channels. For example, although we now better understand the nature of stability criteria for bedforms than we did two decades ago (see Southard, 1971; Yalin, 1977) we remain unable to predict the contribution of bedform roughness to k_s in equation (2) (see Simons & Sentürk, 1977, for a review of many such attempts including the pioneering work of Einstein & Barbarossa, 1952). It appears that our ability to do this is some way off, depending as it likely does on a far more comprehensive understanding of macroturbulence structure (particularly at high discharges) than we can claim at present (see Jackson, 1975b, 1976).

Resistance to flow, however, is clearly also dependent on many factors other than grain and form roughness, not the least of these being the influence of channel bends. But here again, although the general role of planform in determining flow resistance has been recognized for some time (see Bagnold, 1960 and Leopold *et al.*, 1960) specification of the exceedingly complex nature of bend flow remains beyond our present capability (see Callander, 1978).

Sediment transport, the third process equation, is of course the subject of a vast literature. Although any sort of review of this literature is obviously beyond the scope of this commentary, it is useful to consider the nature of a few general problems.

The role of sediment transport in determining channel morphology is set in context by equation (3)

$$\frac{\partial z}{\partial t} = \frac{\partial G_s}{\partial x}. \tag{3}$$

Here the rate of change in bed elevation on the left of equation (3) is dependent on downstream changes in the sediment transport rate. The transient solution of equation (3) is the solution of the degradation/aggradation problem and has been attempted by several researchers (e.g. see Tinney, 1962, Culling, 1960, Scheidegger, 1965, Devdariani, 1967 and Soni, 1981). Central to all these solutions is the specification of the sediment transport function.

Most of the widely used equations designed to predict the important bedload component of G_s in equation (3) are a variant of the early Du Boys relation in which the rate of bedload transport is expressed as a simple function of the excess shear stress beyond that necessary to initiate sediment movement. These Du Boys-type equations can be expressed in terms of threshold velocities or discharges, but all require specification of the critical flow conditions for sediment movement. Several important equations of this type bear the names of their originators (Shields, Schoklitsch and Meyer-Peter and Müller) and are the subject of discussion in most reviews (see Raudkivi, 1967, A.S.C.E. Task Committee, 1971a, b, Graf, 1971, Bogardi, 1974 and Simons & Sentürk, 1977).

The only significant major departure from the mean tractive force concept used by Du Boys is the work of Einstein (1950). His partly stochastic approach, developed from concepts of fluid turbulence, assigned a probability for movement to each bed particle, thus avoiding the use of a threshold entrainment condition. Einstein's 'bedload function' and functionally similar equations (see discussion by Chien, 1956) proposed by others such as Kalinske (1947) and Graf & Acaroglu (1968) are discussed in the reviews cited above.

It should be noted that, although most of these sediment transport equations have a general form based on theoretical considerations, all include important empirically determined coefficients largely based on data from laboratory flumes. This fact immediately raises the question of the appropriateness of these laboratory-determined constants in a field situation. In most circumstances such questions are simply resolved by field observation, but unfortunately this case is further complicated by problems of measurement. Except on very small rivers or at unusually elaborate control structures, it is extremely difficult to obtain an accurate measurement of bedload transport. Our present instrument technology is simply inadequate in most cases. Nevertheless, it seems likely that the several orders of magnitude range in bedload-transport estimates provided by the general-use equations (see Vanoni, 1975) implies that many of the assumed constants are in fact unspecified variables.

One of the primary culprits is the universal assumption that the immersed weight of a bed particle alone determines the force necessary to initiate its movement (or determines its probability of movement in the stochastic models). It may be a reasonable assumption in a laboratory flume with spherical bed particles but it is a highly dubious one in many natural channels. For example, it ignores armouring effects and sheltering effects of imbrication in gravel-bed streams and ignores the effects of organic mats that can form a protective coating during low flows in sand-bed channels.

The bed armour problem is really part of the broader fundamental question of whether equilibrium sediment transport conditions are commonly attained in natural rivers. All sediment transport equations attempt to predict the maximum sediment transport rates given certain fluid dynamic conditions. That is, they assume that sediment supply is not a factor limiting sediment transport. There is more than a little evidence to suggest, however, that this simply is not true. The hysteresis effect in sediment concentration associated with the initial flushing of sediment during passage of a flood wave is a widely recognized phenomenon. Variations in the supply of freeze–thaw debris are known to produce seasonal variations in sediment transport rates in some rivers (see Nanson, 1974). An example of the analogous effect at a geomorphic time-scale is provided by Church & Ryder (1972). They describe the Canadian landscape as one of declining intensity of geomorphic activity since immediate deglaciation. Although sediment supply from the freshly glaciated land surface to rivers is unlikely to have been a factor limiting sediment transport rates in immediate deglacial times, much of this glacial material has now been flushed from the geomorphic system and the sediment supply rate has declined. This fact together with the tendency for Canadian gravel-bed rivers to develop armoured beds, suggests that many of them may well be transporting sediments at a rate which is supply-limited. In the case of supply-limited transport rates, sediment yields from drainage basins must be known if these rates are to be estimated. It is sufficient to say that, in this realm, our understanding at best is qualitative and completely inadequate for the task at hand. Perhaps I should leave the final remarks on sediment transport to Simons & Sentürk (1977) who conclude that 'The mechanics of sediment transport is so complex that it is extremely unlikely that a full understanding will ever be obtained. A universal sediment transport equation is not and may never be available' (p. 644).

The equations for closure, as indicated earlier, are not self-evident but certainly must include (in addition to the three so far considered) a governing equation for lateral movement of channels. Such movements include the complex sets of sinuous motions that we collectively term meandering.

Meandering is a significant process at both engineering and geomorphic time-scales. Lateral migration is sufficiently rapid on many large rivers that it can result in the channel shifting sideways by a complete channel width in less than 20 years. It therefore is a process that can change mean water-surface slopes rapidly (within the limits set by valley slopes) and it can cause local catastrophic changes through the formation of cut-offs. Meander development also introduces considerable flow distortion at bends which must be incorporated into any comprehensive flow resistance equation. It is also a significant factor governing local sediment supply.

Unfortunately, meandering is also a process about which we have a great deal to learn. Although some general empirical relationships among discharge and various planform elements of meandering channels have been known for some time (see Leopold *et al.*, 1964) the underlying causes and controls of these regularities are yet to be identified.

Empirical studies of planform geometry apart, research into meandering has been conducted largely independently on three closely related topics: the cause of meandering, the nature of bend flow, and the nature of lateral migration.

The cause of meandering, a fundamental and long-standing question in fluvial studies, remains unanswered to this day. Earlier work, in which helical flow is featured as an important process, is briefly reviewed by Leopold *et al.* (1964) and more recent developments are discussed by Callander (1978). Currently in favour is the case for dynamic instability of the alluvial channel-bed. Stability analyses have been used to argue (for example, see Callander, 1969, Engelund & Skovaard, 1973 and Parker, 1976) that, because the mobile bed of a straight channel is unstable, any small perturbation will be amplified to produce the pattern of alternating pools and riffles that Friedkin (1945) and many others subsequently have noted are the precursors to meandering. Although these rather complex mathematical models have certain basic elements in common (periodic functions for the initial perturbation in which the wavelength of maximum amplification is assumed to be the corresponding meander wavelength) they employ a variety of other assumptions to close the set of flow equations. In spite of the variety of specific solutions, all succeed in predicting wavelengths similar to those found in natural channels; it is not clear whether this is a weakness or a strength of this approach! Certainly the general nature of the approach is appealing; the periodic function does not specify the process of perturbation and therefore can easily accommodate meandering tendencies in a variety of media. Parker (1976) argues that, although perturbations in sediment transport are necessary to form meanders in rivers, variations in Coriolis acceleration serve the same purpose in oceanic currents, heat differences in meltwater streams, and surface tensions in water threads on glass surfaces. This generality at the same time leaves stability analyses open to the criticism of being 'black box' solutions to complex problems.

Other explanations of meandering proposed in the last couple of decades are less general than the stability analyses but are just as lacking in specific process information. For example, Shen & Komura (1968) and Quick (1974) appeal to a periodic reversal in vorticity while Yalin (1977) argues that meander wavelength scales with the largest macroturbulent eddy that will fit in the channel. The debate continues.

Once a channel bend begins to develop, for whatever reason, the flow is characterized by the interaction of two sets of forces: those demanding the conservation of angular momentum (free vortex flow) and those promoting the lateral transfer of momentum (helical flow). The resultant pattern of bend flow is directly related to erosion and deposition in the channel and is the primary control on the rate and pattern of lateral migration.

Theoretical studies of bend flow, in which solutions have been sought for the continuity and momentum equations for turbulent flow, have led to successful prediction of its general character. Early research is reviewed by Rozovskii (1957) and later contributions are discussed by Callander (1978). Recent advances have improved the theory to the point that it can provide realistic descriptions of the three-dimensional flow field (including the correct sense of helical flow and secondary flow strength of the right order of magnitude), of the bed shear-stress distribution through the bend, and of the general bed configuration. It has clearly been one of the more successful areas of theoretical analysis although the theory certainly is not without rather significant limitations. For example, it only applies to bends with a single helical cell and, perhaps more importantly, it does not apply to bend flow in which there is flow separation.

Although Yen (1975) and Allen (1977) have offered general analytical comment on bed erosion, a theoretical model of lateral migration is not available.

The pace of experimental research on bend flow has increased markedly since the classic works of Mockmore (1943) and Rozovskii (1957), to peak in the mid-1970s. Some of the more notable contributions are reviewed by Callander (1978). Although these experimental studies provide test data for the theoretical models and have added to our understanding of bendflow processes, they have not yielded a quantitative empirical framework for erosion prediction. Indeed, in this context many of them may have provided misleading results: few experimental studies are of live-bed conditions and most adopt constant-width channels with rectangular or trapezoidal sections, a simple geometry which likely represents a significant departure from reality.

There remains a gap between experimental studies of this fixed-bend type and those that have attempted to characterize the initiation and growth of meander bends (see Friedkin, 1945, Wolman & Brush, 1961,

Hickin, 1969, 1972 and Schumm & Khan, 1972). The general pattern of alternate bar formation, flow deflection, and the alternating bank erosion that leads to development of a sinuous channel has been known for some time. But no experiment has related the development of this sinuous deformation of the channel alignment to the changes in the velocity and shear distributions in the flow. Furthermore, most experiments have been conducted in non-cohesive sediments, producing pseudomeanders (see Wolman & Brush, 1961) rather than analogues to natural meanders with pools and riffles. The more realistic results of Schumm & Khan's (1972) experiments using a kaolinite/sand sediment have highlighted this distinction in meander type.

There are still fewer studies of flow in real river bends. The remarkable exposition of fluvial processes on the River Klarälven by Sundborg (1956) might have established an early trend, but it has taken until the 1970s for his lead to be followed. Jackson (1975a) measured primary and secondary flow velocities around several bends of the Lower Wabash River in order to understand better the pattern of point bar sedimentation there. Bridge & Jarvis (1976) and Bridge (1977) have described primary and secondary flow velocities around a bend of the River South Esk in Scotland. They extended the theoretical concepts of Rozovskii (1957), Allen (1970a, b, 1971) and Engelund (1974, 1975) to develop a three-dimensional model of bend sedimentation. In a field study specifically designed to yield information on the hydraulic control of lateral migration rates Hickin (1978) obtained measurements of primary and secondary flow velocities through a continuous series of bends on the gravel-bed Squamish River in British Columbia.

These studies confirm the main theoretical and experimentally based generalizations about bend flow but they also point to important factors ignored in this earlier work. Principal among these is the interaction of flow in adjacent bends and the inevitable occurrence of separation zones (see also Woodyer, 1975, Taylor, Crook & Woodyer, 1971, Leeder & Bridges, 1975 and Hickin, 1979). Unfortunately, these are factors of great significance to the lateral migration process and will have to be understood before an appropriate model can be developed.

The reality is that a universal channel migration model probably never will be available. It is after all just another version of the sediment transport problem, further complicated by the exceedingly more complex flow and entrainment conditions involved.

These circumstances make actual measurements of the process vital to our ability to forecast channel changes.

There have been very few direct field surveys of channel migration; Lewin (1977) provides a brief review of some recent efforts. The successive surveys of Watts Branch, Maryland, first reported by Wolman (1959) and later by Leopold (1973) provide a well-known example of this type of direct monitoring. Other direct observations of channel shifting have been reported by Coleman (1969), Daniel (1971), Hughes (1977), Lewin & Brindle (1977) and Hooke (1977, 1980).

Direct monitoring of channel changes at monumented transects is a recent development and one which, because of the considerable maintenance effort required and the uncertainty of obtaining useful results, is rarely sustained for more than a few years. It follows that the method has been most useful for describing channel migration processes at relatively short time-scales of a few days to just a few years at most. Although this type of direct measurement provides valuable information on the process of lateral migration, the sampling time is far too short to provide a reliable estimate of migration rate for time-scales of the order of decades or longer.

The great majority of lateral migration measurements, however, have been obtained indirectly from serial cartography and aerial photography of channels. This type of historical record generally extends the period of observation to a few decades in the case of aerial photographs and up to a century or so in the case of cartography. Clearly, its usefulness will vary considerably from place to place. In areas with a long cultural history the cartographic record is likely to be useful (at least five surveys over the last 130 years are generally available in Britain, see Lewin & Hughes, 1976) but it is minimally useful in much of the New World. The detailed record of channel pattern for parts of the Mississippi–Missouri system in the United States provides a notable exception to this latter generalization.

Systematic photogrammetric surveys are generally restricted to the post-war period although specific projects may have been undertaken in the first half of this century in some areas. For example, a typical maximum time interval for the photographic recording of channel migration in most of Canada and in much of the United States is about 30 years. Except in special cases, most areas are not likely to be aerially photographed any more frequently than once every five or ten years; this practical limit of coverage defines the short-term limit of resolution of channel migration processes by this method.

Channel migration rates for time periods from about 100 to 500 years have been obtained from the analysis of flood-plain vegetation successions (for example, see Everitt, 1968 and Hickin & Nanson, 1975). Nanson & Beach (1977) also report associated rates of vertical flood-plain accretion based on an analysis of wood-cell structure in cottonwoods. The lower limit of resolution of flood-plain dating by dendrochronology is set by variability in tree-colonization times (probably uniformly small for most species) and subsequent survival rates; the length of record is obviously dependent on the life of the tree.

Few generalizations can be made from these recent field studies that have not already been stated in the classic work of Leopold *et al.* (1964). Perhaps one of the more significant developments has been the willingness to recognize the wide variety of bend evolution other than simple expansion to a sine-generated form and its subsequent downvalley translation. A number of case-studies have stressed the complexity of combined expansion, rotation, and translation in bend development (for example, see Daniel, 1971; Handy, 1972; Chitale, 1973; Hooke, 1977). A very common meander form is the complex compound meander loops described by Brice (1974), and Hickin (1974). Hickin & Nanson (1975) confirmed that migration rates at the apex of bends on the tortuous Beatton River in British Columbia are dependent on channel curvature, which plays an important but not precisely identified role in governing the development of complex meander lobes.

Another noteworthy development in recent studies of river planform is the increased recognition of the importance of cutoffs in meander kinematics (see Kondrat'yev, 1968, Kulemina, 1973 and Mosley, 1975a, b). Until the last few decades our preoccupation with 'equilibrium' forms has led to the dismissal of cutoffs as transient disturbances of the fluvial system. In fact, cutoffs form frequently enough on many rivers that certain reaches may remain in this transient state permanently!

Some conclusions

It is clear from this brief review exercise that in each process case considered, our knowledge is inadequate to define a set of general governing equations that together would constitute a theory of equilibrium channel morphology at an engineering time-scale. For certain purposes, however, the continuity equation for rivers may be relatively well defined and the resistance equation may be an adequate approximation. There also remains some prospect that further significant advances will be made in our understanding of the lateral shifting of rivers. But the existing sediment transport equations are simply inadequate in almost all cases. Not only do they apparently inaccurately describe the mechanics of sediment movement, but it appears likely that the flow-limited equilibrium sediment transport rate they are designed to predict is both rare and insignificant in shaping real channels. It seems that there may be a far better prospect of predicting sediment transport rates from channel morphology rather than vice versa! Certainly we must conclude that a process-based theory for predicting channel changes simply does not exist.

This general conclusion, of course, is hardly a novel one. It was this same conclusion reached a century ago that persuaded river engineers to develop the empirical sets of norms that we have come to know as regime 'theory' and hydraulic geometry.

REGIME THEORY AND HYDRAULIC GEOMETRY

Empirical studies have a long tradition both in the earth sciences and in engineering research. The engineering work on the stable alluvial irrigation canals in India marks an appropriate beginning to such river studies in the modern era. Extending the early observations and ideas of R. G. Kennedy, Gerald Lacey (see the review by Mahmood & Shen, 1971) developed a set of empirical relationships describing the morphology of stable alluvial canals for a limited range of discharges, water-surface slopes, and of a sediment coefficient (now known as the Lacey silt factor). Although these regime relationships adequately describe the geometry of Punjab canals, they have not found a similar degree of success in wider applications. The coefficients of Lacey's empirical equations clearly reflect fortuitous constancy in certain parameters which apparently are important variables in other fluvial contexts. For example, E. W. Lane was prompt in pointing out in 1937 that the Lacey silt factor does not appear to account adequately for variations in boundary materials and sediment loads. Subsequently, Inglis (1949) and perhaps the best known of contemporary regime-method engineers, Tom Blench (1966), have attempted to incorporate sediment load parameters into the regime equations. Nevertheless, they remain empirical equations which will only apply well to the limited set of conditions from which they were derived. A discussion of the relation between regime theory and critical tractive force theory is provided by Henderson (1961); a general review of regime equations is given by Mahmood & Shen (1971).

Regime theory is a very significant contribution to the study of channel adjustment for a variety of reasons. It continues to be a widely used design tool obviously judged by many engineers to be the best technique available for predicting stable channel size and form. Also, it rests firmly on the concept of channel equilibrium, one which we have noted is fundamental to any discussion of river channel changes. Perhaps most important is the fact that, with the engineering steady-state concept of equilibrium relaxed (quasi-equilibrium), regime theory provided the foundation and inspiration of the general geomorphic discussion of river form in the 1950s that to earth scientists became known as the hydraulic geometry of stream channels.

The hydraulic geometry concept is discussed in some detail by Leopold *et al.* (1964), reviewers who contributed much to its development. Briefly, it is the set of power relations between discharge and principally the width, mean depth, and mean flow velocity, but also the water-surface slope, boundary roughness and sediment load, of the natural channel. Of the two basic types of hydraulic geometry, at-a-station and downstream, it is the latter that has application to problems of river channel change. At-a-station hydraulic geometry in most cases simply describes the way in which changing discharge fills the effectively rigid-boundary channel that has been moulded by an earlier and larger discharge. The downstream hydraulic geometry, on the other hand, describes how the fully adjusted channel form accommodates changes in a competent bankfull (or near bankfull) discharge; it is in this 'downstream' context that hydraulic geometry and regime theory are essentially one and the same view of channel equilibrium.

It is important, however, that we do not lose sight of the fact that the hydraulic geometry simply describes what is there. It has little or no theoretical significance. Even the power function form ascribed to the relations is done as much for convenience as for any other reason. There is no one hydraulic geometry and it is most useful for engineering purposes when it is developed for an environmentally homogeneous region (for example, see Bray, 1975, 1979).

Similar comment can be offered on the empirical relationship, $l = cq_{bf}^{\frac{1}{2}}$, widely acknowledged to exist between average meander wavelength (l) and bankfull discharge (q_{bf}). The fact that meander geometry appears to scale with channel width and with the square root of discharge was recognized by a number of researchers including Inglis (1949), Friedkin (1945), and Leopold & Wolman, (1960), but it was George Dury who provided the most comprehensive treatment of its significance as an indicator of environmental change. In a series of papers during the 1950s and 1960s he exploited the meander wavelength/discharge relationship to estimate the discharge reductions and associated Quaternary climatic changes implied by regional underfitness of rivers (see the review by Dury, 1977a). But even at this level of generalization many researchers have not been persuaded that the meander wavelength/discharge relation is sufficiently well defined to warrant these deductions about the former regional climatic and hydrologic environment.

Much of the criticism has focused on three issues. The first concerns the adoption of a single summary discharge variable to characterize the formative flow and is a concept central to both downstream hydraulic geometry and Dury's meander planform analysis. The problem is detailed here in the earlier discussion of the continuity equation.

The second and similar issue concerns the appropriateness of the simple average meander wavelength as the index of meander scale. It has been argued that the variation in meander wavelength in a given reach of river does not represent normally distributed excursions from some mean wavelength but rather reflects the interaction of several fundamental wavelengths related to discharge variability. It was this view of meander planform that led Speight (1965) to apply spectral analysis to meander arrays in order to express the contribution of the assumed component harmonics to the resultant waveform. Other researchers have pursued this line of enquiry (for an example of recent work see Ferguson, 1975, 1976) although the collective results so far have been less than conclusive.

The third issue concerns the general question of the influence of boundary materials on channel form and pattern. For example, Dury (1977a) recounts his case for there being no difference between bedrock and alluvial meanders, and yet this seems contrary to the principle that boundary material influences the width of a channel which in turn is a scaling factor for meander wavelength (Shahjahan, 1970). Although boundary material variation usually is statistically unimportant in these bivariate relationships in which discharge ranges over several orders of magnitude, it becomes increasingly more important as the discharge range contracts. In the limit, knowing the discharge allows a mean morphology to be specified, but the wide scatter of possibilities about this mean condition impose no better than an order-of-magnitude level of resolution.

It was on the assumption that much of this scatter could be attributed to variation in boundary materials that Schumm based his empirical studies of channel morphology, discharge, and bed and bank materials in the 1960s (see the review by Schumm, 1977). In a sense Schumm's work marks a return to the approach advocated by the regime theorists who earlier attempted to incorporate sediment type into their models. He collected and analysed data on flow and channel characteristics from rivers on the Great Plains of the United States and on the Riverine Plains of eastern Australia, yielding the following regression equations.

$$w = 37Q^{0 \cdot 38}/M^{0 \cdot 39} \tag{4}$$
$$\bar{d} = 0 \cdot 6 M^{0 \cdot 34} Q^{0 \cdot 29} \tag{5}$$
$$F = w/\bar{d} = 255/M^{1 \cdot 08} \tag{6}$$
$$M = 55/Q_{\mathrm{bs}} \tag{7}$$
$$S = 60/M^{0 \cdot 38} Q^{0 \cdot 32} \tag{8}$$
$$\lambda = Q_{\mathrm{ma}}^{0 \cdot 48}/M^{0 \cdot 74} = 1890\, Q_{\mathrm{m}}^{0 \cdot 34}/M^{0 \cdot 74} \tag{9}$$
$$P = 0 \cdot 94 M^{0 \cdot 25} \tag{10}$$

in which Q = discharge (ft³ sec⁻¹), Q_{m} = mean annual discharge (ft³ sec⁻¹), Q_{ma} = mean annual flood (ft³ sec⁻¹), Q_{bs} = bed load as a percentage of total load, w = channel width, $\bar{d}$ = channel mean depth (ft), S = water-surface slope, λ = meander wavelength (ft), P = sinuosity and M = the percentage of silt-clay in the perimeter of the channel and an index of sediment load.

Equations (4) and (5) include a sediment parameter which statistically explains much of the residual variation in channel width and depth among the rivers in question left unexplained by variations in discharge. Similar improvement in the definition of the 'meander law' is achieved by taking account of sediment type in equation (9).

For the rivers from which these data were obtained, this closed set of equations provides an empirical model for predicting channel changes. If the direction of these relationships is universal, and if bedload transport (Q_{s}) is inversely related to the sediment parameter, M, equations (4)–(10) imply that

$$Q \propto \frac{w, d, \lambda}{S} \tag{11}$$

and

$$Q_{\mathrm{s}} \propto \frac{w, \lambda, S}{d, P}. \tag{12}$$

These general relationships lead to the following set of channel-change associations (see Schumm, 1977)

$$Q^+ \propto w^+, d^+, \lambda^+, S^- \tag{13}$$
$$Q^- \propto w^-, d^-, \lambda^-, S^+ \tag{14}$$
$$Q_{\mathrm{s}}^+ \propto w^+, d^-, \lambda^+, S^+, P^- \tag{15}$$
$$Q_{\mathrm{s}}^- \propto w^-, d^+, \lambda^-, S^-, P^+ \tag{16}$$
$$Q^+ Q_{\mathrm{s}}^+ \propto w^+, d^\pm, \lambda^+, S^\pm, P^-, F^+ \tag{17}$$
$$Q^- Q_{\mathrm{s}}^- \propto w^-, d^\pm, \lambda^-, S^\pm, P^+, F^- \tag{18}$$
$$Q^+ Q_{\mathrm{s}}^- \propto w^-, d^+, \lambda^\pm, S^-, P^+, F^- \tag{19}$$
$$Q^- Q_{\mathrm{s}}^+ \propto w^-, d^-, \lambda^\pm, S^+, P^-, F^+. \tag{20}$$

Equations (13)–(20) describe a number of possible morphological responses to a variety of changes in the discharge of water and sediment, some more likely than others. Santos-Cayade & Simons (1973) and Schumm (1977) provide a discussion of circumstances in which these changes might occur. Clearly, even for this qualitative scheme, many of the changes are indeterminate, particularly for the more realistic cases, because the magnitude of opposed responses is unspecified. In such circumstances it is very tempting to solve this problem by making the completely unwarranted assumption that equations (4)–(10) are universal. Herein lies the fundamental limitation of this empirical approach: the lack of field data.

Given the difficulties associated with the process-oriented approach discussed earlier, this empirical approach deserves far more attention than it has received. Oddly enough, it is almost the only study of its kind in spite of the fact that it was completed over two decades ago. There seems to be an inexplicable aversion on the part of earth scientists and engineers to repeat the field experiments of others.

The danger of generalizing from particular empirical norms is well illustrated by considering one of Schumm's central relationships: that between the ratio width/depth of a channel and the weighted percentage silt–clay content of the bed and bank materials (equation 6). This relationship is based on 69 channel sections on about 30 rivers in the American Midwest. Today we still have little idea whether or not this relationship applies to other Midwestern rivers or to any in regions beyond. But of course we are very willing to assume that it does. We also seem to have some sort of intellectual defence mechanism that prevents us from recognizing the order-of-magnitude scatter in this particular relationship. For example, it can hardly be comforting for an engineer to know that boundary materials with 10% silt–clay are associated with channels having width/depth ratios that commonly vary between 10 and 60! Perhaps it is even more disturbing that at least one independent field check of this relationship (65 river cross-sections from the Namoi–Gwydir distributary system of eastern Australia) has shown that equation (6) does not apply there at all (Riley, 1975).

These observations do not invalidate the empirical approach but they do point to the need for a more comprehensive data set than is available at present. The norms would be much more useful as a planning and design tool if data were collected and analysed for geologically and hydrologically homogeneous areas. Such a geographic exercise would strengthen the order in particular hydraulic geometries while recognizing the variety in general. The regional regime relations for Albertan gravel-bed rivers developed by Bray (1975, 1979) is a good example of this type of exercise.

Probably the single most important limitation of the approach of hydraulic geometry, one that it shares with the theory of equilibrium channel morphology as previously discussed, is that it only describes static equilibria. The process of channel adjustment from one equilibrium state to another is generally assumed to be instantaneous; transient states are not accommodated although they appear in the hydraulic geometry as scatter in the data (Knighton, 1975, 1977). In other words, rates of change in channel morphology remain largely unspecified. Even if a new equilibrium state can be specified, it is not possible to predict how long it will take to reach it.

Related to this problem of specifying rates of channel change is that of accounting for the effects of large floods. Extreme events, part of the normal process system, may cause significant departures from equilibrium channel morphology that may persist for long periods of time (relaxation time). The Schumm & Lichty (1963) study of the Cimarron River in south-western Kansas provides a well-known example of this behaviour. They reasoned that a large flood in 1914, which severely eroded the floodplain, caused a considerable influx of coarse sediment, resulting in the widening and steepening of the channel between 1914 and 1942. Since that time the channel has slowly worked towards re-establishing its pre-flood morphology ($w/\bar{d}$ halved between 1939 and 1954). Burkham (1972) describes a similar history for the Gila River in Arizona. Schumm (1977) cites several other examples of flood-related channel adjustments.

There may be many more examples of flood-dominated channel morphology that have gone unrecognized. Our recent thinking in this respect has been greatly influenced by the notion that a large proportion of the work done by rivers is effected during events of modest magnitude and relatively high frequency (Wolman & Miller, 1960; Gupta & Fox, 1974) although the question recently has been reconsidered by Baker (1977), and Dury (1977b, 1980). It seems likely that many rivers confirm this conventional view at geomorphic time-scales but, like the Cimarron River, are characterized by long periods of flood-related transient behaviour at engineering time-scales (see also Mosley, 1975b, Stevens, Simons & Richardson, 1975 and Graf, 1981).

Another important limitation of both the theoretical and empirical approaches to channel changes is the existence of non-linearities in channel adjustment. Some of these non-linearities in relatively low flows (for example, the onset of sediment transport) and in intermediate flows (for example, bedform changes) have been described (see Culbertson & Dawdy, 1964 and Richards, 1973) but there may also be as important ones at the often little measured but very significant high flows (for example, those related to the possible onset of general vortex shedding from the boundary). There is a need to collect detailed data sets that will allow these types of non-linearities in the hydraulic geometry to be distinguished from the background noise.

A far more profound set of non-linearities, however, is embodied in the concept of *complex response* in channels (Schumm, 1977).

COMPLEX CHANNEL RESPONSE

These channel responses include all those which are more complex than a simple direct cause-and-effect relationship. They involve the ideas of system thresholds, of episodic development of the fluvial system (Schumm, 1973, 1976; Church, 1980; Howard, 1980), and more recently the perhaps deservedly beleaguered notions of catastrophe 'theory' (see Graf, 1979). These channel responses are governed by an exceedingly complex and poorly understood set of dynamic relationships which have been largely discussed at a conceptual rather than operational level (Hey, 1979).

An interesting and probably very significant example of complex channel response is the experimental work of Parker conducted in the rainfall erosion facility at Colorado State University (see Schumm, 1977). Parker measured sediment yields from an experimental basin immediately following simulated channel rejuvenation by a lowering of base level. As the erosional knick point moved headward through the drainage network, sediment yield at first increased and then declined (as valley storage capacity increased) and subsequently increased again as major tributaries became rejuvenated, only to decline again, and so on. Sediment yields fluctuated over time because of variation in sediment supply relative to storage capacity. A consequence of this fluctuation is that downstream reaches of the experimental channel were at times aggrading (high sediment supply) and at others incising (low sediment supply), all in response to a single lowering of base level while all other external factors (climate/discharge) remained constant. Schumm & Parker (1973) believe that such a complex-response mechanism goes far in explaining the lack of correlation among terraces between and even within valleys in the American south-west. The conventional assumption of externally governed and therefore regionally in-phase terracing no longer appears to be necessary. Also, there is no reason to suppose that such a process is not significant across all time-scales of channel change. Again we see evidence that significant channel change is governed by sediment supply, strengthening an earlier argument made here in the context of sediment transport theory. It would seem judicious, however, that these ideas receive far more experimental confirmation that they currently enjoy before being embraced as matters of principle.

Another of the more important ideas of complex channel response is the exceeding of critical limits or thresholds that can lead to a complete transformation of river morphology (river metamorphosis: Schumm, 1969). An oft-cited example of this process is that implicit in the various channel-pattern stability domains defined in the slope-discharge plane by Lane (1957), Leopold & Wolman (1957) and Ackers & Charlton (1971). These observations suggest that, at a constant discharge, if channel slope is slowly increased, straight channels will be the stable pattern at first but will give way to meandering, and finally braided channels represent the stable pattern at high slopes. At the limits of the domains, very small changes in slope can alter the stable pattern from, for example, meanders to braids. Schumm & Khan (1972) experimentally confirmed this sequence of transitions although there is no agreement on the precise location of the domains on the slope-discharge plane. It is likely that this simple bivariate plot shows association rather than process, and that a more consistent differentiation of pattern would be achieved in a scheme including boundary materials and sediment transport rates (see Henderson, 1961).

It is a curious fact that field confirmations of the original Leopold & Wolman (1957) slope–discharge relationship for braids and meanders are not available in the literature. In this case, as in too many others, early ideas remain speculative because of the failure of earth scientists to follow through with the collection of appropriate field evidence. This has not prevented, however, some of these ideas becoming the foundation of further deductions.

Schumm (1977) has suggested that the many documented cases of channel straightening and widening in the United States are responses to increased peak flows and sediment loads that likely accompanied settlement during the nineteenth century. The interpretation although reasonable does tend to be rather circular, however, because there are no sediment transport data for these rivers. The same limitation applies to Schumm's (1968) interpretation of channel changes on the Murrumbidgee River in New South Wales, Australia. The patterns of change are evident in the landscape but the record of process is largely lost and must be reconstructed by means such as that offered by equations (4)–(20). Obviously such evidence cannot then be used to confirm the process–form relations!

Almost all major natural changes in channels occur so slowly that even if the morphologic record is well documented, the process record is invariably inadequate. But there now may be an exception to this rule. Man, by his activities in and about rivers, has created something of a laboratory of accelerated river change. For this reason man's impact on rivers is of scientific as well as of immediate practical interest and has been the subject of much recent enquiry.

HUMAN-INDUCED CHANNEL CHANGES

The study of the impact of human activity on river channels has recently been reviewed extensively by several authors (for example, see Gregory, 1976a, b, 1979 and Park, 1977, 1981). For this reason it will best serve the present purpose simply to consider some type examples of this research, noting general progress and future prospects.

Studies of human-induced channel change represent an extension of the empirical approach to river behaviour although there are very few actual detailed observations of channel change recorded. Those that are available are concerned either with the consequences of direct modifications such as dam construction, flow diversions, and channel stabilization, or with the general but less direct effects of land-use change.

The effects of dams

By far the largest set of observations exists for the long-recognized downstream effects of dams, particularly the rapid channel degradation associated with the abstraction of sediment load. Many of these published studies have recently been reviewed by Makkavayev (1972) and Petts (1979). Dam closure typically results in substantially reduced flood peaks, increased base flows and, in the case of large reservoirs, almost complete abstraction of sediment load. General consideration of flow and sediment transport continuity (equations 1 and 3) suggests that this should lead to downstream channel contraction and degradation until bed armour or energy slope reduction stabilizes the channel. The quality of these changes is generally confirmed by empirical studies although the degree of response is highly varied. Other effects such as changes in the regime of river ice and the rejuvenation of downstream tributaries (and consequent alluvial fan deposition in the main channel) are discussed by Kellerhals & Gill (1973).

Quantitative prediction of degradation below dams has been attempted by Tinney (1962), Komura & Simons (1967) and Halen, Shindala & Denson (1970), but inevitably there is much discussion (see Herbertson *et al.*, 1968) about the appropriate sediment transport equation to cast into differential form. Nevertheless, some success has been achieved in several cases in the United States although the lack of further testing leaves the general reliability of these semi-theoretical approaches unspecified.

Channel changes upstream of dams are less well documented, but are generally recognized to involve reservoir infilling and aggradation upstream to the back-water limit.

Most empirical studies of degradation involve a simple before-and-after comparison of channel form generally over a period of 5–10 years. Far fewer studies involve regular monitoring of channel change although this is the type of data that is clearly needed. We know from the longer-term behaviour of rivers that the channel adjustment process is commonly complex and may take many decades to be achieved. A 5–10 year observation period may just be the introduction to a much longer and more complicated

story in many cases. Experience with the American Vigil Network indicates that long-term monitoring of channel changes such as that currently being done on the Peace River below the Bennet Dam in British Columbia (by M. Church, personal communication) is an essential research strategy if the process of channel change is to be better understood.

Interbasin river diversions

Similar comment might be offered, of course, in other cases of human-induced channel change, including the converse of the dam closure situation: the inter-basin river diversion. The type of diversion of interest here is one in which a relatively large flow is diverted across a divide and allowed to flow into a dry valley or relatively small stream in which it develops a new larger river channel. Very little is known about the effects of such diversions because they are few in number (although often grand in scale), located in isolated and often uninhabited country, and are generally rather recent in construction. Kellerhals, Church & Davies (1979) have described eleven Canadian cases of interbasin river diversions and they may well represent the majority available anywhere for establishing useful precedents. They classify the diversions into three types: bedrock-controlled diversion routes, steep diversion routes in unconsolidated materials, and alluvial diversions. In the case of the first type (for example, the Churchill River diversion in Manitoba) channel changes are restricted to the flushing out of pockets of unconsolidated sediments and the erosion of weak bedrock. In the second type (for example, the Ogoki River diversion in Ontario) the diversion route is characterized by an expected rapid increase in channel capacity through channel widening and deepening accompanied by the export of large amounts of sediment. Alluvial diversions (for example, the Nechako–Kemano diversion in British Columbia) may remain little altered apart from moderate channel enlargement if the diverted flow is not significantly greater than the higher flood flows of the receiving stream. A most important conclusion of their study is that a detailed knowledge of all materials that might conceivably become exposed to erosion is an essential prerequisite to prediction of the final morphology of a diversion channel. This is particularly important, of course, in determining if and when bed armour will develop.

It is important to note, however, that most of the diversions considered by Kellerhals *et al.* (1979) are less than 20 years old. Yet even in the case of the

34-year-old Ogoki River diversion, the receiving Little Jackfish River has not yet completely stabilized. Clearly there are many more lessons to be learned from their continued surveillance.

Channelization

This term refers to the modifications made to channels in order to improve their navigability, lessen flooding, improve land drainage, or to make them more stable. Alteration of channels is common in all countries but in the United States it has been practised on a massive scale involving over 13,000 river km of levées and floodwalls and a similar additional length of channel 'improvement'. Many of these projects are summarized and analysed in a report for the Council on Environmental Quality (1973). Although many channelization projects are clearly successful, unfortunately it is also the case that many of the 'improvements' involved changes to channel forms that we now know are unstable. For example, many rivers were straightened and thus steepened, causing downstream channel aggradation, bank failure and greatly increased sediment transport rates that exported the problems further downstream.

The classic example of this type of problem is the Mississippi River, probably the most manipulated large river in the world. The Middle and Lower Mississippi River provide an instructive contrast in the effectiveness of differing management strategies.

Stevens, Simons & Schumm (1975) describe some of the man-induced changes in the Middle Mississippi (between St Louis and Cairo) that so dominate the present channel character. The Corps of Engineers is charged with developing and maintaining a navigation channel while at the same time providing flood protection. As a result, most of the Middle Mississippi is lined with Corps of Engineers mainline levées and much of the channel has been contracted to about two-thirds of its former (1888) width by 150 km of dyking system. The channel alignment is maintained by 200 km of bank revetment. It is possible that this engineering scheme simply hastened nature's work; the channel of 1888 was likely overwide, reflecting the effects of earlier extreme floods. In any event the minimum navigation channel is being largely maintained now and the flood hazard has been markedly reduced.

A similar engineering programme was developed by the Corps of Engineers for the Lower Mississippi River (downstream of Cairo) but with one important difference: the channel was straightened by artificial cutoffs (see Winkley, 1977). The cutoff programme was a controversial one. On the one hand it seemed to be a means of improving the navigation channel and of reducing flood stages while on the other there were fears that local increases in slope would cause instability. The outcome of the debate was that the cutoff programme was approved in 1929. The present natural length of the Lower Mississippi River, about 1740 km, appears to have been stable for the last 1000 years (Fisk, 1952), fluctuating between 1830 and 1680 km in response to a natural cutoff frequency of about 14 per 100 yr. Natural relaxation times following a cutoff seem to be 30–80 yr (Winkley, 1977). Between 1929 and 1942 the artificial cutoff rate was increased to eight times the natural rate. In consequence, the river length was shortened to about 1530 km, mean water-surface slope was increased by 12%, while slopes locally were increased up to 20 times the pre-cutoff magnitudes.

The integrated effect of the downstream migration of the sediment slugs produced by each of these cutoffs has been channel widening, bar formation, and increasing bank instability in the southern portion of the Lower Mississippi (i.e. it is tending to braid). In spite of an accelerated maintenance programme depths in the navigation channel are less now than they were 90 years ago. Winkley (1977) estimates that the cutoff channels will not stabilize until the end of the century. The general aggradation problems will presumably persist well into the twenty-first century.

The effects of land use

It has long been recognized that the character of the flood hydrograph is sensitive to land use within the originating catchment (see the reviews in Chow, 1964). For example, the tendency for reductions in infiltration capacity, for whatever reason, to reduce lag times and increase flood peaks is one which is incorporated into most general flood forecasting models (the Stanford Watershed model provides a well-known example). A considerable literature of hydrologic case-studies designed to confirm or better define these relationships among land use and flood characteristics has accumulated during the last few decades, and many of these are discussed in recent general accounts such as those by Gregory & Walling (1979) and Dunne & Leopold (1978).

It has also been known for some time that basin sediment yields are strongly influenced by land use. Indeed, a distinction is made between 'geologically normal' and the usually land-use-related 'accelerated' rates of hill-slope erosion which can catastrophically increase the sediment supply to rivers.

Because channel size and form is largely the product of sediment supply and discharge, physical reasoning dictates that land-use change will promote channel change. Perhaps because it is both obvious and simple, this deduction has rarely been systematically tested although it has been confirmed often enough by casual observation. It is an unfortunate oversight and a potentially rewarding research area because the effects of land-use change on river morphology may provide a useful analogue of the response of channels to natural changes in the general environment.

The few studies of land-use-related channel change that are available are of two types. They either describe the history of channel adjustment to changing land-use or they describe the spatial variation in supposedly equilibrum land-use/channel morphology relations. Because of the difficulty of recognizing such an equilibrium and of controlling for other potentially confounding conditions (for example, see Church & Mark, 1980 for a discussion of scale problems in geomorphology) the former approach may at first appear to be preferable. Direct monitoring of a site history, however, has two distinct disadvantages. First is the time constraint common to all direct monitoring and previously discussed in the context of meandering. Secondly, study areas are necessarily small, and channel responses may be strongly influenced by site-specific characteristics that make generalization difficult.

Channel responses to two types of land use conversion have received most attention because the changes involved are widespread and extreme and therefore more easily measured: those related to deforestation and those related to urbanization. Although there is a small literature on the particular effects of mining on rivers (see Park, 1981) they will not be included in the present discussion. However, the general discussion of channel response to changes in discharge and sediment supply to follow is directly applicable to cases of mining in or near river channels.

The impact of forestry practices on hydrology and channel morphology is not well understood in detail although there now are many case-studies (of the historical type) described in the literature (see Sopper & Lull, 1967, Jeffrey, 1970, Bell, Brown & Hubbard, 1974 and Janda *et al.*, 1975). Perhaps few areas display such extreme responses to deforestation as are commonly observed in the Pacific Northwest of North America. Logging practices in Oregon, Washington and British Columbia often involve clearcutting of the coastal forest on steep slopes of weak volcanics and glacial debris. Usually within a year of logging, slope failures abound, runoff and sediment yields markedly increase and much of the pre-logging drainage pattern on the upper slopes is obliterated by mass movement and devastating debris torrents (Rothacker, 1970; Harr, Harper & Kaygier, 1975). The surviving channels are choked with huge volumes of sediment and logging slash, an important source of which is the spoil from logging road construction (Swanson & Dyrness, 1975). Replanting of these logged areas, however, will stabilize slopes and reduce erosion, sediment yield and runoff within a few years as the regrowth becomes established (for example, see Berndt & Swank, 1970).

The effects of partial vegetation removal on river flow in less extreme geomorphic environments are not well defined, with studies often yielding rather different results for different areas (see Hibbert's inconclusive 1967 review of 39 case-studies). The results of more recent studies are just as variable, however, and point to the prevailing sensitivity of channels to local conditions. Although the balance of studies seems to favour the notion that vegetation removal will increase water yields to some (unpredictable) degree, the effect on flood peaks varies widely. Recent work stresses that, in small catchment studies, factors such as the area of logging roads, compaction of road surfaces, treatment of slash, variation in type and timing of storms, seasonality, etc., can dominate the effects of even extensive vegetation removal (see Ziemer, 1981). It is hardly surprising that comparisons among basins lead to inconclusive results. There is general agreement, however, that sediment yield is strongly dependent on the degree of disturbance to the soil (by skidding, hauling, road construction, etc.) and not on the removal of trees *per se*. For this reason vegetation removal at one site may cause channel enlargement and incision (Carson & Tamm, 1977) while at another it will cause aggradation (Orme & Bailey, 1970, 1971; Janda *et al.*, 1975).

There are no comparative (ergodic) studies of channel response to vegetation changes. Spatial variation in vegetation is typically associated with geologic, geomorphic, edaphic and climatic changes and it is not possible to isolate the precise influence of one from the others. In any case, adequate information on channel geometry, as noted earlier, is simply not available for analysis.

There are, however, a large number of U.S.D.A. studies on runoff and sediment yield for various agricultural and grazing intensities in the United

States; these observations form the basis of the Universal Soil Loss Equation. In detail they show great variation in the degree of change in runoff and sediment yield from one land-use type to another, but the direction of change is well established. Runoff and sediment yields are high from cultivated and heavily grazed rangeland and relatively low for forests and ungrazed rangeland; all yields increase with increases in precipitation.

Langbein & Schumm (1958) and Rango (1970) analysed sediment yields on a regional basis and concluded that as precipitation increases, they increase rapidly from zero to a temperature-dependent maximum and subsequently decline because of the corresponding increase in the protective vegetation cover.

Although these runoff and sediment-yield data are rather variable and inadequate for precision modelling, together they do suggest qualitative hydrologic and sedimentologic changes associated with vegetation change. Unfortunately, however, channel changes are extremely sensitive to the balance between discharge and sediment supply, and these data often do not permit the balance or the quality of channel changes to be specified.

The effects of urbanization on rivers have also been the subject of intensive investigation over the last few decades, but once again far more is known about urban hydrology (see Leopold, 1968; Douglas, 1976) than is known about the morphological response of rivers in urban areas (Graf, 1975, 1976).

The increase in flood peaks and decrease in lag times associated with the spread of impervious surfaces and development of efficient sewerage systems in urbanized areas have been recognized for many years. The results of many historical and comparative studies during the last two decades provide the basis for simple empirical models for flood-peak prediction (Carter, 1961; Demster, 1974) and for general urban hydrology models such as the Stormwater Management Model in the United States and the Road Research Laboratory Hydrograph Model in Britain (Larson, 1972; Roesner, Kibler & Monser, 1972; Aitken, 1973). In general these studies show that urbanization typically will double flood peaks and halve lag times, although Esprey & Winslow (1974) and Hollis (1975) note that such changes depend on the recurrence interval in question (for example, the relative effects appear to decline as the recurrence interval increases). They are obviously also dependent on the character of urbanization (area of roads and parking lots, detached or integrated housing, etc.; see

Dunne & Leopold, 1978) and on the efficiency of the designed drainage system (see Beard & Chang, 1979).

In contrast to the case of logging, the effect of urbanization of flood peaks is relatively direct and straightforward. The impact of urbanization on sediment yields, however, appears very similar to, but far less extreme than, that observed in logged areas where typically sediment is initially mobilized in great amounts and then declines in yield as secondary vegetation becomes established. In the urbanization case the initial construction period is commonly characterized by increases in sediment concentration and yield by respectively one and two orders of magnitude (for example, see Wolman & Schick, 1967, Walling & Gregory, 1970, Guy & Jones, 1972 and Walling, 1974). After the construction phase sediment yields appear to decline with the decreasing availability of sediment sources to levels possibly as low as those associated with forested basins (Wolman, 1967; Wolman & Schick, 1967).

As before, physical reasoning might lead us to deduce that urbanization will initially result in heavy sedimentation and contraction of stream channels and in subsequent channel enlargement to accommodate the new regime of reduced sediment yields and increased flood peaks. Indeed, there is some empirical evidence to support this sequence of events (Wolman, 1967; Graf, 1976). The historical study of Watts Branch in Maryland (see Leopold & Emmett, 1972) shows that urban encroachment on to the headwaters there has caused a progressive decrease in channel size in response to increased suspended sediment loads since observations began in 1953. The rate of channel contraction markedly increased after 1960 (Leopold, 1973) until, by 1972, the channel had been reduced to about 80% of its 1953 capacity. The frequency of flooding has doubled and Watts Branch presumably remains in a transient phase of adjustment to the altered hydrologic regime. Unfortunately this case-study appears to be one of a kind.

There is other evidence available, however, in the form of comparative studies of urban versus 'natural' channels. The relation of the enlargement ratio urban/natural channel size (scaled by drainage area) to land-use was first systematically examined by Hammer (1972) for 78 small watersheds near Philadelphia. He found that the enlargement ratio was high for areas with sewered streets and for those with major impervious areas such as parking lots, and that it was considerably smaller for areas with unsewered streets and for those impervious areas with detached houses. His data suggest that there is at least a 4-year lag between urban development and channel response and that channels in areas of streets and detached housing have curiously contracted to their former pre-urban dimensions by the time they exceed 30 years in age. Hammer's (1972) enlargement ratios accord with Leopold's (1968) average annual flood ratios for corresponding land-use types.

Similar studies have been repeated in England by Hollis & Luckett (1976), and Park (1977) with similar but somewhat more variable results than those reported by Hammer (1972).

One of the important facts emerging from this type of study of urban channel enlargement is that the results of comparative analyses include considerable 'noise' related to variations in lag time (response and relaxation times). Lag times are likely to be dependent on both the scale of the system (drainage area) and stream power (for example, Hammer's data indicate that channel slope is an important statistical determinant of the amount of channel enlargement) among other things. But we simply do not know at present. This, of course, is yet another compelling argument for the need of further historical studies such as that on Watts Branch. It is also important that observations not be restricted to a single reach; degradation in one reach presumably implies aggradation somewhere downstream. To this end of characterizing changes within the entire drainage network, Gregory (1977a) suggests that changes in the channel network volume might better reflect the integrated channel response.

CHANNEL CHANGES IN GEOMORPHIC TIME

It is not the purpose of this section to review the vast array of case-studies describing rivers displaying evidence of long, varied and unique histories. Instead, it will very briefly consider some of the problems and prospects of identifying and interpreting channel changes that have taken place prior to the period of recorded observations.

The primary evidence of channel changes in geomorphic time is the morphology and stratigraphy of alluvial valley fills. Detailed description of this evidence remains essential to any analysis of fluvial history although, with the recent emphasis on contemporary process studies, such detailed field descriptions seem to be less common now than they were in the past. Three well-known examples, amongst many others, characterize this type of study: Fisk's Holocene history of the Mississippi Valley,

Dury's study of regional underfitness of rivers, and Leopold & Miller's study of the alluvial chronology of valleys in the American South-west. Fisk (1952) described some 15,000 years of Mississippi River aggradation based on a knowledge of the detailed stratigraphy of the Mississippi Valley set in the general context of the late Quaternary climate and sea-levels established elsewhere. His reading of the stratigraphic record suggested to him an early deglacial of very high sediment supply in braided channels (extensive deposits of sands and gravels) giving way to a meandering Mississippi as sediment supply, sediment calibre and valley slopes declined to the present. Schumm (1968) cites Fisk's work as a confirmation of the concept of river metamorphosis. Dury (see his brief review, 1977a) identified and surveyed former large channels buried by Holocene alluviation in England, France, Wisconsin and Australia in order to test his theory that misfit streams are widespread and of climatic origin. He used drilling and seismic surveys to delineate the bed profiles of buried rock-cut channels and stratigraphic, palynological and carbon-dating evidence to reconstruct the infilling chronology from the time of channel abandonment some 9000–11,000 yr BP. He went on to use general hydraulic geometry and regional climatic norms to retrodict Holocene hydrologic and climatic conditions. Leopold & Miller (1954) and others (see Hadley, 1960) surveyed valley fills in Wyoming and, using stratigraphic, pedologic, carbon-dating, fossil and archaeological evidence, developed chronologies involving major alternating phases of degradation and aggradation since late Pleistocene times. Although the general chronology is neither complete nor entirely synchronous at all sites, an idealized valley schema involves three terraces in three sedimentary units (see Haynes, 1968). The oldest dates from late Pleistocene (fossil evidence) and was formed by channel incision during an arid phase (the upper terrace sediments include dune sands and calcic palaeosols). Further alluviation and subsequent incision produced another terrace about 600–800 yr BP in similar climatic circumstances. A final major period of deposition followed, and ended with the initiation of modern gullying in 1880–1890 and the formation of the most recent terrace. The reason for the most recent period of gullying is unclear, but likely reflects general climatic shifts and in some places overgrazing of domestic animals. Vigil Network observations suggest that many of these valleys have been in a depositional phase during the second half of this century (Leopold & Emmett, 1972; Emmett, 1974) when the region generally has been experiencing a cooler and wetter climate. The record is by no means clear, however, because of the complicating effects of human activity (see Knox, 1977) during this century.

Each of these studies shows how careful reading of the sedimentological record can yield two types of information about channel changes. The first is the actual chronology of change and the second is the process information implied by the observations. Amid the controversy that so often surrounds the interpretation of this process information, the value of the long-term record of channel behaviour *per se* sometimes seems to be overlooked. Only in this longer-term context can we develop a realistic perspective of present and future channel behaviour in the short term.

Much of the controversy surrounding the interpretation of alluvial chronology results from a certain circularity of argument. For example, in the development of alluvial chronology for the American South-west (and for the Riverine Plains in eastern Australia; see Schumm, 1968) there has been a tendency to evaluate chanel changes in terms of *a priori* models of climatic change and river response and then to argue for those assumptions from the field evidence. This problem, of course, simply points to the need for an independent climatic and hydrologic record against which channel changes can be evaluated.

In this context it is appropriate that more attention is now being given to the potential of extracting such information from the vegetation record. In particular, recent developments in the analysis of palynological (Bryson & Kutzbach, 1974) and dendrochronological (Fritts, 1976; Shroder, 1980) records appear respectively to provide some access to Holocene and detailed access to the last few centuries of climatic and hydrologic conditions. Two recent examples from North America illustrate the potential of these developments.

Mathewes & Heusser (1981) present 12,000 years of mean annual precipitation and mean July temperature records implied by the column of fossil pollen sampled from a lake in south-west British Columbia. Transfer functions for converting pollen frequencies to climatic parameters were adopted from Heusser, Hausser & Streeter (1980), who sampled modern pollen data from 180 coastal sites from the Aleutian Islands to California. They related the frequency of four pollen factors to the modern (1968–1977) July temperature and annual precipitation records at 43 meteorological stations to yield a pair of regression equations used as transfer functions in the Mathewes & Heusser (1981) lake study. Eight radiocarbon dates and a Mazama ash layer provide chronostratigraphic control of the lake core.

In a methodologically similar study, Duvick & Blasing (1981) cored about 100 trees in central Iowa and determined a 300-year chronology of tree-ring width indices in accordance with procedures proposed by Fritts *et al.* (1971) and Fritts, Lofgren & Gordon (1979). They calibrated this record with statewide precipitation records for the period 1920–1979 and tested the resulting regression equation by predicting the actual record for the period 1874–1919. As a result, the tree-ring model was shown to be capable of explaining about 55% of the variance in the test record. This level of explanation makes the model a reliable indicator of wet and dry periods in central Iowa dating from the year 1680.

These are but two of a growing number of such proxy climatic records, and their reliability and range will presumably increase as the data base expands.

CONCLUDING REMARKS

A theme woven into much of this discussion is that river morphology, particularly from the engineer's point of view, may generally be a non-equilibrium property. The literature is replete with examples of rivers for which the significant formative processes are those only viewed properly at a geomorphic time-scale. Perhaps most rivers are dominated by transient behaviour, never fully adjusting to such events as major floods, climatic shifts, and those step function effects known as river metamorphosis.

Recently Burkham (1981) assessed the ability of our fluvial science to respond to the information needs of environmental law, and concluded that its dominating characteristic is uncertainty. It seems likely that an appreciation of the scope of this uncertainty will only come from studies of the long-term river record. One of the more significant gaps in our knowledge is the nature of the river record during the last several hundred years. This would seem to signal a new emphasis on the perspective and techniques of the Quaternary geologist. Perhaps Nanson's (1980) study of the Beatton River in British Columbia is a good example of the approach that should be taken when examining river activity. He combined a survey of contemporary channel form and process, tree-ring analyses of lateral and vertical channel movement and detailed stratigraphic and sedimentological analyses, to develop a record of channel change for the last 400 years. In this case, as in so many others, the character of the present channel was found to be dominated by processes that only display their full range of behaviour in geomorphic time.

REFERENCES

ACKERS, P. & CHARLTON, F.G. (1970) The geometry of small meandering streams. *Proc. Inst. civ. Engrs, Suppl.* **XII**, pap. 7328S, 289–317.

ACKERS, P. & CHARLTON, F.G. (1971) The slope and resistance of small meandering channels. *Proc. Inst. civ. Engrg Suppl.* **XV**, pap. 7362S, 349–370.

AITKEN, A.P. (1973) Hydrologic investigation and design in urban areas—a review. *Aust. Wat. Res. Coun. Tech. Pap.* **5**. Canberra, Australia.

ALLEN, J.R.L. (1970a) A quantitative model of grain size and sedimentary structures in lateral deposits. *Geol. J.* **7**, 129–146.

ALLEN, J.R.L. (1970b) Studies in fluviatile sedimentation: A comparison of fining upwards cyclothems, with special reference to coarse-member composition and interpretation. *J. sedim. Petrol.* **40**, 298–323.

ALLEN, J.R.L. (1971) Rivers and their deposits. *Sci. Prog.* **59**, 109–122.

ALLEN, J.R.L. (1977) Changeable rivers: some aspects of their mechanics and sedimentation. In: *River Channel Changes* (Ed. by K.J. Gregory), pp. 15–45. Wiley, Chichester.

American Society of Civil Engineers, Task Committee on Preparation of Sedimentation Manual (1971a) Sediment transportation mechanics. H. Sediment discharge fomulas. *J. Hydraul. Div. Am. Soc. civ. Engrs* **97**, 523–567; with discussion **97**, 1573–1576; **98** (1972), 284–290, 388–397, 1869–1872.

American Society of Civil Engineers, Task Committee on Preparation of Sedimentation Manual (1971b) Sediment transportation mechanics. Fundamentals of sediment transportation. *J. Hydraul. Div. Am. Soc. civ. Engrs* **97**, 1979–2022.

BAGNOLD, R.A. (1960) Some aspects of river meanders. *Prof. Pap. U.S. geol. Surv.* **282E**.

BAKER, V.R. (1977) Stream-channel response to floods, with examples from central Texas. *Bull. geol. Soc. Am.* **88**, 1057–1071.

BEARD, L.R. & CHANG, S. (1979) Urbanisation impact on streamflow. *J. Hydraul. Div. Am. Soc. civ. Engrs* **105**, 647–659.

BELL, M.A.M., BROWN, J.M. & HUBBARD, W.F. (1974) Impact on harvesting on forest environments and resources. *For. Tech. Rep.* **3**. Canadian Forestry Service, Department of the Environment, Ottawa. 237 pp.

BENSON, M.A. & THOMAS, D.M. (1966) A definition of dominant discharge. *Bull. int. Ass. Sci. Hydrol.* **11**, 76–80.

BERNDT, H.W. & SWANK, G.W. (1970) Forest landuse and streamflow in central Oregon. *U.S. Dep. Agric., Pacific Northwest For. Range Exp. Stat. Res. Pap.* **PNW-93**.

BLENCH, T. (1966) *Mobile-Bed Fluviology.* University of Alberta Technical Services, Edmonton. 300 pp.

BOGARDI, J. (1974) *Sediment Transport in Alluvial Streams.* Akademiai Kiado, Budapest. 826 pp.

BRAY, D.I. (1975) Representative discharges for gravel-bed rivers in Alberta, Canada. *J. Hydrol.* **27**, 143–153.

BRAY, D.I. (1979) Estimating average velocity in gravel-bed rivers. *J. Hydraul. Div. Am. Soc. civ. Engrs* **105**, 1103–1122.

BRIDGE, J.S. (1977) Flow, bed topography, grain size, and sedimentary structure in open channel bends: a three dimensional model. *Earth Surf. Processes*, **2**, 401–416.

BRIDGE, J.S. & JARVIS, J. (1976) Flow and sedimentary processes in the meandering River South Esk, Glen Clova, Scotland. *Earth Surf. Processes*, **1**, 303–336.

BRICE, J.C. (1974) Evolution of meander loops. *Bull. geol. Soc. Am.* **85**, 581–586.

BRYSON, R.A. & KUTBACH, J.E. (1974) On the analysis of pollen-climate canonical transfer functions. *Quat. Res.* **4**, 162–174.

BURKHAM, D.E. (1972) Channel changes of the Gila River in Safford Valley, Arizona 1846–1870. *Prof. Pap. U.S. geol. Surv.* **655-G**.

BURKHAM, D.E. (1981) Uncertainties resulting from changes in river form. *J. Hydraul. Div. Am. Soc. civ. Engrs* **107**, 593–610.

CALLANDER, R.A. (1969) Instability and river channels. *J. Fluid Mech.* **36**, 465–480.

CALLANDER, R.A. (1978) River meandering. *Ann. Rev. Fluid Mech.* **10**, 129–158.

CARLSTON, C.W. (1965) The relationship of free meander geometry to stream discharge and its geomorphic implications. *Am. J. Sci.* **263**, 864–885.

CARSON, M.A. & TAMM, S.W. (1977) The land conversion conundrum of eastern Barbados. *Ann. Ass. Am. Geogr.* **67**, 185–203.

CARTER, R.W. (1961) Magnitude and frequency of floods in suburban areas. *Prof. Pap. U.S. geol. Surv. In: Short Papers in the Geologic and Hydrologic Sciences*, **421-B**, B9–B11.

CHIEN, N. (1956) The present status of research on sediment transport. *Trans. Am. Soc. civ. Engrs* **121**, 833–868.

CHITALE, S.V. (1973) Theories and relationships of river channel patterns. *J. Hydrol.* **19**, 285–308.

CHOW, V.T. (Ed.) (1964) *Handbook of Applied Hydrology.* McGraw-Hill, New York. 1480 pp.

CHURCH, M. (1980) Records of recent geomorphological events. In: *Timescales in Geomorphology* (Ed. by R.A. Cullingford, D.A. Davidson and J. Lewin), pp. 13–29. Wiley, Chichester.

CHURCH, M. & MARK, D.M. (1980) On size and scale in geomorphology. *Prog. Phys. Geogr.* **4** (3), 324–390.

CHURCH, M. & RYDER, J.M. (1972) Paraglacial sedimentation: a consideration of fluvial processes conditioned by glaciation. *Bull. geol. Soc. Am.* **83**, 3059–3072.

COLEMAN, J.M. (1969) Brahmaputra River: channel process and sedimentation. *Sedim. Geol.* **3**, 129–239.

Council on Environmental Quality (1973) *Report on Channel Modifications.* Three volumes. Washington, D.C.

CULBERTSON, J.K. & DAWDY, D.R. (1964) A study of fluvial characteristics and hydraulic variables, Middle Rio Grande, New Mexico. *Wat. Supply Pap. U.S. geol. Surv.* **1498-F**.

CULLING, W.E.H. (1960) Analytic theory of erosion. *J. Geol.* **68**, 316–344.

CULLINGFORD, R.A., DAVIDSON, D.A. & LEWIN, J. (1980) *Timescales in Geomorphology.* Wiley, Chichester. 360 pp.

DANIEL, J.F. (1971) Channel movement of meandering Indiana streams. *Prof. Pap. U.S. geol. Surv.* **732-A**.

DAVIES, T.R. & SUTHERLAND, A.J. (1980) Resistance to flow past deformable boundaries. *Earth Surf. Processes*, **5**, 175–179.

DEMSTER, G.R. (Jr) (1974) Effects of urbanization on floods in the Dallas, Texas Metropolitan Area. *Wat. Res. Inv. U.S. geol. Surv.* 60–73. NTIS, Washington.

DEVDARIANI, A.S. (1967) The profile of equilibrium and a regular regime. *Soviet Geogr.* **8**, 168–183.

DOUGLAS, I. (1976) Urban hydrology. *Geogr. J.* **142**, 65–72.

DUNNE, T. & LEOPOLD, L.B. (1978) *Water in Environmental Planning*. Freeman, San Francisco. 818 pp.

DURY, G.H. (1961) Bankfull discharge—an example of its statistical relationship. *Bull. Int. Ass. Hydrol.* **6**, 48–55.

DURY, G.H. (1977a) Underfit streams: retrospect, and prospect. In: *River Channel Changes* (Ed. by K.J. Gregory), pp. 281–293. Wiley, Chichester.

DURY, G.H. (1977b) Peak flows, low flows, and aspects of geomorphic dominance. In: *River Channel Changes* (Ed. by K.J. Gregory), pp. 61–74. Wiley, Chichester.

DURY, G.H. (1980) Neocatastrophism? A further look. *Prog. Phys. Geogr.* **4** (3), 391–420.

DURY, G.H., HAILES, J.R. & ROBBIE, H.B. (1963) Bankfull discharge and magnitude–frequency series. *Aust. J. Sci.* **26**, 123–124.

DUVICK, D.N. & BLASING, T.J. (1981) A dendroclimatic reconstruction of annual precipitation amounts in Iowa since 1680. *Wat. Resour. Res.* **17** (4), 1183–1189.

EINSTEIN, H.A. (1950) The bedload function for sediment transportation in open channel flows. *Tech. Bull. U.S. Dep. Agric.* **1026**.

EINSTEIN, H.A. & BARBAROSSA, N.L. (1952) River channel roughness. *Trans. Am. Soc. civ. Engrs* **117** (Paper 2528), 1121–1132.

EMMETT, W.W. (1974) Channel aggradation in western United States as indicated by observations at Vigil Network sites. *Z. Geomorph.* **21**, 52–62.

ENGELUND, F. (1974) Flow and bed topography in channel bends. *J. Hydraul. Div. Am. Soc. civ. Engrs* **100**, 1631–1648.

ENGELUND, F. (1975) Instability of flow in a curved alluvial channel. *J. Fluid Mech.* **72**, 145–160.

ENGELUND, F. & SKOVGAARD, O. (1973) On the origin of meandering and braiding in alluvial streams. *J. Fluid Mech.* **57**, 289–302.

ESPREY, W.H. & WINSLOW, D.E. (1974) Urban flood frequency characteristics, *J. Hydraul. Div. Am. Soc. civ. Engrs* **100**, 279–292.

EVERITT, B.L. (1968) Use of the cottonwood in an investigation of the recent history of a floodplain. *Am. J. Sci.* **266**, 417–439.

FERGUSON, R.I. (1975) Meander irregularity and wavelength estimation. *J. Hydrol.* **26**, 315–333.

FERGUSON, R.I. (1976) Disturbed periodic model for river meanders. *Earth Surf. Processes* **1**, 337–347.

FISK, H.N. (1952) Mississippi River Valley geology relation to river regime. *Trans. Am. Soc. civ. Engrs* **117**, 667–689.

FRIEDKIN, J.F. (1945) A laboratory study of the meandering of alluvial rivers. *U.S. WatWays Eng. Exp. Stat. Rep.* 40 pp.

FRITTS, H.C. (1976) *Tree Rings and Climate*. Academic Press, London. 567 pp.

FRITTS, H.C., BLASING, T.J., HAYDEN, B.P. & KUTZBACH, J.E. (1971) Multivariate techniques for specifying tree-growth and climate relationships and reconstructing anomalies in palaeoclimate. *J. appl. Met.* **10**, 845–864.

FRITTS, H.C., LOFGREN, G.R. & GORDON, G.A. (1979) Variations in climate since 1602 as reconstructed from tree rings. *Quat. Res.* **12**, 18–46.

GRAF, W.H. (1971) *Hydraulics of Sediment Transport*. McGraw-Hill, New York. 513 pp.

GRAF, W.L. (1975) The impact of suburbanisation of fluvial geomorphology. *Wat. Resour. Res.* **11**, 690–692.

GRAF, W.L. (1976) Streams, slopes and suburban development. *Geogr. Annlr* **8**, 153–173.

GRAF, W.L. (1979) Catastrophe theory as a model for change in fluvial systems. In: *Adjustments of the Fluvial System* (Ed. by D.D. Rhodes and G.P. Williams), pp. 13–32. Kendall/Hunt, Dubuque.

GRAF, W.L. (1981) Channel instability in a braided, sand bed river. *Wat. Resour. Res.* **17** (4), 1087–1094.

GRAF, W.H. & ACAROGLU, E.R. (1968) Sediment transport in conveyance systems (Part I). *Bull. Int. Ass. Scient. Hydrol.* **13**, 20–39.

GREGORY, K.J. (1976a) Changing drainage basins. *Geogr. J.* **142**, 237–247.

GREGORY, K.J. (1976b) Drainage basin adjustments and man. *Geogr. Pol.* **34**, 155–173.

GREGORY, K.J. (1977a) Stream network volume; an index of channel morphometry. *Bull. Geol. Soc. Am.* **88**, 1075–1080.

GREGORY, K.J. (Ed.) (1977b) *River Channel Changes*. Wiley, Chichester. 448 pp.

GREGORY, K.J. (1979) River channels. In: *Man and Environmental Processes* (Ed. by K.J. Gregory and D.E. Walling), pp. 123–143. Folkestone, Dawson.

GREGORY, K.J. & WALLING, D.E. (Eds) (1979) *Man and Environmental Processes*. Folkestone, Dawson. 276 pp.

GUPTA, A. & FOX, H. (1974) Effects of high magnitude flows on channel form: a case study in the Maryland Piedmont. *Wat. Resour. Res.* **10**, 499–509.

GUY, H.P. & JONES, D.E. (1972) Urban sedimentation—in perspective. *J. Hydraul. Div. Am. Soc. civ. Engrs* **98**, 2099–2166.

HADLEY, R.F. (1960) Recent sedimentation and erosional history of Fivemile Creek, Fremont County, Wyoming. *Prof. Pap. U.S. geol. Surv.* **352A**.

HALES, Z.L., SHINDALA, A. & DENSON, K.H. (1970) Riverbed degradation prediction. *Wat. Resour. Res.* **6**, 499–509.

HAMMER, T.R. (1972) Stream enlargement due to urbanization. *Wat Resour. Res.* **8**, 1530–1540.

HANDY, R.L. (1972) Alluvial cut-off dating from subsequent growth of a meander. *Bull. geol. Soc. Am.* **83**, 475–480.

HARR, R.D., HARPER, W.C. & KAYGIER, J.T. (1975) Changes in storm hydrographs after road building and clear-cutting in the Oregon Coast Range. *Wat. Resour. Res.* **11** (3), 436–444.

HARVEY, A.M. (1969) Channel capacity and the adjustment of streams to hydrologic regime. *J. Hydrol.* **8**, 82–98.

HAYNES, C.V. (1968) Geochronology of late-Quaternary alluvium. In: *Means of Correlation of Quaternary Successions; Proc. VII Inqua Congress* (Ed. by R.B. Morrison, and H.E. Wright), Utah, **8**, 591–631.

HENDERSON, F.M. (1961) Stability of alluvial channels. *J. Hydraul. Div. Am. Soc. civ. Engrs* **87**, 109–138.

HERBERTSON, J.G., PETERS, J.C., BOWLER, R.A., GILL, M.A. DE VRIES, M. & EGIAZAROFF, I. (1968) Discussion of "River-bed degradation below dams" (by S. Komura and D.B. Simons). *J. Hydraul. Div. Am. Soc. civ. Engrs* **94**, 589–598.

HEUSSER, C.J., HAUSSER, L.E. & STREETER, S.S. (1980) Quaternary temperatures and precipitation for the northwest coast of North America. *Nature*, **286**, 702–704.

HEY R.D. (1978) Determinate hydraulic geometry of river channels. *J. Hydraul. Div. Am. Soc. civ. Engrs* **104**, 869–885.

HEY, R.D. (1979) Dynamic process-response model of river channel development. *Earth Surf. Processes*, **4**, 59–72.

HIBBERT, A.R. (1967) Forest treatment effects on water yields. In: *Forest Hydrology* (Ed. by W.E. Sopper and H.W. Lull), pp. 527–543. Pergamon Press, Oxford.

HICKIN, E.J. (1969) A newly identified process of point bar formation in natural streams. *Am. J. Sci.* **267**, 999–1010.

HICKIN, E.J. (1972) Pseudomeanders and point dunes—a flume study. *Am. J. Sci.* **272**, 762–799.

HICKIN, E.J. (1974) The development of meanders in natural river-channels. *Am. J. Sci.* **274**, 414–442.

HICKIN, E.J. (1978) Mean flow-structure in meanders of the Squamish River, British Columbia. *Can. J. Earth Sci.* **15** (11), 1833–1849.

HICKIN, E.J. (1979) Concave-bank benches on the Squamish River, British Columbia, Canada. *Can. J. Earth Sci.* **16** (1), 200–203.

HICKIN, E.J. & NANSON, G.C. (1975) The character of channel migration on the Beatton River, Northeast British Columbia, Canada. *Bull. geol. Soc. Am.* **86**, 487–494.

HOLLIS, G.E. (1975) The effect of urbanisation on floods of different recurrence intervals. *Wat. Resour. Res.* **11**, 431–435.

HOLLIS, G.E. & LUCKETT, J.K. (1976) The response of river channels to urbanization—two case studies from south east England. *J. Hydrol.* **30**, 351–363.

HOOKE, J.M. (1977) The distribution and nature of changes in river channel patterns. In: *River Channel Changes* (Ed. by K.J. Gregory), pp. 265–280. Wiley, Chichester.

HOOKE, J.M. (1980) Magnitude and distribution of rates of river bank erosion. *Earth Surf. Processes* **5**, 143–157.

HOWARD, A.D. (1980) Thresholds in river regime. In: *Thresholds in Geomorphology* (Ed. by D.R. Coates and J.D. Vitek), pp. 227–258. Allen & Unwin, Boston.

HUGHES, D.J. (1977) Rates of erosion on meander arcs. In: *River Channel Changes* (Ed. by K.J. Gregory), pp. 193–205. Wiley, Chichester.

INGLIS, C.C. (1949) The behaviour and control of rivers and canals. *Res. Publs centr. Irrig. Bd India*, **13**, two volumes.

JACKSON, R.G. II (1975a) Velocity-bed form-texture patterns of meander bends in the lower Wabash River of Illinois and Indiana. *Bull. geol. Soc. Am.* **86**, 1511–1522.

JACKSON, R.G. II (1975b) Hierarchial attributes and a unifying model of bed forms composed of cohesionless material and produced by shearing flow. *Bull. geol. Soc. Am.* **86**, 1523–1534.

JACKSON, R.G. II (1976) Sedimentological and fluid dynamics implications of the turbulent bursting phenomena in geophysical flows. *J. Fluid Mech.* **77**, 531–560.

JANDA, R.J., NOLAN, M.K., HARDEN, D.R. & COLEMAN, S.M. (1975) Watershed conditions in the drainage basins of Redwood Creek, Humboldt County, California, as of 1973. *Open File Rep. U.S. geol. Surv.* No. 75568.

JEFFREY, W.W. (1970) Hydrology of landuse. In: *Handbook on the Principles of Hydrology* (Ed. by D.M. Gray), pp. 13.1–13.57. National Research Council of Canada.

KALINSKE, A.A. (1947) Movement of sediment as bed load in rivers. *Am. geophys. un.* **28** (4), 615–620.

KELLERHALS, R., CHURCH, M. & DAVIES, L.B. (1979) Morphological effects of interbasin river diversions. *Can. J. civ. Engrs* **6** (1), 18–31.

KELLERHALS, R. & GILL, D. (1973) Observations and potential downstream effects of large storage projects in Northern Canada. *Proc. 11th Cong. int. Comm. Lge Dams*, Madrid, Spain, pp. 731–754.

KIRBY, M.J. (1977) Maximum sediment efficiency as a criterion for alluvial channels. In: *River Channel Changes* (Ed. by K.J. Gregory), pp. 429–448. Wiley, Chichester.

KNIGHTON, A.D. (1975) Variations in at-a-station hydraulic geometry. *Am. J. Sci.* **275**, 186–218.

KNIGHTON, A.D. (1977) Short-term changes in hydraulic geometry. In: *River Channel Changes* (Ed. by K.J. Gregory), pp. 101–119. Wiley, Chichester.

KNOX, J.C. (1977) Human impacts on Wisconsin stream channels. *Ann. Ass. Am. Geogr.* **67**, 323–342.

KOMURA, S. & SIMONS, D.B. (1967) River-bed degradation below dams. *J. Hydraul. Div. Am. Soc. civ. Engrs* **93**, 1–14.

KONDRAT'YEV, N.Ye. (1968) Hydromorphological principles of computations of free meandering. 1. Signs and indexes of free meandering. *Soviet Hydrol.* **4**, 309–335.

KULEMINA, N.M. (1973) Some characteristics of the process of incomplete meandering of the channel of the.Ob'River. *Soviet Hydrol.* **6**, 518–534.

LANE, E.W. (1937) Stable channels in erodible material. *Trans. Am. Soc. civ. Engrs* **102**, 123–194.

LANE, E.W. (1957) A study of the shape of channels formed by natural stream flowing in erodible material. *M.R.D. Sediment Series*, No. 9, U.S. Army Engineering Division, Missouri River, Corps of Engineers, Omaha, Nebraska.

LANGBEIN, W.B. (1964) Geometry of river channels. *J. Hydraul. Div. Am. Soc. civ. Engrs* **90**, 301–312.

LANGBEIN, W.B. & LEOPOLD, L.B. (1964) Quasi-equilibrium states in channel morphology. *Am. J. Sci.* **262**, 782–794.

LANGBEIN, W.B. & SCHUMM, S.A. (1958) Yield of sediment in relation to mean annual precipitation. *Trans. Am. geophys. Un.* **39** (6), 1076–1084.

LARSON, C.L. (1972) Using hydrologic models to predict the effects of watershed modification. *Nat. Symp. Watershed Trans., Am. Wat. Resour. Ass*, pp. 113–117

LEEDER, M.R. & BRIDGES, P.H. (1975) Flow separation in meander bends. *Nature*, **253**, 338–339.

LEOPOLD, L.B. (1968) Hydrology for urban land planning—a guidebook on the hydrologic effects of urban landuse. *Circ. U.S. geol. Surv.* **554**.

LEOPOLD, L.B. (1973) River channel change with time: an example. *Bull. geol. Soc. Am.* **84**, 1845–1860.

LEOPOLD, L.B., BAGNOLD, R.A., WOLMAN, M.G. & BRUSH, M. (1960) *Prof. Pap. U.S. geol. Surv.* **282-D**.

LEOPOLD, L.B. & EMMETT, W.W. (1972) Some rates of geomorphological processes. *Geogr. Pol.* **23**, 27–35.

LEOPOLD, L.B. & MILLER, J.P. (1954) A postglacial chronology for some alluvial valleys in Wyoming. *Prof. Pap. U.S. geol. Surv.* **1261**.

LEOPOLD, L.B. & WOLMAN, M.G. (1957) River channel patterns: braided, meandering and straight. *Prof. Pap. U.S. geol. Surv.* **282-B**.

LEOPOLD, L.B. & WOLMAN, M.G. (1960) River meanders. *Bull. geol. Soc. Am.* **71**, 769–794.

LEOPOLD, L.B., WOLMAN, M.G. & MILLER, S.P. (1964) *Fluvial Processes in Geomorphology*. Freeman, San Francisco. 522 pp.

LEWIN, J. (1977) Channel pattern changes. In: *River Channel Changes* (Ed. by K.J. Gregory), pp. 167–184. Wiley, Chichester.

LEWIN, J. & BRINDLE, B.J. (1977) Confined meanders. In: *River Channel Changes* (Ed. by J.K. Gregory), pp. 221–233. Wiley, Chichester.

LEWIN, J. & HUGHES, D. (1976) Assessing channel change on Welsh rivers. *Cambria* 3, 1–10.

MADDOCK, T. (1970) Intermediate hydraulics of alluvial channels. *J. Hydraul. Div. Am. Soc. civ. Engrs* 96, 2309–2323.

MAHMOOD, K. & SHEN, H.W. (1971) Regime concept of sediment-transporting canals and rivers. In: *River Mechanics* (Ed. by H.W. Shen), chapter 30. Fort Collins, Colorado.

MAKKAVAYEV, N.I. (1972) The impact of large water engineering projects on geomorphic processes in stream valleys. *Soviet Geogr.* 13, 387–393.

MATHEWES, R.W. & HEUSSER, L.E. (1981) A 12,000 year palynological record of temperature and precipitation trends in southwestern British Columbia. *Can. J. Bot.* 59 (5), 707–710.

MOCKMORE, C.A. (1943) Flow around bends in stable channels. *Trans. Am. Soc. civ. Engrs* 109, 335–360.

MOSLEY, M.P. (1975a) Meander cut-offs on the River Bollin, Cheshire in July, 1973. *Revue Géomorph. dyn.* 24, 21–31.

MOSLEY, M.P. (1975b) Channel changes on the River Bollin, Cheshire, 1872–1973. *East Midland Geogr.* 6, 185–199.

NANSON, G.C. (1974) Bedload and suspended-load transport in a small, steep, mountain stream. *Am. J. Sci.* 274, 471–486.

NANSON, G.C. (1980) Point bar and floodplain formation of the meandering Beatton River, northeastern British Columbia, Canada. *Sedimentology*, 27, 3–29.

NANSON, G.C. & BEACH, H.F. (1977) Forest succession and sedimentation on a meandering-river floodplain, northeast British Columbia, Canada. *J. Biogeogr.* 4, 229–251.

ORME, A.R. & BAILEY, R.G. (1970) The effect of vegetation conversion and flood discharge on stream channel geometry: the case of Southern California watersheds. *Proc. Am. Ass. Geogr.* 2, 101–106.

ORME, A.R. & BAILEY, R.G. (1971) Vegetation conversion and channel geometry in Monroe Canyon Southern California. *Yb. Ass. Pacif. Coast Geogr.* 33, 65–82.

PARK, C.C. (1977) Man-induced changes in stream channel capacity. In: *River Channel Changes* (Ed. by K.J. Gregory), pp. 121–144. Wiley, Chichester.

PARK, C.C. (1981) Man, river systems and environmental impacts. *Prog. Phys. Geogr.* 5 (1), 1–31.

PARKER, G. (1976) Cause and characteristic scales of meandering and braiding in rivers. *J. Fluid Mech.* 76, 457–480.

PETTS, G.E. (1979) Complex response of river channel morphology subsequent to reservoir construction. *Prog. Phys. Geogr.* 3, 329–362.

PICKUP, G. & RIEGER, W.A. (1979) A conceptual model of the relationship between channel characteristics and discharge. *Earth Surf. Processes*, 4, 37–42.

PICKUP, G. & WARNER, R.F. (1976) Effects of hydrologic regime on magnitude and frequency of dominant discharge. *J. Hydrol.* 29, 51–75.

QUICK, M.C. (1974) Mechanism for streamflow meandering. *J. Hydraul. Div. Am. Soc. civ. Engrs* 100, 741–753.

RANGO, A. (1970) Possible effects of precipitation modification on stream channel geometry and sediment yield. *Wat. Resour. Res.* 6, 1765–1770.

RAUDKIVI, A.J. (1967) *Loose Boundary Hydraulics*. Pergamon Press, Oxford. 331 pp.

RICHARDS, K.S. (1973) Hydraulic geometry and channel roughness—a non-linear system. *Am. J. Sci.* 273, 877–896.

RILEY, S.J. (1975) The channel shape–grain size relation in eastern Australia and some palaeohydrologic implication. *Sedim. Geol.* 14, 253–258.

RHODES, D.D. & WILLIAMS, G.P. (1979) *Adjustments of the Fluvial System*. Kendall/Hunt, Dubuque. 372 pp.

ROESNER, L.A., KIBLER, D.F. & MONSER, J.R. (1972) Use of storm drainage models in urban planning. *Nat. Symp. Watershed Trans., Am. Wat. Resour. Ass.* pp. 400–405.

ROTHACKER, J. (1970) Increases in water yield following clearcut logging in the Pacific Northwest. *Wat Resour. Res.* 6 (2), 653–658.

ROZOVSKII, I.L. (1957) *Flow of Water in Bends of Open Channels*. Academy of Sciences of the Ukrainian SSR, Kiev. 233 pp.

SANTOS-CAYADE, J. & SIMONS, D.B. (1973) River response. In: *Environmental Impact on Rivers* (*River Mechanics III*) (Ed. by H.W. Shen), pp. 1–25. Colorado State University, Fort Collins.

SCHEIDEGGER, A.E. (1965) On the dynamics of deposition. *Bull. int. Ass. Sci. Hydrol.* 10, 49–57.

SCHUMM, S.A. (1968) River adjustment to altered hydrologic regimen—Murrumbidgee River and palaeochannels, Australia. *Prof. Pap. U.S. geol. Surv.* 598.

SCHUMM, S.A. (1969) River metamorphosis. *J. Hydraul. Div. Am. Soc. civ. Engrs* 95, 251–273.

SCHUMM, S.A. (1973) Geomorphic thresholds and the complex response of drainage systems. In: *Fluvial Geomorphology* (Ed. by Marie Morisawa), pp. 299–310. Publications in Geomorphology. State University of New York, Binghamton.

SCHUMM, S.A. (1976) Episodic erosion, a modification of the Davis cycle. In: *Theories of Landform Development* (Ed. by W.N. Melhorn and R.C. Flemal), pp. 69–85. Publications in Geomorphology. State University of New York, Binghamton.

SCHUMM, S.A. (1977) *The Fluvial System*. Wiley, New York. 338 pp.

SCHUMM, S.A. & KHAN, H.R. (1972) Experimental study of channel patterns. *Bull. geol. Soc. Am.* 83, 1755–1770.

SCHUMM, S.A. & LICHTY, R.W. (1963) Channel widening and floodplain construction along Cimarron River in southwestern Kansas. *Prof. Pap. U.S. geol. Surv.* 352-D.

SCHUMM, S.A. & LICHTY, R.W. (1965) Time, space and causality in geomorpholoy. *Am. J. Sci.* 263, 110–119.

SCHUMM, S.A. & PARKER, R.S. (1973) Implications of complex response of drainage systems for Quaternary alluvial stratigraphy. *Nature*, 243 (128), 99–100.

SHAHJAHAN, M. (1970) Factors controlling the geometry of fluvial meanders. *Bull. int. Ass. Sci. Hydrol.* 15 (3), 13–24.

SHEN, H.W. & KOMURA, S. (1968) Meandering tendencies in straight alluvial channels. *J. Hydraul. Div. Am. Soc. civ. Engrs* 94, 997–1016.

SHRODER, J.F. (1980) Dendrogeomorphology; review and new techniques of tree-ring dating. *Prog. Phys. Geogr.* 4 (2), 161–188.

SIMONS, D.B. & SENTÜRK, F. (1977) *Sediment Transport Technology*, chapter 6. W.R.P., Fort Collins.

SONI, J.P. (1981) Unsteady sediment transport law and prediction of aggradation parameters. *Wat. Resour. Res.* 17 (1), 33–40.

SOPPER, W.E. & LULL, H.W. (Eds) (1967) *Forest Hydrology*. Pergamon Press, Oxford. 813 pp.

SOUTHARD, J.B. (1971) Representation of bed configurations in depth–size diagrams. *J. sedim. Petrol.* 41, 903–915.

SPEIGHT, J.G. (1965) Meander spectra of the Anabunga River. *J. Hydrol.* **3**, 1–15.

STEVENS, M.A., SIMONS, D.B. & RICHARDSON, E.V. (1975) Non-equilibrium river form. *J. Hydraul. Div. Am. Soc. civ. Engrs* **101**, 557–566.

STEVENS, M.A., SIMONS, D.B. & SCHUMM, S.A. (1975) Man-induced changes of Middle Mississippi River. *J. WatWays Harb. Div. Am. Soc. civ. Engrs* **101**, 119–133.

SUNDBORG, ÅKE (1956) The River Klarälven: a study of fluvial processes. *Geogr. Annlr* **38**, 127–316.

SWANSON, F.J. & DYRNESS, C.T. (1975) Impact of clear-cutting and road construction on soil erosion by landslides in western Canada Range, Oregon. *Geology*, **3** (7), 393–396.

TAYLOR, G., CROOK, K.A.W. & WOODYER, K.D. (1971) Upstream-dipping foreset cross-stratification: origin and implications for palaeoslope analyses. *J. sedim. Petrol.* **41**, 578–581.

TINNEY, E.R. (1962) Process of channel degradation. *J. geophys. Res.* **67** (4), 1475–1480.

VANORI, V.A. (Ed.) (1975) *Sedimentation Engineering.* American Society of Civil Engineers Task Committee for the preparation of the manual of Sedimentation of the Sedimentation Committee of the Hydraulics Division, New York. 745 pp.

WALLING, D.E. (1974) Suspended sediment production and building activity in a small British basin. *Publ. int. Ass. Sci. Hydrol.* **113**, 137–144.

WALLING, D.E. & GREGORY, K.J. (1970) The measurement of the effects of building construction on drainage basin dynamics. *J. Hydrol.* **11**, 129–144.

WILLIAMS, G.P. (1978) Bank-full discharge of rivers. *Wat. Resour. Res.* **14**, 1141–1154.

WINKLEY, B.R. (1977) *Manmade Cutoffs on the Lower Mississippi River: Conception, Construction and River Response.* Potamology Section, U.S. Corps of Engineers, Vicksburg, Mississippi. 213 pp.

WOLMAN, M.G. (1959) Factors influencing erosion of a cohesive river bank. *Am. J. Sci.* **257**, 204–216.

WOLMAN, M.G. (1967) A cycle of sedimentation and erosion in urban river channels. *Geogr. Annlr* **49** A, 385–395.

WOLMAN, M.G. & BRUSH, L.M. (1961) Factors controlling the size and shape of stream channels in coarse, noncohesive sand. *Prof. Pap. U.S. geol. Surv.* **282-G**, 28 pp.

WOLMAN, M.G. & MILLER, J.P. (1960) Magnitude and frequency of forces in geomorphic processes. *J. Geol.* **68**, 64–74.

WOLMAN, M.G. & SCHICK, A.P. (1967) Effects of construction on fluvial sediment, urban and suburban areas of Maryland. *Wat. Resour. Res.* **3**, 451–464.

WOODYER, K.D. (1968) Bankfull frequency in rivers. *J. Hydrol.* **6**, 114–142.

WOODYER, K.D. (1975) Concave-bank benches on the Barwon River, New South Wales. *Aust. Geogr.* **13**, 36–40.

YALIN, M.S. (1977) *Mechanics of Sediment Transport.* 2nd ed. Pergamon Press, Oxford, 290 pp.

YANG, C.T. (1976) Minimum unit stream power and fluvial hydraulics. *J. Hydraul. Div. Proc. Am. Soc. civ. Engrs* **102**, 919–934.

YANG, C.I., SONG, C.C.S. & WOLDENBERG, M.J. (1981) Hydraulic geometry and minimum rate of energy dissipation. *Wat. Resour. Res.* **17** (4), 1014–1018.

YEN, B.C. (1975) Spiral motion and erosion in meanders. *Proc. 16th int. Congr. Ass. Hydraul. Res.* **2**, 338–346.

ZIEMER, R.R. (1981) Storm flow response to road building and partial cutting in small streams in Northern California. *Wat. Resour. Res.* **17** (4), 907–917.

Spec. Publs int. Ass. Sediment. (1983) **6**, 85–95

Distribution of main flow velocity in alternating river bends

H. J. GELDOF*† *and* H. J. DE VRIEND*

* *Delft University of Technology, Department of Civil Engineering, Delft, The Netherlands
and* † *University of Utrecht, Geographical Institute, Utrecht, The Netherlands*

ABSTRACT

Measurements of the depth-averaged main flow velocity in two consecutive sharply curved short bends in the river Dommel, The Netherlands, are compared with results of a mathematical model. The computed main velocity distribution agrees rather well with measured one in the larger part of each bend. Near the bend exits, however, deviations occur which indicate that secondary flow convection should be included in the mathematical model.

INTRODUCTION

The flow and the bed formation of curved alluvial rivers are complicated phenomena. They are not easily explained (Callander, 1978) and are still more difficult to model mathematically (De Vriend, 1981b; Flokstra & Koch, 1981; Koch & Flokstra, 1981). The most striking feature of the essentially three-dimensional flow field in a bend is the helical streamline pattern, in the simplest case with the flow near the water surface directed outwards and the velocity vector near the bottom directed inwards. This induces transverse shear stresses on the bed, causing a transverse component of sediment transport leading to the formation of a point bar near the inner bank and a pool near the outer bank.

This paper deals with the velocity field in a curved stream, assuming the bed to be fixed and the discharge to be constant. The main velocity distribution in this type of flow is analysed using a mathematical model and measurements of the longitudinal velocity component in two consecutive bends of the river Dommel, a small sand-bedded tributary of the Meuse,

0141-3600/83/0106-0085 $02.00

MEASUREMENTS

Location

The bed geometry and flow velocity measurements were performed in May 1980, in a 285 m long section of the river Dommel, approximately 3 km downstream of its crossing with the border between Belgium and The Netherlands (Fig. 1).

Measuring procedure

The geometry of the river channel in the study area was surveyed before and after the velocity measurements (i.e. in the periods 1–5 May and 27–28 May) by levelling along 52 traverses (Fig. 4).

In 23 cross-sections, the longitudinal velocity component, or rather the horizontal velocity component perpendicular to the cross-section, was measured (12–30 May) with a propeller-type current meter (Ott C–2, propeller diameter 0·030 m, pitch 0·100 m), mounted on a movable bridge across the river (Fig. 2). The number of measuring points in a section ranged from 61 to 104, depending on cross-sectional geometry. They were arranged along at least 10 equidistant vertical lines. At each point the velocity was measured during 3 × 30 sec and averaged.

The water level was measured twice a day at four point gauge stations in the study area (Fig. 4). The

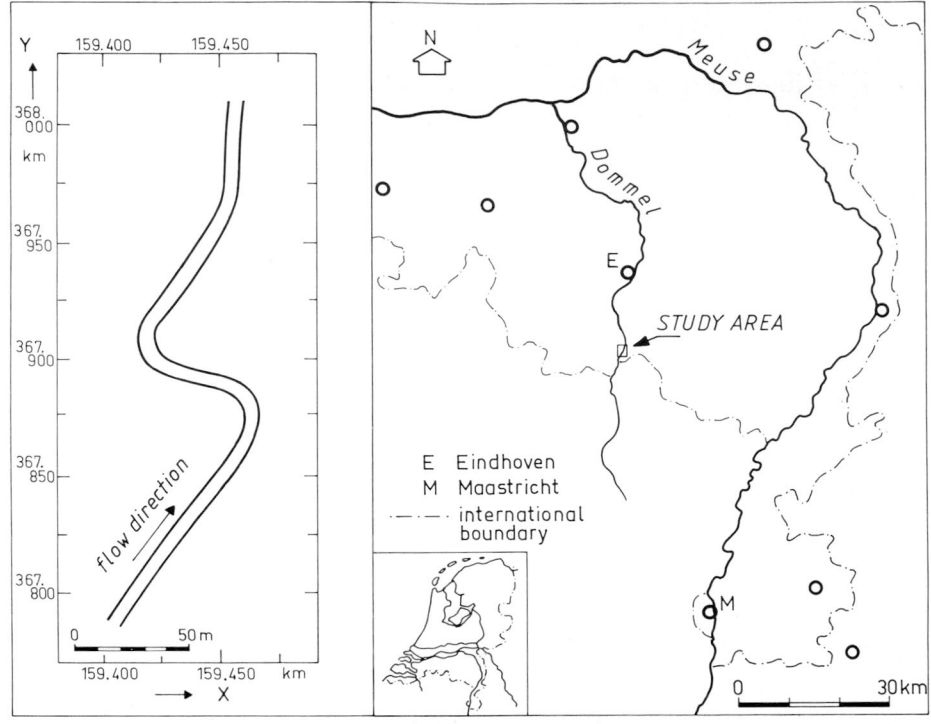

Fig. 1. Location of study area.

Fig. 2. Upstream view of straight reach between the two bends, with movable bridge and current meter.

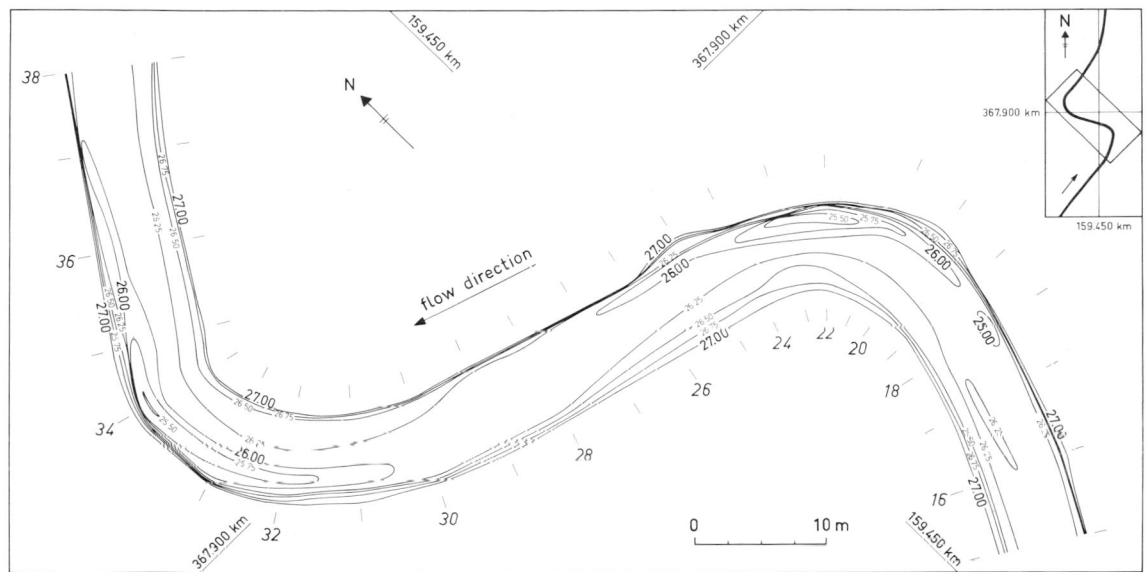

Fig. 3. Contour map of the bed level in the central part of the study area. Bed level indicated in metres above Agreed Ordnance Datum; contour interval 0·25 m. Contours above A.O.D. +27 m are omitted. Surveyed by levelling, 1–5 May 1980.

stations were situated in such a way that the longitudinal slope of the water surface could be determined separately for the upstream straight reach, the bend reach, and the downstream straight reach.

In the cross-sections where velocity measurements were made the transverse slope of the water surface was determined using a point gauge mounted on the movable bridge.

PRESENTATION OF RESULTS

Channel geometry

When comparing the results of the levellings at the beginning and at the end of the survey period, mostly small and unsystematic differences were found. Therefore, the channel geometry is assumed to have remained unaltered.

In addition to the channel alignment outlined in Fig. 1, the channel geometry in the bend section is depicted in further detail in Fig. 3. This contour map shows that the bends are rather sharp and give rise to pronounced deformations of the channel profile, with high point bars and deep pools. Besides, the banks turn out to be quite steep, sometimes almost vertical.

The upstream and downstream reaches, which have not been included in Fig. 3, have steep banks as well, with an almost horizontal bottom in the straight parts

and a slightly distorted profile in the mild bend half-way along the downstream reach (see Fig. 4).

Flow conditions

Throughout the survey period, flow conditions were almost constant. According to the registrations of a Rijkswaterstaat float gauge station located approximately 150 m upstream of the study area, the variation of the daily mean water level amounted to about 0·05 m; using the provisional stage–discharge relation for this station the daily mean discharge varied between 1·00 and 1·22 m³ sec⁻¹. This corresponds reasonably well with the variation of the discharge as derived from the velocity measurements (using the velocity–area method; Anon., 1979) from 1·21 to 1·53 m³ sec⁻¹.

The fall of the water level over the survey section (between point gauge stations 1 and 4, marked in Fig. 4) also showed no important variations: from 0·13 to 0·15 m, with a mean value of 0·145 m. For a discharge of 1·27 m³ sec⁻¹, a water surface width of 6·10 m and a water depth of 0·50 m, this corresponds with a Chézy factor of about 30 m$^{\frac{1}{2}}$ sec⁻¹.

During the period of measurement the vegetation in the channel increased, mainly on the banks. The channel bottom remained almost free from vegetation, except for a few places in the upstream reach and a hump (area 0·5 × 1 m²) near station 30, just before the

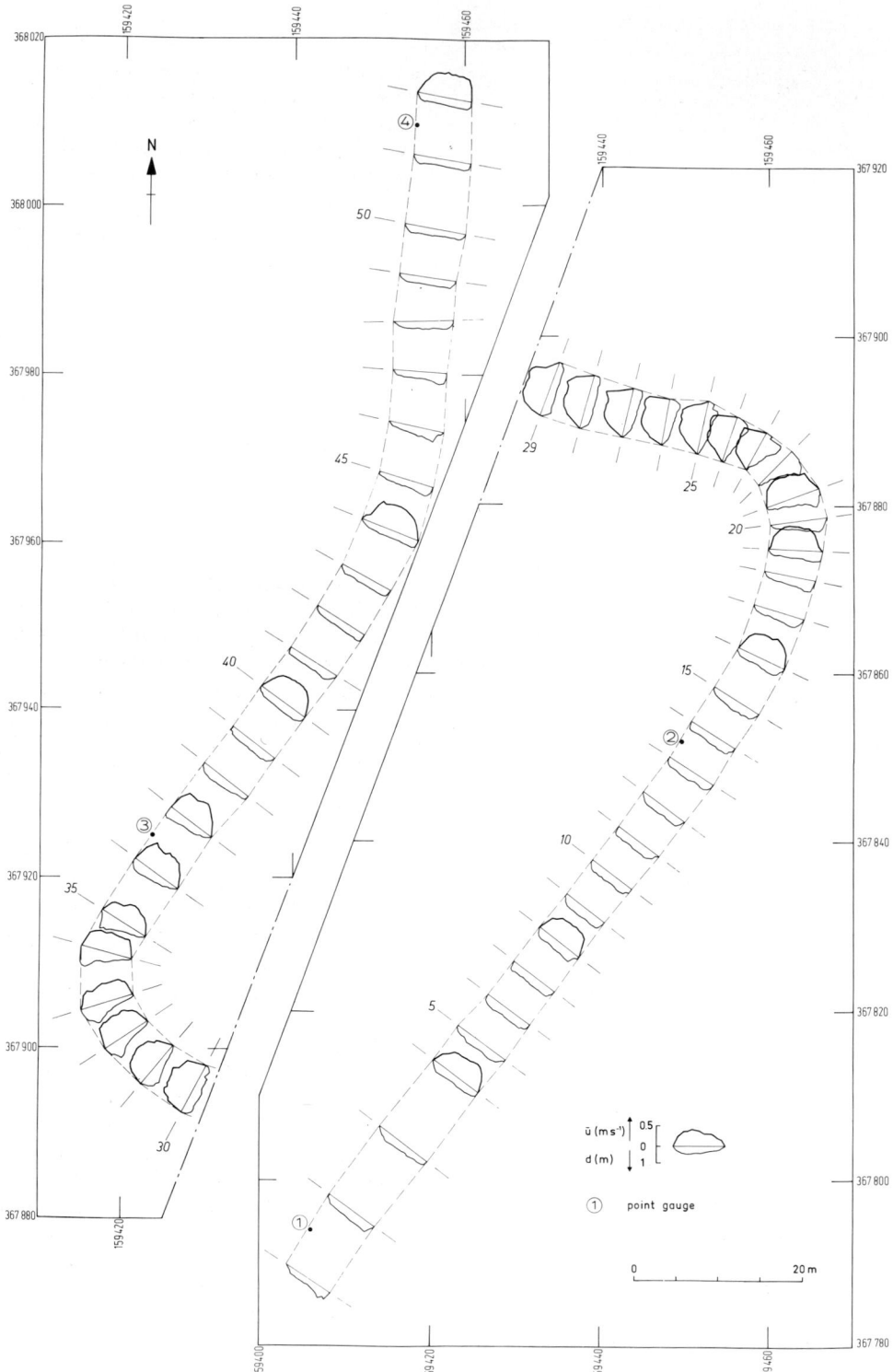

Fig. 4. Water depth and depth-averaged main flow velocity, measured 12–30 May 1980.

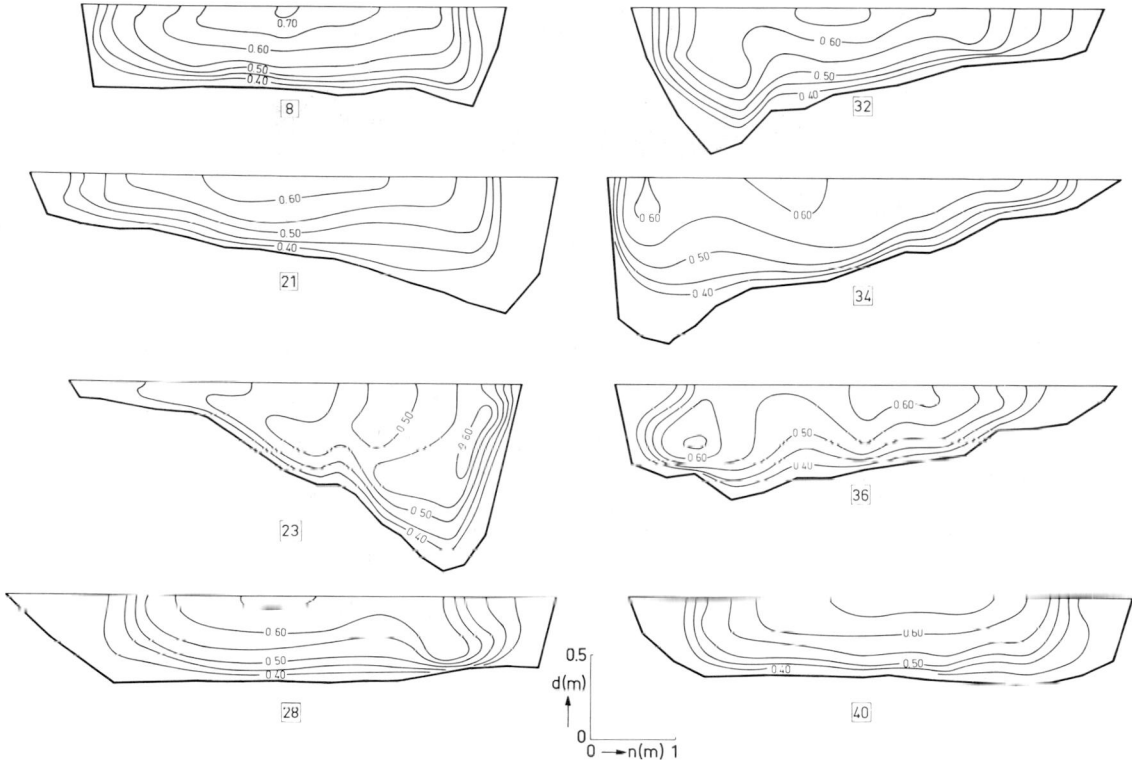

Fig. 5. Downstream view of main flow isovel pattern in eight cross-sections. Isovel interval 0·05 m sec⁻¹.

entrance of the second bend. This bed irregularity may have influenced the flow pattern in the second bend considerably.

Main flow velocity distribution

In general, the longitudinal velocity and the main velocity in a curved river will not coincide exactly (Hickin, 1978). As long as the main flow direction does not deviate too strongly from the channel axis, however, these velocities have almost the same magnitude. Therefore, it is assumed that the observed longitudinal velocity is equal to the main flow velocity.

The results of the velocity measurements are represented in Fig. 4, showing the depth-averaged main velocity distribution after reduction to the reference discharge $Q_0 = 1·27$ m³ sec⁻¹, and in Fig. 5, giving the main flow isovel pattern in eight selected cross-sections.

The principal features of the effect of the bends on the velocity distribution are the same as observed in various other field studies (Jackson, 1975, Bridge & Jarvis, 1976, Hickin, 1978, Dietrich, Smith & Dunne, 1979, also see Leopold, Wolman & Miller, 1964) and

in numerous laboratory experiments with curved flumes, either rectangular (e.g. Siebert, 1980; De Vriend, 1981b) or non-rectangular (Rozovskii, 1961; Yen, 1967; De Vriend & Koch, 1978). In general, the position of the velocity maximum tends to move towards the inner bank upon entering the bend. Subsequently, it gradually shifts outwards, approaching the outer bank at the bend exit. In addition to this horizontal redistribution, the vertical profile of the main flow velocity is considerably deformed as the velocity maximum gradually shifts away from the water surface (Fig. 5) through the bend.

If the bend is followed by a long straight reach, the deformations of the main velocity distribution gradually vanish. If another bend follows at a short distance, however, its flow pattern will be influenced. In the case of opposite bends, this implies that the inward skewing of the main velocity distribution is enhanced, whence it takes longer for the velocity maximum to reach the outer bend (see also Callander, 1978). This may also have consequences for the bend configuration: the point bar near the inner bank will not be as high as in a bend preceded by a long straight reach.

The results of the velocity measurements in the Dommel suggest that the influence of the first bend vanishes before the flow enters the second bend. It should be noted, however, that this picture may have been influenced by the vegetation-covered bed irregularity near station 30, the more so since the point bar in the second bend is much lower than the one in the first bend, even though the two bends have almost the same radius of curvature.

Transverse slope of the water surface

As stated above, the transverse fall of the water level was measured using a point gauge mounted on a movable bridge. This method turned out to be rather sensitive to small fluctuations of the water level: the measured data are widely scattered. Consequently, they do not yield reliable information on the longitudinal distribution of the transverse fall, but the maxima in the bends can still be estimated. After reduction to the reference discharge of $1 \cdot 27$ m³ sec^{-1}, using a water surface width of $6 \cdot 10$ m and a mean water depth of $0 \cdot 50$ m, these maxima amount to $0 \cdot 009$ m for the first bend and $0 \cdot 010$ m for the second one.

MATHEMATICAL SIMULATION

Mathematical model

In order to gain a better insight into the physical process causing the observed redistribution of the depth-averaged main flow velocity, the flow has been simulated mathematically, using a computational model that holds good under the following conditions (cf. Kalkwijk & De Vriend, 1980): (a) the flow is shallow and gently curved, (b) the flow is mainly controlled by friction, (c) the longitudinal component of the velocity is predominant, (d) the convective interaction between the main and the secondary flow is unimportant, (e) the Froude number is so small that a 'rigid-lid' approximation can be applied to the water surface (i.e. the surface is left free in the pressure gradients, but not in the water depth). The model solves the depth-averaged balance equations for mass and momentum, with the vertical distribution of the main flow velocity assumed invariant throughout the flow.

Formulated in the 'channel-fitted' coordinate system (s, n), with s parallel to the channel axis and n perpendicular to it, the system of depth-averaged equations finally becomes:

continuity:

$$\frac{\partial}{\partial s}(\bar{u}\,d) + \frac{\partial}{\partial n}(\bar{v}\,d) + \frac{\bar{v}\,d}{r_s} = 0; \tag{1}$$

s-momentum:

$$\frac{1}{d}\frac{\partial}{\partial s}(\bar{u}^2\,d) + \frac{1}{d}\frac{\partial}{\partial n}(\bar{u}\,\bar{v}\,d) + 2\frac{\bar{u}\,\bar{v}}{r_s} =$$

convective main flow inertia

$$-\frac{1}{\rho}\frac{\partial P}{\partial s} - \frac{g}{C^2}\frac{\bar{u}^2}{d}; \tag{2}$$

longitudinal bed
pressure friction
gradient

n-momentum:

$$-\frac{\bar{u}^2}{R_s} = -\frac{1}{\rho}\frac{\partial P}{\partial n} \tag{3}$$

centripetal   transverse
acceleration   pressure
gradient

with: $\bar{u}, \bar{v}$ = depth-averaged velocity components in
           the s- and n-direction, respectively,
    P = total pressure,
    d = local water depth,
    r_s = local radius of curvature of the s-lines,
    R_s = local radius of curvature of the stream-
         lines of the depth-averaged flow field,
    g = acceleration due to gravity,
    C = Chézy factor,
    ρ = mass density.

Channel schematization

Although, in principle, the mathematical model applies to a much wider range of channel configurations, the version of the computer code available at present is restricted to channels of constant width, with a longitudinal axis consisting of a limited number of straight lines and circular arcs. Therefore, the channel planform has been schematized, fixing the width at $6 \cdot 10$ m and dividing the longitudinal axis into 10 sections, according to Table 1.

Channel bed undulations (mostly ripples, locally dunes), with an estimated maximum height of about $0 \cdot 1$ m, caused significant deviations from the mean bed topography. The bedforms were not observed systematically, nor was their effect eliminated by taking the average of a large number of levellings at each point. Hence the irregularities have to be considered as random scatter of the data without any physical meaning. Therefore it would not be realistic to introduce the raw bed-level data into the computation.

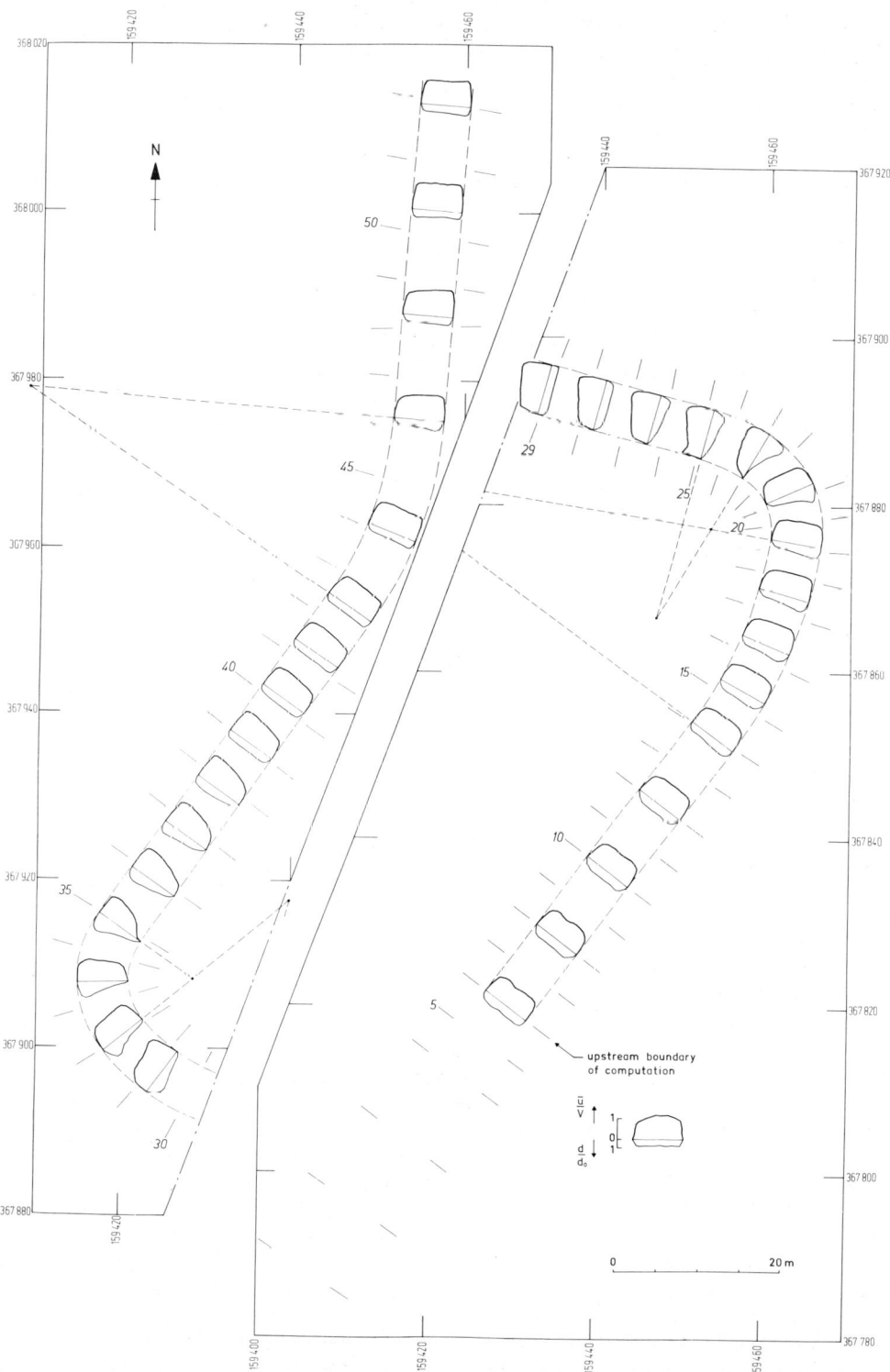

Fig. 6. Schematized water depth and computed depth-averaged main flow velocity.

Table 1. Schematization of the channel planform

Section no.	Centre of curvature		Arc properties of the channel axis			Computational grid	
	X (m)	Y (m)	R_c (m)	Φ_{tot} (°)	L_{tot} (m)	Steps	Δs (m)
1	—	—	∞	0	40·60	12	3·38
2	159,410·6	367,884·1	53·50	27·2	25·40	8	3·18
3	159,453·3	367,877·2	10·20	70·2	12·50	4	3·13
4	159,446·7	367,866·4	22·90	16·2	6·50	2	3·25
5	—	—	∞	0	20·10	6	3·35
6	159,439·9	367,917·9	25·30	−34·9	15·40	5	3·08
7	159,428·6	367,908·4	10·50	−75·3	13·80	4	3·45
8	—	—	∞	0	47·10	14	3·36
9	159,408·5	367,978·9	46·00	30·3	24·30	7	3·47
10	—	—	∞	0	37·50	12	3·13

Another reason for schematizing the bed configuration is the obvious necessity to reconcile the transverse channel profile with the planform schematization discussed before. Therefore the wet perimeter of the channel has been schematized by eye before being used in the computation (see Figs 4 and 6, keeping in mind that the cross-sections of the measuring grid and the computational grid are not exactly coincident).

In accordance with the 'rigid-lid' approximation in the mathematical model, the water surface is taken as horizontal in each cross-section. Its longitudinal slope in the channel axis is assumed to correspond with the overall mean energy slope $(5\cdot2 \times 10^{-4})$, with $z_s = \text{A.O.D.} + 26\cdot620$ m at point-gauge station 3.

The influence of the bedforms is incorporated in the Chézy roughness factor C, estimated from the energy losses measured in the straight reaches. Spatial variations of C due to variations in the water depth are taken into account through the White–Colebrook formula for a rough bed, keeping the roughness height constant. Hence

$$C = C_0 + \frac{\sqrt{g}}{\kappa} \ln (d/d_0) \qquad (4)$$

in which $C_0 = $ Chézy factor for the reference depth d_0

$$(C_0 \simeq 30 \text{ m}^{\frac{1}{2}} \sec^{-1}),$$

$\kappa = $ Von Kármán constant $(\simeq 0\cdot4)$.

Finally, it is pointed out that, as long as only the distribution of the main velocity is concerned and not its absolute magnitude, the mathematical model can be formulated in terms of dimensionless variables, such that the magnitude of the discharge is not relevant (neither the Froude number nor the Reynolds number figure in this model).

Computational results and discussion

Figure 6 gives an overall impression of the model results for the cross-sections that are nearest to the ones of the velocity measuring grid. At first glance, the calculated depth-averaged main velocity field presents the same features as the measured one (Fig. 4): in the first part of each bend the flow tends to shift towards the convex bank; subsequently, it gradually moves outwards, to yield an outward-skewed distribution of $\bar{u}$ near the bend exit and in the straight reach downstream of each bend. (The anomalous peak in $\bar{u}$ near the convex bank at the end of the first bend (station 25), is probably caused by the application of a computational grid that is too coarse for the very shallow flow there.)

On closer inspection, however, significant differences become apparent (Fig. 7). Since lateral diffusion of momentum is disregarded in the model, the predicted influence of the banks on $\bar{u}$ does not extend far enough into the central part of the cross-section, the more so as the vegetation on the banks may have caused a much higher local roughness than estimated in the model schematization.

Secondly, even when limiting consideration to the regions where the influence of the banks seems negligible, the skewness of the $\bar{u}$-distribution in the last part of the first bend is overestimated by the model. When comparing the central part of cross-sections 23 and 24 (Fig. 7), the skewness of the computed velocity distribution remains almost unaltered, whereas the slope of the regression line through the measured data decreases by a factor of 2·3. As the channel geometry hardly changes between these two cross-sections (Fig. 3), this decrease seems rather surprising. But it is not accidental, since it also occurs to an even more marked

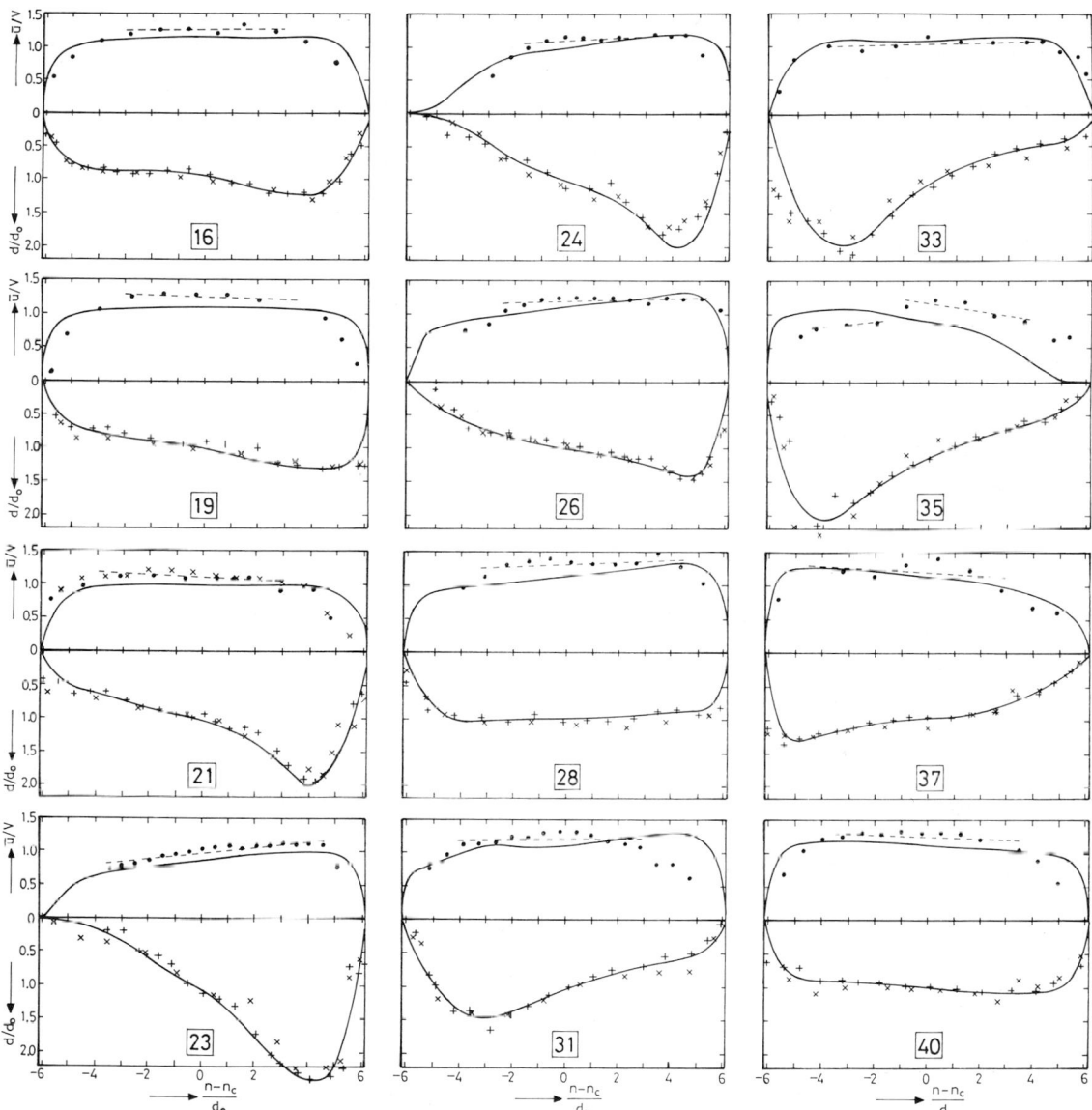

Fig. 7. Comparison between measured and computed depth-average main flow velocity. $\bar{u}/V$-plots: ●, ×, measured data; ------, linear regression line (for central region only); ———, computed. d/d_0-plots; +, ×, measured data; ———, model schematization; [99], cross-section.

degree at the end of the second bend (Fig. 7: section 35 *vs.* section 33).

As this loss of skewness occurs only in the last part of each bend, it seems obvious to attribute it to the secondary flow, which according to the isovel patterns in Fig. 5 has an appreciable effect on the main flow there (cf. Bathurst, Thorne & Hey, 1979). If the secondary circulation consists of a single cell, with the velocity directed towards the outer bank near the

surface and towards the inner bank near the bottom, its convective transport of main flow momentum in the transverse plane enhances the outward skewing of $\bar{u}$ (Kalkwijk & De Vriend, 1980; De Vriend, 1981a). If a second cell with an opposite circulation is formed near the surface in the outer bend, however, the velocity maximum will tend to shift back inwards to the boundary between the two cells.

Though not investigated systematically, the existence of this second cell was confirmed by occasional observations of floating material during the survey. Moreover, the deformations of the main flow isovels near the outer bank (Fig. 5) suggest that a double-cell configuration of the secondary flow does occur in the last parts of the bends (also see Bathurst *et al.*, 1979).

This double-cell configuration provides a qualitative explanation of the observed changes of $\partial \bar{u}/\partial n$ between stations 23 and 24 and between stations 33 and 35. Mathematical models describing the formation of the second cell, however, are extremely complicated and necessarily three-dimensional (De Vriend & Koch, 1981). Hence a quantitative verification of the above hypothesis is still beyond reach.

CONCLUSIONS

In summary, the comparison of the model results with the measured data leads to the conclusion that, for this steep-banked channel with its sharply curved, but rather short bends, the model without secondary flow convection works reasonably well in the straight reaches and in the larger part of each bend. As soon as the influence of this convection becomes important, i.e. in the downstream parts of the bends, the model overestimates the outward skewing of the main flow velocity distribution. The influence of this shortcoming still lingers in the straight reaches following the bends. Consequently, the description of the interaction between the velocity distributions in the two consecutive bends is not entirely correct.

This implies that the mathematical model used here will not be quite suitable as a component of a predictive mathematical model of river bend morphology. On the other hand, the foregoing makes clear that, in the specific type of bends considered here, the role of the secondary flow in the redistribution of the main velocity is not as important as has been suggested (Dietrich *et al.*, 1979). This provides the possibility of using this fairly simple model as a tool for the physical analysis of the main velocity redistribution in this kind of river bend.

ACKNOWLEDGMENTS

J. S. L. J. van Alphen, P. M. Bloks, P. Hoekstra and H. J. Winteraeken, students in physical geography at the University of Utrecht, performed the measurements as part of their field training. Their diligence and enthusiasm have contributed essentially to the success of the field campaign.

We gratefully acknowledge the technical and field assistance by the staff of both the Laboratory of Fluid Mechanics, Delft University of Technology, and the Physical Geography Laboratory, University of Utrecht.

The kind permission of the 'Waterschap De Dommel', Boxtel, for executing the measurements is much appreciated. Rijkswaterstaat, Arnhem, is thanked for providing equipment and float gauge station data.

Finally, we wish to thank Mr D. C. Post and the drawing office of the Department of Civil Engineering, Delft University of Technology, for the skilful preparation of the figures.

REFERENCES

ANON. (1979) *Liquid flow measurement in open channels—velocity-area methods.* I.S.O. 748–1979 (E), 23 pp.

BATHURST, J.C., THORNE, C.R. & HEY, R.D. (1979) Secondary flow and shear stress at river bends. *J. Hydraul. Div. Am. Soc. civ. Engrs* **105**, 1277–1295.

BRIDGE, J.S. & JARVIS, J. (1976) Flow and sedimentary processes in the meandering River South Esk, Glen Clova, Scotland. *Earth Surf. Processes,* **1**, 303–336.

CALLANDER, R.A. (1978) River meandering. *Ann. Rev. Fluid Mech.* **10**, 129–158.

DE VRIEND, H.J. (1981a) Velocity redistribution in curved rectangular channels. *J. Fluid Mech.* **107**, 423–439.

DE VRIEND, H.J. (1981b) *Steady flow in shallow channel bends. Comm. Hydraul. Dep. civ. Engrs* **81–3**, 260 pp. Delft University of Technology (also: doctoral thesis, Delft University of Technology).

DE VRIEND, H.J. & KOCH, F.G. (1978) *Flow of water in a curved open channel with a fixed uneven bed. Delft Hydraul. Lab. & Delft University of Technology. T.O.W. Rep. R657–VI/M*1415-*II,* 15 pp.

DE VRIEND, H.J. & KOCH, F.G. (1981) *Fully three-dimensional computations of steady turbulent flow in curved channels with a rectangular cross-section. Delft Hydraul. Lab. & Delft University of Technology T.O.W. Report R657–X/R*1631 29 pp.

DIETRICH, W.E., SMITH, J.D & DUNNE, TH. (1979). Flow and sediment transport in a sand bedded meander. *J. Geol.* **87**, 305–315.

FLOKSTRA, C. & KOCH, F.G. (1981) Numerical aspects of bed level predictions for alluvial rivers. In: *Proc. int. Conf. Numerical Modelling of River, Channel and Overland Flow for Water Resources and Environmental Application,* I.A.H.R./I.I.A.S.A./W.M.O., Bratislava, Section 2.1, 12 pp. (also: *Delft Hydraul. Lab. Publ. no.* **258**, 1981, 12 pp.).

HICKIN, E.J. (1978) Mean flow structure in meanders of the Squamish River, British Columbia. *Can. J. Earth Sci.* **15**, 1833–1849.

JACKSON, R.G. (1975) Velocity–bed form–texture patterns of meander bends in the lower Wabash River of Illinois and Indiana. *Bull. geol. Soc. Am.* **86**, 1511–1522.

KALKWIJK, J. P. Th. & DE VRIEND, H.J. (1980) Computations of the flow in shallow river bends. *J. Hydraul. Res.* **18**, 327–342 (also: *Comm. Hydraul. Dep. civ. Engrs* **80–1**, 25 pp. Delft University of Technology).

KOCH, F.G. & FLOKSTRA, C. (1981) Bed level computations for curved alluvial channels. In: *Proc. XIXth Congress I.A.H.R.* New Delhi, **2**, subject A(d), paper 16, 357–364. (also: *Delft Hydraul. Lab. Publ. no.* **240**, 1980, 7 pp.)

LEOPOLD, L.B., WOLMAN, M.G. & MILLER, J.P. (1964) *Fluvial Processes in Geomorphology.* Freeman, San Francisco. 522 pp.

ROZOVSKII, I.L. (1961) *Flow of Water in Bends of Open Channels.* Israel Program For Scientific Translations, Jerusalem. 233 pp.

SIEBERT, W. (1980) *Strömungskarakteristiken in einem Kanal mit 180°—Krümmungen.* Dissertation, University of Karlsruhe. 233 pp.

YEN, C.L. (1967) *Bed configuration and characteristics of subcritical flow in a meandering channel.* Ph.D. Thesis, University of Iowa. 52 pp.

APPENDIX: LIST OF SYMBOLS

A.O.D.	Agreed Ordnance Datum (N.A.P.)	—
b_0	water surface width of schematized channel	m
C	Chézy factor	$m^{\frac{1}{2}} sec^{-1}$
C_0	Chézy factor for $d = d_0$	$m^{\frac{1}{2}} sec^{-1}$
d	local water depth	m
d_0	reference water depth	m
g	acceleration due to gravity	$m sec^{-2}$
L_{tot}	length along axis of schematized channel section	m
n	horizontal coordinate perpendicular to the channel axis, positive from left to right bank	m
n_c	n-coordinate of channel axis	m
P	total pressure	$kg\ m^{-1}\ sec^{-2}$
Q	water discharge	$m^3\ sec^{-1}$
Q_0	reference discharge	$m^3\ sec^{-1}$
r_s	local radius of curvature of the s-lines	m
R_c	radius of curvature of axis of schematized channel	m
R_s	local radius of curvature of streamlines of depth-averaged flow field	m
s	horizontal coordinate parallel to axis of schematized channel, positive downstream	m
$\bar{u}$	depth-averaged flow velocity component in the s-direction (main flow velocity)	$m\ sec^{-1}$
$\bar{v}$	depth-averaged flow velocity component in the n-direction (main flow velocity)	$m\ sec^{-1}$
V	reference velocity $\left(\equiv \dfrac{Q_0}{b_0 d_0} \right)$	$m\ sec^{-1}$
X	abscissa of translated national coordinate system	m or km
Y	ordinate of translated national coordinate system	m or km
z_s	local water surface level	m
κ	Von Kármán constant	—
ρ	mass density	$kg\ m^{-3}$
Φ_{tot}	angle of circular arc of schematized channel section, positive for bends turning leftwards	degrees

Spec. Publs int. Ass. Sediment. (1983) **6**, 97–106

Changing size distribution of suspended sediment in arid-zone flash floods

L. E. FROSTICK*, I. REID* and J. T. LAYMAN†

* *Birkbeck College, University of London, U.K. and † La Sainte Union College, Southampton, U.K.*

ABSTRACT

Water and sediment discharge are analysed in detail over several floods in a moderate-sized (7 km^2) ephemeral stream network in Kenya's northern arid zone. Negligible interception by sparse desert scrub and measured low surface infiltration capacities dictate overland flow within minutes of rainfall onset and result in rapid stream response to multi-celled rainstorms (lag response averages 34 min). Surface wash samples, collected in troughs sunk into a variety of slopes, indicate slope sediment yields up to 0·219 kg m^{-2} for single flood events. The contributions of six major tributaries are marked by clearly distinguished discharge pulses in the main channel. These have significant consequences for the size of suspended sediment (a parameter hitherto largely ignored) which responds sympathetically to water discharge pulses. Suspended sediment concentration rapidly achieves a peak with the passage of the flood bore over the dry channel bed and declines from values as high as 15,800 mg l^{-1} in direct relationship with declining water discharge. The average particle-size of suspended sediment declines from 46 to 2 μm, though it is shown that clay-size particles are less responsive to changes in flow. Comparison between expected concentrations in size classes above 63 μm using Laursen's semi-empirical equation, and observed concentrations yields a high correlation coefficient, $r = 0·88$, but the relationship is poor for silt and clay fractions. These are derived almost entirely from surface wash and constitute 56% of annual suspended sediment discharge. Size analyses of flash-flood suspended sediment provide an important and unique insight into the hydrodynamics of arid-zone ephemeral streams, giving a running commentary on the flexible boundary separating bedload and suspended load.

INTRODUCTION

The relationship between the total concentration of suspended sediment and water discharge is well documented for temperate and tropical humid zones where perennial or seasonal river flow makes a sampling programme that covers a range of discharges a feasible proposition (Bauer & Tille, 1967; Douglas, 1968, 1973; Müller & Forstner, 1968; Walling, 1974). The same cannot be said of arid zones. Here, the intermittent and unpredictable pattern of short-lived flash floods peculiar to such areas makes installation of a gauging station expensive . The passage of four or five floods per year dictates a slow rate of data acquisition. Yet such data are crucial to the construction of general models of fluvial sediment transport, since the unvegetated arid landscape sheds

the water of characteristically intense storms in a manner that produces rapid stream responses and record levels of suspended sediment concentration (Bondurant, 1951; Beverage & Culbertson, 1964; Gerson, 1977).

Notwithstanding the wealth of information relating total concentrations of suspended sediment to water discharge, little is known about the changing size distribution of suspended sediment if only because the comparatively low concentrations characteristic of perennial streams preclude the collection of adequate samples. Fleming & Poodle (1970), sampling at an unspecified depth but over a wide range of discharges for the River Clyde, Scotland, declare no increase or decrease in mean size, although they consider this to reflect a poor range of bed material size available rather than the hydraulic competence of the stream.

0141-3600/83/0106-0097 $02.00

Colby & Hembree (1955) indicate a steady median particle size over a wide range of discharge through a channel constriction on the Niobrara river, Nebraska, though at the upper extreme of discharge median particle size falls by a factor of 10 due to the surface wash of fine material from the contributing slopes of the catchment following exceptional rainstorms.

Other reports of suspended sediment size distribution reflect the difficulties of data collection, and use only a limited number of samples either to illustrate the balance of forces that distribute large and small particles within the flow (Nelson & Benedict, 1951; Nordin, 1963; Colby, 1963), or to draw attention to what may be typical size curves of suspended sediment (Beverage & Culbertson, 1964).

Data reflecting changes in the size distribution of suspended sediment load under the unsteady flow conditions of a flood rather than under the steady flows of controlled discharges are needed in particular to verify and refine the predictive equations of Einstein (modified by Colby & Hembree, 1955), Lane & Kalinske (1939) and Laursen (1958). Arnborg, Walker & Pieppo (1967) and Rapp *et al.* (1972) provide useful glimpses of changing particle size over flood-waves and these serve to contradict the general statements of Fleming & Poodle (1970) and Colby & Hembree (1955) which are, after all, based on the data of sampling programmes not orientated to establish the detailed changes that occur during the passage of single floods.

The infrequency of flood flows in arid-zone ephemeral streams focuses attention on single events. Data collected in Kenya's arid north give a detailed picture of the changing size distribution of suspended sediment, and provide both an empirical verification of those size classes whose concentration is susceptible to successful prediction and a clear indication of the limitations of such prediction where considerable quantities of sediment are derived from sources other than the channel bed.

DRAINAGE BASIN CHARACTER AND FLOOD HYDROLOGY

The data are provided by a gauging station established specifically for the early equinoctial rain season of 1979 on the 7 km² Il Kimere catchment east of Lake Turkana in Kenya's arid Northern Province (Fig. 1). The long-term record indicates an average annual rainfall of 308 mm, though recent climatic shifts suggest increasing desertification. The area is uninhabited and now forms part of the Sibilot National Park. The vegetation is a scanty desert scrub that offers little protection against rain-splash erosion, though along the major drainage lines a riparian growth of larger acacias is supported by perched water in the channel-fill (Frostick & Reid, 1979).

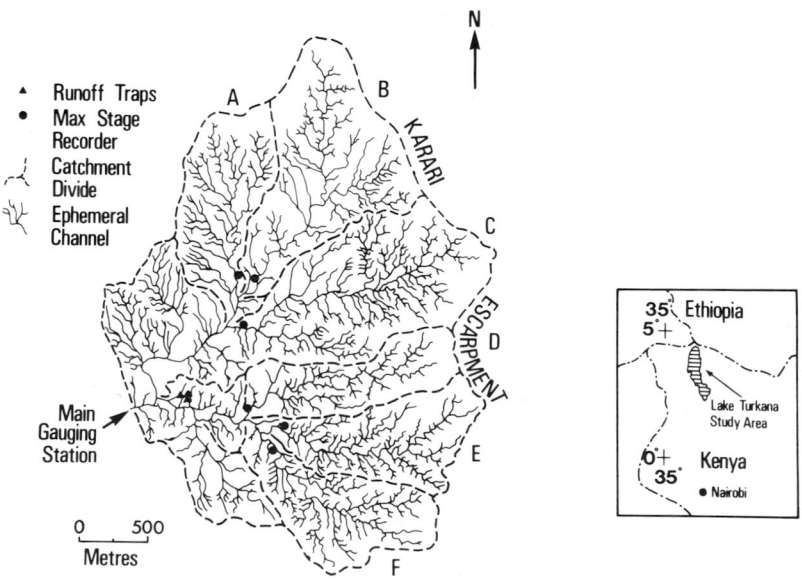

Fig. 1. Index maps of Il Kimere drainage basin.

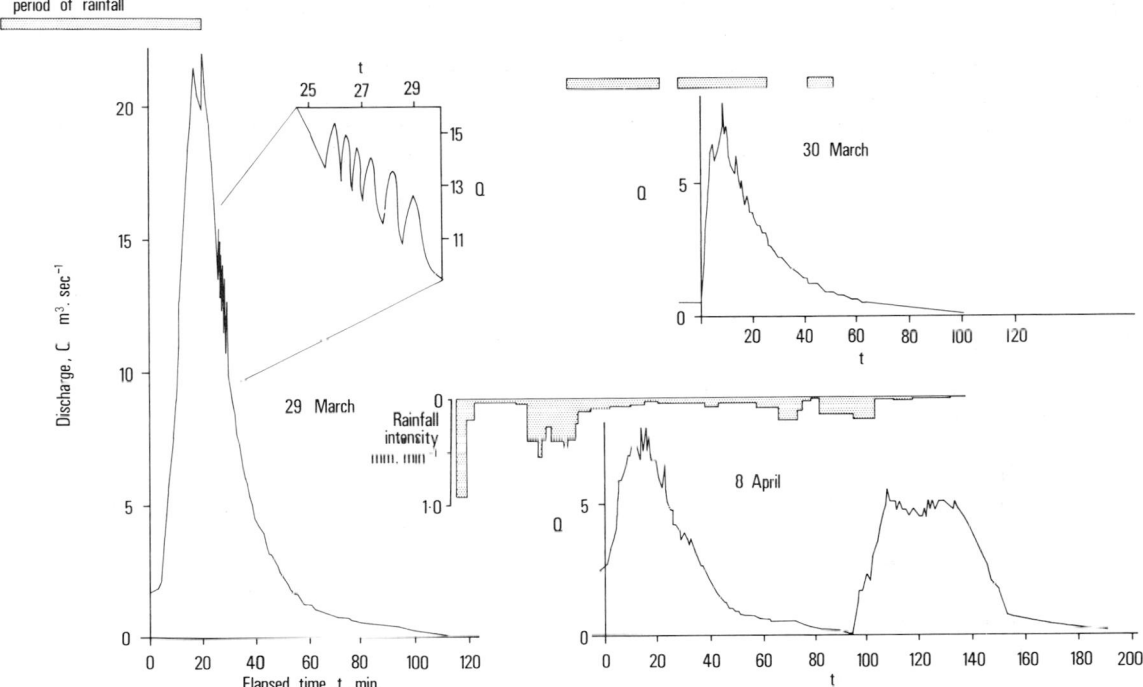

Fig. 2. Il Kimere flood hydrographs, equinoctial rains 1979. Time axis origin refers to arrival of flood bore at gauging station.

Infiltration capacities of 18–72 mm h^{-1}, affected as they are by rainbeat-induced surface crusting, are low compared with the intensity of the rainfall from thunderstorm cells that move in discrete but apparently random tracks over the landscape. As a consequence the onset of overland flow rapidly follows the beginning of rainfall, with water observed to discharge to first-order streams within 10 min. Such immediate shedding of water dictates a high drainage density and brings about a rapid response in the main channel. The lag between rainfall centroid and peak discharge for the three floods that constituted the Il Kimere's half-yearly discharge of 1979 was only 35, 34, and 34 min in each case (Fig. 2). Water arrives at a gauging point as a bore passing over an essentially waterless channel-bed (Leopold & Miller, 1956; Frostick & Reid, 1979). Transmission losses to the porous channel-fill may be substantial (Renard & Keppel, 1966; Burkham, 1970; Butcher & Thornes, 1978) but observations in the Il Kimere channel suggest that direct-contributing low-order streams and rills play an important part in reducing such losses by saturating the channel-fill before the arrival of the main channel bore.

Once the bore has arrived, peak discharge is quickly achieved. The times of rise for the Il Kimere hydrographs range narrowly between 4 and 16 min (Fig. 2). The catchment continues to remain responsive to the changes in rainfall intensity that occur as ancillary storm cells pass over. This can be seen clearly by comparing the double-peaked plateau of the second stream rise of 8 April with the rainfall pattern after an elapsed time of 58 min (Fig. 2).

Of considerable importance to the sedimentary dynamics of ephemeral streams is the control that the drainage network exercises over water discharge. Elsewhere in the area Frostick & Reid (1977, 1979) have shown that staggered tributary discharges produce a series of discharge pulses (but not the translatory waves of Leopold & Miller, 1956, and Renard & Keppel, 1966) that determine the deposition of channel-fill laminae. The Il Kimere has six major tributary basins disposed in two groups (Fig. 1). The hydrographs of Fig. 2 reflect either grouped contribution as in the double peak of the 29 March, or staggered individual basin contributions as illustrated by the six pulses of both the 30 March flood and the recession limb of 29 March (inset). Discharge pulses play an important role in turbulent suspension.

EXPERIMENTAL PROCEDURES

A straight reach of the Il Kimere was selected for gauging. The channel-bed is characteristically a plane-bed sand-fill up to 2 m thick. At the gauging station the bed is incised 2 m below the flood plain and has a width of 11 m. To accommodate sampling at flood peak, a rope bridge was slung from bank to bank. Samples of suspended sediment were taken at the mid-channel position over the entire flood wave using a USDH 48 depth-integrating sampler. Sampling interval ranged between 2 and 20 min depending upon changes in water stage. In three successive floods the numbers of samples collected were 19, 12 and 23. Attempts to determine water velocity using current meters were frustrated by the amount of submerged organic debris carried by the stream (a problem referred to by Nordin, 1963, in a similar environment) and so recourse was made to the Manning equation for open-channel flow. A check on the appropriate value of the roughness coefficient using partially submerged floats and correcting for the logarithmic nature of the velocity profile gave a mean value of

0·021 that varied little with discharge but which falls within the range quoted by Nordin (1963) for flash floods on the Rio Puerco, New Mexico.

Six troughs set into typical slopes and designed to collect overland flow and surface wash provide complementary data on sediment sources outside the stream channel.

Samples were passed through 0·45 μm pore diameter membrane filters and sized using a combination of Coulter Counter, sedimentation and sieve techniques.

TOTAL CONCENTRATION OF SUSPENDED SEDIMENT

Suspended sediment concentration achieves peak values within minutes of initial hydrograph rise (Fig. 3). This is not surprising considering the energy grade-line at the bore front and the turbulence induced by the continuous overriding of the retarded lower part of the flow. The combined effect is most clearly shown in the flood of 30 March, where

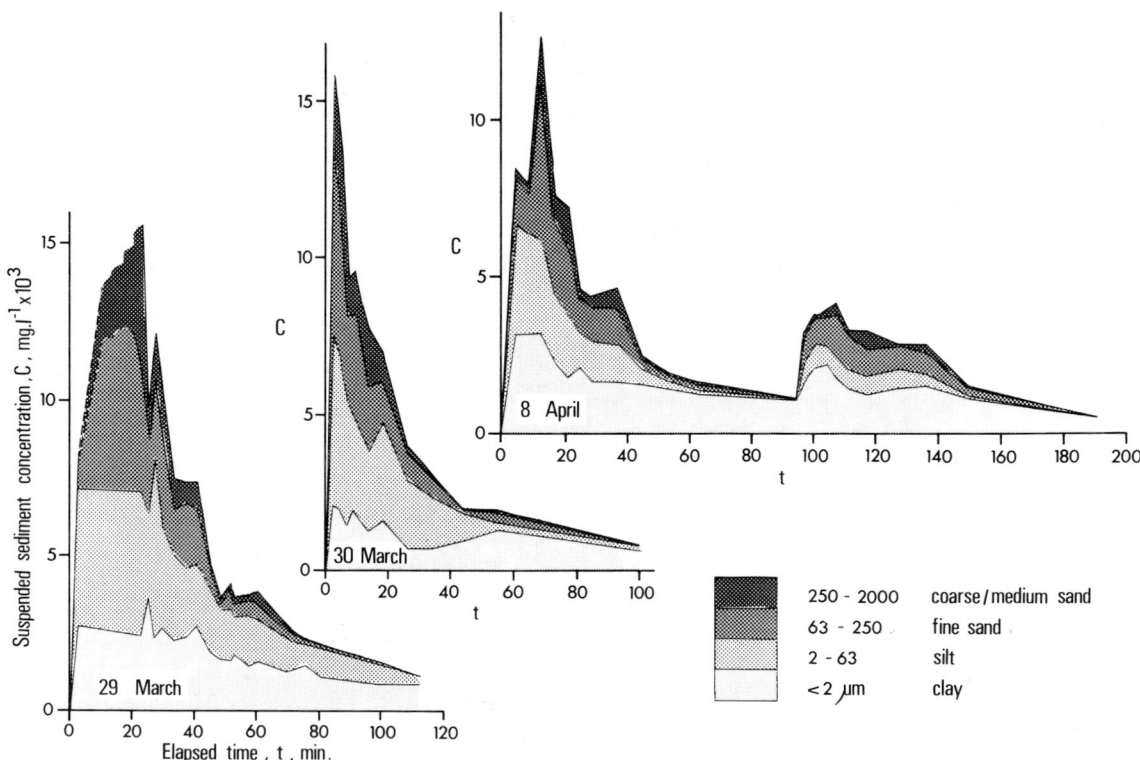

Fig. 3. Concentration of suspendend sediment by size fraction through three flash floods.

maximum concentration is associated with the bore and not with peak flow which occurs at *c.* 8 min. Renard & Laursen (1975) illustrate a similar relationship for one flood on Walnut Gulch, Arizona. However, peak concentration of suspended sediment is not necessarily a bore phenomenon. Although technical difficulties meant a gap in sampling between 2 and 22 min on 29 March, it is clear that during this flood the sediment concentration is related more closely to water discharge, since concentration rises after the bore has passed.

This leads to one important consideration of the part played by the pulses in discharge that are produced by the staggered delivery of water from individual tributary basins. Each of the primary and secondary peaks in suspended sediment concentration other than those associated with flood-bores is synchronous with a momentary increase in water discharge (Fig. 3 *vide* 29 March at $t = 27$ min, 8 April at $t = 12$). Imeson's (1977) suggestion that similar fluctuations result from changes in rain-splash disturbance in or near the channel of his small Ardennes catchment is inappropriate for the much larger Il Kimere system. Observations of the size of suspended sediment involved in the fluctuations indicate that any explanation must account for the disturbance of the coarsest bed particles (up to 2 mm) and these are unlikely to be affected by rain-splash. Instead, it is appropriate to envisage either a wave of sediment-laden water travelling downchannel from the tributary confluence, or a momentary increase in turbulence that forces local bed material into suspension. It is likely that the two mechanisms act

sympathetically and that, while a higher concentration of the fine material is carried forward with the water wave, settling velocity dictates a more local origin for the enhanced concentration of the coarser fraction.

Combining the data of the three Il Kimere floods and plotting total concentration of suspended sediment as a function of water discharge (curve A, Fig. 4) provides a general relationship in keeping with other arid-zone streams (Table 1). High intercept values and low exponents distinguish these from perennial rivers at both high and low latitudes.

There is evidence of a progressive impoverishment of the suspended sediment concentration with successive floods as was suggested for the N. Qishon, Israel by Negev (1969) and for a small Devonshire catchment in south-west England by Walling (1974). The values of the constant *a* in the relationship $C = aQ^b$ where C is sediment concentration and Q water discharge, are successively 3246, 2417 and 2017 and provide grounds for some suspicion that the sediment available for transport is being depleted.

SIZE OF SUSPENDED SEDIMENT

Many factors influence the suspension of bed material. The lift exerted on a small bed particle by the fluid stream may exceed both its submerged mass and any other restraining forces and so lead to suspension. Particles can also be observed to launch themselves off bed roughness elements. But the mechanism thought to be of primary importance in the suspension of bed material is the momentary penetration of the

Table 1. Relationship between total concentration of suspended sediment, C, and water discharge, Q, $C = aQ^b$

Reference	River	Environment	Const. a	Exp. b
Frostick, Reid & Layman (this paper)	Il Kimere, N. Kenya	Arid	2570	0·512
Nordin & Beverage (1965)	R. Grande, N. Mexico	Arid	100	0·700
Negev (1969)	N. Qishon, Israel	Arid	4217	0·159
Nordin (1963)	R. Puerco, N. Mexico	Arid	80,000	0·200
Müller & Forstner (1968)	Alpenrhein, Austria	Temperate humid	0·004	2.200
Jordan (1965)	Mississippi, Missouri	Temperate sub-humid	0·01	1·600
Fahnestock (1963)	White R., Washington	Temperate humid	40	2.500
Bauer & Tille (1967)	Helbe, German D.R.	Temperate humid	31	1·391

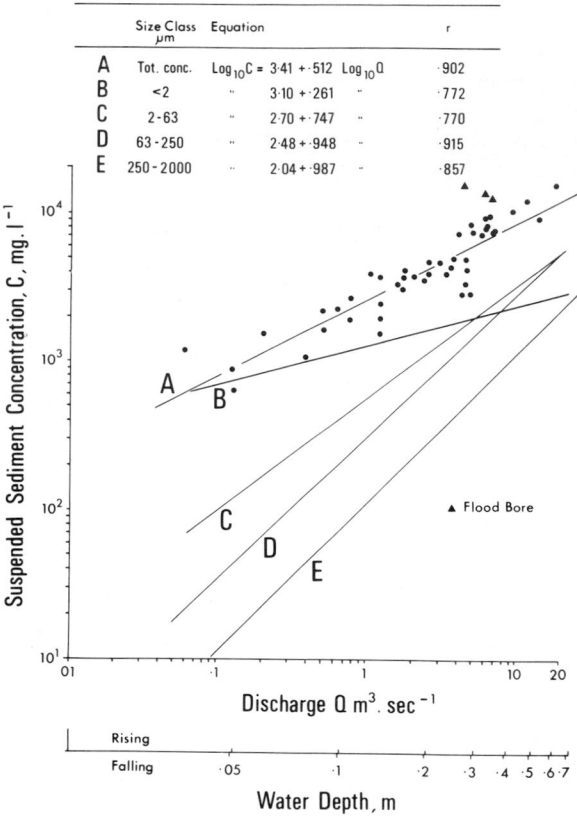

Fig. 4. Least-square relationship between suspended sediment concentration (total and size classes) and water discharge for grouped flood data, Il Kimere 1979.

boundary sublayer by turbulent eddies (Sutherland, 1967). Since the frequency and intensity of the burst and sweep associated with turbulence varies directly with stream flow (Offen & Kline, 1975) an increase in the size of suspended particles is a corollary of increasing fluid velocity provided that a full range of particle sizes is available on the bed or supplied to the stream from adjacent slopes. The Il Kimere yields a set of relationships which substantiate this tenet (Fig. 4). Suspended sediment concentrations for four separate size fractions (clay, silt, fine sand and medium-coarse sand) are plotted against discharge. The least-squares regression lines adopt a disposition relative to each other that clearly demonstrates the dependence of suspended sediment size on flow characteristics.

In fact an examination of individual size-distribution curves shows a relationship between mean fluid velocity (U) and the maximum size of the suspended material (D_{max}) that holds good throughout the course of three successive floods and thereby suggests an identification of the flexible boundary between bedload and suspended load: when $U < 1\cdot0$ m sec^{-1}, $D_{max} < 500\ \mu$m; when $1\cdot7 > U \geqslant 1\cdot0$ m sec^{-1}, $1000 > D_{max} > 500\ \mu$m; and when $U \geqslant 1\cdot7$ m sec^{-1}, $2000 > D_{max} > 1000\ \mu$m.

Figure 5 illustrates the changing size-distribution of selected samples taken during the recession limb of the 29 March flood and reveals clearly the important part played by overland flow in delivering fine sediment to the stream. An inspection of the bed material curve shows only $0\cdot8\%$ less than 63 μm. Yet the size distribution of suspended material close to peak discharge ($U = 2\cdot63$ m sec^{-1}) indicates nearly 50% in this

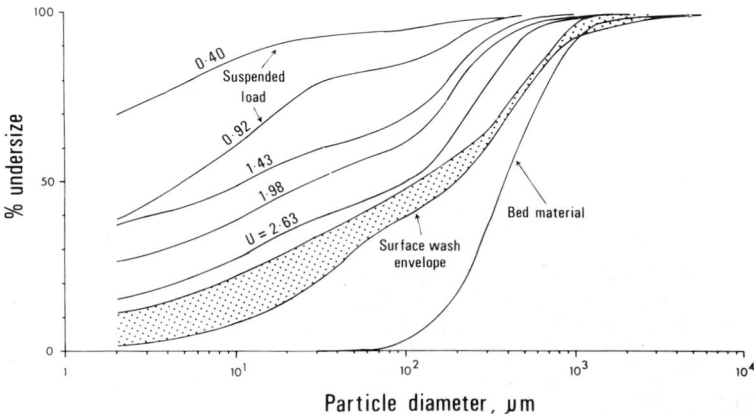

Fig. 5. Particle-size distribution of channel-bed sediment, contemporary surface-wash material, and suspended sediment for selected samples from flood peak and recession limb of 29 March flood.

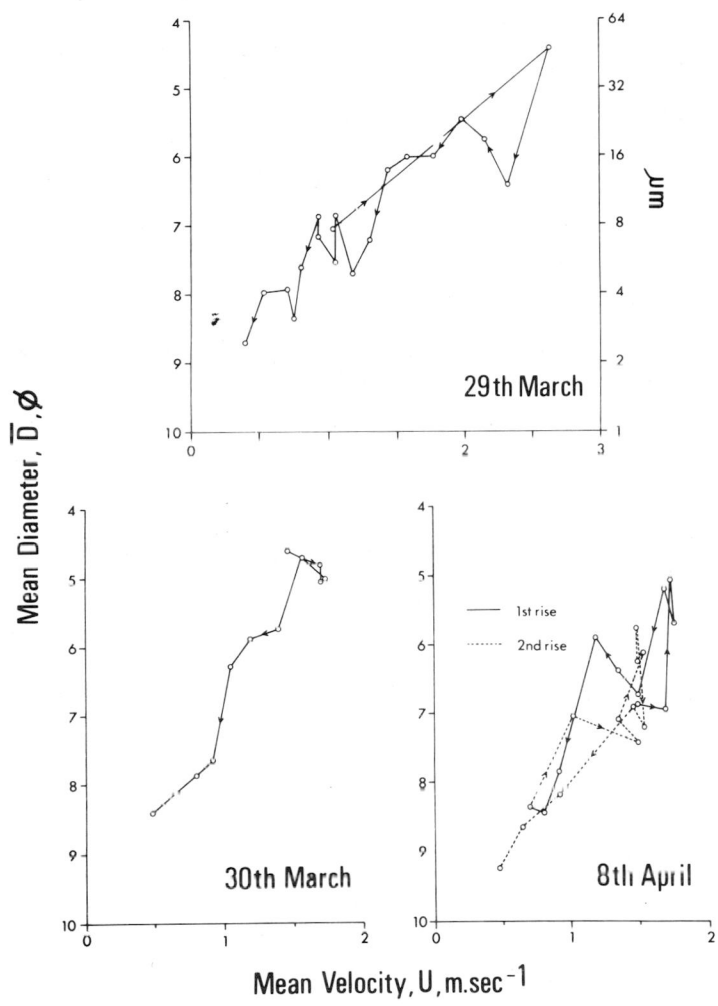

Fig. 6. Mean particle size of suspended sediment $\bar{D}$, and mean water velocity, U, for three flash floods.

category, most of which must be derived from the slopes of the drainage basin. In fact the envelope enclosing the size distributions of surface wash material caught in the slope traps shows clearly that overland flow is capable of supplying all sizes of material suspended in stream flow. As discharge wanes, the suspended sediment is inevitably dominated increasingly by the fine material (< 63 μm) contributed by overland flow (Fig. 3) so that, at a mean velocity of 0·4 m sec^{-1}, the inflection in the size-distribution curve (Fig. 5), which differentiates bed material load and wash load clearly at values of $U \geqslant 1·43$, is no longer even weakly represented. Approximately 46% of the suspended material comes from surface wash at flood peak ($U = 2·63$

m sec^{-1}). This increases steadily to 95% near the end of the recession limb on the flood hydrograph ($U = 0·4$ m sec^{-1}).

With surface wash containing high proportions of silt and clay (34–43%) and a sand bed readily available for scour by the wide but shallow flows characteristic of ephemeral streams, there is a large size range of material available for suspension. Because of this, the Il Kimere displays a direct relationship between mean particle size and mean velocity (Fig. 6) and provides a significant contrast with both the River Clyde (Fleming & Poodle, 1970) and the Niobrara river (Colby & Hembree, 1955). But the slope of the least-squares relationship for each flood indicates a subtle shift in the nature of the sediment carried by the stream as the rainy season progresses:

29 March $\bar{D}_\phi = 8\cdot898 - 1\cdot551U$, $r = -0\cdot901$;
30 March $\bar{D}_\phi = 9\cdot955 - 3\cdot090U$, $r = -0\cdot949$;
8 April $\bar{D}_\phi = 10\cdot175 - 2\cdot474U$, $r = -0\cdot856$;

where $\bar{D}_\phi$ is mean particle diameter in phi units and U is mean velocity, m sec^{-1}. Reference to the relative concentration of particles in different size classes (Fig. 3) shows a greater contribution of silt and clay during both the rising and falling limbs of the 29 March hydrograph. It is clear that readily available fines from both the contributing sideslopes of the drainage basin and the channel (blown in between rain seasons) are flushed during this first flood. There is no significant difference in the particle-size–velocity relationship for the two succeeding floods, suggesting that this flushing of fines is accomplished quickly in this altogether harsh environment.

Comparison with Laursen equation

Laursen (1958) provides a semi-empirical equation for the computation of suspended sediment concentration by size class, C':

$$C' = p\left[\frac{D}{y}\right]1\cdot167\left[\frac{\tau_0'}{\tau_c} - 1\right]f\left[\frac{\sqrt{\tau/\rho}}{w}\right],$$

where p is the fraction of bed material in the size class, y is water depth, τ_0' is boundary shear associated with the bed particles, τ_c is the critical shear of particles diameter D, f is the Darcy–Weisbach resistance coefficient, τ is boundary shear, ρ is fluid density, w fall velocity of particles diameter D, D is mean diameter of the size class. Taking five size classes *above* 63 μm for the selected samples of the 29 March flood depicted in Fig. 4, and regressing observed against expected concentration using the Laursen equation provides a remarkably good fit ($r = 0\cdot88$, sig. level 0·001, Fig. 7). Below 63 μm the prediction falls down for the simple reason clearly shown in Fig. 4, and well documented as a criticism of all theoretical or semi-empirical equations (Colby & Hembree, 1955; Mao & Rice, 1963), that substantial quantities of fine sediment contributed by surface wash are not represented in the bed material. Because of this, the regression of observed (C) on predicted (C') concentrations for *all* size classes gives a poor correlation coefficient $r = 0\cdot692$.

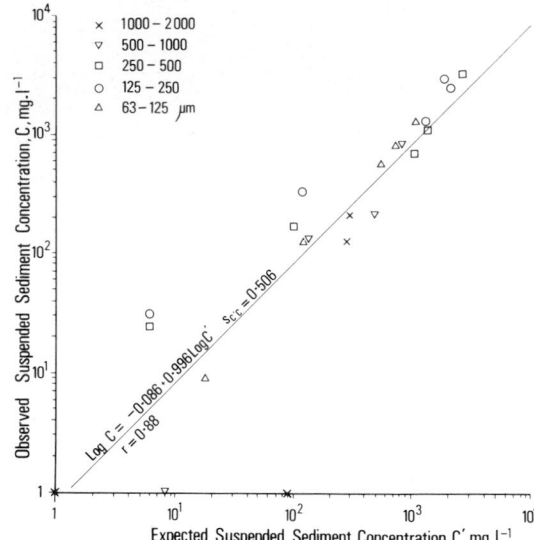

Fig. 7. Regression of observed sand-sized suspended sediment concentration (by classes), C, on expected concentration, C', derived from Laursen's equation (see text for equation).

Sediment yield

An annual value of suspended sediment discharge can be derived for the Il Kimere by assuming six floods per year (three per rainy season) of similar size to those documented (Table 2). Of considerable interest is the fact that Langbein & Schumm's (1958) generalized empirical relationship between precipitation and sediment yield gives, for the Il Kimere's annual runoff of 16 mm, a yield value of c. 302 tonnes km^{-2}, which is remarkably close to the observed 173 tonnes km^{-2}.

All size classes of suspended sediment have been shown to vary directly with water discharge. This wealth of size-distribution data provides a basis for the computation of sediment discharge by size class (Table 2), and indicates that the clay and silt fractions (< 63 μm) that make up an insignificant part of the channel bed material actually constitute 56% of the total basin sediment yield. This adds a significant degree of refinement to the identification of the sediment sources exploited by storm events in arid zones.

CONCLUSION

The Il Kimere provides a useful and clear-cut relationship between the size distribution of its

Table 2. Annual yield of suspended sediment by size class, Il Kimere

Size class (μm)	< 2	2–63	63–250	250–2000	Total
Yield (tonnes km^{-2})	41·9	54·4	45·2	31·1	172·6
Fraction	0·24	0·32	0·26	0·18	1·0

suspended sediment load and storm-flow characteristics. This is undoubtedly attributable in part to the rapid rainfall-runoff response of semi-arid landscapes, and in part to the wide range of sediment size made available by a system where overland flow provides both coarse and fine material and where the sandy channel-fill is left intact between rainy seasons and does not suffer the winnowing processes characteristic of perennial flows.

The detailed sampling programme on the Il Kimere provides the first rating curves of suspended sediment concentration by size class. These may prove to be of general value to sediment transport studies in semi-arid environments where the delivery of different sizes of sediment may dictate, for example, the useful life of irrigation reservoirs. It also provides the first data base for testing the efficiency of predictive semi-empirical formulae (in this case Laursen's) under the unsteady flow conditions of flood events in natural channels. It is clear that the dependence of such formulae upon identifiable and obvious sources of sediment (i.e. the stream bed) limits their applicability to overall suspended sediment concentrations. But Laursen's equation predicts well for those size classes (fine to coarse sand) most likely to be derived from bed material.

The fact that the sandy channel-fill of ephemeral streams contains little silt and clay is extremely useful in pinpointing source areas of suspended sediment in the drainage basin. Silt and clay are derived largely from valley slopes and pass directly out of the catchment with little storage in the channel-fill between floods. In contrast the most important immediate source of sand is undoubtedly scour of the channel bed. Because of its higher fall velocity it moves in a series of storage-transport steps that are inversely proportional in magnitude to particle size.

A partition of sediment yield by size class suggests some source limitation on clay particles since they contribute a smaller fraction than either the silt or the fine–medium sand. This must reflect the extra resistance they offer to rain-splash erosion through cohesion, rather than their surface exposure. Never-

theless fine particles of less than 63 μm, including the clay fraction, provide the major contribution to total basin suspended-sediment yield.

ACKNOWLEDGMENTS

The work was financed by the Natural Environment Research Council of the United Kingdom. We are grateful to Richard Leakey and Frank Fitch for logistical support in the field.

REFERENCES

ARNBORG, L., WALKER, H.J. & PIEPPO, J. (1967) Suspended load in the Colville River, Alaska, 1962. *Geogr. Annlr.* **49 A**, 131–144.

BAUER, L. & TILLE, W. (1967) Regional differentiations of the suspended sediment transport in Thuringia and their relation to soil erosion. *Publs int. Ass. Sci. Hydrol.* **75**, 367–377.

BEVERAGE, J.P. & CULBERTSON, J.K. (1964). Hyperconcentrations of suspended sediment. *J. Hydraul. Div. Proc. Am. Soc. civ. Engrs* **90**, HY 6, 117–128.

BONDURANT, D.C. (1951) Sediment studies at Concash River in New Mexico. *Trans. Am. Soc. civ. Engrs* **116**, 1283–1295.

BURKHAM, D.E. (1970) Depletion of stream flow by infiltration in the main channels of the Tucson basin, Southeastern Arizona. *Wat.-Supply Pap. U.S. geol. Surv.* **1939-B**, 36 pp.

BUTCHER, G.C. & THORNES, J.B. (1978) Spatial variability in runoff processes in an ephemeral channel. *Suppl. Z. Geomorph.* **29**, 83–92.

COLBY, B.R. (1963) Fluvial sediments—a summary of source, transportation, deposition and measurements of sediment discharge *Bull. U.S. geol. Surv.* **1181-A**, 47 pp.

COLBY, B.R. & HEMBREE, C.H. (1955) Computations of total sediment discharge Niobrara River, near Cody, Nebraska. *Wat.-Supply Pap. U.S. geol. Surv.* **1357**, 119 pp.

DOUGLAS, I. (1968) Erosion in the Sungei Gombak catchment, Selangor, Malaysia. *J. trop. Geogr.* **26**, 1–16.

DOUGLAS, I. (1973). Rates of denudation in selected small catchments in eastern Australia. *Univ. Hull Occ. Pap. Geogr.* **21**, 127 pp.

FAHNESTOCK, R.K. (1963) Morphology and hydrology of a glacial stream—White River, Mount Rainier, Washington. *Prof. Pap. U.S. geol. Surv.* **422 A**, 69 pp.

FLEMING, G. & POODLE, T. (1970) Particle size of river

sediments. *J. Hydraul. Div. Proc. Am. Soc. civ. Engrs* **96**, HY 2, 431–439.

FROSTICK, L.E. & REID, I. (1977) The origin of horizontal laminae in ephemeral stream channel-fill. *Sedimentology*, **24**, 1–9.

FROSTICK, L.E. & REID, I. (1979) Drainage-net control of sedimentary parameters in sand-bed ephemeral streams. In: *Geographical Approaches to Fluvial Processes* (Ed. by A.F. Pitty), pp. 173–201. GeoAbstracts, Norwich.

GERSON, R. (1977) Sediment transport for desert watersheds in erodible materials. *Earth Surf. Processes* **2**, 343–361.

IMESON, A.C. (1977) Splash erosion, animal activity and sediment supply in a small forested Luxembourg catchment. *Earth Surf. Processes* **2**, 153–160.

JORDAN, P.R. (1965) Fluvial sediment of the Mississippi river at St Louis, Missouri. *Wat.-Supply Pap. U.S. geol. Surv.* **1802**, 89 pp.

LANE, E.W. & KALINSKE, A.A. (1939) The relation of suspended to bed material in rivers. *Trans. Am. geophys. Un.* **20**, 637–641.

LANGBEIN, W.B. & SCHUMM, S.A. (1958) Yield of sediment in relation to mean annual precipitation. *Trans. Am. geophys. Un.* **39**, 1076–1084.

LAURSEN, E.M. (1958) The total sediment load of streams. *J. Hydraul. Div. Proc. Am. Soc. civ. Engrs* **84**, HY 1, 1530–36.

LEOPOLD, L.B. & MILLER, J.P. (1956) Ephemeral streams—hydraulic factors and their relation to the drainage net. *Prof. Pap. U.S. geol. Surv.* **282-A**, 36 pp.

MAO, S. & RICE, L. (1963) Sediment transport capability in erodible channels. *J. Hydraul. Div. Proc. Am. Soc. civ. Engrs* **89**, HY 4, 69–95.

MÜLLER G. & FORSTNER, U. (1968) General relationship between suspended sediment concentration and water discharge in the Alpenrhein and some other rivers. *Nature*, **217**, 244–245.

NEGEV, M. (1969) Analysis of data on suspended sediment discharge on several streams in Israel. *Isr. Min. Agr. Wat. Comm. Hyd. Serv., Hydr. Pap.* **12**, 27 pp.

NELSON, M.E. & BENEDICT, P.C. (1951) Measurement and analysis of suspended sediment loads in streams. *Trans. Am. Soc. civ. Engrs* **116**, 891–918.

NORDIN C.F. (1963) A preliminary study of sediment transport parameters, Rio Puerco near Bernardo, New Mexico. *Prof. Pap. U.S. geol. Surv.* **462-C**, 21 pp.

NORDIN, C.F. & BEVERAGE, J.P. (1965). Sediment transport in the Rio Grande, New Mexico. *Prof. Pap. U.S. geol. Surv.* **462-F**, 35 pp.

OFFEN, G.R. & KLINE, S.J. (1975). A proposed model of the bursting process in turbulent boundary layers. *J. Fluid Mech.* **70**, 209–228.

RAPP, A., AXELSSON, V., BERRY, L. & MURRAY-RUST, D.H. (1972) Soil erosion and sediment transport in the Morogoro River catchment, Tanzania. *Geogr. Annlr*, **54 A**, 125–155.

RENARD, K.G. & KEPPEL, R.V. (1966) Hydrographs of ephemeral streams in the Southwest. *J. Hydraul. Div. Proc. Am. Soc. civ. Engrs* **92**, HY 2, 33–52.

RENARD, K.G. & LAURSEN, E.M. (1975) Dynamic behaviour model of ephemeral stream. *J. Hydraul. Div. Proc. Am. Soc. civ. Engrs* **101**, HY 5, 511–528.

SUTHERLAND, A.J. (1967) Proposed mechanism for sediment entrainment by turbulent flows. *J. geophys. Res.* **72**, 6183–6194.

WALLING, D.E. (1974) Suspended sediment and solute yields from a small catchment prior to urbanization. In: *Fluvial Processes in Instrumented Watersheds* (Ed. by K.J. Gregory and D.E. Walling). *Spec. Publs Inst. British Geogr.* **6**, 169–192.

Spec. Publs int. Ass. Sediment. (1983) **6**, 107–119

Trapping and tracing: some recent observations of supply and transport of coarse sediment from upland Wales

B. ARKELL*, G. LEEKS†, M. NEWSON† and F. OLDFIELD*

*Department of Geography, University of Liverpool, Liverpool L69 3BX and
†Institute of Hydrology, Staylittle, Llanbrynmair, Powys, Wales*

ABSTRACT

The work reported is part of a project to investigate the relationship between upland land-use, sediment yield and river engineering problems in mid-Wales. Observations of bedload trapping in the forested uplands showed that the erosion of open drainage systems has increased yields; periods of supply-limited and transport-limited yield have been identified.

This paper describes the application of a new magnetic tracing technique to the study of sediment movement in the uplands, from the uplands into the piedmont zone, and from shoal to shoal in that zone. The technique, based on the enhancement of the magnetic susceptibility of the natural bedload provides an effective tracer material which can be detected in low concentrations, and in all particle size ranges. Special field and laboratory techniques associated with the method are described.

In rapidly eroding forest drainage ditches, tracing with particular size ranges from small shoals to bedload traps is improving both the theoretical and technical aspects of the technique. In these source areas, tracer recovery rates of 63% are demonstrated. Downstream from such source areas, the use of the natural bedload in tracing allows replication or replacement of local structure; armour or paving layers can be reproduced. Preliminary observations of the complexities of bed sediment movement at sites on the Rivers Wye, Severn and Llwyd are described.

INTRODUCTION

The long-term study of hydrology and fluvial processes on Plynlimon, an upland massif in central Wales, has been conducted by the Institute of Hydrology since 1968 (Newson, 1979). The instrumentation of two experimental catchments was specifically designed to investigate the effects of land-use change and land management, measuring sediment loads as well as basic hydrological variables. However, the small upland channels in which sediment loads have been measured are of little direct relevance to river engineering problems and recently the Ministry of Agriculture, Fisheries and Food has sponsored an extension of the study into the piedmont reaches (Newson, 1981) of the rivers of mid-Wales. It is in this zone that changes in inputs of water or sediment from the smaller upland catchments could produce costly alterations to the natural regime of erosion and deposition. Piedmont reaches are also used by the water industry in flow regulation for water supply and flood protection.

This paper is a progress report on a cooperative venture by the Institute of Hydrology and the University of Liverpool (Department of Geography) to develop a tracing procedure for coarse bedload as a back-up to trapping on small rivers and as an alternative on larger rivers.

DRAINAGE DITCHING AND INCREASED SEDIMENT YIELDS FROM SMALL CATCHMENTS

It has been generally assumed by British water engineers that trees bind the soil, preventing erosion

0141-3600/83/0106-0107 $02.00

and consequent reservoir sedimentation (Ministry of Health, 1948; Rodwell, 1936). Whilst it is true that in many climatic and soil zones a tree cover reduces sediment yields, the particular case of commercial afforestation in the wet uplands of Britain is rather different. The semi-natural pasture cover over wet, organic soil horizons produces a cohesive surface seldom broken by the characteristically low rainfall intensities. However, if the surface is broken artificially much less cohesive glacial and periglacial sediments are exposed and erosion is rapid. Just such a rupture can occur when the ground is drained by open ditching, as is the case prior to afforestation.

Since 1973 the Institute of Hydrology has measured bedload yield in concrete traps (capacity $10 \cdot 5$ m³) on two sub-catchments within the Plynlimon experimental catchments, the Cyff ($3 \cdot 13$ km², pasture) and the Tanllwyth ($0 \cdot 89$ km², forested). The major sediment movement in these channels is in the gravel and coarser size ranges, up to boulders weighing 20–100 kg each (Fig. 1).

From an initial comparison of the yields of the Cyff and Tanllwyth the greater erosional activity in the latter was obvious (Painter *et al.*, 1974). In eight years of sediment trapping on the Cyff and Tanllwyth the yield of the forested Tanllwyth was approximately five times that of the pastured Cyff. Similar ratios between ditched and unditched catchments were established at Lake Vyrnwy (Cownwy and Marchnant), Llyn Clywedog and the Nant y Moch intakes (Newson, 1980). Because the drainage ditches in the Tanllwyth were the clear sources of the extra sediment a detailed erosion survey was made of the two ditch systems, and three smaller bedload traps were installed on them.

These traps were operated between February 1976 and February 1981 to identify the processes of erosion, storage and sediment transport in ditch systems in an attempt to improve drainage design. It

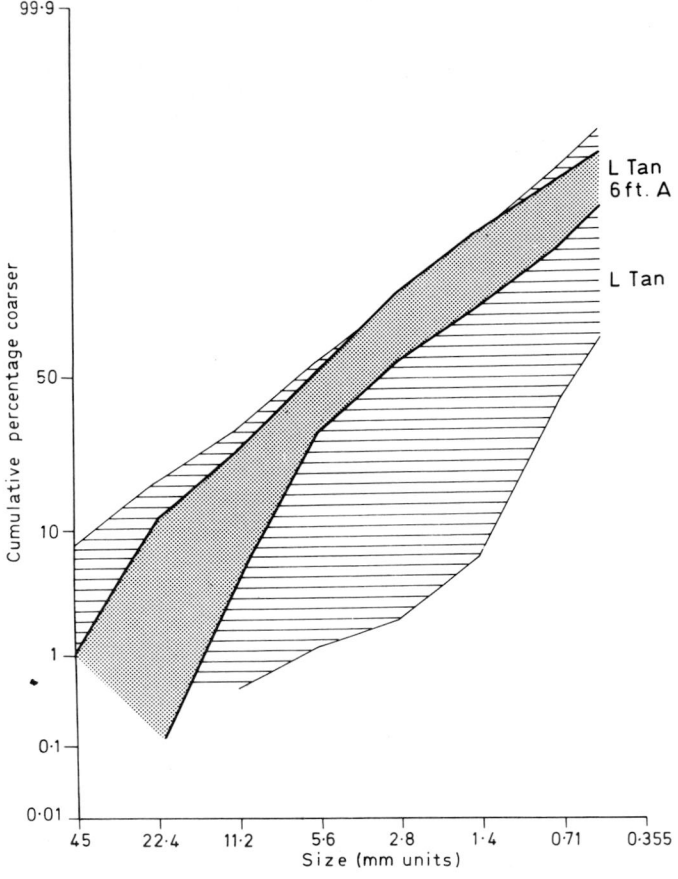

Fig. 1. Particle-size distribution curves for the two forest drainage ditches used for tracing. The range of sediment size over the year study period is plotted, illustrating those occasions when only fines are transported.

is only through knowledge of the dynamics of these processes that predictions of the downstream impact of ditch erosion may be made.

Yields of sediment from the Tanllwyth systems varied between 3·5 and 20·5 times those of the grass-covered Cyff catchment.

Figure 2 shows the relationship between bedload yield and the instantaneous maximum flow, measured at a small weir downstream from each trap. The deviation in the relationship shows the wide variation in the work achieved by different floods, and in particular events in which there was no bedload movement. The implication of this is that supply-limitation of material occurs in these and perhaps

other upland channels (see Newson, 1980) and indeed, data from the main traps on the Cyff and Tanllwyth also illustrate supply and transport-limited phases of channel activity.

These results should not be unexpected; hysteresis and supply-limitation phenomena are commonly seen with suspended sediments (Walling, 1974; Heidel, 1956; dissolved load: Hem, 1970; Hendrickson & Krieger, 1960), and it seems reasonable to expect similar phenomena with bedload transport. However, whilst similar phenomena have been identified elsewhere (Emmett, 1976; Andrews, 1979) very little account is taken in conventional calculations of bedload yield. As Emmett points out: 'At high values of stream

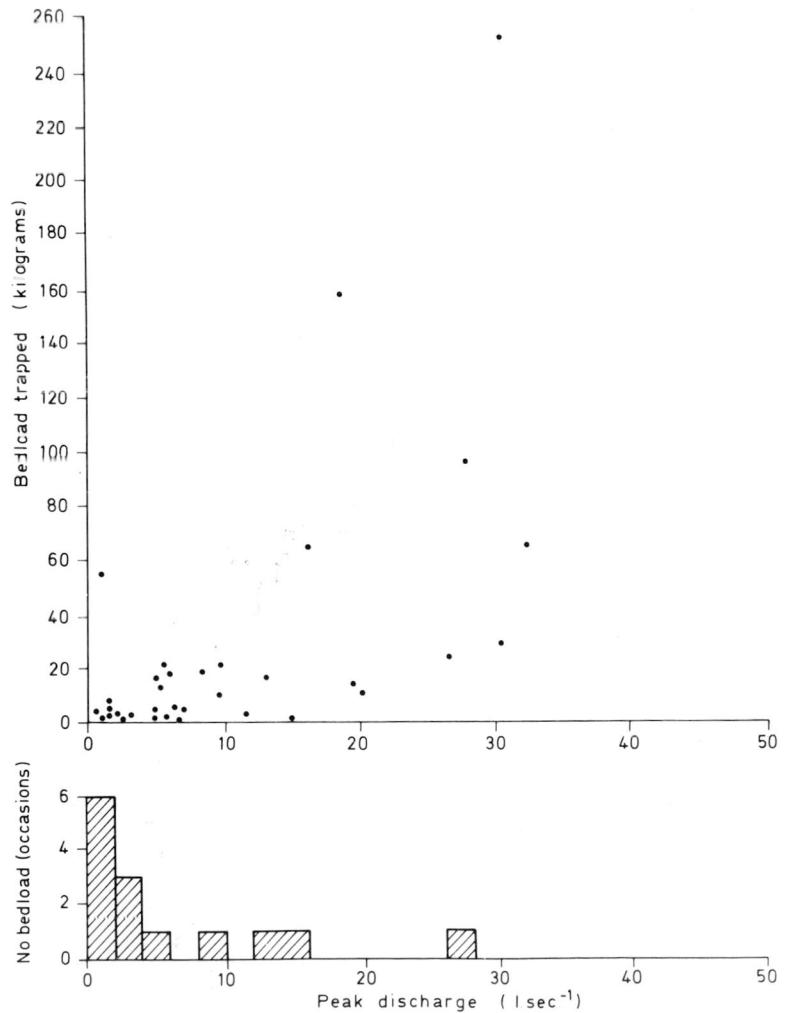

Fig. 2. Relationship between peak discharge and the weight of bedload trapped for the drainage ditch LTan 6 ft A (above). The lower group shows the number of occasions in which no sediment was trapped related to peak flow classes.

flow...bed-load transport rates are correlative with a predictable proportion of stream power expenditure', but for a range of floods and even within floods predictions based on bedload equations become tenuous. This may be particularly so in cases of large short-term, variations in stream discharge. In Plynlimon flow ranges of more than three orders of magnitude are common, with flood peaks reached in less than an hour. Estimations of bedload yield must take into account the supply of material to the channel and its availability for transport. With these controls in mind, the rate of sediment transport is dependent not only upon the momentary discharge, but also on the past history of the channel (Bogen, 1980). It is important to examine the frequency and magnitude of floods over short time periods (seasonal), controls on channel storage, and changes in sources-generating areas.

Since sediment supply is extremely difficult to measure, answers to these problems lie in the analysis of multi-variate time series, serial autocorrelations, or in the use of tracing techniques to investigate transport mechanisms in isolation from supply mechanisms.

MAGNETIC TRACING

The attempt to use magnetically enhanced natural bedload as a tracer arose from observations of forest fires and the persistence of fire-induced magnetic minerals in forest topsoils. Subsequent to the major forest fire in the Gwydyr Forest of North Wales in 1976, work by Rummery *et al.* (1979) and Rummery (1981) showed that it was possible to assess the downstream movement of material derived from the fired area by carrying out magnetic measurements on sediment samples taken from small stream shoals.

Subsequent studies by the Liverpool group (see Oldfield, Thompson & Dickson, 1981) showed that the enhanced magnetic characteristics of the forest-fired sediments could be reproduced within the natural shale bedload from the Tanllwyth system by laboratory furnace treatment and, as with the fired sediments, could be enhanced sufficiently for use as a tracer. Initial field trials indicated the potential of the technique, showing that even with commercially available sensors (sensitive, highly tuned metal detectors): (a) low concentrations of artificially magnetized material could be identified in trapped bedload; and, (b) qualitative field monitoring of loss from 'seeded shoals', of gain in downstream storage sites, and of the location of the bedload during transit was feasible. Since 1980, more detailed assessments of the magnetic tracing technique have been carried out, examining sediment transport systems in eroding forest ditches and in larger reaches downstream from the experimental catchment. This work has followed substantial development of the technology associated both with the laboratory enhancement of the magnetic characteristics of the bedload, and also with its detection in the field.

Tracer methodology

Optimal laboratory treatment (Oldfield *et al.*, 1981) was used to achieve the maximum possible enhancement of the magnetic susceptibility for large amounts of all sizes of material. From uniformly low values (varying between 0·07 and 0·14 × 10^{-6} m^3 kg^{-1}), heat treatment is capable of producing enhancement in the Plynlimon shales up to a factor of 300. Furthermore, by altering the range of variables used, different magnetic signatures can be induced in the same material for use in different traces (within the same reach). Table 1 illustrates the consistency of enhance-

Table 1. The natural and enhanced susceptibility characteristics of Plynlimon bedload

Size range (mm)	Mass susceptibility ($\times 10^{-6}$ m^3 kg^{-1})		Mass susceptibility for fine gravel at low concentrations		
	Natural	Enhanced			
			% concentration		
22·4–44·5	0·098	11·8 ± 3·0			
11·2–22·3	0·112	12·1 ± 2·1			
5·6–11·1	0·122	11·0 ± 0·6	2	5	10
2·8– 5·5	0·143	21·3 ± 1·5	0·73	1·38	2·48
1·4– 2·7	0·144	16·3 ± 0·7	0·59	1·14	2·47

Note: the modal value of mass susceptibility for the Plynlimon catchment soils is approximately 0·35 × 10^{-6} m^3 kg^{-1}, ranging from 0·06 10^{-6} m^3 kg^{-1} for upland peaty soils to 3·72 10^{-6} m^3 kg^{-1} for mature woodland soils.

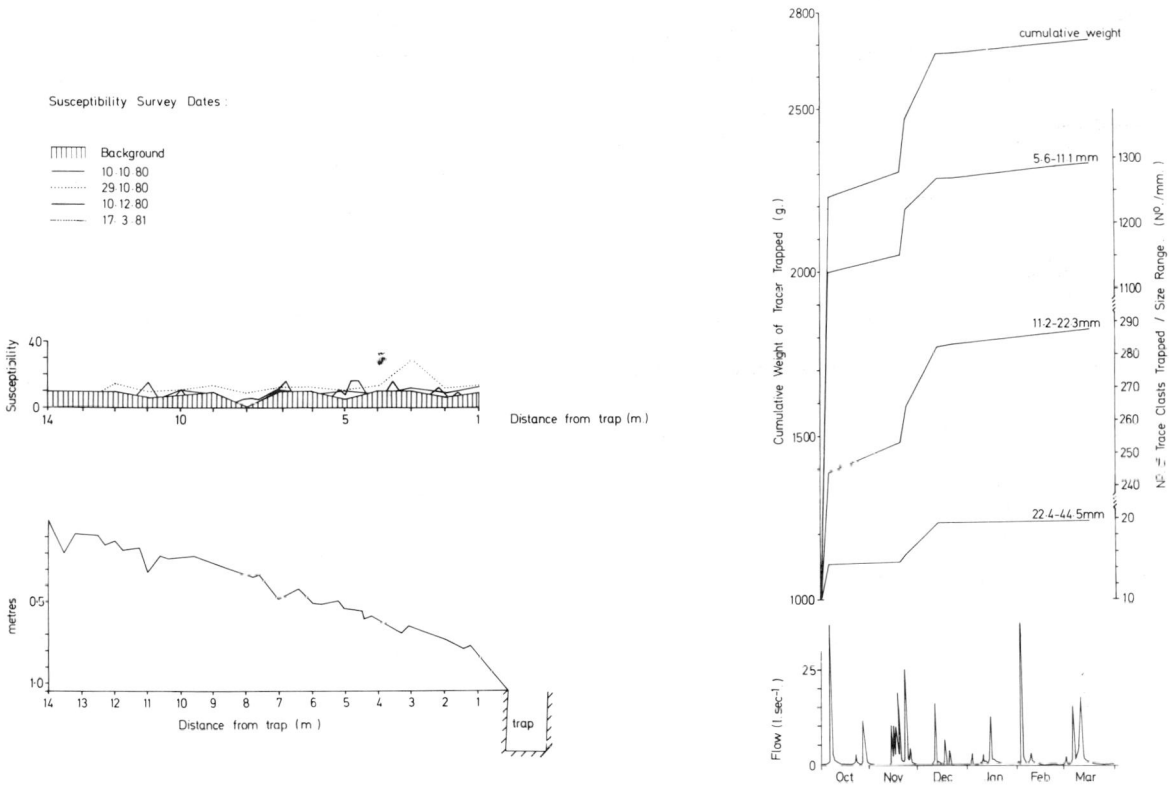

Fig. 3. Tracer movement and recovery in the forest drainage ditch LTan 6 ft A. Fixed-point surface susceptibility variations (top left) are plotted above the longitudinal profile of the study reach; and (right) hydrograph–tracer yield relationships.

ment throughout single size ranges used and the degree of enhancement as compared to the natural bedload and typical soils/source materials in the catchments. Detection of the tracer after emplacement was carried out in four ways, first on the basis of colour. An unforeseen effect of the laboratory processes was the formation of a pink pigmentation on the surface of the newly magnetized material. Thus in the field, initial visual recognition of the pink tracer (Munsell 10R 5/8) against the grey colour (9·5YR N4/0) of the natural shale bedload was often used as an indication of the presence of tracer.

The direct assessment of the movement of tracer in the field by magnetism was carried out in three ways.

(i) Variations in surface susceptibility were measured with a submersible search coil. As with a metal detector, this instrument responds to changes in magnetic flux around the coil sensor created by changes in the volume of magnetic material present; a digital display of susceptibility is produced. However, because the loop is open (by design) and therefore measures variable volumes, its readings are dimensionless and only a semi-quantitative estimate of presence and of the volume of tracer at a point could be made. Actual weights or volumes of tracer were determined by the methods outlined below.

(ii) For material coarser than 4 mm, identification by colour or surface susceptibility was subsequently confirmed by susceptibility sensing using a hand-held ferrite probe. All material separated in this manner was subject to further analysis of weight, size and shape.

(iii) For finer material than in (ii), obviously a different rapid field approach had to be developed. For this, a sensor was designed to measure the mass susceptibility of dried sieve samples. (The susceptibility per unit weight measured within an enclosed loop sensor—units $m^3\,kg^{-1}$.) By comparing the mass susceptibility of a sieved sample against a standard calibration curve of mass susceptibility versus percentage concentration of tracer for each size range used in the trace, the total weight of tracer was estimated.

B. Arkell et al.

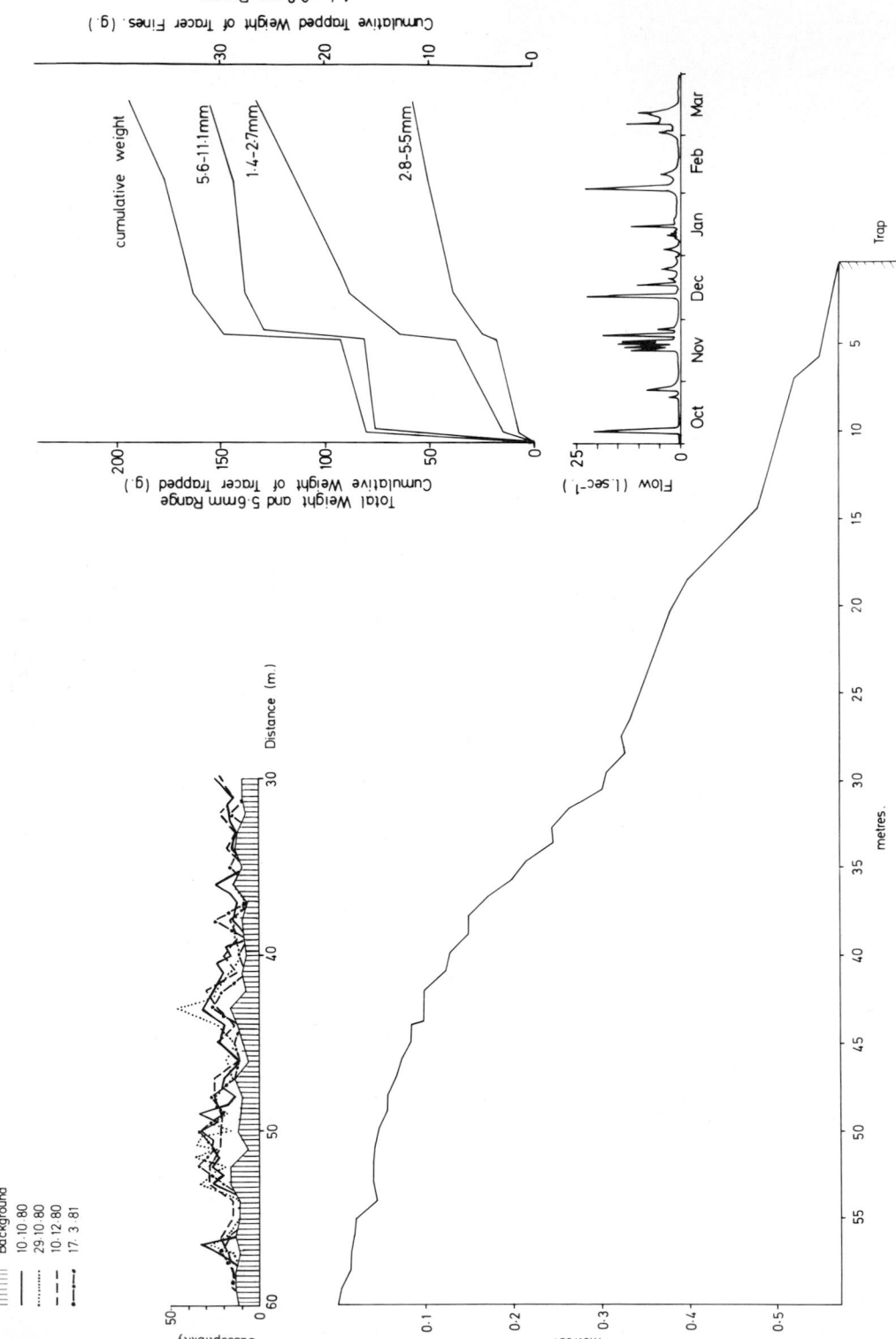

Fig. 4. Tracer movement and recovery in the forest drainage ditch LTan. Fixed-point surface susceptibility variations (top left) are plotted above the longitudinal profile of the study reach; and (right) hydrograph-tracer yield relationships.

MAGNETIC TRACING WITHIN DITCH SYSTEMS

Two experimental lengths of eroding forest ditch in the Tanllwyth system were chosen for study.

(i) LTan 6 ft A: an extremely high-energy system, geometrically simple and similar to a laboratory flume; the trace was carried out in 15 m of ditch upstream from a bedload trap.

(ii) LTan: a low-energy system, sinuous in plan and profile and eroding in places to rough outcrops of bedrock within the upper 30 m of the experimental reach.

The differences in energy between the systems enabled study of the efficiency of the technique in tracing two particle size ranges: in the LTan, finer particle sizes between 1·4 and 11·1 mm and in the LTan 6 ft A coarser material between 5·6 and 44·5 mm. In each ditch a substantial shoal was chosen for the trace, removed and dry sieved for size, then the specific ranges to be monitored experimentally were enhanced and replaced within the sample. The shoals were reconstructed within each ditch at their original position and left to stabilize under low flows. A series of fixed survey points was set up at 1 m intervals along each experimental reach and a background survey of surface susceptibility was carried out with the search coil at each survey point between the shoal and bedload trap.

To date (1 June 1981), the fixed surveys of surface susceptibility have been repeated and the bedload traps emptied and analysed on seven occasions following each major flood subsequent to shoal emplacement.

RESULTS

A summary of the data to 1 June 1981 is given in Figs 3 and 4, showing for each ditch the surface susceptibility change related to the topographic profile of the ditch, and the cumulative weight of tracer trapped related to the storm hydrograph for the study period. The cumulative plots of trapped magnetized material exhibit remarkably similar features. For both ditches the trapping rate has been influenced by three main periods of floods, separated by low recovery rates for more moderate events.

The results from LTan 6 ft A (Fig. 3) indicate that the system is now largely clear of tracer; the cumulative weight of tracer trapped has levelled out, and surface susceptibility variations within the ditch indicate a return to background levels. Separation of the magnetic clasts from the trapped bedload following each event has yielded 63% of the original weight of tracer emplaced in the system, and 40% of this was derived from the first and largest flood during the study period. The recovery for each size range used in the trace is given in Table 2.

The topographic differences between the two ditches give rise to obvious differences in delivery of sediment most probably due to the very different channel storage characteristics. For LTan 6 ft A, the steep, smooth U-shaped profile throughout offers little opportunity for discrete shoaling, and tracer material identified by surface susceptibility measurements appeared spread thinly throughout the channel length, particular concentrations lying in the lee of larger clasts.

In LTan, on the other hand (Fig. 4), the opportunity for shoaling is illustrated by the surface susceptibility plots and topographic profiles. From the first post-emplacement survey, high concentrations of

Table 2. The total recovery of magnetic tracer to 1 June 1981 against the original weights per size range on emplacement

Size range (mm)	Total weight emplaced (g)	Total weight recovered (g)	% recovery
		LTan 6 ft A	
5·6–11·1	1750	1241	71
11·2–22·3	1400	992	71
22·4–44·5	1400	634	45
Total	4550	2867	63
		LTan (see text)	
1·4– 2·7	300		
2·8– 5·5	600		
5·6–11·1	900		

tracer are identified, associated with the development of distinct shoal units downstream from the emplacement zone in areas of reduced channel gradient or in association with bedrock obstructions in the channel. Clearance of the emplacement zone in the first flood event is followed by a build-up into three main units shown in Fig. 4; subsequent floods influenced exchange between shoals, building up into a major shoal unit at 43 m upstream from the trap. Measurements of surface susceptibility revealed a number of magnetic clasts coarser than those emplaced for the trace, and which would appear to be remnants from the initial trials outlined above. That material was removed from the channel and analysed in the laboratory. As indicated earlier, differentiation of the material from the two traces is possible using their definable magnetic 'signatures'; the lower enhancement characteristics of material remaining from the first trace were apparent from susceptibility measures taken with the hand-held ferrite probe (mass susceptibility values of approx. $6{\cdot}97 \times 10^{-6}$ m^3 kg^{-1}, see Table 1). Further studies on the trapped material indicated similar characteristics to these, and whilst no material coarser than 5·6 mm was recovered in the bedload traps, low concentrations of finer tracer material were recovered and these are plotted in Fig. 4. Though on a much smaller scale, trapping concentrations of this material in response to flooding were very similar to those reported for LTan 6 ft A. From these inferences, the present trace material would appear to be still held in storage further up-channel, explaining the consistent concentrations identified by surface susceptibility scans. Further studies are now in progress to analyse concentrations of the different trace materials when mixed together. For these reasons, no estimates of percentage recovery are yet made.

Since the major reason for tracing was the inability of trap and flow records to predict transport conditions, the trace results were interpreted in terms of peak flow measurements. Whilst it can be generally seen from Fig. 5 that there is a relationship between peak flow and trapping concentration, the relationship is not a simple one, and indeed resembles the complexities already discussed with reference to Fig. 2. Although the concentration of tracer within the system will diminish through time anyway (there is only a finite quantity of tracer to be recovered), both ditch systems illustrated a broad correlation between bedload yield and stream power (interpreted from peak discharge). However, this was interrupted by periods where frequent moderate flows and even

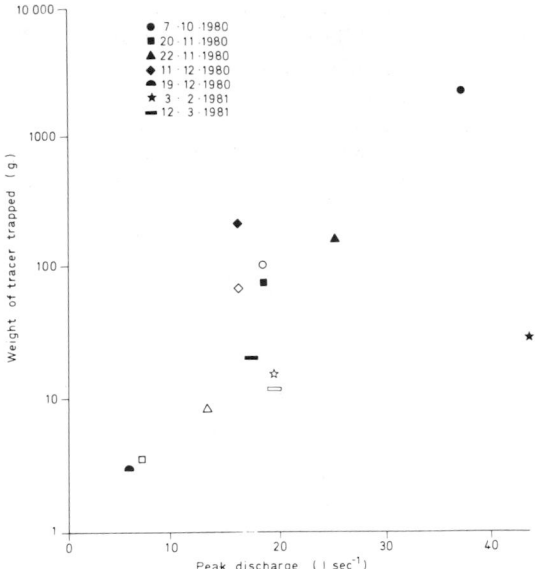

Fig. 5. The relationship between trapping concentration of the tracer to peak discharge for each flood event (cf. Fig. 2). Open symbols relate to the drainage ditch LTan, and closed symbols to the ditch LTan 6 ft A.

larger events (subsequent to the main release of tracer) released low concentrations or no tracer material at all. Exhaustion of the tracer supply from LTan 6 ft A is shown by the range of plots through the last two events monitored.

DOWNSTREAM TRACING WITH THE MAGNETIC TECHNIQUE

Downstream from the Plynlimon catchment, three sites are being used to assess the practicability of the tracer and to study sediment routing systems in different channel situations. To ensure that as far as possible the tracer behaved in the same way as the host sediment, material was taken prior to the trace from an upstream shoal, enhanced, and emplaced at each site in early summer 1980 so that at each site a period of 'acclimatization' was achieved before the first major flood event. The three chosen sites are as follows.

(i) *Morfodion* (*River Severn*)—the river at this point is approximately 30 m wide with a 500 m flood plain. Movement from storage is being monitored from tracer material placed in a trench cut normal to flow through a shoal projecting downstream into the river from the left bank. Special attention was paid to the detailed shoal stratigraphy through the trench to a depth of approximately 30 cm. Sediment size was recorded and each size range replaced accordingly with its magnetic counterpart, keeping to the structure of the shoal as far as possible. In an attempt to replicate existing surface characteristics, an armour layer of coarse magnetic clasts was added over the seeded trench.

(ii) *Dolydd (River Llwyd)*—tagged material was inserted in a pool beneath a road bridge over the Llwyd to monitor the movement of bedload passing under the bridge and sediment routing through shoal channels immediately downstream.

(iii) *Cefn Brwyn (River Wye)*—material was emplaced in a riffle immediately downstream of the Institute of Hydrology compound Crump weir, allowing loss and subsequent downstream movement to be monitored without the additional complication usually found in natural channels of bedload addition from upstream. The immediate downstream areas show little bedform development, with bare rock at or very near the surface. For both the Cefn Brwyn and Dolydd traces, the sites were analysed prior to emplacement for surface and subsurface size distribution, and the tracer was added accordingly.

A sampling strategy comprising a three-tier system of magnetic and topographic monitoring was set up for an objective survey at this scale to recognize sediment movement, throughput between storage, or net changes in channel form following each flood. The system comprised the following:

(i) A close grid cover, established between permanent marker pegs with tapes, at the input site. Here, both magnetic and topographic surveys revealed the amount of material which has left the site.

(ii) At intervals in the first 100 m downstream of the input site transect lines were established for detection of short distance movements.

(iii) For longer-distance movement, sampling with a 'Wolman One Hundred' approach was used for surface susceptibility of individual bedforms, with topographic surveys undertaken by repeat photography rather than levelling as in (i) and (ii).

RESULTS

The main phase of tracer movement at each site began in autumn 1980. The preliminary observations of two traces at Morfodion and Cefn Brwyn are presented in Figs 6, 7 and 8.

Whilst sorting of the tracer through successive floods is evident for all three sites, the scale of sediment movement differs considerably. At Morfodion, movement has been very localized and confined to the shoal/pool vicinity. Observations of the movement of tracer show a trend towards the pool and towards the distal end of the shoal; a sequence of coarse clasts moving diagonally towards the pool whilst progressively finer material is observed towards the distal end of the shoal on the bottom of the steep slope into the pool, and in the interstices between and below larger armour clasts. The gradual reduction of susceptibility over the original trench site through successive events (Fig. 7), from topographic survey, seems due to the burial and dilution of tracer by the addition of material from upstream. Susceptibility readings indicated burial and removal of tagged materials towards the steep slope and into the adjacent pool. From repeat topographic surveys, 17 m³ of material is calculated to have moved at the site (to 1 June 1981) although the net movement is only 3·3 m³ (gain). Thus movement of the tracer into the adjacent pool seems likely, since the original volume of tracer emplaced was only 0·1 m³. These inferences have been verified using a power auger to lift several 1 × 0·2 m cores of sediment from key locations on the shoal and analysing the volume of magnetic material.

The Cefn Brwyn and Dolydd traces have revealed a greater overall distance of tracer movement. The first major flood event for both experiments occurred in October 1980, moving much of the original volume of tracer. Repeat surveys through successive floods show surface accumulations of tracer finer than 11·2 mm downstream and burial of the coarser clasts upstream. Subsequent higher floods were observed to re-expose this sediment, allowing further transport. Much of the material detected downstream of the emplacement sites was finer than 45 mm, suggesting that the bulk of the coarse magnetic clasts are still within the body of the main channel within or adjacent to the emplacement site.

This aspect of burial and re-exposure has been discussed in a number of previous studies, and was one of the main aims of these experiments to test the efficacy of the instrumentation under these conditions. The effective measuring depth of the Search Coil at this stage of development is approximately 5 cm to a maximum of 10 cm, dependent upon the size of tagged clast, its geometric relation to the Search Coil, the depth of overburden, and particle-size variations in the bed armouring. Figure 8 shows the two traces the relationship between surface susceptibility readings and the presence of tagged material at or

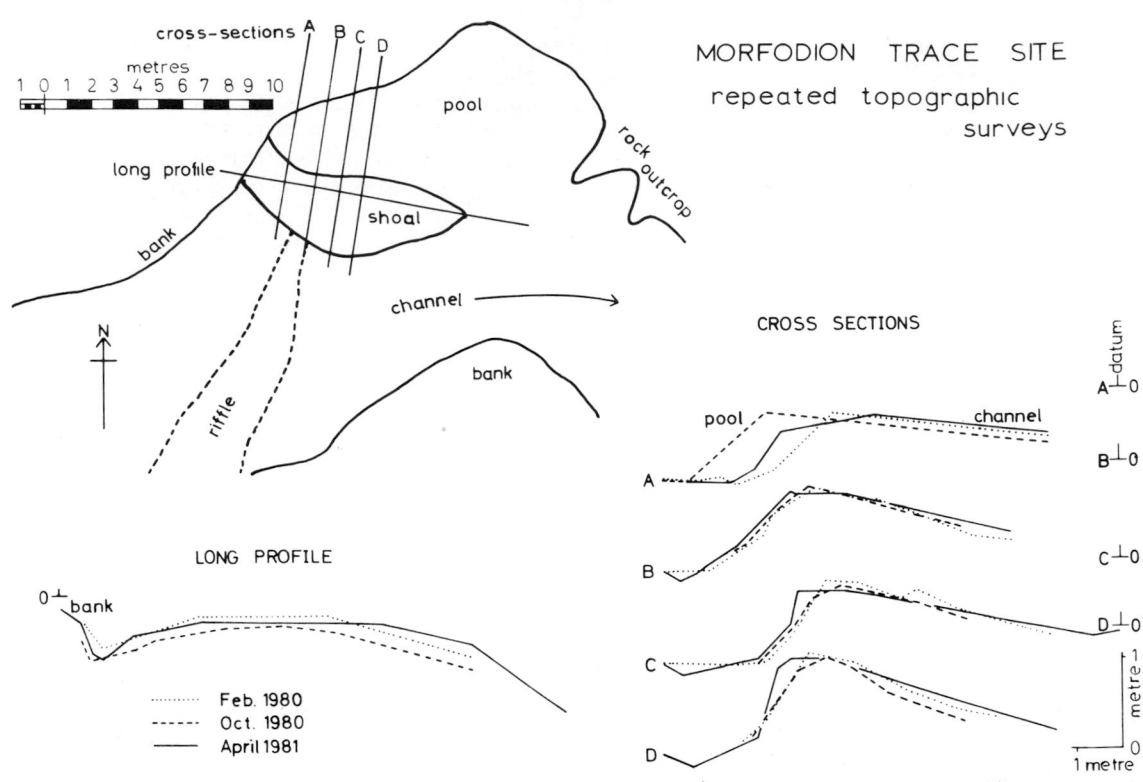

cross-sections A B C D

metres
1 0 1 2 3 4 5 6 7 8 9 10

MORFODION TRACE SITE

repeated topographic
surveys

pool

rock outcrop

long profile

bank

shoal

channel →

N

bank

riffle

CROSS SECTIONS

A─datum
A⊥0

pool   channel

A

B⊥0

B

C⊥0

LONG PROFILE

C

D⊥0

1

0⊥
bank

metre

.... Feb. 1980
------- Oct. 1980
——— April 1981

D

metre

0
1 metre

Fig. 6. See opposite for caption.

beneath the surface, identified in the manner described above. Though the relationship is strong, the geometric constraints make surface susceptibility readings an unreliable direct estimate of tagged clast size.

Channel cross-sectional variation in movement and deposition is also apparent. Zones of high concentration of susceptibility were observed within the main channel of Cefn Brwyn, associated with the channel margins and zones of reduced velocity upstream of a bend in the channel which causes upwelling and deep scour.

Both sites showed less change in topographic form at the emplacement sites than were observed at Morfodion, with the Dolydd site accumulating 0·9 m³ (with gross change of 3 m³) and at the Wye at Cefn Brwyn eroding by 2·1 m³.

DISCUSSION

The application of the magnetic tracing method to the systems described above has illustrated the overriding importance of channel storage on sediment transport systems in gravel-bed streams. Even within small channel systems such as those of the forest drainage ditches, storage of sediment ensures that the output is not solely governed by stream power. This added to the control of sediment availability by weathering further confirms the irregularity of coarse, upland sediment systems, at least at moderate flows. If the routing of sediment through rough upland channels is similar to that of flood waves themselves (see Beven, Gilman & Newson, 1979) we may well need to take a storage-based approach to field studies. We may also find convergence towards a more regular, more linear system at high rates of output.

Similarly, within the downstream traces storage opportunity and control are illustrated. The very slow release of material from the Morfodion shoal contrasts with the relatively rapid transport of material at Cefn Brwyn and the Dolydd, where transport is towards storage sites. This complicates the use of the technique to estimate rates of output as one would with trapping. It also stresses the role of floods once more; it is at high flows that the geometry of many storage sites is produced.

The current drawbacks of the method as illustrated

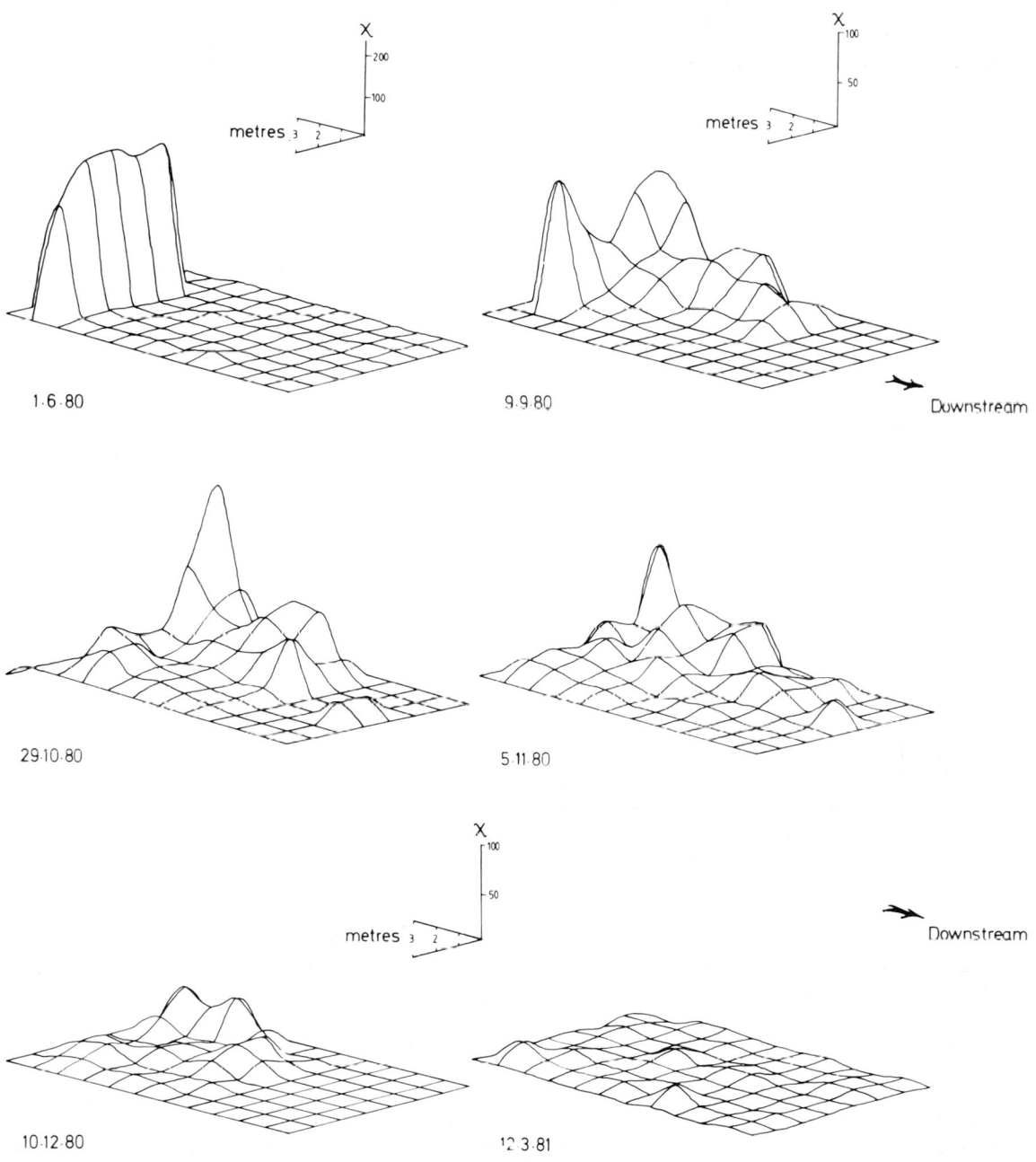

MORFODION : POST FLOOD SOLID ISOPLOT OF MEAN SURFACE SUSCEPTIBILITY. (X).

Fig. 6. Morfodion, River Severn. An illustration of the type of plot that can be created from downstream tracing studies, showing: opposite page, top, shoal sketch plan with repeat topographic surveys; and, above, movement of the tracer by isoplots of mean surface susceptibility m^{-2}.

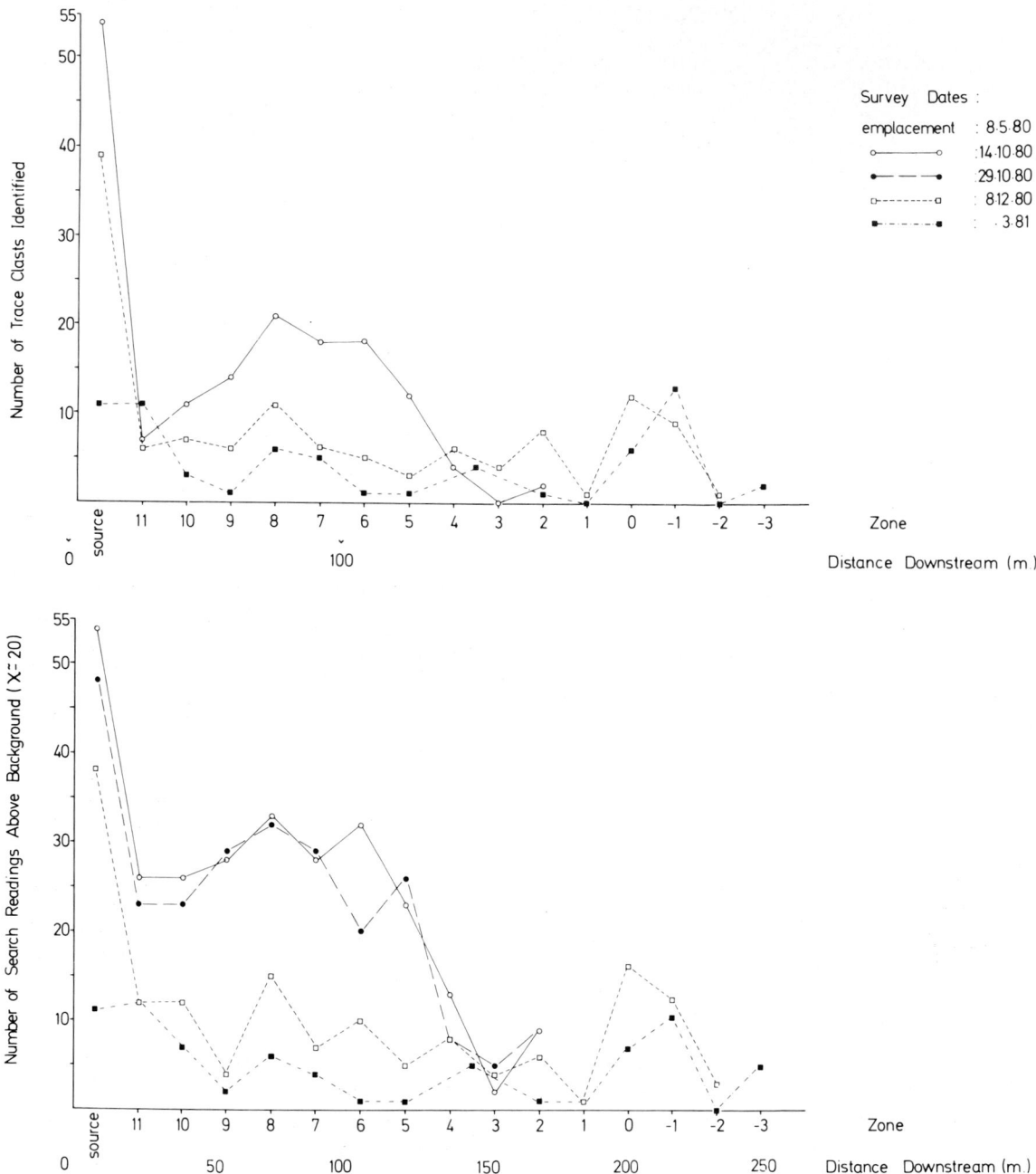

Fig. 7. Cefn Brwyn, River Wye. The relationship between surface susceptibility (below) and identifiable tracer material (above) plotted against distance downstream from the emplacement site.

above—of scale of operation, burial of tracer, dilution and fully quantitative estimates of tracer volume—show signs of being overcome with further research and development. Of particular importance is the performance of instrumentation and its sensitivity to changes in the geometry of the magnetic material being measured. Variations in surface susceptibility show that the field search coil, in its present form, may not detect small quantities of tracer *in situ* below a coarse armour layer; or that a small magnetic clast close to the sensor rim may increase the signal by as much as a larger pebble in the centre of the sensor coil or some way from its perimeter. The scope for progress here is largely technical and depends upon the development of the instrumentation using much larger inductance coils, or paired coils, such as those used in underground pipe detectors.

CONCLUSIONS

The results described above outline the potential of the magnetic tracing technique, offering a cheap, harmless, non-destructive system effective in tracing across a broad range of particle sizes within the same system. The major advance has been that of developing a method which is semi-quantitative, encouraging both extensive and intensive use of tracing in gravel-bedded streams where field experiment is sadly under-utilized. It provides an alternative to the use of portable bedload sampling devices such as the Helley Smith, which require subjective on-site decisions as to when and where to use such samplers, when the very nature of discontinuous bedload movement through a cross-section (depending itself on possible unstable bedforms) can lead to considerable inaccuracies when using samplers from bridges. In comparison to the many other tracing methods, magnetic enhancement offers an effective tracer for all sizes of material. The process of enhancement is cheap, rapid and effective, requiring from a few minutes to a few hours oven time, and all the instrumentation cited is commercially available at present. The technique has potential for use in a variety of situations other than those described here.

ACKNOWLEDGMENTS

This study was part of a project supported by the Ministry of Agriculture, Fisheries and Food and by a CASE studentship awarded by the Natural Environment Research Council. The authors wish to thank Dr A. M. Harvey, University of Liverpool, for many helpful comments and discussion and, for instrument design, Mr G. Bartington of Conservation Instruments, 33 The Green, Charlbury, Oxford.

REFERENCES

ANDREWS, E.D. (1979) Scour and fill in a stream channel, East Fork River, Western Wyoming. *Prof. Pap. U.S. geol. Surv.* **1117**.

BEVEN, K., GILMAN, K. & NEWSON, M.D. (1979) Flow and flow routing in upland channel networks. *Hydrol. Sci. Bull.* **24**, 303–325.

BOGEN, J. (1980) The Hysteresis effect of sediment transport systems. *Norsk Geog. Tidsskr.* **34**, 45–54.

EMMETT, W.W. (1976) Bedload transport in two large, gravel-bed rivers, Idaho and Washington. *Proc. 3rd Federal Inter-Agency sediment. Conf.* pp. 4–101, 4–113.

HEIDEL, S.G. (1956) The progressive lag of sediment concentration with flood waves. *Trans. Am. geophys. Un.* **37**, 56–66.

HEM, J.D. (1970) Study and interpretation of the chemical characteristics of natural water. *Wat.-Supply Irrig. Pap., Wash.* **1373** (2nd ed.).

HENDRICKSON, G.E. & KRIEGER, R.A. (1960) Relationship of chemical quality of water to stream discharge in Kentucky. *21st. int. geol. Congress Rep., Part I. Geochemical Cycles*, pp. 66–75.

MINISTRY OF HEALTH (1948) *Report of the Gathering Grounds Committee.* H.M.S.O., London.

NEWSON, M.D. (1979) The results of ten years' experimental study on Plynlimon, mid-Wales and their importance for the water industry. *Jl. Instn Wat. Engrs* **33**, 321–333.

NEWSON, M.D. (1980) The erosion of drainage ditches and its effect on bed-load yield in mid-Wales: reconnaissance case studies. *Earth Surf. Processes*, **5**, 275–290.

NEWSON, M.D. (1981) Mountain streams, In: *British Rivers* (Ed. by J. Lewin), pp. 59–89. Allen & Unwin, London.

OLDFIELD, F., THOMPSON, R. & DICKSON, D.P.E. (1981) Artificial enhancement of stream bedload; a hydrological application of superparamagnetism. *Phys. Earth planet. Int.* **26**, 107–124.

PAINTER, R.B., BLYTH, K., MOSEDALE, J.C. & KELLY, M. (1974) The effect of afforestation on erosion processes and sediment yield. *Publs int. Ass. Sci. Hydrol.* **113**, 62–67.

RODWELL, J.A. (1936) Pre-storage problems. *Trans Inst. Wat. Engrs* **41**, 52.

RUMMERY, T.A., BLOEMENDAL, J., DEARING, J.A. & OLDFIELD, F. (1979) The persistence of fire-induced magnetic oxides in the soils and lake sediments. *Annals Géophys*, **35**, 1–5.

RUMMERY, T.A. (1981) *The effects of fire on soil and sediment magnetism.* Unpublished Ph.D. Thesis. University of Liverpool.

WALLING, D.E. (1974) Suspended sediment and solute yields from a small catchment prior to urbanisation. In: *Fluvial Processes in Instrumented Watersheds* (Ed. by K. J. Gregory and D. E. Walling). *Spec. Publs Inst. Geogr.* **6**, 169–192.

Spec. Publs int. Ass. Sediment. (1983) **6**, 121–132

Meander changes in relation to bend morphology and secondary flows

J. M. HOOKE* *and* A. M. HARVEY†

Department of Geography, Portsmouth Polytechnic, U.K., and
†Department of Geography, University of Liverpool, U.K.

ABSTRACT

The changes in meander patterns since 1840 have been identified from maps, air photographs and field survey for a 14 km reach of the River Dane in Cheshire, England, with a mobile meandering channel. An important type of change is the development of asymmetric lobes and double-headed bends. Plots of path length against curvature show groupings for three major types of movement: migration, growth and lobing. Plots of path length and curvature also reflect the pool-riffle to bend relationships. Beyond critical combinations of path length and curvature it appears that secondary flows can no longer be sustained and new riffles develop leading to new bend forms. This provides a mechanism for explaining the historical changes but suggests that no equilibrium form is evident on this channel in the study period and that a progressive model of meander development is appropriate.

INTRODUCTION

Most studies of river meandering tend to focus either on the properties of the pattern as a whole or on the morphological, hydraulic and sedimentological properties of individual bends. Few studies relate these approaches to progressive change in meander form. Work on pattern includes studies of sinuosity (Schumm, 1963; Brice, 1964) and meander spectra and direction variance (Speight, 1965; Ferguson, 1975, 1977a) and of bend characteristics, particularly sine-generated forms (Langbein & Leopold, 1966), and the relationship to discharge and other factors (Leopold & Wolman, 1957, 1960; Carlston, 1965), though Ferguson (1977b) has compared spectra of streams at different dates. Bedform and pool and riffle relationships to meanders are well known (Leopold & Wolman, 1960; Harvey, 1975; Richards, 1976a, b; Lewin, 1976) but most work in this field has been concerned primarily either with the detailed behaviour of individual bends or with the gross characteristics of the pool and riffle, meander relationships. Keller (1972) has considered progressive changes in these

relationships. Study of secondary flows through pools and riffles and meander bends provides the basis for some understanding of processes within meanders (Hey & Thorne, 1975; Bridge & Jarvis, 1976; Hickin, 1978), but again this work tends to stress short-term relationships.

Studies of the longer-term development of meanders have mostly been concerned with rates of meander development and migration (Daniel, 1971; Hickin, 1977; Hughes, 1977); with meander development and floodplain formation (Bluck, 1971; Lewin, 1978) or with changing morphology of the bends themselves (Brice, 1973, 1974; Hickin, 1974; Hooke, 1977; Hickin & Nanson, 1975). However, few of these studies can be related to short-term analyses of the morphometric or hydraulic characteristics of bends. Exceptions include the work of Kondrat'yev (1968) on meander deformations and the work of Thorne & Lewin (1979) linking bank processes to meander changes.

The purpose of this paper is to clarify the nature of the relationships between pattern change, bend morphology and bedforms in a mobile meandering reach. Bend morphology and channel behaviour over the period since 1840 have been examined on a 14 km

0141-3600/83/0106-0121 $02.00
© 1983 International Association of Sedimentologists

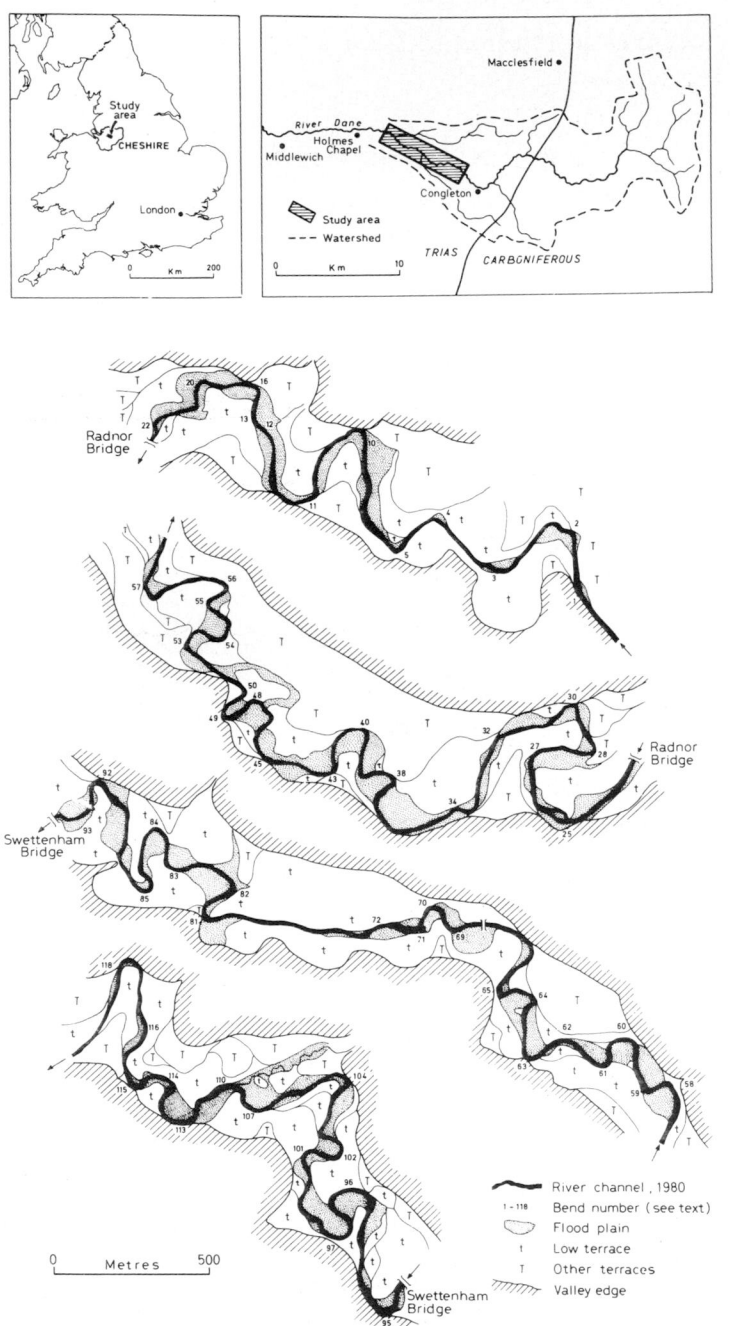

Fig. 1. Location and morphology of the study reach on the River Dane, Cheshire.

study reach of the River Dane in Cheshire (Fig. 1) and the relationships of these changes to pool and riffle and secondary flow characteristics have been investigated.

The River Dane, at the study reach, drains an area of 152 km² with a relief range of 45–550 m. It has a mean annual flow of 62 m³ s⁻¹ at Congleton (Fig. 1) and through the study reach has channel widths of *c*. 10 m. The river sediments range from silt to sand, gravel and cobbles and are derived from the Carboniferous sandstones and grits underlying the Pennine, upper part of the catchment and from the Triassic rocks of the Cheshire plain, mainly Keuper sandstones near Congleton and marls below Radnor Bridge (Fig. 1). These rocks are overlain by Pleistocene glacial and fluvioglacial deposits. The valley is approximately 350 m wide and trenched about 30 m below the surrounding area. Land use on the valley floor is mainly pasture.

The present valley has developed since *c*. 15,500 BP (Johnson, 1969) through a succession of Holocene terrace stages. The earlier terraces (T on Fig. 1) down to *c*. 4 m above the modern river are all floored by bedrock but the youngest terrace formation (t on Fig. 1) involved a trenching to below present river level, downstream of Radnor Bridge (Fig. 1), then aggradation to *c*. 3 m above the present river. Since this aggradation the youngest terrace has been trenched and the modern floodplain formed, most of it since 1840 and predominantly by lateral accretion associated with meander migration and development. Where the present river impinges on either the valley wall or the older terraces meandering is confined (Lewin & Brindle, 1977), but where it impinges on floodplain or the youngest terrace meandering is free.

HISTORICAL CHANGES

Historical information on the course of the River Dane was obtained from maps and air photographs. The earliest large-scale maps available are tithe maps dating from around 1840 and mapped on a parish basis at scales of 1/2376 and 1/4752 (Hooke & Perry, 1976). Ordnance Survey County Series maps at 1/2500 are available for *c*. 1870 and *c*. 1910. The course in 1947 was mapped from aerial photographs and the 1968 course is from O.S. National Grid Series 1/2500 maps. The 1980 course was produced by field mapping of changes from the 1968 O.S. 1/2500 map base. The courses at these six dates were superimposed at 1/2500 scale to examine changes over this time period (Fig. 2).

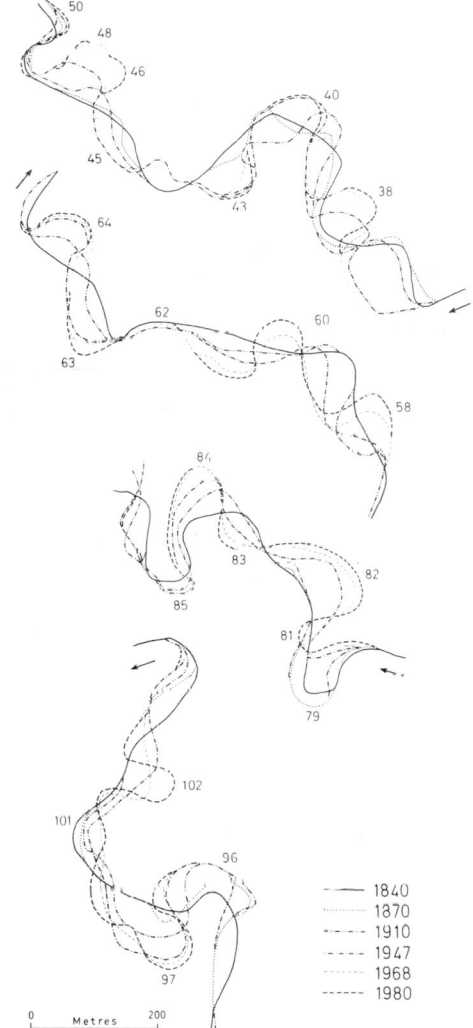

Fig. 2. Historical sequence of bend development on selected reaches of the River Dane.

In a preliminary analysis individual bends were identified and numbered (Fig. 1). The sequences of movement were elucidated by examining and coding the directions and types of changes in apex position and shape and in the limbs of each bend. This revealed a wide range of bend behaviour which was grouped into eight major types. It became apparent that many bends showed a tendency to increase in complexity over time, often developing asymmetric lobes in the apex region which continued to grow to form double-headed bends with a central inflection. Subsequently many of these lobes continue to develop as separate bends. For the sum of all bend movements between each date the following proportions for the major groups were found: simple migration, down-valley movement dominant (14%); confined migration, with part of the bend remaining stable (11%); growth, increase in amplitude dominant (15%); lobing, including double-heading (5%); new bends (5%); retraction and cutoff (7%); complex changes, islands and small irregular movements (18%) and stable bends (24%). However, approximately 24% of non-stable bends have exhibited lobing or double heading at some time in their development since 1840. It is the major process whereby new bends develop.

To facilitate more detailed analysis and to obtain morphological data, the channel centre lines were digitized using a 10 m step distance (approximately 1 width) to produce point coordinates. A computer program was used to calculate differences in angles (curvature) of path orientation at each digitized point. The positions of inflection points were located where curvature changed direction, with the program eliminating very short, low-curvature wobbles less than 30 m (approximately 3 channel widths) in length. For each bend between inflection points meander parameters were computed, including direct distance, path length, sinuosity, total curvature, maximum point curvature and mean curvature.

Mean values of meander characteristics for the course were calculated for each date. The number of bends increased from 77 in 1840 to 89 in 1910, and to 98 in 1980. Mean path length per bend decreased from 149 m in 1840 to 136 m in 1910 but has changed little since. Mean direct length has decreased from 113 to 94 m in 140 yr and mean bend sinuosity has increased from 1·30 to 1·40. The mean of maximum curvatures also increased (from 0·44 to 0·52 radians/10 m) as did total bend curvature (from 1·95 to 2·47 radians). Overall, the total path length increased from 13·13 km in 1840 to 14·28 km in 1980. Throughout this paper curvature is expressed in radians, either for the bend as a whole or for given lengths of channel. Radius of curvature can be derived from $r = (L/\theta)$ where r is radius (m), L is length of channel (m) and θ is curvature (radians).

Within the study reach a number of stable sections were identified, most of them short, but including two longer stable reaches; a rock-controlled reach between bends 3 and 5, and a low-sinuosity reach between bends 72 and 77. Between the stable sections mobile reaches were identified and path length and sinuosity were calculated for these at each date (Table 1). Most mobile reaches show increases in path length and sinuosity over time with particularly large increases through bends 57–61 and 78–85. Overall, development and growth of bends is much greater than retraction and cutoff.

Table 1. Total values of path length and sinuosity in individual reaches

Reach bend nos.	Dominant movements	Reach path length (m)						Reach sinuosity					
		1840	1870	1910	1947	1968	1980	1840	1870	1910	1947	1968	1980
	Mobile reaches												
6–10	Complex	621	—	638	730	612	647	2·64	—	2·61	3·04	2·56	2·74
12–22	Complex migration	872	—	908	865	992	853	1·98	—	2·10	1·99	2·24	1·98
23–30	Migration, retraction, lobing	1033	1085	1092	971	1077	1048	3·57	3·56	3·66	3·51	3·54	3·72
34–49	Migration growth	1197	1239	1315	1362	1456	1492	1·57	1·64	1·74	1·77	1·88	1·92
52–56	Growth, migration	463	451	405	402	504	539	1·50	1·57	1·53	1·67	2·12	2·18
57–61	Migration, growth	304	—	289	426	445	515	1·17	—	1·24	1·50	1·56	1·82
63–65	Migration, growth	342	—	376	475	475	487	1·28	—	1·37	1·76	1·75	1·77
65–72	Cutoff	772	—	811	858	806	768	1·23	—	1·30	1·37	1·28	1·23
78–85	Growth	741	786	869	910	949	986	1·48	1·57	1·74	1·82	1·90	1·97
85–94	Complex migration, retraction	625	644	637	539	596	594	1·37	1·42	1·40	1·18	1·31	1·31
96–103	Growth	648	649	727	862	846	1004	1·72	1·87	2·17	2·57	2·59	3·07
105–115	Complex	899	629	970	1041	1064	1099	1·53	1·36	1·61	1·76	1·83	1·89
	Stable reaches												
3–5	Stable	573	—	614	575	802	651	1·32	—	1·34	1·30	1·38	1·32
72–77	Stable	443	443	444	459	464	460	1·06	1·06	1·06	1·10	1·11	1·10

Data from short stable sections at bends 1, 2, 11, 23, 31–33, 50–51, 56, 62, 66, 85, 95, 104 omitted.

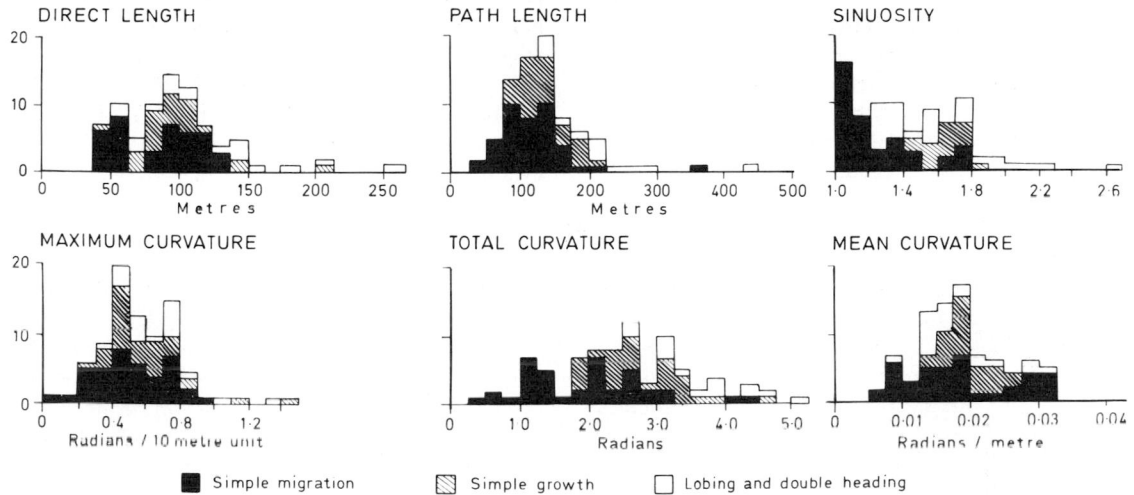

Fig. 3. Frequency distributions of bend characteristics for selected bend movements. Channel width is *c.* 10 m. Maximum curvature, in radians $(\theta)/10$ m unit, can be converted to equivalent radius of curvature (r) by $r = (10/\theta)$.

Table 2. Means and standard deviations of bend characteristics in relation to selected dominant types of movement

Dominant type of movement	Direct length (m)		Path length (m)		Bend sinuosity		Max curvature (rad/10 m)		Total curvature (rad)		Mean curvature (rad)	
	$\bar{X}$	σ	$\bar{X}$	σ	$\bar{X}$	σ	$\bar{X}$	σ	$\bar{X}$	σ	$\bar{X}$	σ
Migration	89·8	30·8	111·9	39·9	1·25	0·25	0·533	0·236	1·95	0·81	0·0184	0·008
Growth	94·7	24·5	140·0	39·7	1·49	0·25	0·594	0·188	2·75	0·64	0·0204	0·004
Lobing	121·5	53·7	219·0	91·3	1·85	0·38	0·645	0·290	3·80	0·99	0·0192	0·007
Double-heading	142·3	33·9	263·0	63·3	1·86	0·28	0·603	0·184	4·05	0·54	0·0156	0·004
Retraction and cutoff	75·5	19·6	99·2	31·6	1·30	0·20	0·593	0·266	1·99	0·64	0·0207	0·005

On this table data for simple and confined migration groups have been combined: lobing and double-heading groups treated separately.

In order to test whether there is a relationship between the type of change and the morphology of a bend, the characteristics of bends exhibiting simple types of migration, growth and lobing have been plotted as frequency histograms in Fig. 3. These represent values at the beginning of the period of change. Table 2 gives the means and standard deviations of meander parameters according to type of movement.

The distributions of meander characteristics for different types of movement overlap but show some distinctive groupings. Migrating bends tend to have relatively short direct lengths and path lengths; maximum and mean curvature also tend to be low though the scatter is wide. Most significantly, migrating bends have low sinuosity and low total curvature. Growth bends have slightly higher direct lengths but markedly higher path lengths, sinuosity and total curvature. Maximum and mean curvature also tend to be slightly higher. The data ranges are generally more limited than for migrating bends and the growth bends tend to be of higher amplitude and more sinuous. Comparison of growth bends with those that develop lobes and double-heads shows that the latter groups have significantly higher values of path lengths, sinuosity and total curvature, especially the double-headed bends. Minimum values of path length and curvature may tentatively be suggested by the histograms. Bends which retract or cut off (Table 2) tend to have short direct lengths and path lengths, low–moderate sinuosity, moderate–high maximum curvature and a comparatively high mean curvature.

Map examination of retracting bends suggests that many are influenced by activity up- or downstream, some becoming confined against bedrock, or are eliminated by rapidly migrating or growing bends from upstream. Several occur in the zones upstream of bridges. Stable bends encompass a wide range of values. The few free meanders which are stable are low-amplitude bends, as in reach 72–77. Other bends stabilize because they reach bedrock or other restrictions.

Individual parameters alone do not adequately characterize meander forms but combinations of characteristics help to distinguish different types of bend development. The combination of curvature and path length in a bend is important both on theoretical and empirical grounds. Figure 4 is a graph of maximum curvature per 30 m channel length, derived from the maximum three adjacent curvature values, against path length of a bend, between the defined inflection points, with the selected simple movements as in Fig. 3 distinguished. Maximum three rather than single maximum curvature values have been used to overcome the influence of digitizing distance, though a plot of the single values produces a similar distribution.

In Fig. 4 migrating bends cover a wide range of maximum curvature but there appears to be an upper limit to path length, tentatively suggested by the line. The growth bends tend to have higher curvature and

path lengths and there is a zone of growth beyond the upper limit of migration. Some new bends which grow would, however, plot in the region of low curvature and short path length. More importantly, a comparison between growth and lobing bends indicates that there is a critical combination of path length and maximum curvature beyond which a bend tends to develop a lobe and eventually a second bend. Retracting bends, though not plotted on Fig. 4 for clarity, overlap with the migration and growth distribution. Thus a bend which plots in the central region may exhibit one of several types of behaviour. Examination of individual cases on the maps and in the field suggest that much depends on the presence of restrictions and on the influence of bends up- and downstream. The three true cutoffs, bends 67, 79 and 90 plot close together with moderately high curvature (*c.* 1·6 rad/30 m) and path lengths of about 140 m; in each case further growth was limited and movement from upstream into the bends was causing tightening. Stable bends plot with a wide distribution on Fig. 4 but include both low-amplitude free bends as in reach 72–77 and those which stabilize against a restriction. Very few freely growing and migrating bends appear to reach a form at which they become stable.

Other curvature characteristics may also be important, including curvature distribution within the bend. Detailed analysis is not included here, but a crude classification according to the number of

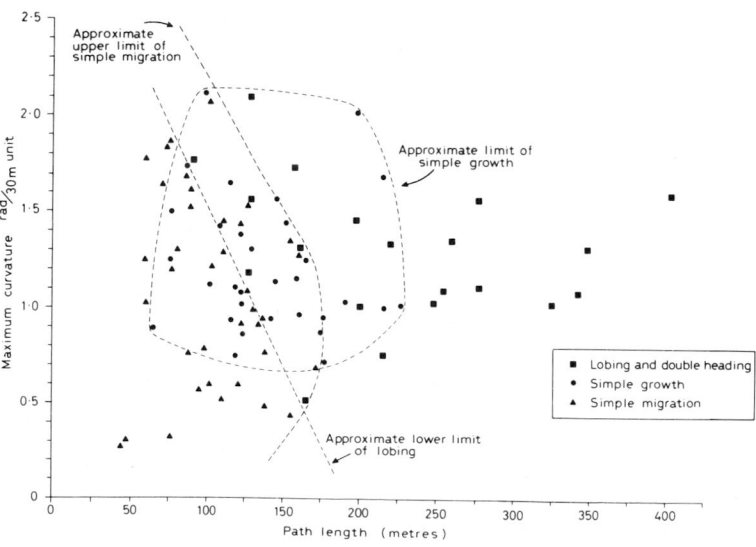

Fig. 4. Path length and curvature relationship for selected bend movements. Channel width *c.* 10 m. Maximum curvature expressed in radians (θ)/30 m length can be converted to equivalent radius of curvature (r) by $r = (30/\theta)$.

curvature peaks reveals a much higher proportion of simple curvature-distribution bends tending to migrate (26%) than to grow (13%).

Examples of individual bend behaviour demonstrate these trends. Bends 38, 39, 60 and 64 (Fig. 2) have migrated great distances, mostly with a change in form due to unequal limb movement. Rapid migration has rarely continued throughout the 140-yr period due to the restricted floodplain. After migration bends may grow, stabilize or retract. Bends 40–42, 46–48 and 97–101 (Fig. 2) illustrate growth sequences. In 1840 bend 40 was a large square-headed bend representing the lobing stage, which by 1870 had developed into two separate bends, 40 and 42, and these continued individual growth and migration to 1910. By 1948 one large bend was present again, due to migration and expansion of 40 and elimination of 42 by migration and retraction into 43. An upstream lobe then developed on bend 40 by 1968 but retracted by 1980. The bend is now tightening but growth is restricted by bedrock in the terrace base. Bend 48, which is not constrained by bedrock, shows the simpler, earlier part of the sequence, with a combination of migration and growth from 1840 to 1947. During this period sinuosity increased from 1·05 to 1·28 and total curvature from 1·09 to 2·40 radians. From 1947 the bend grew rapidly, increasing in sinuosity to 1·75 and total curvature to 3·72 radians by 1968. By 1980 it had a distinct double-head with a central inflection. Flow characteristics in this bend are discussed below. Bends 97–101 demonstrate the complete growth sequence. In 1840, 97 was a large square bend extending from present 97 to 101. An upstream loop began to develop by 1870 and by 1910 had so grown that new bends 100 and 101 formed on its downstream end. The upstream loop (97) continued to grow rapidly to 1947, expanding its head, on which a second inflection developed by 1968 (bends 98 and 99). These have continued to develop since then.

This type of growth behaviour has not been widely recognized elsewhere although Lewin (1972), Brice (1974) and Hickin (1974) recognize complex forms of meanders. Further study is needed to elucidate and explain the distribution of these growth processes, but the mechanism of change can be clarified by examining the characteristics of pools and riffles and secondary flows in relation to bend morphology in the present channel.

THE MODERN CHANNEL

During field mapping of the 1980 channel pools and riffles were identified and located. Pools were defined as the central part of the deep channel areas and riffles as shallowings, usually marked by broken water, between pools. Secondary flow characteristics were also observed in many reaches by mapping surface-convergent and -divergent flow and zones of upwelling and descending vortices, and by detailed surveys in selected bends using velocity measurements and streamer observations. These observations confirm other work on secondary flows (Hey & Thorne, 1975) with a tendency for surface convergence from riffle into pool and divergence from pool into riffle and with strongest cell development in areas of tightest curvature.

To examine the relationships between pools and riffles and bend morphology, inter-riffle units, defined between riffle crests, were classified according to their number in a bend and whether riffles were at inflection points and pools at bend apices (in phase). Meander bends, defined between inflection points, were similarly classified according to number and position of riffle-pool sequences (Table 3). Each category was further subdivided into rock-constrained and free segments of the channel.

Figure 5 shows histograms of direct and path

Table 3. Classification of inter-riffle units and meander bends

	Classification of inter-riffle units
R_I	Inter-riffle unit in phase with meander bend
R_M	Major inter-riffle unit of multiple riffle unit bend
R_X	Minor inter-riffle unit within multiple riffle unit bend or associated with minor wiggle
R_Z	Inter riffle unit totally out of phase with meandering pattern
	Classification of meander bends
M_{1A}	Meander bend with one-to-one in-phase relationship with pool-riffle sequence
M_{1B}	Meander bend with one inter-riffle unit but at least partially out of phase
M_2	Meander bend with multiple riffle units
M_O	Meander bend totally out of phase with pool-riffle sequence including those without riffles

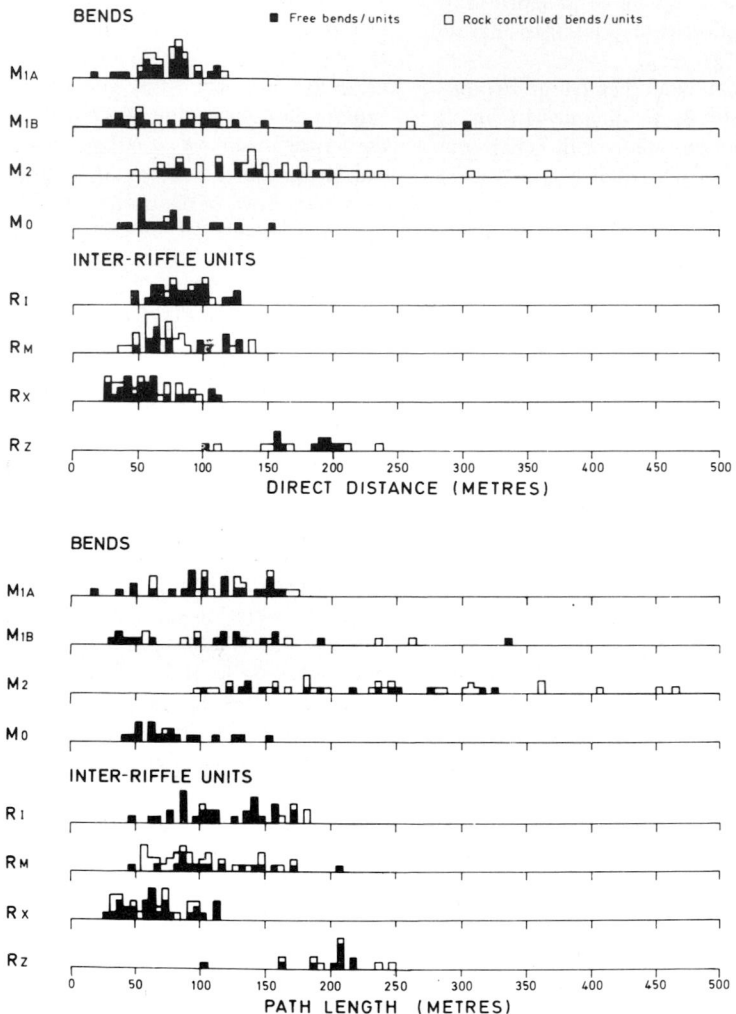

Fig. 5. Histograms of direct and path lengths in 1980 of bends and inter-riffle units classified according to Table 3.

distances for each of these groups. A general association between riffle spacing and meander half-wavelengths is well known, but apart from Keller's (1972) work little is known of the range of behaviour. For direct distance, Fig. 5 reveals little variation between most classes; as can be expected, multiple unit bends (M_2), out-of-phase single-unit bends (M_{1B}), and out-of-phase riffle units (R_Z) have a much wider range than other categories. Minor inter-riffle units (R_X), which can be interpreted as the most recently developed forms, show a smaller range. Rock confinement seems to have little influence on these direct measurements except at the extremes of

classes M_2, M_{1B} and R_Z. However, no single preferred riffle spacing is apparent and even the simple categories have riffle spacing ranging between 3 and 10 channel widths.

When path lengths are considered the data ranges increase in relation to the sinuosity of the channel. Expectedly, categories R_X and M_0 show a similar range as before, with a maximum for R_X of $c.$ 115 m, but all other categories show an increase and a wider scatter of points. The implications appear to be that a one-to-one in-phase relationship is only sustainable up to a maximum path length; categories R_I and M_{1A} have maximum path lengths of $c.$ 160 m and category

M_2 (multiple unit bends) has a minimum approximately twice that of category R_I. This suggests in a manner similar to that proposed by Keller (1972) that beyond a maximum path length, here represented by R_I units, new riffles begin to form. This is presumably because associated regular secondary flow structure can no longer be sustained.

It may be expected that the threshold path length beyond which new riffles begin to develop would also be related to tightness of curvature, since this influences strength of secondary flows. Figure 6 (left) shows inter-riffle path lengths plotted against maximum 30 m curvature for each inter-riffle unit class. Plotting positions for free class R_I units can be included within an approximate envelope curve which represents the upper limit for path length for simple one-to-one riffle units. This limit appears to increase with curvature up to c. 1·6 radians/30 m and then to decrease. Class R_X, immature units, plot within the curve but cluster in the lower part. The majority of class R_M free units also plot within the curve. Units plotting outside include class R_Z, those totally out of phase with the meandering pattern and often in irregular low-sinuosity channel segments, and some rock-controlled units where adjustment is limited.

Figure 6 (right) shows a similar graph plotted for bend classes. A similar envelope curve encloses the free M_{1A} points. Most of the free M_{1B} points plot within the envelope but the few exceptions include irregular, at least partially out-of-phase bends. Class M_O bends, those totally out of phase, include the low-sinuosity channel reaches such as bends 72–77 (Fig. 1) and plot in the bottom left of the graph. Multiple riffle unit bends, class M_2 (both free and rock-controlled), plot outside this curve, suggesting that their path lengths have developed beyond the limit for single inter-riffle units. If the changes in the most recent period, 1968–1980, are considered then in Fig. 6 (right) the most active bends would plot in the upper right part of the graph.

The relationship between pool and riffle characteristics and bend morphology can be used to help understand the historical changes and suggests a mechanism by which changes take place. Figure 4 shows that the development of lobes and double-heads involves extension of path length, and comparison with Fig. 6 (right) indicates that this is associated with the formation of additional new riffles beyond a critical path length and curvature combination. It is also associated with a change from single to double or complex curvature distribution in the bend.

A general model of meander development on the Dane can be proposed on the basis of these relationships. Bend development appears to be related to the formation of pools and riffles through secondary flows (Hey & Thorne, 1975), which focus scour in zones of surface convergence and descending flow and deposition in zones of divergence and

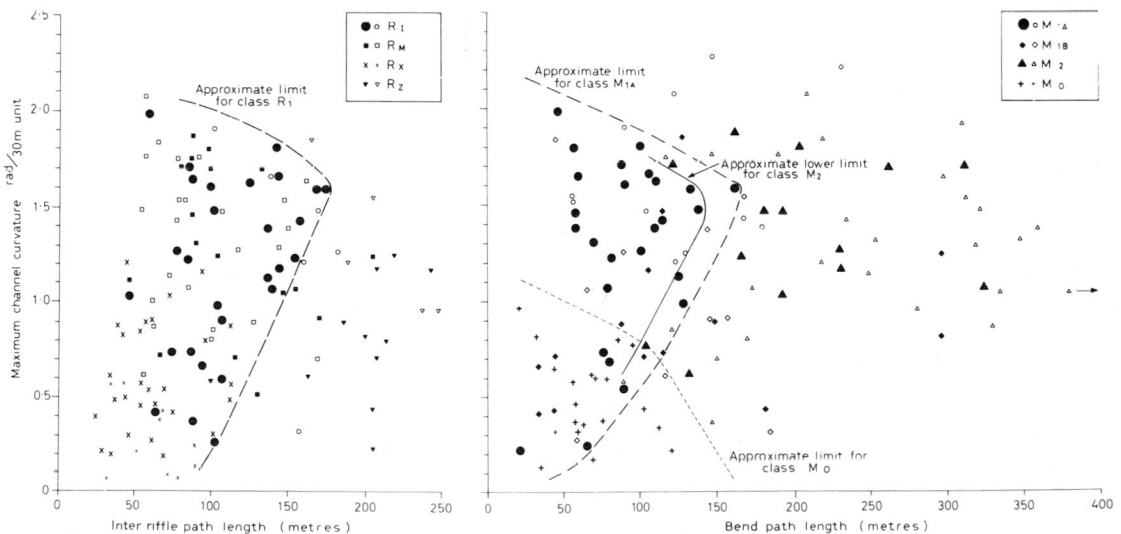

Fig. 6. Path length and curvature relationship for bends and inter-riffle units in 1980, classified according to Table 3. Heavy symbols indicate free bends/units: light symbols indicate rock-controlled bends/units. Channel width c. 10 m. Maximum curvature expressed in radians (θ)/30 m length can be converted to equivalent radius of curvature (r) by $r = (30/\theta)$.

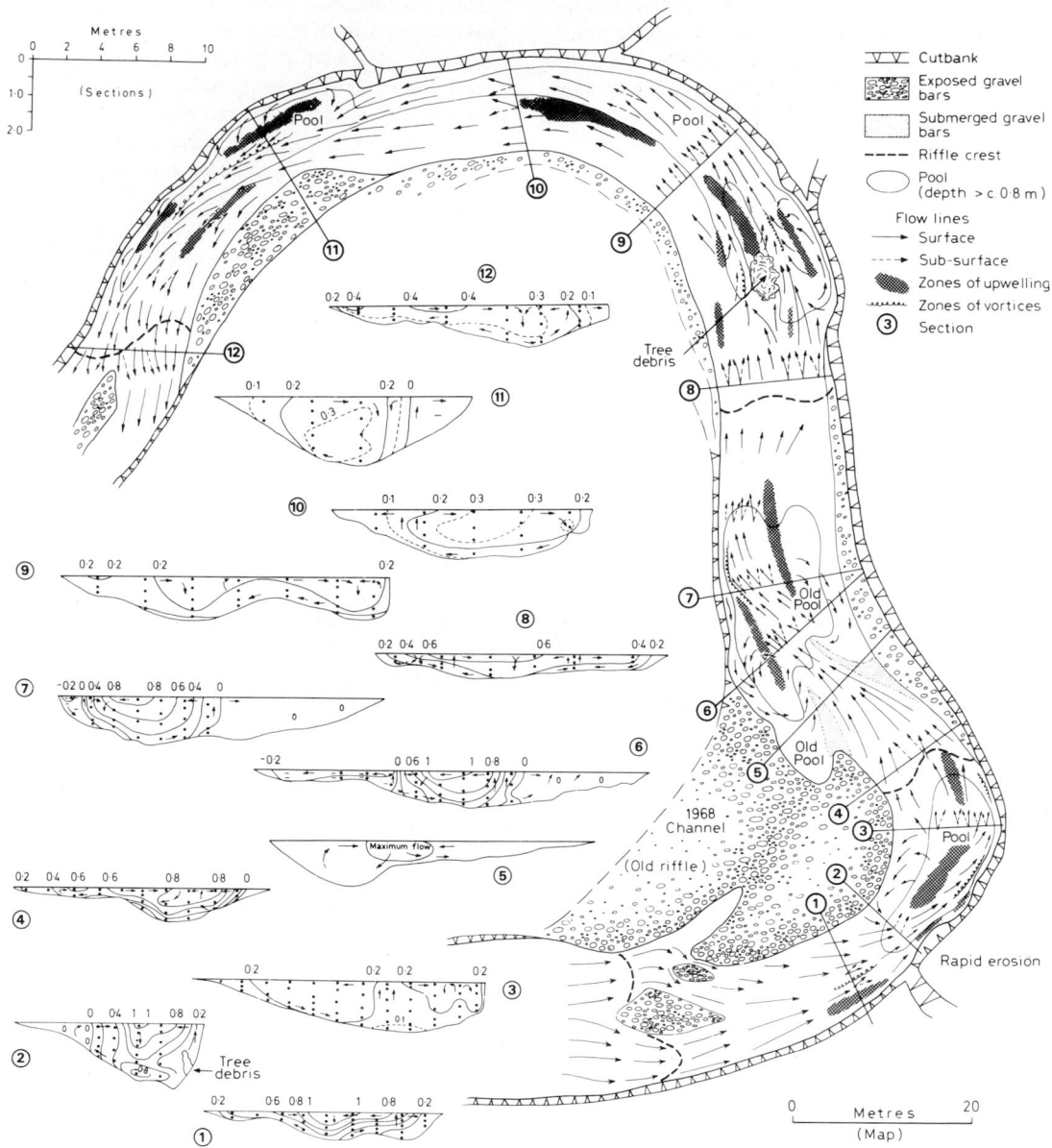

Fig. 7. Morphology, secondary flows and cross-section isovels through bends 46–48. Surveyed April 1981. Velocities on cross-sections in m s^{-1}.

upwelling (see below). Initially bends of low, single-peaked curvature and short path length will have a tendency to migrate, often with differential limb migration, increasing maximum curvature and path length. More rapid development may follow, with growth becoming dominant, and as path length increases there may be a tendency for single-peaked

curvature to be replaced by a double peak on either side of the bend apex. If path length continues to increase beyond the threshold length a one-to-one relationship with pools and riffles may no longer be sustainable. A new riffle may form, perhaps where the secondary flow begins to die away, leading to lobing and eventually to double-heading with each lobe then behaving as a separate bend. The crucial part of this sequence seems to be the development of secondary riffles and the lobing tendency, both of which tend to occur above the threshold path length. Double heading may occur as a result of these trends in free meanders but will be accentuated if growth causes impingement of the meander apex against bedrock in the valley side or at the base of the older terraces. This differs from Keller's (1972) model in that major path length increase is not in the limbs of the meanders but in the apex region.

Verification of this general model of bend development can be approached through consideration of the secondary flows, pools and riffles and morphological sequence in one freely meandering area, bends 46–48 (Fig. 7), that has double-headed between 1968 and 1980. The historical development of the bend (Fig. 2) has been discussed earlier. In 1947 the path length (100 m) and maximum curvature (1·11 rad/30 m) plot on Fig. 6 (right) in a position which would indicate a one-to-one bend. The curvature distribution is double-peaked and the position on Fig. 4 would suggest likely growth. By 1968 growth caused increases in length (to 175 m) and curvature (to 1·32 rad/30 m) and the bend would plot on Fig. 6 (right) outside the one-to-one envelope. Curvature distribution was multi-peaked and plotting position on Fig. 4 would indicate growth or lobing. It is likely that a second riffle unit had formed and this is confirmed in the field (Fig. 7); the position of the 1968 channel is still visible, with a former riffle being present in the old channel and a former pool identifiable in the present channel between sections 4 and 6. By 1980 erosion had continued and a new lobe had indeed formed on the upstream end, making the bend clearly double-headed. The increased path length (to 185 m) and maximum curvature (to 1·47 rad/30 m) confirm a plotting position well outside the one-to-one envelope and within the lobing and double-heading area. The constituent bends, new lobe 46 and the downstream remnant 48, plot back in the one-to-one envelope and within the migration or growth areas, and the interlobe wiggle (bend 47) plots in the lower left position of Fig. 6 (right) as a new bend. The secondary flows in these bends (Fig. 7) have been measured in the field. They illustrate adjustment to the channel change and indicate zones of present erosion and deposition. Within the new pool between sections 2 and 3 there is very strong secondary circulation, convergent near the outer bank, and active erosion on the outer bank. The new riffle at section 4 shows divergence then convergence into the pool at section

7. This riffle extends downstream from section 5, and gravel shoals can be seen to be infilling the former pool of the 1968 channel. The central part of this shoal is scoured by strong convergent descending flow (section 6). The remnants of the old pool are occupied by almost stationary water and are accumulating sand, especially on the left. The lower bend shows the expected secondary circulation with weak divergence upstream of the riffle at section 8, marked divergence at the lower riffle (section 12) and transverse flow with outer bank convergence throughout the long pool (sections 9–11). The secondary circulation is most pronounced at either end of this pool and weakest in the middle.

CONCLUSION

Analysis of the historical data has shown the major types of change which have taken place on the River Dane and the sequences of development in individual bends. A characteristic of this river is the rapid growth of certain bends to a stage where a second inflection is formed and a second bend develops on the head of the original. Analysis of the morphological properties of these bends reveals that this happens after a critical path length is reached. Investigation of the present distribution and characteristics of pools and riffles and secondary flows indicates the mechanism of change and provides some explanation of the historical sequences. Beyond a critical path length, which also depends on curvature of the bend, the regular secondary cell structure can no longer be sustained, the pattern breaks down and an extra pool and riffle develops. These then create their own secondary flow patterns of convergence and divergence, which accentuate the incipient bend form by associated erosion and deposition. Classic bend forms are amongst the most unstable on this river, tending to be the ones which will grow rapidly and develop lobes.

There appears to be an intrinsic threshold in this system (Schumm, 1979). Bends at all stages in this sequence can be found at any time in the 140 yr period and so the changes cannot be wholly explained by a specific extrinsic alteration at some time. Obviously not all bends exhibit this behaviour but many of the irregular or anomalous changes can be explained by rock control, valley constrictions or island development and irregular deflection of flow by fallen trees. Further work is needed on analysis of curvature characteristics in bends and their influence on flow circulations and on the mechanism of new riffle formation at threshold conditions. The relationship of rates of change to bend morphology also needs to be examined further. However, the assumption of an equilibrium form does not seem applicable on the River Dane over the past 140 yr. A progressive, dynamic model is more appropriate.

ACKNOWLEDGMENTS

We wish to thank Robert Perry and Computing Services, Manchester Polytechnic, for computing assistance, Alan Thompson and Tim Rummery for field assistance, and the staff of the photographic section and drawing office of the University of Liverpool, particularly Sandra Mather, for producing the diagrams. J.M.H. is grateful to the British Geomorphological Research Group for a grant towards air photographs and fieldwork costs.

REFERENCES

BLUCK, B.J. (1971) Sedimentation in the meandering River Endrick. *Scot. J. Geol.* **7**, 93–138.

BRICE, J.C. (1964) Channel patterns and terraces of the Loup Rivers in Nebraska. *Prof. Pap. U.S. geol. Surv.* **422D**, 41 pp.

BRICE, J.C. (1973) Meandering pattern of the White River in Indiana—an analysis. In: *Fluvial Geomorphology* (Ed. by M. Morisawa), pp. 179–200. Binghamton, New York.

BRICE, J.C. (1974) Evolution of meander loops. *Bull. geol Soc. Am.* **85**, 581–586.

BRIDGE, J.S. & JARVIS, J. (1976) Flow and sedimentary processes in the meandering River South Esk, Glen Cova, Scotland. *Earth Surf. Processes*, **1**, 303–336.

CARLSTON, C.W. (1965) The relation of free meander geometry to stream discharge and its geomorphic implications. *Am. J. Sci.* **263**, 864–885.

DANIEL, J.F. (1971) Channel movement of meandering Indiana streams. *Prof. Pap. U.S. geol. Surv.* **732A**, 18 pp.

FERGUSON, R.I. (1975) Meander irregularity and wavelength estimation. *J. Hydrol.* **26**, 315–333.

FERGUSON, R.I. (1977a) Meander sinuosity and direction variance. *Bull. geol. Soc. Am.* **88**, 212–214.

FERGUSON, R.I. (1977b) Meander migration: equilibrium and change. In: *River Channel Changes* (Ed. by K.J. Gregory), pp. 235–248. Wiley, Chichester.

HARVEY, A.M. (1975) Some aspects of the relations between channel characteristics and riffle spacing in meandering streams. *Am. J. Sci.* **275**, 470–478.

HEY, R.D. & THORNE, C.R. (1975) Secondary flows in meandering river channels. *Area*, **7**, 191–195.

HICKIN, E.J. (1974) The development of meanders in natural river channels. *Am. J. Sci.* **274**, 414–442.

HICKIN, E.J. (1977) The analysis of river planform responses to changes in discharge. In: *River Channel Changes* (Ed. by K.J. Gregory), pp. 249–263. Wiley, Chichester.

HICKIN, E.J. (1978) Mean flow structure in meanders of the Squamish River, British Columbia. *Can. J. Earth Sci.* **15**, 1833–1849.

HICKIN, E.J. & NANSON, G.C. (1975) The character of channel migration on the Beatton River, N.E. British Columbia. *Bull. geol. Soc. Am.* **86**, 487–494.

HOOKE, J.M. (1977) The distribution and nature of changes in river channel patterns; the example of Devon. In: *River Channel Changes* (Ed. by K.J. Gregory), pp. 265–280. Wiley, Chichester.

HOOKE, J.M. & PERRY, R.A. (1976) The planimetric accuracy of tithe maps. *Cartog. J.* **13**, 177–183.

HUGHES, D.J. (1977). Rates of erosion on meander arcs. In: *River Channel Changes* (Ed. by K.J. Gregory), pp. 193–205. Wiley, Chichester.

JOHNSON, R.H. (1969) A reconnaissance survey of some river terraces in part of the Mersey and Weaver catchments. *Mem. Proc. Manchr lit. phil. Soc.* **112**, 1–35.

KELLER, E.A. (1972) Development of alluvial stream channels: a five-stage model. *Bull. geol. Soc. Am.* **83**, 1531–1536.

KONDRAT'YEV, N.YE. (1968) Hydromorphic principles of computations of free meandering. *Soviet Hydrol.* **4**, 309–335.

LANGBEIN, W.B. & LEOPOLD, L.B. (1966) River meanders—theory of minimum variance. *Prof. Pap. U.S. geol. Surv.* **422H**, 15 pp.

LEOPOLD, L.B. & WOLMAN, M.G. (1957) River channel patterns: braided, meandering and straight. *Prof. Pap. U.S. geol. Surv.* **282B**, 39–73.

LEOPOLD, L.B. & WOLMAN, M.G. (1960) River meanders. *Bull. geol. Soc. Am.* **71**, 769–794.

LEWIN, J. (1972) Late-stage meander growth. *Nature Phys. Sci.* **240**, 116.

LEWIN, J. (1976) Initiation of bedforms and meanders in coarse grained sediments. *Bull. geol. Soc. Am.* **87**, 281–285.

LEWIN, J. (1978) Meander development and floodplain sedimentation: a case study for mid-Wales. *Geol. J.* **13**, 25–36.

LEWIN, J. & BRINDLE, B.J. (1977) Confined meanders. In: *River Channel Changes* (Ed. by K.J. Gregory), pp. 221–233. Wiley, Chichester.

RICHARDS, K.S. (1976a) Channel width and the riffle-pool sequence. *Bull. geol. Soc. Am.* **87**, 883–890.

RICHARDS, K.S. (1976b) The morphology of riffle-pool sequences. *Earth Surf. Processes*, **1**, 71–88.

SCHUMM, S.A. (1963) Sinuosity of alluvial rivers on the Great Plains. *Bull. geol. Soc. Am.* **74**, 1089–1100.

SCHUMM, S.A. (1979) Geomorphic thresholds: the concept and its applications. *Trans Inst. Br. Geog.* **N.S.4**, 485–515.

SPEIGHT, J.G. (1965) Meander spectra of the Angabunga River, Papua. *J. Hydrol.* **3**, 1–15.

THORNE, C.R. & LEWIN, J. (1979) Bank processes, bed material movement and planform development in a meandering river. In: *Adjustments of the Fluvial System* (Ed. by D.D. Rhodes and G.P. Williams), pp. 117–137. Kendall-Hunt, Dubuque.

Spec. Publs int. Ass. Sediment. (1983) **6**, 133–143

Lateral accretion of fine-grained concave benches on meandering rivers

GERALD C. NANSON

University of Wollongong, P.O. Box 1144, Wollongong, N.S.W. 2500, Australia

KENNETH PAGE

Riverina College of Advanced Education, P.O. Box 588, Wagga Wagga, N.S.W. 2650, Australia

ABSTRACT

Not all laterally accreting facies are point bar in origin. On confined meandering rivers where concave benches are commonly associated with rapid channel migration, appreciable amounts of fine sand and mud can be deposited by within-channel lateral accretion of concave benches. Concave benches develop against the upstream limb of the concave bank of abruptly curving meander bends, and are formed of mainly fine suspended load deposited within the channel below the riffle in the upstream end of the pool. Erosion of the upstream limb of the convex bank widens the channel, producing a zone of expanded flow which facilitates flow separation near the upstream limb of the opposite concave bank. A platform of sand in the form of a longitudinal-shaped bar is deposited in this zone, followed by further aggradation with fine sand, mud and organic matter. Even when fully formed, at high flow the concave bench remains isolated from the rest of the floodplain by a secondary channel around the margin of the original concave bank of the main channel. With the continued downvalley migration of the meander bend, another concave bench is formed, and this process continues until eventually a new floodplain surface is locally created by the lateral accretion of these benches.

INTRODUCTION

It has been widely, and usually correctly, assumed that the floodplains of actively migrating meandering rivers are formed largely from laterally accreted and relatively coarse-grained point-bar deposits (e.g. Allen, 1965; Blatt, Middleton & Murray, 1972; Reineck & Singh, 1980). Although relatively small but varying contributions have been acknowledged as coming from flood basin, channel fill, levee and crevasse-splay deposition, a largely ignored process has been that of concave bench deposition. Carey (1963, 1969) described concave bank sedimentation in abrupt angle bends on the lower Mississippi River, and referred to it as eddy accretion. Woodyer (1975) introduced the term *concave-bank bench* for fine-grained deposits on the concave bank of tightly

curving bends on the Barwon River in New South Wales, Australia, and Taylor & Woodyer (1978) and Woodyer, Taylor & Crook (1979) have briefly described them in the context of other types of benches formed along that river. Lewin (1978) has described fine-grained sedimentation on the up-stream concave bank of the River Rheidol in Wales, in both confined and unconfined bends, but he did not note any particular depositional morphology. We have studied these benches on the Murrumbidgee River of New South Wales, and suggested they be termed simply *concave benches* (Page & Nanson, 1982).

A concave bench is a slightly crescent-shaped bench tucked in against the upstream limb of the concave bank (Figs 1, 2 and 3). They are slightly lower than the general floodplain surface and often have a secondary channel around the margin of the original convex bank (Figs 1 and 3). They occur in a wide

0141-3600/83/0106-0133 $02.00

Fig. 1. An aerial view of a mature concave bench on the Murrumbidgee River. The bench is the large forested 'island' separated from the original concave bank of the river by a narrow secondary channel. Flow is from left to top.

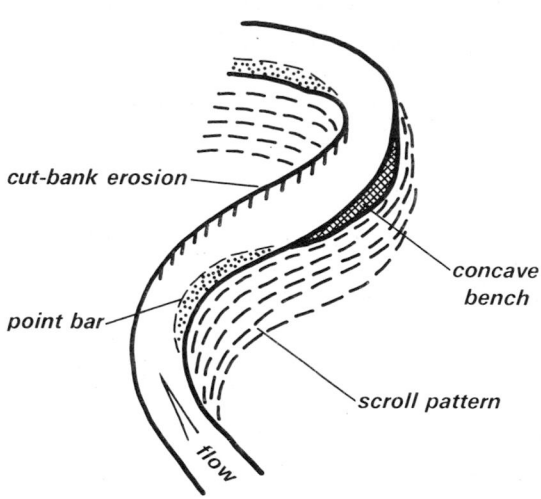

Fig. 2. A schematic diagram of a concave bench and point-bar on a channel migrating towards the top of the figure.

variety of meandering river regimes ranging from the largely suspended load streams of eastern Australia to the coarse gravel-bed and glacial-fed Squamish River of British Columbia, Canada (Hickin, 1979). In each case, however, they form from the within-channel deposition of mostly suspended fine sand, mud and organic matter; hence they dramatically contrast the

generally coarser point-bar and other within-channel sediments. This paper will demonstrate that concave-bench sediments can form very extensive depositional units of lateral accretion that are very different in character when compared to point-bar sediments.

CONCAVE-BENCH FORMATION

The Murrumbidgee, near Wagga Wagga, New South Wales, is a suspended load and sand and gravel bedload river with an average discharge of 120 m³ s⁻¹, an 80 m wide channel 6–7 m deep, and with a slope and sinuosity of 0·0003 and 2·3, respectively. The upper half or more of the floodplain thickness is dominated by cohesive fine sand, silt and clay, whereas the lower half is medium to coarse sand with some basal gravel.

Concave benches are located along the Murrumbidgee River on the upstream limb of concave banks in relatively abrupt meander bends, where the radius of curvature to mean channel width ratio is less than about 2·3 (Fig. 1). They are usually downstream of the tail of a point-bar, and the scroll pattern on the floodplain sometimes indicates that the entire point-bar and concave-bench complex has migrated with the downvalley sweep of the meander belt (Fig. 2).

Sequential aerial photographs (1944 and 1971) of

Fig. 3. A concave bench on the Fontas River, British Columbia. The cambered cross-section slopes gently towards the water, and more steeply towards the secondary channel seen adjacent to the original concave bank of the river. The boat and figure are at the downstream end.

concave benches at various stages of development along the river have permitted an interpretation of the sequential development of concave benches along the Murrumbidgee (Page & Nanson, 1982). The first stage of development is when cutbank erosion widens the channel. Usually the point-bar starts to prograde, leaving a zone of expanded flow downstream and adjacent to the upstream limb of the concave bank (Fig. 4A). A separation zone forms here, and during times of flood we have observed strong upstream secondary flow along the concave bank caused by the super-elevation of water near the bend axis of this tightly curving bank.

The second stage occurs with the deposition of a longitudinal-shaped bar of medium sand within the relatively quiet waters of the separation zone (Fig. 4B).

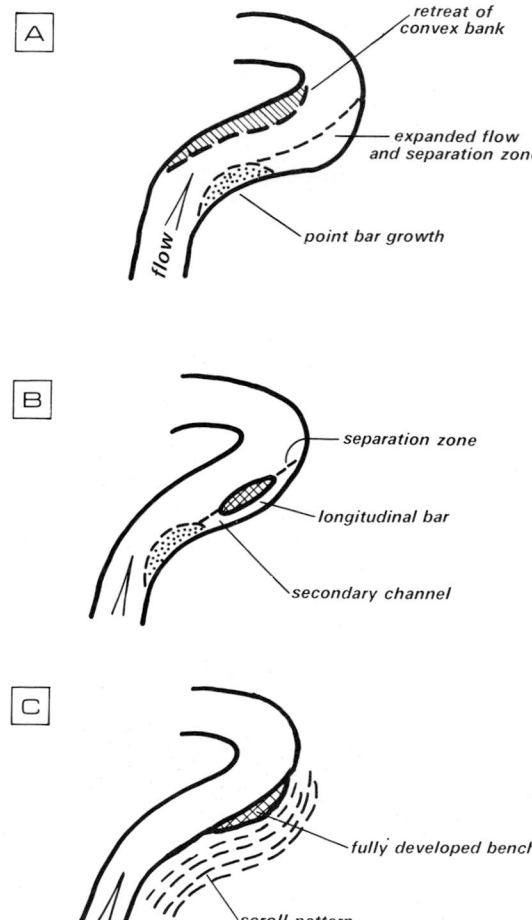

Fig. 4. A sequence of concave bench and point-bar migration.

Deposition is probably concentrated initially near the separation boundary but then gradually extends the bar towards the concave bank. The bar consists of a tabular cross-laminated coset of well-sorted medium sand with foresets dipping perpendicular to the channel axis and towards the concave bank (Fig. 5B). It has a gentle channelward-sloping stoss face and an abrupt avalanche face roughly parallel to the concave bank (Fig. 5A). By failing to migrate completely to the concave bank a secondary channel results which is retained through to the final stage of bench construction (Figs 1 and 3).

This longitudinal-shaped bar appears to act as a nucleus for further deposition. The third-stage of bench evolution is the gradual aggradation of the bar with fine sand and silt to a level of less frequent inundation, thereby allowing the establishment of vegetation. Sedimentation then continues with an increasing amount of mud until the bench attains a level just below the general floodplain surface. The result is a fully developed bench joined both to the recently formed point-bar and to the adjacent floodplain (Fig. 4C).

With continued erosion of the opposite cutbank this sequence of bar-then-bench formation appears to repeat itself as the channel migrates, in a manner similar to the lateral accretion of point-bar complexes. The result is that of floodplain alluvium deposited both by point-bar and concave bench accretion (Fig. 4C).

BENCH AND FLOODPLAIN SEDIMENTOLOGY

The sedimentology of Gobbagumbalin bench on the Murrumbidgee River has been described by us in detail elsewhere (Page & Nanson, 1982), but will be briefly summarized here for the purpose of facies comparison with adjacent point-bar deposits (Fig. 6).

Well-sorted medium sand forms a basement deposit in this bench and probably represents the longitudinal-shaped bar of medium sand seen during the initial stages of bench formation elsewhere on the river. Overlying this is a variable thickness of fine sandy mud and muddy sand which generally fines upwards. Mean sizes range from 250 to 8 μm, with between 10 and 40% of each sample being less than 2 μm. Consequently, they are much finer than any other type of within-channel deposit on the Murrumbidgee. Although laminated, there are very few visible flow structures (Fig. 7); those that are present are

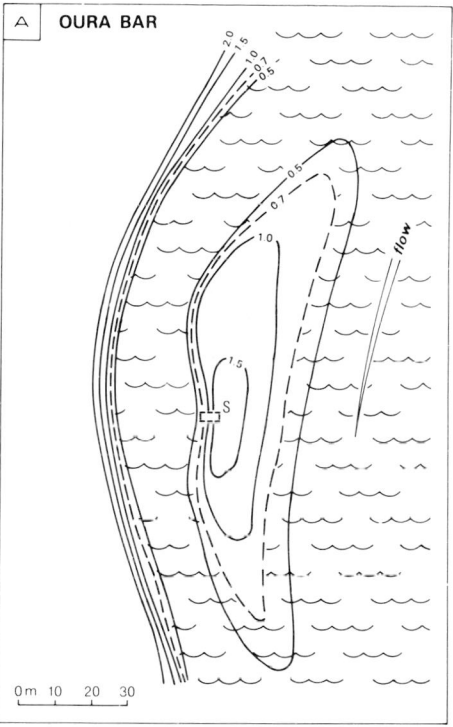

Fig. 5. A longitudinal-shaped bar on the Murrumbidgee River. 'S' marks the location of the photograph showing well-defined foresets from a tabular cross-laminated set directed towards the secondary channel adjacent to the convex bank (left). The ruler is 30 cm.

Spec. Publs int. Ass. Sediment. (1983) **6**, 145–154

Alluvial cutoffs in Wales and the Borderlands

G. W. LEWIS *and* J. LEWIN

Geography Department, University College of Wales, Aberystwyth SY23 3DB, U.K.

ABSTRACT

An extensive survey of 964 km of river valley identified 145 river channel cutoffs. Examination of historic map and air photograph evidence allows analysis of the development pattern and timing of many of the cutoffs, whilst results from additional ground survey of the sedimentary infill are also presented for 92 of the cutoffs. Most cutoffs (55%) are of simple chute type, 16% involve simple neck cutoffs, and 13% involve multiloop forms. 'Mobile bar' cutoffs account for 11%. The geographical distribution of cutoffs and cutoff types is discussed. Forty-five per cent of identified cutoffs can be dated to within approximately the last 100 yr, while infill sedimentation rates average 0.015 m yr^{-1}.

INTRODUCTION

Many studies, in contrasting fluvial environments, have shown that alluvial cutoffs are very widely created by actively migrating river channels, and that the areas of dead channel so produced form important sites for the subsequent deposition of fine sediment. In Wales and the Borderlands a survey of historical maps and air photographs of 40 actively migrating channel reaches showed that no fewer than 17 involved cutoffs at some stage in around the past 200 yr. Such activity has been documented in case-studies elsewhere in Britain (Johnson & Painter, 1967) whilst the geometry and impact of cutoffs, both artificial and natural, have been examined in detail for a number of major world rivers (Matthes, 1948; Kulemina, 1973; Weihaupt, 1977; Ching-shen, Shao-Chuan & Wen-Chung, 1978). The literature also suggests that there is a considerable variety in field cutoff types ranging, for example, from the breaching of very narrow necks on single meander loops (Mosley, 1975), through to much longer flood chutes developed across point bar complexes (Handy, 1972), and to new channels that may short-circuit whole groups of individual meander loops (Kulemina, 1973). It is, however, not generally clear from the

0141-3600/83/0106-0145 $02.00

literature how common each of such varied cutoff types is in any particular fluvial environment, how frequently cutoffs take place, where along a river they are likely to occur, and how quickly they may fill with sediment. This paper reports on an extensive survey of Wales and the Borderlands which provides some answers to such questions.

STUDY AREA AND METHODS

964 km of floodplain on 14 larger rivers in Wales and the Borderlands have been surveyed (Fig. 1). For the most part these rivers flow in narrow (<2 km) incised and glaciated valleys in Palaeozoic bedrock. Valley floors may be underlain by considerable depths of generally coarse, Pleistocene sediments, although in places drainage adjustment and incision during the Pleistocene may have produced reaches in which a confined channel is developed directly on bedrock. Bedrock valley cutoffs are found on the Teifi (Jones, 1965), Dee and Wye, but it is cutoffs developed on alluvial fills that are considered here. A previous study (Lewin, Hughes & Blacknell 1977) showed that 25% of a random sample of river reaches in the study area showed active channel migration at rates ranging from 0.1 to 5.5% of channel area per year. The remaining

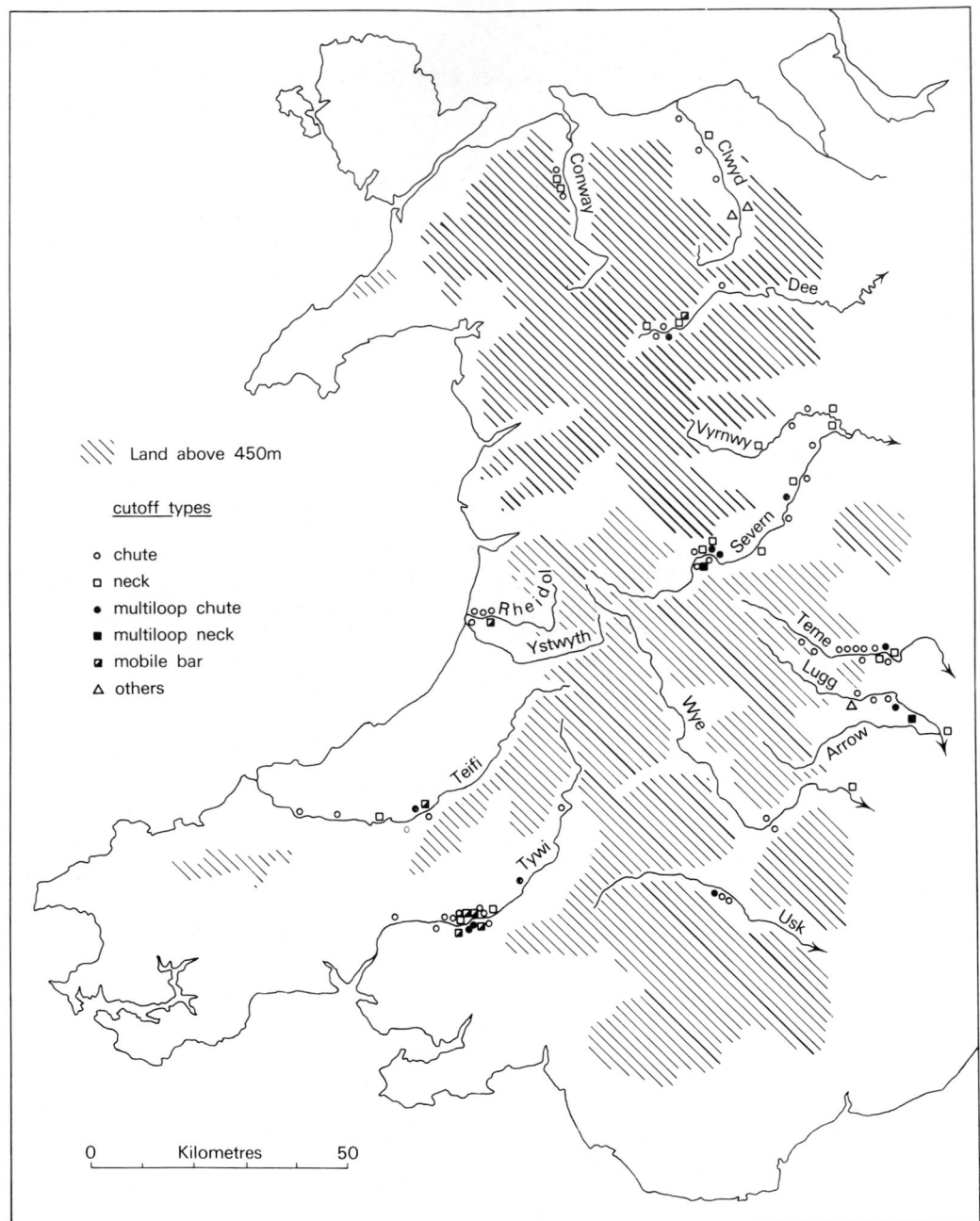

Fig. 1. Distribution of cutoffs in the study area.

Fig. 2. Part of the Rheidol floodplain (NGR SN6380) photographed in June 1970 after a prolonged dry spell. The crop markings seldom have any topographic expression and reflect sedimentation patterns on the active floodplain within the past few centuries.

75% showed no measurable change for periods of up to 78 yr.

Cutoffs were initially identified on air photography as elongated arcuate or sinuous depressions (dry or water-filled) on the floodplain surface with dimensions approximately those of the present river at each location. One hundred and forty-five cutoffs were located, though this must be a conservative estimate of those actually present. It does not include features where only one edge of a channel (often marked by a property boundary) remains distinct, where there is a possibility of confusion with point bar swales, or where sedimentation has completely filled the former channel. In fact crop marks on air photography obtained under dry summer conditions suggest that cutoff infill sediments are more significant for surface floodplain sedimentation than generally available evidence is able to reveal (Fig. 2). Analysis of such surface patterning under favourable conditions in the study area suggests that channel fills may amount to around 12% of surface floodplain sediments, though these may have little topographic expression and thus would not be counted in the survey reported here.

A second phase in the investigation involved examination of all the historical maps and air photographs available for each of the identified cutoff sites. The sources available included:

(1) Tithe maps, dating from the 1840s.

(2) First edition 1:10,560 Ordnance Survey maps, 1880s.

(3) Second edition 1:10,560 Ordnance Survey maps, early 1900s.

(4) Provisional edition 1:10,560 Ordnance Survey maps, generally 1930s.

(5) R.A.F. photography, late 1940s and early 1950s.

(6) Various commercial air photography from the 1960s and 1970s.

The location of the channel, with the cutoff as it developed or existed, was plotted on a common base for each site, thus allowing approximate dating of form developments. Clearly the historical evidence is not ideal, for dating precision is determined by the closeness in time of the bounding surveys. This may be as little as 6 yr or as much as 60, with an average of 35 yr. Again, map sources may show the development of general features such as the location and length of the chute and cutoff channel, blocking-off or plugging of the upstream end of the cutoff from the main river, and the subsequent diminution in size of the cutoff lake or its replacement by marshy ground. But availability of even this level of detail does depend (particularly for the Tithe surveys) on the quality of the map sources involved, whilst comparison between selective cartographic and unscaled but non-selective photographic information has its problems.

The third and final phase of the work here reported was a field visit to each site, which in particular involved survey of cutoff dimensions and the depth of cutoff fill, involving augering to obtain the average

Table 1. Surveyed rivers, cutoffs and range of historical sources

River	River length (km) Total	River length (km) Studied	Number of Cutoffs	Range of sources	Number of cutoffs historically dated
Arrow	50·1	50·1	—	1904–1946	—
Clwyd	53·8	53·8	6	1839–1979	3
Conwy	46·2	46·2	4	1839–1974	2
Dee	116·0	59·2	7	1839–1961	4
Lugg	86·6	55·0	7	1902–1946	5
Rheidol	32·9	32·9	5	1845–1973	3
Severn to Vyrnwy confluence	95·7	95·7	14	1840–1975	6
Teifi	107·8	107·8	7	1840–1946	3
Teme	127·4	73·9	12	1904–1973	9
Tywi	107·1	107·1	19	1839–1971	10
Usk	127·7	69·3	3	1838–1972	1
Vyrnwy	61·7	61·7	5	1840–1947	1
Wye	238·2	114·2	3	1841–1946	—
Ystwyth	37·2	37·2	—	1845–1975	—
Total		964·4	92		47

Table 2. Cutoff types

	Frequency (%)	Area (m²)		Length/width ratio	
		Mean	σ	Mean	σ
1 Chute	55	6320	7750	16·3	9·3
2 Neck	16	8360	5690	15·3	8·1
3 Mobile bar	11	3060	1410	16·9	7·4
4 Multiloop chute	12 ⎫	9450	6720	20·6	7·7
5 Multiloop neck	1 ⎭				

Remaining 5% due to artificial straightening or extensive channel realignment.

depth of fine sediment above coarse bed material. Some of the initially identified sites proved unsatisfactory, either because of non-availability of historical source materials or because of inaccessibility or doubtful site identification in the field, and the discussion that follows is based on the total of 92 sites which were examined in the field (Table 1).

CUTOFF TYPES

Historical evidence and site inspection suggested that the cutoffs studied may be divided into five types, three involving single-arc channels and two being multiloop versions of single-arc process types (Table 2). Although use of the historical material necessitates some subjective judgements because of the time lapses involved, neck cutoffs are identified as those in which adjacent channel reaches were probably less than a channel width apart at the time of breaching: chute cutoffs are those in which a much longer breach channel was created. On this basis the overall majority are of the chute type.

Chutes may occur on the inside of meander beds in a wide variety of positions and channel curvatures. For example, cutoffs may be located between cross-over points along the axis of bends of circular arc or tangentially connecting outer bends of meander loops. Alternatively cutoffs may be positioned in chord locations within loops. Depending on the radius of curvature (in channel widths) at the time of cutoff, the cutoff channel itself may also vary in its relative surface area by comparison with that of truncated point bar sediments with which it was associated. Figure 3(A) shows some possible cutoff locations, together with relative areas of cutoff (c) and point bar (p) sediments for an axial cutoff with meander radius of two channel widths. The cutoffs analysed here tend to occur at low radius of curvature (Fig. 3B), and

are dominantly situated within one channel width of a tangential position connecting the outer bends of meander loops (i.e. if bend radius is 2 channel widths (w), then the cutoff may be expected at a distance 2 w from the bend axis on the side away from the loop apex). Cutoff location in relation to meander radius is plotted in Fig. 3(C), together with curves showing the theoretical relative surface areas of point and cutoff sediments. In field situations, meander loops do not regularly conform to circular arcs, so that relationships between chute location, bend curvature, and the relative size of cutoffs and point bars may be complex.

Figure 4 shows examples of chute and neck cutoffs involving both single and multiple loops: on the river Severn reach (A), a neck cutoff formed shortly prior to 1948, with two additional chute cutoffs by 1975; on the River Teme there was one recent neck cutoff in 1902, with a small chute cutoff at the downstream end of the reach by 1948. However, this river (reaches B and D) does show both how rapid and how complex loop evolution may be, and although cutoffs can be identified, these are but one form of channel transformation which the relatively long time interval between available surveys makes it difficult to disentangle. A further point is that some of these changes are artificial, if also temporary and ineffectual (e.g. the right-angle bend at the upstream end of reach D in 1948 was designed to follow a new field boundary). In recent years (generally after the historical sources used here) river and water authorities have undertaken more systematic channel realignment, and our figure of 7% to include cutoffs from this cause is probably conservative. A final example from the River Teifi (C) shows the problem: the two quasi-straight sections of double channel in 1906 appear artificial, but only one of them proved successful as a permanent diversion of the stream course.

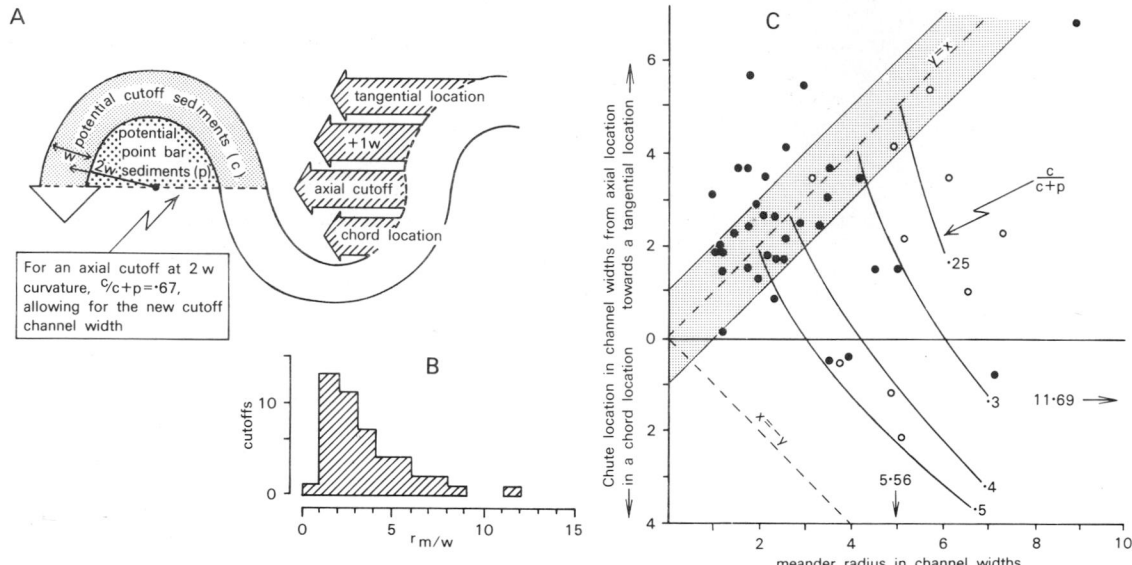

Fig. 3. Relationships between cutoff location, bend curvature, and the relative significance of cutoff infill sediments. (A) Possible cutoff location and the relative importance of cutoff sediments (c) by comparison with point bar sediments (p) for the case of an axial cutoff at meander radius of 2 channel widths. (B) Cutoff frequency and bend radius (in channel widths for the field sites. (C) Bend radius and cutoff location. The plotted lines show the relative proportion of cutoff sediments and $y = x$ and $x = -y$ are limits for tangential and chord cutoffs for circular arc bends. Surveyed loops that are manifestly complex are shown as open circles. Note that for sharper bends, the surface area of cutoff sedimentation becomes relatively significant, and that cutoffs are clustered in the shaded area which is ± 1 channel width from a tangential location.

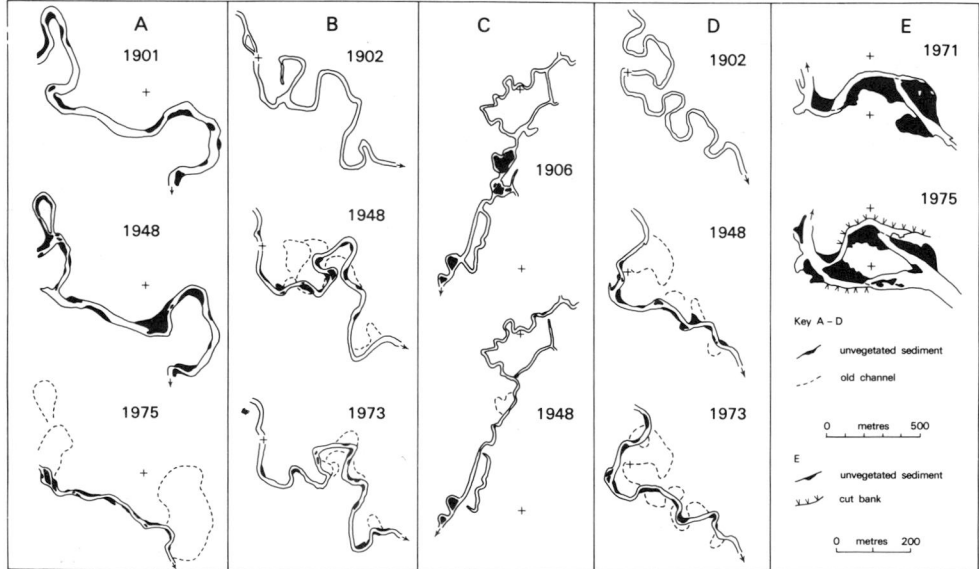

Fig. 4. Examples of cutoff types in the study area: (A) R. Severn at Penstrowed (SO 065923); (B) R. Teme at Nacklestone (SO 422719); (C) R. Teifi at Maes Llyn (SN 688632); (D) R. Teme at Hollybush (SO 411731); (E) R. Tywi near Llandovery (SN 748323).

An additional cutoff type, here termed a 'mobile bar cutoff' and not normally identified in the literature, is worthy of some discussion. Mobile, lobate bar forms are quite commonly found in the gravel and cobble bed sediments of the study area (Lewin, 1978). These have divergent flow across the bar front at high flows, with chute, pool, and cutbank arcs offset downstream. Cutbank development, largely on one side of the channel or the other, is enhanced by the development of dissecting channels through the bar front or side. These may thus be orientated either obliquely or in line with the general downstream flow. Bar mobility, or high-flow re-formation of the bar and the subsequent creation of a new dissecting channel, may lead to the blocking-off and abandonment of former marginal pools and cutbank arcs. The result may be the creation of a smaller (see Table 2), arcuate dead channel of lower curvature than that found in chute or neck cutoffs and, as such, is probably more characteristic of braided, mobile-bar river environments. Sediment occlusion of channels is also a mechanism for the production of dead sloughs. Figure 4(E) shows an example on the River Tywi. A major lobate bar with dissecting channel produced pool and cut-arc development on the right bank of the river. By 1975, this lobe had migrated downstream, leaving a progressively abandoned and occluded dead channel infilling with fine sediment behind it. Some complication is added by a new channel on the left bank so that the whole of the right bank complex may subsequently be abandoned.

CUTOFF DISTRIBUTION

The distribution of cutoffs (Fig. 1) is interesting. Some rivers (Usk, Wye) have few manifest cutoffs on their floodplains especially because the areal extent of the latter is restricted so that the opportunity for both channel migration and alluvial cutoff develop-

ment is limited. Others show a marked bunching of cutoffs somewhere in the middle course of the river (Rheidol, Severn, Teme, Tywi). To judge from the presence of a wide alluvial floor and high rates of channel change elsewhere along these valleys, this is an adjustment to process rather than opportunity. For example, rates of channel change within the zone of cutoff development on the Tywi (Fig. 1) are around $2-4\%$ yr^{-1} with bankfull streampower (Bagnold, 1977) being less than 500 W m^{-1} for a 35–40 m channel. Upstream of this reach, for some 10 km, rates of change can be $4-6\%$ yr^{-1} with a bankfull streampower of over 500 W m^{-1} for a 29–33 m channel of steeper gradient. There thus appears to be a division within the middle reaches of the river between an upstream section of lower sinuosity, steeper gradient, and more rapid channel change and bar migration, and a lower section in which extending and translating loops—and cutoffs—are much more common. Table 3 summarizes cutoff numbers and frequencies according to channel type for the whole survey area. Cutoff size is very variable, but a plot of catchment area against cutoff size (Fig. 5) again shows the relatively small number of cutoffs in the lower river reaches where channel migration rates are small, and the highly varied size of cutoffs in the middle reaches. Multiloop cutoffs appear to have relatively high length/width ratios (Table 2), but the other types are all rather similar.

There is also some suggestion in the data that the type of cutoff present may relate to floodplain slope, as was suggested by Tower a considerable time ago (1904). Translation, he suggested, would be more common on steeper slopes and loop expansion on gentler ones. Thus neck cutoffs may be found at low gradients, but with chute cutoffs presumably occurring with translating steep-gradient loops as well. As plotted in Fig. 6, the data do lend some support to this, for neck and multiloop cutoffs tend to be found only on low gradients, though somewhat surprisingly this also applies to the mobile bar cutoffs.

Table 3. Frequency of surveyed cutoffs by channel type

	Length surveyed		Cutoffs		Cutoff frequency along the valley	
	(km)	(%)	(no.)	(%)	km^{-1}	In channel widths
Confined	231	23·9	2	2·2	0·009	0·0001
Low sinuosity	491	50·8	35	38·0	0·071	0·0011
High sinuosity, active	130	13·5	46	50·0	0·355	0·0058
High sinuosity, stable	113	11·8	9	9·8	0·080	0·0017

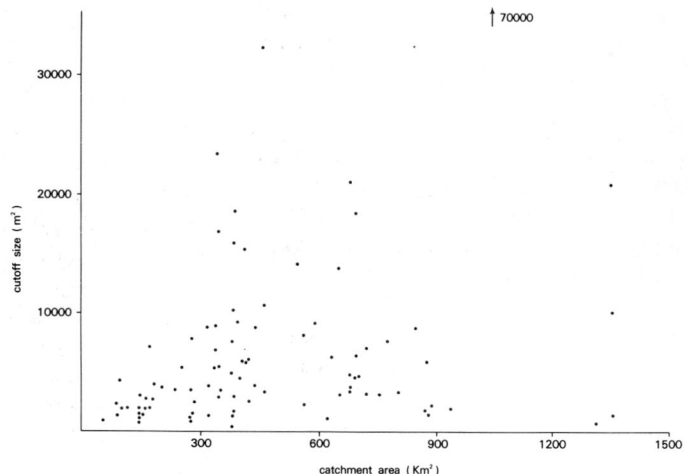

Fig. 5. Cutoff size and catchment area.

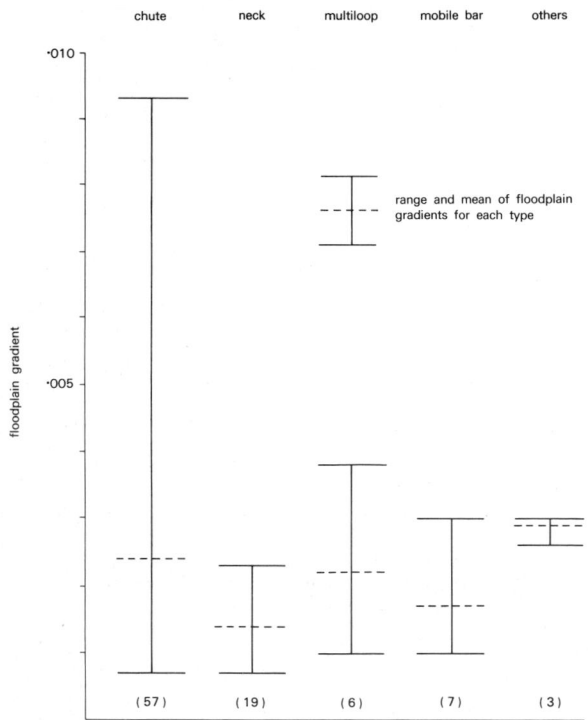

Fig. 6. Cutoff type and floodplain gradient. Numbers of each type for which gradient was measured are shown at the bottom of the figure.

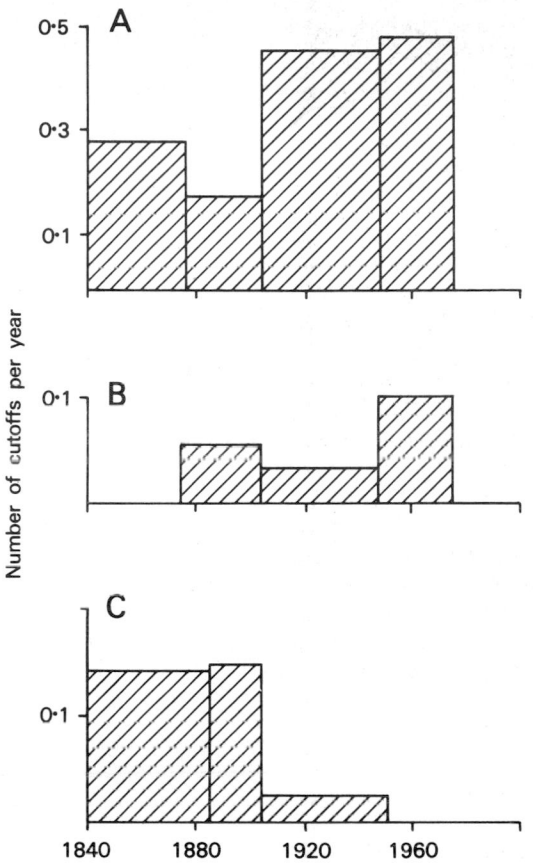

Fig. 7. Frequency of cutoffs for historically datable time periods: (A) all sites; (B) R. Severn; (C) R. Tywi. No data available before 1839.

CUTOFF DATING AND INFILL RATES

Figure 7 shows the number of cutoffs per year for all sites and for the Rivers Severn and Tywi individually. Of the 92 cutoffs studied in the field, 47 (51%) date from the periods when historical maps are available (Table 1), 45% from the last 100 yr. The data suggest that for close on 1000 km of river, cutoffs occurred at a rate of one every 5 yr in the period 1880–1900, and approaching every other year from 1950–1970. But for individual rivers the rate is very variable, so that most dated cutoffs identified on the Tywi are from the nineteenth century, but from this century on the Severn. Particularly in view of the small number of dated cutoffs for individual rivers, it is difficult to see any systematic explanation for such variations. It should also be remembered that the geographical distribution of cutoffs along rivers is sporadic and that cutoff frequency within reaches where cutoffs do actually occur is much higher than these overall averages.

Maximum infill rates for fine sediment (not including bed sediment plugs at cutoff extremities) vary from 0·003 to 0·071 m yr^{-1} (mean = 0·015, $\sigma = 0\cdot012$, $n = 92$). Seven out of the nine infill rates over 0·03 m yr^{-1} date from 1946 and after, so there must be the implication that recent cutoffs infill fastest, presumably whilst they are still in close relationship with the present river. Infill rates show no clear relationship to river or cutoff size. Again, such rates will not be maintained indefinitely. Once the cutoff is filled (and thus not one of the sample discussed here) it must aggrade or scour as the floodplain surface in general, where rates of overbank sedimentation appear to be lower than for cutoffs, to judge by the thickness of overbank fines in the study area at sites resedimented in the past century. Observed rates of cutoff infill are closely comparable with others that have been given, as conveniently summarized by Bridge & Leeder (1979, table 1), especially where reported aggradation rates have been averaged over a number of years.

CONCLUSIONS

The extensive type of documentary and field study reported here allows some new conclusions concerning the type, formation, frequency, location and infill rates for cutoffs in Wales and the Borderlands to be reached. At the same time, the exercise has its limitations, both in that it cannot reveal the 'true' proportion of cutoff fill sediments present (Fig. 2 suggests that these may be considerable), and because process detail may not be obtainable through the sources used. Further work is being undertaken both at type sites of recent or imminent cutoffs, and on the detailed nature of infill sediment.

ACKNOWLEDGMENT

G.W.L. is grateful for an NERC Research Studentship.

REFERENCES

BAGNOLD, R.A. (1977) Bed load transport by natural rivers. *Wat. Resour. Res.* **13**, 303–312.

BRIDGE, J.S. & LEEDER, M.R. (1979) A simulation model of alluvial stratigraphy. *Sedimentology*, **26**, 617–644.

CHING-SHEN, P., SHAO-CHUAN, S. & WEN-CHUNG, T. (1978). A study of the channel development after the completion of artificial cutoffs in the middle Yangtze River. *Scientia sin.* **XXI**, 783–804.

HANDY, R.L. (1972). Alluvial cutoff dating from subsequent growth of a meander. *Bull. geol. Soc. Am.* **83**, 475–480.

JOHNSON, R.H. & PAINTER, J. (1967) The development of a cutoff on the River Irk at Chadderton, Lancashire. *Geography*, **52**, 41–49.

JONES, O.T. (1965) The glacial and post-glacial history of the Lower Teifi valley. *Q. Jl geol. Soc. Lond.* **121**, 247–281.

KULEMINA, N.M. (1973) Some characteristics of the process of incomplete meandering of the channel of the upper Ob' river. *Soviet Hydrol.* **6**, 518–534.

LEWIN, J. (1978) Meander development and floodplain sedimentation: a case study from mid-Wales. *Geol. J.* **13**, 25–36.

LEWIN, J., HUGHES, D. & BLACKNELL, C. (1977) Incidence of river erosion. *Area*, **9**, 177–180.

MATTHES, G.H. (1948) Mississippi river cutoffs. *Trans. Am. Soc. civ. Engrs* **113**, 1–39.

MOSLEY, M.P. (1975) Meander cutoffs on the River Bollin, Cheshire, in July 1973. *Rev. Géomorph. dyn.* **XXIV**, 21–31.

TOWER, W.S. (1904) The development of cutoff meanders. *Bull. Am. Geog. Soc. N.Y.* **36**, 589–599.

WEIHAUPT, J.G. (1977) Morphometric definitions and classifications of oxbow lakes. Yukon River Basin, Alaska. *Wat. Resour. Res.* **13**, 195–196.

Spec. Publs int. Ass. Sediment. (1983) **6**, 155–168

Anastomosed fluvial deposits: modern examples from Western Canada

DERALD G. SMITH

Department of Geography, The University of Calgary, Calgary, Alberta T2N 1N4, Canada

ABSTRACT

Facies of two Canadian modern anastomosing river systems are discussed, the upper Columbia and lower Saskatchewan Rivers, which occur in intermontane and plains settings respectively. Both systems contain aggrading, multiple, low gradient, sand bed channels with adjacent splay, levée, and shallow wetland environments, all aggrading in accordance with channel sedimentation. While aggrading cross-valley alluvial fans or subsidence tend to control sedimentation rates in intermontane valleys, basin subsidence and/or regional tilting controls deposition rates in plains settings.

Sedimentation style in the upper Columbia River valley (120 × 1·5 km) consists of low-sinuosity, stable channels, depositing multi-storied channel sands and numerous sandy crevasse splay deposits. Channel deposits form as sand stringers laterally contained by deposits of levée silt and lacustrine mud. Aggrading at an average rate of 60 cm per 100 years over the past 2500 years, the anastomosing system is very dynamic, exhibiting avulsions and channel fills.

Deposition in the lower Saskatchewan River valley (120 × 80 km) a much wider basin with slower aggradation rates (29 cm 100 yr^{-1}, C-14 date on peat buried beneath a levée) results in laterally extensive sheets of overbank levée deposits of fine sand which grade into even more laterally extensive thick deposits of peat. With time, some dominant channels become highly sinuous, thus causing increased flow resistance, major avulsions upriver and eventual channel filling and abandonment. Facies differences of the two anastomosed river systems are believed to be caused by both the rate of sedimentation and width of the sedimentary basin.

GENERAL BACKGROUND

Over the last few decades, our understanding of different fluvial systems and associated sediment facies has resulted in the meandering, braided and several deltaic (river, tidal, and wave-dominated) depositional models. While these models have become the standard 'frameworks' for comparison and interpretation of ancient fluvial rock sequences, recent field evidence suggests that the anastomosed river model should be added to the list. The processes, morphology, associated deposits and sedimentology of anastomosed rivers are very different when compared to meandering and braided systems.

Two recent papers by Smith & Smith (1980) and Smith & Putnam (1980) best describe the characteristics of modern and ancient anastomosed rivers and associated deposits. They described anastomosed rivers as rapidly aggrading, stable, multiple, interconnected, low-gradient, low-sinuosity, laterally confined sand or gravel bed channels. Six sedimentary environments and associated textural facies are recognized as common in anastomosed systems: *channel* and *crevasse splay* deposits of sand or fine gravel, *levée* deposits of sandy silt, *lacustrine* deposits of mud, *marsh* deposits of mud and organic material, and *bog* deposits of peat. In short, anastomosed systems are rapidly aggrading low-energy channel and wetland complexes.

An understanding of anastomosed river systems and associated deposits is paramount to the geologist exploring for oil, gas and coal (Smith & Putnam, 1981). Because of the limitations of drill core, logs and seismic data of subsurface fluvial rock sequences, estimation of reservoir geometries and locations of future drill holes can be greatly affected by understanding the facies characteristics of anastomosed deposits (Putnam & Oliver, 1980). Also, because of the fact that rapid aggradation is common in modern anastomosing fluvial systems preservation potential should be excellent and there should exist in some thick rock sequences multiple confined channel and sheet sandstone reservoirs, as well as laterally adjacent multiple seams of coal with considerable lateral extent.

Modern examples of anastomosing fluvial systems are not common because of the rather unusual combination of geomorphic, tectonic and climatic conditions required. First, to achieve rapid continuous aggradation, either the sedimentary basin must subside or the local base level controlling the basin must rise. Subsequently, the anastomosed river is always trying to catch up and attain an equilibrium graded condition by rapid sedimentation. Thus, anastomosing rivers have very low gradients and little excess energy, resulting in channel sedimentation and lacustrine deposition in the wetlands. Smith & Putnam (1980) point out that the optimum structural settings for anastomosis would be tectonically active intermontane basins and molasse foreland plains, where subsidence is induced by mountain building. They believe that during Mesozoic and early Cenozoic time tectonic instability and climatic stability were optimum conditions for the deposition of thick accumulations of anastomosed fluvial sediments in and adjacent to the Rocky Mountains. Secondly, sediment influx into the basin must be great enough to maintain an alluvial plain, otherwise the subsiding basin would become a sea, open estuary or lake. Thirdly, regional climate must be wet enough in magnitude, frequency and duration to form well-developed channels and the growth of vegetation to help stabilize channel banks and levées. However, a wet climate may have to be reconsidered in view of ephemeral flows forming anastomosed systems in the arid interior of Australia reported by Rust (1981). Finally, it is likely that some modern anastomosed systems have been modified beyond recognition by farming and urban land use.

In spite of the suspected modification by man of some anastomosing river systems in active subsiding basins, some rivers still exhibit anastomosing-like morphology (multiple channels and associated wetlands): the lower Mississippi River in the Yazoo Basin, U.S.; the upper Nile River, S. Sudan; the middle Niger River, Mali; the lower Yangtze River, China; and the Ob River, W. Siberia, U.S.S.R. While all of these areas display anastomosed morphology, none has been described sedimentologically. In Canada, the upper Columbia and lower Saskatchewan Rivers (Figs 1, 2 and 3) and the Magdalena River in Colombia, South America, have been interpreted as rapidly aggrading anastomosed river systems. Aggradation in both Canadian rivers is due to a rising base level downriver (cross-valley alluvial fan and isostatic rebound), while basin subsidence is cited as the cause of aggradation in the Magdalena (Proyecto Colombo-Holandes de la Cuenca Magdalena-Cauca, 1977).

Most of the data and discussion in this paper are

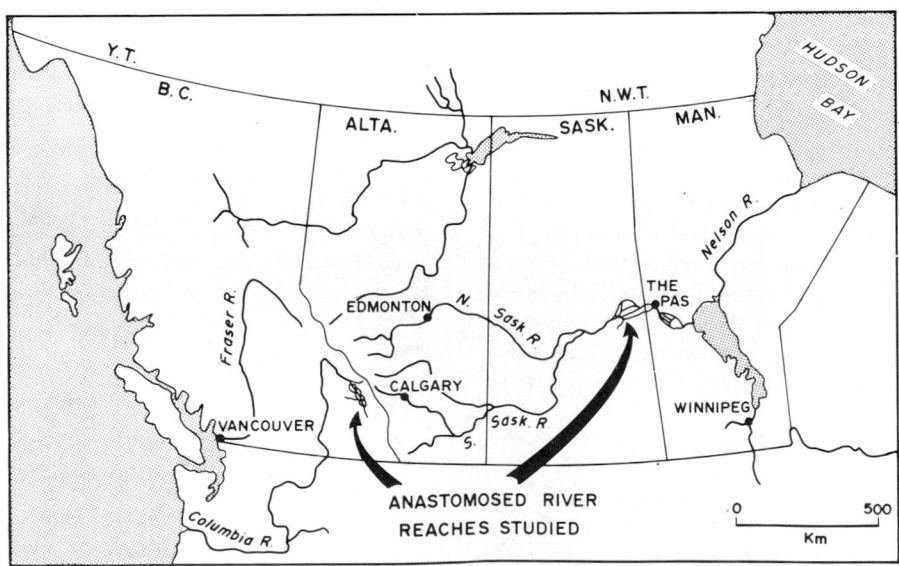

Fig. 1. Locations of the Columbia and Saskatchewan anastomosed channel reaches in western Canada.

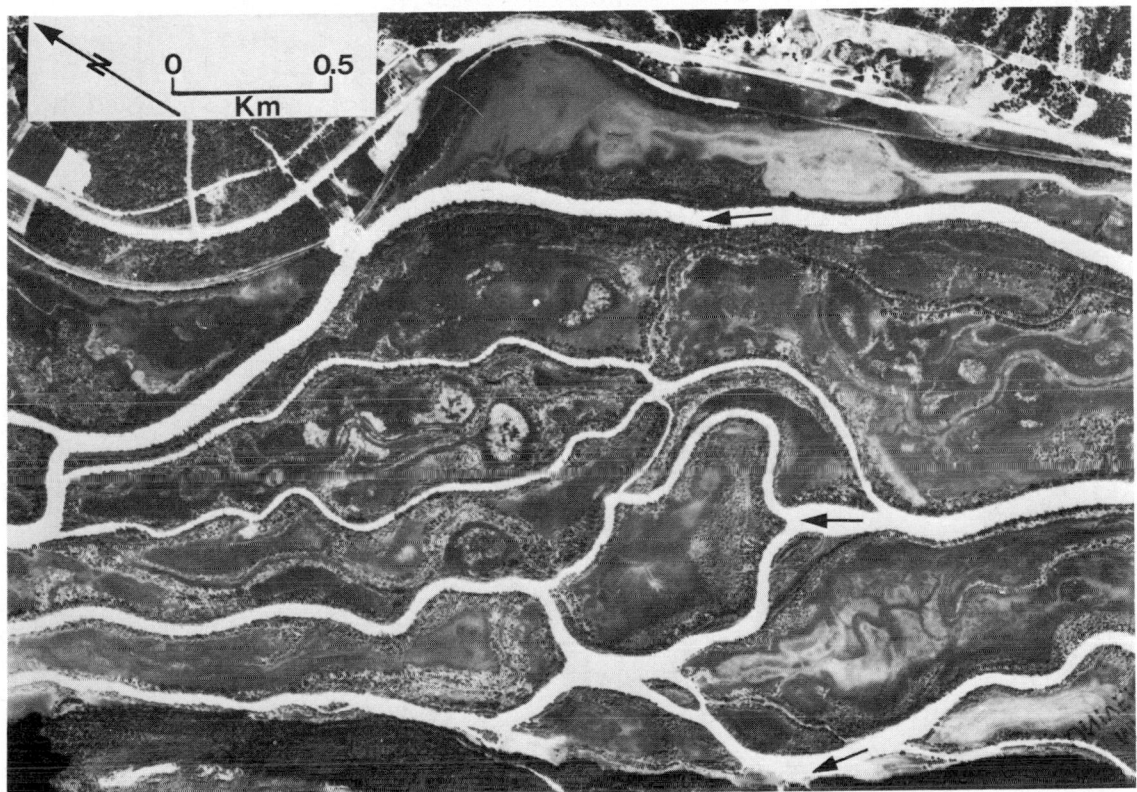

Fig. 2. Anastomosed reach of the Columbia River showing multiple, low-sinuosity, interconnected channels located 47 km south-east upriver from Golden. (Photo courtesy of B.C. Lands and Forest, B.C. 7883 No. 070.)

about the upper Columbia River. The lower Saskatchewan and Magdalena Rivers are compared with the Columbia in the latter part of the discussion.

Upper Columbia River

Located on the floor of the Rocky Mountain Trench between Radium Hot Springs and Golden, B.C. (Fig. 1), the north-west-flowing upper Columbia River (mean elevation 790 m) represents the most accessible known example of an anastomosed fluvial system in Canada. The channels and wetland complex (120 × 1·5 km) occupy a narrow intermontane valley-bottom flanked on the north-east by the Beaverfoot and Brisco Ranges (up to 2728 m elevation) of the Rocky Mountains and on the south-west by the glacier-capped Purcell Mountains (up to 3457 m elevation).

Within the 120 km of anastomosed channels, the morphologically and sedimentologically most spectacular reach is located between the communities of Spillimacheen and Nicholson (Fig. 2), a distance of 55 km. The deposits are supplied by large volumes of mixed-load sediment transported by snow and glacier meltwater in the Bugaboo Creek and Spillimacheen River tributaries. Downriver, both communities of Nicholson and Golden are located on large aggrading cross-valley alluvial fans fed by Canyon Creek and the Kicking Horse River; the latter is most active. Fan sedimentation from these tributaries keeps the anastomosed channels and wetlands active by raising local base level with time. It is doubtful if tectonic subsidence is active in the valley at present.

Flow discharge of the upper Columbia River is unmodified by man and fluctuates in accordance with rainfall and snow melt runoff. The drainage area above Nicholson, is 6600 km², with the average maximum and minimum monthly discharges of 325 m³ sec⁻¹ in July and 23 m³ sec⁻¹ in February respectively (Water Survey of Canada, 1980).

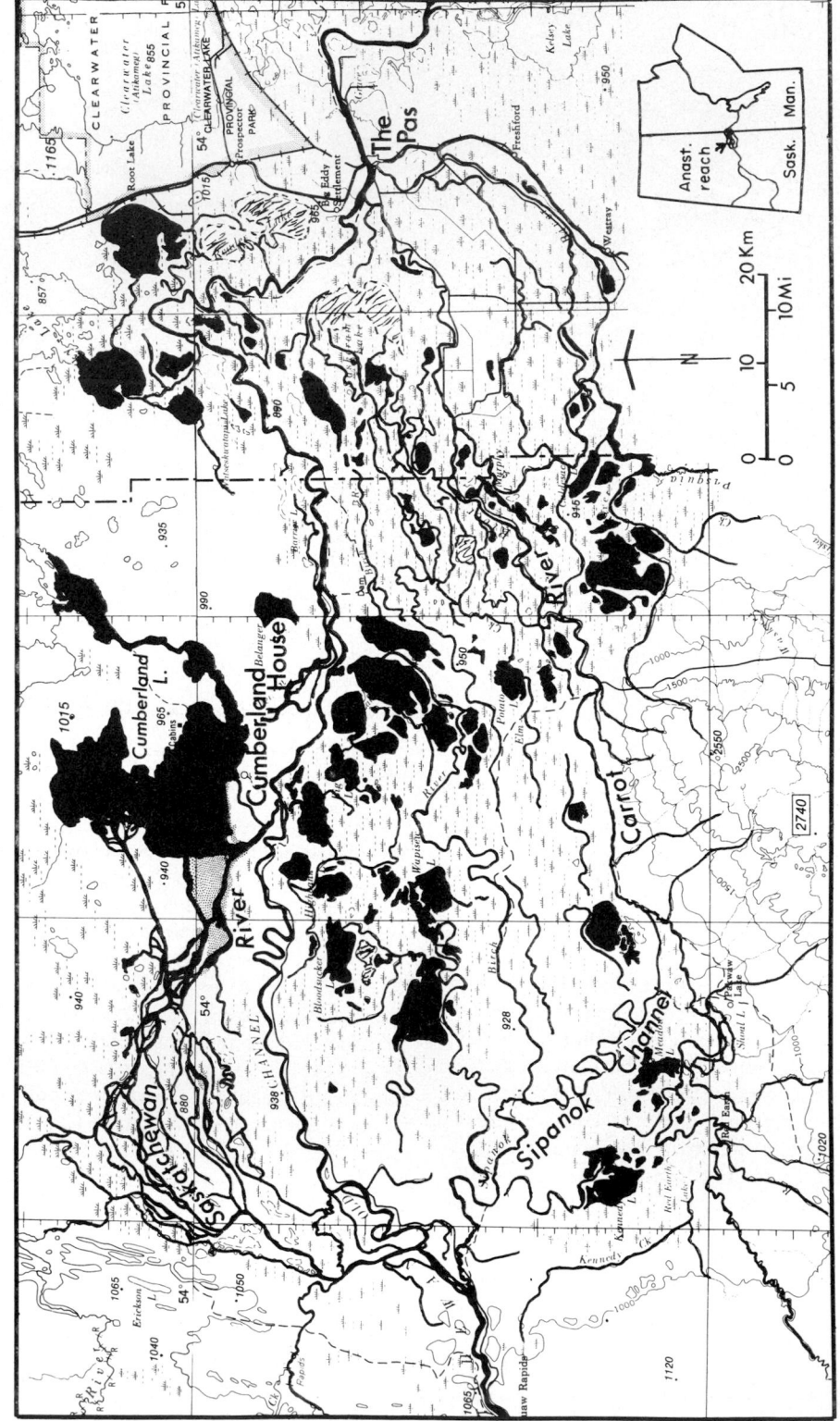

Fig. 3. Anastomosed channel complex of the lower Saskatchewan River.

ANASTOMOSED SEDIMENTARY ENVIRONMENTS AND SEDIMENTS

The same six anastomosed sedimentary environments and associated textural facies observed in the three small-sized mountain rivers by Smith & Smith (1980) occur in the larger-sized upper Columbia, lower Saskatchewan and Magdalena Rivers. These environments include three channel-related facies: channel, crevasse-splay and levée; and three wetland facies; lacustrine, marsh and peat bog.

Channels

The gross morphology and channel pattern distinguish anastomosing from meandering and braiding rivers. Anastomosing systems exhibit stable, low-sinuosity (initially), multiple, confined channels while highly sinuous, single channels are common to meandering rivers and unstable, shallow, multiple channels characterize braids (Table 1). The American Geological Institute Glossary of Geology describes 'braided' and 'anastomosed' as synonymous. However, in this paper and in previous works by Schumm (1968), Smith & Smith (1980), Smith & Putnam (1980) and Rust (1981), the term 'anastomosed' is defined very differently. The unmistakable differences in fluvial styles are compared in Tables 1 and 2. Because of the dissimilar characteristics between braided and anastomosed styles, only meandering is compared in Table 2.

Anastomosing processes are primarily related to the

Table 1. Distinctive river patterns and associated morphologies

Braided river	Meandering river	Anastomosed river
1 Unstable shallow braided channels	1 Highly sinuous single channel	1 Stable low-sinuosity multiple channels
2 Braid bars	2 Scroll bars	2 Prominent levées
3 Wide braid belt	3 Oxbow lakes	3 Extensive wetlands
	↓	↓ ↓
	Average meandering river Beatton River	Lower Saskatchewan River Cumberland Marshes / Upper Columbia River Radium to Golden

Morphologic characteristics (see Table 2)

Table 2. Detailed geomorphic comparison of a typical meandering river with the lower Saskatchewan and upper Columbia Rivers

	Morphology	Meandering river — Average meandering river Beatton R., NE, B.C.	Anastomosed rivers studied	
			Lower Saskatchewan River (Cumberland Marshes)	Upper Columbia River (Radium–Golden)
1	Gradient	Medium (30 cm km^{-1})	Low (12·2 cm km^{-1})	Low (9·6 cm km^{-1})
2	Sinuosity	High (2·1)	Medium (1·4)	Low (1·16)
3	Aggradation rate	Low (± 5 cm 100 yr^{-1}, est.)	Medium (5–30 cm 100 yr^{-1})	High (30–100 cm 100 yr^{-1})
4	Levées	Rare	Prominent	Prominent
5	Scroll bars	Common	Few	Rare
6	Oxbows	Common	Rare	Absent
7	Abandoned channels	Absent	Common	Common
8	Avulsions	Never	Few	Common
9	Crevasse splays	Absent	Few	Common
10	Meander belt	Wide	Confined	Absent
11	Shallow lakes	Rare	Common	Common
12	Peat bogs	Rare	Common	Few

Fig. 4. Anastomosed fluvial system in Columbia River valley, located 25 km upriver and south-east of Golden, B.C. Levée in foreground has been breached by a large crevasse, forming a major avulsion sand splay (light-coloured deposit) in the mid-distance. At lower right, a sand channel-fill is forming upriver from a log jam while downriver the channel is infilling with mud. At upper right, an inactive splay has been overgrown by trees and willows.

lateral stability of channels, a function of both very low channel gradients and cohesive fine-grained bank sediment and overhanging bank vegetation, which reduces flow velocity and erosion along banks. While Smith (1976) and Smith & Smith (1980) earlier emphasized bank vegetation as being an important factor in maintaining channel anastomosis, recent field work on the upper Columbia and lower Saskatchewan Rivers and by Rust (1981) suggests that channel stability is more related to low channel gradients and cohesiveness of bank sediment and not vegetation. Gradients and sinuosities in the Saskatchewan channels average 12·2 cm km^{-1} and 1·4, respectively, while in the Columbia they average 9·6 cm km^{-1} and 1·16 (Table 2). When these data are compared with the meandering Beatton River (Table 2) studied by Nanson (1980), anastomosed rivers appear to have little excess energy for lateral erosion and deposition, which may account for the low sinuosities.

In spite of the stable channel behaviour, anasto-

mosed systems are laterally dynamic in the form of major avulsions and channel filling. An avulsion of 1882 on the Saskatchewan River (Peel, 1972) is over 50 km long with several 'new' channels down-cutting through avulsion-splay sand and underlying wetland deposits. The former dominant channel (Old Channel) is filling (primarily narrowing) with sand and muddy sand throughout the entire 50 km reach (Fig. 3). In the Columbia River, a recent avulsion (about 1965), which started as a crevasse-splay, has now enlarged the crevasse to 40 m wide the avulsion channel scouring through the sand splay and cutting into underlying lacustrine mud deposits (Fig. 4).

Underlying muds and peats into which avulsion channels incise tend to confine the lateral freedom of the river, enhancing channel stability (Fig. 5). On the lower Saskatchewan River, a peat layer located on lacustrine mud and beneath 6·6 m of avulsion splay sand and levée deposits marks the approximate time of avulsion and channel incision at about 2000 yr BP (Fig. 5). Recently (last 100 yr), the Old Channel has

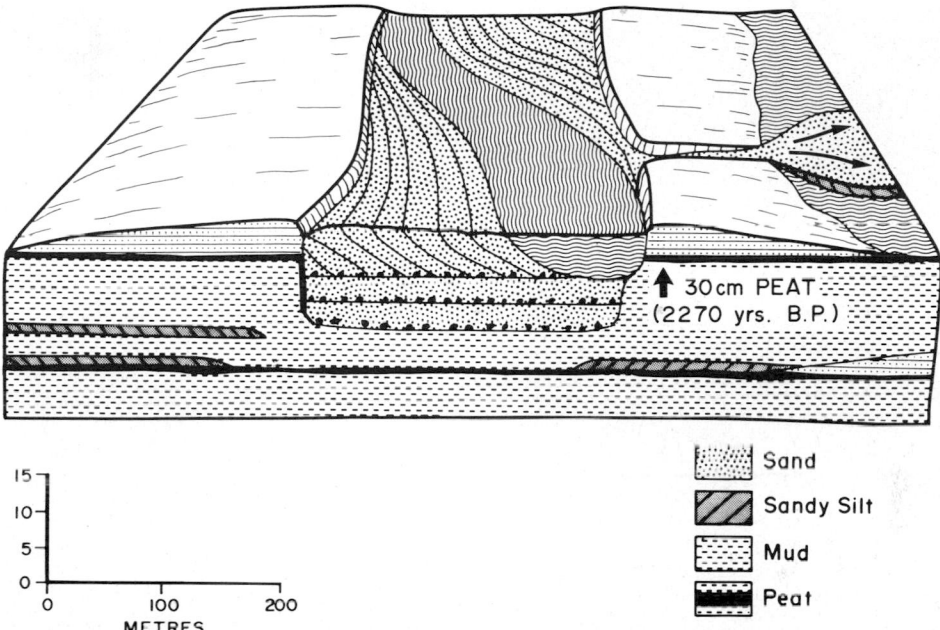

Fig. 5. Schematic reconstruction of the Old Channel infilling in the lower Saskatchewan River. Note how the channel incised 'locking' itself into cohesive lacustrine and peat deposits. The lower channel fill and levées were deposited during 2000 yr of aggradation time. Also note how the channel width decreases (infilling) forming meander scrolls (side bars) in accordance with reduced discharge caused by an avulsion upriver in the last 100 yr. The drill hole on the right bank corresponds to hole No. 5 in Fig. 10.

been infilling, caused by a major avulsion 40 km upriver (Fig. 3).

Oxbow lakes and mud-filled oxbows, common to meandering rivers, are absent in anastomosing systems. There is obviously a transition of morphology between meandering and anastomosing. Some channels through the Saskatchewan River wetlands are classified as close to the meandering transition, because some channels exhibit highly sinuous meanders, while in the Columbia Valley there is a near absence of meandering (Fig. 6).

Bedforms in both the Columbia and Saskatchewan channels are dominated by sand waves (Fig. 6) in river depths varying from 2 to 15 m (indicated by echo-sounding equipment). Trenches excavated into channel fills during low water and drill cores show planar tabular cross-bed sets. Within the crossbeds, multi-storied fining-upward textural sequences are spaced about 1 m apart. The multi-storied sequences are interpreted as flood cycle deposits during the channel aggradation and final infilling phases (Fig. 6). In the Saskatchewan channels, mean sediment size is medium sand, while due to the mountainous proximity coarse sand to granules dominate the Columbia channels.

Anastomosing channel deposits are characteristically thick (5–15 m), narrow, interconnected stringers of sand (Figs 2 and 4), and when buried they are encased laterally by sandy-silt levée deposits and with mud or peats both above and below. The dramatic lateral and vertical facies changes make the channel deposits unique when compared to other river styles which deposit laterally extensive sheets of point bar and braid bar sands. Thickness of many Columbia anastomosed channel deposits exceeds 10 m, with most fills resting on scoured mud or peat bases. The remarkable thickness of active channel deposits is supporting evidence for vertical aggradation rather than lateral accretion as the dominant sedimentation pattern.

In straight river reaches (Fig. 5) channel aggradation deposits are vertical, but narrow point bars (side bars) form on the inside of some low sinuous channels (Fig. 6). Beneath most low-sinuosity channel bends a diagonal cross-sectional channel-fill geometry is a result of both very slow lateral accretion rates (estimated at 1 cm yr^{-1} based on 2 years of measurement) and channel aggradation 0·5 cm yr^{-1} (calculated from buried volcanic ash 15 m beneath the surface dated at 2500 yr BC).

Fig. 6. A channel fill of sand in an anastomosed channel. Note the sand waves on the channel bed. A lacustrine environment is located to the left, and marsh on the right side of the photo.

Crevasse splays

Sandy crevasse splays, which prograde into wetlands, are common features of anastomosing systems. These form when overbank flow cuts a small channel (crevasse) through a levée, then transports and deposits bedload as a lobate sheet of sand (splay) into a wetland environment.

Crevasses (levée breaks) are erosional features which often initially form in beaver drag trails located between the river and wetland. If the levée is relatively narrow, a high gradient flow will quickly sluice a deep crevasse which will allow the transport of channel sand into the wetland basin. Crevasse gradients are controlled by the distance downriver where both the wetland flow returns to the river and wetland water levels are equal to the river stage. The greater the distance between crevasse and location of return flow, the greater the crevasse flow gradient and the better the chance of a larger splay deposit.

Each sheet-like splay deposit of sand has a complex sedimentation history; however, they generally follow a consistent pattern of a coarsening-upward then fining-upward. As the splay progrades into the wetland, a basal unit of coarse silt and very fine sand up to 20 cm thick is deposited from suspended load well beyond the prograding front of sheet sand directly on lacustrine, marsh or peat deposits (Fig. 4). Next, organic waterlogged vegetal litter (up to 10 cm thick) transported as bottom load, is deposited in the 'quiet water' at the toe of a foreset slope of thick prograding sand. The organic litter bed is buried between the underlying silty fine sand and overlying coarse sand. Then prograding medium to coarse sand as bedload in the form of current ripples, dunes and transverse bars deposits the main splay sand sheet.

Many splays usually stop enlarging when the crevasse becomes plugged by a log jam, but sometimes the splay becomes so large that flow resistance increases with decreased flow gradient and transport velocity. In either case flow velocity and sedimentation rates decrease as do sediment volume and grain size which results in an upper splay fining-upward texture trend. The crevasse then fills with sand and the splay is capped with fine-grained mud from overbank suspended load. Finally, vegetation invades the splay, which is topographically favourable for tree and shrub growth (Fig. 4). Thus, large splays 'grow, live and die' over a period of perhaps 100 yr. Field evidence in the Columbia River suggests that some splays become inactive because of the high ratio of suspended to bed load. Because of lacustrine sedimentation and rapid invasion of aquatic plants ahead of the prograding sand sheet, the rate of hydraulic efficiency is reduced, and thus sand deposition is sometimes terminated. In such cases, the sand sheet usually thins as it prograges on to a thickening lacustrine deposit. However, most splays become inactive by loss of hydraulic efficiency due to log jams and/or decreased flow gradients.

The geometry and sedimentology of the splay environment are the most variable of the six anastomosed depositional environments studied in the Columbia Valley. The planform is usually lobate, with several sublobe extensions along the distal margin. The aerial extent of the larger splays varies from 0·3 to 1 km². Proximal deposits are consistently about 2–3 m thick, and thin distally to about 0·3 m at a distance of 0·5–1·6 km from the crevasse. Grain-size distribution at the splay base consists of laminated coarse silt and fine sand (20 cm) overlain by an organic litter bed (10 cm) which is generally overlain by a thick sequence (1–2·5 m) of coarsening-upward medium grain to granule-sized sediment with current ripples and a few high-angle cross-bed structures. Thin (1–3 mm) organic lenses in ripple troughs are abundant throughout. Grain size in the upper (0·5–1 m) splay deposit fines up into overbank mud, on which vegetation thrives, extensively rooting the upper splay deposits.

Vegetated inactive splay deposits (Fig. 4) are encased in mud or peat, except at the crevasse where ground water can move freely between the channel and splay deposits. Cores from a vibra-corer indicate that multiple stacked splay sequences were rather common, usually separated by 1–3 m of mud.

Levées

Prominent levée deposits which flank the channels stand out morphologically because they are so densely forested (Figs 4 and 6). Levée widths and heights vary from 50 and 2·5 m respectively in the Columbia Valley to 1 km and 4·0 m respectively in the lower Saskatchewan River. Sediments consist of laminated (1 cm thick) fine sand and silt with occasional organic lenses. Laminations conform with the surface topography. A dense growth of willow and other shrubs overhand the channel margins, protecting the banks, in part, from river erosion. Deposition processes, morphology and sedimentology of levées generally lack variability and complexity. Laterally, grain size tends to decrease with increased distance from the channel bank. Vertically, levées (overbank deposits) do not exhibit grain-size trend except near the levée base. Here the levée rests on a sandy splay deposit, which represents the initial avulsion-channel-forming stage (see hole no. 5, Fig. 10).

Lakes

Extensive (up to 50 km², Saskatchewan River) shallow (1–3 m deep) lakes are common to anastomosed systems (Fig. 3). The Lakes are irregular to round in shape. The large number suggests that lake sedimentation lags behind that of channel and levée deposition (left side, Fig. 6). Because of the low elevation of the lakes relative to the channels, lakes are likely to be future avulsion-channel routes. Most lakes contain some laminated clay and silty clay, although most deposits are non-laminated, suggesting bioturbation by bottom-feeding waterfowl and ungulates (moose). Lakes in close proximity to channels tend to contain clay–silt deposits. As with levées, lacustrine environments lack variability.

Marshes

Backswamps referred to by Smith & Smith (1980) will be termed marshes here. In the Columbia River, marsh deposits consist of bioturbated and sometimes thin laminated organic and clastic mud (silty mud and muddy silt) deposits. Marshes do not drain seasonally, and thus maintain a dense aquatic flora. However, marshes occasionally receive floodwater charged with suspended load. Hence marsh deposits are mixtures of clastic sediment and organic black-coloured mud.

Peats

Extensive bogs (30 km²) of varying thickness up to 3·5 m and still active have been observed in the Saskatchewan marshes (Dirschl, 1972; D. Phillips and N. D. Smith, personal communication). In the Columbia valley, less extensive and thinner (maximum 1·3 m) peats have been encountered.

AGGRADATION RATES OF ANASTOMOSED SYSTEMS

Three modern large-sized anastomosed river systems may be compared; the lower Saskatchewan River in a plains setting, and the upper Columbia and Magdalena Rivers located in intermontane basins—the latter is located in Colombia, South America. The upper Columbia in British Columbia, Canada, is the most understood; the Saskatchewan is currently (1981–82) under study (N. D. Smith and T. Cross); abundant information is available about the Magdalena, mainly from a published paper by Tanner (1974) and 17 unpublished volumes of a consulting report by Proyecto Colombo-Holandes de la cuenca Magdalena-Cauca (1977).

Aggradation rates in the Columbia River were determined from a buried layer of volcanic ash which has been identified as Bridge River tephra from a vent located about 160 km north of Vancouver. From other localities, organic material adjacent to the tephra layer has been dated at 2500 yr BC using C-14 dating. In the Columbia valley, the tephra is buried at an average depth of 15 m beneath the surface. Based on the tephra, an average aggradation rate of 60 cm $(100 \text{ yr})^{-1}$ has been calculated for the anastomosed channel and wetland deposits of the Columbia.

In the Saskatchewan anastomosed deposits, a 50 cm thick peat was C-14 dated at 2270 yr BP, buried 6·5 m beneath a levée and avulsion splay deposit (Fig. 10, hole no. 5). While an aggradation rate of 29 cm $(100 \text{ yr})^{-1}$ can be calculated, the rate is too high and is not an average, because levée (alluvial ridge) deposition occurs at a greater rate as compared to the wetlands and the basin average. Perhaps 15 cm $(100 \text{ yr})^{-1}$ is a more realistic estimate of aggradation for the Saskatchewan. Also the 2270 yr date is an average for 50 cm of peat (i.e. at 25 cm into the peat) thus 25 cm of deposition years must be subtracted.

In the Magdalena channels and wetlands of Colombia, South America, sixteen C-14 dates from buried material recovered from drill cores from depths of 28·5 m below the surface (Proyecto Colombo-Holandes de la Cuenca Magdalena-Cuaca, 1977) resulted in an average aggradation rate of 38 cm $(100 \text{ yr})^{-1}$ during the last 7500 yr. Basin subsidence is cited as the main cause of continued aggradation.

In all three depobasins sedimentation rates are episodic in different parts of the basin. For example, for many years some areas of the Saskatchewan system aggrade no faster than the production rate of peat, while in other areas in and near active channels

aggradation is in accordance with overbank deposition. Both areas may be only several kilometres apart, but one is aggrading at perhaps 8 cm $(100 \text{ yr})^{-1}$ while the other deposits at 30 cm $(100 \text{ yr})^{-1}$.

While all material used for dating was recovered from drill cores from levées, average aggradation rates can be extended for all sedimentary environments over the entire basin with some consideration for depth and knowledge of the system. Shallower material will be less reliable, as was discussed in the Saskatchewan, while deeper material should be more reliable, as in the Magdalena. Because of the fact that all dates came from levée environments it could be argued that levée sedimentation is greater than average and calculated average sedimentation rates are too high. It should also be pointed out that anastomosed systems are dynamic, and it is unlikely that any channel will survive much longer than 2000 yr (i.e. Old Channel, Saskatchewan River). Therefore, sedimentation rates for the Magdalena spanning 7500 yr of time are probably the most reliable of the three river systems.

Rapid sedimentation rates are the norm for anastomosed river systems and vary between 60 and 15 cm $(100 \text{ yr})^{-1}$. Assuming that an average sedimentation rate of 30 cm $(100 \text{ yr})^{-1}$ in some ancient anastomosed fluvial rock sequences were to occur uninterrupted over a 1,000,000 yr time period, 3000 m of sediment would accumulate, not considering compaction. Such sedimentation rates seem remarkably high. Perhaps the anastomosed style of river behaviour and deposition is short lived, representing an extreme in sedimentation and subsidence. Thus it would seem reasonable to expect the occurrence of ancient anastomosed river deposits to be associated with periods of nearby active mountain building and erosion, provided that mountain orogenesis and subsidence of adjacent depobasins are synchronous in time.

ANASTOMOSED FACIES ASSOCIATION

Facies associations in the subsurface of anastomosing river deposits were determined from cores taken by a vibra-corer, split spoon sampler and soil auger. Results from the analysed cores reflect the same sharp facies changes and associations in the vertical sequence that one sees horizontally at the surface.

A series of core logs taken from the Columbia anastomosed floodplain at Harrogate, B.C., in cross-section (H ×) shown in Fig. 7, illustrate the variety of facies (i.e. sand splays and peats) with depth.

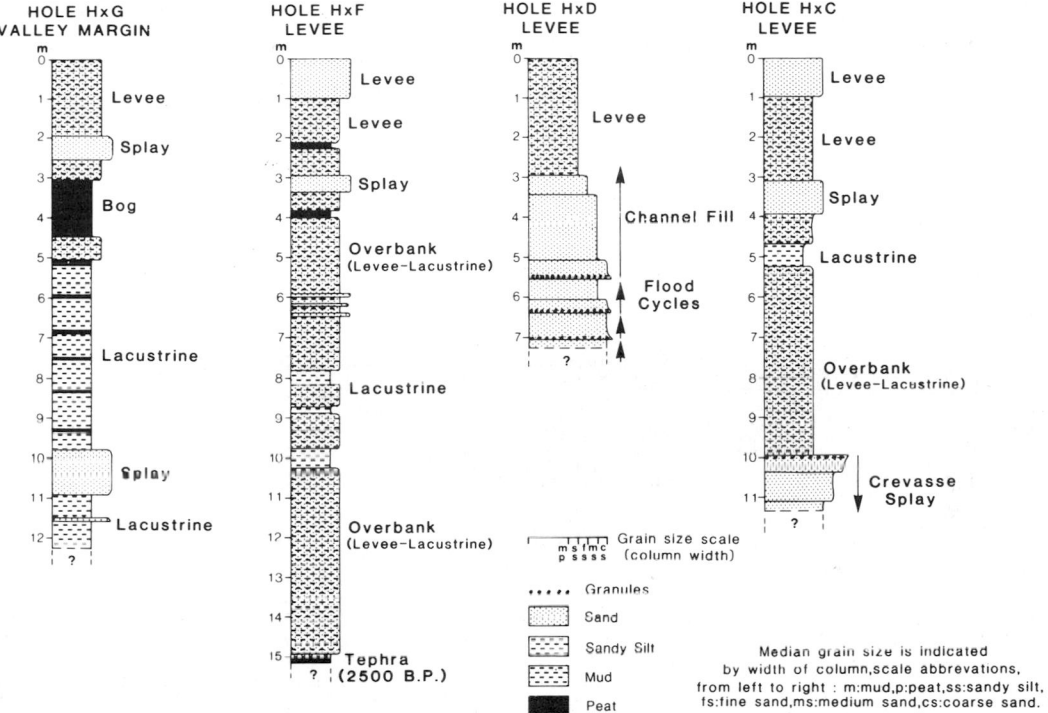

Fig. 7. Four core logs showing typical anastomosed facies changes and interpreted sedimentary environments in the Columbia River. Locations of the logs are indicated by corresponding letters on Fig. 8.

The four core logs labelled with capital letters (G, F, D and C) shown in Fig. 7 are located and interpreted along with other log data in Fig. 8. Of particular interest is hole H × D, which has a channel fill beginning at 3 m deep and extending to an unknown depth. It is interpreted as a channel which had slowly migrated laterally while at the same time aggrading, thus depositing a diagonal-shaped sand body (Fig. 8). The expected thick point-bar deposits common in meandering rivers are generally rare to absent, resulting in bed deposits as a channel fill. Such fills are multi-storied textural cycles with stacks of planar tabular crossbeds.

The other three holes (G, F and C) show sand splay facies averaging about 1 m thick. Thicknesses of most active splays deposits vary between 3 and 0·3 m from proximal to distal sites, respectively.

Peats in the Columbia valley are rare, perhaps due to the narrow valley width restricting the space for extensive wetlands. In the lower Saskatchewan anastomosed deposits, peats are common, ranging up to 3·5 m thick (Dirschl, 1972) and are laterally extensive; here, however, aggradation is slower (estimated at 15 cm $(100 \text{ yr})^{-1}$).

The anastomosed facies associations of the Columbia valley are shown schematically in longitudinal profile in Fig. 9. Note how the buried channel sands and laterally adjacent levée deposits are encased in wetland deposits. A three-dimensional view would show multiple, interconnected, low-sinuosity stringers of channel sand with occasional sheets of splay sand extending laterally.

In a more laterally extensive sedimentary basin such as the lower Saskatchewan, the same facies associations occur, except that peats are more laterally extensive and thicker (Fig. 10). The greater accumulation of peat in the Saskatchewan basin as compared to the Columbia is interpreted as a response to slower sedimentation rates and the fact that extensive peat bogs become isolated from the active channels for perhaps thousands of years. Such a fluvial style would satisfy all of the necessary conditions for thick peat accumulation and burial. Such a style might also help to explain the deposition of some thick non-marine coal sequences in the ancient rocks (i.e. Flores, 1979).

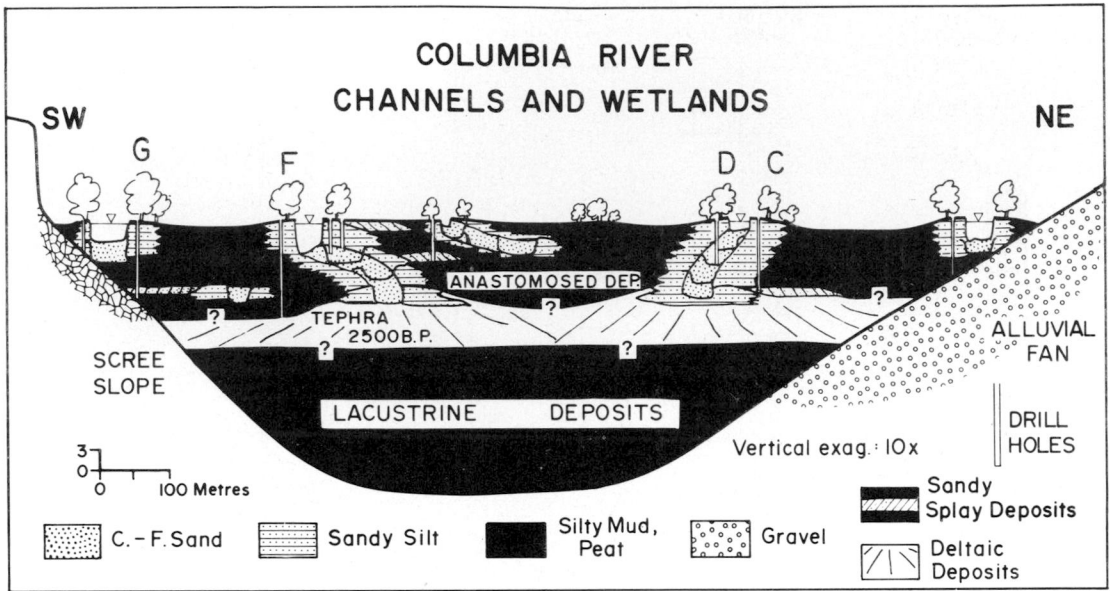

Fig. 8. A reconstruction of anastomosed subsurface deposits from a valley cross-section located near Harrogate.

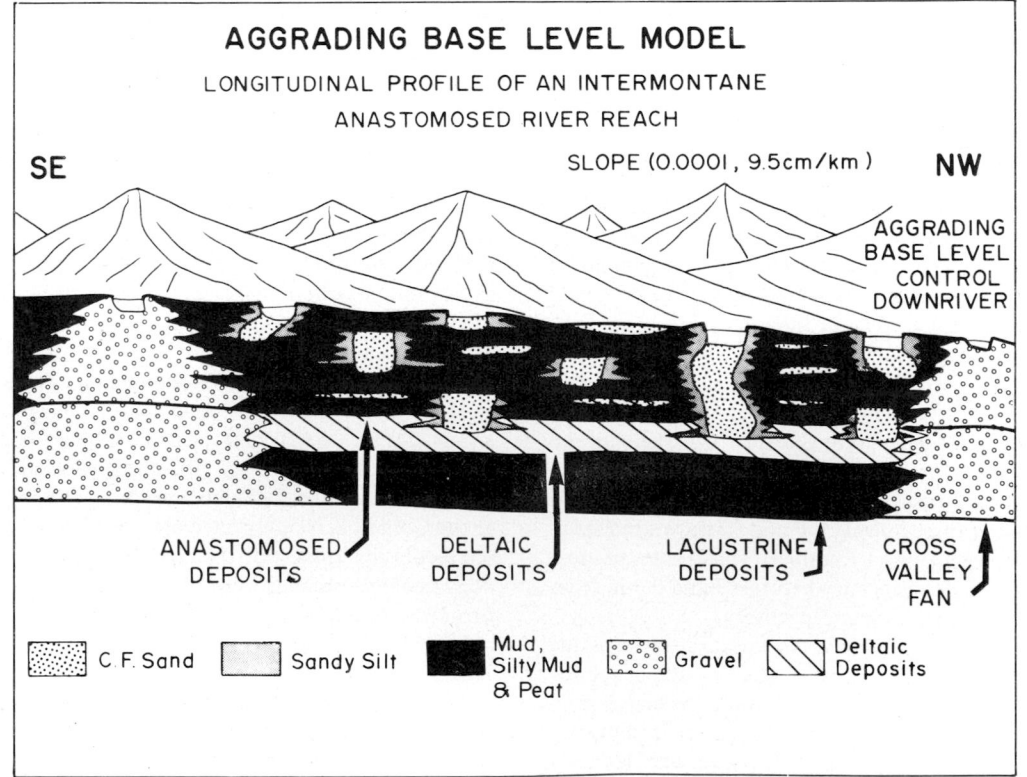

Fig. 9. Schematic diagram of the longitudinal profile of the Columbia River deposits based on 60 drill-hole core logs. Note the channel and levée deposits encased in mud.

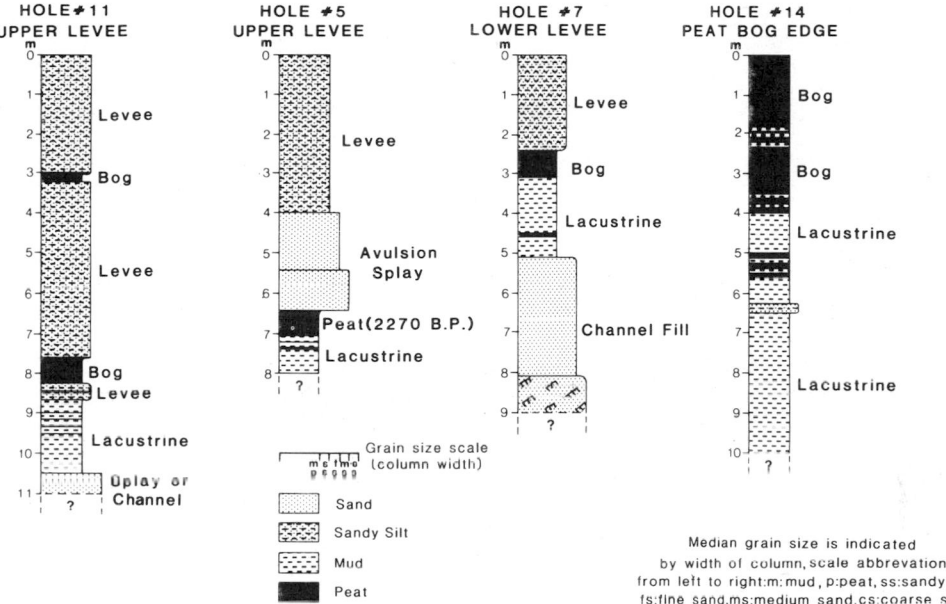

Fig. 10. Four core logs from the lower Saskatchewan River anastomosed channel reach. Note the abundance of peat in each log. Hole No. 5 is located on the right bank in Fig. 5.

CONCLUSIONS

Based on the association between surface morphology and subsurface deposits, aggrading anastomosing fluvial systems are distinctive, displaying a unique assemblage of sedimentary environments and textural facies. A comparison with meandering and braiding river systems shows that anastomosing systems have a relatively high percentage (60–90) of associated wetland (lacustrine marsh and peat) to channel deposits.

Geomorphically, anastomosing systems compared to meandering and braiding have lower channel gradients and sinuosities, higher aggradation rates, prominent levées, few scroll bars, no oxbows and many avulsions, crevasse splays and abandoned channels.

The channel and splay sand facies associated with wetland deposits are very similar to low-energy river-dominated deltas and could easily be interpreted as such. With this in mind, the evolving development of the anastomosed river deposition model may help to explain some anomalies in ancient rock sequences such as meandering river and deltaic-like rock sequences located in structural settings lacking former water bodies or tectonic stability. The model may also help to explain the formation of some non-marine coal sequences.

ACKNOWLEDGMENTS

Encouragement by and discussions with N. D. Smith, A. Miall, P. Putnam, D. Long, R. McLean, P. Glaister, M. Lerand, T. Cross, B. Rust, D. Cant, F. Hein and many others to continue field research on anastomosed river systems with the goal of developing a depositional model are gratefully appreciated. Help by field assistants R. Goreckie, O. Holdenrich, A. Quin and M. Kavanagh during the summers of 1980 and 1981 and drafting by M. Styk is also appreciated. Support was kindly provided by the National Science and Engineering Research Council of Canada, Esso Resources Canada, Shell Resources Canada and Gulf Resources Canada.

REFERENCES

DIRSCHL, H.J. (1972) Geobotanical processes in the Saskatchewan River delta. *Can. J. Earth Sci.* **9**, 1529–1549.

FLORES, R.M. (1979) Coal depositional models in some Tertiary Cretaceous coal fields in the U.S. Western Interior. *Org. Geochem.* **1**, 225–235.

NANSON, G.C. (1980) Point bar and floodplain formation of the meandering Beatton River, northeastern British Columbia, Canada. *Sedimentology,* **27**, 3–29.

PEEL, B. (1972) *Steamboats on the Saskatchewan.* Prairie Books, Saskatoon, Saskatchewan, Canada. 238 pp.

Proyecto Colombo-Holandes de la Cuenca Magdalena-Cauca (1977) Himat, Bogota, Colombia. 17 v.

PUTNAM, P.E. & OLIVER, T.A. (1980) Stratigraphic traps in channel sandstones in the Upper Mannville (Albian) of east-central Alberta. *Bull. Can. Petrol. Geol.* **28**, 489–508.

RUST, B.R. (1981) Sedimentation in arid-zone anastomosing fluvial system: Cooper's Creek, Central Australia. *J. sedim. Petrol.* **51**, 745–755.

SCHUMM, S.A. (1968) Speculations concerning paleohydrologic controls terrestrial sedimentation. *Bull. geol. Soc. Am.* **79**, 1573–1588.

SMITH, D.G. (1976) Effect of vegetation on lateral migration of anastomosed channels of a glacier meltwater river. *Bull. geol. Soc. Am.* **87**, 857–860.

SMITH, D.G. & SMITH, N.D. (1980) Sedimentation in anastomosed river systems: examples from alluvial valleys near Banff, Alberta. *J. sedim. Petrol.* **50**, 157–164.

SMITH, D.G. & PUTNAM, P.E. (1980) Anastomosed river deposits: modern and ancient examples in Alberta, Canada. *Can. J. Earth Sci.* **17**, 1396–1406.

SMITH, D.G. & PUTNAM, P.E. (1981) Anastomosed river deposits: modern and ancient examples in western Canada. *Reservoir, Can. Soc. Petrol. Geol.* (abstract), **8**, 1–2.

TANNER, W.F. (1974) The incomplete flood plain. *Geology*, **2**, 105–106.

WATER SURVEY OF CANADA (1980) *Historical Streamflow Summary of British Columbia*. Environment Canada, Ottawa. 861 pp.

Spec. Publs int. Ass. Sediment. (1983) **6**, 169–180

Pattern of instability in a wandering gravel bed channel

MICHAEL CHURCH

*Department of Geography, The University of British Columbia, Vancouver,
British Columbia V6T 1W5, Canada*

ABSTRACT

Bella Coola River drains 5450 km² (7% glacierized) of the Coast Mountains and Fraser Plateau of British Columbia, whence early summer nival floods and autumn rainstorms may yield flows greater than 1000 m³ sec⁻¹. The river flows over early post-glacial alluvial cobble gravels. Sediment entrained from the river banks and delivered from tributaries is stored in the channel principally in 'sedimentation zones' where the river is laterally unstable. These are connected by stable, cobble-paved 'transport reaches'. Overall morphology is that of a 'wandering gravel river'.

A sequence of maps beginning in the late nineteenth century shows that the river has become more stable and that the locus of lateral instability in the lower course has progressed downstream from near 25 km *ca.* 1900 to near the mouth today. One of two mechanisms might be invoked to explain the effect:

(i) introduction of unusual volumes of sediment into the main channel by the erosion of Neoglacial (eighteenth- and nineteenth-century) moraines of alpine glaciers, now becoming exhausted;

(ii) recent progradation of the alluvial fan at Nusatsum River constricting the main channel and blocking sediment transfer downstream.

The pattern is disturbed by sediment yield to the main channel from occasional extreme floods in tributaries.

INTRODUCTION

We are used to the notion that alluvial rivers modify their channel pattern by bank erosion and deposition of sediment in channel bars. In most rivers, the process appears to proceed essentially continuously downstream so long as the channel remains unconfined, and to be reasonably orderly. Variations in alluvial materials and in hydrology modulate the channel pattern from reach to reach. This paper describes a cobble-gravel alluvial channel whose behaviour departs from this orderly pattern.

Bella Coola River sporadically transports a moderate bedload volume which is deposited locally along the channel in 'sedimentation zones'. Here the river is laterally unstable: it is subject to avulsion and tends to occupy serially several flood channels. By comparison, other reaches appear to be stable in a single channel. The river is not evidently aggrading or degrading. The possibility for this type of behaviour was brought to attention by Leopold & Wolman (1957). Channels of this character appear to be common in upland and mountain valleys, particularly ones that have been glaciated relatively recently. They have been nominated 'wandering gravel channels' by Neill (1973).

SETTING AND HYDROLOGY

Bella Coola River drains 5450 km² of the central Coast Mountains and northern Fraser Plateau of British Columbia (Fig. 1). The river flows 65 km west to the sea from the confluence of its major tributaries, Atnarko and Talchako Rivers. In this study the lowest 25 km will be examined (Fig. 2).

The granitic, metasedimentary and volcanic rocks have been heavily glaciated, producing deep, straight troughs between high, craggy ridges. Bare rock faces

0141-3600/83/0106-0169 $02.00

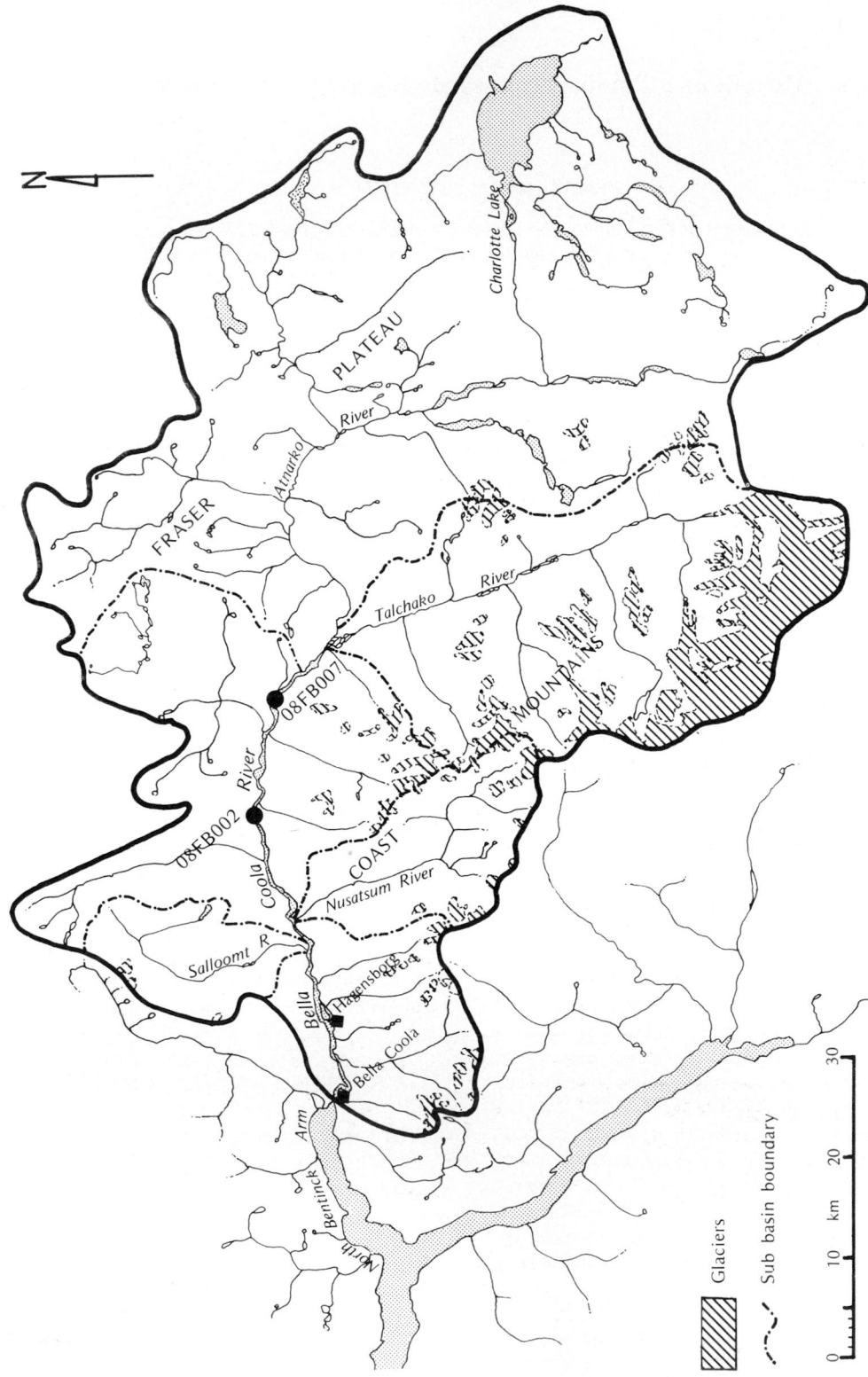

Fig. 1. Bella Coola River basin. The designations 08FB indicate gauging stations of the Water Survey of Canada.

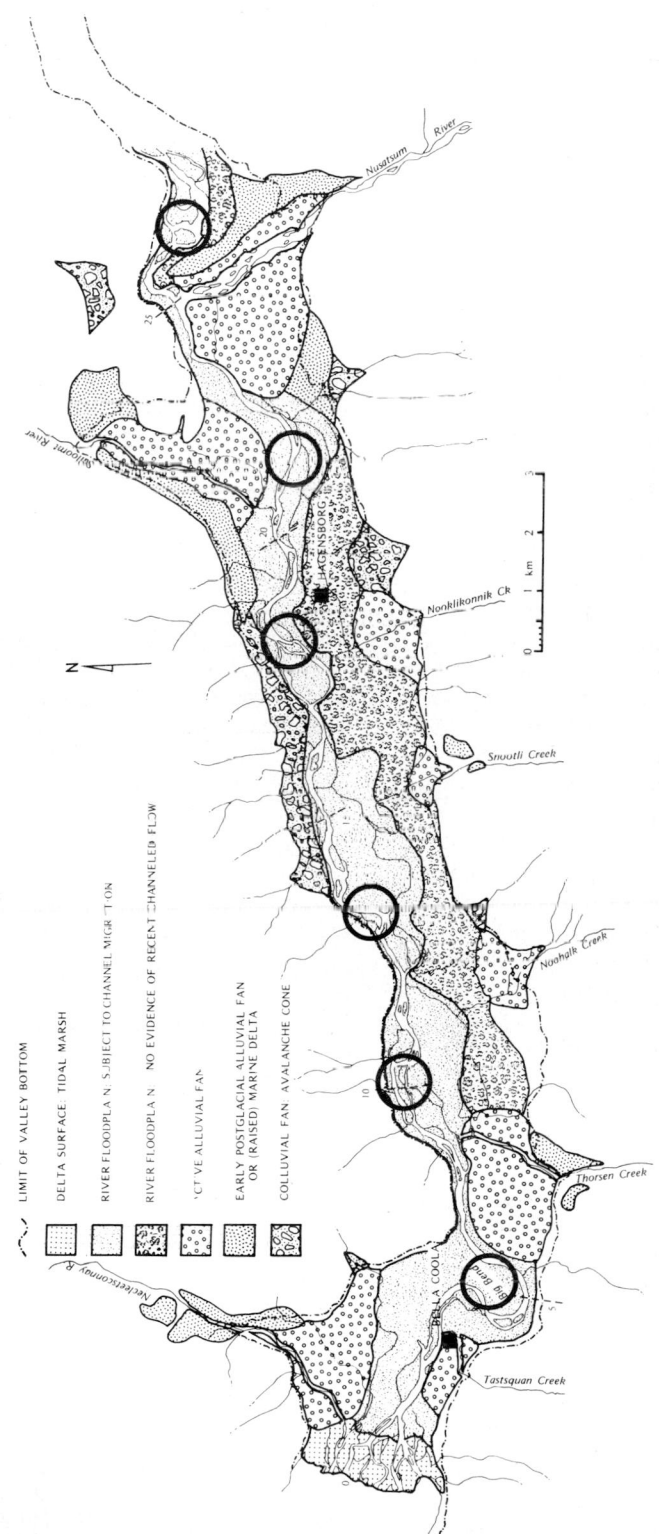

Fig. 2. Morphological map of the study reach. The circled areas indicate the main sites of historical instability of the river: the numbers along the river course indicate kilometres upstream from the river mouth.

are common, with moraine, boulder-gravel alluvial fans and cobble-gravel river deposits in the valleys. The area was deglaciated about 11,000 years ago and there was marine inundation of the valley. However, present sea-level was reached about 7500 yr BP (Andrews & Retherford, 1978). Most of the alluvial sedimentation occurred shortly after deglaciation. Despite the relative stability of the surface environment during the past 9000 years, sufficient coarse material is mobilized—from exposed Pleistocene deposits, from more recent, Neoglacial moraines, and from present-day mass wasting—to maintain considerable sediment transport and permit local channel aggradation. Unlike rivers in less active sedimentary environments in British Columbia, no post-glacial degradation has occurred.

At present, 6·9% of the basin is glacier covered. Virtually all the ice lies within Talchako River drainage and covers one-third of the area of that basin, where the highest peaks exceed 3000 m.

Lying on the east side of the Coast Mountains, the basin experiences a steep decline of precipitation from west to east—from 2250 to 300 mm according to valley climate stations. Prolonged, high, snowmelt flows occur every summer, frequently augmented by rain. In optimum conditions, about 40 mm per day of water equivalent may be melted from the snowpack in the latitude of Bella Coola basin, but because of the large elevation range and variable snowpack condition, not all of the basin is apt to contribute so much runoff at once. Autumn and early winter floods derive from heavy rainstorms, sometimes combined with melting snow in a saturated basin. Such floods are brief but intense. The highest flow on record is estimated at 1050 m³ sec⁻¹ daily mean flow on January 23, 1968, at gauging station 08FB002 just above the study reach. Climatic data indicate that more severe conditions may occur, and two greater floods are known, from 1934 and 1936.

THE CHANNEL PATTERN AND ITS RELATION TO SEDIMENT MOVEMENT

Bella Coola River typically exhibits a wide, shallow, cobble-gravel channel with irregular channel pattern (Figs 2 and 3). A sequence of cross-sections surveyed in the vicinity of Hagensborg (Tempest, 1974) yields the following mean geometry:

bankfull area (A): 391 m² (range: 180–868 m²),
water surface width (w_s): 157 m² (range 67–219 m),
mean depth ($A/w_s = d_*$): 2·5 m,
width–depth ratio (w_s/d_*): 72 (range: 31–165).

These data correspond to a flow of about 565 m³ sec⁻¹ at the gauge.

The natural river banks consist of alluvial gravels overlain by 0–2 m of overbank sand, or of lag cobbles or non-alluvial materials: in recent years, guide banks and dykes have been constructed in several reaches in an effort to promote stability.

The channel exhibits a pool–riffle sequence, the riffles being defined by alternating diagonal bars or by medial bars in the centre of the channel. Average riffle separation in the 25 km study reach is 5·8 w_s, close to the modal value for gravel rivers. The river pattern alternates between a single, irregularly sinuous channel and reaches that are braided or consist of several flood channels with islands.

Figure 4 illustrates the long profile of the river: the average gradient is 0·0025. Also marked on this diagram are the locations of the main alluvial fans in the lower valley and the unstable reaches with multiple channels. The association is obvious, though not invariable. Coarse sediment accumulates in the laterally unstable reaches, which have been called 'sedimentation zones' (Church & Jones, 1982). In lower Bella Coola River, they have a near-regular spacing of about 4 km (Fig. 2). Bed material in the sedimentation zones consists of a wide range of materials from medium sand up to cobbles, only locally well sorted and evidently periodically mobile. By comparison, the connecting single-channel reaches are floored by imbricated cobbles that form a stable surface, with minor accumulations of mobile sediments in bars.

HISTORICAL CHANGES ALONG THE RIVER

Vertical stability

Local changes in bed elevation accompany accumulation and removal of sediment in river bars and in the main sedimentation zones along the river. However, the overall vertical stability of the river requires further consideration.

Three pieces of evidence suggest that the channel has remained essentially at the same elevation for a considerable period. An old fish weir at km 20·3 was noted by Tempest (1974) as evidence that the river had not changed its vertical position there for 40 years. In the stable reaches, bars are paved by large, tightly imbricated, iron-stained cobbles. These stones have been neither disturbed nor buried for many years. Finally, in some places on the adjacent floodplain,

Fig. 3. Illustrations of channel morphology in Bella Coola River. The photographs were made on April 21, 1977; $Q \sim 40$ m^3 sec^{-1}. (A) View upstream at Big Bend showing the effect of rapid channel migration (left to right) following the September 1973 breakthrough (see also Fig. 6). Note the large volume of coarse sediment, much of which was previously present in the meander point bar, and extensive wood debris. In the background is the stable Reserve Reach, which is guided by a right-bank bedrock spur at the upstream end (photo by S. O. Russell). (B) View of the sedimentation zone upstream from the confluence with Thorsen Creek. The braided/anastomosed habit, complex medial bars and prominent wood debris are all typical (photo by S. O. Russell).

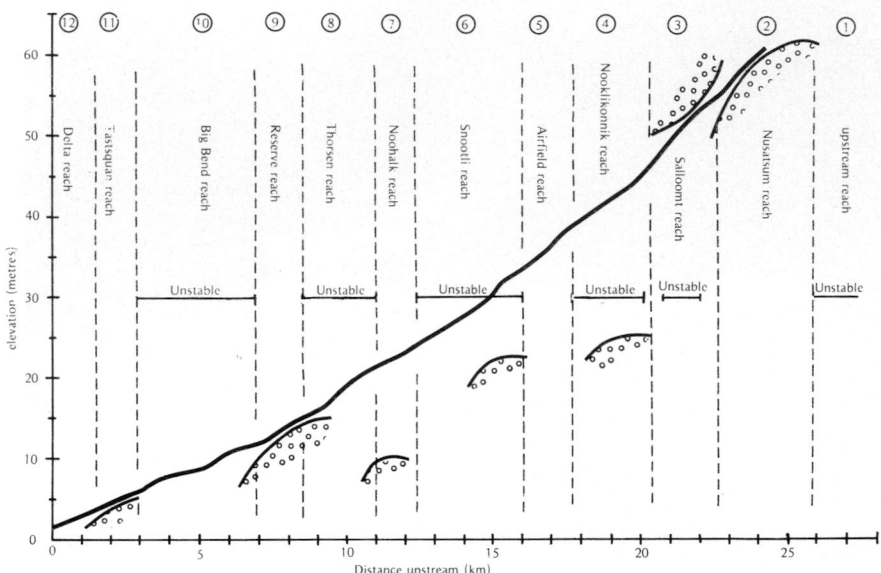

Fig. 4. Water surface profile for $Q \sim 100$ m³ sec⁻¹ (from Tempest, 1974). Reach limits (see also Fig. 5) and tributary alluvial fan positions are shown. The distance of the fan symbol from the profile is intended to give a qualitative impression of how closely the fan approaches (hence may constrict) Bella Coola River.

long-abandoned channels still flood at high water: neither aggradation nor degradation has isolated them completely, although they have filled up with sand.

Lateral stability

The first survey of the river was the legal survey of 1889–93. The river banks were accurately determined. Though features within the channel zone were mainly sketched, the expert planimetric drawing is easily sufficient to infer the condition of the river. For the topographic map of 1905, details of the river appear to have been less expertly recorded. Air photographs are available for many years from 1944, with the first complete air survey in 1946.

The legal survey and air photographs from selected years were used to construct a sequence of planimetric maps showing the position of the river channel. Plotting from air photos was done using a 'Sketch-master' viewer. The initial plots were approximately rectified by plotting on to a series of planimetric control points abstracted from the 1:50,000 topographic map of 1966. Channel displacements cannot be measured precisely, but superposition of the maps shows that known stable banks (bedrock) register satisfactorily from map to map, so that river behaviour

can be assessed qualitatively over 90 years and in detail for the last 35.

A selection of the 12 maps is shown in Fig. 5. In general, the river today appears to have fewer multi-channel reaches than in any previous record. This seems mainly to be the result of increased lateral stability in this century, although the training works near Hagensborg may have had some effect. A more detailed view is revealing.

The 1893 map shows that upstream from Nusatsum River (reach 1) the river was generally confined to a single channel. However, by 1946 there were multiple channels here. Nusatsum River itself appears to have been subject to considerably widening in the interim as well. Since 1946 there has been a tendency for Bella Coola River to settle into two or three main channels about stable channel islands in this reach.

Reach 2 is confined between bedrock and the Nusatsum River fan and has remained stable throughout the record. Reach 3, opposite Salloomt River fan, which appears to have been active in 1893 and still in 1946, has markedly stabilized in recent years.

Reach 4, opposite Nooklikonnik Creek, flowed in two channels about a large island in 1893. During the ensuing half-century one channel became dominant:

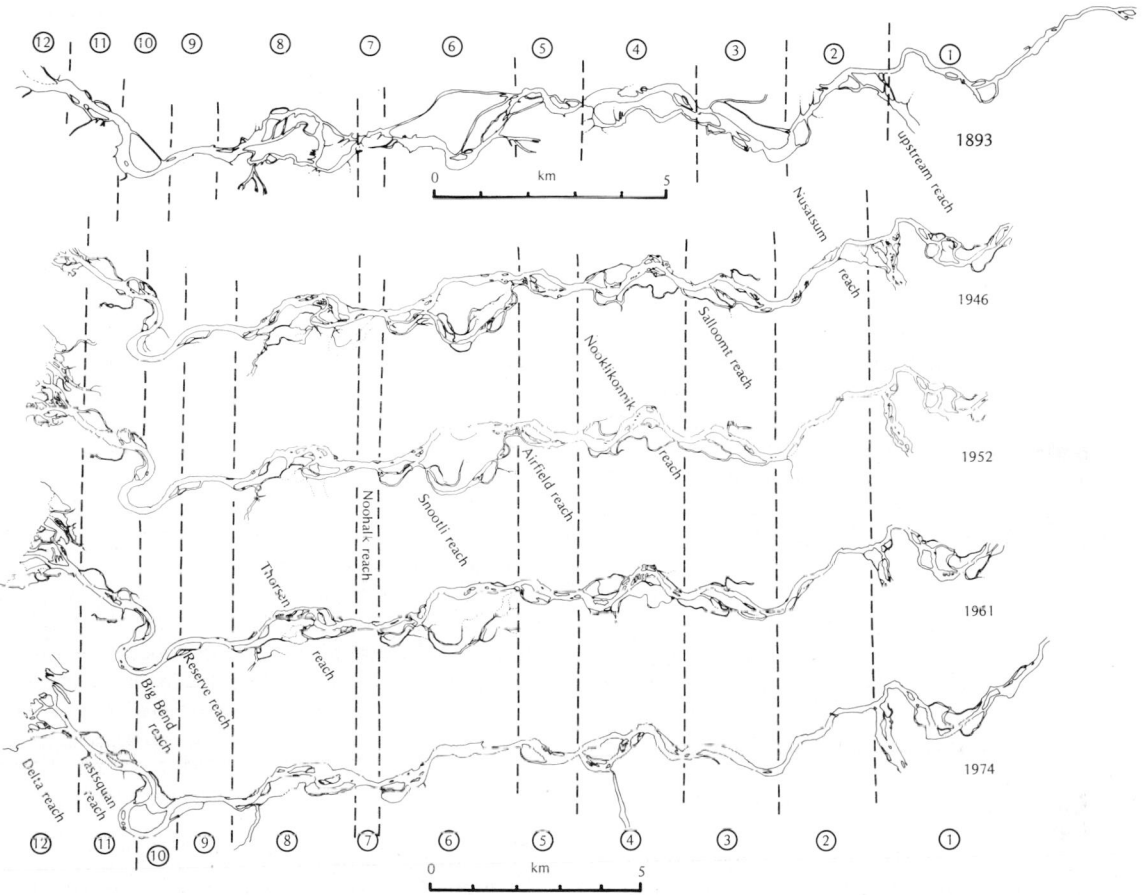

Fig. 5. Successive maps of lower Bella Coola River derived from cadastral survey (1892) and photogrammetric plotting. Note that the 1892 map is presented at a slightly larger scale than the others.

however, the changed alignment shows that lateral migration by bank erosion was proceeding. In the decade 1946–54 the channel remained stable, but by 1961 renewed deposition at the head of the reach prompted the river to reoccupy old flood channels. The 1965 flood produced a major northward shift in the river. Throughout this history, there has been extensive sediment storage in the reach: none the less, it appears that in the late 1950s and 1960s the volume increased substantially.

Reach 5 (Airfield reach) is a short section of channel between Nooklikonnik and Snootli (6) reaches that has remained substantially unaltered throughout the record. However, noticeable widening and occupation of two channels occurred during the period in the 1940s and 1950s when a stabilizing tendency was most

evident in Nooklikonnik reach (4). Construction of guidebanks to protect Bella Coola airport has influenced the river in recent time.

Reach 6 was a zone of channel braiding in the late nineteenth century that was still active in 1946. Since 1950 it has 'healed' steadily and is today relatively stable on the north side of the valley. A small dam constructed in a flood channel has played some role in this stabilization.

Reach 7 (Noohalk reach) is, again, a short, stable reach that leads into reach 8, upstream from Thorsen Creek fan. The fan prominently blocks the main valley. Here the trend of river behaviour is similar to that in reach 6 but is not so far advanced. Extensive braiding and deposition still occur at the downstream end of this reach. Downstream from Thorsen Creek is a long, stable reach (9) where the alluvial fan holds the river against bedrock on the north side of the valley.

Reach 10, upstream from Tastsquan Creek fan— which also blocks the valley— is known as 'Big Bend'. The bend grew considerably between the turn of the century and 1946, and then remained more or less stable until September 1973, when it was cut off during a major flood. Since then, the river has extended rapidly part way back toward its old course and a large volume of sediment remains stored on the point bar. The bend is now very tight on the distal limb and further instability may be anticipated here.

Reach 11, opposite Tastsquan Creek, has remained essentially stable throughout the record. It is at present shoaling and beginning to attack the left bank, which has been rip-rapped to protect Bella Coola village. The head of tidewater occurs in this reach. Delta reach (12) has changed extensively during the period of record, but was not a subject of this study.

PATTERN OF INSTABILITY

Figure 5 reveals an interesting pattern. In 1893 the river flowed in multiple channels nearly everywhere downstream from Nusatsum River, but appears to have been especially unstable in Thorsen reach (8). By 1946 the channel pattern had become considerably less complex downstream through reach 5, yet in reach 1 substantial channel division had appeared which persists today. During the first half of the twentieth century the main channel shifted from the south to the north side of the channel zone in Snootli (6) and Thorsen (8) reaches and considerable extension of the Big Bend occurred. After 1946 we observe (directly, from frequent air photography) steady reduction in channel complexity downstream from Snootli reach, so that by 1974 substantial channel division remained only downstream from Thorsen reach (8) and upstream from Nusatsum River (reach 1). It should be realized that, after 1968, river training has influenced the pattern, especially near Hagensborg.

Two especially interesting developments appear in 1974. Big Bend (reach 10) was breached in late 1973

(Fig. 6) after a period of relative stability. A large volume of bed sediment has accumulated here and is slowly moving downstream into Tastsquan reach (11). which has also been stable for a long time. Reach 9, upstream, has remained stable throughout the record, so that additional sediment has moved directly down from the region of Thorsen Creek, at least.

Nooklikonnik reach (4) underwent a substantial change between 1961 and 1974 (Fig. 7), the main channel shifting to the north side. The effect can be related directly to delivery of coarse sediment from Nooklikonnik Creek, which itself enlarged its channel rapidly during the late 1960s. The major floods of 1965 and 1968 delivered much of this material following rock-slides in the upstream watershed in the early part of the decade.

Some further features of the pattern of instability bear notice.

(i) Reaches 2, 5, 7, 9 and (heretofore) 11 have remained stable throughout the century-long period of record.

(ii) The main loci of instability lie upstream of or opposite the major tributary alluvial fans. The locations of lateral instability are coincident with the 'sedimentation zones' where bedload sediment is stored for more or less protracted periods. Stabilization of such a reach coincides with removal of much of the potentially mobile material (or at least with its incorporation into floodplain adjacent to a relatively clean channel).

The detailed pattern of instability appears, then, to be conditioned by tributary junctions which determine positions of preferrred sediment accumulation. However, the pattern is neither invariable nor strongly correlated with tributary size and activity: in the lower course of the river the sedimentation zones are spaced relatively evenly.

That the bulk of periodically mobile sediment is caught in several discrete storage zones along the channel indicates that bed sediment is supply-limited in this river, for otherwise it would be continuously distributed along the course. This factor—a common feature of gravel rivers in the present epoch—is an important determinant of channel behaviour.

To a degree, 'supply limitation' may indicate

Fig. 6. Big Bend and Tastaquan reaches of Bella Coola River. (A) July 31, 1948. $Q \sim 320$ m^3 sec^{-1}, showing the configuration that was maintained without substantial change until September 1973. Photo from Canada Department of Energy, Mines and Resources, National Air Photo Library; A11564-301. Crown copyright reserved. (B) June 14, 1974. $Q \sim 300$ m^3 sec^{-1}, showing the breach of September 1973 and the plug of coarse sediment that prompted the cutoff. The 1979 alignment is superimposed. In Tastsquan Reach the river has begun to attack the left bank in front of Bella Coola village. This bank is now rip-rapped to protect the settlement. Photo from British Columbia Ministry of the Environment, Air Surveys Division; BC5593-262. Copyright reserved.

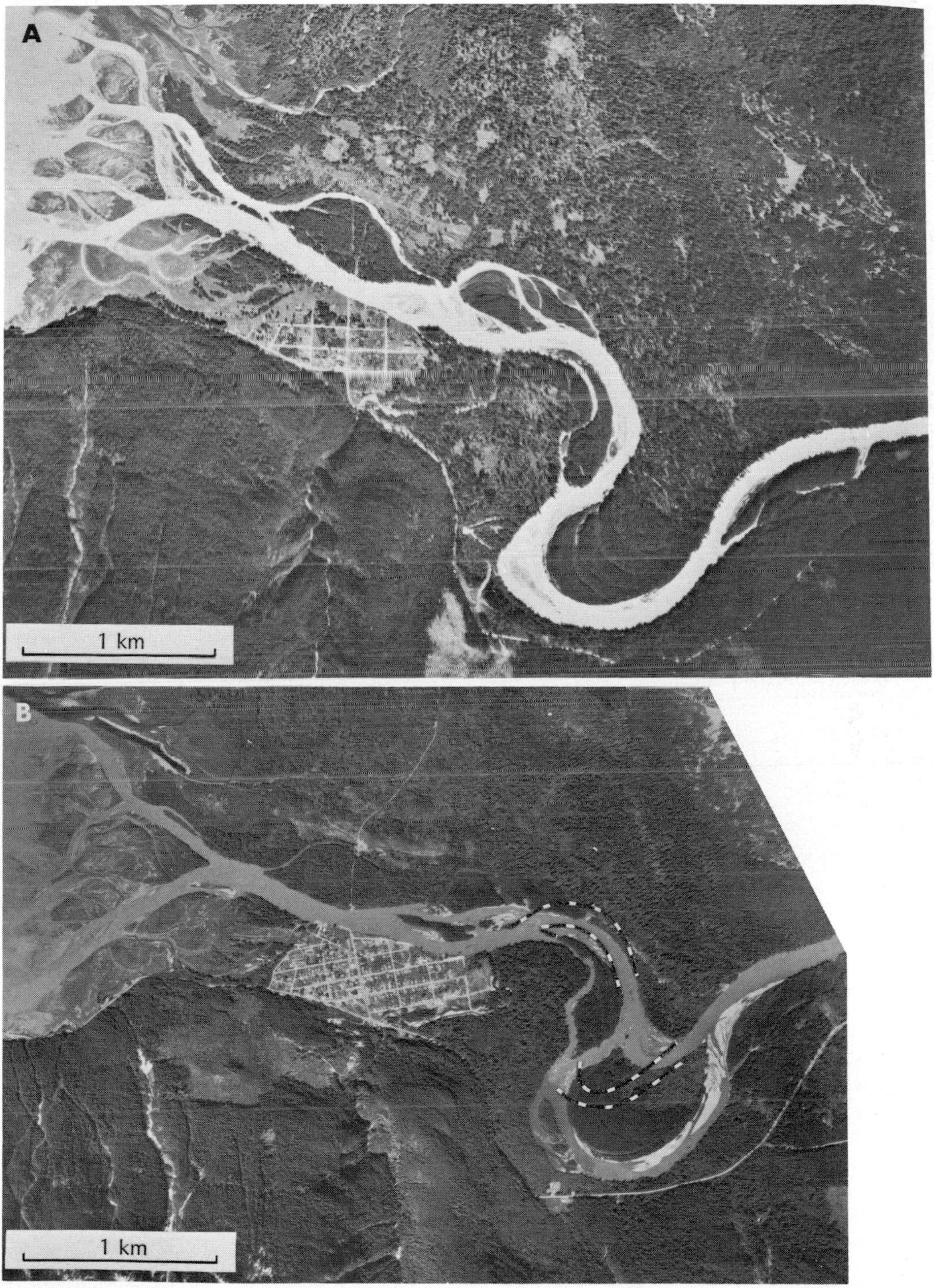

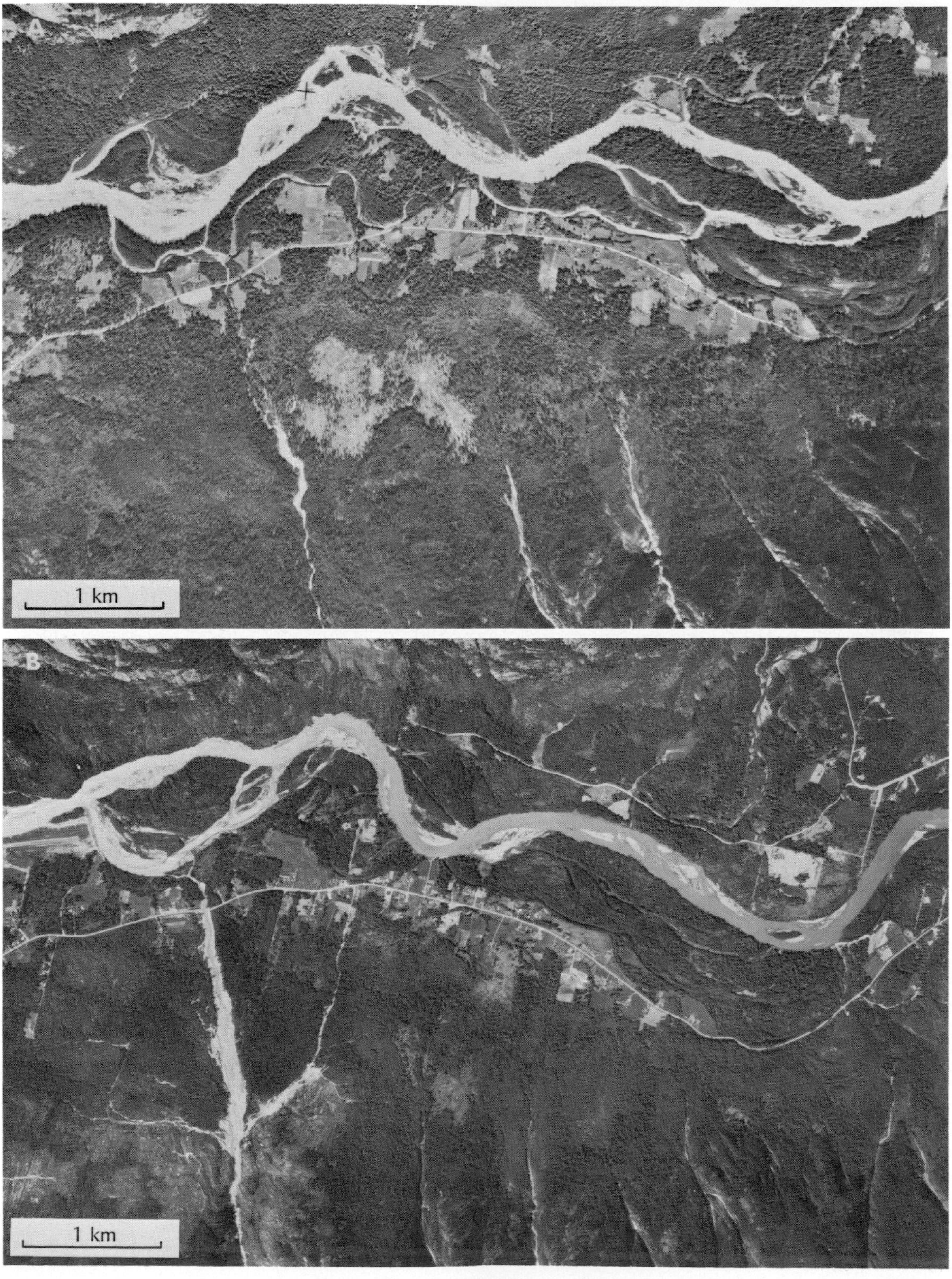
A
1 km
B
1 km

merely the readiness with which sediment may be mobilized from bars in the sedimentation zones. Bar-surface armour and vegetation protects material for a period before a particular flood alters the channel alignment locally and initiates serious attack on the sediments. This appears to be the history of the recent destabilization of Big Bend reach: the September 1973 flood may simply have breached some intrinsic threshold (in the sense of Schumm, 1977). It follows that much of the 'mobile' sediment now exposed there (Fig. 3A) may have been present for a long time—sediment flux through the reach may be very much smaller than storage volume. These considerations do not, however, explain the long-term change in level of fluvial activity that appears to be occurring along the channel.

CAUSES OF CHANGE

The underlying trend in river behaviour has been towards greater stability in this century. Furthermore, the trend may be two centuries old. During his 1793 descent of the river to the Pacific Ocean, Alexander Mackenzie noted of the lower course: 'The navigation of the river now became more difficult, from the numerous channels into which it was divided, without any sensible diminution in the velocity of its current.'

From the pattern of periodic bed sediment movement and storage described above, and directly observed in recent years in the Big Bend reach, I infer that diminution in coarse sediment transport is the cause for this. This diminution appears to be related to reduced supply of coarse sediments to the lower course of the river. This may result either from reduced supply from the tributaries, or from reduced passage of sediment down the main stem.

The tributaries—especially those draining the glacierized mountains south of the Bella Coola River—are short and steep. Sediment moves relatively quickly through the channel system once mobilized at source, as the recent example of Nooklikonnik Creek demonstrates, hence we must consider sources of sediment supply. There are two major possibilities in

contemporary time: (i) erosion of alpine Neoglacial moraines; (ii) response to human activities—particularly roadbuilding and logging—in this century. Neoglaciation culminated rather late in this region (AD 1840–1900) but was certainly well under way during the eighteenth century (Ryder, Thomson & Alley, 1982). On the other hand, whereas settlement by Europeans began around 1885, it was not until after 1950 that extensive mechanized logging began in the mountains. In consideration of this, the substantial Neoglacial moraines appear to be the major likely source of material.

Sediment transfer along the river may be affected by hydrological changes. Changes significant for sediment transport are apt to be associated with the frequency of major rainstorms. The hydrological record does not permit evaluation of changes on a time-scale of one or two centuries. There is regional evidence that the first fifty years of the twentieth century were rather more dry than the last thirty, but flood records are not clearly divisible, nor is the behaviour of the river obviously associated with such a schedule.

Map evidence indicates that the major tributary, Nusatsum River, may influence sediment transfer into the study reach. The main, long-term source of sediment for lower Bella Coola River is the relatively heavily glacierized Talchako drainage basin. The appearance that sediment has been accumulating upstream from the mouth of Nusatsum River since 1900, along with the historically confirmed increase in activity of this tributary, suggest that the downstream progress of bed sediment may be blocked by recent progradation of the tributary alluvial fan. Nusatsum River basin is glacierized, so that Neoglacial sediment yield from this tributary could account for this outcome.

DISCUSSION

Bella Coola River is a 'wandering gravel bed river': that is, it exhibits an irregular pattern of channel instability. Although a single dominant channel is

Fig. 7. Nooklikonnik and Salloomt reaches of Bella Coola River. (A) July 31, 1948, showing a late stage of stabilization in Salloomt reach with only minor flood channels still active beyond the main channel. Photo from Canada Department of Energy, Mines and Resources, National Air Photo Library; A11564-316. Crown copyright reserved. (B) June 14, 1974, showing the single active channel in Salloomt Reach. This channel is, however, wider than in 1948 and exhibits a prominent diagonal bar and small channel island. Old flood channels, now completely vegetated, still pass water during high floods. At Nooklikonnik Creek, sediments delivered from the tributary and from the main channel have forced the river to the north side of the valley. Photo from British Columbia Ministry of the Environment, Air Surveys Division; BC 5593-258. Copyright reserved.

everywhere evident, the river consists of a sequence of braided/anastomosed reaches connected by relatively stable, single-thread reaches. The former hold a large portion of the sporadically mobile bed sand and gravel, hence they are termed 'sedimentation zones'. They occur at points of locally reduced gradient and often are associated with tributary alluvial fans that constrict the valley floor. The localization of deposition implies that movable bed sediment is supply limited.

Within the last century (or more) there has been a tendency for reduced lateral instability along the river. This appears to be connected with diminution of bed sediment supply. One of two mechanisms might explain the effect:

(i) decline in the transport of unusual volumes of sediment introduced into the channel by the erosion of Neoglacial (eighteenth- and nineteenth-century) moraines of alpine glaciers;

(ii) recent progradation of the large alluvial fan at Nusatsum River constricting the main channel and blocking sediment transfer from upstream.

The first proposal appeals to sediment supply, the second to local channel behaviour. They now form hypotheses for further research.

The stabilizing tendency has propagated downstream from near Salloomt River to Big Bend—about 15 km—during the past 30 years, for an average rate of about $0 \cdot 5$ km yr^{-1}. This might represent the rate of evacuation from the system of the tail of a diffusing slug of mobile sediment. The rate value compares reasonably with computed values for downstream transfer of a bed sediment 'wave' in the much smaller Tamaki River, New Zealand, given by Mosley (1978) as $1 \cdot 75$ km yr^{-1} for the peak concentration and about $1 \cdot 0$ km yr^{-1} (from his fig. 7) for the tail, on the assumption of about 12 sediment-moving events per year. However, diffusion must be limited in rivers of Bella Coola type since sediment is reconcentrated in storage areas that are themselves sometimes stable for periods comparable with the transit time for sediment along many kilometres of channel.

Much of the material undoubtedly orginated about 100 km upstream in the relatively heavily glacierized Talchako basin. We realize from the example of Nooklikonnik Creek that delivery time of coarse sediment to the main channel, once mobilized from source, is rather short—of the order of 10 years. A transfer rate of $0 \cdot 5$–$1 \cdot 0$ km yr^{-1} for sediment in the main channel would suggest that the material was first mobilized one or two centuries ago—that is, within the period of the last Neoglacial maximum.

Some interesting implications for the interpretation

of coarse alluvial sediments may be drawn from this study. The sedimentary association of the wandering gravel bed river is bar gravels—including point-bar and medial bar configurations—overlain by sand with minor sedimentary structures, of the order of 1 m thick, occasionally cut by channel fills of sand or silt. Wood may be rather common. Inasmuch as the channel type occurs in conditions of supply-limitation of mobile bed materials, it is apt to cap aggradational sequences of fluvial gravels and to indicate the end of that episode. In Bella Coola valley, this facies represents the entire Holocene portion of late Quaternary sedimentation.

ACKNOWLEDGMENTS

Dr S. O. Russell, Department of Civil Engineering, University of British Columbia, and J. S. Hart have contributed ideas and observations to this report. The initial reconnaissance of Bella Coola River was undertaken during work for Sigma Engineering Ltd, Vancouver. Further study was supported by the Natural Science and Engineering Research Council of Canada. Charles Tremewen drew the figures. This paper is a contribution to International Geological Correlation Programme Project 158A, 'Palaeohydrology of the temperate zone during the last 15,000 years'.

REFERENCES

ANDREWS, J.T. & RETHERFORD, R.M. (1978) A reconnaissance survey of late Quaternary sea levels, Bella Bella/Bella Coola region, central British Columbia coast. *Can. J. Earth Sci.* **15**, 341–350.

CHURCH, M. & JONES, D.P. (1982) Channel bars in gravel-bed rivers. In: *Gravel-bed Rivers: Fluvial Processes, Engineering and Management* (Ed. by R. E. Hey, J. C. Bathurst and C. R. Thorne). Wiley, Chichester. In press.

LEOPOLD, L.B. & WOLMAN, M.G. (1957) River channel patterns: braided, meandering and straight. *Prof. Pap. U.S. geol. Surv.* **282-B**, 39–85.

MOSLEY, M.P. (1978) Bed material transport in the Tamaki River near Dannevirke, North Island, New Zealand. *N. Z. J. Sci.* **21**, 619–626.

NEILL, C.R. (1973) *Hydraulic and morphologic characteristics of Athabasca River near Fort Assiniboine.* Alberta Research Council, Edmonton. *Hway River Engng Div. Rep.* **REH/73/3**, 23 pp.

RYDER, J.M., THOMSON, B. & ALLEY, N.F. (1982) *Neoglacial chronology for the southern Coast Mountains of British Columbia. MS submitted for review.*

SCHUMM, S.A. (1977) *The Fluvial System.* Wiley (Interscience), New York. 388 pp.

TEMPEST, W. (1974) *Bella Coola River: flood and erosion control.* British Columbia Department of Lands, Forests and Water Resources, Water Resources Service, Water Investigations Br., File 0254122 no. 11, 83 pp. (limited circulation report).

Spec. Publs int. Ass. Sediment. (1983) **6**, 181–193

Bar development and channel changes in the gravelly River Feshie, Scotland

R. I. FERGUSON* *and* A. WERRITTY†

*Department of Environmental Science, University of Stirling, Stirling FK9 4LA and
†Department of Geography, University of St Andrews, St Andrews KY16 9AL, U.K.

ABSTRACT

Unconfined reaches of the River Feshie in the Cairngorm Mountains have active low-sinuosity moderately divided patterns in which glacial outwash gravel is reworked by floods of up to 100 m³ sec⁻¹ on a gradient of 0·009. Alternate bars of diagonal or lateral form are characteristic; they may develop from lobate or elongate medial bars, and advance episodically during floods. A model involving diagonal-bar progradation and consequent bank erosion opposite accreting bar margins, interrupted by avulsion within the channel by chute incision or from the channel by ponded-pool overflow, is used to explain surveyed changes in one sub-reach since its initiation by avulsion in 1976. The main oblique bar front advanced 160 m by 1981 and its exposed portion changed from medial to lateral to medial through avulsion around or across the bar. The channel widened by 136% from 1977 to 1981 with bank erosion locally exceeding 10 m yr⁻¹. Bar progradation into sloughs, and sheet transport over vegetated floodplain, led to coarsening-upward sedimentation with a falling-stage veneer of finer deposits on bar tops and lee margins. The behaviour of this wandering gravel river resembles laboratory pseudo-meandering and may represent a balance between divergent meandering and braiding tendencies.

INTRODUCTION

The distinction between meandering and braiding in rivers with gravel or cobble beds is not sharp. Many actively changing pebbly rivers have moderately divided channels of low to medium sinuosity which combine features of both meandering and braiding, rather than closely resembling textbook examples of one pattern or the other. Such rivers have generally wide shallow channels, flanked and locally divided by expanses of bare gravel, but lack the degree of channel division characteristic of archetypal pro-glacial braided rivers. They have a well-defined pool–riffle sequence with a meander-like alternation of riffle orientation, but erosional banks are not always or exclusively found on the outside of bends and channel sinuosity is lower than in freely meandering rivers.

Rivers with transitional channel patterns of this type are common in piedmont situations in most non-tropical environments. They are generally regarded as braided rather than meandering, but we follow C. R. Neill and M. Church in recognizing them as a separate class of 'wandering gravel rivers' (Church, Moore & Rood, 1981). This term emphasizes the activity of such channels, but we are unaware of any time-lapse studies of their erosional or depositional development except for the work of Bluck (1976, 1979, 1982) and ourselves (Werritty & Ferguson, 1980; Werritty, 1982). We report here on five years of observations of bar development and channel change in a wandering reach of the gravelly River Feshie in Scotland.

STUDY AREA

The Feshie is a right-bank tributary of the Spey draining 240 km² of the western Cairngorm Mountains in the Scottish Highlands (Fig. 1). Most of its drainage basin is underlain by Moinian schists and lies at 700–1000 m above sea-level, but the north-east part on the Cairngorm granite batholith rises to 1265 m.

0141-3600/83/0106-0181 $02.00

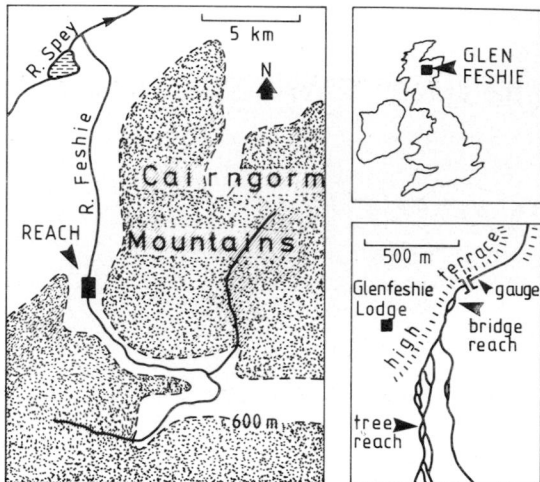

Fig. 1. Location of Glenfeshie, study reach, and sub-reaches studied in detail.

The basin is bisected by a deep glacial trough through which the river flows westwards before turning north at about 400 m above sea-level into the wider outwash-mantled lower part of Glenfeshie. Till is restricted to the valley sides, but there is some blanket peat at 600–800 m. The sparse pine woods of the lower glen give way above c. 500 m to *Calluna–Empetrum* heath, alpine grassland, and almost bare frost-shattered regolith on the highest ground.

The lower course of the river is confined locally by bedrock, and more extensively by post-glacial terraces, but in three reaches it is free to migrate laterally and is actively reworking outwash gravels. In these reaches the Feshie is one of Britain's most braided rivers, and historical maps and air photographs show that its divided low-sinuosity pattern has undergone frequent changes in channel alignments, degree of sinuosity and division, and extent of bare gravel (Werritty & Ferguson, 1980). Attention is restricted in this paper to the uppermost braided reach near Glenfeshie Lodge (Fig. 1).

Active low-sinuosity channel patterns in Britain are associated with high stream power (Ferguson, 1981), which the Feshie possesses by virtue of its combination of steep gradient (averaging 0·009 in the study area) and frequent large floods. A gauging station operated from 1951 to 1974 by the Department of Agriculture and Fisheries for Scotland (subsequently Scottish Development Department), near the mouth of the catchment, recorded a mean flow of 8 m³ sec⁻¹ but floods of up to 200 m³ sec⁻¹. In order to relate channel

changes to flood events, the University of St Andrews installed an autographic water-level recorder in 1978 at a partly confined section at the foot of the study area (Fig. 1). Discharges up to 20 m³ sec⁻¹ have been current-metered from an adjacent bridge to establish a well-defined rating which has been used to extrapolate flood discharges. The 110 km² catchment above this site has a mean flow of between 3 and 4 m³ sec⁻¹ and a minimum of less than 1 m³ sec⁻¹. Channel capacity varies from point to point and over time, but bankfull discharges are generally reckoned to lie between 20 and 30 m³ sec⁻¹. Contrary to the usual assumption that overbank flooding occurs only about once a year, 51 floods in the first three years of record exceeded 20 m³ sec⁻¹ and 16 exceeded 30 m³ sec⁻¹. The three largest each exceeded 100 m³ sec⁻¹. Most floods occur after prolonged frontal rainfall in autumn or winter, but some follow convective storms in summer. Diurnal snowmelt peaks are common in spring and reached 35 m³ sec⁻¹ in May 1979. The river is very flashy by British standards. One summer flood rose to over 100 m³ sec⁻¹ in under 2 hr and was over within 24 hr.

With this hydrologic regime and the relatively steep slope, the river is actively reworking the non-cohesive sediments of its valley floor. Bedload transport is considerable and extensive during major floods and leads to the episodic development of gravel-bar complexes accompanied by channel incision and infill of 0·5 m yr⁻¹ or more and bank erosion that can exceed 10 m yr⁻¹.

BAR TYPES

The 1 km long study reach contains extensive unvegetated gravel sheets not only: (1) as mid-channel bars and (2) alongside low-flow channels but also (3) overlying vegetated floodplain surfaces. The frequency of depositional and erosional activity means that most gravel accumulations have a compound history, with multiple episodes of aggradation and often some erosional trimming. As a result bars differ widely in morphology and are not easily classified.

Mid-channel bars

Most mid-channel bars in the Feshie are elongated in the downchannel direction and lack prominent distal slip faces, at least when examined at low flow; they are longitudinal in the classification of Miall (1977). Three subtypes can be recognized according to flow pattern over the bar at normal discharges.

(a) Flow is radial over the curving sides and front of *lobate* bars (Smith, 1974; equivalent to 'linguoid' of Church & Gilbert, 1975 and Bluck, 1979).

(b) *Longitudinal* bars (Smith, 1974; Church & Gilbert, 1975; equivalent to 'medial' of Bluck, 1979) have a bigger and more elongated exposed surface with flow separating off both sides of the bar head into flanking chute or riffle channels. If these channels have erosional slip faces the bar tail may be trimmed to a point.

(c) *Diagonal* bars (Smith, 1974; Church & Gilbert, 1975) are also elongated but are asymmetric with flow off one side only of the bar head in one or more riffles. According to the number and position of riffles, and whether the riffle-free side of the bar is bounded by a channel or attached to the floodplain, separate areas of exposed gravel may be attached to either bank or neither, but they are all part of a single functional unit.

Lateral bars

Lateral bars in the Feshie are compact or elongated accumulations attached on one side to a floodplain, either merging into it or lying under an erosional bank. They are often attached only proximally, with a distal slough into which the bar grows, and may have originated as mid-channel forms. Once attached, they undergo superficial change and may be enlarged by the accretion of mid-channel bars. These bars resemble the 'lateral bars' of Bluck (1976, 1979), whose studies included a lower reach of the Feshie. They may represent bank-attached diagonal bars whose riffles have degraded. They often lie on the inside of gentle channel bends and are morphologically transitional to point bars with chutes, clear examples of which are absent in the study reach although one exists just downstream at a confined meander bend.

Overbank bars

Gravel deposits are exceptionally found in overbank locations. A tabular overbank gravel bar some 30 m long, 2–4 m wide and 0·4 m thick was observed in 1977 on top of the grassy floodplain at the eroding margin of a recently formed channel; it had disappeared through bank erosion six months later. Thinner but much more extensive gravel sheets were subsequently deposited over other vegetated areas during major floods when overbank flows diverged from the lowflow channel.

General characteristics

Despite their diversity of form and location relative to the low-flow channel, gravel bars in the Feshie share many common sedimentological characteristics which have also been described from a lower region of the river by Buck (1978). Well-defined foresets are rare, and bars consist predominantly of massive or crudely horizontally bedded cobble sheets, clast-supported but with a matrix of gravel and sand. Surface imbrication is almost universal and generally well developed in the often platy schist clasts. Bars on which surface size distributions have been measured show consistent downbar fining and some lateral sorting, but the larger bars often have on their surface small gravel or cobble sheets of contrasting grain size. These features are usually only a few grain layers thick, with well-defined lobate slip-face fronts. They resemble the supra-platform unit bars described by Bluck (1982) from other low-sinuosity rivers in Scotland and perhaps the 'gravel sheets' of Hein & Walker (1978), but from a morphological viewpoint we regard them not as primary units but as, perhaps, the equivalent in gravel rivers of the sand dunes commonly found on top of transverse sandy braid bars and meander point bars. Smaller-scale superficial structures include cobble clusters and sand, gravel or even cobble tails behind beached turf blocks, tree trunks or large clasts. The distal and slough-side margins of laterally attached bars are usually decorated by finer-grained sediments either deposited directly from falling floods or washed out from the bar framework to form miniature deltas.

DIAGONAL BAR EVOLUTION: THE BRIDGE REACH

Diagonal flow over the longitudinally aligned edge of an elongated gravel sheet is widespread in the Feshie and seems to be a normal stage in the evolution of mid-channel unit bars into the extensive compound features that eventually accrete to the floodplains. The characteristics of this 'diagonal' stage in bar development are well illustrated by observations over 5 years of a sub-reach of the Feshie illustrated in Fig. 2. This 'bridge reach' extends above and below the wooden bridge by which our gauging station is sited (Fig. 1) and is bounded on its left by a 10 m high terrace.

Throughout the 1976–1981 study period a pair of characteristic diagonal bars has been present in this

Fig. 2. Views looking up bridge reach: (A) at low flow in June 1979 showing alternate diagonal bars; (B) in moderate flood (c. 20 m³ sec⁻¹) in April 1979 showing drowning out of riffles; (C) at low flow in May 1981 showing downstream migration of upper bar and lower riffle, and potential for avulsion.

reach, and older ground and aerial photographs show similar features, although not always in identical positions. The 1979 view (Fig. 2A) is typical in that at low discharge a prominent riffle flows transversely from an upper channel (left in photograph) to a lower one, over a distinct longitudinally aligned step in the bed. Except where it is crossed and dissected by the riffle this step is exposed as a strip of gravel running down and obliquely across the channel, but this exposed bar is merely the leading edge of a much larger gravel sheet that extends downstream and laterally from a pool in the upper channel. The complete functional unit thus consists of proximal pool, partially submerged bar and riffle.

In the background of Fig. 2(A) the prominent diagonal-bar assemblage is overlapped by the less well-defined oblique front of another, higher unit facing the opposite side of the channel. The two bar fronts form an obtuse-angled zigzag down the centre of the channel and are crossed in a mirror-image zigzag by the two low-flow riffles. Successive bar–pool–riffle units are thus laterally reversed with their proximal pools staggered on alternate sides of the

channel and their riffles slanted in opposite directions, to form a meandering thalweg within the essentially straight channel.

At low or medium flows, as illustrated in Fig. 2(A), the water surface is much steeper in riffles than pools and there is a streamwise non-uniformity of flow marked by vertical, and usually also lateral, divergence and consequent deceleration out of each riffle into the head of a pool, from which flow converges and accelerates towards the next riffle. Velocities in pools are usually low, but riffles may be competent at relatively low discharges, and the bar face beneath them is often dissected into distinct chutes.

In floods (Fig. 2B) the water surface profile becomes more uniform as riffles are drowned out and the river tends to straighten. Pool heads become deeper and steeper, but riffles lose their identity as water depth increases and more and more of the flow straightens out along bar tops instead of slanting over the side face of the bar. In floods larger than that shown in Fig. 2(B) it is likely that the low-flow pattern of convergence into, and divergence out of, riffles is reversed with flow diverging over bar tops and margins and reconverging into the deeper channels of the low-flow pools (compare Figs 3A and 3B).

This reversal of hydraulic conditions remains unquantified but could explain the kinematics of diagonal-bar development. The bridge-reach bars advanced intermittently a total of about 50 m between 1976 and 1981 with maximum advances over intervals during which major floods occurred. The dip direction of surface imbrication suggests downstream advance with some radial divergence, entirely consistent with a tendency for competent high flows to diverge out of pools over bar tops rather than converging through a riffle as at low flow. Sediment could then be transported out of proximal pools and deposited on distal bar margins (Fig. 3B, C). Pool–riffle evolution in gravel rivers has previously been discussed in terms of flow convergence and divergence by Church & Gilbert (1975), and radial dispersal of sediment from chutes to lobes has been identified as a fundamental braiding mechanism in laboratory experiments by Ashmore (1982) and Southard *et al.* (1981). In the case of diagonal bars it must additionally be recognized that the distal part of the bar is only submerged at high discharges, and if flow is then divergent the bar will prograde in each competent flood and remain dormant in between; this episodic advance is just what is observed.

Progradation of diagonal bars has several consequences. If the low-flow riffle also migrates downstream, the former pool head will be abandoned as a slough ('inner-channel' of Bluck, 1976, 1979) which is liable to be infilled by the oblique onlap of the upstream bar front as described by Bluck (1979). Secondly, lateral aggradation of the bar face tends to constrict the channel alongside which may therefore erode, particularly opposite the riffle where a current jet is directed towards the bank. Riffle migration may thus cause a wave of bank erosion travelling downchannel. This has been observed on both sides of the bridge reach, although the left-bank terrace inhibits erosion and has led to higher-than-normal velocities in the 'pool' on this side.

Two other potential consequences of diagonal-bar progradation are especially important because they interfere with its continued occurrence. First, distal elongation of the bar–pool–riffle unit increases the scope for avulsion of the main riffle. Riffle migration along the elongated side of the bar is a normal consequence of progradation as already described, but it is also possible for a chute to be incised down the middle of the bar or through its shorter, previously riffle-free margin (Fig. 3D, 'chute'). The previous riffle may survive or may dry up at normal flows with a transfer of activity to the new chute. Incision of such a chute in the upper of a pair of alternate diagonal bars has the additional effect of bypassing the lower bar (Fig. 3D, 'bypass') which, deprived of some or all of the flow through its proximal pool, may accrete to the bank on this side and cease to function and evolve as previously.

By May 1981 (Fig. 2C) the bridge reach was at the stage where chute incision seemed likely either through the left (terrace) margin of the upper bar into the slough, thus bypassing the lower bar unit, or along the top of the lower bar beneath the bridge. A major flood in September 1981 in fact triggered this second possibility, as well as causing tens of metres of progradation of the lower bar, so that in November

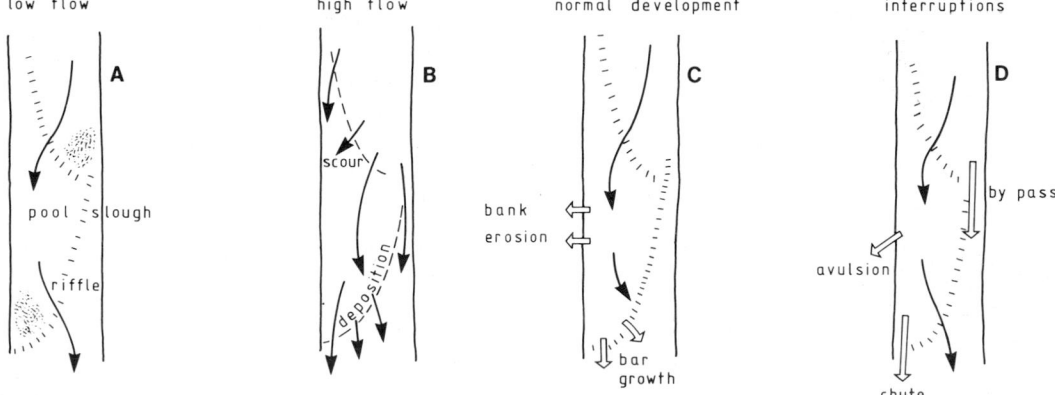

Fig. 3. Schematic diagrams of flow over alternate diagonal bars and consequent morphologic development: (A) meandering thalweg at flow with convergence over bar faces; (B) divergence and deposition on bar face at high flow; (C) normal pattern of episodic progradation in floods and consequent bank erosion; (D) possible interruptions due to avulsion within channel or from it. See text for details.

1981 the bridge spanned a shallow bar-head divergence zone feeding both the original riffle (now below the bridge) and the new chute nearer the true right bank.

The second way in which normal bar development may be interrupted is by true avulsion, away from the channel altogether. This may occur if bar-margin deposition leads to sufficient ponding of the proximal pool for it to overflow on the side away from the bar and form a distributary over the floodplain (Fig. 3D, 'avulsion'). If this were to happen on the right bank of the bridge reach (Fig. 2C, middle left) diversion would be into a tributary channel that rejoins the main channel almost immediately; but elsewhere the floodplain is traversed by inactive former main channels that do not necessarily recombine for several hundred metres and could be reactivated in what Church & Gilbert (1975) termed 'secondary anastomosis' and Werritty & Ferguson (1980) discussed as 'switching'.

In summary, observations of the bridge reach are compatible with: (1) episodic progradation of asymmetric longitudinal bars because of reversal during floods of the normal pattern of flow convergence over a diagonal riffle on the elongated side of the bar; and (2) possible interruption of this 'normal' development by avulsion, either within the channel by chute incision or away from it because of ponding.

BAR DEVELOPMENT AND CHANNEL CHANGES: THE TREE REACH

A combination of 'normal' progradation and bank erosion with interruptions by chute formation and avulsion also provides a framework for understanding the evolution of another sub-reach studied in more detail. This 'tree reach' (Fig. 1) is near the western edge of the active area and its west-bank floodplain is grass-covered throughout, with isolated mature pine trees that are distinguishable on aerial photographs made in different years. Analysis of these and old maps (Werritty & Ferguson, 1980) reveals repeated switching of the river between different channel alignments in this reach (Fig. 4). The present alignment dates only from the start of the study period: subsequent surveys therefore trace the complete evolution of this channel and its bars.

This reach was initially surveyed in June 1976 by instrumental levelling of five transects across the entire active area. In 1977, 1978 and 1979 this was repeated and supplemented by tacheometric plane-table mapping. From April 1979 to April 1981 a 160 m length of the reach was surveyed more extensively by repeated levelling at 2 m intervals along nine parallel sections, each 60 m long, between benchmarks 20 m apart along the channel. Scour or fill to a depth greater than about 10 cm could thus be distinguished from stage-dependent changes in exposed bar area as a result of differences in discharges or drawdown following scour in adjacent channels.

1976–1977

The principal changes in the tree reach between the first and second surveys were avulsion of the river from its previous course by overflow from one bar–pool–riffle unit, and progradation down the new channel of the next unit upstream with consequent bank erosion.

Fig. 4. Tree reach showing course of river in different years. Dates in brackets indicate former distributaries, other dates indicate former main channels. Photograph shows low flow, right to left, in April 1980. Area mapped and illustrated in Figs 5–10 extends from fallen tree (left of centre) to right margin.

These changes are mapped in Fig. 5, in which D indicates the shallow 1976 distributary flowing over the grassy floodplain on the left bank of the original channel (C), on the outside of the pool alongside a lateral bar. By June 1977 the distributary had incised sufficiently to divert all the normal flow of the river, and its downstream continuation had been abandoned in favour of a grassed-over palaeochannel further west (P).

Meanwhile, the oblique face of the next bar platform upstream had prograded (AA) and become exposed as a longitudinally aligned, but functionally diagonal bar with the upper channel incised behind it (S) into the former lateral bar. The low-flow riffle in 1977 was supplemented at above-average discharges by a second riffle (R) across the head of the diagonal bar into the left-bank channel, along which substantial bank erosion occurred (BB) presumably in response to the migration of the bar and its main riffle.

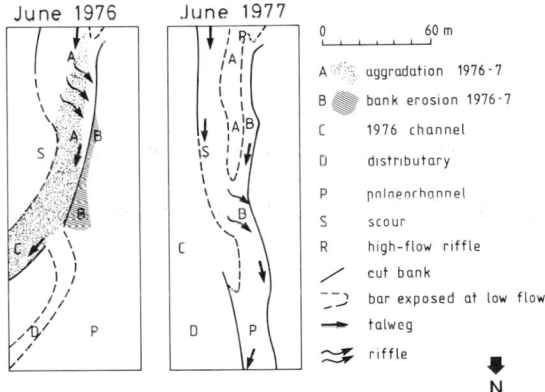

Fig. 5. Bar development and channel change in tree reach, 1976–77. Labels A, B, etc. are at same positions in both maps and mark changes discussed in text.

1977–1978

Complex changes occurred between June 1977 and August 1978 through a combination of 'normal' progradation of, and bank erosion below, the diagonal bar and avulsion both within the channel and from it, the former neutralizing the latter.

Aggradation on the back of the diagonal bar (Fig. 6, A on section and map) forced the upper channel to migrate away and caused a shallow distributary (D) to form by overflow into the 1976 channel. The bar itself advanced further downstream (AA) but an increasing proportion of the flow began to bypass it

following incision (S) of the riffle over the bar head. This prevented permanent avulsion at D back into the 1976 course and led to substantial widening of the left-bank channel by bank erosion (B) of up to 10 m. A small lateral bar formed in an embayment further down this bank.

1978–1979

A major flood of over 100 m³ sec⁻¹ in November 1978 transformed the appearance of the channel by a combination of bar progradation and consequent channel changes. The diagonal bar advanced 20–30 m (Fig. 7, A3) and became a mid-channel feature through incision of a distal chute (S next to A3). Further upstream a new left-bank lateral bar (A1) was transported into the reach from a tributary channel, blocking the head of the former main channel and diverting most of the flow into the upper, right-bank channel which consequently scoured (S next to A1). This led in turn to abandonment of the distributary on this side except during high snowmelt flows in the spring of 1979, which also caused further progradation of the lower bar and the formation of a subsidiary lobe in its lee (A2). The progradation at A3 in and after the main flood was accompanied by bank erosion of up to 10 m along a 100 m length of the left bank (B).

1979–1981

Another major flood one week after the 1979 survey reached almost as high a peak discharge as in November 1978 but was of much shorter duration. Perhaps for this reason it had less effect on the channel, its most spectacular consequence being the deposition of an extensive cobble sheet (Fig. 8) on top of the grass- and heather-covered right-bank floodplain downstream of the 1976 course, which was probably the source of the sediment. By 1981 the former soil and plant cover had become a decimetre-thick 'palaeosol' sandwiched between cobble layers.

Changes within the channel during and after this flood consisted essentially of progressive bar aggradation and consequent bank erosion opposite (Fig. 9). A lobate mid-channel accumulation in the wide shallow channel at the top of the reach advanced into the pool below and emerged as a longitudinal bar (A1) between an aggrading riffle over the head of the left-lateral bar and an incising and migrating chute on the right (S by A1). A side lobe of the lateral bar prograded briefly in 1979 (A2) and the main mid-channel bar aggraded on both sides (A3) and

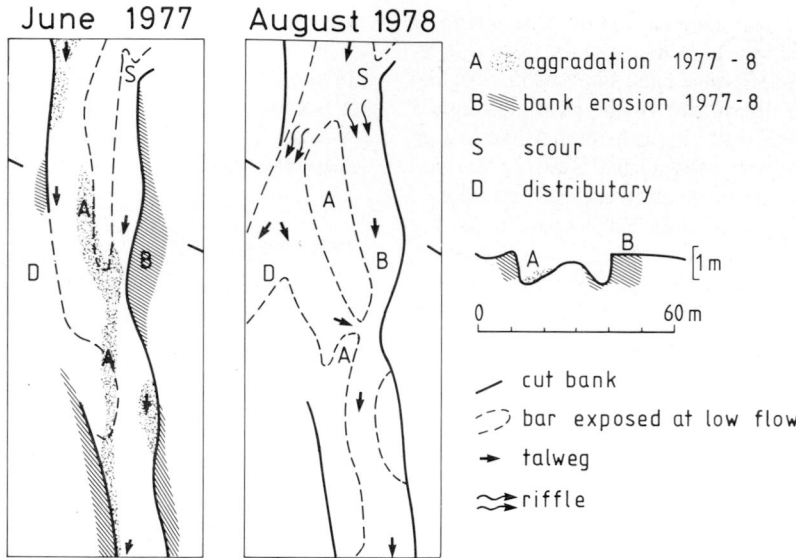

Fig. 6. Bar development and channel change in tree reach, 1977–78. Labels A, B, etc. are at same positions in both maps and mark changes discussed in text.

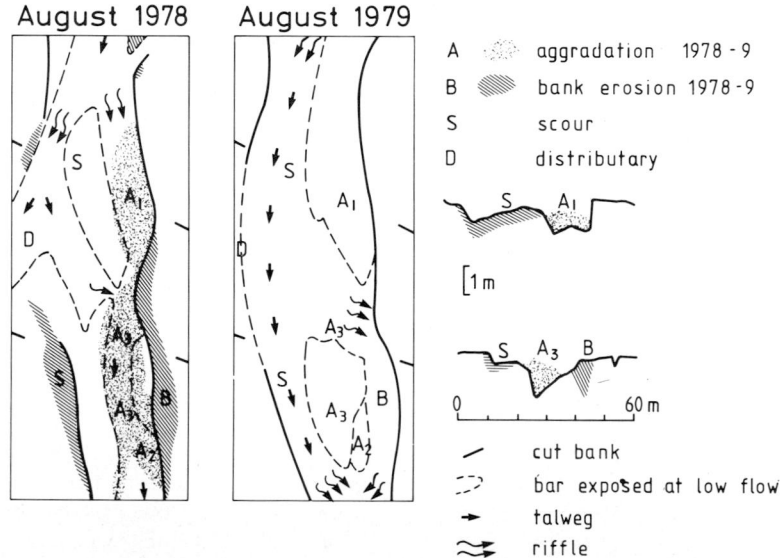

Fig. 7. Bar development and channel change in tree reach, 1978–79. Labels A, B, etc. are at same positions in both maps and mark changes discussed in text.

began to migrate out of the surveyed area. The accretion on its back forced the active upper channel to migrate away, incise (S by A3), and erode a 60 m stretch of the right bank by up to 12 m (B). This incision was accompanied by the development of a new riffle (R2), and the combination of upper-channel

incision and bar aggradation led to abandonment of the previous riffle (R1) at normal flows, so that the lower channel became a slough and this bar and the younger lateral bar further up the left bank became a continuous area of exposed gravel.

The next major flood, in September 1981, continued

Fig. 8. Cobble sheet deposited on grass- and heather-covered lower right bank of tree reach in flood of August 1979. Scale is 1 m long and flow is away from camera.

this pattern by causing further progradation of the distal bar, accretion on its back, and consequent right-bank erosion, but the simultaneous advance of the longitudinal bar at the head of the reach (Fig. 9, A1) led to deflection of its left-hand riffle, submergence of most of the left-lateral bar, and reactivation of the 1979 riffle (R1) and left-bank slough.

Rate of development

The evolution of the tree-reach channel since its creation by avulsion in 1976 is summarized in the maps of Fig. 10. The main accretionary front within this reach had migrated about 160 m downstream by 1981. Its oblique alignment persisted throughout but its surface expression as exposed longitudinal, diagonal or lateral bars altered repeatedly because of changes in the relative importance of the upper and lower channels. Submerged in 1976, it emerged as an elongated proximally attached diagonal bar in 1977, a less elongated distally attached diagonal bar in 1978, a broader mid-channel bar in 1979, and a laterally attached bar in 1981. Minor bars appeared and disappeared below this main accretionary front over the years and two new large bars, one lateral and the second longitudinal, entered the upper part of the reach in 1978 and 1979 and occupied substantial parts of it by autumn 1981.

Annual re-surveys of individual cross-sections repeatedly showed both scour and fill within one section over one year, sometimes by as much as 1 m in a floodplain whose total relief is little over 2 m. Both

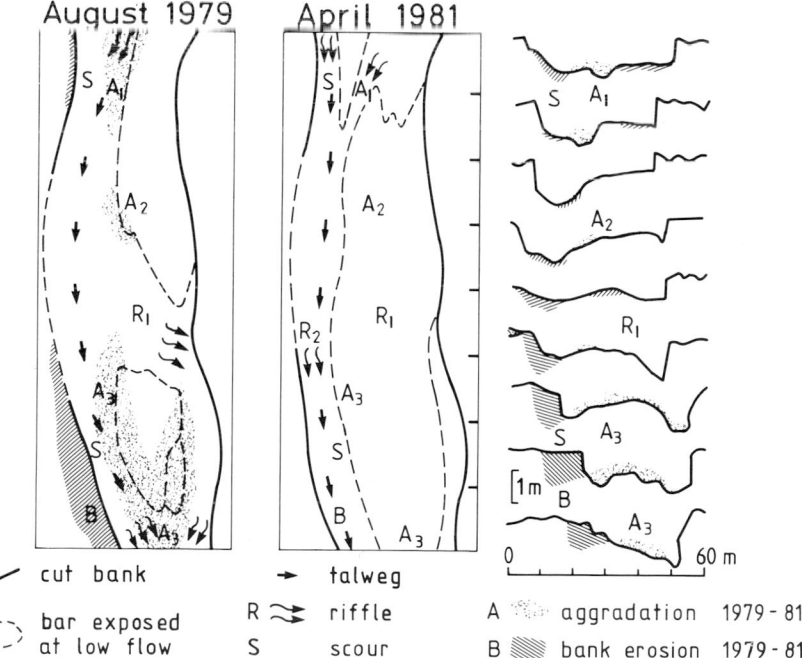

Fig. 9. Bar development and channel change in tree reach, 1979–81. Labels A, B, etc. are at same positions in both maps and mark changes discussed in text.

diagonal bars is unlikely to persist for long. Accretion at the distal margin of a bar alters the non-uniform pattern of flow over the bar in future floods, through changes in both flow direction (convergence–divergence in the horizontal plane) and depth (convergence–divergence in the vertical plane). Such changes must alter the spatial pattern of competence and capacity for bedload transport, with consequent feedback to the future evolution of bar and channel form. Feedback is initially through the constriction of channels alongside accreting bars; the resulting acceleration of flow typically leads to channel scour and bank erosion at locations that migrate as the bar does. This is positive feedback, since channel widening encourages further accretion. But as bars develop and migrate the feedback through flow changes is increasingly likely to be negative, through either chute incision and within-channel avulsion or pool ponding and overbank avulsion. These erosional interruptions to bar progradation can convert medial bars to lateral ones by bypassing their proximal pools, or lateral bars to medial ones by chute incision. The former possibility is regarded by Bluck (1979, fig. 11) as the usual mode of gravel-floodplain construction, but he has also reported cases from Iceland of lateral-bar dissection and detachment (Bluck, 1974, p. 539).

Since the downstream migration of alternate diagonal bars can be interrupted, and in more than one way, the wandering gravel river is not in a steady state, does not evolve progressively in any single way, and does not undergo any regular cycle of development. However, in our experience the potential for avulsion within or beyond the channel can be recognized beforehand even though prediction of these events is uncertain and of the longer-term future almost impossible. It may be that changes follow some kind of Markovian scheme with constant transition probabilities; as is well known to sedimentologists, this can generate characteristic, but not invariable, longer-term cycles of change.

That channel changes are stochastic not deterministic is perhaps inevitable given the rather random magnitude and frequency of competent floods. A large flood followed by a small one need not have the same consequences as the same floods in reverse order, or two floods of different size altogether. Minor floods also play a major role in superficial reworking of bars that are already the product of multiple episodes of aggradation and possibly erosion in bigger floods. Sedimentary structures are correspondingly varied, with a spatial and temporal mixture of fining-upwards deposition in falling stages on bar platforms and lee

margins (Bluck, 1976) and coarsening-upwards deposition during floods as bars prograde into slough pools (Bluck, 1979), or gravel sheets are transported over vegetated floodplain surfaces (as described above and illustrated in Fig. 8).

Our discussion is based on observations over a period long enough to encompass very considerable bar development and channel change, and using surveying methods that measure true aggradation and incision as well as possibly stage-dependent changes in exposed bar outline. Our conclusions must nevertheless remain tentative until replicated over a longer period and in other wandering gravel rivers. A firmer understanding of the processes involved is also required. Bar progradation and avulsion within and from the channel are discussed here in a purely kinematic sense without backing from indirect or direct assessment of local shear stresses through tracer pebble studies, *in situ* competence experiments, or measurement of velocity profiles. These constitute the next stage of research on the Feshie. Similar case studies of bar and channel change, combined with process measurements, in other rivers would be cf great value in improving understanding of a widespread and neglected class of gravelly channels.

CONCLUSIONS

(1) We interpret the evolution over 5 years of two sub-reaches of the Feshie in terms of one basic process, progradation of alternate diagonal bars, that can have varied consequences including bank erosion and avulsion both within the channel and from it. These kinematic 'processes' are explained as feedbacks from strongly non-uniform convergent–divergent flow, through sediment transport, to bar form; but detailed hydraulic measurements to confirm this remain to be done.

(2) The observed changes and the inferred basic process have affinities with the early stages of meandering in straightened gravel rivers and of both 'pseudomeandering' and braiding in coarse-bed laboratory experiments. We speculate that alternate diagonal bars are the common starting point from which divergent feedback mechanisms generate meandering or braiding patterns, unles they almost balance, giving the intermediate 'wandering gravel river' pattern of the Feshie.

(3) The varied consequences of diagonal-bar progradation mean that the wandering gravel river is not in a steady state and neither evolves in any simple way

nor undergoes any regular cycle of development, though it may undergo Markovian cycles. Most bars visible at any one time consequently have compound histories and need not have originated in their present positions with respect to the low-flow channel. Sedimentation is predominantly of horizontally bedded imbricated gravel, but with a mixture in space and time of coarsening-upwards and fining-upwards behaviour.

ACKNOWLEDGMENTS

We thank the Glenfeshie Estate and Nature Conservancy Council for permission to work in the study area, and the Estate for access by a private road. The Universities of St Andrews and Stirling, and the British Geomorphological Research Group, contributed to fieldwork expenses. Surveying was greatly assisted by many undergraduate and graduate students from St Andrews and Stirling and by unwary visiting geoscientists. We have profited from discussions in the field with participants in B.G.R.G., International Fluvial Conference, and private field trips and wish to acknowledge in particular helpful discussions with Brian Bluck, Mike Church and John Lewin.

REFERENCES

ASHMORE, P.E. (1982) Laboratory modelling of gravel braided stream morphology. *Earth Surf. Processes*, **7**, 201–225.

BLUCK, B.J. (1974) Structure and directional properties of some valley sandur deposits in southern Iceland. *Sedimentology*, **21**, 533–554.

BLUCK, B.J. (1976) Sedimentation in some Scottish rivers of low sinuosity. *Trans. Roy. Soc. Edinb.* **69**, 425–456.

BLUCK, B.J. (1979) Structure of coarse-grained braided stream alluvium. *Trans. Roy. Soc. Edinb.* **70**, 181–221.

BLUCK, B.J. (1982) Texture of gravel bars in braided streams. In: *Gravel-bed Rivers: Fluvial Processes, Engineering and Management* (Ed. by R.D. Hey, J.C. Bathurst and C.R. Thorne). Wiley, Chichester, in press.

BUCK, S.G. (1978) *The sedimentology of a coarse bed-load braided river, the R. Feshie, Inverness-shire, Scotland.* Unpublished M.Sc. Thesis. University of Reading, U.K. 120 pp.

CHURCH, M. & GILBERT, R. (1975) Proglacial fluvial and lacustrine environments. In: *Glaciofluvial and Glaciolacustrine Sedimentation* (Ed. by A.V. Jopling and B.C. McDonald). *Spec. Publs Soc. econ. Paleont. Miner.* **23**, 22–100.

CHURCH, M., MOORE, D. & ROOD, K. (1981) *Catalogue of Alluvial River Channel Regime Data.* Department of Geography, University of British Columbia.

FERGUSON, R.I. (1973) Channel pattern and sediment type. *Area*, **5**, 38–41.

FERGUSON, R.I. (1981) Channel forms and channel changes. In: *British Rivers* (Ed. by J. Lewin), pp. 90–125. Allen & Unwin, London.

HEIN, F.J. & WALKER, R.G. (1977). Bar evolution and development of stratification in the gravelly, braided, Kicking Horse River, British Columbia. *Can. J. Earth Sci.* **14**, 562–570.

HICKIN, E.J. (1969) A newly identified process of point bar formation in natural streams. *Am. J. Sci.* **267**, 999–1010.

HOOKE, J. (1980) Magnitude and distribution of rates of river bank erosion. *Earth Surf. Processes*, **5**, 143–157.

LEOPOLD, L.B. & WOLMAN, M.G. (1957) River channel patterns: braided, meandering and straight. *Prof. Pap. U.S. geol. Surv.* **282-B**.

LEWIN, J. (1976) Initiation of bedforms and meanders in coarse-grained sediment. *Bull. geol. Soc. Am.* **87**, 281–285.

LEWIN, J., HUGHES, D. & BLACKNELL, C. (1977) Incidence of river erosion. *Area*, **9**, 177–180.

MIALL, A.D. (1977) A review of the braided-river depositional environment. *Earth Sci. Rev.* **13**, 1–62.

PARKER, G. (1976) Cause and characteristic scales of meandering and braiding in rivers. *J. Fluid Mech.* **76**, 457–480.

PARKER, G. & PETERSON, A.W. (1980) Bar resistance of gravel-bed streams. *J. Hydraul. Div. Am. Soc. civ. Engrs* **106**, 1559–1573.

SMITH, N.D. (1974) Sedimentology and bar formation in the upper Kicking Horse River, a braided outwash stream. *J. Geol.* **82**, 203–223.

SOUTHARD, J.B., SMITH, N.D., DRAKE, T.G. & KUHNLE, A. (1981) Field and laboratory studies of braiding in shallow gravel-bed streams. In: *Abstracts: Modern and Ancient Fluvial Systems: sedimentology and processes.* University of Keele, U.K.

STEBBINGS, J. (1964) The shape of self-formed model alluvial channels. *Proc. Inst. civ. Engrs* **25**, 485–510 (with discussion, **26**, 225–232).

WERRITTY, A. & FERGUSON, R.I. (1980) Pattern changes in a Scottish braided river over 1, 30 and 200 years. In: *Timescales in Geomorphology* (Ed. by R. Cullingford *et al.*), pp. 53–68. Wiley, Chichester.

WERRITTY, A. (1982) Stream response to flash floods in upland Scotland. In: *Proceedings of the IGU Commission on Field Experiments in Geomorphology* (Ed. by D.E. Walling and T.P. Burt). Geo-books, Norwich, in press.

WOLMAN, M.G. & BRUSH, L.M. (1961) Factors controlling the size and shape of stream channels in coarse noncohesive sand. *Prof. Pap. U.S. geol. Surv.* **283-G**.

Spec. Publs int. Ass. Sediment. (1983) **6**, 195–206

Morphology and sedimentology of a sinuous gravel-bed channel system: lower Babbage River, Yukon coastal plain, Canada

D. L. FORBES*

University of British Columbia, Vancouver

ABSTRACT

The lower Babbage River is a non-braided sinuous gravel-bed stream of the poorly documented class-5 meandering channel facies assemblage (Jackson, 1978). It exhibits a low-Arctic nival runoff regime. The low-flow channel has a sinuosity of 1·89, a mean slope of $7·4 \times 10^{-4}$, mean width 62 m, mean maximum depth 1·36 m, and a variable width/depth ratio $24 \leqslant b/h_* \leqslant 226$. The sinuosity and mean width of the channel zone are respectively 1·3 and 222 m. The channel is dominated by lateral or point bars, diagonal bars, and a prominent pool-riffle sequence. The alluvial sediments are described using a modified version of the lithofacies classification proposed by Miall (1978). Imbricate gravel of lithofacies Gm ($12 < d_{50} < 20$ mm) forms an armour layer on channel-floor and bar-top surfaces. Bar-body deposits of sandy gravel ($6 < d_{50} < 12$ mm) include planar cross-stratified gravel (Gp) and massive or crudely stratified units (Gm). Bar-top sediments include gravel sheets and armour surfaces (Gm), shallow scours (Ge), and scour fills or veneers of sand and mud. Trough cross-stratified scour- or channel-fill gravels (Gt) are plausible but unconfirmed additions to the lithofacies set. Parallel-ridge gravels (Gr) are found only in reaches subject to overflow icing. Bar-tail sediments include planar cross-stratified gravels (Gp) and planar, trough and ripple cross-stratified sands (Sp, St, Sr). Channel-zone gravel bodies may exhibit planar bases and convex tops, with total width of 100–200 m and maximum thickness greater than 3 m. Overbank sediments, including peat (C), ice (I), stratified silts and sands (Fl and Sh: $0·01 < d_{50} < 0·3$ mm), and organic muds (Fsc), may exceed 8 m in thickness and cover a much greater area than any of the channel facies in the lower valley.

INTRODUCTION

The character of deposition in non-braided sand-gravel and gravel streams has attracted increased attention during the past few years. Recent contributions include reports by Arche (1981), Bluck (1971, 1976), Church & Kellerhals (1978), Gustavson (1978), Jackson (1976, 1978), Lewin (1972, 1976), Martini (1977), Nijman & Puigdefabregas (1978), Ori (1979, 1982) and Ori & Ricci Lucci (1981). It has become clear that, just as braided channel systems exhibit a wide variety of morphological features and associated

depositional sequences, so a range of river types and a number of distinctive lithofacies sets can be recognized in non-braided streams.

Jackson (1978) proposed a five-part classification of meandering channel lithofacies, with coarse-grained channel systems grouped under class 5. Deposits of such streams remain poorly documented. Non-braided channel systems that may be assignable to class 5 include, in addition to the Little Wind and other Wyoming rivers (Jackson, 1978; Leopold & Wolman, 1957), the Rio Grande in Jamaica (Forbes, unpublished observations, 1978), the Nueces in Texas (Gustavson, 1978), various Welsh rivers (Lewin, 1972, 1976), parts of the Squamish, Peace, North Saskatchewan, Red Deer, and other rivers in western Canada (Kellerhals, Neill & Bray, 1972; Hickin, 1978), and

* Present address: Atlantic Geoscience Centre, Geological Survey of Canada, Box 1006, Dartmouth, Nova Scotia B2Y 4A2, Canada.

0141-3600/83/0106-0195 $02.00

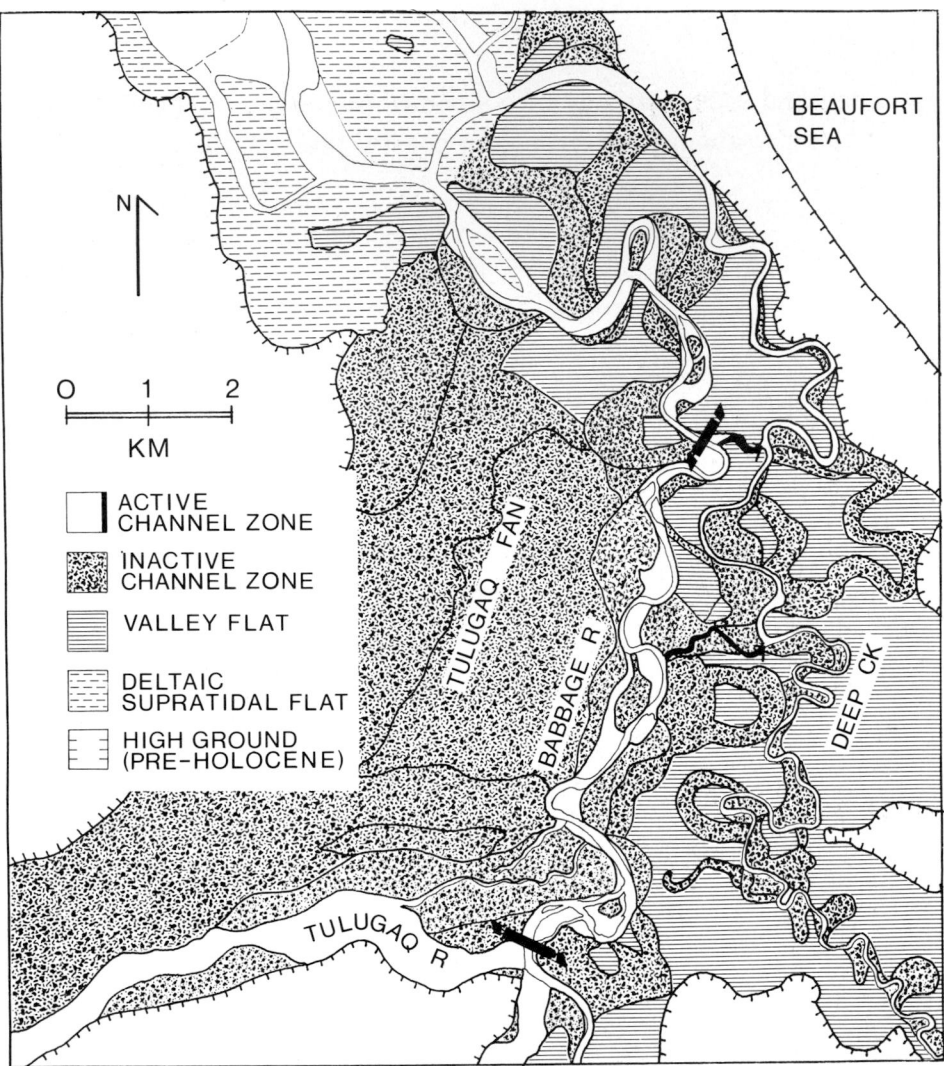

Fig. 1. Distribution of channel-zone facies in the lower Babbage River valley, Yukon Coastal Plain. Inactive channels are those recognizable at the present surface and are mapped from air photographs. Reach described in this paper is delimited by heavy bars at Tulugaq River confluence and at limit of storm-surge backwater. Two distributaries flowing from Babbage River to Deep Creek are shown in black. Note Tulugaq River fan and deltaic facies associated with late Holocene marine transgression.

Quaternary distal fan deposits of the River Reno in Italy (Ori, 1979, 1982). Although the rivers cited above include both high- and low-sinuosity channels ($1\cdot04 \leqslant S < 5$), it is not clear that an arbitrary sinuosity criterion leads to the most appropriate classification. The range of examples does support the suggestion (Jackson, 1978, p. 561) that class 5 streams may occur in a variety of climatic settings.

This paper provides a review of morphological and sedimentological data from the lower Babbage River

(Figs 1 and 2). This non-braided sinuous gravel-bed stream shows many similarities to the class 5 model (Jackson, 1978, p. 563).

General setting

The Babbage River drains some 4200 km² in the northern Yukon Territory (western Canadian Arctic), including parts of the Barn and British Mountains rising to 1300 m elevation and a segment of the Yukon

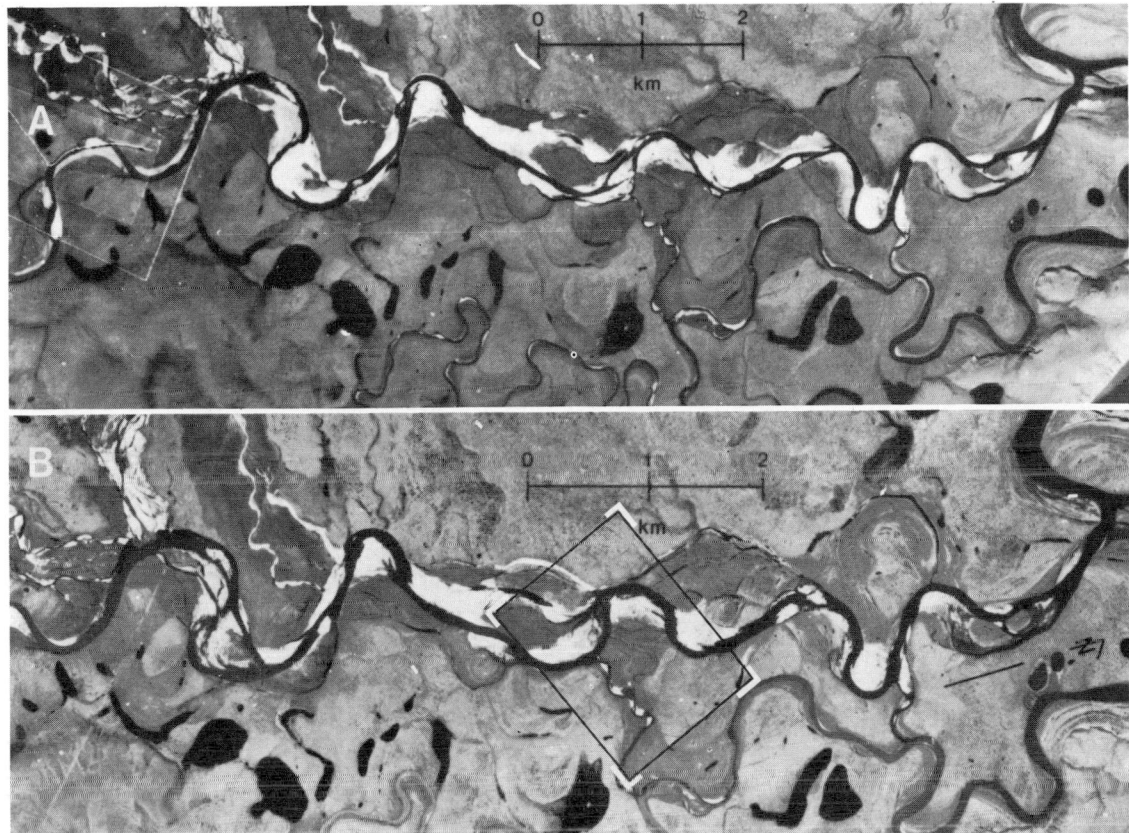

Fig. 2. Lower Babbage River between Tulugaq River confluence and tidal reach. (A) 1954 (part of NAPL A14406-48). (B) 1976 (part of NAPL A24502-170). Rectangle outlines area in Fig. 5.

coastal plain. The river empties into the Beaufort Sea near Kay Point (69° 18′ N, 138° 24′ W), some 100 km west-north-west of the Mackenzie Delta. Source materials include Precambrian metaclastics, Palaeozoic and Mesozoic clastics and carbonates, and Cenozoic clastic sediments, including Tertiary pediment gravels and various Quaternary glacial and non-glacial deposits.

The mean annual near-surface ground temperature is approximately -8.5 ± 1.5 °C (Mackay, 1975) and permafrost extends to depths exceeding 600 m in some places (J. R. Mackay, personal communication, 1981). The mean annual air temperature (screen) is -10.4 °C at Shingle Point, a coastal station near the eastern boundary of the basin, where mean annual precipitation totals 188 mm (76 mm as snow).

RUNOFF AND SEDIMENT TRANSPORT REGIMES

The Babbage River and a small adjacent stream, Deep Creek (Fig. 1), exhibit a low-Arctic nival runoff regime (Fig. 3), characterized by zero flow at the mouth during five to seven months of the year and a major snowmelt runoff flood in June. The channel is ice-covered or, in places, ice-filled during the winter months. A number of springs in the basin continue to flow through the winter, producing extensive icings. Peak discharge during the snowmelt flood in the lower Babbage River (below the Tulugaq River confluence) exceeded 500 m³ s⁻¹ in three out of four years of observation (1974–1977). A second major flood of comparable magnitude, resulting from combined rain and snowmelt runoff, occurred in late June 1976 (Fig. 3). Storm runoff events with peak flow in the range $50 < Q < 100$ m³ s⁻¹ are not uncommon during July, August and September; on the other hand, discharge

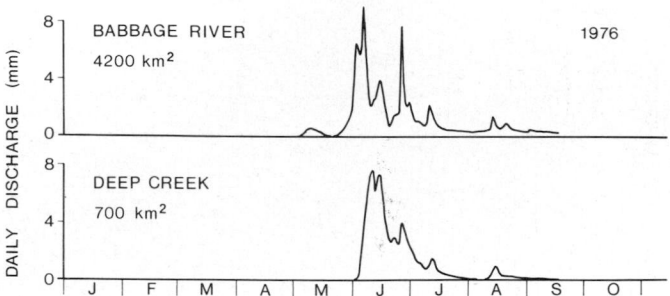

Fig. 3. Specific daily discharge (mm) in Babbage River below Tulugaq River and Deep Creek above Babbage distributaries, January–October 1976. Note prolonged zero flow in winter and major snowmelt runoff events in June; second major flood in Babbage River included both rain and snowmelt runoff.

may decline to less than 10 m³ s⁻¹ during the summer recession in July. Water Survey of Canada records, initiated in 1977 at a section in the upper basin (drainage area 1500 km²), include an unusual summer storm event exceeding the nival flood, with peak flow greater than 380 m³ s⁻¹ (August 1979).

Major sediment transport events are associated with the snowmelt runoff flood and, to a lesser degree, with storm runoff later in the season. Bed-contact transport rates in the Babbage River have been estimated following Bagnold (1966) at high flow and using Helley-Smith sampler data for discharges less than 60 m³ s⁻¹ (Forbes, 1981). Transport rates may approach 100 kg s⁻¹ at high flow ($Q > 500$ m³ s⁻¹) but decline to near zero at low flow ($Q < 50$ m³ s⁻¹). Estimates of suspended sediment transport range from about 10^{-1} kg s⁻¹ at $Q = 10$ m³ s⁻¹ to almost 10^3 kg s⁻¹ at high flow. Estimated annual bed-contact and suspended sediment transport amount, respectively, to 8×10^7 and 2.5×10^8 kg for 1975 and to 7×10^7 and 1.7×10^8 kg for 1976. Although less than 8% of total sediment transport is gravel ($d > 2$ mm), nevertheless the gravel fraction has a very significant effect on the form of the channel and the nature of the depositional record.

CHANNEL MORPHOLOGY OF THE LOWER BABBAGE RIVER

The lower reaches of the Babbage River and Deep Creek occupy a broad valley of probable early Wisconsinan age (Mackay, 1959; Rampton, 1982) in which a diverse body of Wisconsinan and Holocene alluvial sediment has accumulated (Fig. 1). Deep Creek, the Babbage River, and two distributaries flowing to Deep Creek from the Babbage, form an anastomosing channel system (cf. Smith & Smith, 1980) in the lower valley above the delta.

The distribution of Holocene channel sediments and the morphology of the lower Babbage River have been strongly influenced by lateral channel migration, meander cutoff, and avulsion; by sediment discharge from the braided tributary Tulugaq River; and by deposition and channel avulsion on the Tulugaq fan (Fig. 1). At present, the channel is slightly entrenched but is believed to be aggrading downstream from the Tulugaq River confluence.

The present study pertains to a 10 km reach of the Babbage River between the Tulugaq River and the limit of storm-surge backwater (Figs 1 and 2). Within the lower 6·25 km segment of this reach, the channel has a sinuosity of 1·89, a variable width/depth ratio $24 \leqslant b/h_* \leqslant 226$, and a mean slope of $7·4 \times 10^{-4}$ at low flow ($Q \simeq 25$ m³ s⁻¹). The low-flow channel has mean maximum depth $\bar{h}_t = 1·36 \pm 0·15$ m ($0·4 \leqslant h_t \leqslant 3·6$ m) and mean width $\bar{b} = 62 \pm 4$ m ($39 \leqslant b \leqslant 109$ m) ($n = 25$). The mean width of the channel zone, including unvegetated bar surfaces, is $b_f = 222$ m. Further data on a short segment of the reach described here were provided by McDonald & Lewis (1973, pp. 143–149).

The channel is locally divided and is dominated by lateral or point bars, diagonal bars, and a prominent pool-riffle sequence (Figs 4 and 5). The pool-riffle wavelength is typically no longer than one-quarter of the dominant planform wavelength, L_p; the latter is related to the scale of the lateral bars ($\bar{L}_p \simeq 900$ m $\simeq 4b_f$). Although the riffles may function primarily as hydraulic resistance elements, the lateral bars constitute the major gravel storage sites in the fluvial system (cf. Church & Jones, 1982). Riffles occur on diagonal and transverse bars within the low-flow channel (Fig. 6) and at locations where the channel crosses over the

Fig. 4. Lateral bars and major riffle in Babbage River 7 km downstream from Tulugaq River confluence (part of reach illustrated in Fig. 5). (A) August 1974, looking upstream with distributary channel and Deep Creek in distance; note second riffle on diagonal bar in left foreground. (B) August 1975, looking downstream; note chute-channel erosion in middle distance and bar-tail sand in foreground.

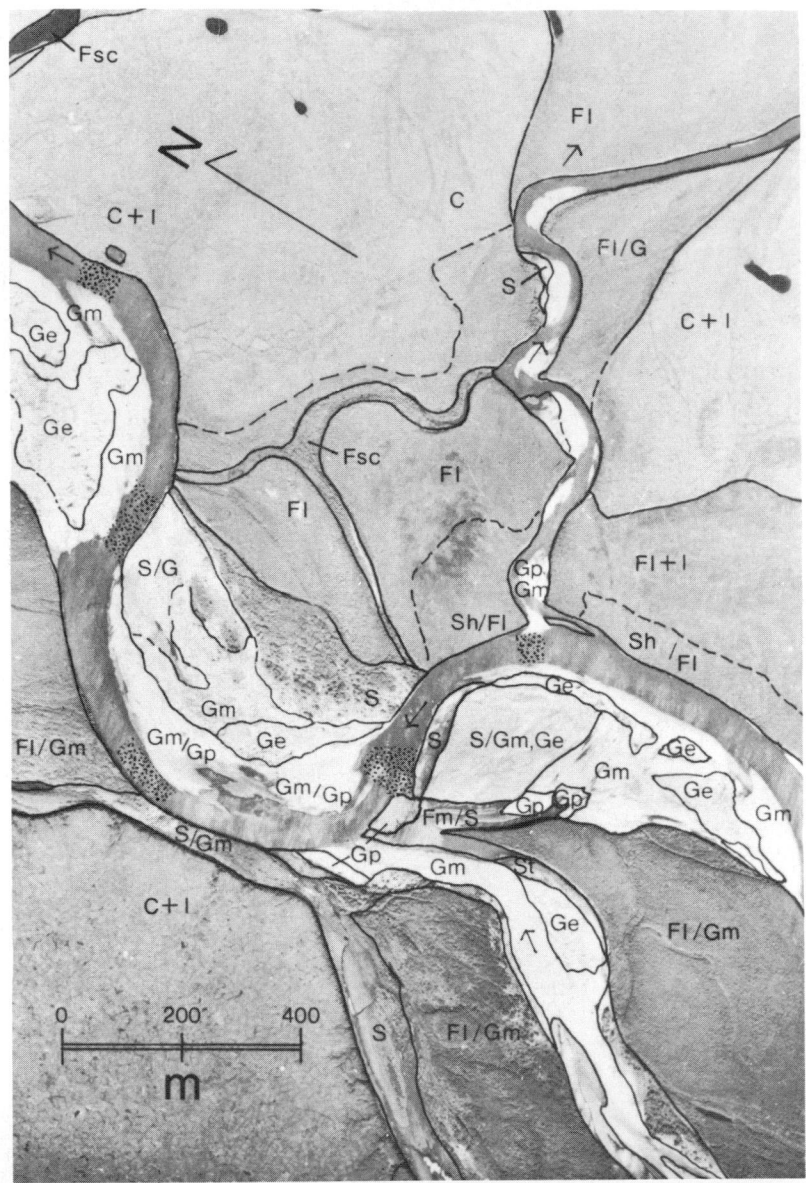

Fig. 5. Surface distribution of lithofacies types in representative area of valley approximately 7 km downstream from Tulugaq River confluence (see Figs 2 and 4 for location). Lithofacies codes as defined in Table 1. Part of NAPL A21826-87, August 1970.

major sediment storage body formed by successive lateral bars (Fig. 5). This sediment body, which dominates the channel zone, has a sinuosity of about 1·3 (see Fig. 2). While the detailed morphology of riffles and chute channels may be altered with each major flood (Fig. 4), substantial changes in the form of the channel zone, including intermittent dissection of lateral bars, occur at longer intervals (Fig. 2).

The style of junction between lateral bars and the valley flat is variable. Some bars slope consistently up to the valley flat, but others (roughly 4 in 10) exhibit an inner cutbank with an adjacent secondary channel (see Fig. 7). Discontinuous inner accretionary banks are present at some sites (Fig. 4A).

A number of distinctive morphological and sedimentary features of the lower Babbage River result

Fig. 6. Transverse bar forming riffle in lower Babbage River about 4 km downstream from Tulugaq River confluence. Flow from left to right. Channel width approximately 80 m; August 1974.

Fig. 8. Distinctive quasi-parallel gravel ridges (lithofacies Gr) associated with icing development in a headwater reach of the Babbage River. Similar features were observed on bar-top surfaces downstream from Tulugaq River in August 1974.

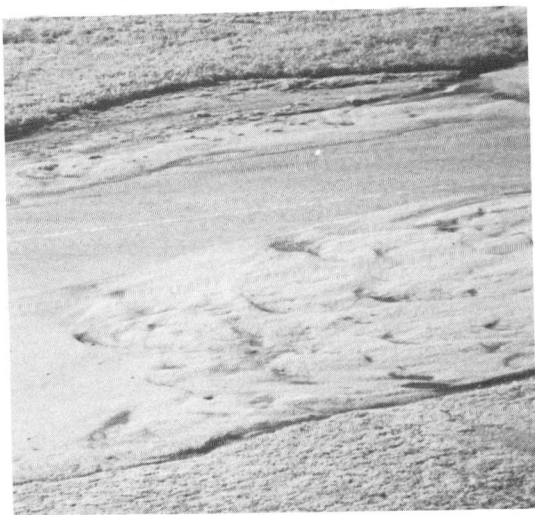

Fig. 7. Babbage River channel 10 km downstream from Tulugaq River confluence (flow left to right). Lateral bar with main channel in distance and secondary channel against inner cutbank in foreground. Note shallow scours (lithofacies Ge) on bar-top surface and sheet gravel (Gm) on proximal bar at left. Width of bar approximately 100 m; August 1975.

from formation and break-up of ice in the channel. Ice jams and bottomfast ice may play a role in the maintenance of active upper bar surfaces and may cause flow diversions leading to channel avulsion. Bottomfast ice formed by progressive downward freezing in

low-flow channels may form a temporary raised channel bed during initial snowmelt runoff in spring (Forbes, 1979), but is not associated with any particular morphological or sedimentary feature. On the other hand, a distinctive morphology characterized by quasi-parallel ridges of order 0·1–1 m in height, with wavelength of order 1–10 m (Fig. 8) is uniquely associated with reaches in which overflow icings develop. Other features directly attributable to ice are minor ice-rafted deposits, and small constructional, scour and ablation features due to ice grounding on bar surfaces (Fig. 9).

LITHOFACIES CLASSIFICATION

Various types of depositional environment generate characteristic deposits having distinctive sets of textural and structural attributes. These lithological characteristics may be summarized in terms of assemblages of standard lithofacies types, each defined by a limited range of particle size, specified compositional properties, structural characteristics or other criteria. A lithofacies classification of this kind, developed by Miall (1977, 1978) for application to braided river deposits, has been adopted for the present purpose in a modified and expanded form (Table 1, Fig. 5). Modifications include the application

Fig. 9. Minor ice-related features on bar-top surfaces of lower Babbage River. (A) Ice-rafted gravel on bar-top 10 km downstream from Tulugaq confluence (cf. Fig. 7); paddle for scale; August 1976. (B) Ice-pushed gravel, ice scour, and ablation pit on proximal bar surface 4·5 km downstream from Tulugaq; spade for scale; June 1976.

of lithofacies code Fsc to massive or finely laminated muds accumulating in valley-flat lakes and ponds and of code C to modern peat. The set of structures associated with lithofacies Fl is expanded to include root channels, desiccation cracks and ice wedges. Explicit differentiation of aqueous and aeolian sand deposits in the lithofacies code is rejected on the ground that the classification should be descriptive rather than genetic. In the absence of suitable descriptive criteria, therefore, wind-blown sands on terrace surfaces are coded Sh (rather than She). Additions to the set of lithofacies previously defined (Miall, 1978, p. 598) include shallow scours in gravel (Ge), parallel-ridge gravel deposits (Gr) in reaches subject to icing, ice (I), and undifferentiated sands and gravels (S and G).

(Although ice is ephemeral in the long-term stratigraphic context, it can be a major determinant of facies geometry and surface morphology in high-latitude systems. Furthermore, major ice bodies may be preserved in the record over intervals of order 10^4–10^5 yr, as in some of the sediments exposed along the central Yukon coast — Mackay, Rampton & Fyles, 1972.)

ALLUVIAL DEPOSITS OF THE LOWER BABBAGE RIVER

Imbricate pebble or pebble–cobble gravel of lithofacies Gm ($12 < d_{50} < 20$ mm) is characterictic of channel-floor and bar-top surfaces, where it forms a thin armour layer (in the sense of Bray & Church, 1980). Maximum clast size ranges from 45 mm to more than 100 mm. Although the armour surface at many channel sections is laterally continuous from the low-flow channel on to the bar, progradation of the bar over channel-floor gravel would generate a three-part sequence analogous to the channel-floor, bar-body and bar-top facies described by Ori (1979).

Bar-body deposits are predominantly sandy gravel ($6 < d_{50} < 12$ mm). They include planar cross-stratified gravel of lithofacies Gp, associated with diagonal and transverse bars (Figs 4, 6 and 10); and massive or crudely stratified deposits of lithofacies Gm, formed by progradation of thin sandy-gravel sheets across the bar surface (cf. Hein & Walker, 1977; Gustavson, 1978). The channel geometry (McDonald & Lewis, 1973, p. 145; Forbes, unpublished data) suggests that preserved bar-body units may have planar bases and convex upper surfaces (cf. Arche, 1981), mean stream-normal width of approximately 160 m, and maximum thickness of 3 m or more.

as the
upstrea
straight
control
channel
transpo
and gra
tempora
reach, v
varies v
flood ti
into thi
bars. A
downstr
This bra
expressi
conditic
There a
grown s
result o
promine
these ac
functior
braided
depende
followin

    Below
sinuous,
upstrear
low run
the strai
river. C
moving
downstr
dominat
end of t
higher v
reversin
runoff 1
pattern i
near the
discharg
in the u
suspende
bedrock
in low-v
(Renwic
substanti
transpor
11).

    The o
meander

Table 1. Lithofacies types of the lower Babbage River

code	Lithology	Structure	Location	Area (%*)
G†	Gravel ($d_{50} > 2$ mm)	Undifferentiated	Channel zone	7
Gm	Gravel ($d_{50} > 2$ mm)	Massive or horizontally stratified, commonly imbricate	Channel-floor; bar-body; bar-top	6
Gp	Gravel ($d_{50} > 2$ mm)	Planar cross-stratified	Riffle; distal bar body (bar-tail)	3
Gt	Gravel ($d_{50} > 2$ mm),	Trough cross-stratified	Bar-top?	?
Ge†	Gravel ($d_{50} > 2$ mm)	Shallow scours with gravel bases and scour fill of Gt?, Sr, Fm	Bar-top	3
Gr†	Gravel ($d_{50} > 2$ mm)	Low laterally extensive quasi-parallel ridges	Bar-top	<1
St	Sand ($0·06 < d_{50} < 2$ mm)	Undifferentiated	Channel zone; overbank	8
Sh	Sand ($0·06 < d_{50} < 2$ mm)	Horizontally laminated	Bar-top; overbank	<1
Sl	Sand ($0·06 < d_{50} < 2$ mm)	Low-angle cross-stratified	Bar-top	<1
Sp	Sand ($0·06 < d_{50} < 2$ mm)	Planar cross-stratified	Bar-tail; channel-fill	<1
St	Sand ($0·06 < d_{50} < 2$ mm)	Trough cross-stratified	Bar-tail; channel-fill	<1
Sr	Sand ($0·06 < d_{50} < 2$ mm)	Ripple cross-stratified	Bar-top; bar-tail; channel-fill; overbank	<1
Fl‡	Sand–silt–mud ($d_{50} < 0·06$ mm)	Finely laminated with roots, desiccation cracks, ice wedges	Overbank	29
Fsc‡	Mud ($d_{50} \ll 0·06$ mm)	Massive or finely laminated	Lake-bottom	1
Fm	Mud ($d_{50} \ll 0·06$ mm)	Massive, desiccation cracks	Bar-top; overbank	<1
C†	Peat	Massive or crudely stratified; may be interstratified with F or S; rare allochthonous clasts	Valley flat (autochthonous); bar-top (allochthonous)	39
It†	Ice	Massive, vertically foliated, or horizontally foliated; may be interstratified with other lithofacies	Valley flat	3?

* Percentage of area covered by each lithofacies type; estimates based on area shown in Fig. 5 (cf. Fig. 2); note that G may include other gravel lithofacies and S other sand lithofacies.
† Newly defined lithofacies type.
‡ Modified definition (cf. Miall, 1978).

PRANDTL, L. (1952) *Essentials of Fluid Dynamics*. Blackie, London. 452 pp..

RENWICK, W.H. & ASHLEY, G.M. (1982) Influence of tidal fluctuations on sediment transport—sources, storages, and sinks in the Raritan River, New Jersey. *Abstr. Prog. geol. Soc. Am.* **14**, Nos 1 and 2, 76.

UNITED STATES GEOLOGICAL SURVEY (1964–1980) Water resources data for New Jersey. *U.S.G.S. Water-data Rep. NJ*-64-1 *to NJ*-80-1.

WRIGHT, L.D. (1977) Sediment transport and deposition at river mouths: a synthesis. *Bull. geol. Soc. Am.* **88**, 857–868.

Spec. Publs int. Ass. Sediment. (1983) **6**, 219–227

High-magnitude floods and stream channel response

AVIJIT GUPTA

Department of Geography, National University of Singapore, Singapore 0511, Singapore

ABSTRACT

This paper discusses two premises: (*a*) large floods may play a bigger role than normally recognized in determining channel morphology, and (*b*) such effects tend to follow a geographical distribution of drainage basins. Post-flood observations of stream channels are common but inductive conclusions are not. Field observations from Jamaica, Maryland, U.S.A. and South-east Asia, and a review of case-studies indicate that high-magnitude floods leave a stable imprint on channel form, if certain conditions are fulfilled. Such conditions involve the presence of a considerable supply of coarse detritus, length of the recurrence interval of large floods, the ratio of peak flood discharge to bankfull flow, etc. Stability of flood-affected forms does not only depend on the erosive, or transporting or depositional ability of the flood discharge, but also on the inability of smaller flows to modify such forms in the inter-flood period. Attempts are made to construct and locate environments where the effect of large floods will be too pronounced to explain the channel entirely in terms of the most commonly used parameter: the bankfull discharge. Under such circumstances channel forms may be visualized as lag effects of a high-magnitude flood being slowly altered in most cases by flows of smaller magnitude. Erosional and depositional effects of such channels may also be preserved in the floodplain alluvium or in terrace deposits, thereby acting as a palaeohydrological indicator.

INTRODUCTION

Those who have seen a major flood travelling down a river, or have visited a channel immediately after the passage of a large flood, are usually impressed by the modifications made by the flow. The wealth of post-flood observations on stream channels is a testimony to this experience. However, existing studies seem to differ from one another regarding the role of major floods on the formation and maintenance of stream channels. The flood effects on certain streams–even if impressive immediately after the flood–often turn out to be transitory, and are soon altered or concealed by the work of more frequent and smaller flows. In contrast large floods on other streams leave impressive and at least quasi-permanent results: channels widened, depositional forms created, bed and bank materials coarsened. It may even be possible in the case of some channels to attribute much of their appearance to the work of large floods.

All this leads to the inference that there are factors which determine the nature of stream channels' response to high-magnitude floods. Certain streams may owe their appearance not only to bankfull discharge but also to the periodic passage of large floods down their channels. Post-superflood occurrence of smaller and commoner floods tends to return the channel to its pre-superflood form, and the duration of time necessary to accomplish this varies from stream to stream. Channels can occur in a condition of disequilbrium to the accepted channel forming discharge at bankfull level. The purpose of this paper is to examine the implications of such inferences.

TRANSITORY FLOOD EFFECTS ON STREAM CHANNELS: A REVIEW

In order to determine the effects of large floods on river channels a selection of case studies on flood-affected channels will be reviewed, starting with

0141-3600/83/0106-0219 $02.00

examples of studies in situations where superflood effects have been transitory.

A carefully recorded story of unusual flooding is available from the upper course of the Patuxent river in Maryland, U.S.A. and its main tributaries for 1971 and 1972. Four large floods occurred in this drainage basin between 3 August 1971 and 24 June 1972, two of which (on 11–12 September 1971 and 21–24 June 1972) had recurrence intervals of near or about 100 yr. As these floods arrived in the middle of a field investigation being carried out in the upper part of the Patuxent Basin, it was possible to examine pre-flood, syn-flood and post-flood channel conditions repeatedly. This happened in an area of broadly undulating landscape dissected by narrow and deep river valleys where the Patuxent flows over a thin valley-fill overlying phyllites and schists of Precambrian age. The average annual rainfall is around 1000 mm with no distinct dry season, soils are moderately deep, and the drainage area is mainly in woodland and farmland.

The details of these flood events have been described elsewhere (Gupta & Fox, 1974). In sum, the effects involved widening of channel, increase in transport competence, clearing the channel of nearly all but very coarse bed material, erosion of channel bars, overbank splays, and destruction of vegetation adjacent to the channel. Within several months following a flood the channels showed signs of returning to a smaller size; fine sand, silt and clay returned to the channels; and bars were rebuilt at their old locations. The only flood evidences that remained were the presence of smaller and younger trees next to the channel in places, the localized occurrence of 0·6–0·9 m scarplets on the floodplain, and equally localized pebble layers found in cutbanks. None of these is particularly distinctive, and apparently in Maryland Piedmont channels are not affected by high-magnitude floods on a permanent basis. Similar results were obtained by an independent study on a 155 km² neighbouring drainage basin (Costa, 1974).

Another example of the transitoriness of the effects of large floods was observed in the channel and on the floodplain of a small, 4·5 km long stream in the secondary forest around the MacRitchie Reservoir in Singapore. The drainage basin in this gently rolling landscape is on granite, and towards the lower part of the basin there is a thin layer of alluvium underlying the channel and the floodplain. The bed materials are mainly coarse sand with granules and a few fine pebbles, whereas the banks are in coarse sand with a top layer of silt. The channel meanders with a low sinuosity, and mid-channel bars are locally formed behind logjams.

Singapore experienced storm rainfall of unusual intensity and amount on 2 December 1978. The raingauge nearest to the channel recorded 360·3 mm, and the highest hourly intensity was 55·2 mm. Approximately one-sixth of the mean annual precipitation was delivered by this single storm. In the resulting flood the water was about 1·2 m above the floodplain, and though in the absence of gauging it was not possible to determine the peak discharge or the duration of overbank flow, from field observation it was found that overbank flooding continued for more than a day.

The effect of this flood again was not impressive. Post-flood observations revealed 8–10 cm thick splays of coarse sand and granules, thin silt layer on the floodplain, and destruction of floodplain vegetation including both shrubs and riverside trees. There were a few cases of mid-channel building of bars associated with trees fallen across this narrow (4–5 m wide) channel. One may thus restrict the role of high-magnitude floods in the small rivers draining the rolling countryside of southern West Malaysia and Singapore to these. There was no change in the overall channel geometry and form, and over time smaller flows have modified whatever flood effect there was, and the channel has returned to its original appearance.

In an often quoted reconnaissance study of flooding in Connecticut, Wolman & Eiler found the extent of valley bottom modification by flood to be surprisingly small and localized. The flood effects listed, however, include a number of boulders, 1·5–2 m in diameter, moved by floodwaters. But outside certain suitable locations, the general effect of about 6 m of water was not pronounced (Wolman & Eiler, 1958).

On the other hand there are many areas of the world where the effects of high-magnitude floods on stream channels have not only been found impressive immediately after the flooding, but where such modifications of channels have persisted over time as normal forms in the channel and valley bottom.

STABLE IMPRINT OF LARGE FLOODS ON CHANNEL FORM: A REVIEW

There are a number of post-superflood case-studies which indicate not only an impressive amount of erosion and deposition carried out by the floodwaters, but also at least the semi-permanence of the

Table 1. Estimated high rainfall eastern Jamaica

Station (north to south)	Approximate location in river basins	Data available from	24 h rainfall with recurrence interval of (mm)			
			2 yr	5 yr	10 yr	25 yr
Buff Bay	Lower Buff Bay	1895	215	290 (264)	320 (315)	380 (378)
Balcarres	Middle Buff Bay	1942	305	445	584	735
Mt Holstein	Middle Buff Bay	1910	280	460 (368)	520 (460)	572 (579)
Cinchona	Headwaters, both Buff Bay & Yallahs	1895	205	330 (297)	420 (353)	635 (442)
Mavis Bank	Middle Yallahs	1924	140	240 (244)	330 (302)	395 (378)
Easington	Lower Yallahs	1895	130	240 (295)	290 (363)	355 (477)

Source: Data processed from *Jamaica Weather Report*.

Note: Evans (1972) has estimated maximum 24 hr rainfall for certain stations in Jamaica. His figures, where available, are shown within parentheses. There are cases of both excellent agreement and wide disparity, but both estimates indicate the expected recurrence of very high rainfall events in eastern Jamaica.

flood-modified valley forms. Two river basins studied in eastern Jamaica provide good examples of such valleys.

A part of the rain received annually in eastern Jamaica may arrive as intermittent downpours of very high magnitude and intensity. Such heavy rainfall is associated with cold fronts travelling from the north, waves in the easterlies, or tropical storms which occasionally develop into hurricanes. Vickers (1967) listed 21 separate instances of at least 400 mm of rainfall in one day in Jamaica in 95 years (1868–1963). The maximum one-day rainfall recorded was 1109 mm which fell at Bowden Pen, Portland on 23 January 1960. In fact this station had 2789 mm of rain in the four days of 22–25 January 1960. Between 5 and 8 November 1909, 2451 mm of rain fell at Silver Hill, St Andrew (Vickers, 1967). In order to determine the expected recurrence of high-magnitude rainfall, data

from six reliable stations in the Yallahs and Buff Bay Basins were analysed following the technique used by Lirios (1969) in eastern Jamaica, based on Gumbel's method of extreme value analysis. This is presented in Table 1, which shows that in this area a 24 h 5 yr recurrence interval rainfall is at least 250 mm; and for the same duration but with a recurrence interval of 10 yr a precipitation of 300–550 mm can be expected. This suggests the occurrence of high-magnitude flood at least once on average in every 10 yr, especially as the figures in Table 1 are for 24 h only.

Quantitative information regarding the magnitude of these floods is, however, scanty. Gauging stations are few in number in eastern Jamaica, and those that exist either have records of limited length or gaps in the record during high floods. An attempt was made to compile a record of major floods, in which water flowed a few metres deep on the floodplain, from field

Table 2. Estimates of major floods on the Yallahs River

Station	Drainage area (km²)	Type	Peak flow m³ s⁻¹	Peak flow m³ s⁻¹ km⁻²	Source
Mahogany Vale	63·5	30 yr recurrence interval	1415	22·3	(1)
		1886 hurricane	2320	36·6	(1) calculated from floodheight assuming n
		1909 hurricane	2320	36·6	(As 1886)
		Probable maximum flood	4250	67	(1)
Middleton Abbey	111·4	15 yr recurrence interval	905	8·1	(2)
		Probable maximum flood	3685	33·1	(2)

Gauge measurement data not available for large floods. Source of estimated discharge:

(1) Howard Humphreys and Sons, 1967–1968, The Yallahs River Water Supply Scheme Feasibility Study for the Government of Jamaica and the Water Commission, vols 1–4. Ministry of Overseas Development.

(2) Harza Engineering Co. and Hue Lyew Chin, 1971, Blue Mountain Water Supply Project Prefeasibility Report, appendix D, Hydrology, June.

Table 3. Ten largest boulders transported in flood

River	Size (m)	Lithology
Big Thompson River	6·98–3·66–3·60	Granite
1976 flood (Shroba *et al.*, 1979)	5·95–2·75–0·92	Pegmatite
	3·75–2·62–1·07	Granite
	2·96–2·53–1·22	Granite
	2·75–2·44–2·23	Granite
	2·75–2·44–0·79	Granite
	3·20–2·23–1·65	Granite
	3·81–2·20–1·83	Granite
	2·90–2·14–1·98	Granite
	3·66–1·74–1·34	Granite
Buff Bay River	5·5–4·9–3·7	Conglomerate
(found in channel or on floodplain)	5·5–4·6–3·4	Conglomerate
	5·2–3·1–1·5	Conglomerate
	4·6–3·1–3·0	Conglomerate
	3·1–3·0–2·3	Limestone
	3·0–3·0–1·8	Conglomerate
	4·4–2·8–2·1	Conglomerate
	4·0–2·8–1·8	Conglomerate
	4·3–2·4–1·8	Conglomerate
	3·1–2·4–2·1	Conglomerate
Yallahs River	4·6–3·4–2·0	Conglomerate
(found in channel or on floodplain)	3·7–3·4–2·8	Conglomerate
	5·5–2·8–1·5	Conglomerate
	3·7–2·8–1·8	Conglomerate
	4·6–2·4–1·5	Conglomerate
	2·4–2·4–1·5	Diorite
	2·1–2·1–1·4	Conglomerate
	2·2–1·9–1·0	Conglomerate
	3·5–1·8–1·8	Conglomerate
	2·4–1·8–1·5	Conglomerate

observations, information gathered from local people, accounts in the *Jamaica Weather Report*, and descriptions in the *Daily Gleaner*. This shows that at least 13 major floods have taken place in the Yallahs channel in this century. Several measurements and estimates of flooding are available (Table 2).

Besides rainfall, it is worthwhile remembering that eastern Jamaica usually has a local relief of 450–600 m, with steep slopes between 20° and 30° coming down to sharply demarcated narrow valley bottoms. Frequent mass movements and high-velocity slope-wash after intense rainfall are accentuated by destruction of natural vegetation. For example, in October 1963, following the precipitation from hurricane Flora, the Yallahs undercut part of its left bank at Mahogany Vale, starting a landslide which brought down approximately 40,000 m³ of greywackes and shales. Such landslides and high-magnitude floods are responsible for the extreme coarseness of the deposit and certain valley forms.

The general nature of the alluvium is very coarse: predominantly pebbles and cobbles with sand and silt in sheltered places. One also finds scattered boulders, much larger than the rest along the stream channels and on the floodplain. Boulders with an intermediate diameter between 1 and 2 m are common, and the measurements of some of the very large boulders are given in Table 3, which also shows boulder measurements from a post-superflood study on the Big Thompson River in Colorado, U.S.A., to indicate that the Jamaican figures are not unique.

Details of transport of these boulders have been discussed elsewhere (Gupta, 1975), but it is obvious that these rivers in flood move very coarse material primarily contributed by landslides, and hence the coarse valley alluvium is not a relict from a past environment. Furthermore, the relatively smaller-sized material—cobbles and pebbles forming the bed, banks and bars—will be carried downstream in the large floods, and deposited at sheltered places during the falling stages, especially as these floods, like the one in October 1963, may run for several days.

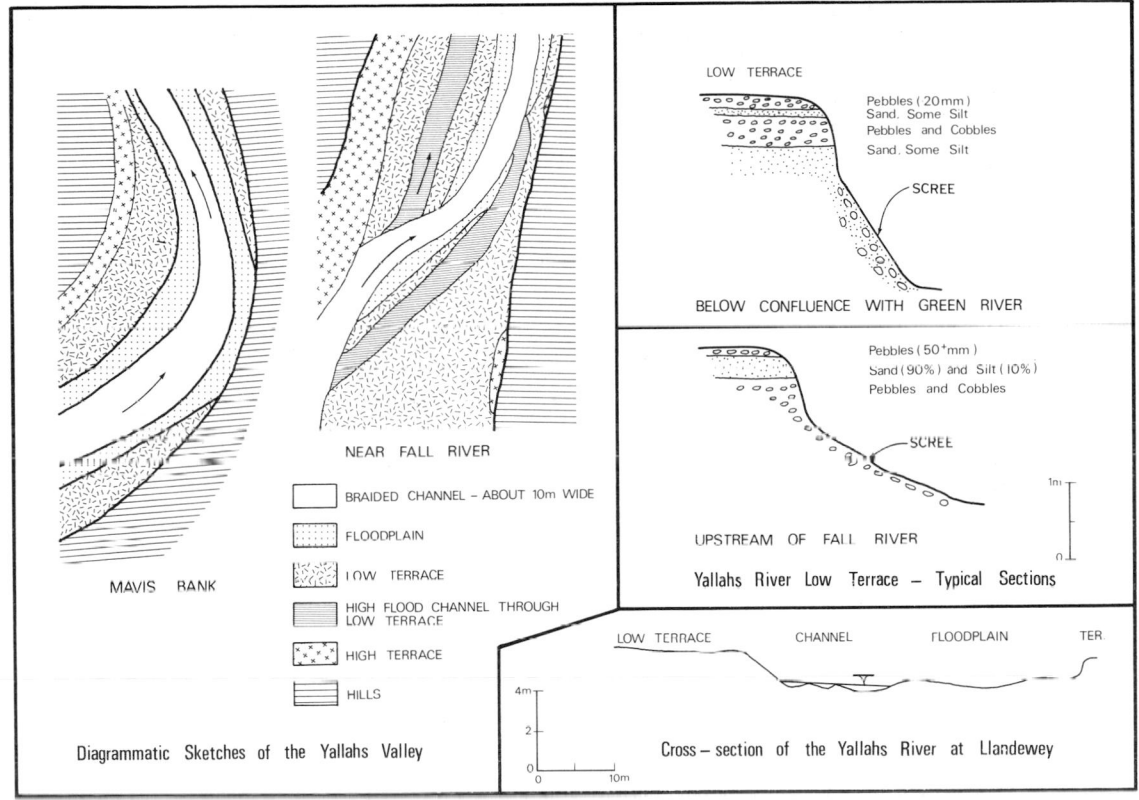

Fig. 1. Examples of sediment and form created by high-magnitude floods in the Yallahs Valley.

Figure 1 shows a section across the Yallahs River which may be taken as representative of the valleys of the mountains in eastern Jamaica. Apart from the channel and floodplain made by coarse detritus, there is a set of low terraces which conspicuously stand about 1·75–3 m above the floodplain. On top of these terraces occur big flood channels and mid-channel bars up to 2·5 m high at places. The terraces are built of subangular to subrounded pebbles, cobbles, boulders and sand. The stratigraphy varies between layers of coarse sand alternating with beds of pebbles, and 1 m of sand overlying several metres of pebbles.

These terraces and the vegetation on top of them are periodically destroyed by superfloods. Rebuilt and modified terraces emerge during the waning stage of the floods, which also leave the stream with a wide channel at the base of the terraces roughly equal to the sum of floodplain and channel widths as shown in Fig. 1. This wide channel at this stage is probably

cleared of all but very coarse cobbles and boulders. Flows of lesser magnitude rebuild the floodplain, shrink the channel to its normal size, and deposit bars inside the channel.

Studies of large floods and flood-modified channels and valleys are common in the literature. Table 4 is a comparative compilation of these findings. In general the major effects on landscape seem to be destruction of vegetation, landslides, deposition of coarse detritus in the valley, post-flood widening of the channels, and in some cases formation of a low terrace which probably is better termed a superfloodplain. The question is whether such modifications are temporary or made to last. In the case of the streams draining the mountains of eastern Jamaica, the superflood effects seem to be permanent. This is probably due to the approximately 10 yr recurrence of superfloods, which ensures the survival of flood-originated sediments and forms.

A. Gupta

Table 4. Stream channel response to high-magnitude floods (measure and permanency)

Source	River	Station	Drainage area (km²)	discharge (m³ s⁻¹)	Unit peak discharge (m³ s⁻¹ km⁻²)	Maximum flood depth (m)	Velocity (m s⁻¹)	Recurrence interval (yr)	Basin physiography	Lithology (basin)	Flood effect along the channel
(A) *Examples of non-effective stream channel response to high-magnitude floods*											
Costa (1974)	Western Run (June 1972–Agnes)	Mouth	155	1076·5 (est.)	6·9	7·9	—	>200	Rolling country. Deep soil, forest, pasture and cropland	Gneiss, schist, marble saprolite	1, 2, 4, 7 (0·7-0·5-0·4) m
Gupta & Fox (1974)	Patuxent River (11–12 Sept. 1971)	Unity	90·2	620	7·0	5·2	—	≫100	Rolling country. Deep soil, forest, pasture and cropland	Phyllites and schist	1, 2, 3, 4, 7, 10 (1·3-1·2-0·7) m
Gupta & Fox (1974)	Patuxent River (21–22 June 1972)	Unity	90·2	410	4·5	4·9	—	>100	Same	Same	1, 2, 3, 4, 7, 10 (1·3-1·2-0·7) m
(B) *Examples of effective stream channel response to high-magnitude floods*											
Baker (1977)	Elm Creek (11 May 1972)	Near confluence	12·5	1130 (est.)	90 (est.)	7	6·4 (est.)	400	Mainly limestone (jointed)		1, 2, 3, 4, 5, 6, 10 (2·5-1·5-0·9) m
Baker (1977)	Blieders Creek (11 May 1972)		39	1370	35	6-9	2·5-3·3 (est.)	400		Mainly limestone (jointed)	1, 2, 3, 4, 5, 6, 7, 10 (2·4-1·1-0·6) m
Glancy & Harmsen (1975)	Eldorado Canyon (14 Sept. 1974)	Below Eagle and Techatticup Washes	59	2153 (est.)	36·5 (est.)		7·6-11·8		Steep mountainous country	Igneous and metamorphic rocks, volcanics	1, 2, 3, 4, 5, 7, 8, 10 (1·8-1·2-0·7) m
Gupta (1975)	Yallahs River (Nov. 1909 flood)	Mahogany Vale	63·5	2320 (est.)	36·6 (est.)				Steep mountainous country. Local relief 450-600 m, slopes 20°-30°. Landslides. Thin soil. Scanty vegetation cover	Conglomerate, greywacke, shale, volcanics	1, 3, 4, 5, 7, (3·7-2·7-1·8) m found at this station at present
Gupta (1975)	Yallahs River (est. 30 yr flood)	Mahogany Vale	63·5	1415 (est.)	22·3 (est.)			30	Same	Same	1, 2, 3, 4, 5, 7, 8, 10 (as above)
Gupta (1975)	Buff Bay river (9 Nov. 1970)	Tranquility	52	498 (est.)	9·6	5 (approx.)		3-6	Similar	Conglomerate, greywacke, shale, limestone, volcanics	2, 4, 5, 6, 7 (b = 0·5 m)
McCain et al. (1979), Shroba et al. (1979)	Big Thompson River (31 July-1 Aug. 1976)	Above Drake	88	800	9·07	2·5	6·7	≫100	Mountainous terrain. Local relief up to 1000 m, slopes 20°-40°, mountain coniferous forest, grassy lowland	Igneous and metamorphics	1, 2, 3, 4, 5, 6, 7, 8, 10 (7·3-7·3-3·6) m
Scott & Gravlee (1968)	Rubicon River (flood surge associated with dam failure: 23 Dec. 1974)	Foresthill				23·8	7 (est.)		Mountainous terrain, steep canyon, altitudinal vegetation zones	Diorite and other igneous rocks and metamorphics	1, 3, 4, 5, 6, 7, 8, 9, 10
Stewart & LaMarche (1967)	Coffee Creek (22-23 Dec. 1964)	USGS gauging station	277	504	1·8		up to 4·5 (est.)	100	Steep mountainous country, mostly pine-fir forest	Igneous and metamorphic rocks. Schists. Surficial deposits. Glacial till	1, 2, 3, 4, 5, 7, (8), 9 (1·9-1·5-1) m

Note: This table has been compiled from a selection of case-studies. Section (A) indicates cases where the flood effects have been transitory, whereas section (B) includes examples of permanent stream channel response to high-magnitude floods.

The data represented have been sorted and generalized, which in spite of due care may give rise to over-simplification or erroneous impression. The table, however, does provide an opportunity for comparative study of large floods. Blank spaces indicate absence of information.

The flood measurements refer to a single station whereas the flood effects refer to the entire channel.

Notation: 1, Widening of channel. 2, Erosion of bars/formation of chutes. 3, Scouring of floodplain. 4, Increase in competence (measurement of the largest boulder moved in flood given within parentheses). 5, Deposition of coarse gravel in channel. 6, Building of transverse gravel waves/gravel bars. 7, Deposition on floodplain, often of coarse gravel. 8, Formation of levee in coarse gravel. 9, Formation of a terrace-like feature. 10, Destruction of vegetation.

STABLE STREAM CHANNEL RESPONSE TO FLOODS: THE NECESSARY CONDITIONS

A large flood almost anywhere in the world will at least temporarily modify stream channels. The stability of the flood-affected forms, however, will depend not only on the flood size but also on the inability of the post-flood range of flows to reverse the effects of the large flood. Permanency of flood effects is possible only under favourable conditions.

Table 4, a compilation of flood data from several case-studies, brings out some of the favourable conditions but is unfortunately somewhat restricted in geographical coverage; and it should be stated that reports of flood affected channels are more widespread than the cases mentioned in Table 4. The basins involved are small in size, and are commonly located in mountainous country with steep slopes of 20° or more. The instability of the slopes combines with the breakage of country rock into large blocks. Baker, for example, shows that well-jointed limestone is partly responsible for coarsening the channel material in floods (Baker, 1977). Conglomerate and limestone supply the Yallahs and Buff Bay Rivers of Jamaica with very coarse gravel in profusion. Suitability of the basin physiography and the size of the flood are undoubtedly important, but the effects will be further enhanced if a periodicity is present in the occurrence of high-magnitude floods, such as once in every 10 or 20 yr. A high ratio between the discharge at flood peak and either the mean annual flood or the bankfull discharge is also one of the conditions favourable for the stability of flood effect. Stability of floodchannel

forms is created when large floods erode, transport and deposit coarse gravel, which remains at best partially modified as the post-flood flows lack the competency to modify forms made of very coarse material. Effects of floods with a recurrence interval of about 100 yr have been described in detail, for example, by Stewart & LaMarche (1967) and Shroba *et al.* (1979). Stewart & LaMarche attributed a large number of valley forms to the superflood, and not to smaller and more common flows. Where high-magnitude floods like these occur with a recurrence interval of only twenty years as in the Caribbean, the effects are obviously even more pronounced. Such flood-potential areas apparently have a geographic distribution, for example, areas with high flood potential in conterminous United States have been mapped by Beard (Baker, 1977).

Figure 2, to illustrate this point further, shows common tracks of large storms in the tropics. In the neighbourhood of such tracks over land one may find examples of potential areas where flood forms persist over time. Similar potential areas may exist in other parts of the world. If suitable basin physiography and geology are also present, river valleys in these areas may display forms which are solely flood-originated. Such forms can be identified in the smaller valleys of the Barisan Mountains in north Sumatra. There are reports of flood-originated forms in the valleys of the eastern Himalayas (Starkel, 1976), in eastern Taiwan (J. Street, personal communication) and probably in the mountains of Papua New Guinea. One may therefore conclude that in certain parts of the world the landscape, at least river valleys, is to some extent determined by high-magnitude floods, and not

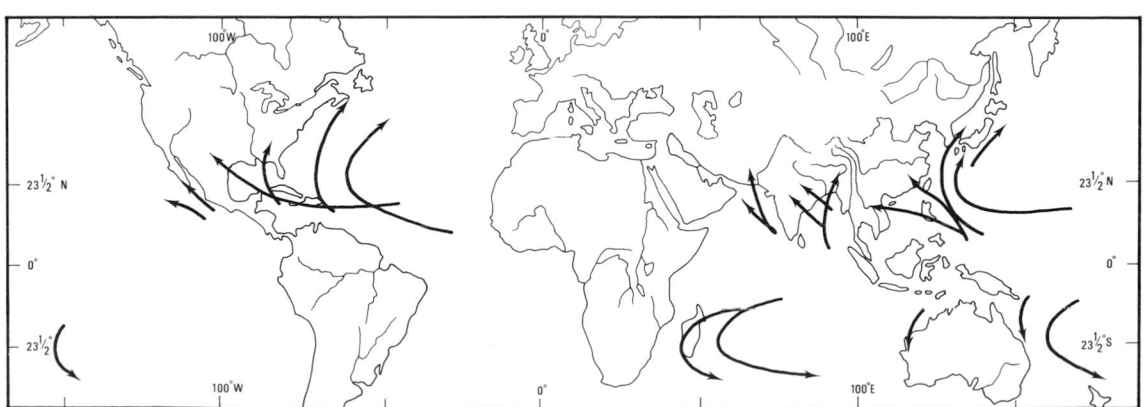

Fig. 2. Examples of potential areas where flood forms may persist over time in the low latitudes. Arrows indicate common tracks of large storms in the tropics.

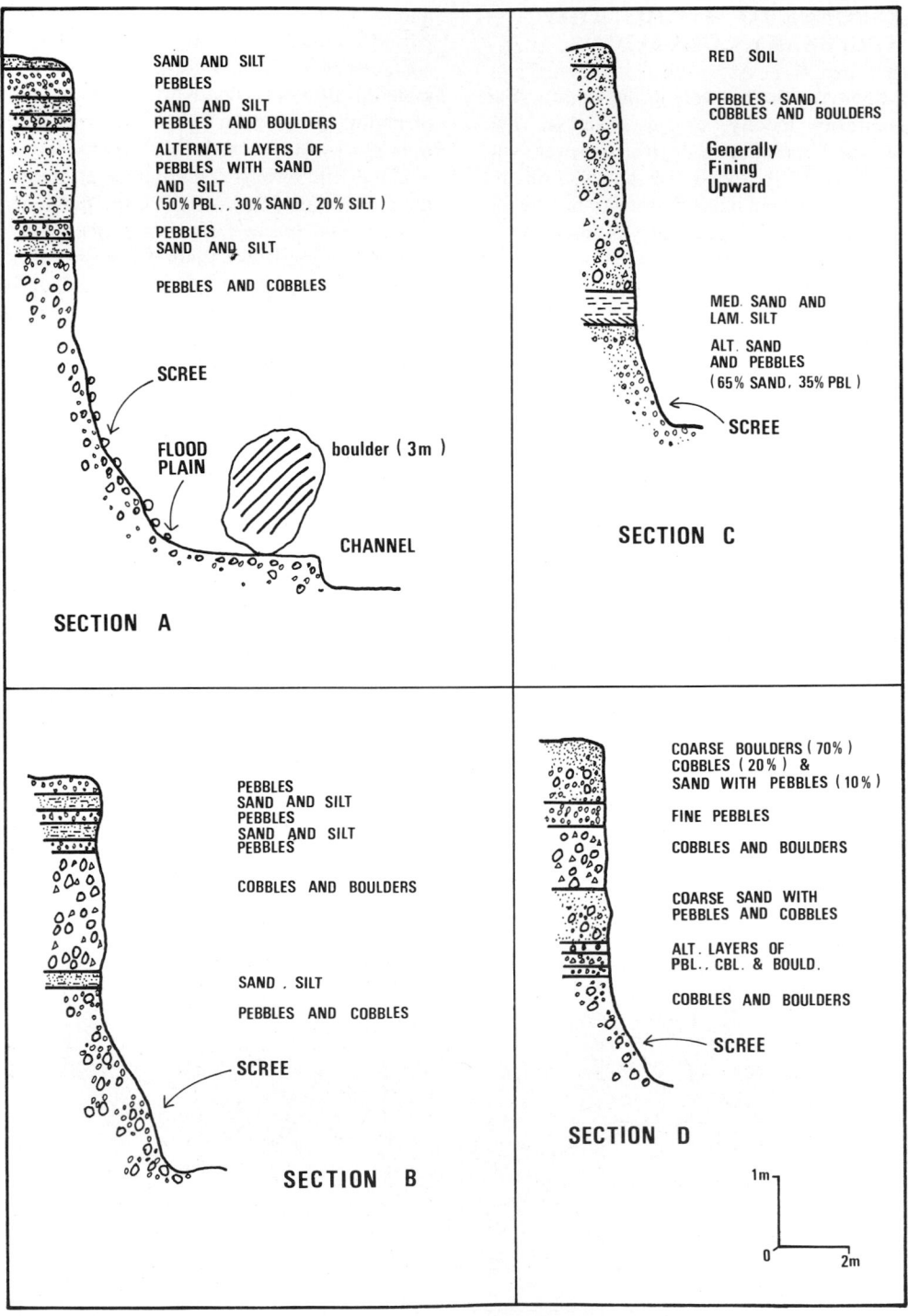

Fig. 3. Sections in the high terrace—the Yallahs Valley.

entirely by more frequent and smaller discharges. Similar conclusions have been reached by geomorphologists for certain parts of the world, e.g. Beatty (1974) in the Great Basin of the Western United States and Selby (1974) in New Zealand; though it should be stated that such reports are often more concerned with the role of mass movements on slopes than with the activity of streams.

CONCLUSION

Large floods with high velocity reshape the channel and the floodplain: an example of a geomorphic process crossing what Schumm (1977) has termed the extrinsic threshold. Smaller flows after the flood gradually reverse the flood effects over time especially if such flows carry a high sediment load. The effect of large floods will be modified by smaller flows like bankfull discharge, but if the deposited flood alluvium is coarse, often the lag effect of high-magnitude floods will persist, and the river in terms of adjustment to bankfull discharge will display nonequilibrium channel forms (Stevens, Simons & Richardson, 1975). Thus the river forms may have to be considered as a function of the sequence of flood events in some cases. If large floods which cross the extrinsic threshold occur periodically, then the valley forms will also be persistent, and such rivers might even display the flood-created forms and the 'ordinary' features together, as in the valleys of eastern Jamaica.

What would be the forms that indicate a river valley distinguished by the passage of a superflood? Most of the features listed at the bottom of Table 4 will be identifiable in the field. In the case of a palaeoalluvium, however, the diagnostic features may be limited to the extreme coarseness of the deposit which may include boulders, and the repeated pattern of episodic flooding. Figure 3, which shows sections from a high and older terrace deposit found in the Yallahs Valley, illustrates this point.

One may therefore put forward a hypothesis that given suitable geology, basin physiography, and a periodicity of superfloods, river channel and valley forms are created and maintained not by flows like bankfull discharge or mean annual flood only, but also by relatively rare high-magnitude floods. This hypothesis has received support from several case-studies on rivers so far. It is worthwhile examining more rivers in whose drainage basins the requisite conditions are present, in order to determine whether it is necessary to modify our understanding of channel formation and maintenance in particular geographical areas.

REFERENCES

Baker, V.R. (1977) Stream-channel response to floods, with examples from central Texas. *Bull. geol. Soc. Am.* **88**, 1057–1171.

Beatty, C.B. (1974) Debris flow, alluvial fans, and a revitalized catastrophism. *Suppl. Z. Geomorph.* **21**, 39–51.

Costa, J.E. (1974) Response and recovery of a Piedmont watershed from tropical storm Agnes, June 1972. *Wat. Resour. Res.* **10**, 106–112.

Evans, C.J. (1972) Estimates of maximum 24-hour rainfall amounts for return periods from 5 to 100 years. *J. Sci. Res. Council, Jamaica*, **3**, 25–45.

Glancy, P.A. & Harmsen, L. (1975) A hydrologic assessment of the September 14, 1974 flood in Eldorado Canyon, Nevada. *Prof. Pap. U.S. geol. Surv.* **930**, 28 pp.

Gupta, A. (1975) Stream characteristics in Eastern Jamaica, an environment of seasonal flow and large floods. *Am. J. Sci.* **275**, 825–847.

Gupta, A. & Fox, H. (1974) Effects of high-magnitude floods on channel form: a case study in Maryland Piedmont. *Wat. Resour. Res.* **10**, 499–509.

Lirios, J.F. (1969) *Rainfall intensity–duration–frequency maps for Kingston and St. Andrew, Jamaica.* Caribbean Meteorological Institute, mimeographed.

McCain, J.F., Hoxit, L.R., Maddox, R.A., Chappell, C.F. & Caracena, F. (1979) Storm and flood of July 31–August 1, 1976, in the Big Thompson River and Cache la Poudre River Basins, Larimer and Weld Counties, Colorado, part A—meteorology and hydrology in Big Thompson River and Cache la Poudre River Basins. *Prof. Pap. U.S. geol. Surv.* **1115**, 1–85.

Schumm, S.A. (1977) *The Fluvial System.* Wiley, New York. 338 pp.

Scott, K.M. & Gravlee, G.C. (1968) Flood surge on the Rubicon River, California – hydrology, hydraulics and boulder transport. *Prof. Pap. U.S. geol. Surv.* **422-M**, 40 pp.

Selby, M.J. (1974) Dominant geomorphic events in landform evolution. *Bull. int. Ass. Engng Geol.* **9**, 85–89.

Shroba, R.R., Schmidt, P.W., Crosby, E.J. & Hansen, W.R. (1979) Storm and flood of July 31–August 1, 1976, in the Big Thompson River and Cache la Poudre River Basins, Larimer and Weld Counties, Colorado, part B—geologic and geomorphic effects in the Big Thompson Canyon area, Larimer County. *Prof. Pap. U.S. geol. Surv.* **1115**, 87–152.

Starkel, L. (1976) The role of extreme (catastrophic) meteorological events in contemporary evolution of slopes. In: *Geomorphology and Climate* (Ed. by E. Derbyshire), pp. 203–246. Wiley, New York.

Stewart, J.H. & LaMarche, V.C. (1967) Erosion and deposition produced by the floods of December 1964 on Coffee Creek, Trinity County, California. *Prof. Pap. U.S. geol. Surv.* **422-K**, 22 pp.

Stevens, M.A., Simons, D.B. & Richardson, E.V. (1975) Nonequilibrium river form. *J. Hydraul. Div. Am. Soc. civ. Engrs* **101**, 557–566.

Vickers, D.O. (1967) Very heavy and intense rainfalls in Jamaica; *Proc. Univ. West Indies Conf. Climatology and Related Fields.* pp. 57–63, Mona. West Indies, 1966.

Wolman, M.G. & Eiler, J.P. (1958) Reconnaissance study of erosion and deposition produced by the flood of August 1955, in Connecticut. *Trans. Am. geophys. Un.* **39**, 1–14.

Spec. Publs int. Ass. Sediment. (1983) **6**, 229–239

Palaeohydrologic analysis of Holocene flood slack-water sediments

VICTOR R. BAKER

Department of Geosciences, University of Arizona, Tucson, Arizona 85721, U.S.A.

R. CRAIG KOCHEL

Department of Environmental Sciences, University of Virginia, Charlottesville, Virginia 22093, U.S.A.

PETER C. PATTON

Department of Earth and Environmental Sciences, Wesleyan University, Middletown, Connecticut 06457, U.S.A.

G. PICKUP

CSIRO, Institute of Earth Resources–Central Australian Laboratory, P.O. Box 211, Alice Springs, N.T. 5750, Australia

ABSTRACT

Estimates of the frequency and discharge of large floods can be refined and extended over the past 10,000 years through the study of slack-water sediments deposited in bedrock canyons and gorges. Slack-water deposits are typically fine-grained sand and silt that accumulate during major floods in protected areas where current velocity is reduced, such as in back-flooded tributary mouths, at channel expansions, and downstream from bedrock spurs. Relatively narrow bedrock canyons in arid, savanna, and semi-arid regions provide excellent areas for estimating flood discharges from the elevation of slack-water deposits because: (1) relatively small discharge increases are accompanied by large increases in river stage, (2) the bedrock canyons provide stable cross-sections for slope-area calculations, and (3) the paucity of vegetation limits sediment bioturbation. The accumulation and preservation of slack-water sediment sequences appear to be controlled by tributary–mainstream junction morphology and by tributary drainage basin efficiency.

Radiocarbon dating of wood, charcoal, buried soils and fine-grained organic detritus in slack-water deposits has been accomplished for our studies in central Texas, western Texas, northern and central Australia. Along the lower Pecos River of western Texas the slack-water sediment stratigraphies record between six and ten major flood events. This record extends back at least 2000 years in some canyons, and as far as 10,000 years at the Arenosa Shelter site. In this region we have been able to apply geological data to extend observational records of major floods to achieve flood-frequency curves over a time base in millennia, rather than in decades. The procedure allows for the realistic assessment of recurrence intervals for large-magnitude events.

INTRODUCTION

Two general approaches have been adopted by researchers investigating fluvial palaeohydrology. One approach estimates average hydrological conditions coincident with the formation of palaeochannels. Such studies involve mapping of modern and ancient channels, sediment analyses, and comparisons of channel morphologies of the fluvial system through time. This information is useful for providing estimates of relatively frequent flow events such as bankfull discharge. The analysis leads to implications

0141-3600/83/0106-0229 $02.00

about climatic change and its effects on the independent fluvial variables of discharge and sediment load. Studies utilizing this approach include Dury (1965, 1976), Schumm (1965, 1968), and Baker & Penteado-Orellana (1977).

Another approach in palaeohydrologic research considers the nature of discrete hydrologic events, such as catastrophic floods (Baker, 1973). River stage is inferred from indirect evidence, and palaeodischarges are calculated. The technique which may provide the most abundant and accurate information in this regard is the study of slack-water sediments, a seldom-used but potentially promising tool (Baker, Kochel & Patton, 1979; Patton, Baker & Kochel, 1979). If a stratigraphic record of flood stage exists, this approach yields not only the magnitude of the event (stage and discharge) but also the frequency of events.

Slack-water deposits are typically fine-grained sand and silt that accumulate rapidly from suspension during major floods in protected areas where current velocity is reduced. Where slack-water sediments accumulate in areas shielded from erosion by subsequent floods, the sedimentation sites continue to receive additional deposits from subsequent floods of equal or greater magnitude. If datable materials are entrapped in these sequences, frequency estimates can be made of the average return periods for the responsible flood events.

Slack-water sedimentation has been recognized in the catastrophic Pleistocene flooding of the Channeled Scabland in eastern Washington (Bretz, 1929, 1969; Baker, 1973; Waitt, 1980). Holocene slack-water deposits have been described along several major rivers, particularly at the mouths of tributary streams: Colorado River in the Grand Canyon (McKee, 1938); Ohio River floods of 1937 (Mansfield, 1938); Connecticut River floods of 1927, 1936 and 1938 (Jahns, 1947), Skagit River floods (Washington) of the 1800s (Stewart & Bodhaine, 1961); in tributary mouths during 1972 Hurricane Agnes flooding in Maryland (Costa, 1974) and in Pennsylvania (Moss & Kochel, 1978); and along various streams in central Texas (Baker, 1975; Patton, 1977) and in central and south-western Texas (Patton et al., 1979; Kochel & Baker, 1982).

SLACK-WATER DEPOSITION

Rivers of the Northern Territory, Australia (Fig. 1), provide excellent examples of slack-water deposition.

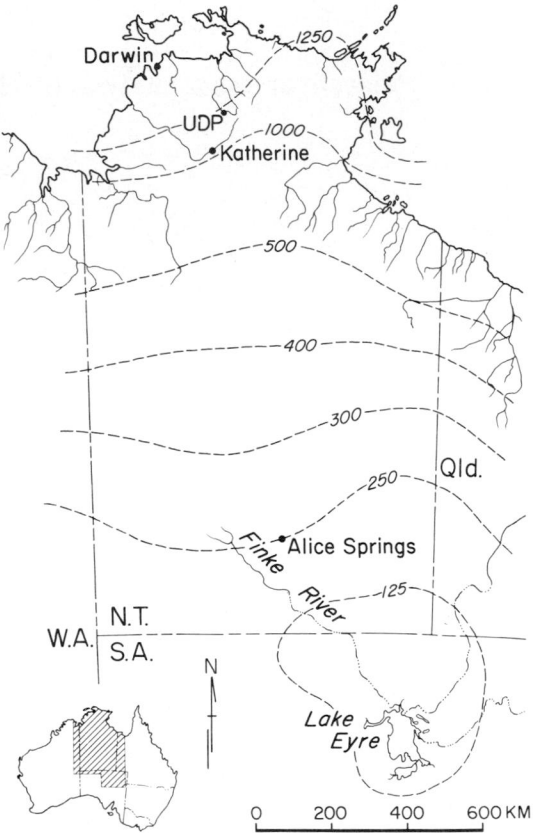

Fig. 1. Location map showing streams studied in Northern Territory, Australia. The dashed lines show mean annual rainfall in millimetres.

Where these rivers become confined by older, resistant sediments (Fig. 2) or by bedrock (Fig. 3), vertical accretion of fine-grained sediment from suspension becomes the dominant mode of flood-plain development. In such rivers the pattern of flood-plain sedimentation is different from the models developed for streams which are unconfined by bedrock and free to migrate laterally. Rapidly migrating streams have relatively uniform bank heights and form their flood plains primarily by lateral accretion of sediment (Wolman & Leopold, 1957). In contrast, those streams confined by bedrock, or which migrate only slowly, have flood plains developed by vertical accretion of flood sediments. This creates a more erratic flood-plain stratigraphy and more variable stream bank heights. Flood plains which form in this manner are the most useful for palaeohydrologic investigations.

By far the most common site for the accumulation

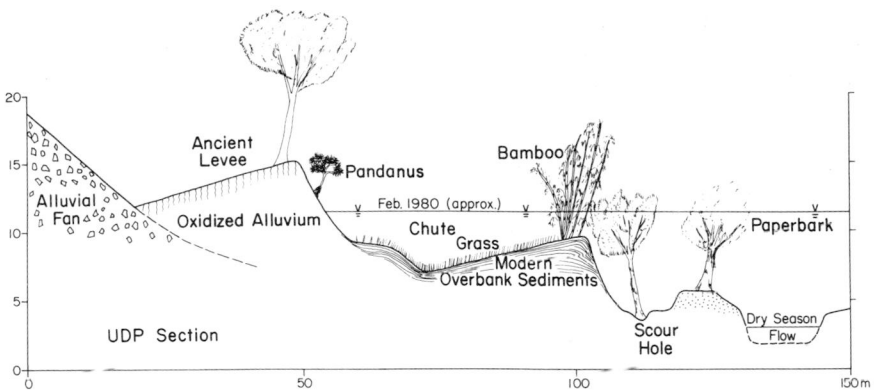

Fig. 2. Schematic cross-section through the northern half of the South Alligator River flood plain at UDP, Northern Territory (see Fig. 1). 0·6–1·0 m of modern overbank sediments have accumulated since 280 ± 100 ^{14}C yr BP (TX-4211).

Fig. 3. Katherine River in Katherine Gorge during wet-season flooding in February 1980. Note slack-water zone at mouth of tributary.

of slack-water deposits is in the mouths of tributaries. During large floods in highly confined bedrock valleys, flood waters may fill the valley from one bedrock wall to the other (Fig. 3). Flooding of the tributaries is typically not synchronous with the passage of the crest down the main stream. Tributaries usually debouch their floods first, and water stage then falls rapidly. Later, the flood waves move down the

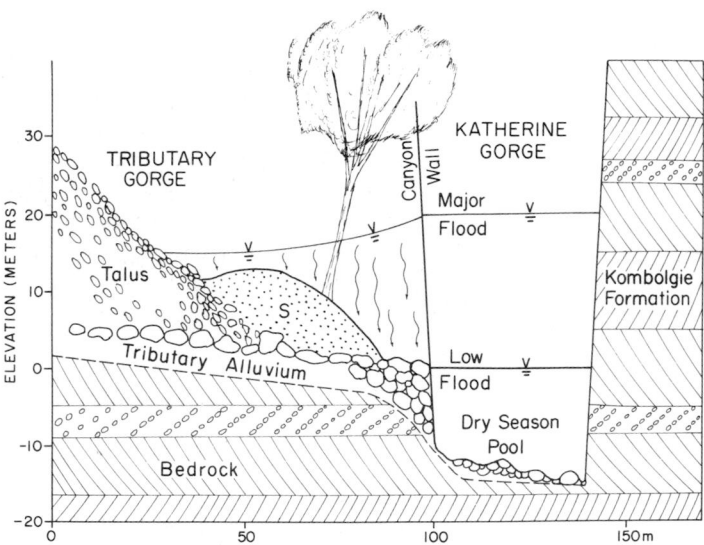

Fig. 4. Schematic illustration of slack-water deposit accumulation (S) in the Katherine Gorge, Northern Territory, Australia.

Fig. 5. Dry-season view of same tributary shown in Fig. 4. The slack-water sand deposit occurs at the mouth of the tributary.

mainstream through portions of the basin not currently in flood. In this manner, backflooding of tributary mouths occurs, and powerful surges of sediment-laden water proceed up the tributary canyons, flooding them up to a stage corresponding to the elevation of the flood crest on the main stream. Figure 4 illustrates one of these reverse surges operating in the mouth of a small tributary to the

Katherine Gorge, near Katherine, Northern Territory. The effect is similar to that of a stilling basin connected to a flume. Velocity in the back-flooded tributary mouth rapidly decreases, and suspended sediment is deposited. If the stage of the flood exceeds the height of previously accumulated slack-water deposits, a new slack-water flood sedimentation unit will be produced (Fig. 5). However, if the stage of the flood fails to

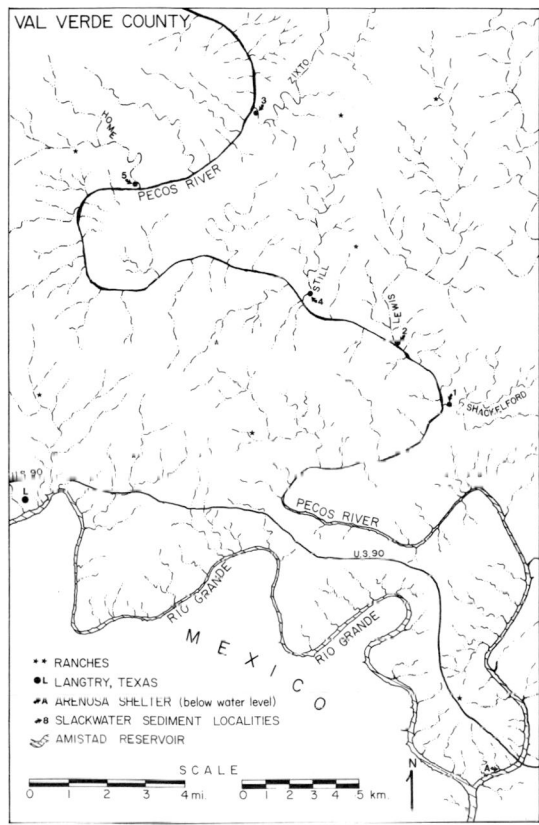

Fig. 6. Location map showing slack-water sediment accumulation sites along the Pecos River in western Texas.

exceed the existing surface of slack-water deposits in the tributary mouths, the new deposit will occur as an inset terrace in the tributary valley.

Mainstream-derived slack-water sediments must be carefully distinguished from tributary alluvium. In tributary mouths along the Susquehanna River, typical slack-water sedimentary sequences for the Agnes event were composed of two units (Moss & Kochel, 1978). The basal layer deposited by the tributary (ranging from a few centimetres to 1 m thick) was composed of coarse-grained gravel and cobbles derived from the basin. The upper layer was composed of fine-grained sand, silt and clay deposited by back-flooding of the Susquehanna River. Proof that these sediments were derived from the Susquehanna was the presence of coal, which does not outcrop in any tributary basins studied. The rivers of the Northern Territory mentioned herein transport loads of granitic quartz, feldspar and mica. Tributaries to

gorges through ranges of sedimentary rocks produce easily distinguishable alluvium.

In the Pecos River canyon of western Texas (Fig. 6) the mainstream sediment is predominantly fine-grained, tan quartz sand (average grain size = 3.5ϕ). Tributary basins are underlain by carbonate rocks. Alluvium derived from local tributary floods sometimes appears interstratified in the quartz slack-water sands and is composed of either limestone gravel and cobbles or grey carbonaceous sand and silt. Low-altitude colour aerial photographs permit a rapid confirmation of the source of the tributary mouth deposits because the Pecos-derived sands appear tan compared with their surrounding background of grey tones characteristic of the limestone bedrock and shallow soils of the region. Grain-size is variable in each flood sedimentation unit and within a single unit, but usually ranges from coarse silt ($4–5\phi$) to medium sand ($2–3\phi$) with a mean of fine sand (3.5ϕ). These sizes are optimum for suspended transport during large floods and for rapid settling from suspension in areas of slack water.

Horizontal laminations are the dominant sedimentary structure in lower Pecos River slack-water sediments. Laminations are in part the result of abrupt grain-size changes (medium sand to coarse silt) and in part due to concentrations of fine-grained organic detritus. In the former case, sand laminae usually range from 0.5 to 2 or 3 cm in thickness, while silt partings are usually only a few millimetres thick and are commonly laterally discontinuous. Likewise, the organic-rich laminae are usually an order of magnitude thinner than interstratified sand laminae and are also frequently discontinuous. Horizontal laminations are usually considered to form as the result of the migration of bedforms (such as small ripples) up to the tributary canyons during rapid influx of back-flood waters. Groups of laminations with variable grain size indicate variations in both the rate of sediment supply and current velocity. Possible explanations for these could be pulsations of water and sediment pouring through the entrance to the tributary canyon from the mainstream.

Second in order of abundance of sedimentary structures are structureless units of fine sand and silt. These units range in thickness from a few centimetres to 1.5 m, usually between 0.5 and 1.0 m. Structureless units are commonly ungraded and contain few or no silt partings or organic-rich partings. All indications suggest that these units represent very rapid deposition from suspension which prevented segregation of grains into clusters and therefore prevented formation

16

Table 1. Recurrence interval calculations for Arenosa Shelter
(from Patton & Dibble, 1982)

Stratigraphic layers	Number of floods	Time interval (yr BP)	River stage (m)	Estimated discharge (m^3 sec^{-1})	Recurrence interval* (yr)
5	3	1970 ± 70–present	14·8	11,320	630–660–680
4			15·4	12,450	950–985–1020
1			24·4	27,735	1900–1970–2040
27, 26A, 26, 24	4	4450 ± 150–4150 ± 150	10·1	5095	0–75–150
23C, 23A	4	4150 ± 150–3350 ± 85	11·0	6095	140–200–250
22B, 21A	2	3220 ± 70–3600 ± 70	11·4	6790	120–190–260
19–11	5	3220 ± 70–2230 ± 80	12·6	8490	170–200–230
8	1	2440 ± 140–1970 ± 110	14·1	10,755	220–470–720

* Recurrence intervals in the upper three strata (5, 4, 1) calculated by the formula: $n + 1/m$, where n = yr, m = magnitude ranking.

Project (now the Texas Archeological Survey) from 1965 to 1968 (Dibble, 1967). Dibble's work revealed an essentially continuous record of occupation in the site over the past 10,000 yr (Dibble, 1965, 1967). Excavation at the site using backhoes provided a detailed view of over 10 m of interstratified alluvium and cultural sediments.

Patton (1977) studied sediment samples, sediment monoliths and photographs from the site, and generated a depositional stratigraphy. Patton & Dibble (1982) note that this analysis represents a minimum number of floods due to intermixing of sediments by cultural activities in the shelter. The 1954 flood sediment (the thickest in the entire section) buried a surface which was about 1400 yr old. Therefore, the minimum recurrence interval for the 1954 flood at Arenosa would be 1400 yr. Patton (1977) and Patton & Dibble (1982) computed discharges for the flood events preserved, and assigned recurrence intervals to the upper three floods using a time-series approach. Table 1 summarizes their estimates on flood frequency at Arenosa Shelter. The time intervals were fixed by 18 radiocarbon dates. Each flood was assigned a rank according to magnitude, and the following formula was used to calculate recurrence intervals:

$$\frac{n+1}{m},$$

where n is the number of years of record bracketed by radiocarbon dates and m is the ranking of the flood. In this manner flood flows were assigned a frequency by grouping floods of similar stage (within a pre-determined range), thus having similar magnitudes, and estimating the probability of a flow equal or greater in magnitude within this interval (Patton & Dibble, 1982). Therefore, the probabilities calculated

by this method of breaking the stratigraphy into time-series are estimates of the probability of the occurrence of a flood of equal or greater magnitude within a given time interval. The resulting analyses (Patton, 1977; Patton & Dibble, 1982) indicated that a lower limit of 2000 yr could be assessed for the recurrence interval of the 1954 flood. Without a doubt, the 1954 flood was the largest to occur during the past 2000 yr at Arenosa. The 1954 flood sediment completed the filling of Arenosa Shelter.

ACKNOWLEDGMENTS

This research was supported by the Division of Earth Sciences, National Science Foundation, NSF Grants EAR 77-23025, EAR 81-00391, and EAR 81-19981. The work in Australia was supported by the Australian-American Educational Foundation and by the North Australian Research Unit, Australian National University. We thank S. Valastro, Jr, for his assistance with the geochronological aspects of our studies. Henry A. Polach of the Australian National University performed the radiocarbon analyses shown in Fig. 8.

REFERENCES

BAGNOLD, R.A. (1966) An approach to the sediment transport problem from general physics. *Prof. Pap. U.S. geol. Surv.* **422-I,** 37 pp.

BAKER, V.R. (1973) Paleohydrology and sedimentology of Lake Missoula flooding in eastern Washington. *Spec. Pap. geol. Soc. Am.* **144,** 79 pp.

BAKER, V.R. (1975) Flood hazards along the Balcones Escarpment in central Texas: alternative approaches to their recognition, mapping, and management. *Bur. econ. Geol. Circ. Univ. Texas, Austin,* **75-5,** 22 pp.

BAKER, V.R. (1977) Stream-channel response to floods, with examples from central Texas. *Bull. geol. Soc. Am.* **88**, 1057–1071.

BAKER, V.R., KOCHEL, R.C. & PATTON, P.C. (1979) Long-term flood frequency analysis using geological data. In: *The Hydrology of Areas of Low Precipitation. Sci. Publ. int. Ass. hydrol.* **128**, 3–9.

BAKER, V.R. & PENTEADO-ORELLANA, M.M. (1977) Adjustment to Quaternary climatic change by the Colorado River in central Texas. *J. Geol.* **85**, 395–422.

BRETZ, JH. (1929) Valley deposits immediately east of the Channeled Scabland of Washington. *J. Geol.* **37**, 393–427, 505–541.

BRETZ, JH. (1969) The Lake Missoula floods and the Channeled Scabland. *J. Geol.* **77**, 505–543.

CHOW, V.T. (1959) *Open-Channel Hydraulics.* McGraw-Hill, New York. 680 pp.

COSTA, J.E. (1974) Response and recovery of a piedmont watershed from Tropical Storm Agnes, June, 1972. *Wat. Resour. Res.* **10**, 106–112.

DIBBLE, D.S. (1965) Bonfire shelter: a stratified bison kill site in the Amistad Reservoir area, Val Verde County, Texas. *Misc. Pap. Texas archeol. Salvage Proj.* **5**, 128 pp.

DIBBLE, D.S. (1967) Excavations at Arenosa shelter, 1965-1966. Preliminary report submitted to the National Park Service by the *Texas Archeological Salvage Project*, Austin, Texas, 85 pp.

DURY, G.H. (1965) Theoretical implications of underfit streams. *Prof. Pap. U.S. geol. Surv.* **452-C**, 43 pp.

DURY, G.H. (1976) Discharge prediction, present and former, from channel dimensions. *J. Hydrol.* **30**, 219–245.

FOLK, R.L. (1974) *Petrology of Sedimentary Rocks.* Hemphill, Austin, Texas. 182 pp.

JAHNS, R.H. (1947) Geologic features of the Connecticut Valley, Massachusetts, as related to recent floods. *Wat.-supply Pap. U.S. geol. Surv.* **996**, 158 pp.

KOCHEL, R.C. (1980) *Interpretation of flood paleohydrology using slack-water deposits, Lower Pecos and Devils Rivers, southwestern Texas.* Ph.D. dissertation. University of Texas, Austin. 360 pp.

KOCHEL, R.C. & BAKER, V.R. (1982) Paleoflood hydrology. *Science*, **215**, 353–361.

MANSFIELD, G.R. (1938) Flood deposits of the Ohio River, January–February, 1937, *Wat.-supply Pap. U.S. geol. Surv.* **838**, 693–736.

MCKEE, E. (1938) Original structures in Colorado River flood deposits of Grand Canyon. *J. sedim. Petrol.* **8**, 77–83.

MOSS, J.H. & KOCHEL, R.C. (1978) Unexpected geomorphic effects of the Hurricane Agnes storm and flood, Conestoga drainage basin, south-eastern Pennsylvania. *J. Geol.* **86**, 1–11.

PATTON, P.C. (1977) *Geomorphic criteria for estimating the magnitude and frequency of flooding in central Texas.* Ph.D. dissertation. University of Texas, Austin. 222 pp.

PATTON, P.C. & BAKER, V.R. (1977) Geomorphic response of central Texas stream channels to catastrophic rainfall and runoff. In: *Geomorphology in Arid Regions* (Ed. by D.O. Doehring), pp. 189–217. Allen & Unwin, Winchester, Massachusetts.

PATTON, P.C. & DIBBLE, D.S. (1982) Archeologic and geomorphic evidence for the paleohydrologic record of the Pecos River in west Texas. *Am. J. Sci.* **282**, 97–121.

PATTON, P.C., BAKER, V.R. & KOCHEL, R.C. (1979) Slack-water deposits: a geomorphic technique for the interpretation of fluvial paleohydrology. In: *Adjustments of the Fluvial System* (Ed. by D.D. Rhodes and G. Williams), pp. 225–252. Kendall-Hunt, Dubuque, Iowa.

SCHUMM, S.A. (1965) Quaternary paleohydrology, In: *The Quaternary of the United States* (Ed. by H.E. Wright, Jr, and D.G. Frey), pp. 783–794. Princeton University Press.

SCHUMM, S.A. (1968) River adjustment to altered hydrologic regimen, Murrumbidgee River and paleochannels, Australia. *Prof. Pap. U.S. geol. Surv.* **598**, 65 pp.

STEWART, J.E. & BODHAINE, G.L. (1961) Floods in the Skagit River basin, Washington. *Wat.-supply Pap. U.S. geol. Surv.* **1527**, 66 pp.

WAITT, R.B. (1980) About forty late-glacial Lake Missoula jökulhlaups through southern Washington. *J. Geol.* **88**, 653–679.

WOLMAN, M.G. & LEOPOLD, L.B. (1957) River floodplains: some observations on their formation. *Prof. Pap. U.S. geol. Surv.* **282-C**, 87–109.

Spec. Publs int. Ass. Sediment. (1983) **6**, 241–250

Present channel stability and late Quaternary valley deposits in northern Mississippi

E. H. GRISSINGER *and* J. B. MURPHEY

U.S. Department of Agriculture, Agricultural Research Service, USDA Sedimentation Laboratory,
P.O. Box 1157, *Oxford, Mississippi* 38655, *U.S.A.*

ABSTRACT

Seven lithologic units have been identified in the Holocene valley-fill deposits of the loess region of northern Mississippi. These units are (1) post-settlement alluvium, (2) meander-belt alluvium, (3) channel fill, (4) massive silt, (5) bog-type materials, (6) unconsolidated grey silt and (7) channel lag deposits. In addition, several pre-Holocene deposits have been sampled, but at this time are poorly defined.

The chronology and relative lithologies of these units from the loess region are consistent with the valley fill deposits over most of northern Mississippi. This valley-fill sequence is comparable with sequences for other sections of the United States and is coherent with the Holocene palaeoclimatic conditions.

The individual lithologic units and their sequences influence present-day channel bed and bank failure mechanics. Failure of the post-settlement and meander-belt alluvium results primarily from gravity stress accentuated by tension crack development. Failure of the massive silt also results from gravity stress but the failure mechanism for this unit is controlled by its distinctive polygonal structure. Exposure of the easily erodible unconsolidated bog-type and channel lag deposits in a bank toe position typically increases rates of failure due to gravity stress. The interactions of these lithologic controls of failure mechanics with channel realignment and entrenchment during historic times has produced the present-day channel morphology. The degree of entrenchment is, in turn, related to the presence or absence of pre-Holocene consolidated sandstones.

INTRODUCTION

Stream channel instability is a serious problem for many areas of relatively high relief bordering the lower Mississippi Alluvial Plain (the Mississippi Delta). Instability problems are especially critical in the northern Mississippi bluff area bordering the Delta due to the erosiveness of the channel bed and bank materials, and in the Delta due to the damage produced by this sediment to channels. As one part of a comprehensive study of these instability problems, possible influences of channel bed and bank materials on failure mechanics have been investigated. This paper summarizes the results concerning: (a) the nature and distribution of valley-fill lithologic units, (b) their relation with palaeoclimatic conditions and (c) the influences of these units on channel morphology and bed and bank stability.

STUDY AREA

The study area is the Yazoo–Little Tallahatchie catchment (YLT) east of the Mississippi Alluvial Plain (Fig. 1). This area is within the East Gulf Coastal Plain Physiographic Province. Physiographic subprovinces include, from west to east: (1) the Loess or Bluff Hills, (2) the North Central Hills, (3) the Flatwoods and (4) the Pontotoc Ridge. These subprovinces reflect stratigraphic controls. The Pontotoc Ridge formed on the Ripley Sand of the Cretaceous Selma Group and the

0141-3600/83/0106-0241 $02.00

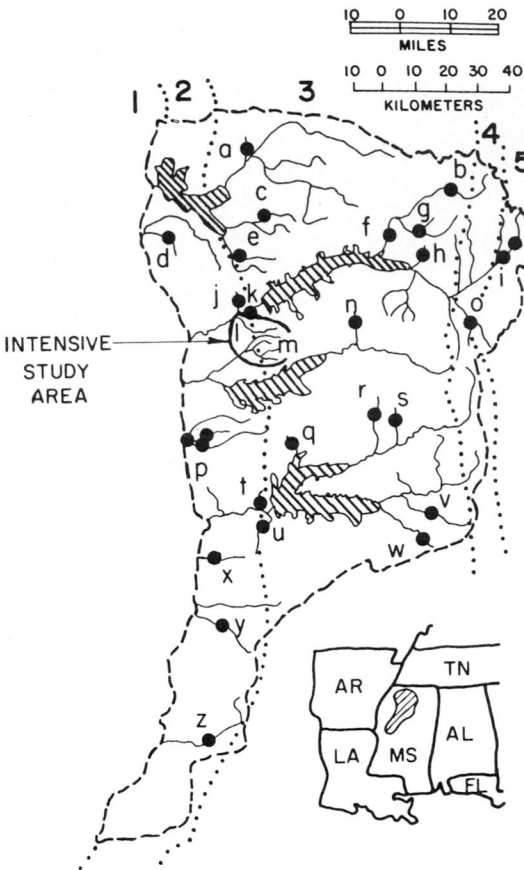

Fig. 1. Yazoo–Little Tallahatchie catchment study area. Physiographic subprovinces are separated by dotted line: (1) Mississippi Alluvial Plain, (2) Loess or Bluff Hills, (3) North Central Hills, (4) Flatwoods and (5) Pontotoc Ridge.

Flatwoods developed on Porters Creek Clay of Palaeocene age. Unconsolidated sands, silts and clays of the Wilcox and Claiborne Groups, both Eocene in age, crop out in the North Central Hills. Both are blanketed by loess and fluvial sands and gravels in the Loess or Bluff Hills Subprovince. The Pleistocene loess thins rapidly from west to east. The sands and gravels are referenced as either the Citronelle or Lafayette Formation and have been decribed as Plio-Pleistocene in age.

For ease of discussion, the study area has been divided into an intensive study are which includes an intensively instrumented catchment and a general study area which includes the remainder of the YLT (Fig. 1). The intensive study area is mainly within the Bluff Hills Subprovince with the easternmost part in the North Central Hills. Field investigations in the general study area were conducted to compare the

chronology and nature of the valley-fill sequence throughout the general study area with comparable data from the intensive study area. At all locations, the valley-fill sequence had been mapped prior to this study as an undifferentiated deposit.

LITHOLOGIC UNITS

Unit differentiation in the intensive study area was based primarily on the properties of outcrop samples. Supplemental information was obtained from borehole samples used to define unit distributions. In the general study area, borehole samples were the primary source of information; outcrop samples were supplemental. Wood or other organic detritus is abundant in these deposits, and 140 samples were dated. Sample collection was restricted to organics that would contribute to definition of the chronology of the valley-fill deposits but was otherwise random. All ages were calculated using the Libby half-life of 5568 years. No correction has been made for variation in atmospheric ^{14}C concentration. Of these 140 samples, 12 have ages greater than 40,000 years Before Present (yr BP), five have ages in the span from 17,000 to 40,000 yr BP, and 123 have ages less than 13,200 yr BP (Fig. 2b). No samples have ages in the span from 13,200 to 17,000 yr BP. Of these latter dates, 64 are for samples from the intensive study area (Fig. 2a) and 59 from the general study area.

Units in the intensive study area

Seven lithologic units have been identified in the intensive study area. An idealized section of these units is presented in Fig. 3 and the age frequency is presented in Fig. 2(a) for the 64 Holocene-age samples. The seven units are: (1) post-settlement alluvium, (2) meander-belt alluvium, (3) channel fill, (4) massive silt, (5) bog-type materials, (6) unconsolidated grey silt and (7) channel lag deposits. In addition, pre-Holocene consolidated sandstones frequently crop out in the channel beds. The units are presented in the following discussion as members or facies within depositional sequences.

Pre-Holocene depositional sequences

The most common pre-Holocene lithologic unit is the consolidated sandstone. Sandstones are usually cross-stratified and frequently contain gravel. All samples have ages greater than 40,000 yr BP. Outcrops are limited in size, rarely exceeding 20–30 m horizontal

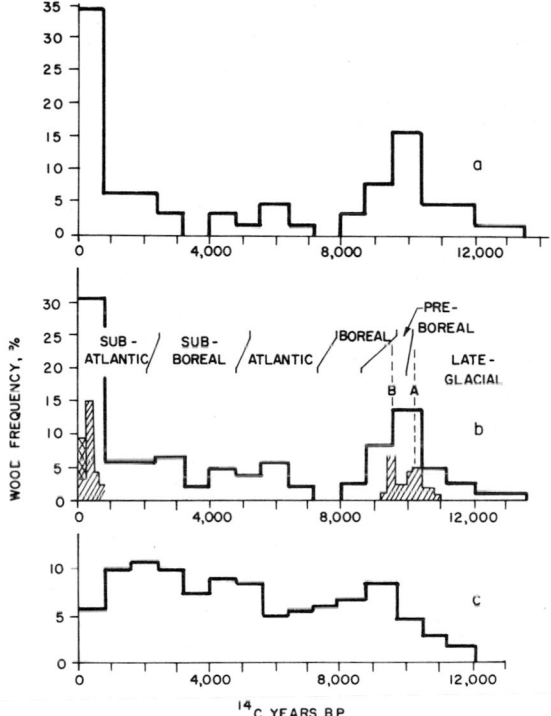

Fig. 2. Histograms of ¹⁴C dates. (a) Sixty-four samples from the intensive study area in 800 yr classes. (b) All samples in 800 yr classes, with 200 yr classes noted by hatching and those too young to date noted by cross-hatching. Diagonal lines represent times of transition between climatic episodes (source: Godwin, 1966). (c) Summary of 815 independent ¹⁴C dates (source: Wendland & Bryson, 1974).

distance, and are typically truncated and disconformably overlain by Holocene valley-fill deposits. Other pre-Holocene deposits range in age from 17,000 to 34,000 yr BP. The depth below floodplain elevation for these samples ranges from 5 to more than 16 m. Pre-Quaternary materials crop out in the upper reaches of several streams. These materials are relatively coarse textured, frequently indurated and cross-stratified, with foreset bed angles indicating flow from west to east. The limited number of pre-Holocene outcrops, however, negates further definition of these depositional sequences.

Early Holocene depositional sequence

This sequence includes the massive silt, channel lag, bog-type and grey silt units. The massive silt overlies any of the other three units in this sequence. Contacts vary from disconformable to fining-upward gradational interfaces. Organics are common in all units except the massive silt deposit, and ages for these organics comprise the early Holocene frequency mode (Fig. 2a). The unconsolidated grey silt, the bog-type and the channel lag units are considered to be facies of one member of this sequence and the massive silt unit to be a separate member.

The massive silt unit is a widespread, predominantly fine-textured valley-fill deposit. It is present in all valleys and frequently exceeds 4 m in thickness. Unit truncation or complete removal by erosion at some valley sites is common and we have not been able at this time to define the longitudinal valley profile

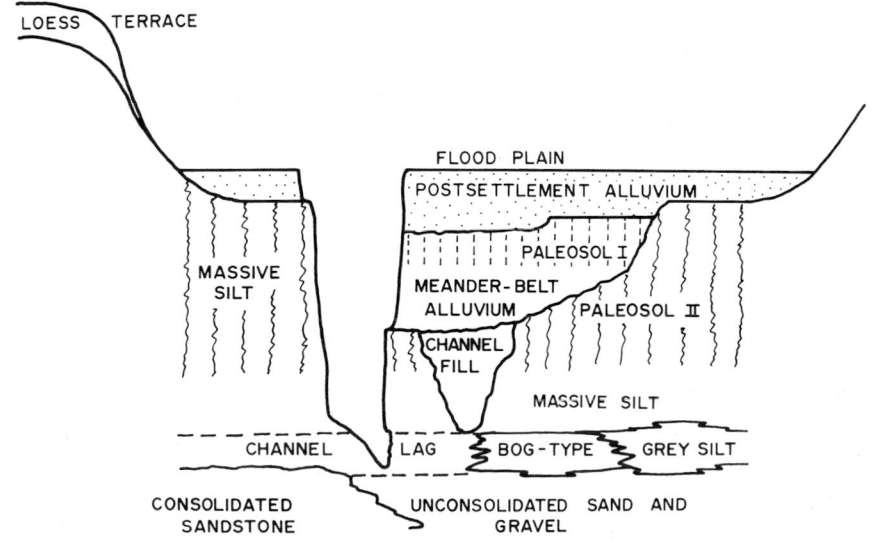

Fig. 3. Idealized section of lithologic units.

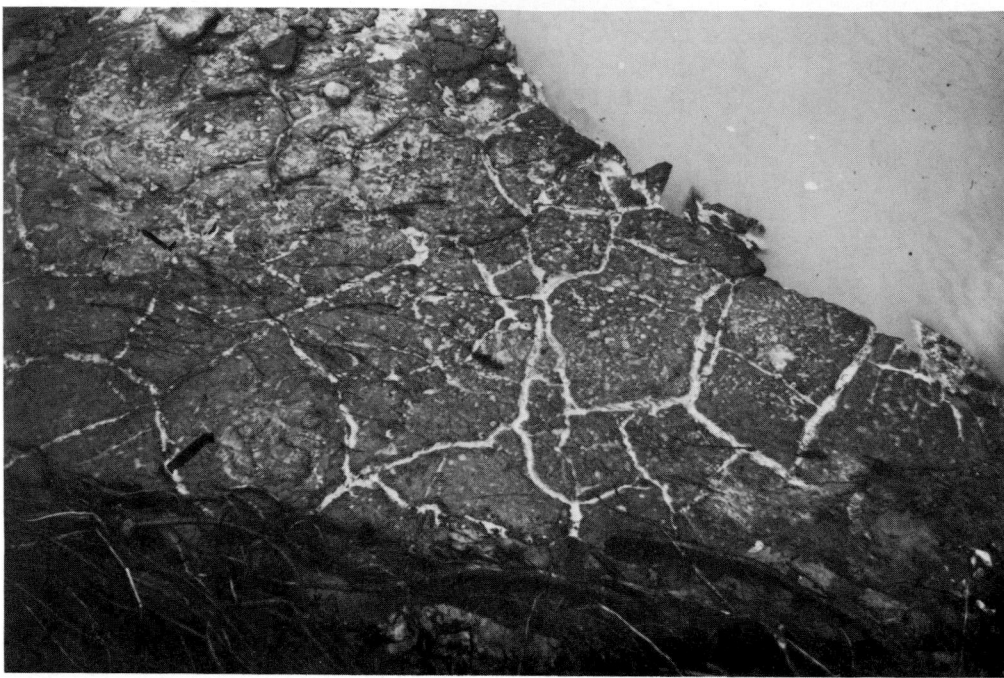

Fig. 4. Polygonal structure typical of palaeosol II weathering in massive silt.

adequately due to this complexity. Bedding is rare and has been observed only in the sandier zone of (gradational) contact with the subjacent member. The weathering profile on the massive silt, termed palaeosol II (Fig. 3), is distinctive. This palaeosol has no A1 horizon, a thick A2 horizon and a dense B2 horizon with a well-developed polygonal structure (Fig. 4). Seams between the polygons are often wider than 2 cm. A distinguishable basal phase of the massive silt is present in several valleys. This phase has more clay than the superjacent (typical) massive silt and is usually separated from it by a sharp textural contact. The polygonal seams, however, are continuous from the massive silt into the basal phase. This basal phase usually occurs as narrow ribbon-like deposits, suggesting that it may be fill in an abandoned stream course.

Some aspects of this early Holocene depositional sequence are ambiguous. The typical massive silt unit contains no large relict channels which could be used as evidence for channelized flow. This unit, however, is coarser textured than typical upland loess deposits and is somewhat younger than the presently accepted age for the youngest loess unit (the late Quaternary Peoria loess). Tributary valley inundation (ponding) due to plugging by trunk stream aggradation, such as described by Pflug (1969) for tributaries in eastern

Brazil, is not reasonable for this study area. On several streams the elevation of the massive silt decreases by more than 30 m over a length of about 16·5 km down-valley, but 5–6 m terraces at the downstream end of the valleys are free of this material. Certainly, ponding depths sufficient to induce deposition in the upper-valley position would also have induced deposition over the terraces present in the lower-valley position. These terraces are capped with Peoria loess, which is not present as a deposit in a floodplain position. We interpret this terrace unit/floodplain unit relation as evidence for valley entrenchment immediately preceding the Holocene. Additional evidence for such entrenchment, presented by Grissinger, Murphey & Little (1982)*, includes (1) the absence of organics with ages in the interval from about 13,200 to 17,000 yr BP, (2) the absence of boreal species in our wood samples, (3) the excessive depth of several of the early Holocene wood samples, (4) the disconformable contact between the Holocene deposits and the > 40,000 yr BP consolidated sandstone and (5) the absence of any evidence for pre-Holocene palaeosols

* This article contains a more detailed description of the lithologic units, their relation with valley-fill sequences in other areas and their agreement with palaeoclimatic conditions. The data base included 60 ^{14}C ages with a maximum datable age of 12,050 ± 180 yr BP.

which would indicate the presence of buried pre-Holocene floodplains. They argued that this depositional sequence of valley erosion → bog-type, grey silt or channel lag deposits → massive silt recorded a condition of decreasing fluvial energy resulting from decreasing pluvial activity and rising base-level controls. In this scenario, the massive silt is a low-energy fluvial deposit but at this time cannot be further defined.

Mid-Holocene depositional sequence

The single lithologic unit in this sequence is the channel-fill unit which was deposited subsequent to a period of stream entrenchment. Ages of these samples form the mid-Holocene frequency mode (Fig. 2a). Sediments are relatively coarse textured, grey coloured and have a poorly defined weathering profile. Bedding ranges from lenticular to festoon cross-stratification but is often difficult to discern. These deposits are present in all valleys but are not widespread.

Late Holocene depositional sequence

This sequence includes two lithologic units, the post-settlement alluvium (PSA) and meander-belt alluvium (MBA), which form the youngest mode of the age frequency (Fig. 2a). PSA sediments cap all floodplain surfaces. They were produced in historic times largely by human activity. Thickness is usually less than 1 m but may locally exceed 3 m. The PSA has well-preserved fluvial bedding ranging from horizontal discontinuous to lenticular to micro-cross-stratification. It is unweathered, with an Ap horizon directly overlying a C horizon.

MBA sediments include both vertical and lateral accretion deposits having a maximum age of about 3000 yr BP. Wood and other organic detritus occur throughout the deposits. These sediments are distributed throughout the valleys, unconformably overlying older deposits. The entrenched streams apparently meandered across the floodplain, eroding older sediments and depositing the MBA sediments in a well-defined meander belt. Where not subsequently eroded, this unit has an A1 horizon which varies in thickness from a few centimetres to more than 25 cm, no A2 horizon and a weakly developed B horizon. This weathering profile has been named palaeosol 1 (Fig. 3). In all cases, fluvial bedding is readily observable but is less well preserved than that in the PSA. Bedding in the vertical accretion deposits is comparable with that of the PSA, whereas festoon cross-stratification is common in the lateral accretion

deposits. These two units can be identified as individual members of this sequence where the palaeosol I profile is intact.

Units in the general study area

Lithologic units in the general study area occur in the same sequences and have age relations identical with comparable units from the intensive study area (Fig. 2a). The two age–frequency distributions were not significantly different at the 95% level, as evaluated by the two-sided Kolmogorov–Smirnov test (T_{max} of 0·153 < critical value of 0·247). The combined age–frequency distribution is presented in Fig. 2(b). The depositional sequences and relative textures of the lithologic units are also comparable. Absolute textures are variable, however, depending on the nature of the source material. Lithologic units in the Flatwoods Physiographic Subprovince are finer textured than comparable units in the intensive study area, reflecting the fine texture of the Porters Creek Clay source material. Other depositional and weathering properties are consistent.

Age-to-depth relations are presented in Fig. 5 to define the nature of the controls of the Holocene valley-fill system further. Fig. 5(a) includes the relation of all ^{14}C samples from the general study area except those for gravel-laden catchments and for catchments adjacent to major drainage ways. This regression equation is

$$D = -1·29 + 0·77 \ln A \qquad (1)$$

where D is the sample depth and A the sample age. The correlation coefficient, $r = 0·79$, is significant at the 99% level. This relation explains 63% of the depth variation; 37% of the variation remains unexplained. A large part of this unexplained depth variation is associated with two secondary relations, one at about 2000–3000 yr BP and the second at about 4000–6000 yr BP. This latter age span is the time of mid-Holocene entrenchment and subsequent deposition of the channel-fill unit. The correlation coefficient for this secondary relation is 0·68, which is not significant at the 95% level ($r_{0·95} = 0·71$). The younger secondary relation is significant at the 95% level, $r = 0·75$. This age range corresponds to the initiation of the late Holocene depositional sequence but is not associated with a differentiable lithologic unit within this sequence.

Figure 5(b) includes comparable age–depth relations for gravel-laden catchments, including Goodwin, Long and Johnson catchments of the intensive study area (noted as m, Fig. 1), for Tillatoba catch-

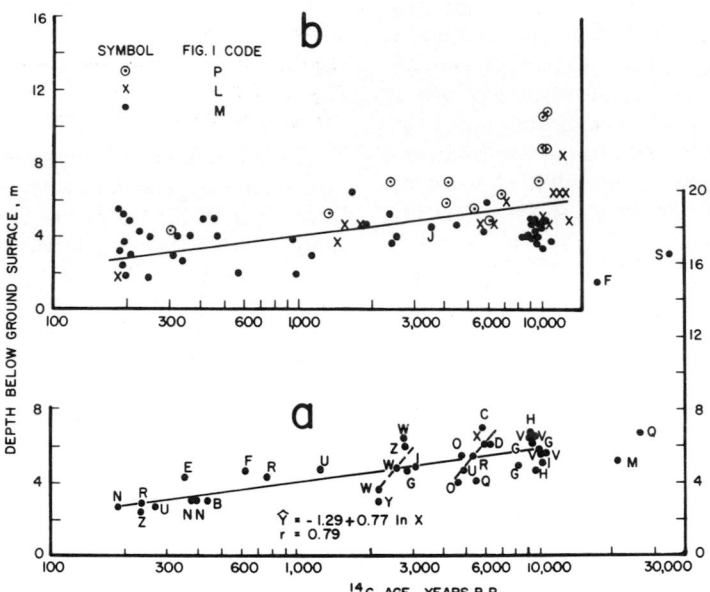

Fig. 5. Age-to-depth relations for valley-fill deposits. (Letters identify sample location, see Fig. 1.) (a) All samples from general study area except those from Tillatoba Creek catchment (noted as p, Fig. 1). (b) All Holocene age samples from the intensive study area and Tillatoba Creek catchment. The regression for (a) is reproduced for comparison.

ment (noted as p, Fig. 1), which is adjacent to a major drainage way, and for Hotophia catchment (noted as 1, Fig. 1), which is both gravel-laden and adjacent to a major drainage way. The correlation coefficient is significant at the 99% level, $r = 0.52$, but explains only 27% of the depth variation. A large part of the unexplained variation is associated with samples from the early Holocene depositional sequence. For Tillatoba catchment, adjacent to a major drainage way, all five samples had greater depths than predicted by equation (1). The probability of this being a random event is only about 3%. For the gravel-laden catchments, 17 out of 18 samples had depths less than those predicted by equation (1). The probability of this being a random event is less than 0.1%.

The two age-to-depth regressions (Fig. 5a, b) reflect two types of controls of the Holocene depositional sequences, i.e. local and regional controls. Local controls, including gravel abundance and relative position in the drainage net, influenced sample depth (Fig. 5b) presumably by regulating the degree of pre-Holocene entrenchment. Regional control is implied by the significance of the age-to-depth regression for all portions of the study area not materially influenced by local controls (Fig. 5a). Additional evidence for regional control is the similar nature and chronology of the individual units and the

repetition of the depositional sequences over the entire study area. The nature and/or chronology are also generally comparable with valley-fill sequences in Iowa (Daniels & Jordan, 1966; Ruhe, 1969), in Missouri (Brakenridge, 1980), in south-western Wisconsin (Knox & Johnson, 1974; Knox, McDowell & Johnson, 1981) and in the south-western United States (Haynes, 1968). More detailed comparisons of these valley-fill sequences have been presented by Grissinger *et al.* (1982).

PALAEOCLIMATIC CONTROLS

The general consistency of the Holocene depositional sequences over a large part of the mid-continental United States suggests that palaeoclimate was the regional control. To evaluate this possibility, the age frequency for this study was compared with one presented by Wendland & Bryson (1974). Their data consisted of dates of discontinuities (within pollen profiles, glacial records, sea-level heights and peat beds) which defined times of palaeoclimatic transition. About 75% of their data pertained to Europe. The age–frequency distribution for the combined data base from northern Mississippi is presented in Fig. 2(b) and that of Wendland & Bryson (1974) in Fig. 2(c). Both distributions are trimodal, with relatively comparable peak ages for the two older modes. These qualitative similarities between the two distributions were noted previously, based on a restricted data base of 57 samples (Grissinger *et al.*, 1982).

The additional data for northern Mississippi permitted more quantitative comparison between these two distributions using the two-sided Kolmogorov–Smirnov test. The distributions (Fig. 2b, c) are significantly different at the 95% level (T_{max} of 0·280 at 990 yr BP > critical value of 0·134). This significant difference resulted from the relatively large number of northern Mississippi samples with ages less than 800 yr BP (Fig. 2b). All of these samples which were datable (excepting the PSA samples) originated in meander-belt deposits and are thus not comparable with the data set used by Wendland & Bryson (1974) to identify times of palaeoclimatic transition. Wendland & Bryson did report a major botanic-geologic discontinuity at 850 yr BP coincident with a change in the Mill Creek culture of Iowa. This change was attributed to changes in atmospheric circulation patterns which may also be responsible for this peak frequency in the northern Mississippi data.

An additional but secondary time of significant difference was established by performing the same test on all samples with ages greater than 800 yr BP. This maximum cumulative difference (T_{max} of 0·304 at 9320 yr BP > critical value of 0·157 at $P = 0·95$) resulted from the larger number of dates in Wendland & Bryson's data between the early and mid-Holocene peak frequencies (at about 7000–9320 yr BP). This significant difference may reflect climatic differences between the two areas or it might have been produced by a diachronous climatic shift which would influence Wendland & Bryson's data base proportionately more than the data base for the much smaller northern Mississippi study area. No other differences between these two distributions were significant. The two early Holocene peak frequencies (200 yr class interval, Fig. 2b) fit the transition times between the Boreal to pre-Boreal and the pre-Boreal to late glacial climatic episodes and agree with times of discontinuities reported by Wendland & Bryson (1974).

The fit between the early and mid-Holocene modes for these two independent frequencies supports the hypothesis of climatic control of valley-fill sedimentary processes. This agreement implies that major climatic changes were generally synchronous. The lack of synchronism in late Holocene times and from about 7000 to 9320 yr BP, however, implies asynchronous change presumably reflecting both diachronous climatic shifts and the influence of local controls. Local controls in each system, for example in our study area the abundance of gravel and the relative position in the drainage net, may be just as significant as regional climatic controls.

CHANNEL STABILITY AND MORPHOLOGY

Many, if not most, streams in the study area were straightened or otherwise modified prior to 1937, the date of the earliest aerial photographic record. In many instances, the positions of the streams in the valleys were changed so that channels were no longer within the late Holocene meander-belt deposit. Entrenchment occurred during this same time period, primarily in the form of knickpoints which proceeded upstream at an alarming rate. For Johnson Creek in the intensive study area, the rate of knickpoint migration averaged 160 m yr^{-1} from 1940 to 1975. Thalweg lowering of 3–4 m was common. Entrenchment is still in progress in many streams of the study area producing widespread problems of bank instability which are financial burdens to both private and public interests.

The two features of channel realignment and entrenchment have changed many of the streams from alluvial channel systems to systems presently controlled by the properties and distribution of mid-Holocene and older deposits. The lifetime of these controls may be insignificant in geologic time but is significant to current channel behaviour, especially to channel instability problems. The types of failure and the processes involved in bank and bed failures are controlled by properties of individual lithologic units and by unit sequences.

Controls of channel bed and bank stability

Bank materials which control stability include post-settlement and meander-belt alluvium, channel fill and the early Holocene massive silt, bog-type, channel lag and unconsolidated silt deposits. Failure of the post-settlement and meander-belt alluvium results from gravity stress, in many instances accentuated by tension crack development. The tension cracks are vertical, and parallel with the banks. Failure of the massive silt also results from gravity stress with failure mass defined by the polygonal structure typical of the palaeosol II-type weathering. The frequency of failure is related to current entrenchment, which has

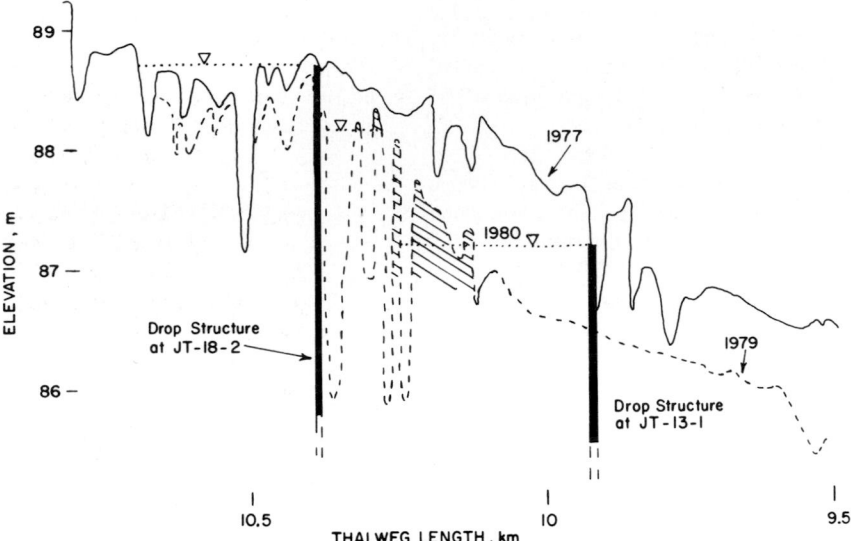

Fig. 6. Entrenchment into and through the massive silt and sill-pool development, Johnson Creek, the intensive study area (Fig. 1). The thalweg profile for 1977 is shown as the solid line and that for 1979 as the dashed line. Drop structures were installed in 1980, and sill destruction between 1979 and 1980 is identified by hatching. Sill materials are the basal phase of the massive silt.

increased bank height and steepness and has exposed the relatively weak bog-type and channel lag deposits in a bank toe position. Materials of both deposits are easily eroded by channelized flow whereas the unconsolidated silt has sufficient cohesion to be relatively stable. Channel-fill materials are also easily eroded by channelized flow. Although failure is primarily gravity controlled, continuing instability is perpetuated by removal of the slough material by channelized flow. The rate of removal is controlled by disaggregation of the slough blocks.

Channels downstream of knickpoints have sand-to-gravel beds and the stability of these beds is primarily dependent on sediment supply to, and transport properties of, the hydraulic system. Two units limit knickpoint migration. Outcrops of the pre-Holocene consolidated sandstone function as bed-control sills and inhibit knickpoint migration. The control life of these outcrops varies with unit thickness and flow conditions; relatively thick outcrops of one or more metres have persisted for several tens of years whereas thinner outcrops have failed in shorter periods of time. The second unit which limits knickpoint migration is the basal phase of the massive silt. The control life of these sills is several years, with failure usually resulting from the development of chutes through the polygonally structured basal phase followed by block disarticulation. An unusual feature of this type of

knickpoint migration is the development of scour holes upstream of the sills (Fig. 6). These scour holes frequently have bed elevations two or more metres below sill outcrop elevations.

Channel morphology

Johnson and Goodwin Creeks occupy adjacent catchments in the intensive study area. These catchments have similar land-use patterns, topography, soils and rainfall. The present channel morphologies, however, are variable, reflecting differences in both the texture of the current bed material and the distribution of valley-fill deposits which control channel adjustment. Thus the morphologies for these creeks cannot be expressed as a singular function of discharge.

Johnson Creek has two functional segments separated by a knickpoint. Channel width-to-depth relations are inconsistent downstream of the knickpoint (Fig. 7b) due primarily to excessive widening where banks are late or mid-Holocene materials. The relations are significant upstream of the knickpoint (Fig. 7a) where bed and bank materials are primarily early Holocene materials. This correlation is positive and significant at the 99% level ($r = 0.69$). Pre-Holocene consolidated sandstones do not crop out in Johnson Creek, and gravel bed material is uncommon

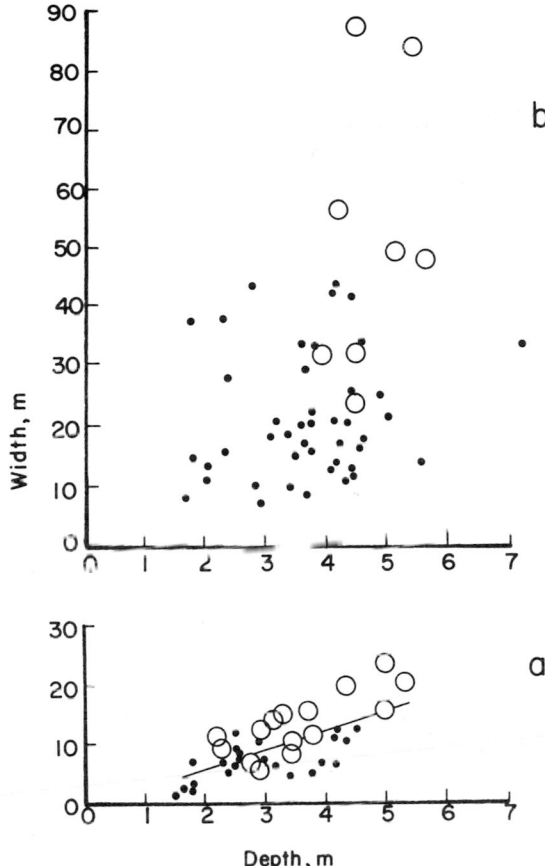

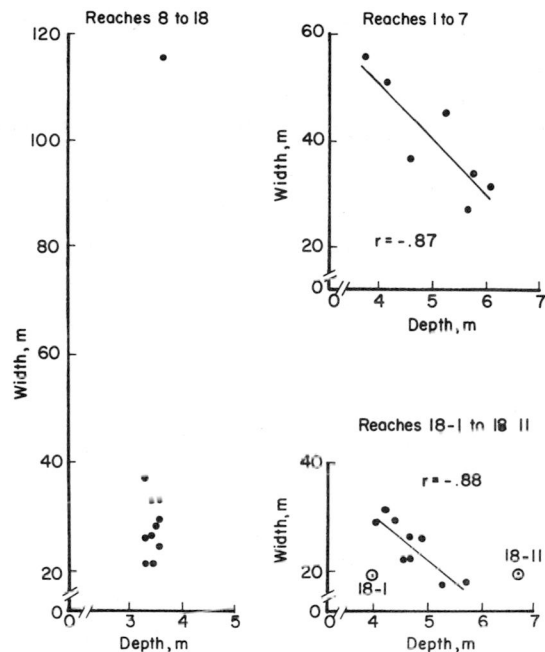

Fig. 8. Average widths and depths for Goodwin Creek segments from 1977 survey data.

Fig. 7. Width-to-depth relations for Johnson Creek upstream (a) and downstream (b) of the knickpoint from stereoscopic analysis of aerial photographs for the years prior to 1968 identified by dots and for the year 1975 identified by circles (Ethridge, 1979).

upstream of the knickpoint. The significant width-to-depth relation reflects simple channel enlargement controlled by the early Holocene massive silt deposit.

Consolidated sandstones crop out throughout the Goodwin Creek channel, and gravel bed material is common. Knickpoint movement and entrenchment on Goodwin Creek have been minor relative to that for Johnson Creek. The Goodwin Creek channel has three functional segments with transition reaches less well defined than for Johnson Creek (reach numbers increase upstream). These transition reaches are defined by a relatively small knickpoint in reach 8 and by the presence of pre-Quaternary bed and bank materials upstream of reach 18-1.

Average width-to-depth relations are presented in Fig. 8. Width and depth are inversely related in reaches 1–7 (downstream of the knickpoint) and

upstream of reach 18-1 where bank and bed materials are pre-Quaternary. Both of these regressions are significant at the 99% level. (Reaches 18-1 and 18-11 were not included; the former is transitional and the latter influenced by road-culvert control.) Depth is constant and independent of width for the middle segment (between reaches 8 and 18) where thalweg elevation is controlled by the massive silt unit and by outcrops of the consolidated sandstone.

For the upstream and downstream segments, the channel width/depth ratios are related to sinuosity (Fig. 9). The correlation coefficient, $r = 0.82$, is significant at the 99% level for the upstream segment but was not calculated for the downstream segment due to the limited number of reaches within each of the two relations for the segment. Reaches 5 and 8 each contain relatively large bendways which formed due to local material controls. Thalweg slopes for these reaches are less than downstream reaches (Fig. 10). (The slope for reach 8 was adjusted to remove the knickpoint drop at the lower end of this reach.) We believe that these two large bendways, at reaches 5 and 8, disrupt the downstream movement of coarse sediment and thus adversely affect downstream bank stability. Thalweg slopes throughout the remainder of Goodwin Creek channel are locally

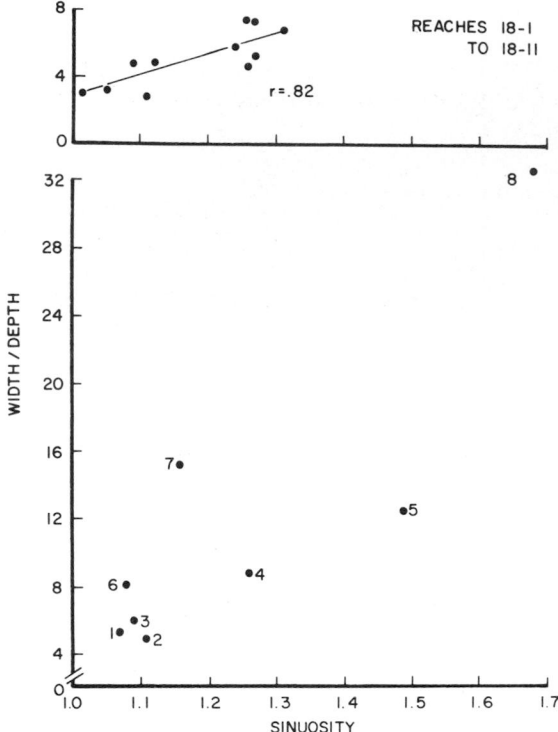

Fig. 9. Width/depth ratios versus sinuosities for the upstream and downstream segments of Goodwin Creek, from 1977 survey data.

controlled by the presence or absence of consolidated sandstone sills.

CONCLUSIONS

The Holocene valley-fill deposits in northern Mississippi have been differentiated into seven litho-logic units which, together with several pre-Holocene deposits, control present-day channel bed and bank stability and morphology. Due to this bed and bank material control of near-term channel behaviour, channel morphology is considered to be a variable independent of present flow conditions. A corollary of this reasoning is that flow properties such as secondary flow, boundary shear stress and total flow resistance are, to some degree, dependent upon channel morphology. In essence, the present massive channel instability problems of the study area result from the immaturity of the present fluvial system; the system will continue to change and eventually become a stable, mature system when channel morphology is dependent on flow properties.

At this time, channels in the study area are not alluvial channels; they are not 'free to adjust' and are not 'composed of material identical with their present

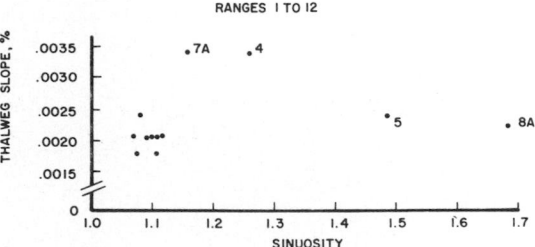

Fig. 10. Thalweg slope versus sinuosity for the downstream segment of Goodwin Creek, from 1977 survey data.

sediment load'. The channel morphologies are controlled by the nature and distributions of the Holocene valley-fill deposits and several older mater-ials; deposition of the Holocene units, in turn, was controlled by palaeoclimatic conditions. Thus, in a broad sense, the present climatic conditions are interacting with Holocene palaeoclimatic conditions to produce the present problems of acute channel instability.

REFERENCES

BRAKENRIDGE, G.R. (1980) Widespread episodes of stream erosion during the Holocene and their climatic cause. *Nature*, **283**, 655–656.

DANIELS, R.B. & JORDAN, R.H. (1966) Physiographic history and the soils, entrenched stream systems, and gullies, Harrison County, Iowa. *Tech. Bull. U.S. Dep. Agric.* **1348**, U.S. Government Printing Office, Washington, D.C. 102 pp.

ETHRIDGE, L.T. (1979) *Photogrammetric interpretation of stream channel morphology, Johnson and Goodwin Creeks, Panola County, Mississippi.* M.E.S. Project, Department of Geology, the University of Mississippi. 24 pp.

GODWIN, H. (1966) Introductory address. In: *World Climate from 8000 to 0 B.C.*, pp. 3–14. Royal Meteorological Society, London.

GRISSINGER, E.H., MURPHEY, J.B. & LITTLE, W.C. (1982) Late-Quaternary valley-fill deposits in north-central Mississippi. *SE. Geol.* **23**, 147–162.

HAYNES, C.V. (Jr) (1968) Geochronology of late-Quaternary alluvium. In: *Means of Correlation of Quaternary Successions. Proc. VII Congr. int. Ass. Quat. Res.* **8**, 591–631.

KNOX, J. C. & JOHNSON, W.C. (1974) Late Quaternary valley alluviation in the Driftless Area of southwestern Wiscon-sin. In: *Late Quaternary Environments of Wisconsin*, pp. 134–162. American Quaternary Association, Third Bien-nial Meeting, University of Wisconsin.

KNOX, J.C., McDOWELL, P.F. & JOHNSON, W.C. (1981) Holocene fluvial stratigraphy and climatic change in the Driftless Area, Wisconsin. In: *Quaternary Paleoclimate*, pp. 107–127. Geo Abstracts, Norwich, England.

PFLUG, R. (1969) Quaternary lakes of eastern Brazil. *Photogrammetria*, **24**, 29–35.

RUHE, R.V. (1969) *Quaternary Landscapes in Iowa*. Iowa State University Press, Ames. 255 pp.

WENDLAND, W.M. & BRYSON, R.A. (1974) Dating climatic episodes of the Holocene. *Quat. Res.* **4**, 9–24.

Spec. Publs int. Ass. Sediment. (1983) **6**, 251–266

Proglacial channel systems: change and thresholds for change over long, intermediate and short time-scales

JUDITH K. MAIZELS

Department of Geography, University of Aberdeen, St Mary's, High Street, Old Aberdeen AB9 2UF, U.K.

ABSTRACT

This paper examines the nature, direction and magnitude of change in channel pattern and form in a series of proglacial drainage systems that have experienced deglaciation or glacial fluctuation over a variety of time-scales. Simple quantitative measures are introduced to indicate the direction and magnitude of channel change, and associated changes in meltwater discharge, through the different terrace sequences. The results indicate that in areas experiencing long-term deglaciation channel systems have changed from those comprising dominantly braided, steeply graded, low-sinuosity, high width-depth ratio bedload channels to those represented by deeper, more sinuous, low-gradient, single-thread channels. In addition, channel systems observed on terrace surfaces are shown to include threshold or transitional channels, themselves closely associated with terrace formation and a change in channel equilibrium from a stable or aggradational to an erosional regime. The major channel changes observed appear to reflect factors which are associated with long-term deglaciation, including a change in water and sediment supply, in the magnitude and frequency of extreme flood events, in regional and local base-level and/or the local exceedance of internal geomorphic thresholds.

INTRODUCTION

This paper examines the nature, direction and magnitude of change in channel pattern and form in three terraced palaeosandur deposits, whose catchments have experienced deglaciation over long-term time-scales. The North Esk palaeosandur deposit in north-east Scotland, for example, dates from the final stages of Older Dryas ice-sheet decay, and its catchment has possessed no glaciers since the end of the Lateglacial. By contrast, sandur deposits investigated in west and south-west Greenland first accumulated during ice-sheet recession *c.* 7000 and 2500 yr BP respectively, while the meltwaters continue to be fed by the ice sheet still active within their catchments. Additional evidence is also briefly described from areas experiencing shorter periods of deglaciation or glacial fluctuation. In the ice-marginal environment important upvalley links exist with the glacier system that controls the supply of meltwater and sediment to the ice-marginal meltwater streams, while predominant downvalley links exist with the proglacial system where many fluid-flow, sediment-transport, channel-form and base-level parameters are all closely interlinked. Hence proglacial channel changes may result from a number of different factors, acting together or independently, and producing simple or complex forms of response. The main factors are: (a) changes in water and sediment supply associated either with a change in ice mass or with extreme hydrologic events; (b) a change in regional or local base-level, which affects channel slope and sediment transport rates, and hence the degree of erosion or aggradation; and (c) intrinsic controls such as the local exceedance of an internal geomorphic threshold. However, changes in both extrinsic and intrinsic controls are only reflected in the channel system once certain critical limits have been exceeded, and some of the problems involved in trying to identify the thresholds for channel change during deglaciation are briefly considered in the final part of this paper.

0141-3600/83/0106-0251 $02.00

METHODS OF PALAEOCHANNEL ANALYSIS

The analysis of palaeochannels is aimed at establishing the nature of both 'directly' and 'indirectly determinate' variables. Directly determinate variables are those that can be established from direct field measurements of palaeochannel parameters, such as channel pattern, form and sedimentary characteristics; indirectly determinate variables are those that can be estimated through analogy with present-day fluid–sediment relations, and include estimates of bed roughness and flow resistance, flow velocity and bankfull discharge. Both these types of determinate variables allow one to estimate aspects of former flow conditions associated with changing channel forms and terrace development, the controls on channel change, and the possible thresholds for channel change. 'Indeterminate' variables, by contrast, are those that cannot be determined in palaeohydrologic analysis using reconstruction techniques available at present, and include the rates of ice ablation, the degree of englacial channel development and meltwater storage, and the rates of glacial and non-glacial weathering and erosion, as examples of possible indeterminate controls on meltwater and sediment supply, respectively.

Surface, morphology, channel pattern and form

Channel systems were first identified from aerial photographs and a base map produced of each area to show main terrace areas and associated palaeochannel courses. Channel pattern and sinuosity were determined directly from the base maps. Channel multiplicity was determined from the mean number of channels per 100 m width of terrace surface, and defined as a simple braiding index. The value of this index allows one to estimate the possible mean and maximum number of channels that might have been present across the whole sandur surface, assuming both that the width of the (active) palaeosandur is known and that channels on the whole (active) palaeosandur surface were operative at one time. Channel systems were then levelled in the field in order to provide information on bankfull channel width and depth, and channel slope, for the different terrace surfaces.

Sedimentology

Sediment sampling of older, vegetated sandur deposits (North Esk) was restricted to natural exposures and quarry sections, and hence is not always representative of certain mapped channel systems. However, in non-vegetated sandur areas sediments were sampled from the bed, banks and intervening bars of individual channels. At each site the intermediate diameter of the 10 largest clasts from the coarsest sedimentation unit was measured. Many workers have used this value as an indicator of peak flow conditions rather than calculating, for example, the 84th percentile of the whole size distribution. However, Limerinos (1970) claims that the use of this latter measure cannot be rationalized in hydraulic terms, but only in terms of empirical results and ease of sampling; similarly, Hey (1979) comments that for natural gravels the roughness height may in fact be best represented by the 99th percentile of the sediment size distribution. Hence it seems that measurement of the largest clasts present on the bed may provide a reasonably reliable estimate of maximum skin resistance. Analysis of sedimentary structures, imbrication, bedforms, stratigraphy, facies types and clast form was also undertaken, and details are given elsewhere (Maizels, 1976, 1980a).

Estimates of palaeohydrologic conditions

A number of methods has been proposed for estimating the palaeodischarge of streams, at least to within an order of magnitude, but very few are directly applicable to the analysis of meltwater stream deposits. These deposits characteristically represent coarse-bed, low-sinuosity, unstable, active multiple-channel bedload streams in which up to 100% of stream flow may be provided by ice melt. This means, first, that proglacial discharge may increase only slightly downstream, through minor flow additions from precipitation, ground water and slope runoff (Fig. 1, model 2), and may even decline where much water infiltrates through permeable outwash deposits. Meltwater discharge would only exhibit a major increase when the stream meets a tributary river, which may be of either non-glacial or glacial origin, its origin in turn affecting the downstream rate of increase in discharge, (i.e. model 2/1 or 2/2 in successive downstream reaches). Secondly, it means that palaeodischarge conditions are more reliably estimated on the basis of flow functions which define the critical flow conditions required to entrain sediment particles of a known size in fully turbulent rough-bed streams, than on the basis of empirical relations such as those between channel form, discharge, catchment area, climate and sediment yield (e.g. see Ethridge & Schumm, 1978).

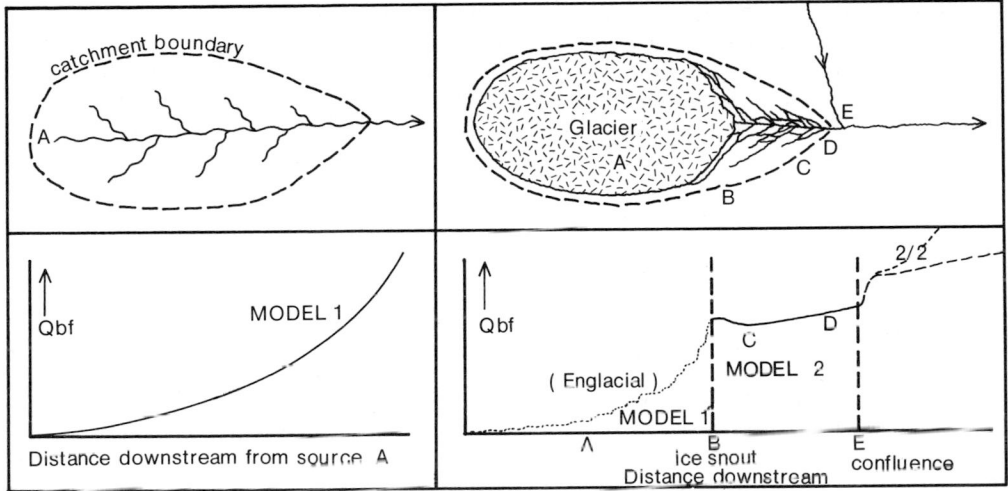

Fig. 1. Downstream changes in stream discharge: comparative models for humid and ice-fed stream systems.

Shields' (1936) entrainment function, for example, has been fairly widely applied in the palaeohydrologic analysis of outwash deposits (e.g. Church, 1970, 1978; Baker, 1974; Cheetham, 1976, 1980; Clague, 1975; Maizels, 1976) and is also adopted in this study, since it enables one to relate particle size directly to the critical force, where particle size exceeds about 8 mm and flow is fully turbulent:

$$\tau_c = C_1 (\gamma_s - \gamma) D, \qquad (1)$$

where τ_c is the critical tractive force, γ and γ_s are the specific weight of water and the sediment particles respectively, D is the clast diameter (m), and C_1 is Shields' coefficient. C_1 is generally taken as 0·06 although this represents an extrapolated value (Shields, 1936) and several workers have since shown that the value of this coefficient may increase to as much as 0·11 with increasingly underloose boundary conditions (e.g. Church, 1970, 1978) or decrease to about 0·03 for uniform sediments (e.g. Cheetham, 1976). Shields' function is of particular value here since it can be used in conjunction with the DuBoys equation for boundary shear stress as the basis for estimates of critical flow depth and, subsequently, of critical flow velocity. Critical flow depth may then be calculated from

$$d = C_2 D S^{-1}, \qquad (2)$$

where C_2 varies from 0·099 when $C_1 = 0·06$, to 0·183 when $C_1 = 0·11$, d is the critical mean flow depth, S represents the energy gradient, γ and γ_s are taken to be constant for a given meltwater deposit. Where measured bankfull depth is taken to represent peak

flow conditions, however, two assumptions must be made: first, that estimates of peak flow represent flow within confined channels, and secondly, that original channel depths have not been modified by subsequent flows before final abandonment. In this study all channel depths used in palaeoflow calculations have been determined using equation (2).

Critical mean flow velocity was initially estimated from Manning's equation, where the resistance coefficient n was determined using Strickler's (1923) function (Maizels, 1982). However, these estimates of resistance were found to underestimate flow resistance significantly, and hence to overestimate mean flow velocity, when compared with estimates using the Darcy–Weisbach friction factor, f. Hey (1979) has demonstrated that f is a function of both relative roughness of the bed and cross-sectional channel shape:

$$\frac{1}{\sqrt{f}} = 2·03 \log \left(\frac{ad}{3·5D} \right), \qquad (3)$$

where a is a function of the ratio between the hydraulic radius, R_h, and maximum flow depth. Hence, both skin and form resistance are significant in coarse-bed, shallow meltwater streams, where width–depth ratios exceed a value of $c.\ 15·0$, and relative roughness exceeds values of $c.\ 0·2$, although also significant in some channels exhibiting values as low as 0·03. Critical mean flow velocity, $\bar{V}_c$, was therefore estimated using the Darcy–Weisbach equation:

$$\bar{V}_c = \left(\frac{8g\ dS}{f} \right)^{\frac{1}{2}}, \qquad (4)$$

where g is the acceleration due to gravity, and d is taken to be equivalent to R_h. Combining (4) with (2) and (3) allows one to express the critical mean velocity solely in terms of D, S and the cross-sectional channel shape:

$$\bar{V}_c = (4\cdot24 C_2 D)^{\frac{1}{2}} \log\left(\frac{C_2 a}{3\cdot5 S}\right). \qquad (5)$$

Palaeodischarge amounts were then estimated from the standard flow continuity equation, i.e.:

$$Q = wd\,\bar{V}_c.$$

PALAEOCHANNEL CHANGES IN PROGLACIAL AREAS

Analysis of palaeochannel changes in proglacial areas has been based on the application of the methods described above to several case-studies of the sandur deposits that have been formed: (a) during deglaciation or glacial fluctuations over different time-scales, and (b) in areas that at the present time are either non-glacial or still retain an ice mass within the catchment.

Case-study 1: long-term deglaciation of a present-day non-glacial catchment: North Esk sandur deposits, North-east Scotland

The River North Esk, and its main tributary, the River West Water, rise in the south-east Grampians, and flow south-eastwards across the Vale of Strathmore to the North Sea (Fig. 2). The Vale of Strathmore was covered by an extensive ice sheet during the Late Devensian, which, between 14,000 and 13,000 yr BP wasted back north-westwards into the Grampian glens from the North Sea coast (Paterson, 1974; Sissons, 1974). The valley glacier phase of deglaciation was accompanied by the deposition of massive sandur sediments extending from the former ice-marginal zone just north of the Highland Boundary Fault zone marking the upland edge, for some 10 km downstream. These outwash deposits have since been dissected to form four main terrace systems. Each terrace system exhibits at many sites intricate and well-developed palaeochannel networks, as well as evidence of localized channel incision and minor terrace development (Maizels, 1982). Hence a 'terrace system' is one comprising many minor surface configurations and irregularities, but such that 'within-terrace' relief (usually $< 1\cdot5$ m) is still smaller than 'between-terrace' relief (usually $> 1\cdot5$ m). In addition, the major terrace bluffs can be traced clearly for several kilometres downvalley (Fig. 2).

Morphology of the terraced sandur deposits

Four main terrace systems, T1 to T4, have been identified both from aerial photograph analysis and in the field. The altitudinal and morphological sequence of terrace systems indicates a south-eastward migration of the North Esk–West Water confluence of 2·8 km through the terrace sequence, accompanied by alternating periods of aggradation and incision. While the highest terrace system, T1, exhibits a relatively steep downvalley gradient, the three lower terraces all follow gentler gradients which, in addition, do not appear to have altered significantly during these three later stages of terrace formation. The present River North Esk, which is incised both into the valley-fill deposits and the underlying Old Red Sandstone bedrock, exhibits a much lower mean gradient. Mean gradient values for individual terrace systems can be misleading, however, since gradients in fact decline downvalley, following an approximate exponential function. Isostatic readjustment does not appear to have increased terrace gradients by any significant amount; an estimated maximum increase of $0\cdot038$ m km^{-1} (Maizels, 1976) lies well within the range of local slope variations (Table 1).

Channel pattern and form

All of the terrace systems, but particularly terrace systems T3 and T4, exhibit complex braided palaeochannel systems at many locations (Fig. 2). These channel systems are characterized by multiple channel networks and intervening longitudinal bars, and by channels exhibiting high width–depth ratios (averaging 75) and low sinuosities. The terrace surfaces are locally incised by relatively well-defined deep, low-gradient, sinuous (SIN $< 1\cdot62$) channels, while any adjacent terrace bluffs follow a series of conspicuous meander scars. These sinuous channels are particularly common on the lowest terrace, T4, and this phenomenon is reflected in the overall increase in the percentage of more sinuous channel reaches (i.e. SIN $\geq 1\cdot1$) through the terrace sequence from T1 to T4 (Table 1). Similarly, the mean braiding index exhibits a marked decline through the terrace sequence as the number of single-thread channels increases. An overall decrease in mean width–depth ratio also accompanies these changes in channel pattern, with the braided channels exhibiting a mean width–depth ratio of 108 (±31), compared with those of the sinuous channels of T4 averaging only 40 (±15).

The channel changes observed through the terrace sequence are further complicated by channel changes

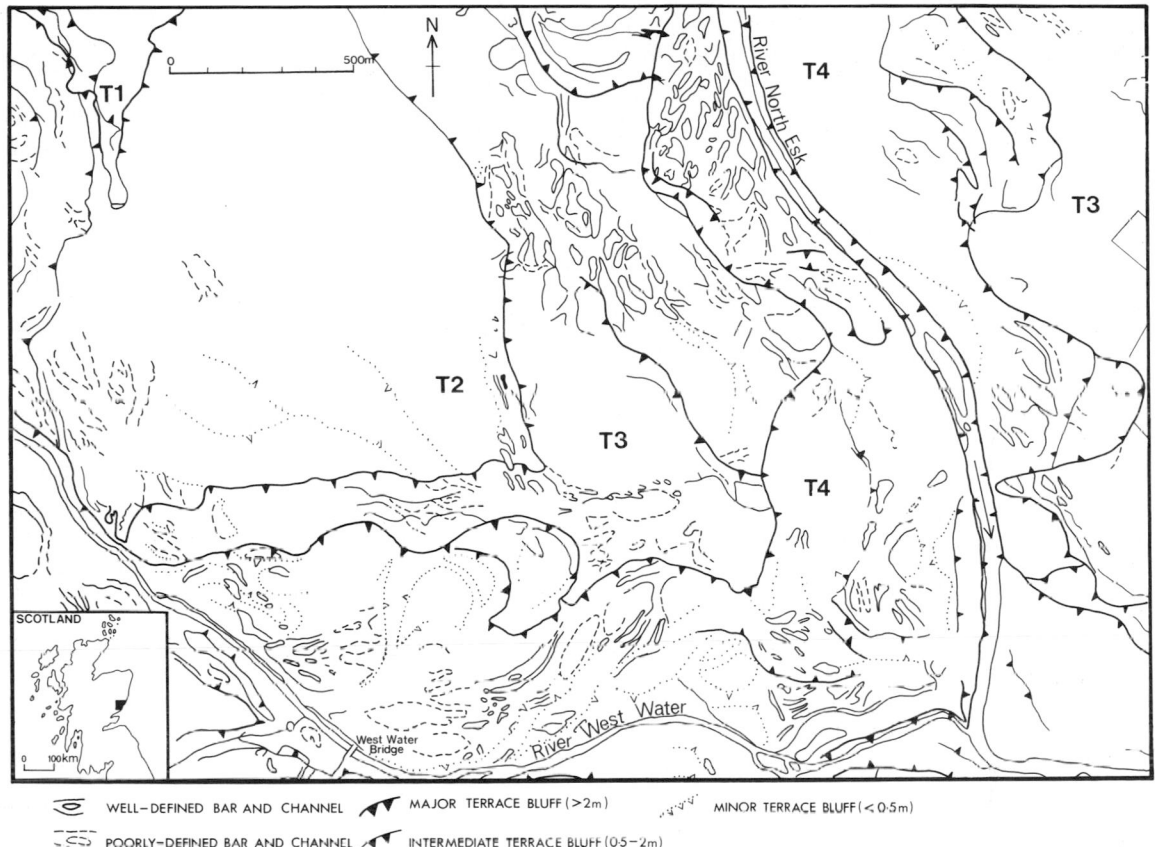

Fig. 2. Examples of palaeochannel systems on the North Esk terrace surfaces, T1 to T4. Low-sinuosity, braided channels can be differentiated from sinuous, single-thread channels, the latter normally associated with conspicuous meander scars.

Table 1. Mean palaeochannel characteristics of the North Esk sandur terraces

	Palaeochannel characteristic			
Terrace	Distal gradient (m km^{-1}) mean SD	Sinuosity mean SD	Percentage channel reaches with SIN $\geqslant 1\cdot10$	Braiding index mean SD
T1	35·7±11·6	1·022±0·025	0·0	5·50±0·71
T2	5·1±3·1	1·019±0·041	7·1	2·88±0·84
T3	5·0±3·9	1·034±0·058	12·5	2·69±1·20
T4	5·1±6·0	1·068±0·120	15·9	2·31±1·19
Present River North Esk	3·7±0·3	1·235		1·05±0·23

that occur in the downstream direction along the terrace surface. An overall increase in channel width and decrease in channel depth and multiplicity were found to occur from proximal to distal zones, associated with declining channel gradients downstream (Maizels, 1982).

Sedimentology

The outwash deposits comprise up to 6 m of massive, coarse, poorly sorted imbricated gravels and cobbles, with isolated lenses of cross-bedded and plane-bedded coarse and medium sands, characteristic of Miall's (1978) *Gm* gravel lithofacies type, and similar to the Scott outwash sediments (facies assemblage GII of Rust, 1978) comprising over 90% gravel content. The sediments are interpreted as accumulating in a proglacial braided stream environment, characterized by rapid deposition in aggrading conditions, and only minor channel scouring.

Estimates of palaeodischarge

Estimates of palaeodischarge are based on field measurements of channel width and depth and on calculations of palaeovelocities using equation (5). Bankfull discharge is calculated as having averaged up to 52 m³ sec⁻¹ per channel, but with a BI = 2·69, between 20 and 60 channels might have existed across the sandur surface. Total bankfull discharge would therefore have reached a maximum of between 1040 and 3120 m³ sec⁻¹. These values are of a similar order to the 1300 m³ sec⁻¹ estimated by Williams & Rust (1969) for the summer discharge of the Donjek Glacier, Alaska, and the 3700 m³ sec⁻¹ for the maximum recorded discharge on the Hoffellssandur by Krigström (1962). Similar calculations for the River West Water indicate an individual bankfull discharge of *c.* 28 m³ sec⁻¹. Together, the North Esk and West Water peak channel flows may well have exceeded 3500 m³ sec⁻¹, compared with the present-day mean annual river discharge of only 15–20 m³ sec⁻¹ and a maximum recorded peak flow of 262 m³ sec⁻¹. Hence a peak discharge decline of between 4·0 and 13·4 times appears to have accompanied the change in channel pattern and form, and the progressive channel incision at the close of the Older Dryas. Estimates of sediment transport have not been attempted, but it seems likely that this too has decreased by at least a similar amount.

Case-study 2: long-term deglaciation of a present-day glacial catchment: (a) Watson River sandur deposits, Søndre Strømfjord, West Greenland

The western margin of the Greenland ice sheet is drained at a latitude of *c.* 67° N by the Watson River, which flows for about 23 km westwards to the head of Søndre Strømfjord. The active meltwater stream is bounded by a series of abandoned sandur terraces, the surfaces of which are traversed by the courses of palaeomeltwater channels. Four main areas of terrace sequences are present within the Watson Valley (Maizels, 1980a, b) of which the most complex sequence occurs in Terrace Area C in the Mt Keglen area. Deglaciation in the Keglen area commenced between 7000 and 6500 yr BP, following the Keglen stage 're-advance' (Weidick, 1968, 1976; Ten Brink, 1975). Maximum deglaciation was attained about 5500 yr BP, followed by a further small re-advance, while isostatic uplift of up to 40 m is believed to have ceased by about 4000 yr BP. During the latter part of the Holocene, only small-scale marginal oscillations have occurred, culminating in the Little Ice Age, or Ørkendalen, re-advance, between 700 and 300 yr BP. The Keglen terrace area comprises eight terrace forms, and although several of these are narrow and fragmented, they each exhibit distinctive evidence of palaeochannel courses (Fig. 3).

Morphology of Keglen terraces

The highest terraces, TI to TIII, provide evidence of ice proximity during the accumulation of the terrace sediments and development of surface channels. The channels exhibit relatively high gradients of 11–26 m km⁻¹, while several kettle-hole depressions within the channel courses are clearly linked to ice-marginal channels on the valley side. These higher terraces are separated by a major bluff, often reaching over 5 m in height, from two series of lower terraces. Terrace system IV is of intermediate altitude (Fig. 4) and is relatively steeply graded (20–42 m km⁻¹), while terrace systems V–VIII are more extensive low-lying terraces with minimum gradients of only 3 m km⁻¹. Blown sand deposits are also extensive on these low terraces, with many of the palaeochannels infilled to depths of up to 1·5 m, choking both their up- and downstream reaches with dune and hummock accumulations. The present river, flowing across a sediment-filled rock basin between two gorge sections, has an overall gradient of 7 m km⁻¹, suggesting that even when taking account of isostatic uplift, gradients have tended to decrease following maximum deglaci-

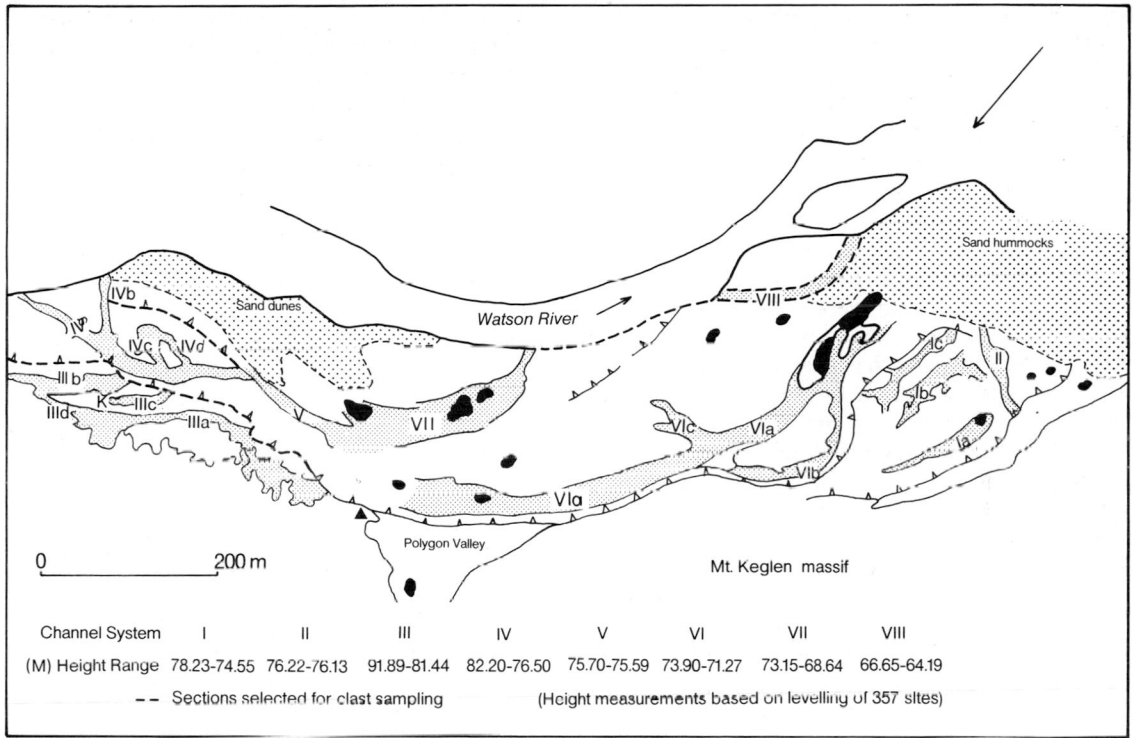

Fig. 3. Classification of palaeochannel systems on the Watson River terraces, Keglen area, West Greenland, based on morphology and levelling.

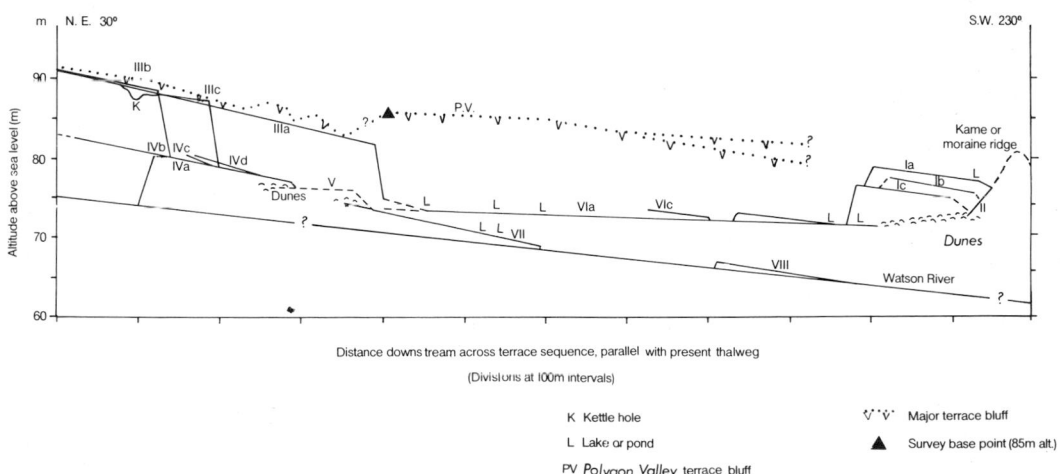

Fig. 4. Height–distance diagram for palaeochannel courses on the Watson River terraces, Keglen area, West Greenland.

ation, and that some stability may have been achieved during the Holocene.

Channel pattern and form

Measures of channel multiplicity and sinuosity indicate that braided, low-sinuosity channels (BI = 3–6; SIN = 1·00–1·05) are confined to the higher and intermediate terraces (TI–V). On the lower terraces (TVI–VII) the channels become significantly more sinuous (SIN = 1·14–1·41), and marked by single-thread courses (BI = 1–2). The present Watson River in the Keglen reach exhibits a sinuosity of only 1·09, but the whole meltwater stream includes braided, straight and sinuous reaches, the latter becoming more dominant in distal zones.

The upper and intermediate braided channel systems are characterized by relatively narrow channels, with mean widths ranging between only 6·1 and 12·7 m. By contrast the mean widths of the younger, lower channels are at least double those of the braided channels, ranging between 23 and 28 m, while the present Watson River extends for some 52 m between its banks. Mean depths of the braided channels rarely exceed 1·2 m, while the broad low-gradient sinuous channels range up to 4 m deep in some reaches. Width–depth ratios show a corresponding decrease through the terrace sequence, from a maximum of 47·5 on TI to 22·1 on TVI.

Sedimentology

The sandur sediments comprise over 18 m of coarse gneissose boulder beds, with clasts averaging 0·24 to 0·35 m in diameter. These boulder beds contain little matrix material and no apparent bedform structures, and are overlain by variable thicknesses of aeolian cover sands. Maximum clast diameter appears to exhibit only a marginal decrease through the terrace sequence, probably reflecting a relatively local clast origin, short distance of transport, and large-scale local reworking of sediment.

Estimates of palaeodischarge

Calculations of palaeodischarge indicate that bankfull flow was several times greater in the broad sinuous channels (up to c. 640 m³ sec⁻¹) than in the individual braided channels of the upper terraces (c. 38–104 m³ sec⁻¹). However, when one takes account of the maximum possible number of channels present across the sandur surface, based on the braiding index, the total palaeodischarge values for the braided channel systems increase to significantly higher amounts than on the lower terraces, namely to between 1092 and 2661 m³ sec⁻¹, compared with values of between 114 and 640 m³ sec⁻¹, respectively. Hence, maximum discharge from the melting ice margin occurred during the early deglaciation period, and has experienced a gradual, although fluctuating, decline through the Holocene. The decrease in discharge, however, is of a smaller order than the decrease estimated for the North Esk catchment, with peak channel flows of the Watson River falling by between about 4·2 and 9·6 times. These lower rates of decrease may reflect the presence of the Watson River as a still actively ice-fed stream.

Case-study 3: long-term deglaciation of a present-day glacial catchment: (b) Narssarssuaq Valley, S.W. Greenland

The Narssarssuaq proglacial valley extends for some 8 km from the snout of the Kiagtut sermia glacier to the estuary of the meltwater stream where it enters Eriksfjord in south-west Greenland (lat. 61° 10′ N). The present meandering Narssarssuaq River (SIN = 1·65) emerges from a proglacial lake and flows south-westwards through a series of relict sandur terraces. Six main sandur terrace systems have been identified (Maizels, 1980b), the oldest apparently dating from the mid-Sub Atlantic ice recession, i.e. c. 2300 yr BP (Weidick, 1963). Of these terraces, Terrace Area T5 (Fig. 5), at an altitude of c. 25 m and c. 2 km from the present ice-snout, presents the most complete sequence of channel changes.

The northern tract of this terrace is traversed by a proximal braided channel system (BI = 4), comprising steeply graded, straight channels and extensive kettle-hole pitting. With increasing distance southwards across the terrace surface to the valley wall, the palaeochannels become increasingly broad, deep and sinuous (Fig. 6), until finally a major incised channel meander is present ('T7' channel) circuiting the back of the terrace at the foot of the valley side wall.

Estimates of palaeodischarge reflect this clear pattern of channel change that is in evidence across the single terrace surface, apparently representing the progressive incision and migration of increasingly sinuous streams during the initial period of deglaciation. Estimated total bankfull discharge amounts exhibit an overall decline through the channel sequence (Fig. 6), with the discharge decrease from channel systems 5/8 to T7 averaging only about 3·4 times during the mid-Sub Atlantic period.

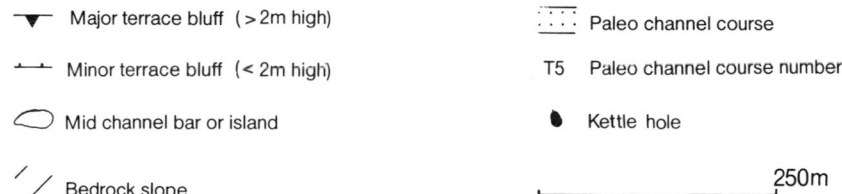

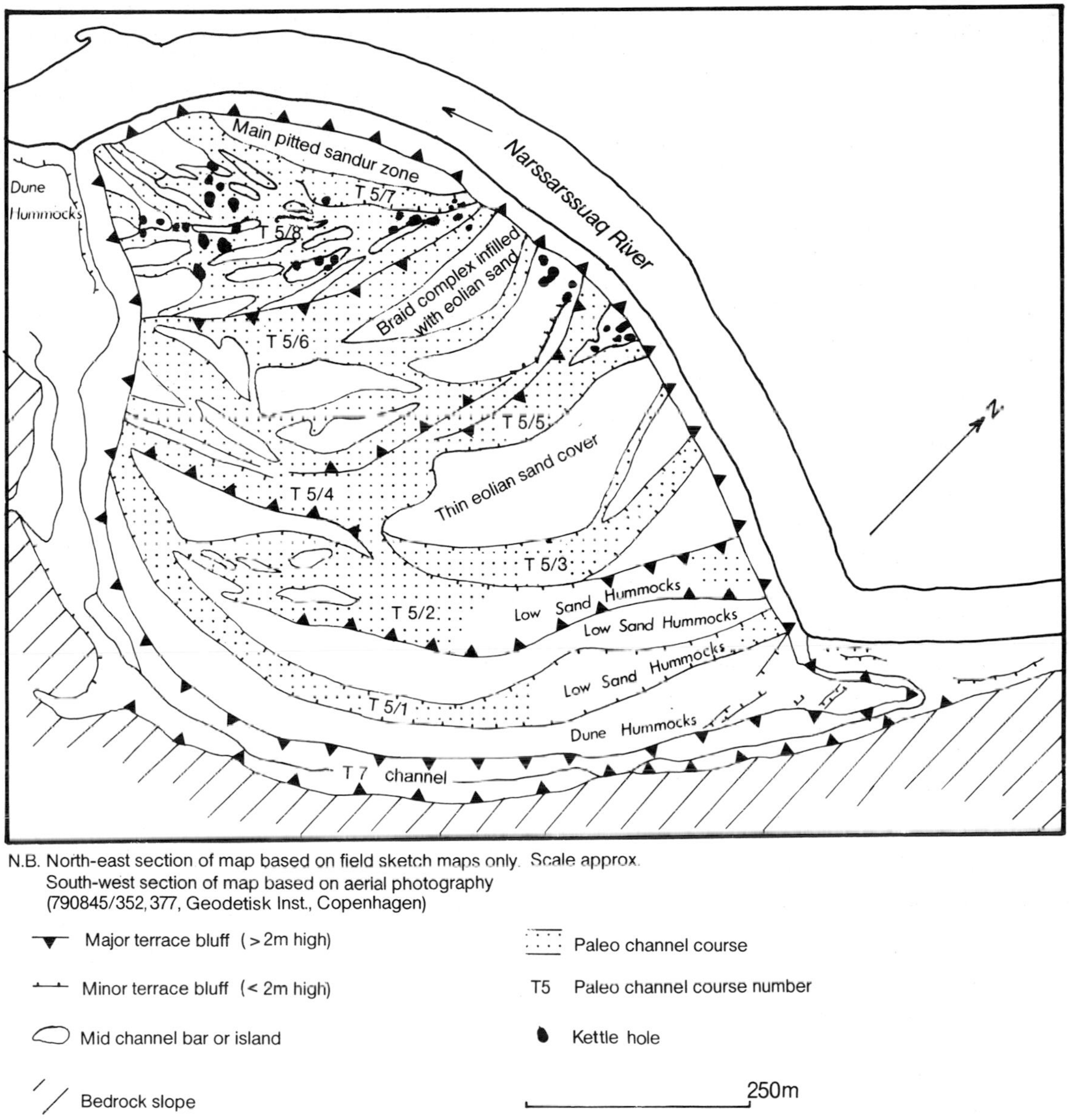
Main pitted sandur zone
Narssarssuaq River
Dune Hummocks
T 5/7
T 5/8
Braid complex infilled with eolian sand
T 5/6
T 5/5
Thin eolian sand cover
T 5/4
T 5/3
Low Sand Hummocks
T 5/2
Low Sand Hummocks
Low Sand Hummocks
T 5/1
Dune Hummocks
T 7 channel
N.
N.B. North-east section of map based on field sketch maps only. Scale approx.
South-west section of map based on aerial photography
(790845/352, 377, Geodetisk Inst., Copenhagen)
Major terrace bluff (>2m high)
Minor terrace bluff (<2m high)
Mid channel bar or island
Bedrock slope
Paleo channel course
T5 Paleo channel course number
Kettle hole
250m

Fig. 5. Palaeochannels on terrace area T5, Narssarssuaq Valley, S.W. Greenland.

Case-study 4: shorter-term deglaciation of a present-day glacial catchment: Nigardsbreen, southern Norway

The valley train of the Nigardsbreen glacier in southern Norway extends for *c.* 2·5 km downstream of a large proglacial lake. Over the past 230 years the Nigardsbreen glacier has receded a distance of almost 4 km from its Little Ice Age maximum of AD 1748, leaving in its wake a series of terminal moraine ridges and intervening outwash deposits. The moraine ridges have been relatively well dated by lichenometric methods (Andersen & Søllid, 1971), while the outwash deposits themselves appear to be associated with particular moraine ridges. These outwash deposits exhibit very distinctive palaeochannel courses and minor terrace development in each inter-moraine tract.

Each of the palaeochannel systems studied in the Nigardsbreen valley exhibits complex, dense, low-

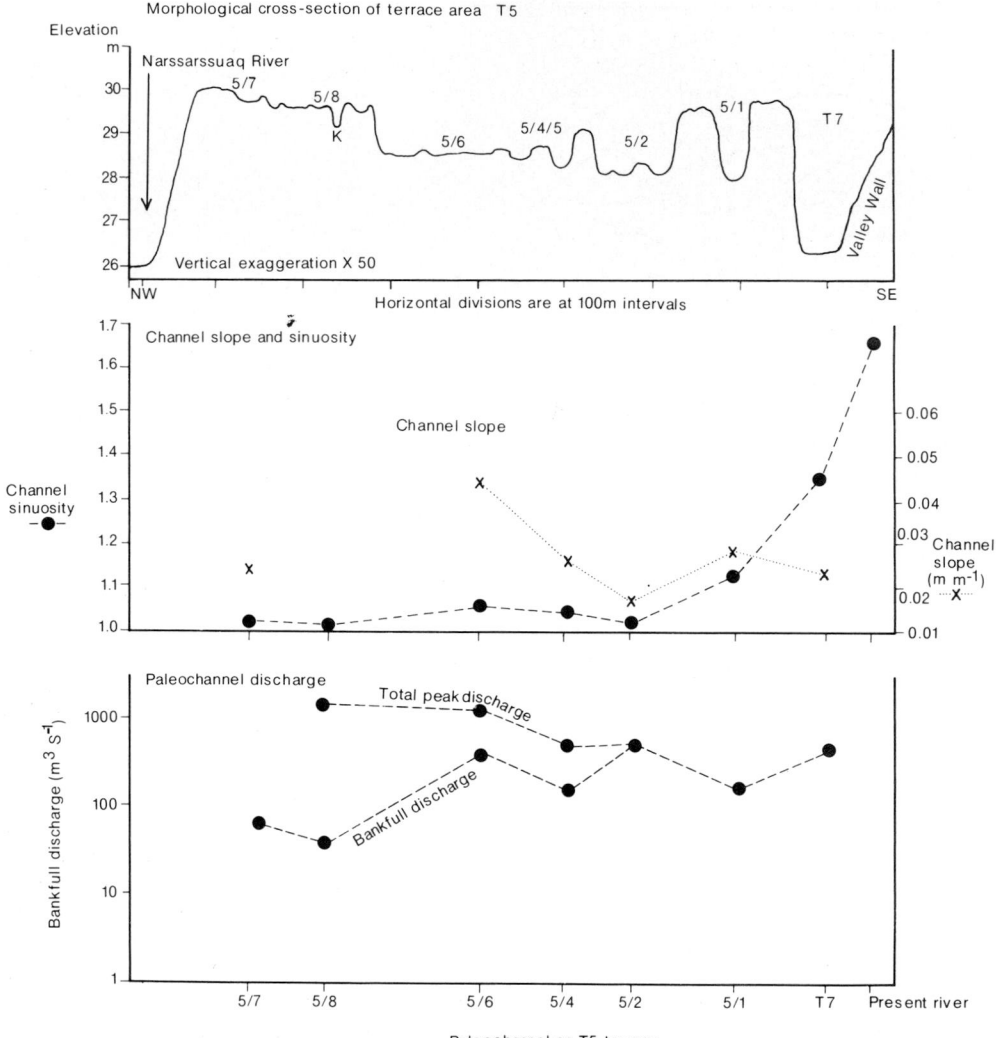

Fig. 6. Changes in channel morphology and palaeodischarge across terrace area T5, Narssarssuaq Valley, S.W. Greenland.

sinuosity, braided channel networks, such that the pattern and form of proximal channels does not appear to have varied significantly over the past 230 yr period of deglaciation. However, the present meltwater stream exhibits both straight, multiple-channel reaches and more sinuous, single-thread reaches, the latter clearly incised into older proglacial deposits. This mixed channel pattern and the extent of recent channel incision suggest that the active meltwater channel may lie at present within the threshold zone for channel change, and is only now exhibiting some significant change in morphology. Although the Nigardsbreen glacier experienced an overall decline in ice mass of 200×10^6 m³ (below 1000 m altitude)

between 1951 and 1974, measurements of water and sediment discharge (Ziegler, 1972; Østrem, Liestøl & Wold, 1977; Kjeldsen, 1980) indicate no comparable trend over this period which might have affected stream runoff patterns. The role of peak flow events, however, remains to be assessed in this context.

Case-study 5: short-term glacial fluctuation of a present-day glacial catchment: Bossons Valley, French Alps

Changes in channel pattern and form, and associated changes in surface morphology, were monitored annually (or bi-annually) over a six-year period, on the Bossons proglacial valley train in the French Alps (Maizels, 1979). The valley train extends for some 200 m in front of the Bossons glacier, and is about 40 m wide, confined between seventeenth-century lateral moraines. Between 1968 and 1975 the Bossons glacier advanced over a distance of *c.* 110 m, and this advance was accompanied both by extensive net aggradation across the whole valley train averaging *c.* 1300 m³ yr⁻¹ and by steepening of mean valley train gradients from 78 to 106 m km⁻¹.

During this glacier advance the proglacial braided channel pattern was maintained, but the density of channels and bars increased significantly as a response to an increase in frequency in localized pockets of sedimentation throughout the channel system. Hence, while channel sinuosity remained between 1·11 and 1·21, the braiding index increased from 17 to 26 and mean bar density trebled from 0·58 to 1·77 bar 100 m⁻². Mean bankfull width and depth both decreased by almost 60%, thus maintaining width–depth ratios to between 20 and 40. Estimates of maximum discharge amounts indicate that these have also remained of a similar order throughout the period (1·33–1·49 m³ sec⁻¹). Thus, while sediment supply appears to have increased substantially over the period, maximum discharge has remained fairly constant, resulting in extensive proglacial aggradation; and despite this change in the balance between water and sediment supply, the braided proglacial channel pattern has not only been maintained but accentuated and reinforced.

DISCUSSION OF RESULTS

Origin of terrace channel systems

Surface palaeochannel systems represent the final channels occupying, and perhaps forming, the uppermost layers of the deposit. They therefore represent either the last channels in an aggradational (or stable) sequence or, if a terrace form is present, the first channels in a degradational sequence. However, where a terrace has been formed, no evidence is preserved of the channel (or channels) that actually performed the incision ('C' in Fig. 7), other than, perhaps, abandoned meander scars. Evidence is only available for the channel system existing immediately prior to, and possibly subsequent to, incision. Since the surface channel systems so closely precede, or even accompany, such a major change in channel stability, these surface channels appear to include 'threshold' or 'near threshold', i.e. transitional, channels (e.g. Falkowski, 1975; Brunsden & Thornes, 1979). Although many of the surface channels may represent

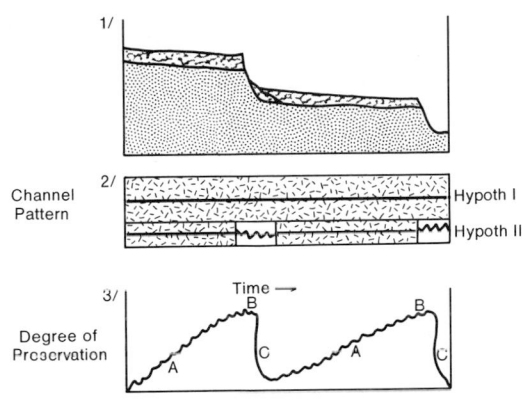

Fig. 7. Possible channel pattern changes during a simple sequence of terrace formation.
(1) Morphology of a simple terrace sequence.
(2) Possible channel changes during terrace formation: *Hypothesis 1* —no change in pattern during incision; pattern remains braided throughout. *Hypothesis 2* braided channel system changes to deep, sinuous, single-thread forms during incision and reverts back to braided channels when incision completed.
(3) Availability of evidence of channel change during the formation of a simple terrace sequence, according to the degree of preservation of the evidence.

threshold or transitional channels, they do not necessarily represent the type of channel that did exceed certain thresholds for channel change and that degraded their beds. Surface channel systems can therefore comprise a complex range of channel types, originating both as 'characteristic form' and 'transitional form' channels. This in turn suggests that not all surface channels ('B' in Fig. 7) can be considered truly representative of the formation of the whole terrace deposit ('A' in Fig. 7).

Surface channels on the Lateglacial terrace sequence of the North Esk sandur deposits, for example, appear to represent both the final aggrading (or stable?) channel system, comprising low-sinuosity, steeply graded multiple channels with high width–depth ratios, and the initial degrading channel system, comprising higher-sinuosity, more gently graded, single channels that are more deeply incised into the terrace surface. Hence both the 'late aggrading' system (Type I, Fig. 8) and the 'early degrading' system (Type II) may be identified as falling within or close to the threshold zone for channel changes. These different channel systems can be distinguished quite clearly according to a number of measured channel parameters, such as sinuosity (Fig. 8), and with

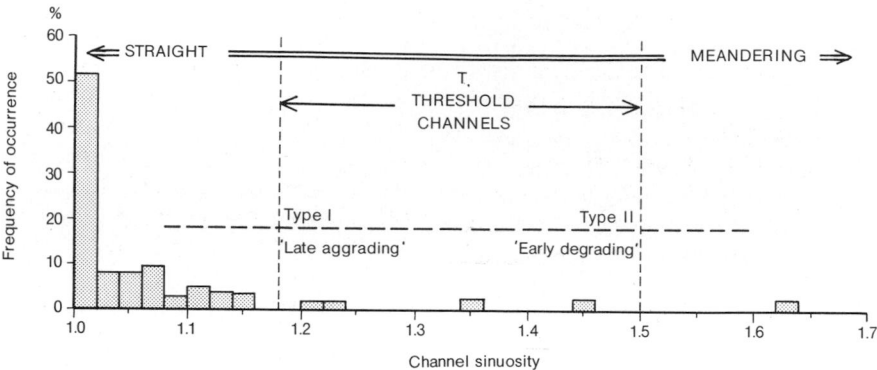

Fig. 8. Variability of palaeochannel sinuosity on the T4 terrace surface of the North Esk sandur deposits, N.E. Scotland.

sufficient data surface channels could be differentiated using some form of discriminant function to help identify the nature of surface stream activity.

The incision of the outwash deposits to form a terrace bluff appears to have occurred through the selective development of deeper channels on the upper terrace surface. The deeper palaeochannels preserved on the terrace surface represent those channels that initially experienced selective development but were then abandoned in favour of a more efficient, more steeply graded (?) channel (or channels) elsewhere. The location of such incising channels may be controlled by the spatial relationship between the selective upstream migration of incising reaches and the initially random flow patterns across the gravel outwash surface. The presence of deep sinuous channels on an otherwise braided terrace surface, at least in the North Esk deposits, suggests that it was this former type of channel that may actually have created the terrace. Each terrace-cutting episode may well have been accompanied by a short-lived change in channel pattern and form (Hypothesis 2 in Fig. 5), suggesting that incision was accompanied by a sufficient change either in meltwater (Q_w) and sediment (Q_s) discharge, or in intrinsic flow conditions, to cause also a change in channel form.

Nature, direction and magnitude of channel changes

In those areas experiencing long-term deglaciation, the evidence suggests that the nature and direction of channel changes appear to follow a broadly similar pattern. The earliest channel systems are characteristically braided, steeply graded, low-sinuosity, high width–depth ratio bedload channels typical of flow through proximal proglacial meltwater systems. The later channel systems appear to include a number of deeper, more sinuous channels until the final observed channel forms are characteristically single-thread, gently graded, higher-sinuosity, lower width–depth ratio, suspended load channels, more typical of distal meltwater systems or non-glacial, humid alluvial catchments. Hence the investigations of the North Esk, Watson and Narssarssuaq channel systems have revealed some major changes in channel patterns and form which appear to be related to a change from an ice-fed to either a humid, or a less dominantly ice-fed, regime. These changes correspond closely to evidence from a number of other channel systems in which major changes in channel pattern, form and stability appear to be associated with long-term deglaciation (e.g. Kozarski & Rotnicki, 1977; Mycielska-Dowgiallo, 1977; Koutaniemi, 1980; Kozarski & Tobolski, 1981; Starkel, 1981; Szumanski, 1981). In the two areas experiencing shorter-term periods of deglaciation or glacial advance, however, no significant changes in channel pattern or form have been identified, although in Nigardsbreen, some transitional forms may be active at the present day.

The change from braided to meandering channel patterns between upper and lower terrace surfaces may also reflect a change from proximal to distal types of channel systems associated with ice recession. The observed channel changes would then not necessarily be a direct response to a changing Q_s/Q_w balance in meltwaters emerging at the ice-margin, but simply to changes in slope and stream competence with increasing distance from the ice-margin. However, since incision has occurred between the development of the two channel systems, it seems likely that a change in the Q_s/Q_w balance at source did occur during ice recession on a sufficient scale to result in the development of degrading channel forms and hence in outwash incision.

The magnitude of the changes observed in these areas is difficult to quantify in meaningful terms, but one simple and useful procedure is to compare the measures of sinuosity (SIN), braiding index (BI), width–depth ratio (w/d), channel slope (S) and maximum estimated bankfull discharge (Q_w) of the earliest channels with those of the latest, or post-glacial, channels. Unfortunately, not all these variables are totally independent of one another; the braiding index is partly a function of channel width, while sinuosity and discharge values are partly dependent on channel slope. Hence, while the partial interdependence of some variables may affect the interpretation of the present estimates of the degree of channel change, it is hoped that in future studies a modified range of independent channel parameters will be introduced. The changes in the five given parameters can be used as indices of the direction of change (increase or decrease) and the proportionate degree of change (expressed as proportionate maximum mean change between the earliest and latest channel system) in each of the channel systems investigated (summarized in Table 2). Clearly, these mean values can only be very approximate at this stage, and should be interpreted with some caution, but they do seem to indicate the likely order of magnitude of change in channel pattern, form and capacity.

The indices of channel change given in Table 2 indicate that the pattern and form of channels in the areas of long-term deglaciation have experienced significantly greater changes than in areas of shorter-term deglaciation or glacial advance. Al-

though these latter areas have also experienced some channel changes they have not been significant enough to produce a change in pattern or form, i.e. threshold conditions for significant channel change have not been met in these areas (and see Fahnestock & Bradley, 1973; Werritty & Ferguson, 1980). The maximum changes recorded for each individual channel system may also be summarized, as in the case of North Esk, using the following simple notation (after Schumm, 1971).

$$Q_w^{-13\cdot4}Q_s^{-?} \rightarrow \text{SiN}^{+1\cdot2}\,\text{BI}^{-5\cdot5}\left(\frac{w}{d}\right)^{-15\cdot9}S^{-9\cdot7}.$$

Maximum observed changes for the three main areas investigated appear to have occurred in the shape of the channel cross-section and in channel slope, such that the smaller discharges were accommodated in fewer, deeper, more sinuous streams. Differences in the degree of change in channel sinuosity, however, do not appear to be particularly significant, supporting the suggestion that the parameters most sensitive to changes in meltwater and sediment supply, for example, are channel form and slope. The maximum discharge decrease has occurred in the North Esk terrace sequence, which is not unexpected since the North Esk is now a non-glacial catchment. The major unknown variable in the equation is Q_s, but estimates of the approximate order of magnitude of sediment discharge may be possible at a later date.

Channel threshold conditions

Characteristics of 'threshold channels' may be defined as those that are intermediate between those

Table 2. Indices of mean channel change in selected proglacial outwash deposits

Outwash deposit	Sinuosity	Channel characteristic			
		Braiding index	Width–depth ratio	Channel slope	Total bankfull discharge[e]
North Esk	+1·21	−5·5	−15·87[c]	−9·65	−13·40
Watson River	+1·38	−6·0	−6·31	−13·53	−9·55
Narssarssuaq River	+1·33	−4·0	−3·34	−3·66	−3·14
Nigardsbreen	[a]+1·02	−1·5	[d]	[d]	−1·04[c]
	[b]+1·11	+1·31	[d]	[d]	
Bossons	+1·06	+1·53	+1·47	+1·48	+1·12[c]

[a] Based on correlation with 1850 and 1909 moraines.
[b] Based on correlation with 1850 outwash and present day.
[c] Estimated using equations (2)–(5).
[d] Field data still to be analysed.
[e] Discharge data available for period 1968–79. Comparisons made between 5 yr means.

of the actively braided and meandering channel systems, and include channels of both Types I and II transitional channel forms. Unfortunately, the deposits investigated here provide few examples of highly sinuous (SIN > 1·5) channels, so that the thresholds for channel meandering remain particularly uncertain. Increases in mean channel sinuosity through the terrace sequences in the North Esk, Watson and Narssarssuaq valleys were from 1·019 to 1·235, 1·022 to 1·410 and 1·008 to 1·345, respectively. These values suggest that the transitional forms occur within the range 1·03 to 1·23+, but particularly around 1·15 to 1·2+, i.e. at a relatively low sinuosity. The braiding index of the sinuous channels lies between 1 and 2, suggesting that a BI value of about 2 may represent a channel system that lies close to the threshold conditions for change. Similarly, a width–depth ratio of between 10 and 20 appears to differentiate the two end-member channel systems. In addition, the observed channel changes suggest that, before reaching a new characteristic form, threshold channels may undergo changes through time of 6–20% in sinuosity, 55–70% in braiding index, 35–70% in channel slope, and up to 75% in total bankfull discharge.

CONCLUSION

Major channel changes from actively braided to meandering habit have been observed in sandur terrace sequences in the three areas that have experienced some degree of long-term deglaciation. These changes may be attributed to any of the following three factors, or combination of these.

(i) Changes in ice mass, which will in turn affect the rates of meltwater and sediment supply to the proglacial stream system, as well as the relative contributions of periglacial and paraglacial sediments during deglaciation. Changes in ice mass appear to have acted as the major control on channel change in the areas of long-term deglaciation.

(ii) Changes in base-level, which affect channel gradients and rates of incision and aggradation. Changes in base-level appear to have had only a minor effect on channel change in the areas investigated.

(iii) Episodic events, the occurrence of which is suggested by the presence of stepped terrace sequences, often with similar channel patterns and forms on both the upper and lower terrace surfaces. Episodic events may include: (a) the periodic exceedance of intrinsic geomorphic thresholds as a complex response to an initial change in the Q_s/Q_w balance or in channel slope (e.g. Schumm, 1980), and (b) a catastrophic flood event or jokulhlaup, which could provide the rapid, large-scale erosive capacity that may be required to dissect an outwash deposit on a massive scale, after which similar equilibrium channel forms are re-established at a lower level. Both these forms of episodic changes in hydraulic conditions may have played an important role in the formation of the observed terrace sequences; however, insufficient evidence is available at present to allow one to confirm the significance of internal hydraulic changes and major flood events on observed channel change and terrace formation.

The observed channel changes appear to have operated through the exceedance of thresholds by transitional channel systems; the threshold levels for change obviously vary in each locality according to local geomorphic, hydrologic, climatologic, vegetation and base-level conditions, but similar threshold zones may be present in many different outwash deposits (Fig. 9). It should prove worthwhile to attempt to identify these threshold zones in other outwash deposits, and to define the limits within which water and sediment flow conditions may change without exceeding any thresholds for change of channel pattern, form and slope. Such studies might allow the prediction of the types of hydrologic conditions that are likely to have produced the observed channel changes over given periods of time or extents of deglaciation, as well as the types of channel change that are likely to occur during particular conditions of glacial fluctuation. The study of channel changes in proglacial areas provides substantial scope for the analysis of the controls and threshold conditions for channel change, since these are so prone to rapid and large-scale changes in water and sediment supply. A particular need is for long-term measurements of water and sediment discharge in areas of stable tectonics in order to isolate the effects of ice-mass changes, intrinsic geomorphic controls, and single hydrological events on the evolution of channel systems.

ACKNOWLEDGMENTS

I would like to thank P. Glennie, S. Kennedy, J. Livingstone, J. McIntosh and members of the Aberdeen University West Greenland (1979) and Norway (1981) Expeditions for field and technical assistance, and to express my thanks to Drs J. Birnie,

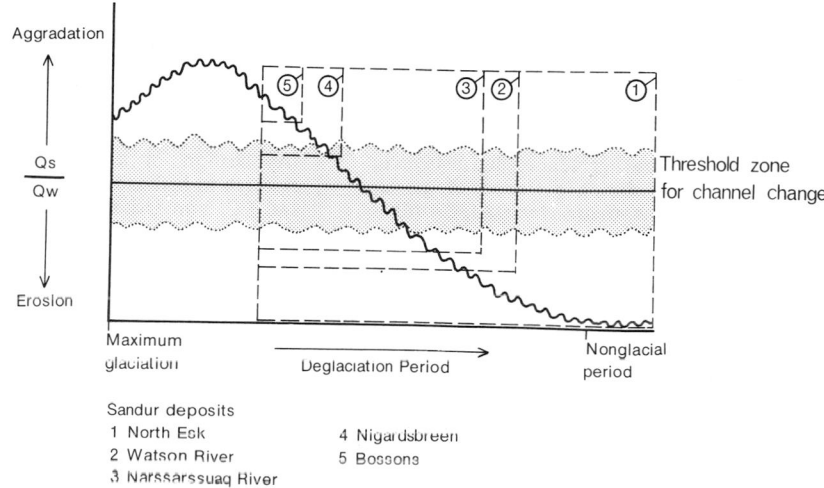

Fig. 9. A model illustrating the possible relation between areas experiencing different degrees of channel change through time and the changing balance of water and sediment supply around a threshold zone for channel change during a period of deglaciation.

C. Clapperton and A. Gemmell, Mr J. Rose, Mr M. Sharp, Drs D. Sugden, R. Ward, A. Werritty and B. Willetts for valuable discussions during the preparation and revision of this paper.

REFERENCES

ANDERSEN, J.L. & SØLLID, J.L. (1971) Glacial chronology and glacial geomorphology in the marginal zones of the glaciers, Midtalsbreen and Nigardsbreen, South Norway. *Norsk geogr. Tidsskr.* **25**, 1–38.

BAKER, V.R. (1974) Paleohydraulic interpretation of Quaternary alluvium near Golden, Colorado. *Quat. Res.* **41**, 94–112.

BRUNSDEN, D. & THORNES, J.B. (1979) Landscape sensitivity and change. *Trans. Instn British Geogr.* **4**, 463–484.

CHEETHAM, G.H. (1976) Palaeohydrological investigations of river terrace gravels. In: *Geo-archaeology: Earth Sciences and the Past* (Ed. by D.A. Davidson and M. Shackley), pp. 335–344. Duckworth, London.

CHEETHAM, G.H. (1980) Late Quaternary palaeohydrology, the Kennet Valley case-study. In: *The Shaping of Southern England* (Ed. by D.K.C. Jones), pp. 203–223. Academic Press, London.

CHURCH, M. (1970) *Baffin Island sandar: a study of Arctic fluvial environments. Old sandar: essays in paleogeomorphology*, pp. 442–493. Unpublished Ph.D. Thesis. University of British Columbia.

CHURCH, M. (1978) Palaeohydrological reconstructions from a Holocene valley fill. In: *Fluvial Sedimentology* (Ed. by A.D. Miall). *Mem. Can. Soc. Petrol. Geol.*, Calgary, **5**, 743–772.

CLAGUE, J.J. (1975) Sedimentology and paleohydrology of Late Wisconsin outwash, Rocky Mountain Trench, southeastern British Columbia. In: *Glaciofluvial and Glaciolacustrine Sedimentation* (Ed. by A.V. Jopling and B.C. McDonald). *Spec. Publs Soc. econ. Paleont. Miner.*, Tulsa, **23**, 223–237.

ETHRIDGE, F.G. & SCHUMM, S.A. (1978) Reconstructing paleochannel morphologic and flow characteristics: methodology, limitations, and assessment. In: *Fluvial Sedimentology* (Ed. by A.D. Miall). *Mem. Can. Soc. Petrol. Geol.*, Calgary, **5**, 703–721.

FAHNESTOCK, R.K. & BRADLEY, W.C. (1973) Knik and Matanuska Rivers, Alaska: a contrast in braiding. In: *Fluvial Geomorphology* (Ed. by M. Morisawa), pp. 221–250. Binghamton, New York.

FALKOWSKI, E. (1975) Variability of channel processes of lowland rivers in Poland and changes of the valley floors during the Holocene. *Biul. Geol.* **19**, 45–78.

HEY, R.D. (1979) Flow resistance in gravel-bed rivers. *J. Hydraul. Div. Am. Soc. civ. Engrs* **105**, HY4, 365–379.

KJELDSEN, O. (1980) Materialtransportundersøkelser i Norske bre-elver. *Norges Vassdrags- og Elektrisitetsvesen.* Rapport Nr. 1-80, Oslo. 43 pp.

KOUTANIEMI, L. (1980) Some aspects of the palaeohydrology connected with the development of the relief in the Oulanka river valley, northeastern Finland, and a review of complementary study concerning IGCP – Project No. 158. *Bull. Ass. fr. Étude Quat.*, 2ème série, **17**, 1–2, 71–75.

KOZARSKI, S. & ROTNICKI, K. (1977) Valley floors and changes of river channel patterns in the North Polish plain during the Late-Würm and Holocene. *Quaest. Geogr.* **4**, 51–93.

KOZARSKI, S. & TOBOLSKI, K. (1981) *Guide-book of excursions.* INQUA Eurosiberian Subcommission for the Study of the Holocene IGCP No. 158. *Symp. Paleohydrology of the Temperate Zone*, Poznan, September.

KRIGSTRÖM, A. (1962) Geomorphological studies of sandur plains and their braided rivers in Iceland. *Geogr. Annlr* **44**, 328–346.

LIMERINOS, J.T. (1970) Determination of the Manning coefficient from measured bed roughness in natural

Facies models

Spec. Publs int. Ass. Sediment. (1983) **6**, 279–286

Basin analysis of fluvial sediments

ANDREW D. MIALL

Department of Geology, University of Toronto, Toronto M5S 1A1, Canada

ABSTRACT

Major fluvial basins may contain hundreds or thousands of metres of fill, accumulated over tens of millions of years. In this time, plate motions may have subjected the basin to complex tectonic changes and carried it through several climatic zones. Analysis of the sediments at the basin level therefore requires us to recognize the effects of a range of sedimentary controls and to use a variety of basin analysis techniques.

This approach resolves into three main areas:

(1) The development of special basin analysis methods for non-marine rocks. Problems of stratigraphy and correlation are particularly acute because of rapid lateral facies changes. Climate is a first-order sedimentary control; its analysis depends on a variety of developing petrological, geochemical and pedological techniques.

(2) Development of models for alluvial architecture which reflect, more fully than at present, the variables of lithosome geometry, interpreted channel geomorphology and channel shifting behaviour.

(3) Exploration of the basin-scale effects of tectonics and climate through the documentation of detailed case-studies. Application of plate tectonic principles should permit the development of generalized structural, stratigraphic and palaeogeographic models as this documentation is improved.

INTRODUCTION

The study of fluvial sedimentology as applied to ancient deposits has led to a preoccupation with the development of facies models to explain relatively small-scale architectural variations, essentially at the outcrop or local level. For example, this seemed to be one of the major themes of the first (1977) fluvial conference at Calgary, as recorded in the proceedings volume (Miall, 1978a). A particularly popular approach for fluviologists has been to focus on vertical profile studies. These are both relatively simple to perform, and have considerable relevance to subsurface exploration work. Such profile studies commonly lead to an emphasis on cyclic sequences, and their interpretation in geomorphological terms. Although this work is far from complete, it is also necessary to pursue the problem of how broader-scale sedimentary controls affect fluvial architecture and basin-fill styles.

As summarized in a recent review (Miall, 1980), cyclic sequences can be developed, modified or confused by the imprint of tectonic events and climatic change, and their analysis then requires a multidisciplinary, basin analysis approach.

At the basin-wide scale the study of fluvial stratigraphy, facies, geochemistry, petrology, palaeogeography and structure have much to contribute to an understanding of regional geological history. Major fluvial basins (e.g. Weimer, 1970; Embry & Klovan, 1976; Galloway, 1981) may contain hundreds or thousands of metres of sediment requiring tens of millions of years for their accumulation. During such extended periods of time the basin may be carried hundreds or thousands of kilometres by plate motions involving convergent, divergent or transform plate tectonics, major orogenic modification and, possibly, the passage through several climatic zones. Our understanding of how these processes affect fluvial rocks requires a great deal of research. Case-studies of

0141-3600/83/0106-0279 $02.00

many basins, of all ages, in a diversity of tectonic and climatic settings are required, in order to provide the necessary data base for generalization and modelling at the scale of sedimentary basins. The application of plate tectonic principles to interpretation of fluvial basins has barely begun (e.g. Miall, 1981a).

Elsewhere (Miall, 1978b), I remarked that the ideal fluvial sedimentologist would be a 'Quaternary sedimentologist with experience in petroleum geology and river engineering, a passion for hydraulics, statistics and scuba diving, a more than passing interest in tectonics and a lot of money for coring equipment'. Whilst preparing this paper for the 1981 Keele conference it became clear that these requirements were already woefully inadequate. To be a fluvial specialist in the 1980s requires a knowledge of biostratigraphy, palaeomagnetism, geochemistry, petrology, pedology, palaeoclimatology and structural geology in addition to those skills listed above. The purpose of this article is to review some recent papers (including those in this book) that touch on this vast area, in an attempt to see in what directions our subject is developing. It was clear at Keele that, far from consolidating our corner of the sedimentological revolution, we are still in a phase of energetically collecting new data and ideas.

BASIN ANALYSIS METHODS

A review of all the techniques used to analyse large-scale fluvial deposits should not be necessary for readers of this book. Methods of lithofacies documentation and stratigraphic mapping, studies of cyclic sedimentation and palaeocurrent analysis are amply described in the recent literature, are generally well understood and are now widely used. I shall focus here on three main topics, the problem of stratigraphic correlation, and the problem of determining palaeoclimate and taphonomy.

Stratigraphic correlation

Fluvial deposits are characterized by numerous lateral facies changes. The fluvial depositional process produces few obvious stratigraphic marker beds and correlation of both surface and subsurface sections may at best, be generalized. Obviously, detailed analyses of alluvial architecture are impossible if the data base consists entirely of poorly correlated vertical sections. Since this is commonly all we have, it is important to explore all the correlation methods

available to us. Four techniques are discussed briefly in this section.

Fluviologists have tended to make little use of biostratigraphy. Fossils are limited mainly to vertebrates, plants and palynomorphs, and these only in Devonian and younger strata. Invertebrates tend to be rare, long-ranging and therefore stratigraphically unhelpful. The same is true of many macroflora. However, there seems to me to be much undeveloped potential in the area of vertebrate palaeontology and palynology. Historically, the science of vertebrate palaeontology has been pursued as much, if not more, by biologists than by geologists. This has led to some lax collecting methods and, sometimes, a startling unconcern for stratigraphy. There is a view amongst some vertebrate palaeontologists that phylogenies can be erected by constructing logical trends in functional morphology, regardless of the relative stratigraphic origins of the specimens in question. However, this is changing, and many vertebrate studies are now carried out in cooperation with stratigraphy using the techniques outlined below (e.g. Johnson et al., 1979; Bishop, 1978; Hay, 1976). However, vertebrates themselves may be of use in refined stratigraphic correlation. A. Matter (pers. comm., 1981) reports that meticulous collecting in the Alpine molasse, Switzerland, is permitting stratigraphic correlations between sections with an accuracy of a few tens of metres. Berggren et al. (1978) discussed the use of vertebrate biostratigraphy of non-marine sediments in refining the chronostratigraphy of the Palaeogene.

Shaw (1964) and Miller (1977) developed a technique of biostratigraphic correlation using graphical methods to compare faunas or floras instead of conventional biozones. The technique, if carefully applied, may yield correlations with accuracies of only a few metres. It is admirably suited to palynological data and therefore should commend itself to use by basin analysts dealing with thick fluvial sequences. Graphical methods are widely used in petroleum exploration but, to my knowledge, have not been used for sedimentological purposes.

Ash beds or bentonites may be useful for correlation, if present. Allen & Williams (1981) discussed the use of ash beds in correlating Devonian sections in South Wales. Beds immediately above or below an ash horizon in separate outcrops can be closely compared to derive a picture of the alluvial architecture at particular instants in time.

A technique finding increasing use in Cenozoic non-marine strata is magnetic reversal stratigraphy. Palaeomagnetic methods have been applied to redbeds and other non-marine deposits for about twenty years (e.g. Picard, 1964). In some research the aim has been to obtain palaeopole positions, but of greater interest here is the potential for developing local reversal sequences and correlating these with the standard reversal stratigraphy of volcanic rocks or oceanic sediments. Johnson *et al.* (1979) reported on studies of the Neogene Siwalik sediments of Pakistan, which formed in an ancestral Indo-Gangetic plain as the Himalayas were uplifted. Closely spaced samples permitted the erection of a reversal stratigraphy, which was correlated with the standard sequence by use of radiometrically dated ash beds and vertebrate fossils. Development of this correlation network has permitted these authors to make detailed calculations of the effects of local tectonics on subsidence, sedimentation and uplift rates.

Unfortunately it is unlikely that the technique will ever be useful for pre-Cretaceous non-marine rocks, because of the increasing complications of diagenetically modified remanence with increasing age. Also, the production of a standard sequence for pre-late Jurassic rocks will be difficult in the absence of old oceanic crust and its overlying sediments.

The last correlation technique to be discussed here is the use of palaeosol horizons. These are time markers which commonly develop over wide areas of a river floodplain during periods of non-deposition. Their potential for correlation purposes has long been recognized (e.g. Allen, 1974) but, in fact, they have been little used. Palaeosols commonly are poorly exposed and tend to be passed over by the non-specialist. Careful geochemical work, however, has shown that individual horizons can be characterized for correlation purposes and much information on palaeoclimate obtained therefrom (Johnson, 1977; Bown & Kraus, 1981).

Palaeoclimate

Climate is a first-order sedimentary control in non-marine environments. It controls the type and density of vegetation, the amount and variability of discharge and weathering characteristics, and hence influences sediment load. Garner (1959) showed that in two fault-bounded basins a few hundred kilometres apart on the west flank of the Andes the basin fills were completely different, largely because of climatic contrasts. One basin, formed under arid conditions, is filled with coarse clastic debris deposited by ephemeral streams, the other contains a dominantly fine-grained fill brought in by perennial streams flowing in a humid, vegetated environment.

Determination of palaeoclimate is not a simple matter in ancient fluvial rocks. Such indications of dryness as desiccation cracks or evaporite crystal casts, or evidence of discharge fluctuations such as reactivation surfaces and clay drapes in sedimentary structures, may reflect only seasonal dryness rather than long-term aridity. Coal can develop in tundra–muskeg environments (Le Blanc Smith & Eriksson, 1979), as well as in temperate and tropical humid settings.

Under certain circumstances palaeosols might be extremely useful as climatic indicators. Leeder (1975) related calcrete development to climatic and floodplain exposure factors. Johnson (1977) and Bown & Kraus (1981) discussed geochemical and pedological characteristics of selected palaeosols, and considered their climatic significance.

Various petrological techniques are also being explored as palaeoclimatic indicators. Dutta & Suttner (1981) discuss three approaches. Where the source area geology remains constant through a period of climatic change, the composition of the first-cycle detritus may change in response to different weathering behaviour. They argue that immature sandstones tend to be produced in arid climates and mature sandstones in humid climates (other factors being equal). Early diagenetic clay minerals are thought to reflect the degree of saturation and pore fluid movement which are, at least partly, a function of climate-controlled meteoric water input. Oxygen isotope data from these clays are used to determine palaeotemperatures. Obviously there are dangers in these approaches. Constancy of rock type in the source area may be difficult to demonstrate; clay minerals are highly susceptible to later diagenetic modification. Nevertheless, there are many possibilities here which seem worth pursuing.

Taphonomy

Many fluvial deposits are rich in vertebrate remains. These range from entire carcasses entombed in muddy backswamp deposits to isolated and abraded bones transported long distances in a channel lag gravel. The state of preservation reflects a combination of the effects of decay and disarticulation, activity of predators, amount of transport and rate of burial. Investigation of these factors is the subject of taphonomy and may be of considerable interest in

19

making detailed environmental interpretations of ancient sediments (e.g. Dodson, 1971; Smith, 1980).

Many experimental and field studies of vertebrate remains in fluvial systems are in progress, including the use of flumes to investigate the hydrodynamic behaviour of bones (Behrensmeyer & Hill, 1980). Whether such work will ever make a useful contribution to fluvial facies and palaeohydraulic studies remains to be seen. As noted above, there are several factors which determine the nature of a given bone locality or bone bed, and sorting these out seems likely to prove a complex task.

FLUVIAL ARCHITECTURE

The architecture of a fluvial sequence refers to the three-dimensional geometry and interrelationships of the deposits of the channel, levee, crevasse, floodplain and other sub-environments of a fluvial depositional system. The term may be used for the meander belt or channel complex of a single river or the interrelationships (stratigraphy and palaeogeography) of the various component rivers and their deposits of an entire basin.

To obtain an understanding of fluvial architecture requires three-dimensional data, and it also assumes a thorough understanding of fluvial geomorphology in the appropriate tectonic and climatic setting. We are only just beginning to put these pieces together.

The vertical profile approach

Three-dimensional data for any stratigraphic unit are rare. Large cliff or roadcut exposures may provide good two-dimensional data (e.g. Campbell, 1976; Horne et al., 1978); closely spaced exploratory wells for coal, uranium, petroleum, etc. may provide the beginnings of a three-dimensional network, but well density and data quality are rarely sufficient for reliable architectural reconstruction. Subsurface data are commonly limited to geophysical logs and a few cores.

Until recently, many sedimentologists have attempted to avoid these problems by focusing on the diagnostic potential of lithofacies assemblages and vertical profiles (e.g. Cant & Walker, 1976; Miall, 1977, 1978c; Jackson, 1978; Rust, 1978; Friend et al., 1976). While this approach has served successfully to categorize various deposit types (e.g. gravelly braided, silty meandering), and certain geomorphic patterns

(e.g. channel aggradation, flood cycles), it has been less successful in predicting lateral facies relationships or channel migration and stacking behaviour.

Vertical profile studies lend themselves readily to statistical data manipulation for the documentation of cyclic depositional patterns (e.g. Cant & Walker, 1976; Miall & Gibling, 1978). One of the most thorough such studies was that by Friend et al. (1976), who recorded all their field data on computer processable forms. They divided their vertical sections into arbitrary 10 m samples and compared their characteristics using factor analysis. Eight sample types, divisible into 32 subtypes, resulted from this work. Each of these is a discrete facies assemblage, requiring its own facies model. Thirty-two subtly different facies models for one fluvial basin is enlarging the facies model approach somewhat beyond its original conception as a simplifying and clarifying exercise. Friend and his coworkers did not, in fact, attempt to explore all these ramifications of their data, but it would seem from examples like this that the vertical profile, cycle model may, in some cases, not help us very much. Miall (1980) reviewed some of the ambiguities arising from the production of similar cycles from different geomorphic or tectonic processes, and the superimposition (nesting) of several cycle types within each other. Schumm (1981) commented on geomorphic controls of cyclic mechanisms, and the several time-scales of cyclic processes which produce cycle nesting.

Unfortunately, the nature of subsurface data is such that subsurface stratigraphers and exploration geologists still rely extensively on vertical profiles in order to subdivide fluvial deposits into different geomorphic categories (e.g. Embry & Klovan, 1976; Galloway, 1981). This is unavoidable, and the only solution is to increase well density to spacings of a few hundred metres or less, so that reliable lateral correlations become possible.

Sedimentological applications of fluvial geomorphology

One of the problems with current architectural analysis is that it is based on an incomplete understanding of modern fluvial processes. However, it is surprising how rapidly our comprehension has improved. Although the most casual observer must have long been aware of different fluvial styles, it was not until about ten years ago that this knowledge began to filter into sedimentology. Visher's (1972) review dealt only with meandering rivers; Miall

(1977) discussed the work then available on four river types. More recent complications are discussed below. Relevant geomorphological work began with that of Schumm (1963) who divided rivers into bed-load, suspended-load and mixed-load channels, and showed how load type controls channel shape. Galay, Kellerhals & Bray (1973) and Mollard (1973) illustrated an almost continuous spectrum of fluvial styles from low to high sinuosity and single to multiple channels. Schumm (1981) illustrated 14 basic channel types.

Even this may not be enough, as began to be apparent at the Calgary Fluvial Symposium in 1977. Baker (1978) showed that many of our concepts of river behaviour are derived from temperate or cold regions and do not apply to tropical humid regions covered by dense rain forests, such as the Amazon Basin. Smith & Smith (1980) discussed how base-level control can cause vertical aggradation, channel stability and the development of anastomosed systems. Long (1978) and Cotter (1978) demonstrated the validity of an earlier suggestion by Schumm (1968) to the effect that vegetation has such an important effect on stabilizing channel patterns that before the evolution of extensive land vegetation in the Devonian fluvial styles may have been entirely different. Miall (1977), Friend (1978) and Tunbridge (1981) emphasized the importance and distinctiveness of the deposits of unchannelized flash floods. Friend (1978) and Parkash, Awasthi & Gohain (1983) discussed terminal fans, which are inland ephemeral distributary systems that deposit their sediment load as the runoff infiltrates into the channel bed. Schumm (1981) discussed the importance of geomorphic thresholds in controlling discontinuous erosional and depositional behaviour.

It is clear from this brief review that fluvial architecture depends on a wide variety of interdependent controls, including channel shape, channel shifting behaviour, load type, discharge variability and subsidence rate. But the present state of the art is one of having more variables than equations to explain them. For the time being, at least, the exercise of making quantitative palaeohydraulic reconstructions from ancient deposits (as described by Leeder, 1973; Ethridge & Schumm, 1978; etc.) should probably be discontinued except, perhaps, for the basic meandering river model of Allen (1963, 1965) on which the procedures are based.

Efforts to solve some of these difficulties include one strand of research which has attempted to quantify all relevant geomorphic processes and simulate architec-ture models using simple calculations (Allen, 1974), or probability models (Leeder, 1978). Later attempts (Allen, 1978; Bridge & Leeder, 1979; Crane, 1981) have employed computers to perform the necessary calculations and draw cross-sections of simulated channel and floodplain geometries.

The importance of lateral control

One of the most important single advances in fluvial sedimentology was the development of the lateral accretion model for point bars by Allen (1963), and its subsequent application to an ancient fluvial deposit (Allen, 1965). Bluck (1979) showed how lateral accretion and lateral channel migration could occur in multiple channel, bed-load rivers. A given deposit may have been produced by vertical aggradation or lateral channel wandering or a combination of both. Distinction of the two types of behaviour may be difficult in vertical profiles alone, although Cant & Walker (1976) discussed the use of detailed palaeo-current data in vertical sections to analyse bar and bedform migration patterns, which may be of some assistance.

Undoubtedly lateral control is the way to tackle this problem. Careful tracing of individual bedding units and entire channel fill deposits can reveal within-channel depositional patterns, channel geometry and shifting behaviour. Increasing recognition of this fact has led to a growing number of publications containing detailed drawings or interpreted photo-mosaics of long cliff or roadside exposures (Campbell, 1976; Ethridge & Flores, 1981; and several papers in this volume). Friend (1983) emphasizes the importance of analysing lithosome geometry and offers a classification of sediment body shape which reflects many aspects of channel behaviour.

There are two important components of such a classification, external shape and internal form. Nami & Leeder (1978), Friend, Slater & Williams (1979), Friend (1982) and others recognized ribbon, sheet and channelized lenticular geometries, which reflect stable or migrating channels, and unchannelized (including crevasse splay) flow. Precisely how a given channel is filled and a deposit is produced can be analysed by carefully examining internal bedforms, bars and scour surfaces and in particular how these components behave when traced laterally. The lateral accretion point bar or epsilon cross-bed model is one of the simplest ways to produce a channel fill. Vertical aggradation and progressive abandonment is another method. Many channels fill by a process combining both these styles, preserving complex internal scour surfaces, multicomponent bar forms and numerous reactivation surfaces (e.g. Haszeldine, 1983), representing both vertical scour and fill, and the lateral growth of side bars or sand-flats.

TECTONIC CONTROL

At a regional or basin-wide scale, contemporaneous tectonics control most broader features of the basin architecture, including position and orientation of the rivers and subsidence rates, which affect local channel deposit density and interconnectedness. It is therefore useful to examine fluvial basins in their tectonic context.

Fluvial basins occur in twelve major plate tectonic settings, a discussion of which has been given elsewhere (Miall, 1981b). Plate setting may affect fluvial styles in various ways. For example fore-arc and back-arc basins may be supplied by vast quantities of volcaniclastic debris during eruptions, producing catastrophic debris flows and damming of channels (Kuenzi, Horst & McGehee, 1979; Vessell & Davies, 1981). Smith & Putnam (1980) suggested that retro-arc or foreland basins may subside more rapidly than the adjacent continental hinterland, providing the necessary downstream base-level control for the development of an anastomosed channel pattern. Cratonic basins may be characterized by slow subsidence rates, permitting extensive lateral channel sweeping and sediment reworking. There is scope for much fruitful research here in relating fluvial behaviour to plate tectonic setting (e.g. Miall, 1981a).

In some basins the fill style is actively modified by syndepositional tectonics, including fold and fault movement (Miall, 1978d). Major phases of alluviation may be related to pulses of uplift, as in the Hornelen Basin, Norway (Steel *et al.*, 1977; Steel & Aasheim, 1978). These can take the form of laterally persistent, fining-upward or coarsening-upward cycles. Active folding within some basins, particularly in foredeep and retro-arc settings, may divert river patterns and lead to the development of intraformational unconformities.

In most modern basins the rivers can be categorized into two types, those flowing perpendicular or transverse to tectonic grain and those paralleling tectonic strike. The latter are usually trunk streams, and include all the world's major rivers. Basins with both longitudinal rivers and transverse tributaries occur along or parallel to tectonic sutures, in rift and intermontane settings and in many retro-arc basins. Coastal plain basins contain only transverse rivers. They occur in some fore-arc and retro-arc basins and on divergent plate margins. Miall (1981b) has used these variations to erect nine basin-fill models based on the presence or absence of a longitudinal trunk river and on coastal sedimentation patterns. An interesting topic for future research will be to determine whether there are any relationships between these models and the various sediment body-shape types identified by Friend (1983).

RESEARCH TRENDS

There is a great need for detailed studies of fluvial architecture in three dimensions, in order to explore fully the range of sediment body-shape variations outlined by Friend (1983). Unfortunately well-exposed, structurally undeformed sections in which such research can be performed are not that common. Of considerable interest would be tight networks of cored holes through the Recent deposits of some modern major rivers on active alluvial plains, e.g. the Kosi, Ganga, etc.

Because vertical profiles are so widely used by subsurface geologists it would be useful to continue the effort to relate such profiles to geomorphic processes. Refinement of dipmeter interpretation techniques may be of considerable interest if they could permit geometric reconstructions of bedforms and bars from vertical sections.

Our efforts to classify and model fluvial sediments on a basin scale are hampered by a lack of thoroughly studied, well-documented case histories. Therefore there is much room for basic, descriptive research. However, the research should be as broadly based as possible because, as I have tried to make clear here and elsewhere, it is only by adopting a multidisciplinary research method that we can hope to understand fully how a particular basin was filled.

REFERENCES

ALLEN, J.R.L. (1963) Henry Clifton Sorby and the sedimentary structures of sands and sandstones in relation to flow conditions. *Geol. Mijnb.* **42**, 223–228.

ALLEN, J.R.L. (1965) The sedimentation and palaeogeography of the Old Red Sandstone of Anglesey, North Wales. *Proc. Yorks. geol. Soc.* **35**, 139–185.

ALLEN, J.R.L. (1974) Studies in fluviatile sedimentation:

implications of pedogenic carbonate units, Lower Old Red Sandstone, Anglo-Welsh outcrop. *Geol. J.* **9**, 181–208.

ALLEN, J.R.L. (1978) Studies in fluviatile sedimentation: an exploratory quantitative model for the architecture of avulsion-controlled alluvial suites. *Sedim. Geol.* **21**, 129–147.

ALLEN, J.R.L. & WILLIAMS, B.P.J. (1981) Sedimentology and stratigraphy of the Townsend Tuff Bed (Lower Old Red Sandstone) in South Wales and the Welsh Borders. *J. geol. Soc. London*, **138**, 15–29.

BAKER, V.R. (1978) Adjustment of fluvial systems to climate and source terrain in tropical and subtropical environments. In: *Fluvial Sedimentology* (Ed. by A.D. Miall). *Mem. Can. Soc. Petrol. Geol., Calgary*, **5**, 211–230.

BEHRENSMEYER, A.K. & HILL, A.P. (Eds) (1980) *Fossils in the Making: Vertebrate Taphonomy and Paleoecology.* University of Chicago Press. 388 pp.

BERGGREN, W.A., McKENNA, M.C., HARDENBOL, J. & OBRADOVICH, J.D. (1978) Revised Paleogene polarity time scale. *J. Geol.* **86**, 67–81.

BISHOP, W.W. (Ed.) (1978) *Geological Background to Fossil Man.* Scottish Academic Press, Edinburgh. 585 pp.

BLUCK, B.J. (1979) Structure of coarse grained braided stream alluvium. *Trans. R. Soc. Edinb.* **70**, 181–221.

BOWN, T.M. & KRAUS, M.J. (1981) Lower Eocene alluvial paleosols (Willwood Formation, northwest Wyoming, U.S.A.) and their significance for paleoecology, paleoclimatology, and basin analysis. *Palaeogeogr. Palaeoclim. Palaeoecol.* **34**, 1–30.

BRIDGE, J.S. & LEEDER, M.R. (1979) A simulation model of alluvial stratigraphy. *Sedimentology*, **26**, 617–644.

CAMPBELL, C.V. (1976) Reservoir geometry of a fluvial sheet sandstone. *Bull. Am. Ass. Petrol. Geol.* **60**, 1009–1020.

CANT, D.J. & WALKER, R.G. (1976) Development of a braided-fluvial facies model for the Devonian Battery Point Sandstone, Quebec. *Can. J. Earth Sci.* **13**, 102–119.

COTTER, E. (1978) The evolution of fluvial style, with special reference to the central Appalachian Paleozoic. In: *Fluvial Sedimentology* (Ed. by A.D. Miall). *Mem. Can. Soc. Petrol. Geol., Calgary*, **5**, 361–384.

CRANE, R.C. (1981) Controls on the architecture of alluvial suites produced by computer-aided theoretical simulations of flood plain processes. *Abstracts: Modern and Ancient Fluvial Systems, Sedimentology and Processes*, p. 26. Keele, U.K.

DODSON, P. (1971) Sedimentology and taphonomy of the Oldman Formation (Campanian), Dinosaur Provincial Park, Alberta (Canada). *Palaeogeogr. Palaeoclim. Palaeoecol.* **10**, 21–74.

DUTTA, P.K. & SUTTNER, L.J. (1981) Fluvial sandstone composition and paleoclimate. *Abstracts: Modern and Ancient Fluvial Systems, Sedimentology and Processes*, p. 34. Keele, U.K.

EMBRY, A. & KLOVAN, J.E. (1976) The Middle-Upper Devonian clastic wedge of the Franklin Geosyncline. *Bull. Can. Petrol. Geol.* **24**, 485–639.

ETHRIDGE, F.G. & FLORES, R.M. (Eds) (1981) *Recent and Ancient Nonmarine Depositional Environments: Models for Exploration. Spec. Publs Soc. econ. Paleont. Miner., Tulsa.* 349 pp.

ETHRIDGE, F.G. & SCHUMM, S.A. (1978) Reconstructing paleochannel morphologic and flow characteristics: methodology, limitations and assessment. In: *Fluvial Sedimen-*

tology (Ed. by A.D. Miall). *Mem. Can. Soc. Petrol. Geol., Calgary*, **5**, 703–721.

FRIEND, P.F. (1978) Distinctive features of some ancient river systems. In: *Fluvial Sedimentology* (Ed. by A.D. Miall). *Mem. Can. Soc. Petrol. Geol., Calgary*, **5**, 531–542.

FRIEND, P.F. (1983) Towards the field classification of alluvial architecture or sequence. In: *Modern and Ancient Fluvial Systems* (Ed. by J.D. Collinson and J. Lewin). *Spec. Publs int. Ass. Sediment.* **6**, 345–354. Blackwell Scientific Publications, Oxford.

FRIEND, P.F., ALEXANDER-MARRACK, P.D., NICHOLSON, J. & YEATS, A.K. (1976) Devonian sediments of East Greenland. *Meddr Grønland*, **206**.

FRIEND, P.F., SLATER, M.J. & WILLIAMS, R.C. (1979). Vertical and lateral building of river sandstone bodies, Ebro Basin, Spain. *J. geol. Soc. London*, **136**, 39–46.

GALAY, V.J., KELLERHALS, R. & BRAY, D.I. (1973) Diversity of river types in Canada. In: *Fluvial Processes and Sedimentation, Proc. Hydrology Symp. Edmonton, National Research Council*, pp. 217–250.

GALLOWAY, W.E. (1981) Depositional architecture of Cenozoic Gulf Coastal Plain fluvial systems. In: *Recent and Ancient Nonmarine Depositional Environments* (Ed. by F.G. Ethridge and R.M. Flores). *Spec. Publs Soc. econ. Paleont. Miner., Tulsa*, **31**, 127–156.

GARNER, J.F. (1959) Stratigraphic-sedimentary significance of contemporary climate and relief in four regions of the Andes Mountains. *Bull. geol. Soc. Am.* **70**, 1327–1368.

HASZELDINE, R.S. (1983) Low sinuosity sandy fluvial deposits, Westphalian B. Coal Measures, NE England. In: *Modern and Ancient Fluvial Systems* (Ed. by J.D. Collinson and J. Lewin). *Spec. Publs int. Ass. Sediment.* **6**, 449–456. Blackwell Scientific Publications, Oxford.

HAY, R.L. (1976) *Geology of the Olduvai Gorge.* University of California Press. 203 pp.

HORNE, J.C., FERM, J.C., CARUCCIO, F.T. & BAGANZ, B.P. (1978) Depositional models in coal exploration and mine planning in the Appalachian region. *Bull. Am. Ass. Petrol. Geol.* **62**, 2379–2411.

JACKSON, R.G., II (1978) Preliminary evaluation of lithofacies models for meandering alluvial streams. In: *Fluvial Sedimentology* (Ed. by A.D. Miall). *Mem. Can. Soc. Petrol. Geol., Calgary*, **5**, 543–576.

JOHNSON, G.D. (1977). Paleopedology of *Ramapithecus*-bearing sediments, North India. *Geol. Rdsch.* **66**, 192–216.

JOHNSON, G.D., JOHNSON, N.M., OPDYKE, N.D. & TAHIRK-HELI, R.A.K. (1979) Magnetic reversal stratigraphy and sedimentary tectonic history of the Upper Siwalik Group, Eastern Salt Range and southwestern Kashmir. In: *Geodynamics of Pakistan* (Ed. by A. Farah and K.A. DeJong), pp. 149–165. Geological Survey of Pakistan.

KUENZI, W.D., HORST, O.H. & McGEHEE, R.V. (1979) Effect of volcanic activity on fluvial-deltaic sedimentation in a modern arc-trench gap, southwestern Guatemala. *Bull. geol. Soc. Am.* **90**, Pt I, 827–838.

LE BLANC SMITH, G. & ERIKSSON, K.A. (1979) A fluvioglacial and glaciolacustrine deltaic depositional model for Permo-Carboniferous coals of northeastern Karoo Basin, South Africa. *Palaeogeogr. Palaeoclim. Palaeoecol.* **27**, 67–84.

LEEDER, M.R. (1973) Fluviatile fining-upward cycles and the magnitude of paleochannels. *Geol. Mag.* **110**, 265–276.

LEEDER, M.R. (1975) Pedogenic carbonates and flood

sediment accretion rates: a quantitative model for alluvial arid-zone lithofacies. *Geol. Mag.* **112**, 257–270.

LEEDER, M.R. (1978) A quantitative stratigraphic model for alluvium, with special reference to channel deposit density and interconnectedness. In: *Fluvial Sedimentology* (Ed. by A.D. Miall). *Mem. Can. Soc. Petrol. Geol., Calgary*, **5**, 587–596.

LONG, D.G.F. (1978) Proterozoic stream deposits: some problems of recognition and interpretation of ancient sandy fluvial systems. In: *Fluvial Sedimentology* (Ed. by A.D. Miall). *Mem. Can. Soc. Petrol. Geol., Calgary*, **5**, 313–342.

MIALL, A.D. (1977) A review of the braided river depositional environment. *Earth Sci. Rev.* **13**, 1–62.

MIALL, A.D. (Ed.) (1978a) *Fluvial Sedimentology. Mem. Can. Soc. Petrol. Geol., Calgary*, **5**, 859 pp.

MIALL, A.D. (1978b) Fluvial sedimentology: an historical review. In: *Fluvial Sedimentology* (Ed. by A.D. Miall). *Mem. Can. Soc. Petrol. Geol., Calgary*, **5**, 1–47.

MIALL, A.D. (1978c) Lithofacies types and vertical profile models in braided rivers: a summary. In: *Fluvial Sedimentology* (Ed. by A.D. Miall). *Mem. Can. Soc. Petrol. Geol., Calgary*, **5**, 597–604.

MIALL, A.D. (1978d) Tectonic setting and syndepositional setting of molasse and other nonmarine-paralic sedimentary basins. *Can. J. Earth Sci.* **15**, 1613–1632.

MIALL, A.D. (1980) Cyclicity and the facies model concept in fluvial deposits. *Bull. Can. Petrol. Geol.* **28**, 59–80.

MIALL, A.D. (Ed.) (1981a) Sedimentation and tectonics in alluvial basins. *Spec. Pap. geol. Ass. Can.* **23**, 272 pp.

MIALL, A.D. (1981b) Alluvial sedimentary basins: tectonic setting and basin architecture. In: *Sedimentation and Tectonics in Alluvial Basins* (Ed. by A.D. Miall). *Spec. Pap. geol. Ass. Can.* **23**, 1–33.

MIALL, A.D. & GIBLING, M.R. (1978) The Siluro Devonian clastic wedge of Somerset Island, Arctic Canada, and some regional paleogeographic implications. *Sedim. Geol.* **21**, 85–127.

MILLER, F.X. (1977) The graphic correlation method in biostratigraphy. In: *Concepts and Methods in Biostratigraphy* (Ed. by E.G. Kauffman and J.E. Hazel), pp. 165–186. Dowden, Hutchinson & Ross, Stroudsburg.

MOLLARD, J.D. (1973) Airphoto interpretation of fluvial features. In: *Fluvial Processes and Sedimentation. Proc. Hydrology Symp. Edmonton, National Research Council*, pp. 341–380.

NAMI, M. & LEEDER, M.R. (1978) Changing channel morphology and magnitude in the Scalby Formation (M. Jurassic) of Yorkshire, England. In: *Fluvial Sedimentology* (Ed. by A.D. Miall), pp. 431–440. *Mem. Can. Soc. Petrol. Geol., Calgary*, **5**, 431–440.

PARKASH, B., AWASTHI, A.K. & GOHAIN, K. (1983) Lithofacies of the Markanda terminal fan, Kurukshetra District, India. In: *Modern and Ancient Fluvial Systems* (Ed. by J.D. Collinson and J. Lewin). *Spec. Publs int. Ass. Sediment.* **6**, 337–344. Blackwell Scientific Publications, Oxford.

PICARD, M.D. (1964) Paleomagnetic correlation of units within Chugwater (Triassic) Formation, west-central Wyoming. *Bull. Am. Ass. Petrol. Geol.* **48**, 269–291.

RUST, B.R. (1978). Depositional models for braided alluvium. In: *Fluvial Sedimentology* (Ed. by A.D. Miall). *Mem. Can. Soc. Petrol. Geol., Calgary*, **5**, 605–625.

SCHUMM, S.A. (1963) A tentative classification of alluvial river channels. *Circ. U.S. geol. Surv.* **477**, 10 pp.

SCHUMM, S.A. (1968). Speculations concerning paleohydrologic controls of terrestrial sedimentation. *Bull. geol. Soc. Am.* **79**, 1573–1588.

SCHUMM, S.A. (1981) Evolution and response of the fluvial system, sedimentological implications. In: *Recent and Ancient Nonmarine Depositional Environments: Models for Exploration* (Ed. by F.G. Ethridge and R.M. Flores). *Spec. Publs Soc. econ. Paleont. Miner., Tulsa*, **31**, 19–30.

SHAW, A.B. (1964) *Time in Stratigraphy.* McGraw-Hill, New York. 365 pp.

SMITH, D.G. & PUTNAM, P.E. (1980) Anastomosed river deposits: modern and ancient examples in Alberta, Canada. *Can. J. Earth Sci.* **17**, 1396–1406.

SMITH, D.G. & SMITH, N.D. (1980) Sedimentation in anastomosed river systems: examples from alluvial valleys near Banff, Alberta. *J. sedim. Petrol.* **50**, 157–164.

SMITH, R.M.H. (1980) The lithology, sedimentology and taphonomy of flood plain deposits of the Lower Beaufort (Adelaide Subgroup) strata near Beaufort West. *Trans. geol. Soc. S. Afr.* **83**, 399–414.

STEEL, R. & AASHEIM, S.M. (1978) Alluvial sand deposition in a rapidly subsiding basin (Devonian, Norway). In: *Fluvial Sedimentology* (Ed. by A.D. Miall). *Mem. Can. Soc. Petrol. Geol., Calgary*, **5**, 385–412.

STEEL, R.J., MAEHLE, S., NILSEN, H., RØE, S.L. & SPINNANGR, Å. (1977) Coarsening-upward cycles in the alluvium of Hornelen Basin (Devonian) Norway: sedimentary response to tectonic events. *Bull. geol. Soc. Am.* **88**, 1124–1134.

TUNBRIDGE, I.P. (1981) Sandy high-energy flood sedimentation—some criteria for recognition, with an example from the Devonian of S.W. England. *Sedim. Geol.* **28**, 79–96.

VESSELL, R.K. & DAVIS, D.K. (1981) Nonmarine sedimentation in an active fore-arc basin. In: *Recent and Ancient Nonmarine Depositional Environments* (Ed. by F.G. Ethridge and R.M. Flores). *Spec. Publs Soc. econ. Paleont. Miner., Tulsa*, **31**, 31–48.

VISHER, G.S. (1972) Physical characteristics of fluvial deposits. In: *Recognition of Ancient Sedimentary Environments* (Ed. by J.K. Rigby and W.K. Hamblin). *Spec. Publs Soc. econ. Paleont. Miner., Tulsa*, **16**, 84–97.

WEIMER, R.J. (1970) Rates of deltaic sedimentation and intrabasin deformation, Upper Cretaceous of Rocky Mountain region. In: *Deltaic Sedimentation Modern and Ancient* (Ed. by J.P. Morgan). *Spec. Publs Soc. econ. Paleont. Miner., Tulsa*, **15**, 270–292.

Spec. Publs int. Ass. Sediment. (1983) **6**, 287–300

Tabular cross-bedding in Messinian fluvial channel conglomerates, Southern Alps, Italy

FRANCESCO MASSARI

Istituto di Geologia dell'Università, Via Giotto 1, 35100 Padova, Italy

ABSTRACT

Channelized fluvial conglomerates of Messinian age are related to a fan-delta system active along an actively subsiding tectonic margin. They can be followed from an active fan-delta zone to an inferred interlobe area, along a belt oriented transversely to the palaeocurrents.

In the active fan-delta area the conglomeratic bodies are complex as a result of vertical stacking of channel deposits, and are dominated by facies Gm (code of Miall, 1978). They grade laterally into uni-storey interlobe sequences isolated in thick mudstones and dominated by facies Gp; a well-developed association of supraplatform facies is also present in the interlobe sequences including sheet sandstones and conglomerates, chute channels and chute bars. These changes are believed to represent a deepening and an increase in sinuosity of the channels, which are inferred to be respectively dominated by longitudinal and transverse bars.

The unimodal orientation of foresets in facies Gp suggests an essentially downstream movement of transverse bars during short-lived and pronounced flood peaks and limited subsequent modification of high-stage features during falling and low stages. The Messinian channel conglomerates are thought to be an ancient counterpart of present-day Nueces River deposits (Gustavson, 1978).

INTRODUCTION

The younger part of the Miocene molasse of the Southern Alps, outcropping in the Eastern Pre-Alps, is a clastic wedge more than 2000 m thick which accumulated from the middle Tortonian to the Messinian along an active tectonic margin. Sedimentation was probably controlled by a hinge line parallel to the regional strike, which allowed rapid downwarping of the basin axis concurrently with the uplift of the Southern Alps. Tectonic control led to a coarsening-upward sequence from prodelta, through deltaic, into subaerial fan deposits.

In the lower part of this sequence two distinct episodes of deltaic progradation may be recognized (Massari, 1978). The older (?late Serravallian–early Tortonian) is only known in the Vittorio Veneto area; the younger (late Tortonian–Messinian) shows a more generalized character and is represented along the

entire southernmost belt of the Eastern Pre-Alps. It is expressed by a stacked series of increasingly proximal mouth-bar sequences, followed by a number of conglomeratic lithosomes organized partly as beach and bay-mouth cycles and partly as channelized sequences. The last mentioned are mostly bounded by interdistributary basin and delta plain (overbank, backswamp and backswamp pond) mudstones comprising a lignite key bed (Fig. 1), and respectively bearing brackish-water molluscs and sparse freshwater and terrestrial gastropods. This study focuses on the features of these conglomeratic lithosomes.

High subsidence and sedimentation rates (of the order of 1000 m My^{-1} during the Messinian) resulted in an alluvial stratigraphy characterized, mostly in the eastern area, by very low interconnectedness, with conglomeratic bodies isolated in thick mudstones. It seems likely that in such a situation the preservation potential was very high.

The conglomerates are for the most part clast-

0141-3600/83/0106-0287 $02.00

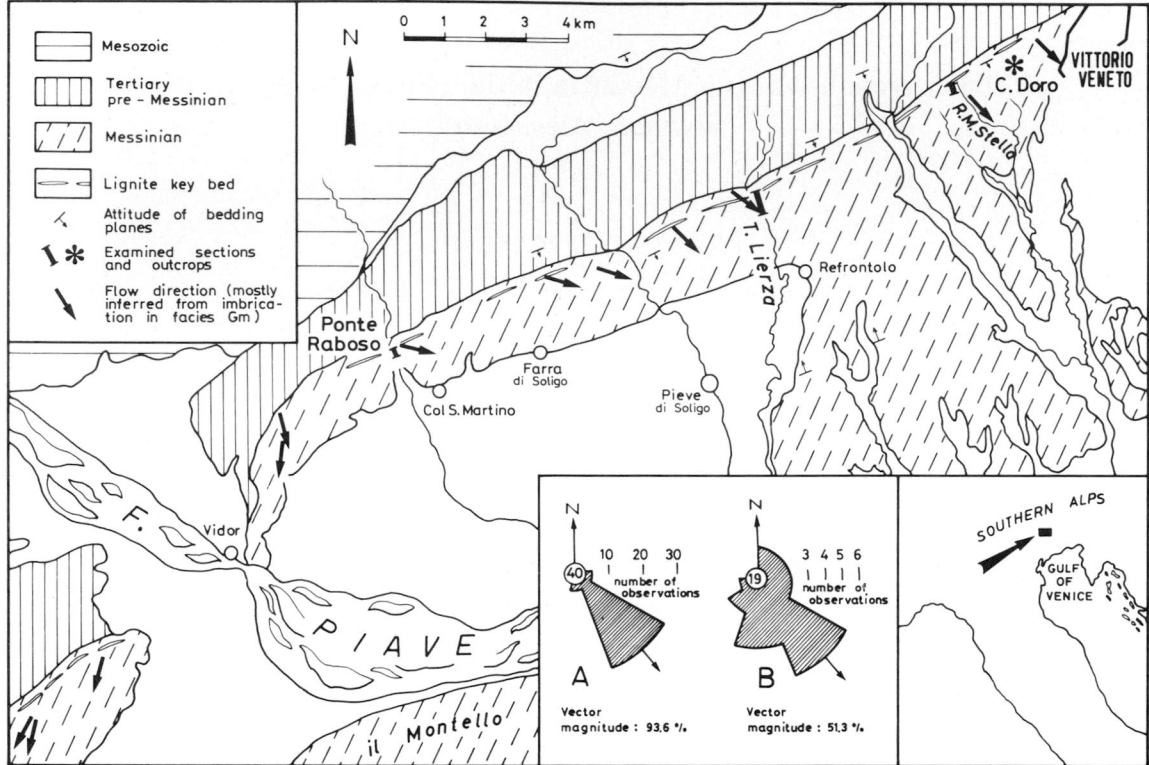

Fig. 1. Simplified geological map of Eastern Pre-Alps between the River Piave and Vittorio Veneto, showing general pattern of palaeocurrents of Messinian conglomerates in the stratigraphic interval studied. A fan-shaped pattern (palaeo-Piave fan-delta) is clearly visible in the western area. Inset: current roses refer only to eastern area (T. Lierza, R.M. Stella and C. Doro sections). (A) shows orientations of trough axes of cross-strata and imbrication in facies Gm; (B) shows cross-set orientations of planar cross-bedding. Arrows indicate vector mean.

supported, with a small amount of fine- to medium-grained sandstone matrix and good rounding of the clasts. Size range extends from the granule to the cobble grade, pebble sizes being dominant. A northern to north-western source may be inferred both by palaeocurrent data and by the south-Alpine affinity of the clast composition (Massari, Rosso & Radicchio, 1974).

The study area is an ENE–WSW trending belt, belonging to the Bassano-Valdobbiadene flexure which marks the transition from the Eastern Pre-Alps to the Veneto plain. Owing to the generally high angles of the structural dip the lithosomes may be traced only along the regional strike, which is approximately transverse to the palaeocurrents (Fig. 1).

FACIES

The code of Miall (1977, 1978) is adopted here, with the additions described below.

Facies Gm (Figs 2, 5 and 7): horizontally bedded pebble–cobble conglomerates, commonly imbricate. They are locally interbedded with impersistent lenticles of sandstones which sometimes contain scattered to aligned pebbles, and in places display a faint plane lamination.

Facies Gt (Figs 2, 5 and 7–9): pebble–cobble framework conglomerates, trough cross-bedded and distinctly stratified, in places interbedded with discrete impersistent lenticles of sandstone. Compared with other facies, this shows poorer sorting, less developed imbrication and weaker fabric. Long-axis fabric commonly displays two modes more or less perpendicular to one another.

Facies Gp (Figs 2 and 5–9): pebble–cobble

framework conglomerates, tabular cross-bedded with dip angle of foresets between 15° and 40°. This facies usually occurs in solitary sets ranging from less than 1 m up to 4·5 m thick. Decrease of set thickness is usually accompanied by improvement of sorting and regularity of bedding, as well as by increase in fabric isotropy. Some foresets show convolute bedding, probably due to slumping or to rapid fluctuations in water stages which would upset grain-to-grain contact pressure (Levey, 1978). The larger sets consist of either planar or tangential foresets (Fig. 7). The former normally show a higher dip angle, are usually finer grained (most commonly in the fine pebble grade) and do not show marked changes in grain size and bed thickness down-dip. In addition, planar foresets often display a normal (rarely a reverse) grading. Thickness of individual foresets ranges from 20 to 70 cm with a mean of about 25 cm. On the other hand, the tangential foresets show a wider range of grain sizes (from sandstones to cobble conglomerates) and are usually thicker (on average 30–40 cm) with down-dip decreasing thickness. Sandy foresets are rare and show a faint lamination parallel to the foreset plane and emphasized by pebble strings. An abrupt lateral transition from tangential to angular-based foresets sometimes occurs, the latter lying conformably to a planar erosion surface truncating the former.

In both types of foresets the dip of the *ab* plane of clasts is usually up the foreset slope, but in places it is downslope; sometimes opposite dips coexist in adjacent parts of the same layer. In addition, high angles of dip (up to 50–60°) are sometimes observed. The long axes of clasts are predominantly oriented parallel to the foreset dip; moreover, in some areas of the sole of foreset beds, the *a*-axes of the clasts are seen arranged with fan-shaped orientations over a range of 35–40°.

Vegetation cover and poor exposures only allow a limited number of observations of set orientation. However, it can be said that the spread of palaeocurrent directions of tabular cross-bedding is apparently higher by comparison with trough cross-bedding and imbrication (Fig. 1). It should be noted that the most divergent orientation with respect to the average flow direction is often displayed by the smaller sets, which are usually finer grained and more distinctly and thinner bedded.

In some cases bundles of foresets are separated by inclined erosion surfaces which may be slightly convex upwards and less steep than the associated foresets (Fig. 7). Intrasets (Collinson, 1968) occur in places between the foresets, most commonly preserved in the lower part of the first-order set. When compared with the first-order foresets, they are finer grained, better sorted and commonly dip in the same direction but at a higher angle. However, some sandy intrasets consist of laminae dipping at a low angle up the foreset slope (Fig. 7).

Other identified facies are: St (trough cross-bedded sandstone), Sh (planc-laminated sandstone), Sr (ripple cross-laminated sandstone), Fl (mudstone or very fine sandstone with plane lamination or delicate and small-scale ripple cross-lamination), Fsc (massive mudstone), Fcf (massive mudstone with fresh water molluscs), C (carbonaceous mudstone).

Three additions to the facies defined by Miall (1977, 1978) are proposed here.

Facies Ge (Figs 2, 5 and 8): scour & fill gravel. This occurs as massive, erosive-based and usually thick beds of clast-supported, pebble–cobble conglomerates often rich in mudstone clasts. It can be considered as the equivalent of facies Sc as defined by Rust (1978) in the sandstones.

Facies Gl (Figs 5 and 7): distinctly stratified and commonly imbricate pebble framework conglomerates, with very low angle of dip (4–6°).

Facies Gu (Fig. 2): fine pebble conglomerates showing undulatory bedding. Wavy bedforms have crests transverse to flow, wavelength of 0·8–2 m and amplitude of 15–30 cm. The conglomerates involved in this structure sometimes contain abundant mudstone clasts. This facies usually occurs on the upper surface of Gm or Ge conglomerate beds.

WESTERN AREA

Changes in lithology and trend of the sequences may be detected within the study area. The conglomerate/mudstone ratio, as well as the mean grain size of conglomerates, decrease eastwards, so that the western area appears characterized by a more proximal sedimentation. The conglomeratic bodies of this area (P. Raboso section, Fig. 2) are dominated by facies Gt, Gm and Ge. In addition facies Gu and Gp rarely occur. A characteristic feature is the rarity to absence of coarse lag gravels at the base of channel sequences.

Interchannel sediments mostly consist of facies Fl, Fcf and Fsc, with rare layers of carbonaceous mudstones and occasional interbeds of fine sandstones.

When conglomerates are underlain by mudstones, the bounding surface usually shows a series of

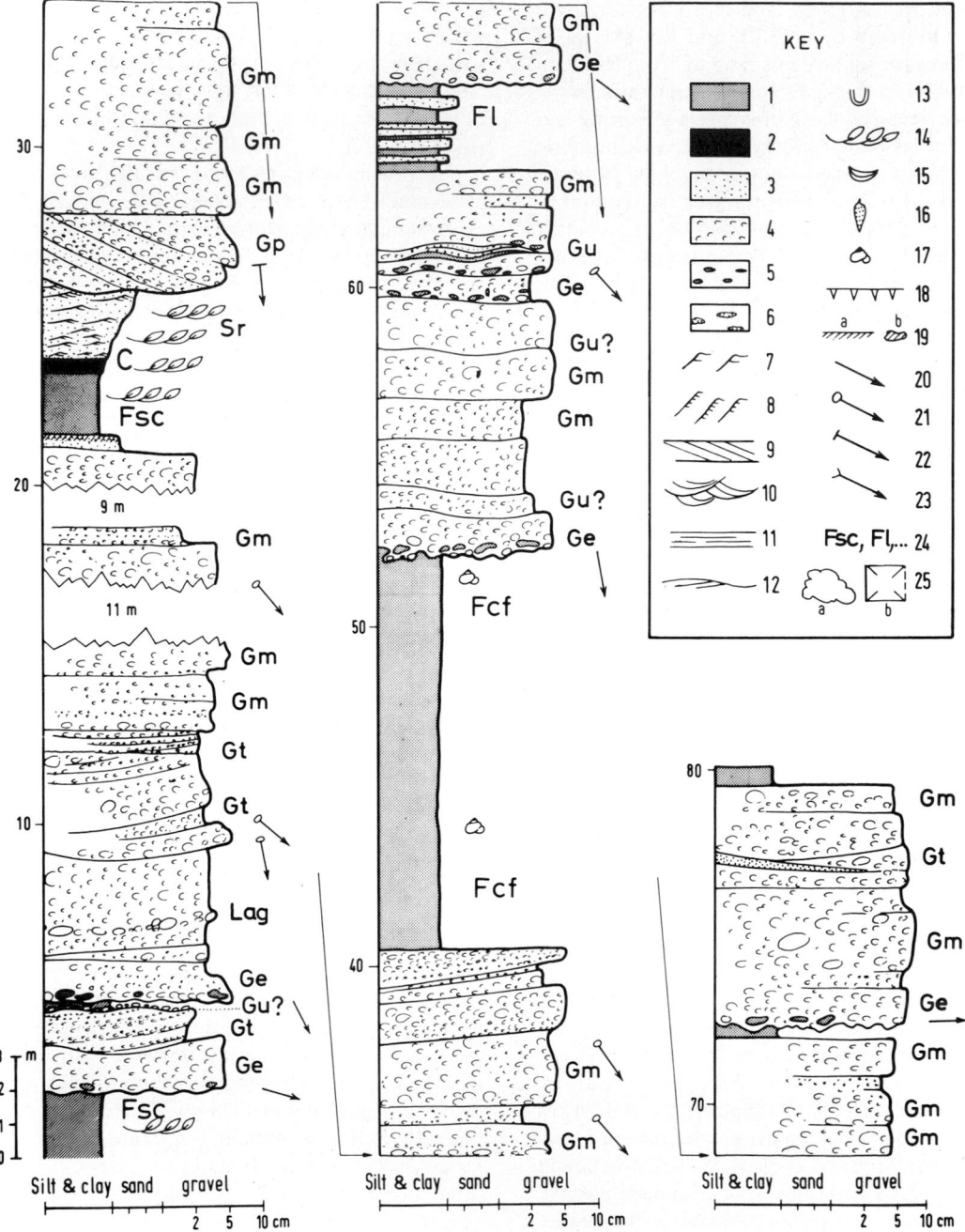

Fig. 2. Stratigraphic log of Ponte Raboso section (western area). (1) mudstone, (2) carbonaceous mudstone, (3) sandstone, (4) conglomerate, (5) mudstone clasts, (6) sandstone intraclasts, (7) ripple drift cross-lamination, (8) climbing ripples, (9) planar cross-bedding, (10) trough cross-bedding, (11) plane lamination, (12) undulatory bedding, (13) bioturbation, (14) plant debris, (15) oyster shells, (16) *Cerithidae* shells, (17) freshwater and terrestrial gastropods, (18) mud cracks, (19) pedogenic carbonate in crusts (a) and nodules (b), 20–23, palaeocurrent directions determined from macrolineation (20), imbrication in facies Gm (21), planar cross-bedding (22) and trough cross-bedding (23), (24) letter code devised by Miall (1978) for facies nomenclature, (25) covered outcrops in field sketches (a) and stratigraphic logs (b).

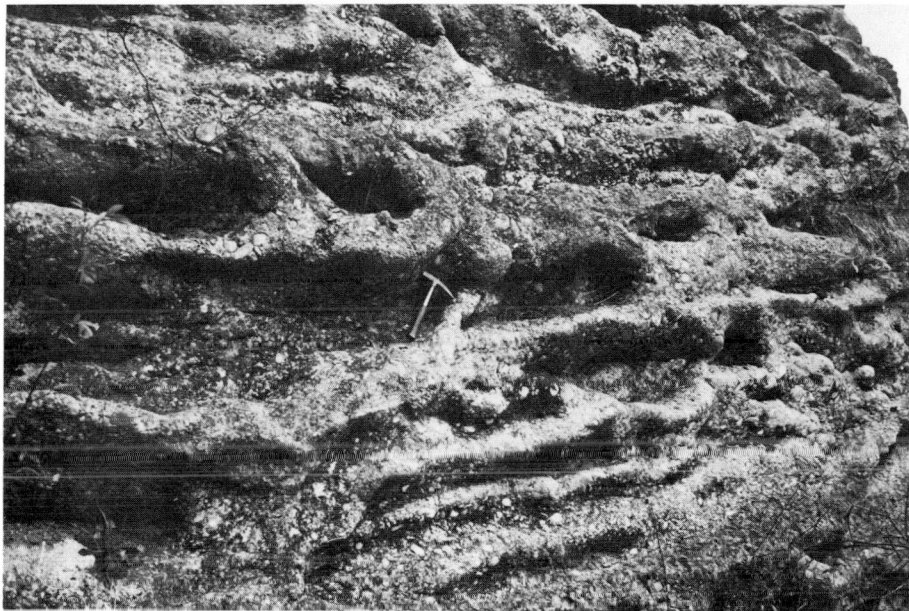

Fig. 3. Moulds of longitudinal furrows and ridges (macrolineation) at the base of a channelized body (Ponte Raboso section).

alternating longitudinal ridges and furrows (Fig 3) which are rounded in cross-section and have comparable dimensions. Similar (but larger scale) erosion features of stream channels incised into fine-grained cohesive sediments were described by Allen (1965) as corrasion furrows and attributed to the corrasive action of the stream load. Friend (1965) reported, in the Devonian of Spitsbergen, similar sole structures ('long welts') ranging from straight and parallel to sinuous and branching, at the base of fluvial channelized sandstone sequences incised into overbank siltstones.

The amplitude of the structure, possibly increased by loading, is irregular, ranging from 3 to 35 cm, spacing of ridges is between 7 and 50 cm. Ridges range from more or less straight and parallel to sinuous. With increase in sinuosity an increase in frequency of coalescing and branching of the ridges may be observed. The larger the scale of the structure, the coarser apparently the grain size of the conglomerates involved and the straighter and more parallel the ridges and furrows. It therefore seems that a relationship exists between these features and current velocity. A similar relationship has been presumed by Dzulynski & Walton (1965) in the ridge and furrow structure associated with turbidites.

The larger clasts are usually concentrated along the furrow bottoms. In addition, the clast long-axes are differently oriented on the flanks of each furrow, diverging at about 35° from the direction of the furrow itself. The macrolineation is therefore orientated approximately along the bisectrix of the angle formed by the average orientations of clast long-axes on the

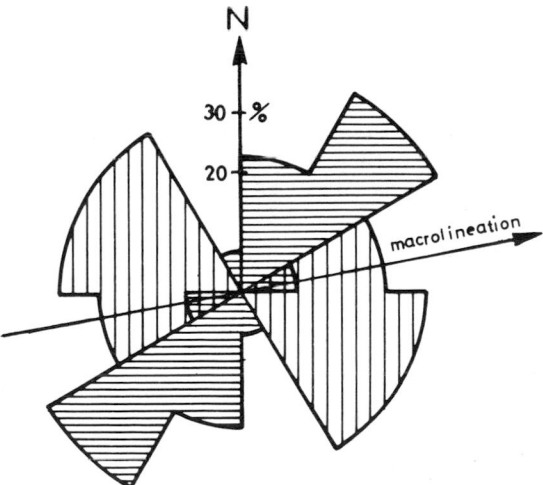

Fig. 4. Diagram of clast long-axis azimuths read on the two flanks of a furrow at the base of a conglomeratic body showing macrolineation (Ponte Raboso section). Number of readings for horizontally hatched group of azimuths, 83; for vertically hatched, 45.

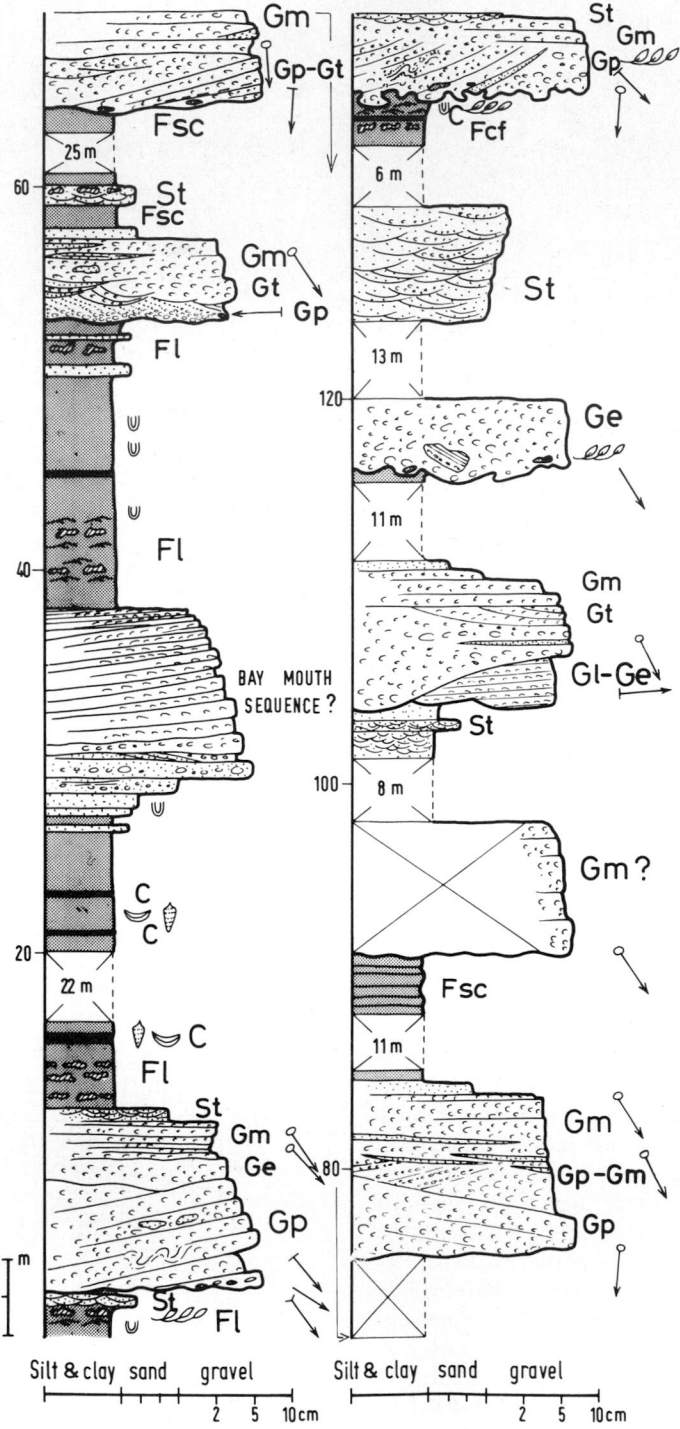

Fig. 5. Stratigraphic log of T. Lierza section. See Fig. 2 for legend.

Fig. 6. The C. Doro channelized body (easternmost area). See Fig. 7 for details.

two sides of the furrows (or ridges) (Fig. 4). The fabric on ridge crests and furrow bottoms on the other hand is more anisotropic, with weak predominance of orientations transverse to flow. Allen (1964) reported similar rose diagrams (though displaying a markedly lower dispersion) for the long-axis microfabric of sand grains involved in primary current lineation. In most cases he identified two modes of grain orientation more or less symmetrical about the lineation.

The structure may be the result of upper flow regime conditions involving longitudinal roll vortices and may be interpreted as a kind of large-scale gravel lineation.

EASTERN AREA

Channel deposits

The channel deposits of the eastern area have been investigated in the T. Lierza (Fig. 5), C. Doro and R. M. Stella sections (Figs 6–9: see Fig. 1 for locations). The coarse lithosomes are tabular in geometry and isolated in a much higher proportion of fine sediments when compared to those of the western area, thus resulting in an alluvial stratigraphy characterized by very low interconnectedness. Facies

Gp is much more common here than in the western area and in most cases is confined to the lower part of the channel sequences. Within this area, the frequency and thickness of facies Gp as well as the regularity of stratification increases eastwards, concurrent with a progressive decrease of mean grain size. A maximum thickness of 4·5 m has been measured in a single set of Gp conglomerates occurring in the easternmost section (C. Doro, Figs 6 and 7).

Conglomerate sequences commonly show a tendency towards ordered organization of the facies and are of the type: Gp(Gt, Ge, Gm, Gl) → Gm(Gt, Gp, Ge, St, Sh). By contrast, other characters fail to show an ordered and systematic trend. Vertical changes in bed thickness and grain size are quite variable, with some sequences displaying a crude fining- and thinning-upwards, mostly detectable in the uppermost part, and others showing an irregular coarsening with confinement of fine gravels and even sandstones in the lowermost part and of coarser sizes at the top.

Most sequences may be subdivided into two distinct levels.

The lower part, encompassing up to two-thirds of the sequence, is most commonly characterized by facies Gp. Foresets are either tangential or planar, the former usually thicker and coarser grained (Fig. 7). The sets of Gp conglomerates are occasionally

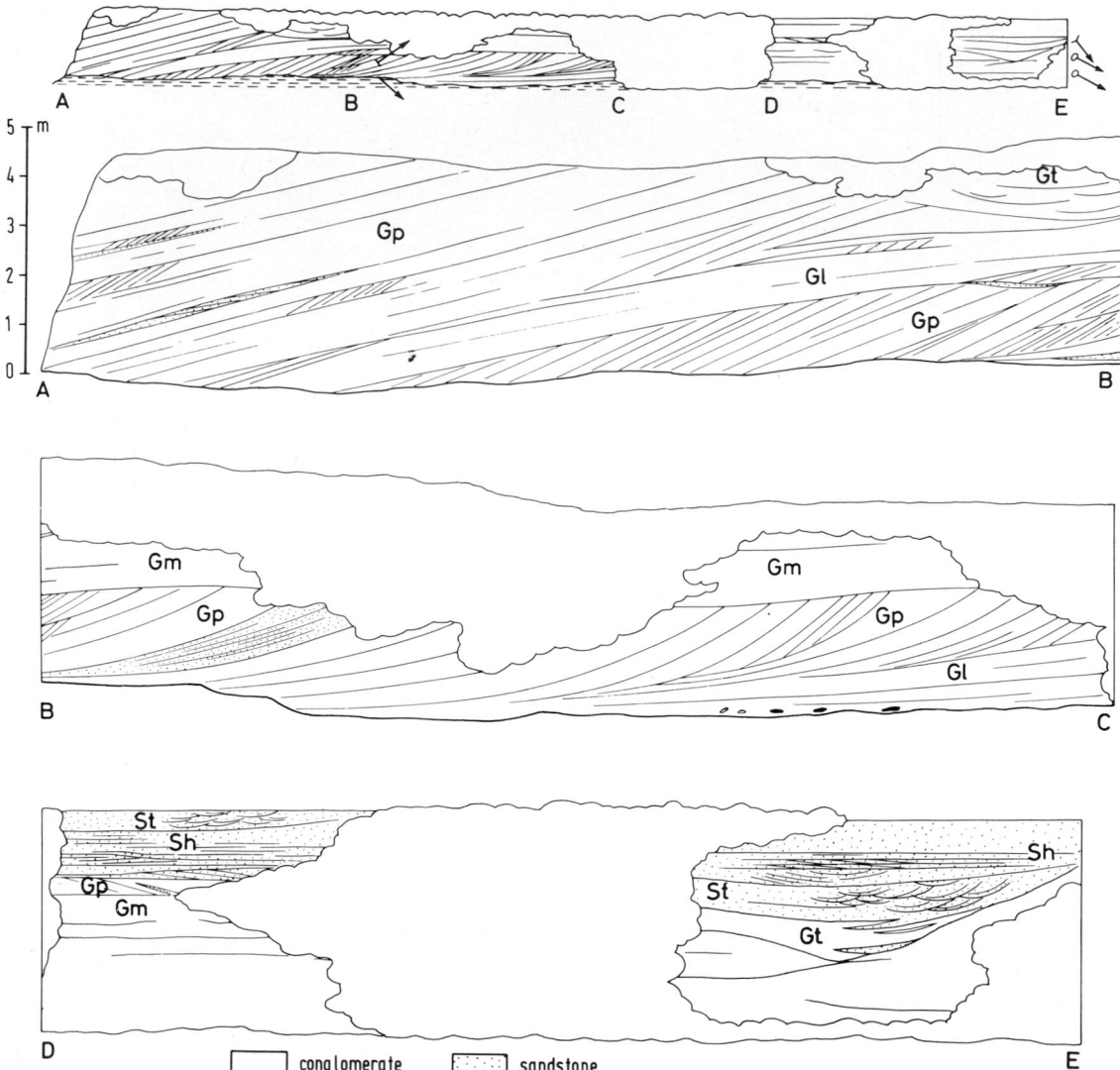

Fig. 7. Field sketch of the C. Doro section (for palaeocurrent symbols see legend of Fig. 2). Note, in the AB stretch, the major convex-up erosion surface, the intrasets, and the very high set of facies Gp on extreme left. See text for interpretation.

subdivided into bundles of foresets by low-angle erosion surfaces, as in the C. Doro section (Fig. 7). Intrasets may occur, mostly in the lower part of the set (Fig. 7, stretch AB), and may noticeably contribute to the fining of sizes in the lowermost part of the sequence.

The spectrum of facies is completed by facies Gm, Gl, Ge and Gt which may occur laterally to facies Gp. Inter-storey erosion surfaces commonly occur within the lower part of the sequence (Figs 7–9). They sometimes show minor scour troughs which may be filled by cross-laminated sand wedges or lenses (Fig.

9A), or small sets of fine-grained Gp conglomerates. In addition, small scour channels with fine-grained filling in places cut Gp conglomerates (Fig. 9B).

The upper part of the sequence may be separated from the lower by an erosion surface and is characterized by great variability of facies:

(a) Thin- to medium-bedded gravelly or sandy layers with more or less distinct horizontal stratification (facies Gm and Sh) (Fig. 7, stretches BC and DE).

(b) Broad, shallow scour channels, more or less symmetrically filled by facies Ge, Gt, St or Sh (Figs 7, 9A, B).

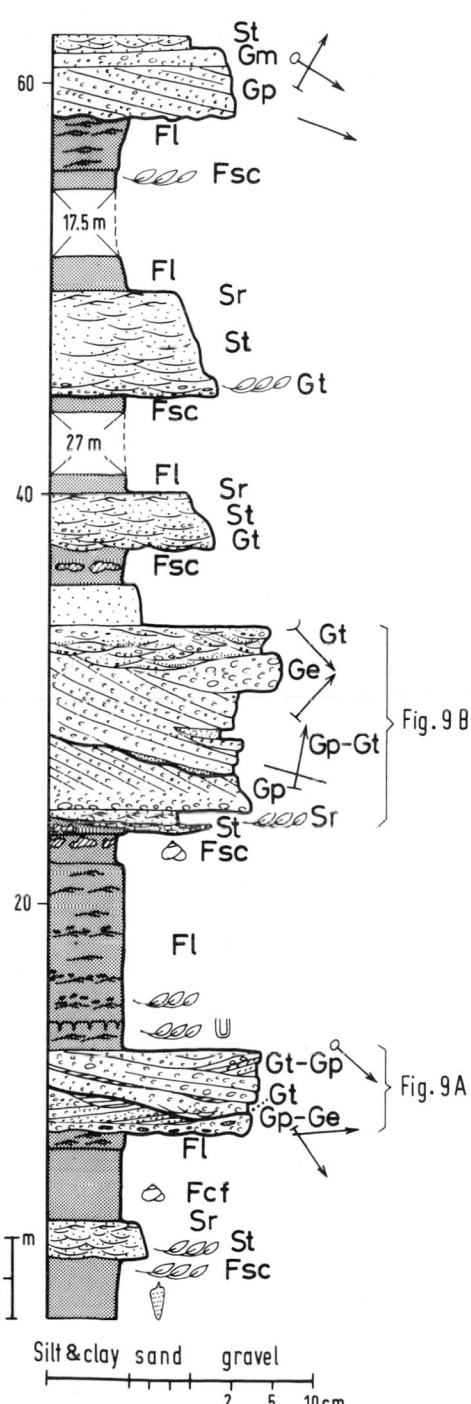

Fig. 8. Stratigraphic log of R.M. Stella section (eastern area). See Fig. 2 for legend.

(c) Planar cross-bedded pebble–cobble conglomerates with slightly convex-up foresets, occurring in sets up to 2 m thick, usually associated with facies group (b) (Fig. 9A).

If underlain by mudstones, the base of the conglomeratic lithosomes occurring in the eastern area commonly shows a macrolineation of the type described above but on a smaller scale relative to the western examples possibly due to finer-grained conglomerates involved in the structure. Moreover, the base of the channelized sequences is also characterized by the absence or rarity of coarse lag gravels.

In addition to major conglomeratic cycles, in the eastern area there are also some channelized units which are dominated by trough cross-bedded sandstones, sometimes with sparse, very fine gravel (upper parts of the sections of Figs 5 and 8). These sequences, up to 6 m thick, are usually fining- and thinning-upwards with decreasing scale of the structures, merging at the top into ripple-drift cross-lamination. Sometimes, however, a coarsening trend occurs (T. Lierza section, upper part of right column).

Interchannel deposits

In the majority of cases, the channelized bodies are underlain either by facies Fl or facies Fsc (Figs 5 and 8). The latter, in addition to the features indicated by Miall (1978), shows common bioturbation, local abundance of comminuted plant debris sometimes surrounded by small pyrite concretions and pedogenic carbonate nodules in some layers. In places, the channelized bodies are underlain by small sequences of massive grey mudstone up to 4 m thick which coarsen upwards into facies Fl (Fig. 8). The latter displays a ripple-drift lamination of upward-increasing scale.

Erosively based units with medium- to small-scale trough cross-bedding grading upwards into ripple cross-lamination sometimes of the climbing type (Fig. 7B at base of channelized sequence) form small fining-upward sequences 40–80 cm thick. These bear in places reworked pedogenic nodules. Other minor units sparsely occurring, interbedded with mudstones, consist of structureless sandstone layers 20–30 cm thick.

INTERPRETATION

As outlined above, some important differences exist between the western and eastern sides of the belt examined.

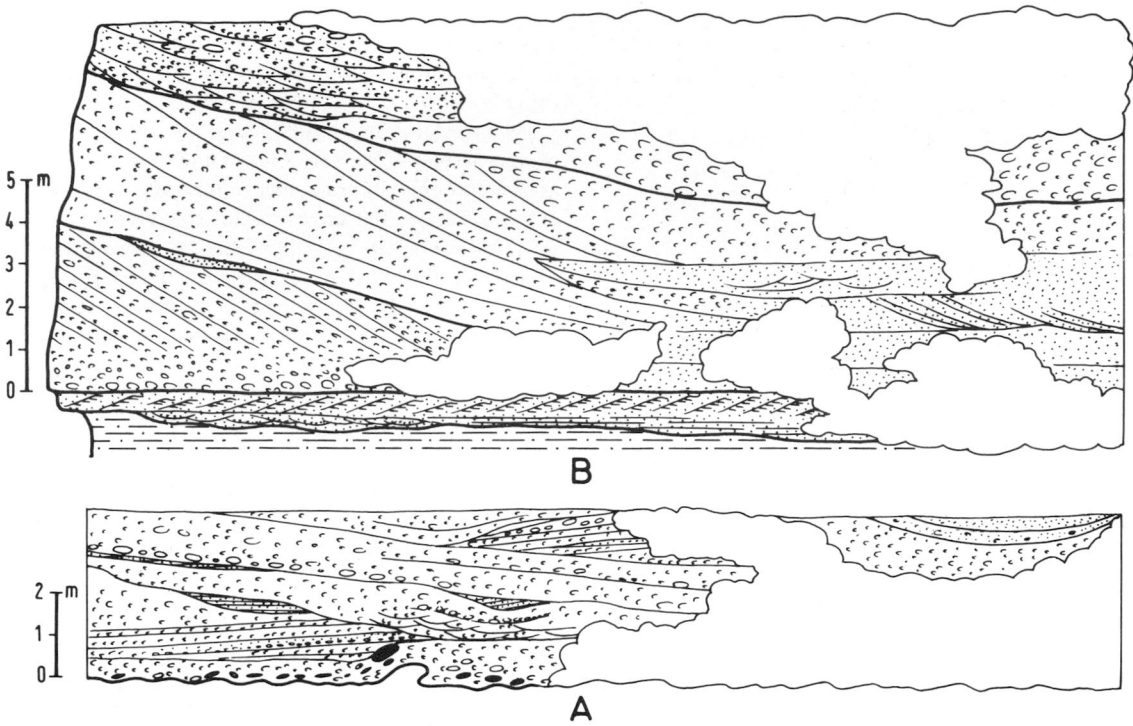

Fig. 9. Field sketch of the first two conglomeratic sequences of Fig. 8 (see Fig. 2 for legend). At the top of sequence A broad symmetrically filled channels are associated with a set of Gp conglomerates (inferred chute channels and chute bar).

Western sequences

Conglomeratic bodies in the western area are probably complex, i.e. they may result from vertical stacking and amalgamation of a number of channel units 2–3 m thick dominated by facies Gm, Gt and Ge (Fig. 2). Though no definite facies trend was identified, the channel sequences may result from the migration of longitudinal bars (facies Gm) on channel deposits (facies Gt and Ge) and are believed to be more proximal in character than those of the eastern area. Facies Gu, though rarely occurring, is exclusive to the western facies association and may record gravel bedforms formed under standing waves.

The observed features suggest a braided system of low-sinuosity channels. The presence of thick, muddy deposits between the conglomeratic bodies seems, however, to imply that the channels were flanked by largely muddy interdistributary and overbank areas, and possibly that there was some confinement by muddy banks limiting the lateral mobility of channels.

Eastern sequences

The changes observed in the eastward direction and particularly the progressive increase in importance

and scale of facies Gp probably reflect a deepening of the channels and decline in width/depth ratio. The eastern sequences (Figs 5–9) are the distal end-member of this continuum of facies associations.

In all examined sections of the eastern area, a tendency towards ordered organization of the facies within the conglomeratic lithosomes is recognized. This suggests that individual lithosomes do not result from stacking of a number of cycles, but correspond to uni-storey sequences in the majority of cases. Scour surfaces within the lithosomes are interpreted as in-channel features rather than major erosion surfaces separating different storeys within complex bodies.

The most striking feature of the eastern sequences is the presence of large-scale sets of planar cross-bedded conglomerates. The relatively wide directional dispersion of this facies, when compared to the low variability of both clast imbrication in facies Gm and trough axes in facies Gt (Fig. 1), suggests that it represents a form of in-channel variance related to differing bedform orientations (Bluck, 1974, 1976).

Although difficulties in determining bedform morphology in the fossil record are well known (Asquith & Cramer, 1975), the hypothesis may be advanced that major sets of Gp conglomerates result from the migration of transverse bars with convex downstream crestlines (Gustavson, 1978). The bars were probably subjected either to downstream migration at the slipface, or to vertical accretion, depending on the flow conditions. During peak discharge stages the bars were probably subjected to both processes, whereas the falling stage was dominated by slipface accretion processes and limited modifications of high-stage features.

Within the sets of facies Gp, bundles of tangential and planar foresets may follow one another without marked changes in orientation. This suggests that the change in geometry may result from a change in the strength of the separation eddy downstream of the bedform crestline. Stronger macroturbulence might thus lead to the formation of tangential foresets, mostly during peak discharge stages, and weaker separation eddies might produce planar foresets, probably mostly during falling stages or medium discharge events (Collinson, 1970). Moreover lateral transition from tangential to planar foresets may be due to a change in the geometry of the leeside separation flow from three-dimensional to essentially two-dimensional. This gives further support to the view that the bedforms involved may have had curved crestlines since, as stated by Allen (1968), a three-dimensional separation flow is produced when crestlines of bedforms are skewed relative to the flow direction.

Active erosion of previously formed bedforms probably took place when the latter were markedly out-of-phase with the flow. This might be expected mostly during falling stages of the river (Collinson, 1970). In the case illustrated in Fig. 7 (stretch AB), the major convex-up low-angle erosion surface may be considered a reactivation surface and may have resulted from washing out of a transverse bar during the falling stage. Under decreasing rate of shearing, deposition on the low-dipping downcurrent side of the washed-out bar could have initiated in the upper flow regime without flow separation, as a consequence of the shallowing of flow depth, and accretion on this side may have formed low-angle cross-strata (facies Gl in Fig. 7, stretch AB). The latter may then have been buried by migration of the avalanche faces of a high-relief transverse bar during subsequent high-stage flow. As a consequence of water stage fluctuations, a composite structure comprising high- and low-angle cross-beds may thus have formed.

The vertical organization of the sequences suggests a topographical differentiation of the system into distinct levels, higher ones being dominated by flood deposits and lower ones recording, in addition to high-energy events, the medium- and low-discharge regimes and particularly the processes operating during the falling stage. This differentiation may therefore reflect the spectrum of processes active at different discharge stages of the river.

Lower part of the sequences. The major sets of facies Gp are interpreted as due to high-stage slip-face accretion on transverse bars, as in the example described by Gustavson (1978) in the Nueces River.

On the other hand, the predominance of finer-grained and most distinctly bedded varieties of facies Gp (including the intrasets) with a relatively wide dispersion of dip directions in the lowermost parts of the sequence and in the lower part of thicker sets, suggests that the deeper parts of the system were the preferred site for slipface accretion during the falling and low-water stages. This may have been achieved either by superimposition and downslope migration of smaller bedforms on transverse bars (leading to formation of intrasets), or by deposition related to components of flow moving along the slipface, or by development of avalanche faces on the downstream margin of riffles, or by breaching of transverse bars as flow stage declined, to form a hierarchy of increasingly smaller gravel lobes as in the case described by Gustavson (1978). The presence of minor scour channels partly cut into major sets of facies Gp and filled with facies St or fine-grained Gt gravels (Fig. 9B) lends support to the last alternative, taking into account the fact that, in the case described by Gustavson, the lobes were fed by minor channels cut into transverse bars and occupied by shallow, turbulent, and fast-flowing waters.

Facies Gt may represent either a high-stage channel-bottom deposit like facies Ge or, if it occurs at higher levels of the sequence, a 'platform' deposit in the sense of Bluck (1971). Thick, pebble–cobble conglomerate units sometimes associated with small sets of facies Gt and bounded at the base by low-angle scour surfaces (middle part of sequence A in Fig. 9) suggest a process of lateral accretion on the platform surface which dips with a very low angle into the pool. Other low-angle scour surfaces occurring within the lower part of the sequences may be interpreted as reactivation surfaces related to water stage fluctuations, as mentioned above.

In the lower part of any sequence, facies Gm may represent either gravel sheets lacking in slipface (Hein & Walker, 1977) or riffle head gravels or thin gravel sheets deposited on the stoss side of some transverse bars, as in the modern Nueces River (Gustavson,

1978). This would explain the lateral transitions between facies Gp and Gm.

Facies Gl may result either from falling-stage accretion on the low profiles of washed-out bars, as mentioned above (Fig. 7, stretch AB), or from the migration of low-dipping downstream side of diffuse gravel sheets, possibly in the riffle reaches of the stream (Rust, 1978). Lateral transitions between facies Gl and Gm seem consistent with this interpretation.

Upper part of the sequences. Layers of horizontally bedded gravels or sands at the top of the sequences suggest that, during peak flood stage, gravelly or sandy sheets may be swept across the higher levels of the system, locally entering the supercritical flow regime. In the modern, low-sinuosity Nueces River (Gustavson, 1978) gravel sheets are mostly deposited at high stage on the upper part of lateral accretion surfaces of gravel meander lobes.

Broad, symmetrically filled, shallow scour channels in places grading laterally to sets of Gp gravels with slightly convex-up foresets (Fig. 9A) may represent respectively chute channels cut into the bar surface during flood stage and chute bars developed at the downcurrent termini of the channels.

It is suggested therefore that a large spectrum of sediments exist in the upper part of the system, which may be compared to the supra-platform of Bluck (1971). The fining or coarsening trend in the upper part of the sequences may reflect the downstream changes of grain size within the supra-platform. If a gradual fining occurs from head to tail of the supra-platform, we may expect relatively coarse grain sizes at the top of the sequence if it ends with bar head gravels, whereas a fining trend may reflect termination in bar tail sediments.

Channels in the eastern area may have been flanked by levees as suggested by the features of the overbank sediments. Facies Fl with ripple drift lamination of increasing scale at the base of some channel sequences may represent levee progradation. Erosively based trough cross-bedded sandstone interbeds, with evidence of waning current, may be interpreted as crevasse channels, and minor sandstone layers may have originated as crevasse splays.

CONCLUSIONS

The changes observed in the textural and sequential characters of the sequences within the study area suggest a predominance of very low sinuosity streams with longitudinal bars in the western area, and an eastward increase in the frequency and height of transverse bars, concurrent with increase of the channel depth and of sinuosity, and decrease in the rate of sediment discharge. However, it should be remembered that the changes observed occur in a direction that is essentially perpendicular to the palaeocurrents, so that they cannot be considered downstream changes. The depositional system was probably characterized by a series of laterally interfingering fan deltas (Massari, Iaccarino & Medizza, 1976). The western area of the belt examined was dominated by the influence of the palaeo-Piave fan-delta (Fig. 1), whereas in the eastern area the much lower conglomerate/mudstone ratio may be related to deposition in an interlobe area. Here rivers may have been deeper, being confined between muddy cohesive and probably vegetated banks; greater flow depth may have favoured the development of foreset slopes and slip-face accretion of thick bars (Hein & Walker, 1977; Rust, 1978). The model presented here seems consistent with that proposed by Hein & Walker (1977), implying that high rate of discharge of coarse gravel will mainly result in a crude horizontal stratification, whereas lower rates of discharge of finer gravels and availability of greater depths result in greater abundance of cross-bedding.

The bedforms of facies Gp may be compared with the transverse bars of the Nueces River (Gustavson, 1978). This is a low-sinuosity meandering stream with a sinuosity index of 1·3 and average stream surface slope of 1·8 m km^{-1}. It transports gravel primarily as bed load and is confined between banks stabilized by thick vegetation. In addition it is characterized by extreme ranges in annual discharge. The transverse bars have straight to broadly curved to sinuous crestlines, nearly transverse to flow direction, and contain planar cross-bedded and horizontally bedded gravel. Transverse bars in the upper Congaree River (Levey, 1978), a meandering stream with a sinuosity index of 1·75, are often located along the margin of point bars closest to the channel thalweg. In the Nueces River they characterize the channel bottom and lower lateral accretion face of gravel meander lobes. As in modern examples, it may be presumed that it is during relatively infrequent, high-magnitude discharges that most significant transport occurs.

Jones (1979) suggested that cross-set directional variance is related to the rate of bedform modification during the falling stage, this in turn depending on the discharge regime of the river. In fact, when there is a lengthy falling-stage period or pronounced low-stage flow, there may be time for substantial modification of high-stage features, this leading to high in-channel variance and commonly to bimodal orientations of the tabular cross-bedding.

The palaeocurrent variance of facies Gp in the study area is believed to be a case of within-channel variance in low-sinuosity streams. Data are too scarce to draw any firm conclusions from the observations, but the absence of bimodality seems to point to a limited modification of high-stage features during falling and low stages. Movement of transverse bars probably took place essentially during flood stages and was predominantly downstream, even if lobate bedforms were involved. In addition to slipface accretion, bar top aggradation was probably also involved. A short-lived period of forward movement and a limited subsequent modification of high-stage features would lead to unimodal cross-set orientations, as in the case described by Jones (1979). This suggests a highly variable discharge regime with a pronounced flood peak and a relatively short, rapidly falling stage without substantial modification of high-stage features.

As stated by Rust (1975), many transverse bars occur in channels which are deep and confined between relatively narrow banks. This may also be the case of the Messinian rivers of the eastern area which may have been unusually deep owing to their confinement between cohesive muddy banks resistant to erosion and probably levéed.

ACKNOWLEDGMENTS

Thanks are due to Mr F. Todesco for the preparation of figures. This work was supported by the Consiglio Nazionale delle Ricerche of Italy, grant no. CT 80 . 02603 . 05.

REFERENCES

ALLEN, J.R.L. (1964) Primary current lineation in the Lower Old Red Sandstone (Devonian), Anglo-Welsh basin. *Sedimentology*, 3, 89–108.

ALLEN, J.R.L. (1965). A review of the origin and characteristics of Recent alluvial sediments. *Sedimentology*, 5, 89–191.

ALLEN, J.R.L. (1968). *Current Ripples*. North-Holland, Amsterdam.

ASQUITH, G.S. & CRAMER, S.I. (1975). Transverse braid bars in the Upper Triassic Trujillo Sandstone in the Texas Panhandle. *J. Geol.* 82, 657–661.

BLUCK, B.J. (1971). Sedimentation in the meandering River Endrick. *Scott. J. Geol.* 7, 93–138.

BLUCK, B.J. (1974). Structure and directional properties of some valley sandur deposits in southern Iceland. *Sedimentology*, 21, 533–554.

BLUCK, B.J. (1976) Sedimentation in some Scottish rivers of low sinuosity. *Trans. Roy. Soc. Edin.* 69, 425–456.

COLLINSON, J.D. (1968). Deltaic sedimentation units in the Upper Carboniferous of northern England. *Sedimentology*, 10, 233–254.

COLLINSON, J.D. (1970). Bedforms of the Tana River, Norway. *Geogr. Annl* 52A, 31–56.

DZULYNSKI, S. & WALTON, E.K. (1965) Sedimentary features of flysch and greywackes. *Dev. Sedim.* 7, Elsevier, Amsterdam, 274 pp.

FRIEND, P.F. (1965). Fluviatile sedimentary structures in the Wood Bay Series (Devonian) of Spitzbergen. *Sedimentology*, 5, 39–68.

GUSTAVSON, T.C. (1978) Bed forms and stratification types of modern gravel meander lobes, Nueces River, Texas. *Sedimentology*, 25, 401–426.

HEIN, F.J. & WALKER, R.G. (1977) Bar evolution and development of stratification in the gravelly, braided, Kicking Horse River, British Columbia. *Can. J. Earth Sci.* 14, 562–570.

JONES, C.M. (1979) Tabular cross-bedding in Upper Carboniferous fluvial channel sediments in the Southern Pennines, England. *Sedim. Geol.* 24, 85–104.

LEVEY, R.A. (1978). Bed-form distribution and internal stratification of coarse-grained point bars, upper Congaree River, S.C. In: *Fluvial Sedimentology* (Ed. by A.D. Miall). *Mem. Can. Soc. Petrol. Geol., Calgary*, 5, 105–127.

MASSARI, F. (1978) High-constructive coarse-textured delta systems, Tortonian, Southern Alps. Evidence of lateral deposits in delta slope channels. *Memorie Soc. geol. ital.* 18, 93–124.

MASSARI, F., IACCARINO, S. & MEDIZZA, F. (1976) Depositional cycles in the Tortonian-Messinian of the Southern Alps (Italy): transition from fan-delta to alluvial fan sedimentation. *Messinian Seminar 2*, Field trip guide book.

MASSARI, F., ROSSO, A. & RADICCHIO, E. (1974) Paleocorrenti e composizione dei conglomerati Tortoniano-Messiniani compresi fra Bassano e Vittorio Veneto. *Mem. Ist. Geol. Min. Univ. Padova*, 31, 20 pp.

MIALL, A.D. (1977) A review of the braided-river depositional environment. *Earth Sci. Rev.* 13, 1–62.

MIALL, A.D. (1978) Lithofacies and vertical profile models in braided river deposits: a summary. In: *Fluvial Sedimentology* (Ed. by A.D. Miall). *Mem. Can. Soc. Petrol. Geol., Calgary*, 5, 597–604.

RUST, B.R. (1975) Fabric and structure in glaciofluvial gravels. In: *Glaciofluvial and Glaciolacustrine Sedimentation* (Ed. by A.V. Jopling and B.C. McDonald). *Spec. Publs Soc. econ. Paleont. Miner.*, Tulsa, 23, 238–248.

RUST, B.R. (1978) Depositional models for braided alluvium. In: *Fluvial Sedimentology* (Ed. by A.D. Miall). *Mem. Can. Soc. Petrol. Geol., Calgary*, 5, 605–625.

Spec. Publs int. Ass. Sediment. (1983) **6**, 301–312

Gravel bars in low-sinuosity streams (Permian and Triassic, central Spain)

AMPARO RAMOS* *and* ALFONSO SOPEÑA†

**Departamento de Estratigrafía, Facultad de Ciencias Geológicas, Universidad Complutense, Madrid 3 and
†Instituto de Geología Económica C.S.I.C., Madrid 3, Spain*

ABSTRACT

The Buntsandstein sequence of the central area of Spain ranges from a variety of fluvial facies, low in the succession, to tidal mudstones and carbonates in the youngest and most eastern areas. This paper is restricted to the conglomerates in the lowest part of the fluvial sequence. These were deposited in bar and channel systems dominated by gravelly, low-sinuosity braided streams. Their distribution reflects both the pre-existing topography and synsedimentary tectonic movements.

Six main facies have been distinguished (in order of decreasing abundance): sheets of massive conglomerates, channel-fill conglomerates, coarse–medium sandstones, lateral accretion conglomerates. tabular cross-stratified conglomerates, fine-grained sediments.

Sheets of gravels, formed as longitudinal bars, were by far the major feature of sediment accumulation. Units of lateral accretion, if not very frequent, are quite important as they probably represent modifications of bars during waning stage.

Two big cycles may be seen in these conglomerates. They correspond to different physiographic models. The lower cycle was mainly formed by smaller channels and bars than was the upper cycle. The lower cycle is mainly formed by channel deposits, whilst in contrast, the upper cycle is mainly made up of longitudinal bars.

The change in sedimentation may be due to tectonic movements that occurred in the basin during Buntsandstein deposition.

INTRODUCTION

The sedimentation of gravelly streams is difficult to study, not only in the ancient record, but also at the present day. The direct observation of clast movement and bedform evolution is particularly difficult because the movement and evolution take place during flood stage with strong currents and normally turbid waters. Moreover, the necessary flow conditions cannot be achieved experimentally in laboratory flumes and it is quite difficult to construct facies models to compare with ancient sediment sequences.

In contrast, grain size, the shapes of bedforms, their relationships and their organization, can sometimes be studied quite well in ancient gravelly deposits These sediments have very complex geometries as has been pointed out by Bluck (1976, 1979), using both ancient

and present-day examples. When working in ancient sediments it is difficult to isolate single depositional events. Very different geometrical bodies may be closely related or they may even belong to a single event; in contrast, similar geometries may have different causes.

A two-stage analysis is presented here as part of a study of 'Buntsandstein' conglomerates in the Iberian Ranges (Central Spain). First, the main facies are described, according to their geometry and internal structure in cross-section. Secondly, a general model is constructed and its significance assessed.

GEOLOGICAL FRAMEWORK

The 'Buntsandstein' sediments in the study area (Fig. 1) are mainly fluviatile conglomerates and

0141-3600/83/0106-0301 $02.00

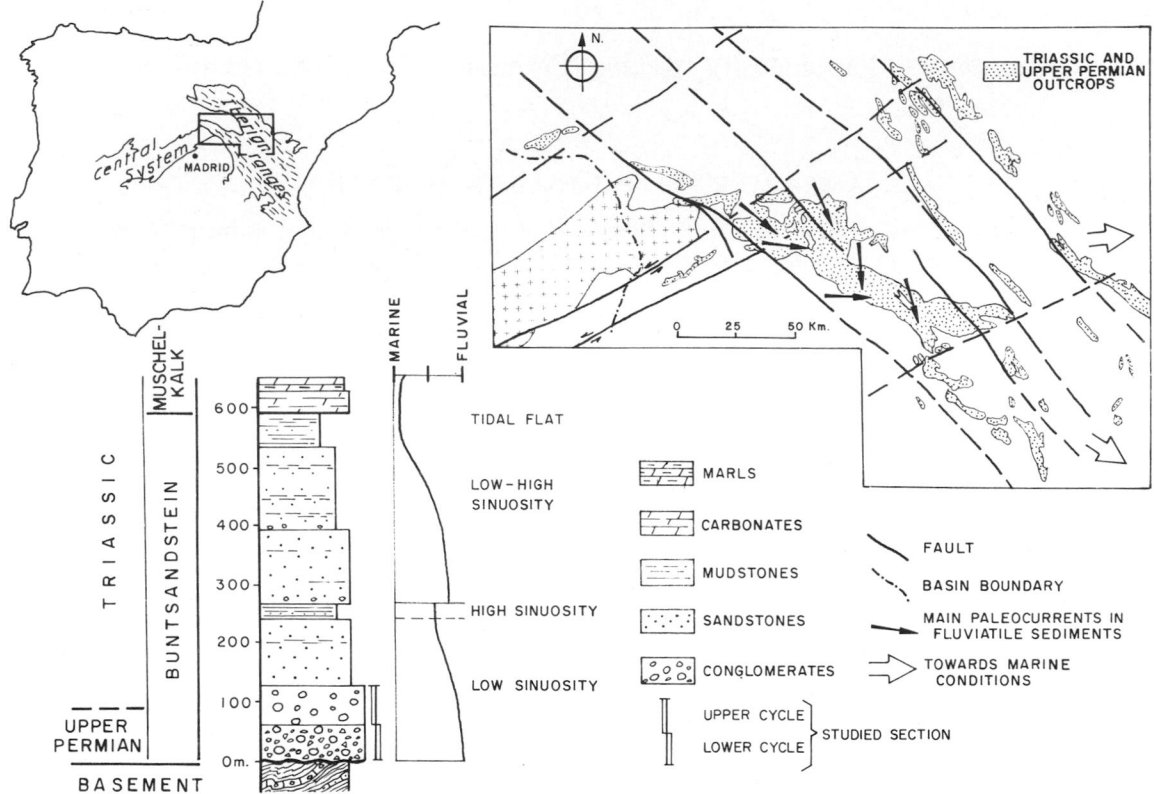

Fig. 1. Location maps and generalized stratigraphic sequence.

sandstones. They form the base of a complete and continuous sequence, from fluviatile sediments below to clastic tidal sediments and carbonate tidal sediments ('Muschelkalk') above. These 'Muschelkalk' facies were the westernmost carbonate sediments of the Tethys sea.

The thickness of the 'Buntsandstein' varies greatly in central Spain and may reach up to 800 m near the Central System. It can be divided into several lithostratigraphic units, whose ages range from Upper Permian (Thuringian) to Middle Triassic (Ladinian) (Ramos, 1979; Sopeña, 1979). The 'Buntsandstein' is mainly composed of sandstone (Fig. 1) but, low in the sequence, there are some conglomerates (this study), mudstones, and a few caliche deposits.

These sediments accumulated in a tectonically active basin. Important strike–slip faults with a NW–SE trend (Fig. 1) developed towards the end of Hercynian movements. Later vertical movements produced a topography of uplifted blocks and basins which filled with 'Buntsandstein' fluviatile sediments.

The fact that some of these movements occurred during the sedimentation is proved by the observation that some faults cut lower 'Buntsandstein' sediments and were covered by upper 'Buntsandstein' sediments.

The lower conglomerates always rest unconformably on a basement of Hercynian metamorphic or Lower Permian sedimentary rocks (Virgili et al., 1980). Their thickness varies from 70 to 180 m. The palaeorelief at the beginning of 'Buntsandstein' probably produced this variation.

The main fluviatile palaeocurrents across the basin were from north to south, or from NW to SE, but there are some local deviations, probably due to small alluvial fans extending from local sources.

FACIES DESCRIPTION

This study is restricted to lower Buntsandstein conglomerates (Fig. 1). More than 40 sections have

FACIES	BEDDING AND SEDIMENTARY STRUCTURES		TEXTURE AND FABRIC	THICK-NESS
SHEETS OF MASSIVE CONGLOMERATES	Massive Imbricated clasts	Ⓐ a	Clast sizes : 5–30 centimetres Rounded – Subrounded clasts Low Sandy matrix proportion	0.5 – 1.5 metres
	Crude flat bedding Imbricated clasts	b		
	Convex upward tops Imbricated clasts	c		
UNITS OF TABULAR CROSS-STRA-TIFIED CON-GLOME-RATES	Tabular cross – stratified	Ⓑ		0.8–1.0 metres
UNITS OF LATERAL ACCRETION CONGLOMERATES	Lateral accretion units with sandstone drapes Imbricated clasts	Ⓒ a	Clast sizes: 3–20 cm. Moderately sorted Sandy matrix	0.6 – 1.8 metres
	Lateral and vertical accretionary surfaces	b		
CHANNEL – FILL CONGLOMERATES	Massive	Ⓓ a	Clast sizes : 3 – 20 centimetres Rounded – subrounded clasts Moderately sorted High sandy matrix proportion	1.0 – 1.8 metres
	Complex – fill Stratified	b		
	Transverse fill Cross stratification	c		
	Multi – storey fill Trough cross – stratification	d		
UNITS OF COARSE – MEDIUM SANDSTONE	Flat or low angle cross – bedded Rare trough cross-bedded	Ⓔ	Coarse – medium grain size	0.5 metres

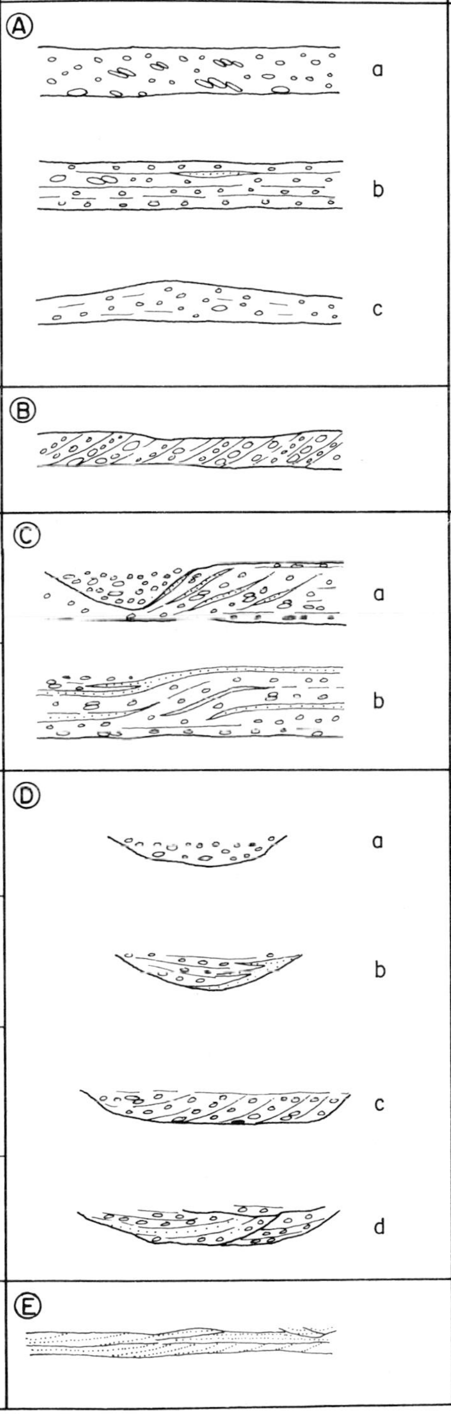

Fig. 2. Characteristics of depositional facies types.

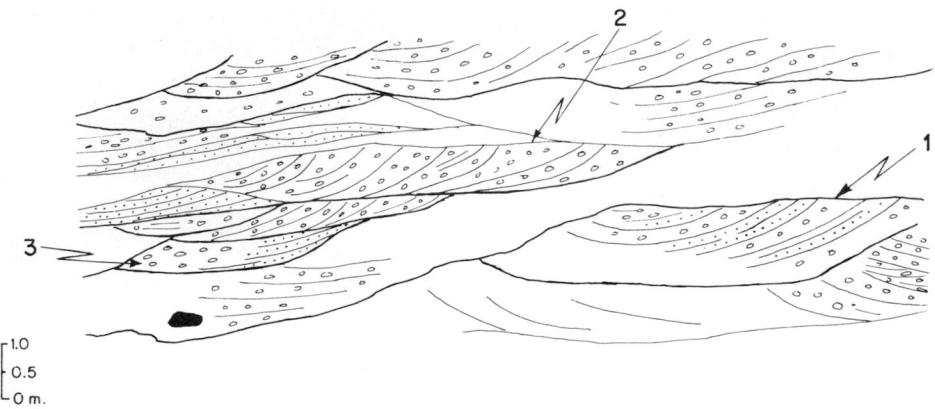

Fig. 7. Field sketch from Fig. 6. (1) Channel fill consisting of trough cross-stratified gravelly sandstones. (2) Channel with a transverse fill consisting of cross-stratified gravels. (3) Channel fill consisting of gravels passing laterally into sandstones.

continuous laterally. The former may be associated with all the other facies. They may be on top of sheets of massive conglomerates, as channel fills or between lateral accretion surfaces. The second subfacies may cover extensive areas and may occur above or below channel-fill conglomerates, sheets of massive conglomerates, or even lateral accretion conglomerates or tabular cross-stratified conglomerates (Figs 3 and 5).

Tabular cross-stratified conglomerates, which are quite rare here, may occur in association with channels.

Finally, the few fine-grained sediments are normally big blocks, always associated with channels or with big scoured surfaces (Fig. 5B).

INTERPRETATION

In our area, sheets of gravels (Fig. 2A) were the major features of sediment accumulation. We think that they formed as longitudinal bars (Fig. 3), similar to those described by many authors in modern gravelly braided streams and their inferred ancient equivalents (Ore, 1964; Williams & Rust, 1969; Smith, 1974; Rust, 1978). Convex-upward surfaces characteristic of these bars have been clearly preserved in many bars in our area.

Sheets of gravel do not normally represent a single depositional event and this is proved by multiple scoured surfaces and thin sandstone beds within them (Fig. 5A).

According to many studies (Leopold & Wolman, 1957; Smith, 1974; Hein & Walker, 1977), longitudinal bars may have been initiated as a diffuse sheet during, or immediately after, flood stage. During flood stage

the stream was strong enough to move even the coarsest clasts. These were deposited when the flow reduced and the morphology of the bed surface (the presence of riffles and pools that were not big enough to influence the clast's movement during flood stage) tended to influence them. The clasts acted as a trap for other clasts, and a bar developed with crude horizontal stratification and imbricate clasts. These characteristics, together with the absence of cross-stratification, suggest that most of the gravel tended to move during high flow stage in sheets, and not by avalanching on the lee sides or lateral flanks of bars (Eynon & Walker, 1974).

During reduced flow, bar movement stopped and its upper surface may even have emerged. Flow was then only active along lateral channels and sometimes also along small cross-over channels on top of the bars. These channels may have been filled with sand, which helps to distinguish individual bars (Fig. 3).

The fact that clasts inside the bars are in contact with each other, that there is little or no matrix (openwork gravels) and that there is often an alternation of different clast sizes, may be the result of high- and low-discharge events (Smith, 1974). During high flows, sands were carried in suspension above gravels. As flow decreased, finer gravels were deposited above the larger ones, and finally sand was deposited, filling pores in the gravels but not reaching the lowest gravel beds. This process produced a lower openwork gravel and an upper, sand-filled gravel.

Clast pavements (Fig. 3B), one clast thick, may be found between two longitudinal bars. They may be the result of flow increases, not strong enough to move the clasts but capable of sweeping the sand matrix from around the clasts, producing a lag between two successive bars, but may also occur after a single flood. Flood gravels tend to have a high matrix content and the falling-stage flow washes out the matrix to give lag sheets.

Upstream imbrication is common inside our longitudinal bars, giving a reliable indicator of palaeocurrents, particularly in the case of larger discoidal clasts (Fig. 4). These larger clasts show a minimum deviation from the trend of the main channel because they are oriented during major floods and resist movement during lower-stage flooding, when normally deviations occur from the general trend (Bluck, 1979). Discoidal clasts have here a preferential tranverse imbrication (in the sense of Harms *et al.*, 1975, p. 137), but cylindrical clasts have a preferential parallel imbrication. An explanation has been offered by Rust (1972a), who suggested that high concentrations of clasts (as in our Buntsandstein), and high velocities favour long-axis orientation parallel to the current.

The accretionary deposits (Fig. 2C) that we find laterally and on top of longitudinal bars (Figs 3 and 5A) may be the modification of these bars during waning flow. Because of the decrease in flow, or perhaps when bars had reached a certain height in relation to the depth of water, few or no clasts could be moved over the bar. Nevertheless, flow was strong enough to move clasts laterally along the bar margins. Similar processes have been described by Costello & Walker (1972) and Smith (1974). According to Bluck (1976) this process is mostly confined to the bar heads (upstream areas, coarse grained, normally with imbricate clasts).

When lateral accretionary processes continued, lateral foresets developed, dipping towards the lateral channels at an angle less than 90° to the general trend as shown by imbrication inside the foresets (Fig. 3). This deviation between structures indicating the flow direction and the bar accretion direction has been recorded by Bluck (1979) in the Tulla River (Scotland).

Vertical accretion occurred on the tops of bars, simultaneously with lateral accretion (Fig. 3). Vertical accretion was always minor in comparison with lateral accretion and was probably controlled by the depth of water. Thus foresets are thinner on top of bars and become thicker laterally and downwards with a maximum towards the base. Sandstone drapes (Figs 3 and 5A) between foresets and alternations of different grainsize foresets may represent changes in flow stage.

Channel deposits (Fig. 2D) may have formed laterally to longitudinal bars. Channel deposits may pass into bars by transition into lateral accretionary deposits (Fig. 3). Sometimes a multi-storey pattern of channels developed (Fig. 2D), with channels cutting into each other both vertically and laterally. This is very characteristic of braided streams (Williams & Rust, 1969; Costello & Walker, 1972). Gravel lag has been recorded in many channels, and may be the result of winnowing of earlier deposits when cut into by a lateral channel.

Mud or fine sand beds that we have found inside multi-storey channels may be the result of the settling of fine sediments in periodically abandoned channels. Channels were sometimes filled with massive gravel (Fig. 2D), making it difficult to recognize whether they have filled during one or more events, but they were more often filled with sets of cross-bedded gravels and coarse sandstones, sub-parallel to the channel margins. These planar cross-bedded gravels and sandstones may be the result of lateral accretion of longitudinal bars, downfilling side channels during the waning stage. This has been recorded in both ancient and recent braided streams by McGowen & Groat (1971) and Fraser & Fishbaugh (1980). Large channels filled with large trough cross-sets of gravel (Figs 2D, 6 and 7) may be the result of bankfull discharge, when the entire valley has acted as a channel (Fraser & Fishbaugh, 1980). Some channels were filled with gravels in the lower part (Fig. 7), the grain size then decreasing laterally, passing into gravelly sandstones (Fig. 7). This may be due to the different stream power in the channel as the depth of water varied. The currents may be powerful enough to move gravels along the channel bed, but only competent to move finer sediments in shallower water.

The planar tabular cross-stratified conglomerates (Figs 2B and 5B) that are rare in our sequences may be due to migration of transverse bars. Similar transverse bars have been described in present-day braided streams (Ore, 1964; Rust, 1972b). According to Hein & Walker (1977) these bars, with well-defined foresets, reflect a lower water and sediment discharge than the longitudinal bars.

Flat- or low-angle-bedded sandstones (Fig. 2F) that can be followed laterally across tens of metres and over different deposits (they may be found simultaneously over several channels or longitudinal or transverse bars) may have been produced during high flood stage, when water was not confined to incised channels. Flood water may then have spread from the main channels across bar surfaces and abandoned channel deposits, flowing with high velocity across all sediments and forming flat- or low-angle-bedded sandstones. Similar processes have been described for

upper and middle fan areas formed by braided streams
(Boothroyd & Ashley, 1975).

Local trough cross-bedded sandstones (Fig. 5A, B)
on top of low-angle or flat-bedded sandstones may be
due to a decrease in energy, causing sands to move as
megaripples and forming trough cross-bedded
sandstones.

Fine-grained sediments (sometimes with biotur-

bation) are very rare in our sediments. As Costello &
Walker (1972) pointed out, in a system where coarse
sediments are predominant, deposition of fine material
will only occur in small, specially sheltered areas on
the floodplain. There were probably not many of these
small areas in our system. Nevertheless, some large
mud or muddy sandstone clasts are associated with
large scoured surfaces which points to the former

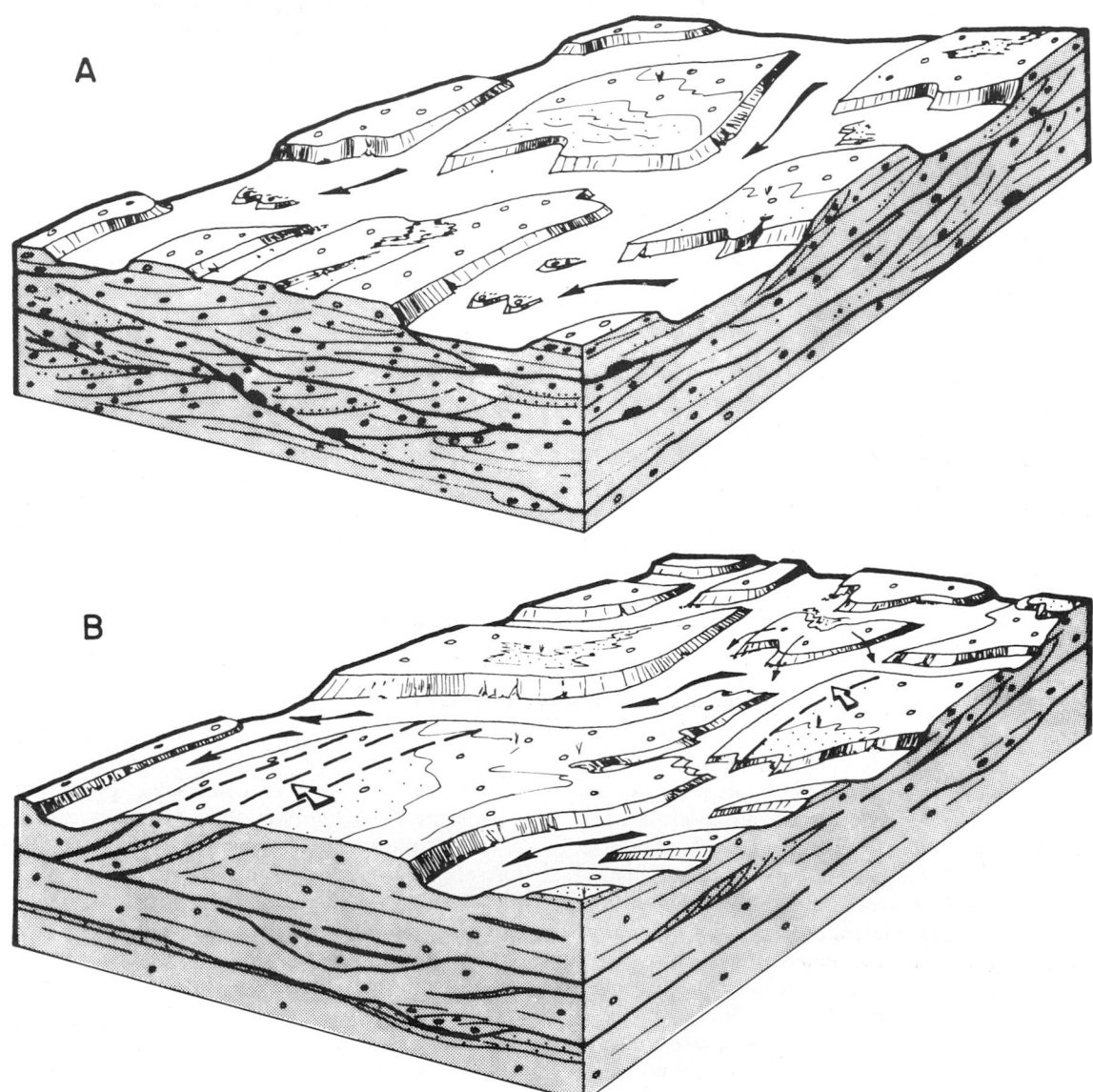

Fig. 8. (A) Lower cycle block diagram. Smaller bars. Higher preservation rate of channel fill. Preservation of some convex-upward
tops of bars. Periodic large scoured surfaces. Very rare overbank deposits (see also Fig. 6). (B) Upper cycle block diagram.
Larger longitudinal bars with lateral growth. Higher channel sinuosity. Preservation of convex-upward top of bars. Periodic
large scoured surfaces. Very rare overbank deposits (see also Fig. 3A).

presence of some of these fine-grained units with a very low preservation rate.

CONCLUSIONS

The main characteristics of conglomerates that form the basal Buntsandstein in the Iberian Ranges (Fig. 1) indicate relatively high-energy streams, with prevalent bedload transport. They are thick gravel deposits and have very few sandstone beds and rare mudstone beds.

According to Bluck (1979), the gravel-to-sand ratio is the most important factor controlling the structure of sediments like these. The ratio has probably influenced the different bedforms in our sediments, producing a complex pattern of bars and channels with very few, or no, overbank deposits.

The high bedload to suspension-load ratio may have favoured the rounding of gravels that can be seen in these sediments. This fact has been pointed out by McGowen & Groat (1971) for the Van Horn Sandstone. However, many clasts must be second cycle, as has been previously shown (Ramos, 1979; Sopeña, 1979), and this also results in a high rounding. These gravels were mainly arranged internally as relatively simple units. They were formed in bars and channels of low-sinuosity braided stream systems.

Two big cycles may be seen in the complete sedimentary sequence of these lower Buntsandstein conglomerates. They differ in the frequency, distribution and preservation of different depositional units, corresponding to different physiographic models (Fig. 8A, B). The change in sedimentation may be related to fluvial system pattern change, due to the tectonic movements that occurred in the basin during Buntsandstein times.

The lower cycle (Fig. 8A) was mainly formed by smaller channels and bars than the upper cycle. Fast migration and fill of channels may have been due to quick and frequent fluctuations, together with slightly higher slope. These features may also have been responsible for the very low preservation rate and for the modifications and scoured surfaces on top of the few bars that have been preserved.

The upper cycle (Fig. 8B) is mainly made up of longitudinal bars with not very complex internal organization. The lateral growth that can be seen indicates higher stability for those bars, together with higher channel sinuosity. According to Smith (1974), these features are characteristic of gravelly streams with distinct high and low discharges.

Nevertheless, the preservation of entire bed forms (e.g. convex-upward tops of bars) together with the conglomerate thickness indicate generally high subsidence related to the vertical movements that have been described above.

No major vertical pattern of grain-size change has been recorded in these conglomerates. The most important vertical features of the two cycles are the periodic large scoured surfaces, each 10–15 m deep, that can be followed laterally across hundreds of metres. They have big blocks from the underlying sediments and may be related to periodic tectonic reactivation.

ACKNOWLEDGMENT

We are very grateful to Dr P. F. Friend (University of Cambridge) for reading the manuscript, patiently correcting the English, and for useful comments to improve it.

REFERENCES

BLUCK, B.J. (1976) Sedimentation in some Scottish rivers of low sinuosity. *Trans R. Soc. Edinb.* **69**, 425–456.

BLUCK, B.J. (1979) Structure of coarse grained braided stream alluvium. *Trans. R. Soc. Edinb.* **70**, 181–221.

BOOTHROYD, J.C. & ASHLEY, G.M. (1975) Process, bar morphology and sedimentary structures on braided outwash fans, North-eastern Gulf of Alaska. In: *Glaciofluvial and Glaciolacustraine Sedimentation* (Ed. by A.V. Jopling and B.C. McDonald). *Spec. Publs Soc. econ. Paleont. Miner.*, Tulsa, **23**, 193–222.

COSTELLO, W.R. & WALKER, R.G. (1972) Pleistocene sedimentology: Credit River, Southern Ontario: a new component of the braided river model *J. sedim. Petrol.* **42**, 389–400.

EYNON, G. & WALKER, R.G. (1974) Facies relationships in Pleistocene outwash gravels, Southern Ontario: a model for bar growth in braided rivers. *Sedimentology*, **21**, 43–70.

FRASER, G.S. & FISHBAUGH, D.A. (1980) Sedimentary structures of the late Wisconsian terraces along the Wabash River. *Great Lake Section Spec. Paper. Soc. econ. Petrol. Mineral Field Trip Guidebk* pp. 59–78.

HARMS, J.C., SOUTHARD, J.B., SPEARING, D.R. & WALKER, R.G. (1975) Depositional environments as interpreted from primary sedimentary structures and stratification sequences. *Soc. econ. Paleont. Mineral. Short Course*, Dallas, **2**, 161 pp.

HEIN, F.J. & WALKER, R.G. (1977) Bar evolution and development of stratification in the gravelly, braided Kicking Horse River, British Columbia. *Can. J. Earth Sci.* **14**, 562–570.

LEOPOLD, L.B. & WOLMAN, M.G. (1957) River channel patterns: braided, meandering and straight. *Prof. pap. U.S. geol. Surv.* **282-B**, 85 pp.

McGOWEN, J.H. & GROAT, C.G. (1971) Van Horn Sandstone, West Texas: an alluvial fan model for mineral exploration. *Rep. Invest. Bur. econ. Geol. Univ. Texas*, **72**, 57 pp.

ORE, H.T. (1964) Some criteria for recognition of braided stream deposits. *Wyoming Univ. Dept. Geol. Contr. Geol.* **3**, 1–14.

RAMOS, A. (1979) Estratigrafía y paleogeografía del Pérmico y Triásico al oeste de Molina de Aragón. *Seminarios de Estratigrafía, Serie Monografías*, **6**, 313 pp.

RUST, B.R. (1972a) Structure and process in a braided river. *Sedimentology*, **18**, 221–245.

RUST, B.R. (1972b) Pebble orientation in fluviatile sediments. *J. sedim. Petrol.* **42**, 384–388.

RUST, B.R. (1978) Depositional models for braided alluvium. In: *Fluvial Sedimentology* (Ed. by A.D. Miall). *Mem. Can. Soc. Petrol. Geol.*, *Calgary*, **5**, 605–625.

SMITH, N.D. (1974) Sedimentology and bar formation in the Upper Kicking Horse River, a braided outwash stream. *J. Geol.* **82**, 205–224.

SOPEÑA, A. (1979) Estratigrafía del Pérmico y Triásico del noroeste de la provincia de Guadalajara. *Seminarios de Estratigrafía, Serie Monografías*, **5**, 329 pp.

VIRGILI, C., SOPEÑA, A., RAMOS, A., HERNANDO, S. & ARCHE, A. (1980) El Pérmico en España. *Revista Española de Micropaleontología*, **XII**, 2, 255–262.

WILLIAMS, P.F. & RUST, B.R. (1969) The sedimentology of a braided river. *J. sedim. Petrol.* **39**, 649–679.

Spec. Publs int. Ass. Sediment. (1983) **6**, 313–321

Coarse-grained meander lobe deposits in the Jarama River, Madrid, Spain

ALFREDO ARCHE

Instituto de Geología Económica, C.S.I.C., Facultad de Ciencias Geológicas, Madrid 3, Spain

ABSTRACT

The sediments of the 18–20 m terrace of the Jarama River have been studied in an area between Velilla de San Antonio and Arganda, near Madrid, and compared with the deposits of the active channel.

Two main vertical sequences are found in the channel sequences of the terraces: one consists mainly of conglomerates with few variations in pebble size and the other is composed of a lower conglomeratic member and an upper sandy member with a sharp or erosive contact. The conglomeratic units can be traced laterally for several hundred metres and have planar bases, convex tops and many internal accretionary surfaces. Lateral migration of meander lobes was the process responsible for the formation of the conglomeratic units. There is a ridge-and-swale topography on top surfaces; varying discharge is the main factor controlling this topography.

Some swales and chute channels receive water from the main channel during moderate floods, plugging them with sand. The sand units have a lenticular geometry, with concave bases and planar tops; they accumulated as transverse and chute bars which accreted longitudinally because of lateral topographic restriction. Overbank deposits overlie these channel sequences and were deposited as crevasse splays, natural levées, suspension deposits and abandoned channel fills. The high proportion of gravel in these deposits can be explained by the presence of thick, erodible conglomerates of Pliocene age at the headwaters of the Jarama River.

INTRODUCTION

In recent years much work has focused on a hitherto neglected category of fluvial sedimentation; that of coarse-grained, bedload rivers of medium to high sinuosity (Bluck, 1971, 1976; Gustavson, 1978; Levey, 1978; McGowen & Groat, 1971; Morton & McGowen, 1980; Nijman & Puigdefabregas, 1978; Ori, 1979). A section of the Jarama River, east of Madrid (Fig. 1) between Velilla de San Antonio and the confluence with the Manzanares River, provides an excellent case-study of this type of sedimentation. Here extensive gravel pits expose the Quaternary sediments of the Jarama River and the present-day processes of sedimentation are not highly modified by man.

The Jarama River has a mean discharge of 11·01 cumecs and a maximum recorded discharge of 175·5 cumecs and a mean slope of 1 m km^{-1} in the studied section (Pelaez *et al.*, 1973).

The sinuosity of the active channel is 1·31 in this section and several mid-channel islands are present (Fig. 2). According to Leopold & Wolman (1957), meandering rivers have a sinuosity of 1·5 or more and a single channel, therefore the Jarama River should be considered as braided because of its low sinuosity and islands, but this value was later considered as a median sinuosity, not a lower limit (Leopold, Wolman & Miller, 1964). The recent deposits consist mainly of meander lobe deposits (Gustavson, 1978), formed by lateral accretion due to the lateral shifting of the active channel (Fig. 2). For these reasons the studied section will be considered to be a medium-sinuosity, bedload-dominated river.

The first sedimentological studies were made by Carrillo (1977) and Ricci Lucchi (1979). Further details are given in Carrillo & Arche (1982).

0141-3600/83/0106-0313 $02.00

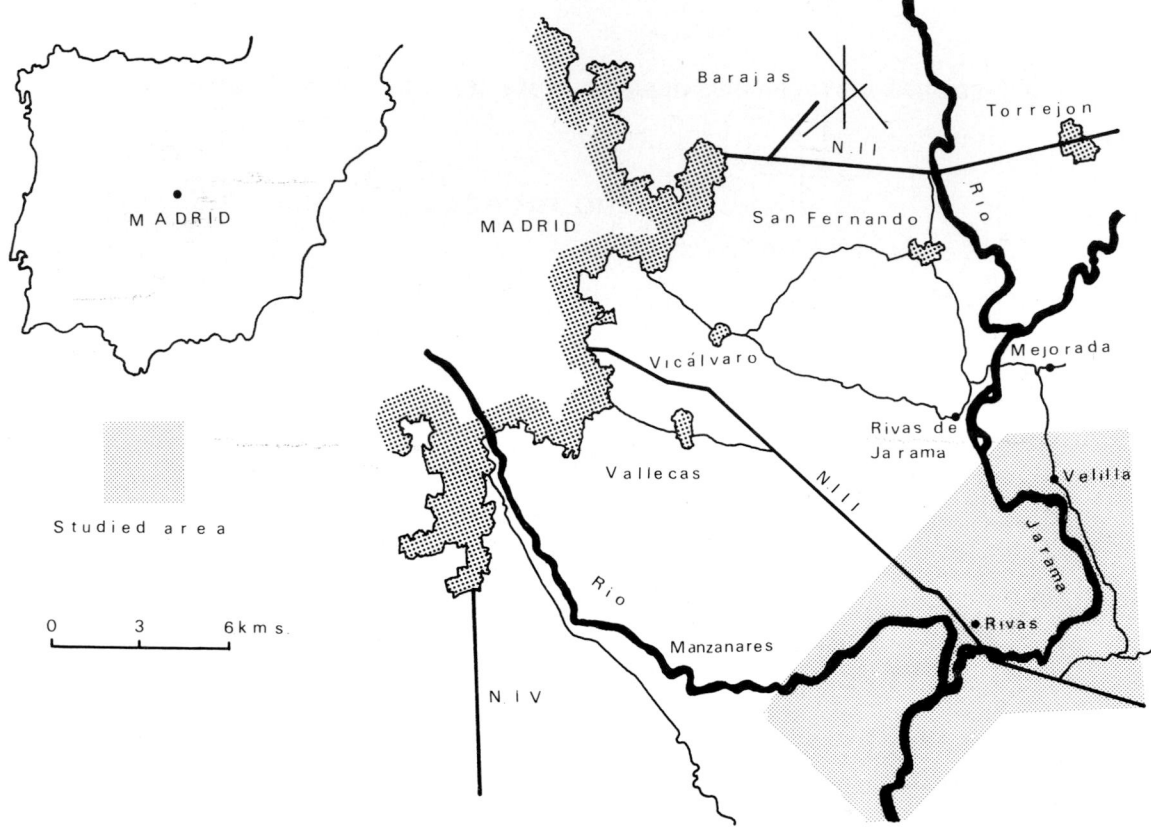

Fig. 1. Geographical setting of the studied area.

THE SEDIMENTS

The sediments of the 18–20 m terrace were observed in the gravel pits and recent deposits in the meanders between Velilla de San Antonio and La Poveda and near the junction with the Manzanares River (Fig. 2).

The sediments of the terraces are composed of sheet-like gravel and sand units forming a multi-storey body (Friend, Slater & Williams, 1979). There are three main sediment types according to grain size which, in decreasing abundance, are gravel, sand and silt and clay.

Gravels

Gravels are composed of subangular to subrounded quartzite pebbles, up to 35 cm in diameter, with a mean size of 8 cm. They form a clast-supported conglomerate with a sandy matrix infilling the voids and locally show imbrication.

Gravels form laterally continuous beds (at least up to 200 m) up to 2·5 m thick with planar, erosive bases and irregular to undulating tops. The dominant internal structure is megacross-stratification (epsilon cross-stratification, Allen, 1963), dipping 10°–20°; some horizontal stratification and massive beds are also found. These surfaces descend to the base of the unit or are restricted to the upper and middle part (Figs 3 and 4, unit B); they sometimes terminate against a channel wall (Figs 3 and 4, unit A, E border) or pass into concave-up channel fills (Figs 3 and 4, unit B). Lenticular, cross-stratified sands commonly occur along the megacross-stratification surfaces, and silt intraclasts, up to 1·2 m long occur along the base.

Gravel imbrication is present, although it is not a very common feature because of the low percentage of discoidal or platey pebbles. The dip of the plane containing the *a-* and *b*-axes of the imbricated pebbles has been measured at six points close to mega-cross-stratification surfaces and between six and 27 measurements were taken at each point. The mean dip

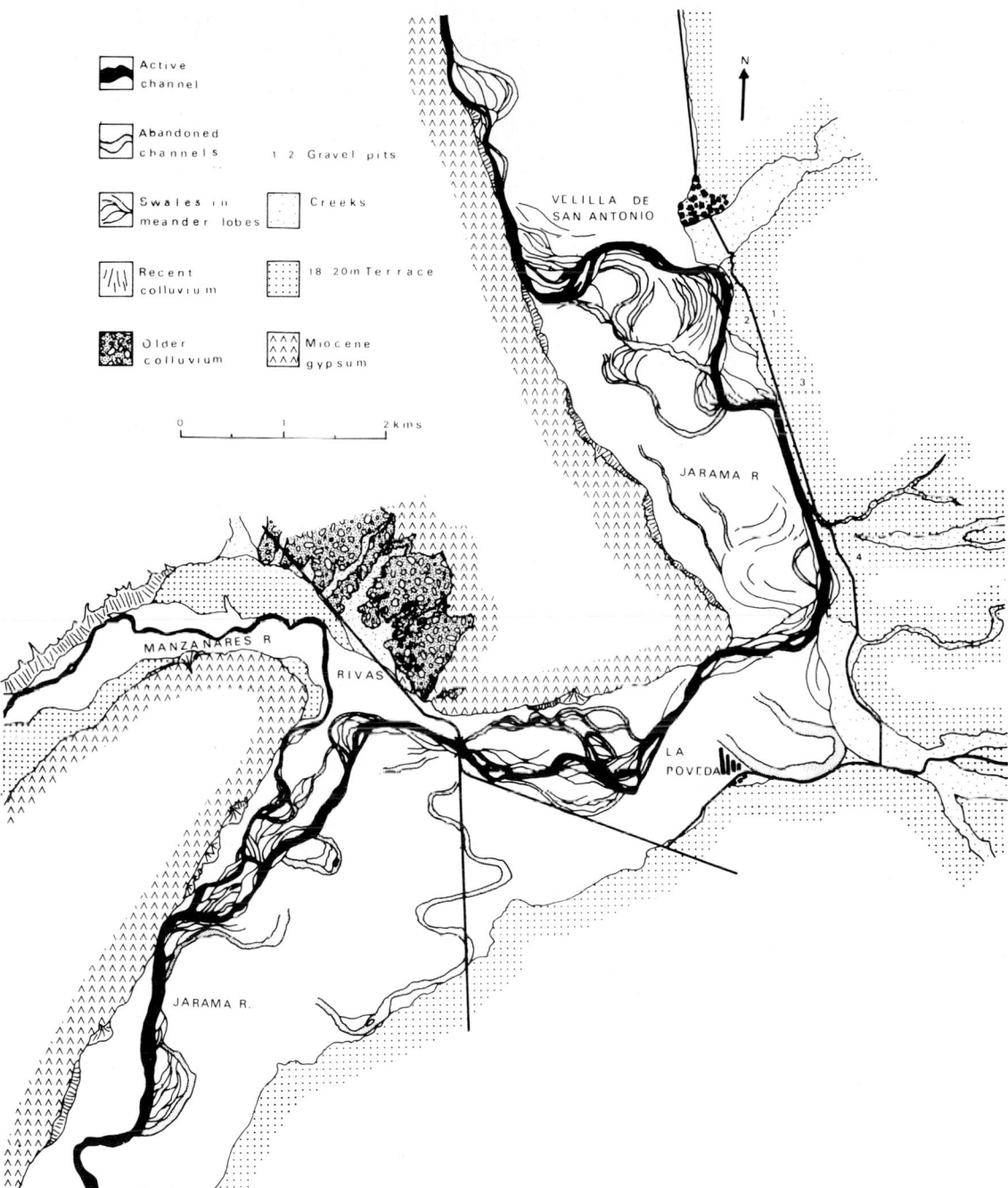

Fig. 2. Geological setting of the studied area and location of the studied gravel pits. The most prominent features of the present-day alluvial plain are also shown.

Fig. 3. Typical aspect of the terrace sediments in gravel pit 4. Six different sequences are exposed. Note megacross-stratification in gravel beds, especially in upper left part. The machine is 4·25 m high.

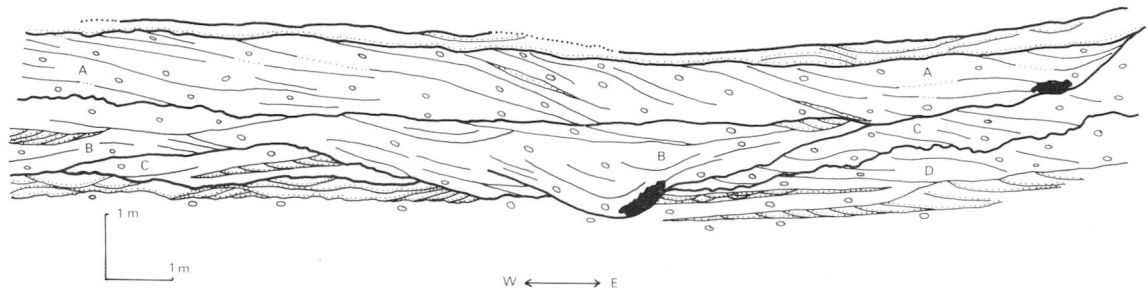

Fig. 4. Detail of Fig. 3 showing the internal structure of the upper part of the exposure.

of the imbrication shows a 60°–90° transverse orientation with the dip direction of the mega-cross-stratification surfaces.

Grain size shows many horizontal and vertical changes, but there is no definite trend.

Sands

Distinct bodies of medium to coarse sand occur as wide (up to several tens of metres) and thick (up to 1·3 m) units with concave, sharp or erosive bases and flat to slightly undulating tops, in addition to the gravel-embedded lenses. Locally they coalesce, but are usually isolated. The main internal structure is trough cross-stratification and sometimes thin silt layers are found along the sets. Several cut-and-fill structures and fluid escape structures are also found. The base of the body can be covered by a thin layer of pebbles and intraclasts.

Palaeocurrents inside sand bodies show little dispersion. Measurements at nine sample points show divergences of between 75° and 90° with the dip direction of the megacross-stratification surfaces. An upwards fining in grain size is common, and climbing ripples and parallel lamination are found occasionally at the top.

Some sand bodies display a very different palaeo-current pattern. Their dimensions are comparable, but a radial pattern and almost opposite directions in contiguous sets are found (Figs 5 and 6).

Silts and clays

Although their volume is very small, their associations are very variable. Three associations are the most frequently found. (1) Thin, repetitive sequences (up to 3 cm) of fine sand, silt and clay with climbing ripples, parallel lamination and massive beds, laterally continuous. These, exceptionally, can be associated with lenticular sandy or gravelly units with erosive bases. (2) Thick (up to 1 m) massive beds of silts and clays containing rootlet levels and plant remains. Some isolated fine sand ripples are occasionally found. (3) A lenticular body filling an abandoned channel, 4 m thick and about 35 m long has been found in gravel pit no. 4, formed by massive silts and clays and some thin, fining-upwards levels of fine sand (Fig. 7).

The recent sediments were observed in several

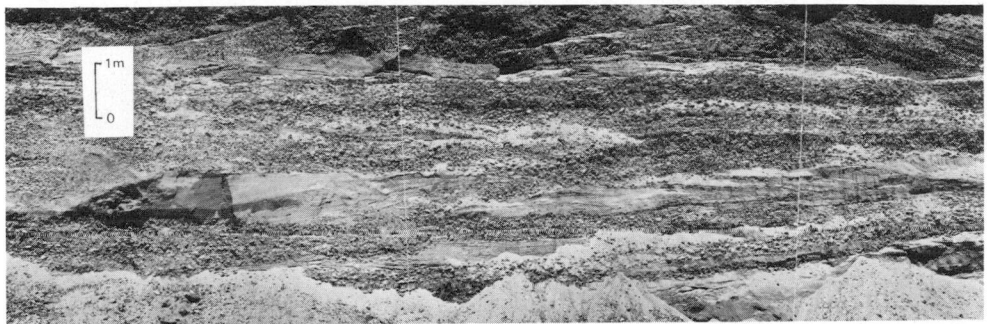

Fig. 5. One of the recognized microdeltas or chute bars in gravel pit 4. Note conglomeratic and silty units on the left.

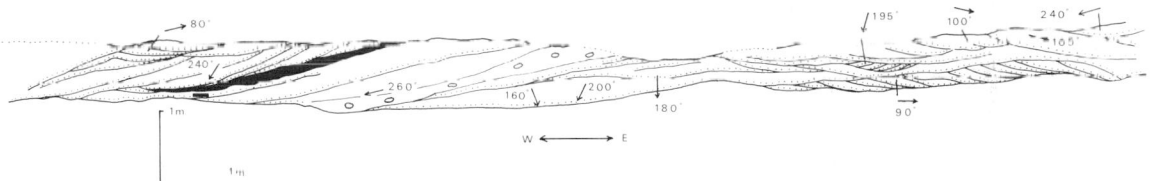

Fig. 6. Palaeocurrents measured in the microdelta or chute bar of Fig. 5.

Fig. 7. Gravel point bar and clay plug infilling an abandoned channel. Gravel pit 4. The channel is about 35 m long.

active point bars near Velilla de San Antonio, La Poveda and near the Jarama and Manzanares River junction. A ridge-and-swale topography (Fig. 2) is developed in some parts, but in others human action has destroyed it beyond recognition by levelling the ground for cultivation. The growth of meander lobes is prominent and several old tight loops are visible in the alluvial plain. Several prominent chute cutoffs have been recognized, separating the meander lobe from the alluvial plain, acting sometimes as a secondary active channel. They cut across the meander lobe and its ridge-and-swale topography.

The active meander lobes are free of vegetation and form point bars. Bar heads are made up of pebble sheets showing good imbrication and, closer to the water, several transverse gravel ridges, up to 45 cm

high with a sandy matrix are found. The gravel clasts have their *a*-axes transverse to the active channel and imbrication planes dip up-current.

Sands are dominant at the bar tails. They appear as: (1) Thin sheets of medium to fine sand covering active point-bar surfaces. Current ripples and flat beds are common surface structures, with current shadows around the scattered pebbles present and some thin, discontinuous mud patches. (2) Medium to coarse sand infilling swales and chute channels, forming transverse bars and superimposed current ripples. Silt and plant remains accumulate at the depressed areas between ripple crests. (3) Chute bars at the bar tails, with an arcuate avalanche face, up to 0·6 m high. Some gravel sets are embedded in the sands. They develop at the end of chute channels cutting across the

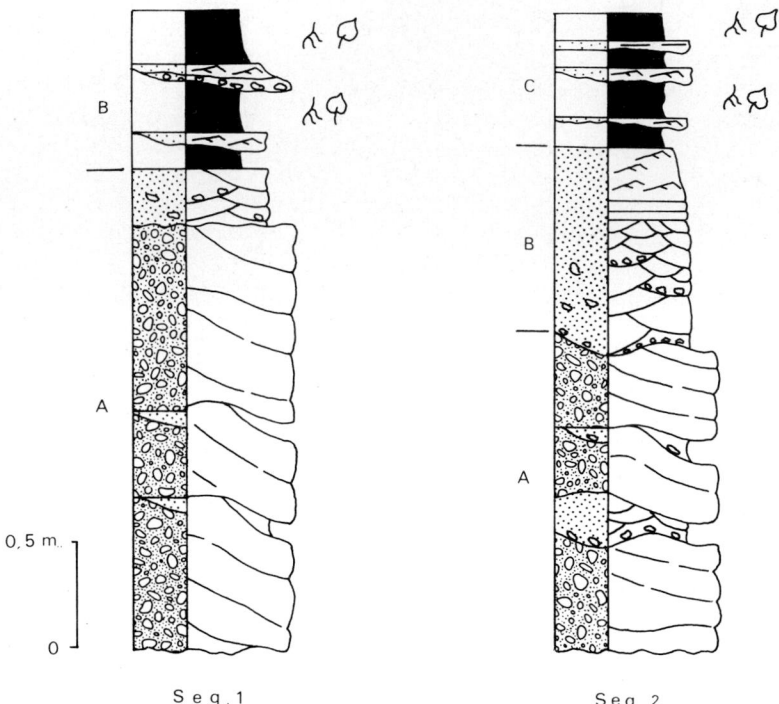

Seq. 1   Seq. 2

Fig. 8. Type sequences for the Jarama River deposits. Sequence 1. Member A: conglomerates with megacross-stratification; the sandy unit can be absent. The bar top is sharp but not erosional. Member B: silts and clays containing plant remains. Sequence 2. Member A: similar to sequence 1; bar top usually erosive. Member B: cross-stratified sands with current ripples on top. Member C: similar to Member B of sequence 1.

meander lobe topography and are transverse to the active channel. Chute channels are flanked by vegetation.

The rest of the alluvial plain is covered by mud, but active cultivation has disturbed the original structures beyond recognition.

SEDIMENTARY SEQUENCES

The sediments in the terrace deposits are organized in two different vertical sequences (Fig. 8).

Sequence 1 has two members: a basal one (Member A) of cross-stratified and flat-bedded gravels at the base; pebbles form a continuous framework, sometimes showing imbrication, and a sand matrix fills the voids. In general gravels are finer at the top, but many complex grain-size variations are found and no channel lag can be observed. Mega cross-stratification is often present.

The top of this member can be flat or slightly undulating but never erosional, and is overlain sometimes by a thin sand layer. The upper member (Member B) is made up of silts and clays.

Sequence 2 has three members: gravels, very similar to sequence 1 (Member A), sands forming large-scale trough cross-stratified units, sometimes tabular, passing sometimes to climbing ripples and parallel lamination to the top. Thin laminae of silt and clays sometimes cover a set. The contact with the conglomeratic member is sharp and erosional (Member B).

Their lower and upper contacts are sharp, and palaeocurrent measurements show either little dispersion (approximately ±30°) or a radial pattern of more than 180° (Figs 5 and 6) with almost opposite directions in contiguous sets. Sands are overlain by silts and clays with a sharp contact (Member C).

DEPOSITIONAL ENVIRONMENTS

The lower members of both sequences are very similar and should represent the same environment. Their major structure, megacross-stratification can be interpreted as lateral accretion surfaces on an active meander lobe, similar to the ones forming in the present active channel. Imbrication of pebbles is almost perpendicular to these surfaces, as well as the palaeocurrent pattern of some of the sands formed along them, embedded in the gravels. The dipping surfaces always begin at the top of the member, but do not always reach the base.

The observation of the modern meander lobes and their sediments show a very similar mechanism of deposition and the orientation of pebbles and sand structures are comparable to the ones found in the terraces. Discharge fluctuations in the actual channel are wide. High water conditions occur every spring, the river running almost bankfull, and floods happen every ten to fifteen years according to available records. Low water conditions occur in summer and autumn and, although complete drying has not been observed nor recorded, less than 0·4 m of water was measured in July 1981 near Velilla after two years of severe drought. Therefore the gravel bodies of both sequences are interpreted as being deposited in coarse-grained meander lobes by lateral shifting of the active channel; and megacross-stratification as lateral accretion surfaces, forming during very low water conditions, although complete drying of the channel was a rare event. Their abundance can be related to great discharge fluctuations as in the present Jarama River, the gravels moving during floods. The lateral shifting of the meanders is obvious in the present alluvial valley (Fig. 2).

The sands, found occasionally on top of the gravels, lie on a flat or concave surface. This surface is sharp but shows no scouring or other erosional marks. They are interpreted as being deposited by waning floods at the downstream margins of the bars; similar sands are found at the bar tails of the actual active lobes. The surface only marks a change in flow conditions.

The ridge-and-swale topography of the top of the sequences cannot be seen in the terrace deposits, because some sequences are eroded, incomplete (Figs 3 and 4) or covered by younger sediments, but when complete their gravel members have a broad, convex top that could be related.

Sequence 1 is then very similar to the lateral bars 1 of Bluck (1976) and comparable to some described examples (Levey, 1978; Gustavson, 1978; Ori, 1979; Jackson, 1978), but contain a much higher percentage of gravel than other examples (McGowen & Groat, 1971; McGowen & Garner, 1970; Morton & McGowen, 1980; Nijman & Puigdefabregas, 1978). It is also similar to the simplest meander lobes of the present river, where no chute channel or bar is present and sands accumulate as sheets at the bar tails.

Sequence 2 has a more complex history, elucidated only after observation of processes within the modern Jarama River. It was observed that sometimes the ridge-and-swale geometry of the surface of the point bars became active in depressed parts, as secondary active channels during moderate floods. Some chute cut-offs are also developing and their chute channels, cutting across the ridges and swales, were also occasionally active. The channels of this network are approximately parallel, but sometimes coalesce. In both cases their small width and depth permitted only weak currents, moving only sands. They form transverse or linguoid bars. The internal structures of the sand member (Member B) of the sequence observed in the terrace—trough cross-stratification and occasional planar cross-stratification passing to current ripples, parallel lamination and thin silt layers at the top—record a decrease in current velocity.

The lenticular geometry and the basal erosional surface covered by pebbles and intraformational clasts can be interpreted as the shape of a channel formed on top of the gravel member, floored by a lag conglomerate. As the channel was infilled, its cross-section became smaller, allowing only weaker currents until only ripples and parallel lamination formed. Finally it became completely plugged and the upper part of the meander lobe was smoothed. Non-channelized silts and clays cover it. This process can be observed in various stages in the present alluvial plain. The chute channels and swales cannot migrate laterally because their banks are composed of gravels and fixed by trees and shrubs so that erosion is impossible by the weak currents. Only a longitudinal infilling is possible. Microdeltas are found in the present Jarama River at the end of chute channels; they have a fan geometry, are oblique to the main river channel and show variable cross-bed directions. The sand bodies with a radial pattern of palaeocurrents found embedded in the conglomerates of the terrace can represent similar deposits. The opposite directions in contiguous sets can be explained as being produced simultaneously in the main channel and the microdelta.

Sequence 2 therefore has a complicated origin and

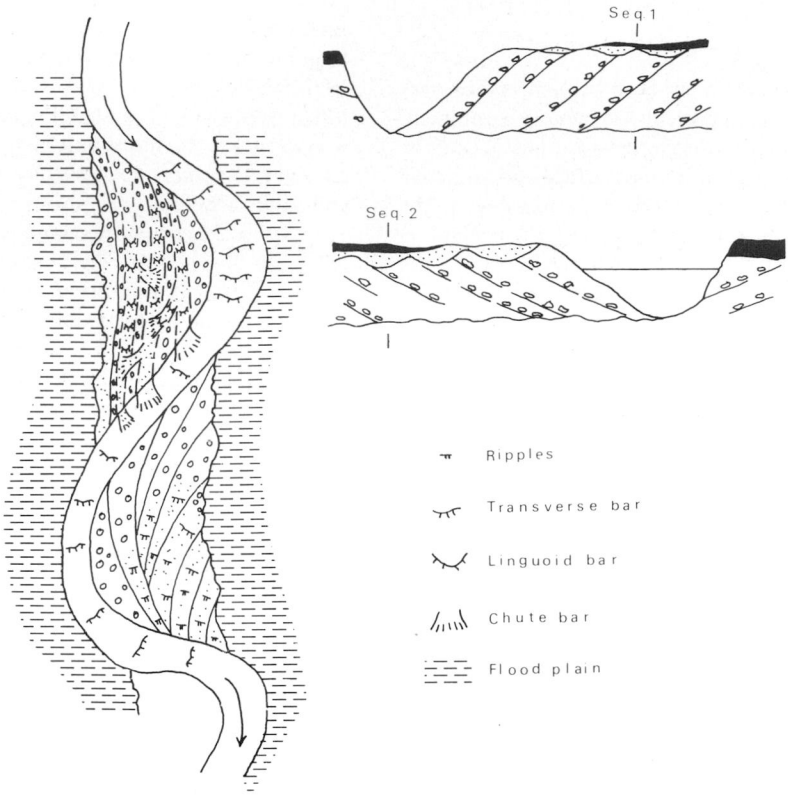

Fig. 9. Reconstruction of the proposed mechanisms for bar growth, inferred from terrace and modern river sediments.

can be compared to the lateral bar 2 of Bluck (1976) and some of the Upper Congaree River point bars of Levey (1978).

Fine-grained members in both sequences have a sharp, non-erosive basal contact and represent several subenvironments.

The thin, repetitive sequences with climbing ripples, parallel lamination and massive silts, containing occasional sandy or gravelly units can be interpreted as crevasse splay deposits or natural levées. The presence of gravels and sands in overbank deposits is rare, but by no means unknown (Ritter, 1975; Smith & Smith, 1980). The massive silt and clays with rootlet levels and leaf and twig impressions represent typical flood-plain deposits, mainly due to fallout from suspension after floods. The fine-grained material shown in Fig. 7 is clearly related to the infilling of a suddenly abandoned channel; fallout from suspension was the dominant process during the formation of this clay plug but occasionally gentle currents introduced detrital material and stirred up the bottom, forming

thin sand-beds. Abandonment of the channel could be caused by meander cut-off or avulsion. Both processes can be recognized in the active alluvial plain today.

CONCLUSIONS

The Jarama River near Madrid is a good example of a medium-sinuosity, coarse-bedload river. Observations in the actual active channel and alluvial plain and in sediments of 18–20 m terrace allow reconstruction of some of the processes acting during their deposition, summarized in the ideal diagram of Fig. 9. The gravel point bars were active during floods, becoming fixed as flow waned. Two different types of point bar formed, one simple, made up only of gravels except for a thin veneer of sand in its bar tail. Absence of sand at the bar head can be due to non-deposition or winnowing by weak currents. The second type is more complex, cut by chute channels and a ridge-and-swale topography. These depressed parts became active during moderate floods, and were plugged by sands. Sand matrix infilling the conglomerate voids was deposited during waning flood at low water levels.

The extremely high proportion of gravels in the deposits of a small river like the Jarama can be explained by the presence of thick, erodible Pliocene conglomerates (the so-called 'rañas') in its headwaters some 60 m north, which act as an important source of clastic sediments.

ACKNOWLEDGMENTS

I would like to thank the firms Aridos de Velilla S.A. and Castiñeira S.A. for granting permission to visit their pits and their personnel for help during fieldwork.

I would like also to thank Dr Phillip Allen who revised a first manuscript of this paper and made very useful comments, Mrs Francisca Rivas who typed the manuscript and Mr José Luis Gonzalez who printed the photographs.

REFERENCES

ALLEN, J. R. L. (1963) The classification of cross stratified units with notes on their origin. *Sedimentology*, **2**, 93–114.

BLUCK, B. J. (1971) Sedimentation in the meandering River Endrick. *Scott. J. Geol.* **7**, 93–138.

BLUCK, B. J. (1976) Sedimentation in some Scottish rivers of low sinuosity. *Trans. R. Soc. Edinb.* **69**, 425–455.

CARRILLO, L. (1977) *Estudio sedimentológico de la terraza media de los ríos Manzanares y Jarama entre Mejorada del Campo y Velilla de San Antonio (Madrid)*. Unpublished Tesis de Licenciatura. Universidad Complutense, Madrid.

CARRILLO, L. & ARCHE, A. (1982) Los depósitos de grano grueso del río Jarama al S. de Velilla de San Antonio. *IX Congreso Español de Sedimentología, Salamanca* (in press).

FRIEND, P. F., SLATER, M. J. & WILLIAMS, R. C. (1979) Vertical and lateral building of river sandstone bodies, Ebro Basin, Spain. *J. geol. Soc. Lond.* **136**, 39–46.

GUSTAVSON, T. C. (1978) Bedforms and stratification types of modern gravel meander lobes, Nueces River, Texas. *Sedimentology*, **25**, 401–426.

JACKSON, R. G. (1978) Preliminary evaluation of lithofacies models for meandering alluvial streams. In: *Fluvial Sedimentology* (Ed. by A. D. Miall). *Mem. Can. Soc. Petrol. Geol., Calgary*, **5**, 543–546.

LEOPOLD, L. B. & WOLMAN, M. G. (1957) River channel patterns, braided, meandering and straight. *Prof. Pap. U.S. geol. Surv.* **282B**, 1–73.

LEOPOLD, L. B., WOLMAN, M. B. & MILLER, J. P. (1964) *Fluvial Processes in Geomorphology*. San Francisco. 479 pp.

LEVEY, R. A. (1978) Bed-form distribution and internal stratification of coarse-grained point bars, Upper Congaree River, S.C. In: *Fluvial Sedimentology* (Ed. by A. D. Miall). *Mem. Can. Soc. Petrol. Geol., Calgary*, **5**, 105–127.

McGOWEN, J. H. & GARNER, L. H. (1970) Physiographic features and stratification types of coarse-grained point-bars: modern and ancient examples. *Sedimentology*, **14**, 77–111.

McGOWEN, J. H. & GROAT, C. G. (1971) Van Horn Sandstone, West Texas: an alluvial fan model for mineral exploration. *Rep. Invest. Bur. econ. Geol. Univ. Texas*, **72**, 1–52.

MORTON, R. A. & McGOWEN, J. H. (1980) Modern depositional environments of the Texas coast. *Gdbk No. 20, Bur. econ. Geol. Univ. Texas, Austin.* 35 pp.

NIJMAN, W. & PUIGDEFABREGAS, C. (1978) Coarse-grained point bar structure in a Molasse-type fluvial system, Eocene Castisent Sandstone Formation South Pyrenean Basin. In: *Fluvial Sedimentology* (Ed. by A. D. Miall). *Mem. Can. Soc. Petrol. Geol., Calgary*, **5**, 487–510.

ORI, G. G. (1979) Barre di meandro nelle alluvioni ghiaiose del Fiume Reno (Bologna). *Boll. Soc. geol. ital.* **98**, 35–54.

PALAEZ, J. R., PEREZ, A., VILAS, L. & AGUEDA, J. A. (1973) Características hidrogeológicas del Cuaternario del río Jarama. *I Congreso Hispano Americano de Geología Económica*, **3**, 513–526.

RICCI LUCCHI, F. (1979) *Sedimentologia, Parte III*. Cooperativa Libraria Universitaria, Bologna. 545 pp.

RITTER, D. F. (1975) Stratigraphic implications of coarse-grained gravel deposited as overbank sediment, Southern Illinois. *J. Geol.* **83**, 645–650.

SMITH, D. G. & SMITH, N. D. (1980) Sedimentation in anastomosed river systems, Examples from alluvial valleys near Banff, Alberta. *J. sedim. Petrol.* **50**, 157–164.

Spec. Publs int. Ass. Sediment. (1983) **6**, 323–336

Coastal alluvial fans and associated marine facies in the Miocene of S.W. Turkey

A. B. HAYWARD*

Grant Institute of Geology, University of Edinburgh, West Mains Road, Edinburgh, U.K.

ABSTRACT

A thick sequence (*c.* 1000 m) of Miocene clastic sediments was derived from a Mesozoic ophiolitic complex during its tectonic emplacement. The upper 350 m of the sequence, the Kasaba Formation of Upper Miocene age, consists of a diverse assemblage of coastal alluvial fan and shallow marine facies.

Alluvial fan deposits in proximal areas consist of dominantly clast-supported massive conglomerates, interpreted as stream deposits, interbedded with subordinate, matrix-supported conglomerate deposited by debris-flows. These pass down palaeoslope into interbedded conglomerates, sandstones and mudstones which form distinct fining-upward units up to 15 m thick. These are considered to have been deposited on an alluvial braid-plain. Pedogenic calcretes and reddened horizons attest to an arid or semi-arid environment.

Shallow marine deposits are the downslope equivalent of the alluvial braid-plain sediments. Patch reefs which developed along the shoreline protected the shoreface from extensive wave and storm reworking. Offshore material finer than pebble gravel was reworked sporadically, probably by storm-induced currents.

The overall depositional setting is interpreted to be a narrow coastal plain where alluvial fans prograded directly into a shallow microtidal sea.

INTRODUCTION

Miocene clastic sediments of the Susuz Dag massif in S.W. Turkey (Fig. 1) document the emplacement of two allochthonous ophiolitic units on to an adjacent carbonate platform. The Miocene succession is approximately 1000 m thick and comprises submarine fan deposits at the base that pass upwards through shallow marine sediments into continental deposits. This paper focuses on the top 350 m of the succession, the Kasaba Formation, of Tortonian (Upper Miocene) age. The sequence is well exposed in sections parallel to the palaeoslope in which the complete transition from an alluvial fan to a shallow marine depositional environment may be studied.

Coastal alluvial fans that pass directly into the sea have been demonstrated both from the Recent, where

they are confined to two-dimensional surface studies (Boothroyd & Ashley, 1975; Boothroyd & Nummedal, 1978; Gwirtzman & Buchbinder, 1978; Wescott & Ethridge, 1980), and from the ancient, where they are generally limited to the two-dimensional analysis of vertical sedimentary sequences (Daily, Moore & Rust, 1980).

This paper provides a model for semi-arid coastal alluvial fans based on lateral and vertical facies transitions observed within the sedimentary sequence.

GEOLOGICAL SETTING

On a regional scale the Tauride Mountains of S.W. Turkey consist of a central para-autochthonous unit, the 'Tauride autochthon' (Brunn *et al.*, 1970, 1971; Dumont *et al.*, 1972), sandwiched between two allochthonous ophiolitic units, the Lycian Nappes (Brunn *et al.*, 1970, 1971; de Graciansky, 1972; Poisson, 1977) and the Antalya Complex (Robertson

* Present address: British Petroleum Co., West Britannic House, London E.C. 2, U.K.

0141-3600/83/0106-0323 $02.00

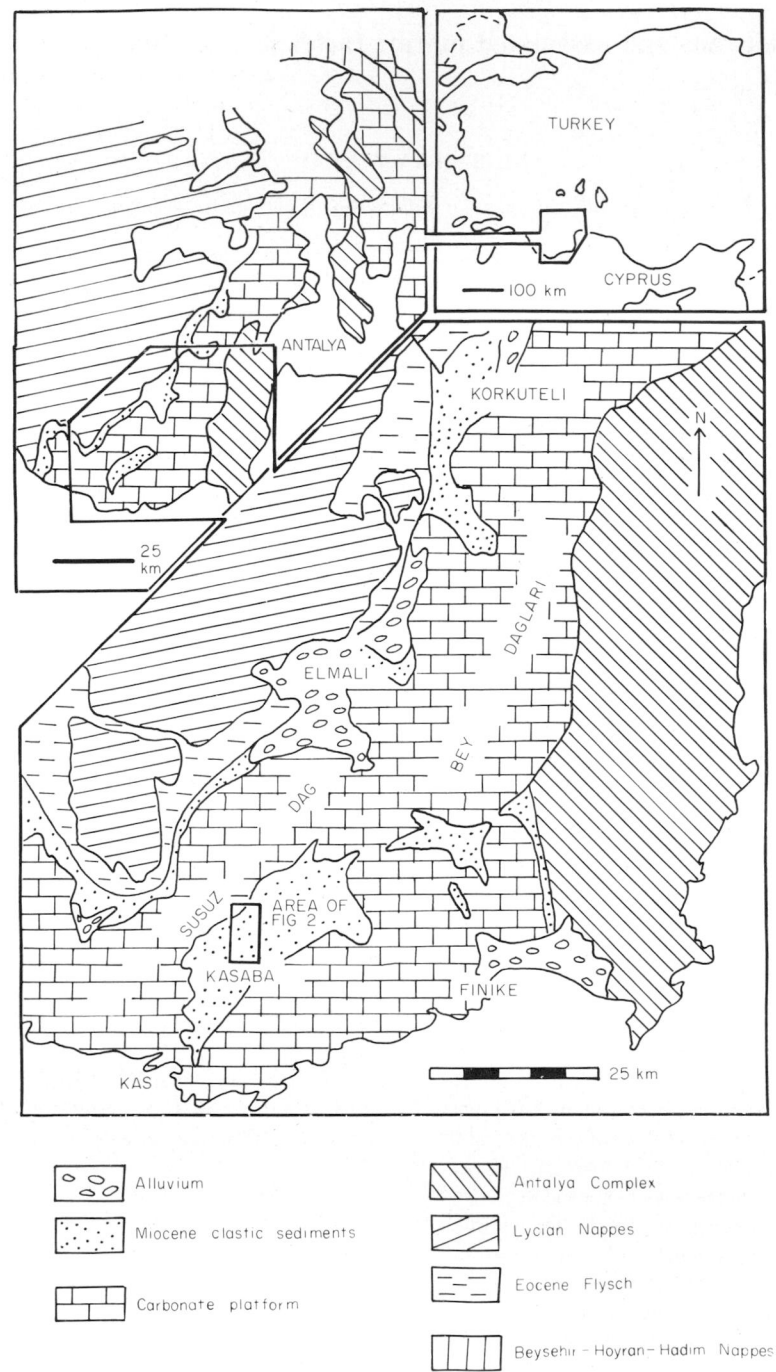

Fig. 1. Location map showing regional tectonic setting of Miocene clastic sediments in S.W. Turkey. Small inset on lower diagram shows location of larger-scale map on Fig. 2.

& Woodock, 1980) (Fig. 1). All of these form an extension of the Alpine orogenic belt into S.W. Turkey (Brinkman, 1976).

The 'Tauride autochthon' comprises dominantly shallow marine sediments of a carbonate platform ranging in age from Liassic to Lower Miocene (Aquitanian) with several non-sequences in the Lower Tertiary (Poisson, 1977; Dumont *et al.*, 1972). Miocene clastic sediments, which include those discussed here, overlie unconformably the carbonate platform rocks.

To the north-west, the Lycian Nappes consist of a series of allochthonous sheets transported from the north-west to the south-east in several phases during the early Tertiary (Brunn *et al.*, 1970, 1971; Graciansky, 1972; Poisson, 1977; Delaune-Mayère *et al.*, 1977). The Upper Miocene deposits described here were derived exclusively from the Lycian Nappes and record their final emplacement from a north-westerly direction on to the carbonate platform.

SEDIMENTARY FACIES

Sedimentary facies associations are used to produce a detailed picture of the depositional environment (cf. Reading, 1978; Walker, 1979). The sedimentary sequences can be subdivided into those deposited in either a continental or a marine environment.

The continental sediments are characterized by calcretes, desiccation cracks and reddened horizons, which are consistent with subaerial exposure in a semi-arid climate. Within the continental sequences two facies associations are recognized. Individual facies are closely comparable to those described by Miall (1977, 1978) and Rust (1978).

The marine deposits contain an abundant macrofauna of bivalves and gastropods, and coral patch-reefs.

Petrographically the sediments of the Kasaba Formation consist of a complete admixture of rock types derived from the Lycian Nappe ophiolitic unit to the north-west. Clast types in the conglomerate consist of moderate to well-rounded (R2–3, Odell, 1977) limestone, dolerite, gabbro, chert and subordinate sepentinite and basalt rock fragments.

The sandstones are generally poorly sorted; rock fragments again predominate. Serpentinite, chert, dolerite, basalt and subordinate quartz and feldspar grains are cemented by carbonate (microsparite). Sandstones from the marine facies association are rich in foraminiferal, algal and other shell fragments.

Continental facies

Conglomerate facies association

This consists of laterally extensive, poorly sorted, reddened conglomerate, with subordinate sandstone, and red and green mudstone. The facies codes used throughout this paper for the description of the continental sediments are those of Miall (1977, 1978).

The conglomerates make up more than 80% of the association (Fig. 2, section a) and comprise poorly to moderately rounded (R 1–3, Odell, 1977) pebbles, cobbles and boulders, with a maximum clast size of 0·80–1·30 m. The conglomerates are dominantly clast-supported, although rare (10%) matrix-supported conglomerates are present. The clast-supported conglomerates can be subdivided into two facies: horizontally stratified or massive (Gm) boulder conglomerate (80%) and volumetrically subordinate trough cross-stratified (Gt) cobble and boulder conglomerate (10%).

In the former, crude stratification is picked out by variations in the average clast size or by high concentrations of larger clasts. Discontinuous clay, silt and planar cross-stratified sandstone lenses are commonly interbedded within the conglomerate so that individual depositional events are difficult to delineate. Contact imbrication of the clast *b*-axes dipping upstream is common.

Trough cross-stratified units between 0·60 and 3·0 m thick infill erosive scours, up to 1·5 m deep and 4–5 m across. The cross-strata are markedly heterogeneous, composed of poorly sorted openwork conglomerate separated by pebbly sandstone. Upwards and laterally, trough cross-sets pass into massive conglomerate of similar grain size. Foreset dips are generally between 10° and 15°, but may be up to 25° in smaller troughs.

Subordinate matrix-supported conglomerates, which vary between 1·0 and 2·5 m thick, consist of clasts of up to 0·8 m in a silty mudstone matrix. Clast percentages range from 5 to 25%. Clasts are aligned parallel to the bounding surfaces or, more rarely, *a*-axes are imbricated up current. Palaeocurrent flow (measured from cross-stratification) was consistently to the south for this facies association (Fig. 3).

Interpretation

The high proportion of conglomerate, the clast size and palaeocurrent directions (Fig. 3) show that this facies association is the most proximal one preserved.

Reddened horizons within the conglomerates and

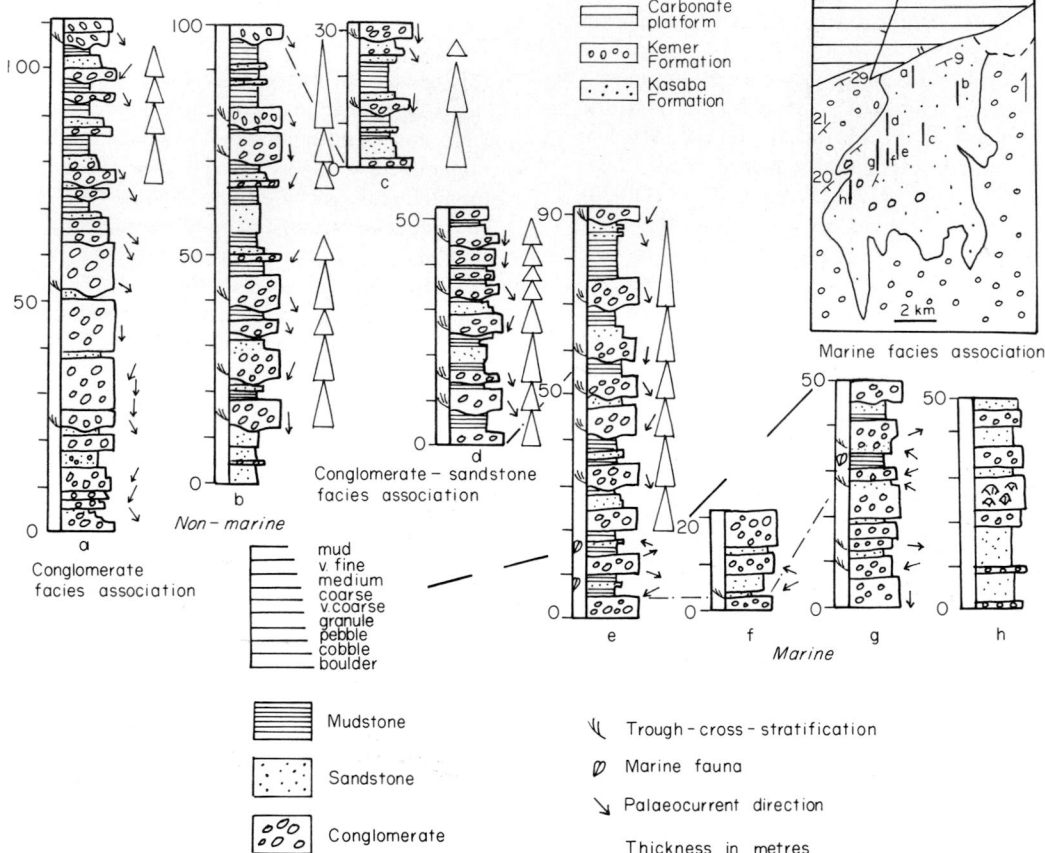

Fig. 2. Sedimentological logs measured in the Kasaba Formation. Note fining-upward cycles in the conglomerate–sandstone facies association and the variation in palaeocurrent orientation between non-marine and marine deposits. Inset is a simplified map of the Kasaba Formation showing location of sedimentological logs and patch-reefs.

rare, laterally discontinuous horizons of red siltstone and sandstone are consistent with subaerial oxidation and suggest subaerial deposition in an arid or semi-arid environment (e.g. Walker, 1967).

Bed lenticularity and poor segregation of sand and gravel is consistent with a fluvial origin (Clifton, 1973). In the massive, clast-supported conglomerate the orientation of clast a-axes is normal to palaeoflow with frequently well-developed b-axis imbrication. This is evidence for both the individual response of particles to a flow mechanism (Harms *et al.*, 1975), and the effect of bedload transport in response to unidirectional flow (Rust, 1972). The absence of cross-stratification indicates that bedform resistance did not cause flow separation and the depth of flow was shallow (Church & Gilbert, 1975; Rust, 1975).

Conglomerates similar to those in S.W. Turkey are

found in modern alluvial environments (Smith, 1970; Rust, 1972; Church, 1972; Miall, 1977, 1978; Rust, 1979) where longitudinal bars are the dominant bedforms (Church & Gilbert, 1975; Collinson, 1978). These bars are formed by unconfined flow and may be stable at maximum flood, or form as a result of decreased strength at falling stage (Leopold & Wolman, 1957; Bluck, 1979). Scouring at the base of individual conglomerate units could be related to local vortices around obstructions or to channels formed by avulsion at high stage (Miall, 1977). Gravel bars in modern braided streams do not generally show cross-stratification (Miall, 1977) and the isolated trough cross-sets are interpreted as scour fill or channel fill features.

The matrix-supported conglomerates have non-erosive bases and frequently show inverse and normal

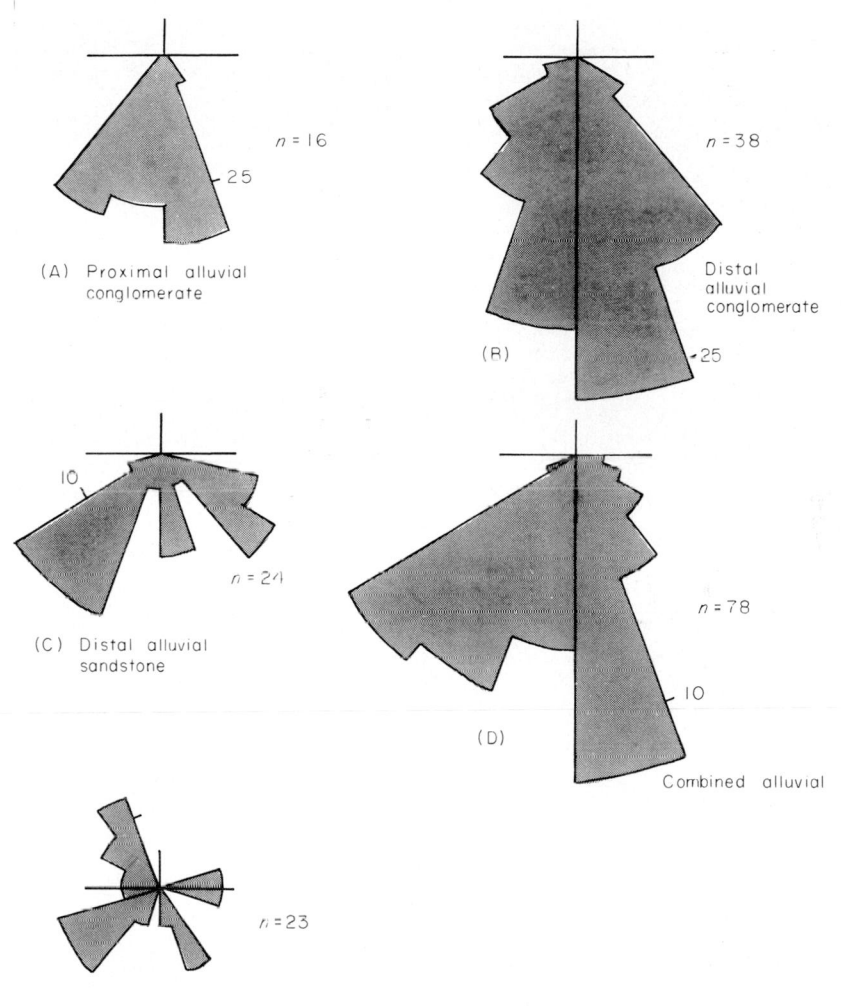

Fig. 3. Palaeocurrent measurements for the different facies associations. (A) imbrication, (B) cross-stratification, (C) cross-stratification, (E) cross-stratification. *N* is a number of readings. Small numbers refer to percentage of total readings.

grading, all of which are consistent with debris flow origin (Enos, 1977). The interbedding of debris-flow conglomerates and stream-deposited conglomerates is consistent with deposition on an alluvial fan (Bull, 1972; Rust, 1978; Daily *et al.*, 1980). The small proportion of debris-flow conglomerates do not suggest a very proximal fan location, and this association is interpreted as being the result of dominantly stream deposition (sheet flood and channel) on the mid to distal parts of an alluvial fan complex.

Conglomerate–sandstone facies association

This association is characterized by well-defined fining-upward units of the order of 10–20 m thick (Fig. 2, sections b, c, d, e). Trough cross-stratified conglomerate (Gt) forms up to 20% of the conglomerate, infilling scoop-shaped scours at the bases of fining-upward units. Laterally and vertically this facies (Gt) passes into massive conglomerate (Gm) of a similar grain size.

Maximum clast size is variable, up to 0·90 m, but is commonly 15–35 cm. Discontinuous lenses, up to 30 cm thick, of claystone, siltstone, cross-stratified

Fig. 4. Conglomerate–sandstone association showing three fining-upward units (photograph taken from top of lowest conglomerate unit). Trough cross-stratified conglomerate (Gt) infills scours at base of conglomerate units. Palaeoflow was into photograph. Note thickness of overbank deposits in lowest f-u unit. Thickness of the middle f-u unit is 7 m.

sandstone and trough cross-stratified conglomerate occur within the massive conglomerates. The most common matrix type in both conglomerate facies is a muddy to silty medium-grained sandstone. The percentage of matrix varies both laterally and vertically; in some places horizons within the conglomerate are matrix-free.

Conglomerate beds have an average thickness of 7 m (approximately half of the thickness of a complete fining-upwards unit). They occur as laterally continuous sheets over 300–400 m, with irregularly scoured erosional bases. On a scale of several hundred metres, bases parallel the underlying sediments (Fig. 4).

Conglomerates make up approximately 50% of the succession (Fig. 2). Palaeocurrent directions measured from trough cross-strata in the conglomerate are consistently south–south-west (Fig. 3); essentially the same as the proximal deposits.

Two types of fining-upward units can be distinguished (Fig. 5). In 60% of the cycles the conglomerates are overlain by either fine to medium massive (Sm) and parallel-stratified (Sh) sandstone or directly by calcrete (Fig. 5B). Red to brown, medium- to fine-grained, massive and parallel-laminated sandstones pass upwards and laterally into massive (Fm) red and green siltstones and claystones with silt

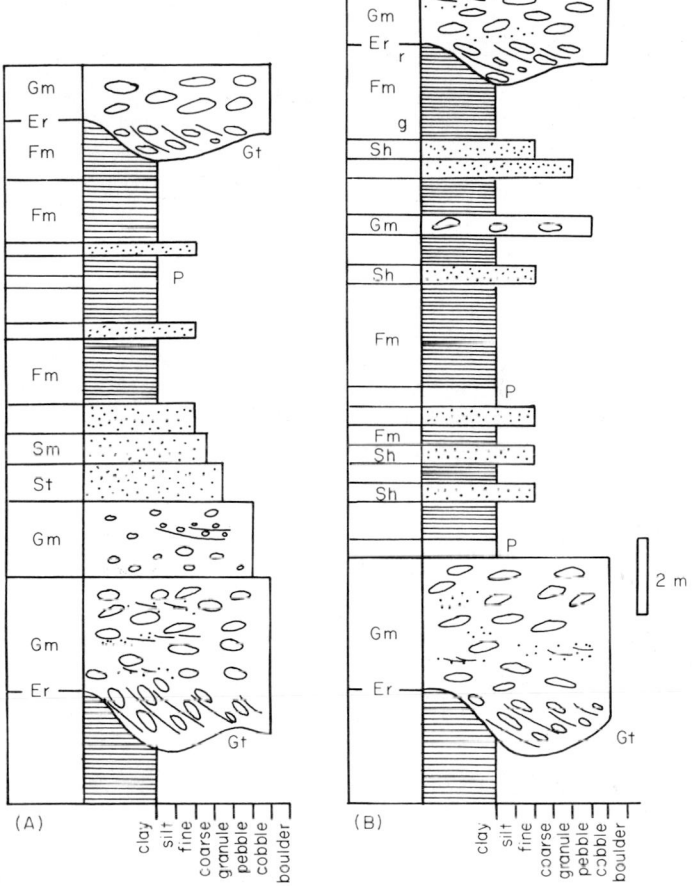

Fig. 5. Detailed sedimentological logs showing the two types of fining-upward cycle: (A) 40%, (B) 60%, recognized in the conglomerate–sandstone association (braidplain). Gm, massive conglomerate; Gt, trough cross-stratified conglomerate; Sh, horizontally stratified sandstone; St, trough cross-stratified sandstone; Sm, massive sandstone; Fm, massive mudstone/siltstone; P, calcrete; Er, erosion surface; r, red; g, green.

laminae and rare pebble conglomerate horizons (up to 40 cm thick). Calcretes (P) form laterally continuous nodular horizons 5–15 cm thick, in which very fine sand- and silt-sized terrigenous grains are dispersed in a micritic cement. Desiccation cracks are associated with calcretes.

In the second type of cycle, conglomerates are overlain gradationally by granule conglomerates and very coarse sandstones (Fig. 5A). Red to brown, poorly sorted granule conglomerate and coarse sandstone form trough-cross-stratified units (Gt, St) comprising shallow troughs, commonly 10–20 cm thick. They are, in many places, interbedded with, and pass laterally into, parallel-stratified medium to coarse-grained sandstone. More rarely tabular cross-stratification occurs in sets 10–15 cm thick, with

foreset dips at shallow (5–10°) angles. The palaeo-current trends for these sandstones (Fig. 3C) are consistently southwards.

The coarse sandstone passes upwards into fine to medium, moderate to poorly sorted, red and brown, massive and parallel-stratified sandstone. Occasional trough cross-stratified sandstone (St), with mudstone rip-up clasts, displays low-angle trough sets less than 10 cm thick. Palaeocurrent trends are similar to those of the coarse sandstone. Sandstones of this type are interbedded with, and pass upwards into, red and green claystone with discontinuous siltstone and rare conglomerate horizons. A preferred facies-transition probability diagram for this association (Fig. 6) emphasizes the cyclic nature of deposition.

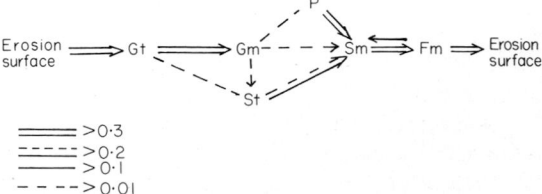

Fig. 6. Preferred facies transition diagram for conglomerate–sandstone facies association. Calculated from 188 transitions after method of Walker (1979). For explanation of symbols see caption of Fig. 5.

Interpretation

The conglomerate–sandstone association contrasts with the previous conglomerate association in the well-developed internal organization exhibited by the fining-upward cycles (Figs 2 and 5). Subaerial exposure is indicated by the occurrence of calcretes and reddened mudstones. The palaeocurrent orientations and smaller mean clast size are all consistent with this association being the distal equivalent of the conglomerate association.

The conglomerates which form the base of individual fining-upward units are related to depositional processes similar to those of the conglomerate association. Trough cross-stratified, clast-supported conglomerates that occur infilling scours above a sharp erosional surface (Fig. 4) at the base of conglomerate units are considered the result of scour and channel-fill. Laterally discontinuous trough sets within conglomerate units are the deposits of migrating bedforms within channels formed at high flood-stage (Martini, 1977; Rust, 1978).

Trough cross-stratified conglomerate in all cases passes transitionally upward into massive conglomerate, suggesting that most of this fluvial conglomerate (facies Gm) was transported as diffuse sheets or within low-relief bedforms (Smith, 1974; Eynon & Walker, 1974; Hein & Walker, 1977). By analogy with modern braided fluvial systems, massive conglomerate probably accreted as planar sheets in the form of longitudinal bars during high flood-stage.

In some instances (Fig. 5A) trough cross-stratified conglomerates pass upwards through massive conglomerate into trough cross-stratified and low-angle cross-stratified coarse sandstone, parallel-stratified sandstone and finally into fine sandstone, mudstone and calcrete.

Trough cross-stratified sandstones are interpreted as dunes formed under lower flow-regime conditions (Harms & Fahnestock, 1965). Low-angle cross-stratified and horizontally stratified sandstones are

formed respectively as very shallow scour fills (Miall, 1977), and as plane beds formed under low flow regime conditions (Harms & Fahnestock, 1965; Harms et al., 1975). This sequence is the result of waning flood and the shallowing of water over actively accreting bars (Williams & Rust, 1969; Miall, 1977, 1978). Mud and fine sand are deposited as the area becomes inactive. Pedogenic calcretes develop as a result of prolonged subaerial exposure.

The calcretes which directly overlie conglomerates are evidence of a depositional break, prior to deposition of the overlying sandstone, mudstone and calcrete triplets. The red and green mudstone and sandstone are characterized by their fine grain size, lack of current structures or only occasional small isolated ripples and rare bioturbation. All these features are consistent with deposition in standing water away from the active area of sedimentation.

The sudden change from conglomerate to mudstone is interpreted as the result of a rapid decline in velocity from high flood-stage and an associated fall in water level. The active area of sedimentation becomes localized down the sides of accreted bars, exposing bar tops and marginal terraces subaerially. Deposition in the inactive areas is primarily by the vertical accretion of fine sediment (Williams & Rust, 1969; Rust, 1979) and washover from the active channel area. In humid regions extensive vegetation develops in this area (Williams & Rust, 1969; Miall, 1977; Boothroyd & Ashley, 1975). In regions with the appropriate arid or semi-arid climate, reddened oxidized horizons and calcretes are formed (Allen, 1965; Rust, 1978). In flood, minor channels may transport sand and gravel across the inactive area (Rust, 1979) producing thin sand and conglomerate lags within the mudstone sequence (Fig. 5A, B).

Superficially this association resembles the middle reaches of the Donjek River in Yukon, chosen by Miall (1977) as the modern type example of a distal gravelly braided fluvial system. Similar sequences have also been described by Rust (1979) from the Carboniferous of eastern Quebec. Alternatively this sequence may be the result of deposition by a meandering stream with coarse gravel and little sand as described by Jackson (1978) from Wyoming. Streams of this nature (facies class 5 of Jackson, 1978) exist near or within mountain ranges and flow on relatively steep slopes. At the present time the deposits produced by this type of fluvial system are poorly documented. Modern examples are characterized by lateral bars rather than point bars (Jackson, 1978) and as a result ancient sequences are unlikely to show any evidence of lateral accretion surfaces, which if preserved are considered diagnostic of finer-grained meandering systems.

One major difference appears to be the proportion of conglomerate and coarse sandstone to mudstone, siltstone and calcrete (overbank sediments). Vertical sequence models, in the literature, for braided fluvial systems (Miall, 1977, 1978; Rust, 1978, 1979) commonly show proportions of greater than 90% active area sediments (cf. 50% in the Kasaba Formation) compared with between 25 and 100% in gravelly meandering systems (Jackson, 1978). However, McLean & Jerzykiewicz (1978) have described a sequence with thick, fine-grained (overbank) sedimentary facies that they attribute to an overriding tectonic control, rather than channel type, and the absolute thickness of fine-grained facies may not be a reliable indicator of channel type.

In conclusion, in the absence of any clear indicators of a meandering channel system such as epsilon cross-stratification and fine-grained channel fills, suggested as diagnostic of meandering fluvial systems (Allen & Friend, 1968), this association is interpreted to have been deposited downslope from the distal alluvial fan sequence (conglomerate association) on a fluvial braid-plain.

Marine association

This comprises interbedded conglomerate, sandstone and mudstone with an abundant marine fauna, and small coral patch-reefs. Within this spectrum of deposits two sequence types are distinguished.

In proximal areas (Fig. 2, sections e, f, g) mudstones are interbedded with conglomerate and sandstone, forming units between 10 and 15 m thick. To the south (Fig. 2, section h) mudstone is absent or subordinate, and conglomerate and sandstone are interbedded with small patch-reefs.

Dark green to buff pebble, cobble and boulder conglomerates have a maximum clast size of 0·50 cm. Cross-stratification is rare within the conglomerates. Where present it consists of poorly defined low-angle (5–7°) cross-stratification in sets 1·0 m thick. However, the conglomerates are generally structureless, with non-erosive bases parallel to the underlying sediment and a random orientation of the clasts (Fig. 7). Green to grey, moderate- to well-sorted trough cross-stratified and parallel-stratified sandstones and granule conglomerates comprise up to 20% of this association (Fig. 2, sections e, f, g). The trough cross-sets are 20–30 cm thick. Burrows within the sandstones consist of U-shaped or vertical tubes which penetrate the top of individual sandstone beds to a depth of 10 cm.

Massive and laminated, dark grey calcareous mudstones form up to 10% of the association (Fig. 2, sections e, f, g). The mudstones are burrowed and contain a marine fauna of dominantly bivalves (*Lutraria ablonga*, *Venus basteroti*) and gastropods (*Turritella turris*, *Bursa marginata*).

Southwards, where mudstone is subordinate or absent (Fig. 2, section g), interbedded patch-reefs occur on depositional slopes of up to 5°. The patch-reefs consist of a central framework, up to 8 m high and 15–20 m across, composed of *in situ* corals against which is banked reef talus breccia that interfingers with the surrounding coarse-grained terrigenous sediment. The latter comprises interstratified pebble-cobble conglomerate horizons, 25–40 cm thick, and coarse pebbly-sandstone horizons 40–60 cm thick. The matrix is generally coarse to very coarse sandstone.

Interpretation

An abundant marine macrofauna of gastropods and bivalves, and small patch-reefs indicate that this sequence was deposited in a fully marine environment.

Conglomerates in the nearshore zone (Fig. 2, sections e, f, g) show no evidence of being reworked by marine processes. The general lack of structure within the conglomerates suggests that they are deposited as poorly sorted sheets by fluvial channels entering a shallow sea.

Offshore, the fine conglomerates and coarse sandstones which occur in association with patch-reefs are segregated into beds of conglomerate and sandstone, suggesting some reworking by wave processes (Clifton, 1973).

The paucity of stratification, absence of bipolar cross-stratification, and ripples, generally considered indicative of tidal currents (Johnson, 1978), suggest that the shoreline was not subject to strong tidal influence. The sporadic occurrence of sedimentary structures indicating marine reworking may suggest that most reworking occurred when storms augmented normal marine processes (Sellwood, 1972).

In summary, the marine sequence was probably deposited in a shallow, partially enclosed, microtidal sea. Patch-reefs developed parallel to the shoreline protecting the nearshore and shoreface from extensive reworking by wave processes. Further offshore storms probably reworked sand and fine conglomerate.

Fig. 7. Laterally continuous conglomerate units of the nearshore marine facies association. Notice the lack of internal organization and non-erosive bases when compared with the alluvial conglomerates. Interbedded mudstones contain an abundant marine fauna of gastropods and bivalves. Central conglomerate unit is 5·50 m thick.

GENERAL MODEL: SUMMARY

There is clear evidence of a N–S alluvial palaeoslope as indicated by the following.

(1) The general decrease in maximum clast size from north to south.

(2) The overall palaeocurrent trend, which is uniformly to the south for the non-marine sequences (Fig. 3).

(3) The increasing marine influence seen in the sediments from north to south.

Facies associations and downslope transitions indicate deposition on an alluvial fan complex passing through a fluvial braid-plain into a shallow sea. The general model is shown in Fig. 8.

The alluvial fan (conglomerate association) with little internal organization passes over a distance of approximately 2 km (Fig. 8) downslope into a sequence with good internal organization (conglomerate–sandstone association) (Fig. 4). In this association fining-upward cycles are characterized by conglomerate units that are laterally continuous up to 400 m down palaeoslope and by thick overbank deposits.

The conglomerate–sandstone association extends for 3 km down palaeoslope before passing into a shallow marine environment (marine association). The transition zone lacks features such as strand lines or low-angle seaward-dipping imbrication in conglomerates, normally associated with beach environments

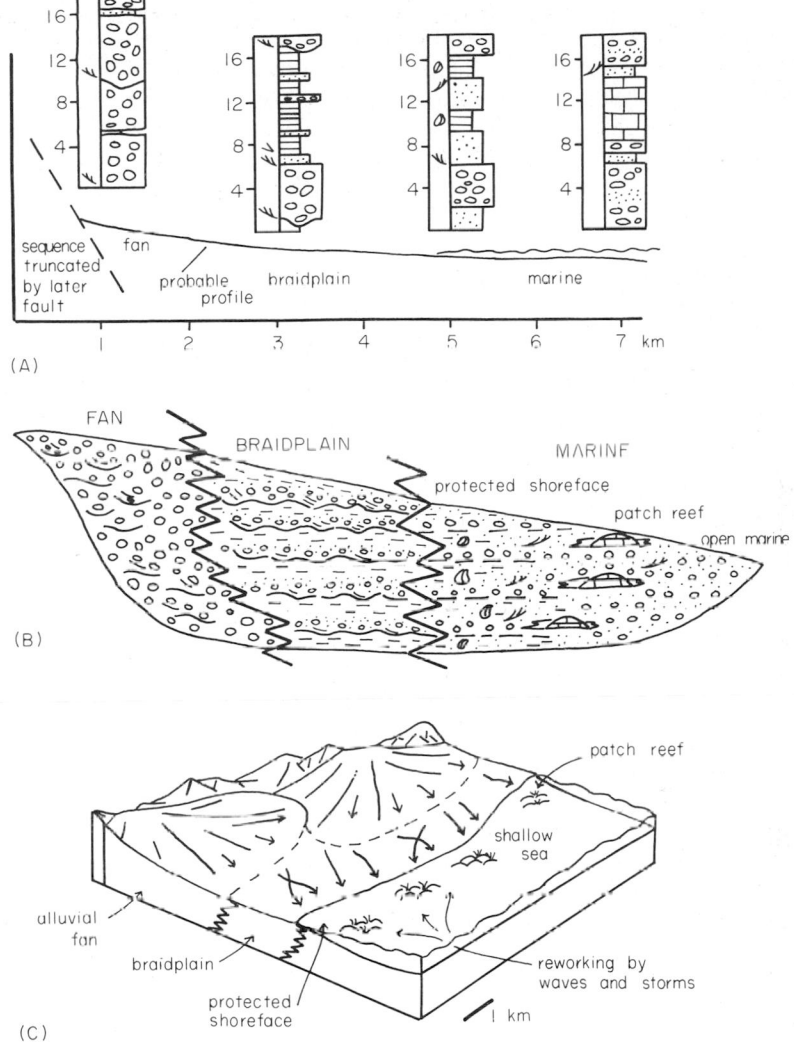

Fig. 8. General depositional model showing down-fan and vertical variations in sedimentary facies and structures. Symbols as on Figs 2 and 5.

(Cailleux, 1945; Bluck, 1967). This is taken to indicate a predominantly low-energy, probably microtidal, marine environment.

The development of patch-reefs parallel to the shoreline partially protected the shoreface from marine processes. The immediate nearshore zone is characterized by very strong fluvial influence, the conglomerates showing no evidence of marine reworking. The lack of large-scale cross-stratification, typically produced when fluvial currents of high velocity expand into standing water (Rust, 1975; Jopling, 1965), probably indicates that the shoreface and offshore slope were shallow.

Palaeocurrents (Fig. 3) in the sandstones of the marine sequence show a wide scatter, in marked contrast to the fluvial sequence. This is possibly the result of storm-induced currents which are unlikely to be consistently unidirectional.

The association of alluvial fans passing directly into a standing body of water has been termed a fan-delta (Gilbert, 1890; Holmes, 1965; Sneh, 1979; Wescott & Ethridge, 1980). However, the term 'coastal alluvial fan' is preferred here, as in many cases terrestrial relief is the major control on sedimentation and the fans frequently show no clear relationship to a marine base level. In Recent examples from the Red Sea there is no break in slope at the sea-level line and geomorphic areas of the 'delta' (delta top, front, etc.) cannot be recognized. Development of coastal alluvial fans requires high relief adjacent to the coastal zone and short-headed, high-gradient bedload streams that remain braided to the coast. Such deposits are relatively rarely described from the geological record although occurrences have been documented by Dabrio (1975) from the Miocene of Spain, by Howell & Link (1979) from the Eocene of South California and by Daily *et al.* (1980) from the Cambrian of South Australia.

Probably the best modern analogue of this sequence in terms of both fan size and climate are the alluvial fans that pass directly into the sea along the Gulf of Aqaba and the Red Sea (Friedman & Sanders, 1978; Gwirtzman & Buchbinder, 1978; Hayward, 1982). Here alluvial fans and braided fluvial systems drain over a coastal plain which varies in width between 1 and 7 km. In particular, the Ras Antatur area of the Gulf of Elat (Gwirtzman & Buchbinder, 1978) includes an alluvial fan that passes downslope into a braid-plain with a regional dip between $0.5°$ and $1°$; this passes directly into the sea with the development of a large fringing reef and associated patch-reefs along the seaward margin.

Coarse-grained clastic wedges of coastal alluvial fans may provide reservoirs for oil and gas originating in adjacent organic-rich mudstones, for example the Brae oilfield in the North Sea. A coarse clastic wedge of sediment shed off the Fladen Ground Spur forms the reservoir. This sequence has recently been interpreted (Harms *et al.*, 1981) as the result of deposition on a coastal alluvial fan.

Discussion of cyclicity within the fluvial sequence

Cyclic alluvial sequences may result from either of two mechanisms (Beerbower, 1964). *Autocyclic* mechanisms require no net change in the total energy and sediment input into a sedimentary system, but simply the redistribution of energy within the system. Examples of this are channel migration, avulsion, crevassing and subsidence due to compaction. Over time, provided no external control is exerted on the system, this will result in equilibrium being approached (Leopold & Wolman, 1957). The cause and effect of this type of cycle have been discussed extensively by Allen (1978) and Bridge & Leeder (1979). *Allocyclic* mechanisms require a change in the energy input into the system. External forces that may be involved include eustatic variations in sea-level, climatic changes and tectonic controls such as irregular elevation of the source area and spasmodic depression of the basin.

The fluvial succession described here shows small-scale fining-upwards cycles that are characterized by strongly erosive bases, with a sharp transition from mudstone to cobble and boulder conglomerate and a very marked asymmetric nature. These features suggest that the cycles are the result of the successive migration of fluvial channels through an area of the floodplain (autocyclic mechanism), as opposed to relative uplift of the source area, providing pulses of coarse sediment (allocyclic mechanism). The latter frequently show a more gradual grain-size transition and are often characterized by symmetrical cycles that coarsen and then fine upwards. An example of this type has recently been described from Norway (Steel & Aasheim, 1978). In this example the cycles are small (~ 10 m thick); they are characterized by a coarsening upwards in proximal areas and coarsening–fining upwards in distal areas. They are attributed to the progradation of the entire fluvial system, controlled by tectonic instability and rapid subsidence of the basin.

The deposition of thick overbank sediments suggests low lateral confinement of the fluvial channel, so that, even in moderate flood conditions, fine-grained material transported in suspension can be widely dispersed over the floodplain in areas away from the main fluvial channel. Preservation of complete cycles formed in this way requires relatively rapid subsidence, preventing reworking of the floodplains and superimposition of channel sandstones and conglomerates (e.g. see models of Bridge & Leeder, 1979). Fluvial sequences with thick overbank sediments have previously been documented from areas of tectonic activity associated with rapid subsidence. In particular McLean & Jerzykiewicz (1978) document a sequence that is primarily controlled by thrust emplacement; subsequent loading and subsidence, in front of the thrust sheets, occurs in response to isostatic adjustment.

In the present area, emplacement of the Lycian Nappes from the north-west in the Lower Miocene (Poisson, 1977) resulted in loading and subsequent subsidence of the carbonate platform on to which the thrust sheets were emplaced. Subsidence slowed during Middle Miocene times as the sedimentary basin filled.

Final emplacement of thrust sheets in the Upper Miocene, during which time the Kasaba Formation was deposited, again brought about rapid subsidence.

CONCLUSIONS

The Upper Miocene (Kasaba Formation) clastic sediments of the Susuz Dag in south-western Turkey document the development of a coastal alluvial fan–braid-plain complex that prograded across a narrow hinterland into a shallow sea.

Alluvial fan deposits in proximal areas consist of dominantly clast-supported conglomerates deposited by sheet floods. These pass down palaeoslope into a sequence comprising interbedded conglomerate, sandstone and mudstone which form distinct fining-upwards cycles. The sequence was deposited on an alluvial braid-plain. Calcretes and reddened horizons attest to an arid environment. Preservation of overbank sediments which form up to 70% of individual cycles is attributed to rapid subsidence associated with large-scale nappe emplacement from the north-west.

Patch-reefs developed on the seaward margin protected the shoreface from extensive reworking by marine processes. Absence of sedimentary structures indicating tidal currents suggests a small tidal range. Offshore material finer than pebble gravel was reworked sporadically by storm-induced currents.

A modern analogue for this environment is the Gulf of Elat in the Red Sea, where arid alluvial fans of comparable size pass directly into the sea. The sedimentary model developed from both lateral and vertical facies transitions may be applicable to other instances of arid alluvial fans prograding directly into the sea.

ACKNOWLEDGMENTS

This work was carried out during the tenure of a Natural Environment Research Council studentship. I am grateful to the Turkish Geological Society (M.T.A.) for providing logistical support in the field. I thank Maureen Fulton for drafting the figures. Drs Alastair Robertson, Bryan Lovell, Dorrik Stow, John Waldron and two anonymous referees made constructive comments of an earlier version of this manuscript.

REFERENCES

ALLEN, J.R.L. (1965) A review of the origin and characteristics of recent alluvial sediments. *Sedimentology*, **5**, 89–191.

ALLEN, J.R.L. (1978) Studies in fluviatile sedimentation: an exploratory quantitative model of the architecture of avulsion controlled alluvial suites. *Sedim. Geol.* **21**, 129–147.

ALLEN, J.R.L. & FRIEND, P.F. (1968) Deposition of the Catskill facies, Appalachian region: with notes on some other Old Red Sandstone Basins. In: *Late Palaeozoic and Mesozoic Continental Sedimentation, Northeastern North America* (Ed. by G. de V. Klein). *Spec. Pap. geol. Soc. Am.* **106**, 21–74.

BEERBOWER, J.R. (1964) Cyclothems and cyclic depositional mechanisms in alluvial plain sedimentation. In: *Symposium on Cyclic Sedimentation* (Ed. by D.F. Merriam). *Bull. Kansas geol. Surv.* **169**, 31–42.

BLUCK, B.J. (1967). Sedimentation of beach gravels: examples from S. Wales. *J. sedim. Petrol.* **37**, 128–156.

BLUCK, B.J. (1979). Structure of coarse-grained braided stream alluvium. *Trans. R. Soc. Edinb.* **70**, 181–221.

BOOTHROYD, J.C. & ASHLEY, G.M. (1975). Processes, bar morphology and sedimentary structures on braided outwash fans, northeastern Gulf of Alaska. In: *Glaciofluvial and Glaciolacustrine Sedimentation* (Ed. by A.V. Jopling and B.C. McDonald). *Spec. Publs Soc. econ. Paleont. Miner., Tulsa*, **23**, 193–222.

BOOTHROYD, J.C. & NUMMEDAL, D. (1978) Proglacial braided outwash: a model for humid alluvial-fan deposits. In: *Fluvial Sedimentology* (Ed. by A.D. Miall). *Mem. Can. Soc. Petrol. Geol., Calgary*, **5**, 641–668.

BRIDGE, J.S. & LEEDER, M.R. (1979) A simulation model of alluvial stratigraphy. *Sedimentology*, **26**, 617–644.

BRINKMAN, R. (1976) *Geology of Turkey.* Enke, Stuttgart. 158 pp.

BRUNN, J.H., DUMONT, J.F., GRACIANSKY, P.C. DE, GUTNIC, M., JUTEAU, T., MARCOUX, J., MONOD, O. & POISSON, A. (1971). Outline of the geology of the western Taurides. In: *Geology and History of Turkey* (Ed. by A.S. Campbell). Petroleum Exploration Society of Libya, Tripoli.

BRUNN, J.H., GRACIANSKY, P.C. DE, GUTNIC, M., JUTEAU, T., LEFEVRE, R., MARCOUX, J., MONOD, O. & POISSON, A. (1970) Structures, Majeures et corrélations stratigraphiques dans les Taurides occidentales. *Bull. Soc. géol. Fr.* **12**, 515–551.

BULL, W.B. (1972) Recognition of alluvial-fan deposits in the stratigraphic record. In: *Recognition of Ancient Sedimentary Environments* (Ed. by J.K. Rigby and W.K. Hamblin). *Spec. Publs Soc. econ. Paleont. Miner., Tulsa*, **16**, 63–83.

CAILLEUX, A. (1945) Distinctions des galets marins et fluviatiles. *Bull. Soc. géol. Fr.* **15**, 375–404.

CHURCH, M. (1972) Baffin Island sandurs: a study of arctic fluvial processes. *Bull. Can. geol. Surv.* **216** 208 pp.

CHURCH, M. & GILBERT, R. (1975) Proglacial fluvial and lacustrine environments. In: *Glaciofluvial and Glaciolacustrine Sedimentation* (Ed. by A.V. Jopling and B.C. McDonald). *Spec. Publs Soc. econ. Palaeont. Miner., Tulsa*, **23**, 22–100.

CLIFTON, H.E. (1973) Pebble segregation and bed lenticularity in wave worked versus alluvial gravel. *Sedimentology*, **20**, 173–189.

COLLINSON, J.D. (1978) Alluvial sediments. In: *Sedimentary Environments and Facies* (Ed. by H.G. Reading), pp. 15–60. Blackwell Scientific Publications, Oxford.

DABRIO, C.J. (1975) La sedimentacion arrecifal Neogena en la region del rio Almonzana. *Estudios Geol.* **41**, 285–296.

DAILY, B., MOORE, P.S. & RUST, B.R. (1980) Terrestrial-marine transition in the Cambrian rocks of Kangaroo Island, South Australia. *Sedimentology*, **27**, 379–399.

DELAUNE-MAYÈRE, M., MARCOUX, J., PARROT, J.-F. & POISSON, A. (1977) Modèle d'évolution mésozoïque de la paléomarge Tethysienne au niveau des nappes radiolaritiques et ophiolitiques de Taurus Lycian, d'Antalya et du Baer-Bassit. In: *Structural History of the Mediterranean Basins* (Ed. by B. Biju-Duvaal and L. Montadert), pp. 79–94. Editions Technip, Paris.

DUMONT, J.F., GUTNIC, M., MARCOUX, M., MONOD, O. & POISSON, A. (1972) Le Trias des Taurides occidentales (Turquie). Définition du bassin pamphylien: un nouveau domaine à ophiolithes à la marge externe de la chaîne taurique. *Z. geol. Ges.* **123**, 385–409.

ENOS, P. (1977) Flow regimes in debris-flows. *Sedimentology*, **24**, 133–142.

EYNON, G. & WALKER, R.G. (1974) Facies relationships in Plesitocene outwash gravels, southern Ontario: a model for bar growth in braided rivers. *Sedimentology*, **21**, 43–70.

FRIEDMAN, G.M. & SANDERS, J.E. (1978) *Principles of Sedimentology*. Wiley, New York. 792 pp.

GILBERT, G.K. (1890) Lake Bonneville. *Mon. U.S. geol. Surv.* **1**.

GRACIANSKY, P.C. DE (1972) *Recherches géologiques dans le Taurus Lycian Occidental*. Thèse. Université de Paris-Sud. 571 pp.

GWIRTZMAN, G. & BUCHBINDER, B. (1978) Recent and Pleistocene coral reefs and coastal sediments of the Gulf of Elat. *Field trip Guidebk*, 10th int. Congr. Sedim., Jerusalem, pp. 163–189.

HARMS, J.C. & FAHNESTOCK, R.K. (1965) Stratification, bedforms and flow phenomena (with an example from the Rio Grande). In: *Primary Sedimentary Structures and their Hydrodynamic Interpretation* (Ed. by G.V. Middleton). *Spec. Publs Soc. econ. Paleont. Miner.*, Tulsa, **12**, 84–115.

HARMS, J.C., SOUTHARD, J.B., SPEARING, D.R. & WALKER, R.G. (1975) Depositional environments as interpreted from primary sedimentary structures and stratification sequences. *Soc. econ. Paleont. Miner.*, Short Course Notes, Dallas, **2**, 161 pp.

HARMS, J.C., TACKENBERG, P., PICKLES, C. & POLLOCK, R.E. (1981) The Brae oilfield area. In: *Petroleum Geology of the Continental Shelf of N.W. Europe* (Ed. by L.V. Illing and G.D. Hobson), pp. 352–357. Heyden & Son, London.

HAYWARD, A.B. (1982) Coral reefs in a clastic sedimentary environment. Miocene S.W. Turkey, Recent Red Sea. *Coral Reefs*, **1**.

HEIN, F.J. & WALKER, R.G. (1977) Bar evolution and development of stratification in the gravelly, braided, Kicking Horse River, British Columbia. *Can. J. Earth Sci.* **14**, 562–570.

HOLMES, A. (1965) *Principles of Physical Geology*. Nelson, London. 1288 pp.

HOWELL, D.G. & LINK, M.H. (1979) Eocene conglomerate sedimentology and basin analysis, San Diego and the southern California Borderland. *J. sedim. Petrol.* **49**, 517–540.

JACKSON, R.G. (1978) Preliminary evaluation of lithofacies models for meandering alluvial streams. In: *Fluvial Sedimentology* (Ed. by A.D. Miall). *Mem. Can. Soc. Petrol. Geol.*, Calgary, **5**, 543–576.

JOHNSON, H.D. (1978) Shallow siliciclastic seas. In: *Sedimentary Environments and Facies* (Ed. by H.G. Reading), pp. 207–258. Blackwell Scientific Publications, Oxford.

JOPLING, A.V. (1965) Hydraulic factors controlling the shape of laminae in laboratory deltas. *J. sedim. Petrol.* **35**, 777–791.

LEOPOLD, L.B. & WOLMAN, M.G. (1957) River channel patterns: braided, meandering and straight. *Prof. Pap. U.S. geol. Surv.* **282**, 39–85.

MARTINI, I.P. (1977) Gravelly flood deposits of Irvine Creek, Ontario, Canada. *Sedimentology*, **24**, 603–622.

MCLEAN, R.J. & JERZYKIEWICZ, T. (1978) Cyclicity, tectonics and coal: some aspects of fluvial sedimentology in the Brazeau-Paskapoo Formations, Coal Valley area, Alberta, Canada. In: *Fluvial Sedimentology* (Ed. by A.D. Miall). *Mem. Can. Soc. Petrol. Geol.*, Calgary, **5**, 441–468.

MIALL, A.D. (1977) A review of the braided river depositional environment. *Earth Sci. Rev.* **3**, 1–62.

MIALL, A.D. (1978) Lithofacies types and vertical profile models in braided river deposits: a summary. In: *Fluvial Sedimentology* (Ed. by A.D. Miall). *Mem. Can. Soc. Petrol. Geol.*, Calgary, **5**, 597–604.

ODELL, J. (1977) Description in the geological sciences and the lithostratigraphic description system. L.S.D. ϕ2. *Geol. Mag.* **114**, 81–114.

POISSON, A. (1977) *Recherches géologiques dans les Taurides occidentales (Turquie)*. Thèse de Docteur des Sciences. Université de Paris-Sud. 795 pp.

READING, H.G. (Ed.) (1978) *Sedimentary Environments and Facies*. Blackwell Scientific Publications, Oxford. 557 pp.

ROBERTSON, A.H.F. & WOODCOCK, N.H. (1980) Strike-slip related sedimentation in the Antalya Complex, S.W. Turkey. In: *Sedimentation in Oblique-Slip Mobile Zones* (Ed. by P.F. Ballance and H.G. Reading). *Spec. Publs int. Ass. Sediment.* **4**, 127–145. Blackwell Scientific Publications, Oxford.

RUST, B.R. (1972) Structure and process in a braided river. *Sedimentology*, **18**, 221–245.

RUST, B.R. (1975) Fabric and structure in glaciofluvial gravels. In: *Glaciofluvial and Glaciolacustrine Sedimentation* (Ed. by A.V. Jopling and B.C. McDonald). *Spec. Publs Soc. econ. Paleont. Miner.*, Tulsa, **23**, 238–248.

RUST, B.R. (1978) Depositional models for braided alluvium. In: *Fluvial Sedimentology* (Ed. by A.D. Miall). *Mem. Can. Soc. Petrol. Geol.*, Calgary, **5**, 605–625.

RUST, B.R. (1979) Coarse alluvial deposits. In: *Facies Models* (Ed. by R.G. Walker). *Geosci. Can. Reprint Series 1*, Geol. Ass. Can. 9–21.

SELLWOOD, B.W. (1972) Tidal-flat sedimentation in the Lower Jurassic of Bornholm, Denmark. *Palaeogeog. Palaeoclim. Palaeoecol.* **11**, 93–106.

SMITH, N.D. (1970) The braided stream depositional environment: comparison of the Platte River with some Silurian clastic rocks, North-central Appalachians. *Bull. geol. Soc. Am.* **81**, 2993–3014.

SMITH, N.D. (1974) Sedimentology and bar formation in the upper Kicking Horse River, a braided outwash stream. *J. Geol.* **82**, 205–223.

SNEH, A. (1979) Late Pleistocene fan-deltas along the Dead Sea Rift. *J. sedim. Petrol.* **49**, 541–552.

STEEL, R.J. & AASHEIM, S.M. (1978) Alluvial sand deposition in a rapidly subsiding basin (Devonian, Norway). *Fluvial Sedimentology* (Ed. by A.D. Miall). *Mem. Can. Soc. Petrol. Geol.*, Calgary, **5**, 605–625.

WALKER, R.G. (Ed.) (1979) *Facies Models. Geosci. Can. Reprint Series 1*, Geol. Ass. Can. 211 pp.

WALKER, T.R. (1967) Formation of red beds in modern and ancient deserts. *Bull. geol. Soc. Am.* **78**, 353–368.

WESCOTT, W.A. & ETHRIDGE, F.G. (1980) Fan-delta sedimentology and tectonic setting—Yallahs Fan delta, southeast Jamaica. *Bull. Am. Ass. Petrol. Geol.* **64**, 374–399.

WILLIAMS, P.F. & RUST, B.R. (1969) The sedimentology of a braided river. *J. sedim. Petrol.* **39**, 649–679.

Spec. Publs int. Ass. Sediment. (1983) **6**, 337–344

Lithofacies of the Markanda terminal fan, Kurukshetra district, Haryana, India

B. PARKASH, A. K. AWASTHI *and* K. GOHAIN

Department of Earth Sciences, University of Roorkee, Roorkee 247672, India

ABSTRACT

The Markanda river is an ephemeral, straight stream which forms a terminal fan in Kurukshetra district, Haryana, India. It has a width of about 80 m and maximum monthly flow volume of 58×10^6 m³. The surface of the fan is marked by a network of abandoned streams forming a distributive pattern. Surficial sediments consist of channel and overbank deposits. The channel sediments generally consist of fining-up sequences (15 cm–2 m thick) starting with an erosional surface, without marked relief, overlain by a sequence of interbedded trough/planar cross-bedded and horizontal bedded sand facies, passing up into cross-laminated and flaser-bedded sand facies. Further up the channel sand may pass up into horizontal laminated silt and massive mud facies. There is a distinct decrease in thickness of channel sands and increase in percentage thickness of horizontal-bedded sand facies from apex to toe of the fan. The overbank sediments consist of thin, climbing ripple laminated sand, trough cross-laminated sand, interbedded mud, silt and sand facies passing into thick (10–80 cm) massive mud facies with rootlets and desiccation cracks. Major deposition in the channels and inter-distributary area seems to have taken place by vertical accretion.

INTRODUCTION

Friend (1978) pointed out that some ancient fluviatile sequences were characterized by features which suggested their accumulation in river systems with lobate topographic form and downstream decrease in river depth. These fluvial formations lacked distinct river incision. Friend (1978) thought that these fluviatile deposits were formed by deposition on terminal fans, such as those described by Mukherji (1976) from the plains of the Sutlej–Yamuna rivers in northern India. Though the morphological aspects of these fans have been described in detail by Mukherji (1975, 1976) little is known regarding their sedimentological characters. Keeping the above in view, this paper describes lithofacies, sedimentary structures and textures of the well-defined terminal fan of the Markanda river in Kurukshetra district of Haryana state, India (Fig. 1).

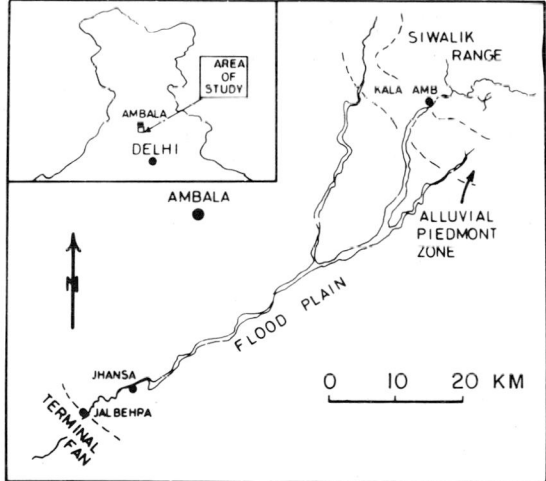

Fig. 1. Location map of the area under investigation.

Physiography

The Markanda river terminal fan covers an area of about 64 km². The whole area is flattish, the highest and lowest points in the area being 825′ (251·7 m) and 805′ (245·5 m) above M.S.L. The Markanda river originates in the outer Himalayan ranges (Siwalik Ranges), traverses through the alluvial piedmont zone in the foothills and forms a floodplain, where it is joined by two small tributaries. Further downstream it forms a terminal fan with an overall concave-up slope as shown by contours (Figs 1 and 3). The total length of the river is 112 km from its source to its terminus. The river bifurcates near Jalbehra village (Figs 1 and 3) at the apex of the fan; the main stream flows south-west and the other small distributary flows towards the south.

The widths of the channels are not uniform along their entire course. Near Jhansa (Fig. 1) the width of the river is 100 m, near Jalbehra about 80 m and it decreases downstream for the main stream. At location T_{22} the channel width is only 25 m. The channel may be incised up to 2 m near the fan apex; bank heights decrease downstream. The older topographic maps show that the larger distributaries become very narrow near the toe of the fan and take a meandering course after leaving the fan-surface.

Both the main channel and the distributary channel

have been restricted by artifical levees in the last 25 years and the fan has been extensively cultivated. However, the older topographic maps and aerial photographs show numerous palaeochannels within the fan, and these indicate repeated bifurcation and multifurcation of the main channel leading to the formation of subfans and a distributive channel pattern (Mukherji, 1976, fig. 2).

Climate and hydrology

The area has a semi-arid type of climate. The average annual rainfall in the area is 53 cm. The maximum rainfall occurs in the months of July, August and September, recording a maximum monthly rainfall of 31 cm.

The discharge of the river is highly variable over the whole year, as can be seen from the flow data given in Table 1 for the period June 1976 to May 1978 for the Kala Amb Project site (Fig. 1), 78 km upstream of the fan apex. The river loses most of its discharge by the time it reaches the fan apex. In fact, except during monsoon months, the river bed is dry in the whole of the fan area.

LITHOFACIES

The Markanda terminal fan has been studied by making trenches and observing bank cuts in the floodplain at 36 locations (Fig. 3). The sedimentary structures and changes in grain size were noted in all cases. Based on sedimentary structures, nine lithofacies are recognized: trough cross-bedded sand, planar cross-bedded sand, horizontal bedded sand, horizontal laminated silt, shallow scours filled with

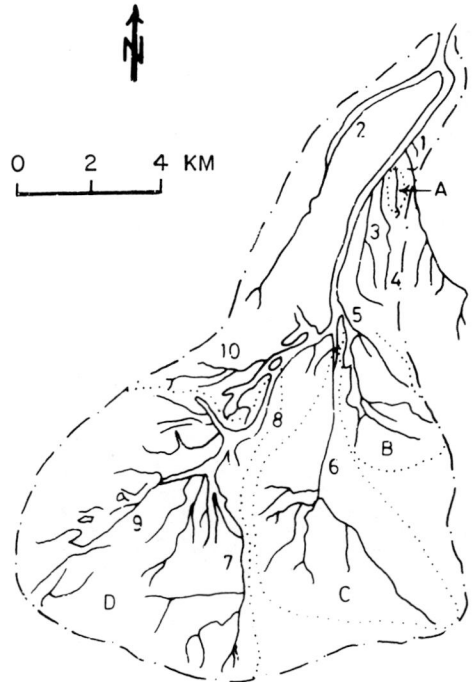

1,2 ...... MAJOR DISTRIBUTARIES

A,B,C,D_ SUBFANS

Fig. 2. Subfans and palaeochannels showing multifurcations on the Markanda terminal fan (after Mukherji, 1976).

Table 1. Monthly stream flow at Kala Amb project site (Fig. 1) $m^3 \times 10^6$

	1976	1977	1978
Jan.	—	—	1·3
Feb.	—	—	1·1
Mar.	—	—	0·9
Apr.	—	—	0·7
May	—	—	0·5
June	—	5·5	—
July	30·1	35·7	—
Aug.	58·2	31·6	—
Sept.	9·0	21·7	—
Oct.	—	7·0	—
Nov.	—	2·6	—
Dec.	—	1·6	—

—, data not available.

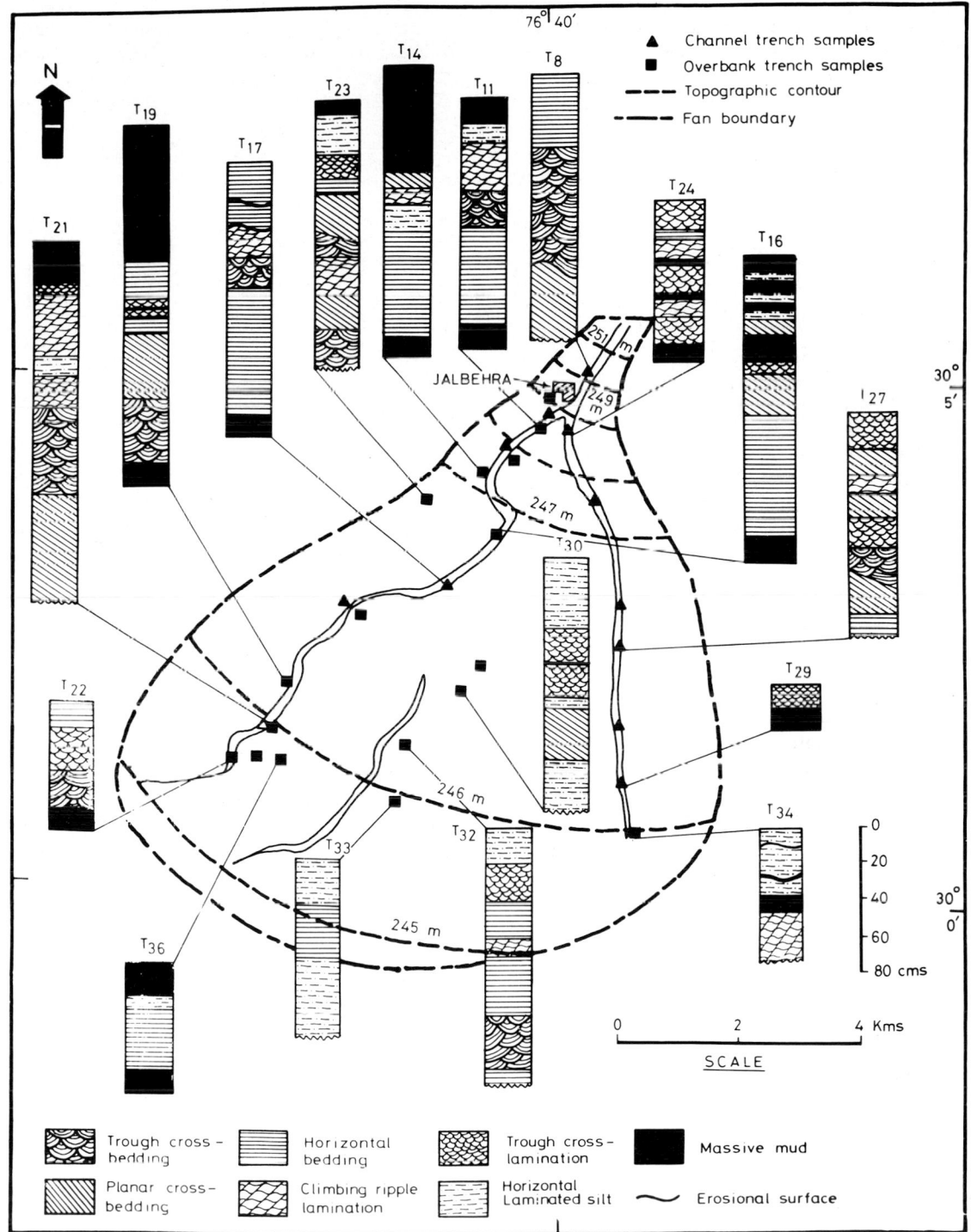

Fig. 3. Vertical sequences of sedimentary structures at various locations on the Markanda terminal fan.

sand, ripple cross-laminated sand, interlayered sand, silt and mud, flaser-bedded sand and mud or silty mud facies. A brief description of these facies is given below.

Trough cross-bedded sand

Cross-bed sets occur singly or in cosets. The facies thickness ranges from 8 to 52 cm. The thickness of the individual sets is variable, the maximum thickness being 26 cm. The base of this facies is generally an erosional surface. Sands are fine grained and well sorted.

Planar cross-bedded sand

Both solitary sets and cosets of planar cross-beds are present. Total thickness varies from 8 to 50 cm. The dip of the foreset laminae ranges from 5° to 32°, low-angle cross-beds occurring rarely. At places these beds are marked by reactivation surfaces (Fig. 4). Grain size and sorting characteristics are similar to those of the trough cross-bedded sand facies.

Horizontal bedded sand

The facies consists of essentially horizontal, laminated units, though thin lenses of cross-laminated sand measuring up to 2 cm are present in places. Laminations are continuous and particularly distinct, generally due to the presence of mica. The facies thickness ranges from 3 to 80 cm (Fig. 5). This facies is sometimes cut by sand dykes. In some places lamination may show slight irregular undulations, due to soft-sediment deformation. This facies is probably a product of deposition under high-velocity conditions in the plane bed phase of the upper flow regime.

Horizontally laminated silt

This facies is characterized by a mean grain size of 0·03–0·06 mm and rarely about 0·10 mm and shows faint, but recognizable horizontal lamination. Rarely, streaks of mud may be present. Rootlets may be present towards the top of the facies. This facies grades up into mud facies at a number of places. The facies is considered to have been deposited from suspension clouds at current velocities below those producing ripples.

Fig. 4. Reactivation surfaces in cross-bedded sand, Section T_{17} (Fig. 3). Reactivation surfaces originated due to fluctuations in flow. Each successive flow was marked by higher discharge and depth as indicated by larger cross-bed set thickness from bottom to top of the coset.

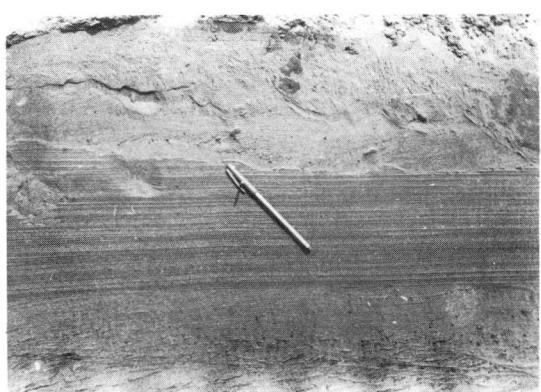

Fig. 5. Distinct erosional surface at the top of horizontal-bedded sand in the middle of the photograph. The bed overlying the erosional surfaces contains mud pebbles.

Scour-fill sand

This facies consists of minor asymmetrical troughs with their larger axes parallel to the flow direction. Generally scours are filled with sands, plant debris and mud pebbles. Faint laminae can be seen in scour-fill sediments. Scour-fill sands may form due to pulsations in velocity during floods. Sometimes flattish erosional surfaces are seen transecting the older sandy facies (Fig. 5) and sediments overlying the erosional surfaces are poorly bedded sands with mud pebbles. A maximum of two such surfaces are observed within the observed sandy horizons; these are encountered only in the upper half of the fan.

Fig. 6. Transitions between types 1 and 2 (lower portion) ripple laminae in drift passing up into type 2 ripple laminae in drift (upper portion). Location 1 km downstream of T_{17}.

Fig. 7. Flaser bedding in top portion of the distributary channel sediment. Section T_{24}.

Ripple cross-laminated sand

This facies comprises a variety of small-scale cross-bedding, namely climbing ripple lamination (Fig. 6) and small-scale planar and trough cross-bedding. Bed thickness of climbing ripple laminated units ranges up to 60 cm. Climbing angles have been recorded up to 21°. The climbing ripple lamination can be classified as ripple laminae in drift (McKee, 1965) and types 1 and 2 ripple laminae in drift (Jopling & Walker, 1968). Both trough and planar cross-laminations occur and facies unit thickness ranges from 6 to 60 cm. This facies is composed of very fine sand to silt, well to moderately sorted.

Flaser-bedded sand

Flaser bedding (Fig. 7) was observed to be closely associated with cross-laminated sand. Mud streaks were observed in troughs of ripples in very fine sand.

Flaser bedding is thought to have been produced during the falling stage of a flood, when fluctuations of flow deposited fine sand/silt and mud alternately.

Laminated sand, silt and mud

This facies is characterized by the presence of inter-layered sand, silt and mud, and coarser layers exhibit faint horizontal lamination. Units can attain a maximum thickness of about 1 m. Convolute bedding was observed in this facies near the toe of the fan.

Mud or silty mud

This facies includes massive mud and silty mud. It exhibits rare faint lamination. The top surface often shows hexagonal desiccation cracks. Rootlets are fairly common. Mud occurs as drapes over undulating surfaces of a few centimetres relief in places.

Other channel features

The present studies were carried out in the months of January and February 1980 when the stream carried no discharge over the fan. Parts of the channel, even near the fan apex, were plastered with mud showing desiccation cracks. Some sand bars with straight crests, trending almost perpendicular to the channel direction and extending over half the width of the channel, were observed. These bars had no well-developed slip faces, heights of 50–80 cm and wavelengths of 15–20 m. The upper surfaces of these bars were in places reworked by wind into tiny ripples. At a few places a thin cover of mud was observed. Trenches in these waves revealed the presence of small planar cross-beds at the base, overlain by trough or climbing ripple lamination with occasional flaser beds, especially towards the top. These bars probably originated as transverse bars and have been modified by the falling stage of flood.

Lateral and vertical distribution of lithofacies

Three subenvironments can be recognized on the Markanda fan, i.e. channel, natural levee and floodplain. Each environment is marked by characteristic sequences of lithofacies. The channel sediments normally overlie older clay beds with an erosional surface without any marked relief, at least over the studied cuts, of a couple of metres in lateral extent. The erosion surface is overlain by interbedded horizontal bedded sand, planar and trough cross-bedded sands and, towards the top of the sequence, trough cross-laminated sand, climbing ripple laminated sand, horizontal bedded silt and flaser-bedded sand are common. These, in turn, may pass up into clay at the top. These sediments show overall fining-up sequences. However, in a few cases, as at sections T_{14}, T_{19}, T_{21}, T_{23} and T_{27}, mainly from the upper half of the fan, multi-storey sequences containing two or three fining-up sequences can be recognized. These sequences may have formed by reoccupation of the abandoned parts of the channels or older channels and renewed sedimentation. The total thickness of the channel sediments decreases from about 2 m at the apex to about 80 cm towards the toe of the fan, and cross-bedded sand facies seem to be replaced by horizontal-bedded sand facies towards the toe of the fan.

The levee sediments comprise climbing ripple and trough cross-laminated sand facies and alternations of sand, silt and mud facies. The floodplain deposits are composed typically of massive mud or silty mud facies and minor amounts of cross-laminated silts. Since both levee and floodplain deposits are deposited by overbank flooding of the fan, these will be treated as overbank deposits in further discussion.

GRAIN-SIZE ANALYSIS

A total of 11 channel bed samples and 77 trench samples were analysed for their grain-size distributions. The analysis was carried out using A.S.T.M. sieves at $\frac{1}{2}\phi$ intervals; the pan fractions with more than 5% of the sample weight were analysed with the help of a Sartorius Sedimentation Balance. The cumulative weight percentage curves for typical samples are shown in Fig. 8.

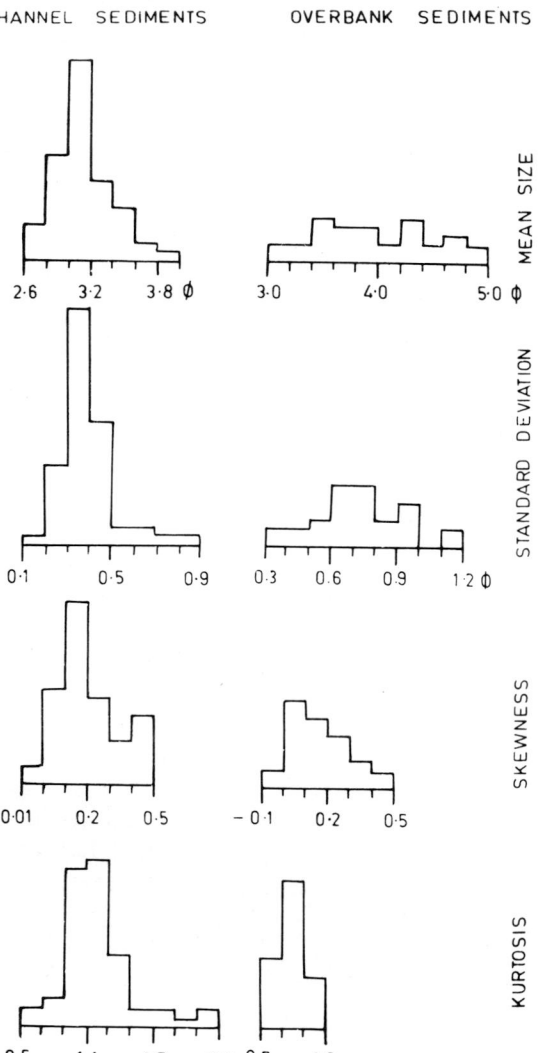

Fig. 9. Histograms showing distribution of grain-size parameters of fan sediments. Numbers of channel and overbank samples used in the above figures are 57 and 31 respectively.

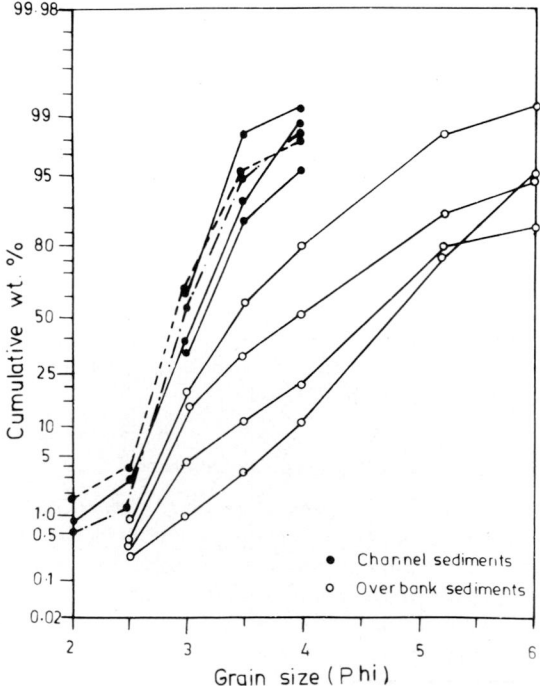

Fig. 8. Grain-size distribution curves for selected fan sediments.

Most of the grain-size curves for channel and overbank sediments are marked by breaks at about 3·5 and 3·0φ respectively, suggesting the presence of two populations. The coarse and fine populations are interpreted as saltation and suspension populations respectively. The overall nature of the curves is similar to those of the Brazos river sands given by Visher (1969).

Variation in statistical parameters

Various grain-size statistical parameters (Folk & Ward, 1957) have been calculated for all samples. Histograms showing distribution of different grain-size parameters of channel and overbank sediments are given in Fig. 9. Channel sediments are fine to very fine sands and are usually very well to well sorted. Their sorting characteristics seem to be more akin to beach sediments than to the fluvial sediments as decribed by Friedman (1967). These exhibit a wide variation in skewness, that is, they are symmetrical to very positively skewed. Overbank sediments are mostly of silt grade, typically moderately sorted and symmetrical to positively skewed. Though both channel and overbank sediment are leptokurtic and mesokurtic, channel sediments tend to be more commonly leptokurtic and overbank sediments tend to be mesokurtic.

Spatial variation

General trends in the variation of different grain-size parameters of the lowermost sandy bed encountered in trenches on the fan surface are shown by contours in Fig. 10. There is a distinct decrease in grain size and a worsening of sorting from apex to toe of the fan. Kurtosis and skewness show a marked change in a narrow zone around the mean size contour of 3·5φ; both increase in value from the apex to this zone and then there is an almost abrupt decrease in their values. Further downslope a slight decrease in their values is observed.

DISCUSSION AND CONCLUSIONS

Observations in the top 2·5 m of fan sediments indicate that the major portion of the fan is underlain by a widespread, and almost continuous layer of channel sediments. Vertical changes in grain size and sedimentary structures in these sands indicate their deposition from decelerating currents. Such sequences have probably resulted from the process of frequent multifurcation of the channel, described by Mukherji (1976), and the filling of these distributaries by waning currents and later their abandonment.

The decrease in depth and width of the main and

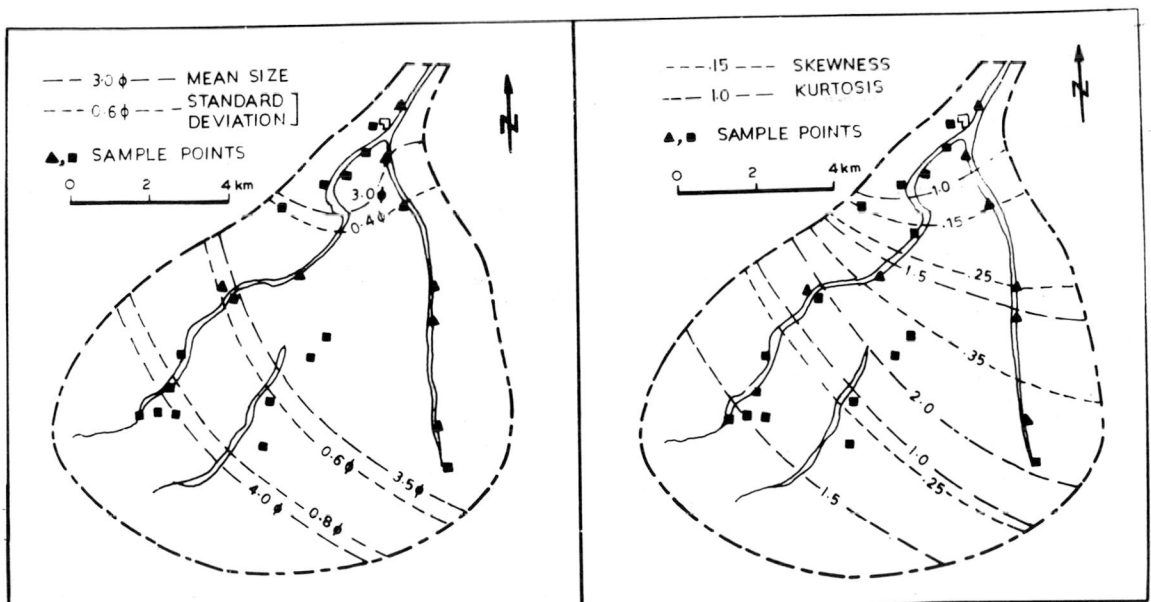

Fig. 10. Contours showing changes in grain-size parameters in the lowermost sandy beds observed in trenches on the surface of the fan.

distributary channels in the downstream direction suggests that the stream loses a fair part of its discharge through infiltration and evaporation, leading to rapid deposition of sediments. Loss in discharge may be due more to lateral infiltration than vertical infiltration because of the presence of the widespread clay layer below the top sandy deposit. Decrease in stream power of the earlier depositing currents is also suggested by the decrease in grain size, the changes in sedimentary structures and decrease in thickness of sandy facies from apex to foot of the fan. Another aspect of the Markanda river worth noting is that it originates in the loosely cemented and highly dissected sediments of the Siwalik Group and carries a considerably larger sediment load during the rainy season. The highly variable discharge characteristics of rivers in the monsoonal region and the large sediment load of the streams during the rainy season, due to the abundant availability of loose detritus in the source area, combined with the loss of discharge due to infiltration, must have been responsible for deposition in midstream, choking the channel and resulting in bifurcation or multifurcation of the channel into distributaries. This process is akin to avulsion in deltas. It also resulted in the formation of subfans on the terminal fan (Fig. 2). The formation of subfans was somewhat similar to the development of supra-fan lobes on the submarine fans described by Walker (1979).

The deposition in the channels has been taking place in two distinct phases. The coarser facies such as planar and trough cross-bedded sands and horizontal-bedded sands seem to have been deposited in the flood stage. The planar and trough cross-beds were formed by deposition during migration of transverse bars and dunes respectively, whereas the horizontal-bedded sand was deposited in the plane-bed phase of the upper flow regime. The finer facies such as ripple cross-laminated sand, laminated sand, silt and mud, flaser-bedded sand and horizontal-laminated silt were deposited during the falling stage of the flood. Sheet flow during the falling stage may have reworked tops of bars to form cross-laminated sand facies as observed in channels at the present. The general deposition in the channels appears to have taken place by vertical accretion and not by lateral accretion on point bars as in the case of meandering streams. The evidence for this is the almost straight nature or low sinuosity of the channels, the absence of point bars and the rare presence (only in the distal portion of the fan) of side bars in the channels, the abundance of cosets of planar and trough cross-beds and the horizontal bedding and lack of the epsilon cross-beds of Allen (1963).

Frequent flooding of the inter-distributary areas is suggested by the presence of fairly thick overbank deposits in these areas. Their preservation potential seems to be as good as that of channel sediments.

The unusually well- to very well-sorted nature of many of the fan sediment samples is probably due to the fact that the source area consists of moderately sorted fluviatile Siwalik sediments which on reworking and transportation have become better sorted.

ACKNOWLEDGMENTS

We are grateful to Drs J. D. Collinson and P. F. Friend, who made many helpful suggestions for the improvement of the manuscript. We are also thankful to Professor R. S. Mithal, Emeritus Scientist, for the encouragement provided during the progress of the work.

REFERENCES

ALLEN, J.R.L. (1963) The classification of cross-stratified units, with notes on their origin. *Sedimentology*, **2**, 93–114.

FOLK, R.L. & WARD, W.C. (1957) Brazos River bar: a study in the significance of grainsize parameters. *J. sedim. Petrol.* **27**, 3–26.

FRIEDMAN, G.M. (1967) Dynamic processes and statistical parameters compared for size frequency distribution of beach and river sands. *J. sedim. Petrol.* **37**, 327–354.

FRIEND, P.F. (1978) Distinctive features of some ancient river systems. In: *Fluvial Sedimentology* (Ed. by A.D. Miall). *Mem. Can. Soc. Petrol. Geol. Calgary*, **5**, 531–542.

JOPLING, A.V. & WALKER, R.G. (1968) Morphology and origin of ripple-drift cross-lamination, with examples from the Pleistocene of Massachusetts. *J. sedim. Petrol.* **38**, 971–984.

MCKEE, E.D. (1965) Experiments on ripple-lamination. In: *Primary Sedimentary Structures and their Hydrodynamic Interpretation* (Ed. by G.V. Middleton). *Spec. Publs Soc. econ. Paleont. Miner.*, Tulsa, **12**, 66–83.

MUKHERJI, A.B. (1975) Geomorphic patterns and processes in the terminal triangular tract of inland streams in Sutlej-Yamuna plain. *J. geol. Soc. India*, **16**, 450–459.

MUKHERJI, A.B. (1976) Terminal fans of inland streams in Sutlej-Yamuna plain, India. *Geomorph.* **20**, 190–204.

VISHER, G.S. (1969) Grain size distributions and depositional processes. *J. sedim. Petrol.* **39**, 1074–1106.

WALKER, R.G. (1979) Turbidites and associated coarse clastic deposits. In: *Facies Models* (Ed. by R.G. Walker). *Geosci. Can. Reprint Series*, **1**, 91–103.

Spec. Publs int. Ass. Sediment. (1983) **6**, 345–354

Towards the field classification of alluvial architecture or sequence

P. F. FRIEND

Department of Earth Sciences, University of Cambridge, Downing Street, Cambridge CB2 3EQ, U.K.

ABSTRACT

Present models of river sedimentation are based largely on a classification of the geomorphology of present-day river channel patterns. The classes are difficult to apply with confidence to ancient sediments, even where good, three-dimensional exposures are available. Although many workers have often referred to the deposits of (a) meandering or (b) braided rivers, present knowledge suggests that these terms should only be used where there is direct evidence of (a) high sinuosity and/or lateral migration or (b) emergent bars (islands).

A broad transport-mode classification can be useful, based on grain-size and distinguishing 'mainly bed-transport', 'mainly suspension', and 'bed-transport and suspension'. A descriptive classification of two- or three-dimensional exposures is proposed, which is based on the shape and interrelations of distinctive sediment bodies, e.g. sheet or ribbon body form, or, more genetically, channel-plugs or bar forms.

Environmental classification tends to concentrate on the degree and mobility of channelization during deposition. Useful classes are: (1) sheet flood, (2) fixed channel, (3) mobile channel. Terms such as meandering or braided may then be suffixed to any of these classes, and some estimation of indices of sinuosity or braiding may locally be possible.

All too often, knowledge of ancient alluvium is limited to more-or-less one-dimensional, 'sequence' information ('vertical' exposure logs, borehole logs). Without two-dimensional data any classification must be limited to transport-mode generalizations, although detailed studies of vertical patterns of palaeocurrent direction, structure and grain size may suggest more precise environmental models.

CLASSIFICATION OF PRESENT-DAY RIVERS

By channel pattern

The most widely used classification of river morphology is the one proposed by Leopold & Wolman (1957), who distinguished the following channel patterns: (1) braided, (2) meandering, (3) straight.

This Leopold & Wolman classification has recently been converted into a fully quantitative classification by Rust (1978), who proposed the use of numerical measures of channel sinuosities and of braiding.

Another channel pattern term, 'anastomosing', has been used in a variety of different ways. Leopold & Wolman (1957) used the term as synonymous with

braided, but Schumm (1968) used it for multiple channel systems that have distinctively stable channels, with low gradients and relatively large amounts of suspended load. Smith (1976) and Rust (1978) have used the term anastomosing in a similar sense to Schumm, for multichannel systems with high channel sinuosities.

By sediment load

Schumm (1968) proposed a classification of stable alluvial channels based on their sediment load: (1) suspended load and dissolved load channels (bed load < 3% of total load), (2) mixed load channels (bed load 3–11% of total load), (3) bed-load (bed-load > 11% of total load).

It is not possible, at the moment, to apply this classification with any quantitative precision to present-day channels. Bed-load is notoriously difficult to measure, and no method has yet been agreed of summarizing the great fluctuations that occur in the sediment loads of most rivers over various time periods. In practice, measures of channel cross-sectional shape and of channel perimeter grain size have been used as indices for this classification, because there is evidence that they vary systematically with the sediment load.

CLASSIFICATION OF ANCIENT FACIES USING GRAIN SIZE

Practical and theoretical attractions

The grain size of a bed of clastic sediment has the great advantage, compared with structural features, that it can be easily estimated, even with incomplete exposures. Grain size can be measured in all outcrops, in spite of diagenetic and weathering effects.

There is a gross correspondence between the grain size of alluvium and the gradient of the river that deposited it (Friend & Moody-Stuart, 1972). Also, a general classification of rivers by the grain size of their alluvium (e.g. Collinson, 1978a) corresponds with differences in their more local morphological features.

Many workers make a distinction between coarser-grained units (generally sand-grade or coarser), and finer units (silt-grade or finer). Sedimentary structures will often confirm that this grain-size distinction corresponds approximately to the distinction between bed-load and suspension deposits.

Schumm (1960) has shown that in many rivers of the American West channel sinuosity is closely correlated with the amount of finer-grained sediment in the banks and bed of the stream channel, and he linked this with his classification (already described) of channels using sediment load. Not only is there field evidence for this correlation, but the effect of the proportion of cohesive sediment on bank strength provides a logical explanation for the corresponding variation in morphology of the channels. This work, therefore, has lent support to the notion of classifying ancient facies using grain size.

Transport-mode classification of vertical sequences

Vertical grain-size profiles, measured either at outcrop or subsurface, provide a direct record of the time-sequence of net accumulation of sediment at one point on the alluvial surface.

It is an important step forward in the analysis of any sequence to apply to it the gross distinction, discussed above, between bed-transport and suspension deposits. Diagnostic sedimentary structures may not be present in many beds, and it may therefore be necessary to use grain-size criteria as an approximate index. Figure 1 illustrates some examples of this approach from my own work on the Devonian strata of Spitsbergen. I have subjectively made the assumption that medium siltstone units are suspension sediments, and that fine sandstone or coarser units are bed-transport units, leaving intermediate grain sizes (coarse siltstone and very fine sandstone) as uncertain. Other workers, using different types of observations on different sediments, may decide on a different criterion. Whatever the criterion, a general classification reflecting gross transport mode can be set up, e.g. (Fig. 1) 'bed transport and suspension deposits', 'mainly bed-transport deposits', and 'mainly suspension deposits'. Other sequences may show a systematic trend, with time, from one of these classes to another, and these might be described as 'suspension to bed-transport sequences', etc.

Problems in environmental classification

A major influence on the development of a classification of alluvial facies has been fig. 35 of J. R. L. Allen's review of 'The origin and characteristics of Recent alluvial sediments' (Allen, 1965). In this figure (my fig. 2), Allen illustrated four 'hypothetical models' 'of common alluvial facies': (1) piedmont formed of alluvial fans, (2) braided stream, (3) low sinuosity stream, (4) strongly meandering stream.

The names of models (2) to (4) correspond clearly with Leopold & Wolman's three classes of channel patterns. Allen's four models specified, in depth below the sediment surface, different geometrical arrangements and overall proportions of: (1) coarse fan, channel bar, or meander belt deposits, (2) mainly top stratum deposits, (3) other facies.

Simplifying these attractive generalizations even further, many subsequent authors (e.g. Friend & Moody-Stuart, 1972) used the overall proportion of coarser-grained sediment to finer-grained sediment as a means of distinguishing braided from meandering stream deposits. Vertical sequences of bed-load type, using the process classification of the previous section and Fig. 1, have tended to be interpreted automatically as braided river deposits, and vertical sequences of bed- and suspended-load type have been interpreted as meandering river deposits. This simplification now appears to have been unjustified in many cases.

It is now clear that there are problems in two types of environmental situations:

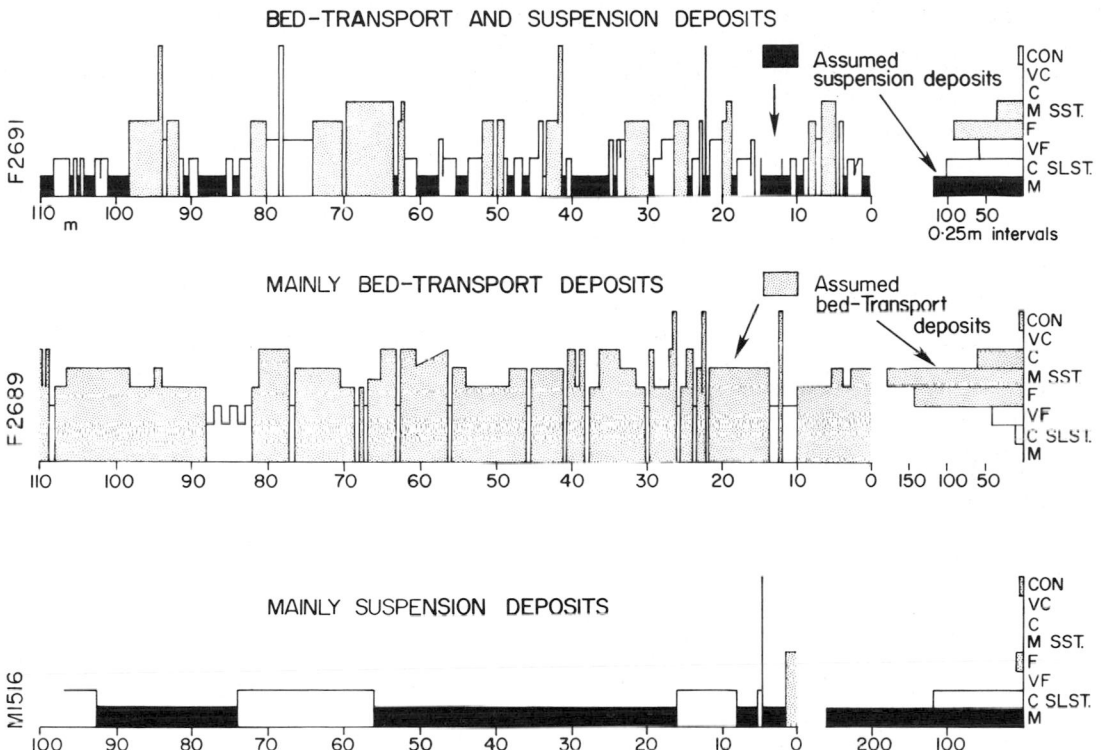

Fig. 1. Three logs from Devonian outcrops in Spitsbergen (for location, etc. see Friend & Moody-Stuart, 1972). Medium siltstone (arbitrarily assumed to be suspension deposits), and fine sandstone to conglomerate (arbitrarily assumed to be bed-transport) are distinguished, and the logs classified accordingly into 'bed-transport and suspension', 'mainly bed-transport', and 'mainly suspension' types. Histograms show, for each sequence, the number of 0.25 m intervals of each grain-size grade.

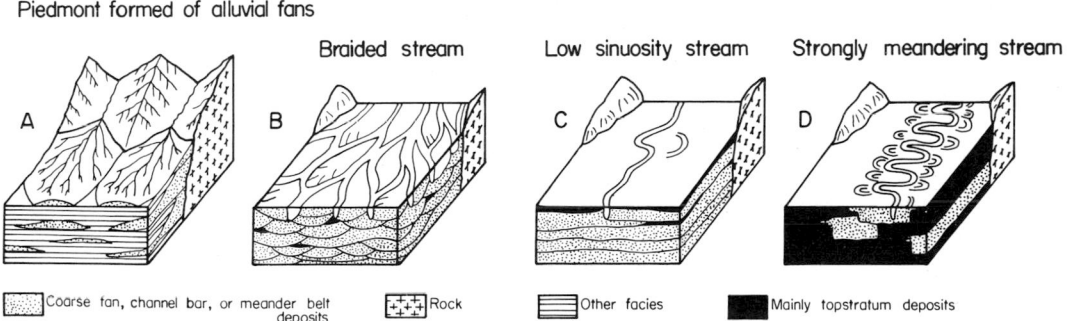

Fig. 2. Redrawn version of the hypothetical models of common alluvial facies (Allen, 1965, fig. 35).

(*a*) *Levées, crevasse splays and fans* (*including terminal fans—Friend, 1978*)

In all these systems, coarse-grained sediment may locally be introduced into an area of fine-grained sedimentation, e.g. the local flood-basin. The introduction may often take place in a relatively short-lived flood event, and in no way is it the result of 'stable' alluvial channel action, as envisaged by Schumm (1963, 1968). No balance need exist between the sediment introduced and the surrounding sedimentation (the 'other facies' indicated by Allen, 1965, fig. 35A, in his model of 'piedmont formed of alluvial fans'), and there is therefore no justification for using the relative proportions of coarse and fine sediment to indicate the sediment load or channel characteristics of the levée, splay or fan.

Many sandstone units previously interpreted, because they occur between intervals of siltstone, as the deposits of high-sinuosity (meandering) streams may be the results of these local introductions, where high sinuosity and relatively deep channels cannot be inferred.

(*b*) *Periods of climatic, tectonic or base-level change*

A number of alluvial basins are now known (e.g. Haszeldine & Anderton, 1980), where the basin-wide extent of sandstone units is clear evidence of basin-wide changes of alluvial style. Such allocyclic (Beerbower, 1964) changes must have been a response to controls external to the immediate processes of local alluvial sedimentation (Miall, 1980). These external controls (climatic, tectonic or base-level variations) must have changed alluvial style by affecting sediment or water input and/or through-flow (Schumm, 1968). In these cases, the overall proportion of fine-grained material in a vertical sequence spanning one or more of these events indicates the time-scale of the variations rather than the properties of the alluvial channels present.

Studies have recently been carried out (Allen, 1978; Bridge & Leeder, 1979), to simulate channel meander-belt sedimentation in a fine-grained alluvial plain setting. These studies have demonstrated various ways in which changes in the basinal subsidence rate and other variables may affect the overall proportion of coarse and fine sediment that accumulates. For example, the overall proportion of coarse-grained sediment may vary up to 100%, depending on the degree to which channel meander belts cut down into their predecessors or are separated by fine-grained over-bank sediments.

(*c*) *Conclusion*

It is my contention, therefore, that the overall proportion of fine-grained sediment in a sequence must not, by itself, be used as an index of channel pattern.

CLASSIFICATION OF ANCIENT FACIES, USING TWO- OR THREE-DIMENSIONAL MACROSCALE FEATURES

Recognition of alluvial bars

Following earlier workers (e.g. Allen, 1966), Jackson (1975) distinguished a hierarchy of scale in sandy bed-forms produced by flow. He classified bed-forms into: (1) *microforms*, typically cm–dm in wavelength, e.g. ripples; (2) *mesoforms*, typically m–10 m in wavelength, e.g. dunes, megaripples; (3) *macroforms*, typically 10 m–10 km in length, e.g. bars, sand waves, deltas. Jackson (1975, p. 1523) stressed the relative permanence of some of the large features: 'The largest bed-forms (macroforms), such as point-bars, respond to the geomorphological regime of the environment and are relatively insensitive to changes in fluid-dynamic regime during an individual dynamic event.'

Recognition of their size, relative to meso- or microform structures, is the key to recognizing surfaces as traces of parts of bars, or whole bars (Fig. 3), in ancient facies. The interpretation of epsilon

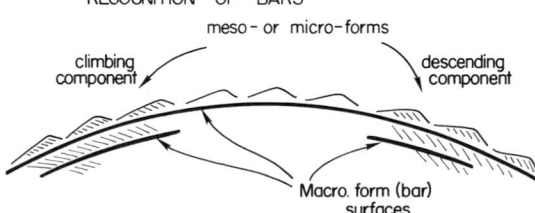

RECOGNITION OF BARS

Fig. 3. Structural features important in recognizing alluvial bars.

cross-bedding (Allen, 1963) as a result of accretion on bar surfaces is an example of this recognition. Epsilon cross-bedding is recognized by the presence of structures generated by smaller bed-forms that migrated along the cross-bedding surfaces, usually obliquely.

Although some river bars may be submerged by river waters throughout their active lives, many are also, at least to some extent, emergent. Evidence for this emergence is surprisingly rarely reported from ancient facies.

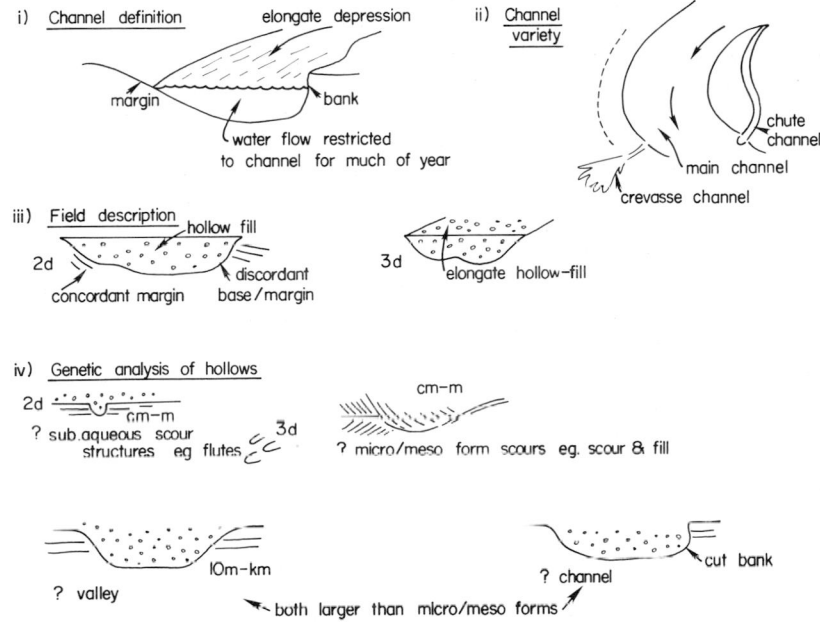

Fig. 4. Channel definition and recognition.

Recognition of channel cut-banks and channels (Fig. 4)

Although some river systems produce sheetflow (at least locally and temporarily), many others are characterized by one or more channels. Channels may be defined as elongate depressions in the alluvial surface, with more or less clearly defined margins or banks between which the river flow is restricted for most of the year. The identification of features of channel forms in an ancient facies is an important step in understanding the system that deposited it. It means that there *were* channels, and implies something about their size.

In many river systems there is more than one channel size, even at the same general locality. For instance, the main channel of a system may contain bars with chute channels, crevasse channels and even distributaries and branch channels.

A macroscale surface, broadly concave upward, in a two-dimensional outcrop face, is the usual evidence of an ancient channel. It seems wise to use a non-genetic term, such as 'hollow' to describe these features, because, although they may be the traces of old channel perimeters, they may also be the traces of subaqueous scour structures. These subaqueous

structures might be either scour-and-fill structures formed between bed-forms, or microscale structures such as flutes, formed at irregularities on the sediment surface. Some large 'hollows' may, at the other extreme, be the traces of ancient valleys, cut by smaller channels, while many hollows may represent only parts of the surface originally formed.

Geometry of sandstone bodies

In alluvial formations consisting of bed- and suspended-load sequences, the geometry of the sand units may be a distinctive characteristic. Friend, Slater & Williams (1979) distinguished 'ribbons' from 'sheets', using a width/depth ratio of 15 to separate the two classes. Some 'ribbons' have 'wings' extending laterally from distinct points of inflection in the basal scours.

Analysis and classification of channels and their behaviour (Figs 5 and 6)

The recognition of channels is a major step in the analysis of an ancient fluvial formation. If channels are not recognized, there may be various explanations, and these are outlined in the next section.

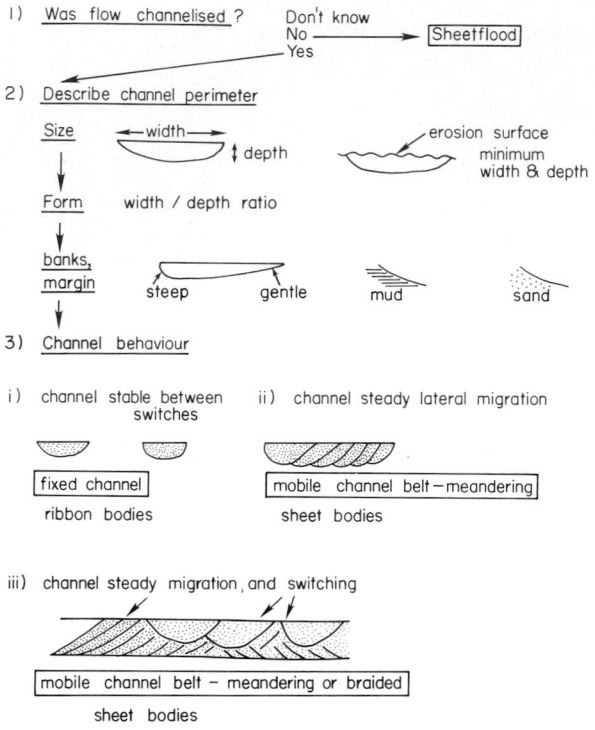

Fig. 5. Key for classification of channel behaviour.

If channel features are recognized, or at least suspected, minimum dimensions can be placed on the channel, and it may be possible to measure the profile of the banks and the nature of the sediment forming the banks. Width/depth ratios can be measured if entire channel perimeters are visible.

As outlined earlier in this paper, most attempts to classify rivers and their sediments have been based largely on the behaviour of the channels present. Figures 5 and 6 illustrate my view of the extent to which a key (Fig. 5) and a classification (Fig. 6) can be usefully specified, in practice, at the moment. The transport mode, based largely on grain size as discussed above, is also built into my classification.

Major classes (Fig. 6) are: (1) *sheetflood*, where the flow was not channelized; (2) *fixed channel*, where the channel was laterally stable between episodes of abrupt switching (e.g. Friend *et al.*, 1979); and (3) *mobile channel belt*, where a channel, or channels, over a period, occupied most sites within the belt. Steady lateral migration of a channel may be distinguished in some cases (e.g. Puigdefabregas & Van Vliet, 1978),

but combinations of migration and switching would be typical of many belts, whether braided or meandering (see Collinson, 1978b, fig. 1B).

Absence of recognizable channel and bar (macroform) features

Failure to recognize ancient channel features may be because of:

(1) *sheetflood* (the flow was not channelized);
(2) *major reworking* (early channel and bar perimeter surfaces may have been largely removed by subsequent channel reworking, so that only late geometries would be preserved);
(3) *low-angle channel perimeters* (channel surfaces were too flat to allow them to be distinguished);
(4) *sediment too uniform* (channel and bar surfaces are not visible because of lack of variation in the sediment);
(5) *outcrops unsuitable* (outcrops may be wrongly oriented, or too small, to allow channel features to be recognized).

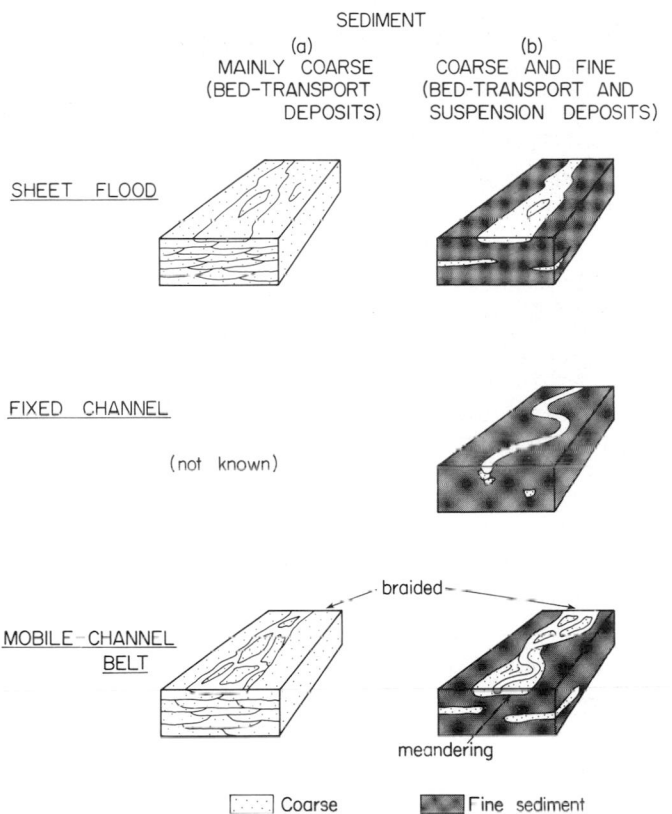

Fig. 6. Models of distinctive types of alluvial architecture and sequence.

Table 1 analyses the eleven descriptions of 'ancient fluvial systems' in the 'Fluvial sedimentology' volume edited by Miall (1978), using the general process classification and then concentrating on the recognition of channel and bar features. If these features are not recognized, then one or more of the above reasons is suggested. If channel and bar features are recognized, then the key described in the previous section is applied.

There is some evidence that recognizable channel features are more rare in older alluvial formations than in younger alluvial formations. This trend has been predicted, and explained as a result of the widespread effects on river regimes that the evolution of sediment-binding vegetation, through the Phanerozoic, might have had (Schumm, 1968).

LIMITATIONS AND DIRECTIONS IN VERTICAL SEQUENCE (1-D) ANALYSIS

All borehole data (whether core log or wire-line log) are essentially vertical sequence, one-dimensional data. So is much reconnaissance outcrop logging, although an experienced sedimentologist will probably detect some macroscale, two- or three-dimensional information, even in rather limited exposures.

Data on patterns of vertical grain-size variation are usually part of this vertical sequence information. Indeed, for descriptive classification, the total proportion of fine-grained sediment in the whole sequence, along with information on the average thickness of coarse and fine intervals, if present, are desirable summary statistics. The second section of this paper has reviewed some of the grave problems in using these gross grain-size distribution statistics to indicate directly the morphological class of the river responsible.

Table 1. Process classification, recognition of macroform structures and environmental classification of certain ancient fluvial facies (Miall, 1978)

(A) *Fluvial facies* (in order of decreasing stratigraphic age)
1   Archaean Moodies Group, Southern Africa (Eriksson, 1978)
2   Proterozoic, various formations, mainly Canadian (Long, 1978)
3   Precambrian Belt Supergroup, U.S.A. (Winston, 1978)
4   Ordovician to Carboniferous, many formations, Central Appalachian U.S.A. (Cotter, 1978)
5   Devonian, Western Norway (Steel & Aasheim, 1978)
6   Permian and Triassic, Ecca and Beaufort Groups, Eastern Karoo Basin, South Africa (Hobday, 1978)
7   Jurassic, Scalby Formation, Yorkshire, England (Nami & Leeder, 1978)
8   Upper Cretaceous–lower Tertiary, Brazeau-Paskapoo Formations, Alberta, Canada (McLean & Jerzykiewicz, 1978)
9   Tertiary, Southern Pyrenees, Spain (Puigdefabregas & Van Vliet, 1978)
10 Eocene Castisent Sandstone Formation, South Pyrenean Basin, Spain (Nijman & Puigdefabregas, 1978)
11 Plio-Pleistocene, Koobi Fora Formation, Kenya (Vondra & Burggraf, 1978)

(B) *Transport mode classification*
   Mainly bed-transport sequences (1, 2, 3, 4, 5, 6, 11)
   Bed-transport and suspension sequences (2, 3, 4, 5, 6, 7, 8, 9, 10)
   Suspension sequences (3, 5)

(C) *Recognition of macroform structures, environmental classification*
(1) No structures, or minor, shallow ones
        ?Sheetflood 3, 5, 11
        ?Major reworking 1, 2, 3, 4, 5
        ?Low-angle channel perimeters 1, 2, 3, 4, 5, 11
        ?Sediment too uniform 1, 2
        ?Outcrops unsuitable 1, 8, 11
(2) Structures described, environmental classification
        Mobile channel belt (*meandering*) 6, 7, 9, 10
        Mobile channel belt (*braided*) 7
        Mobile channel belt 4
        Fixed channel 6, 7

Although this paper has presented, so far, a gloomy view of the value of vertical sequence information in answering environmental questions about ancient river deposits, some methods of analysing these data, that are relatively promising, are outlined below.

(1) Numerical studies of patterns of sequence in grain size and structure within single coarse units. Although general trends of decreasing grain size and structures of decreasing flow strength (e.g. Allen, 1970) do not, by themselves, appear to be diagnostic of a particular river channel pattern, e.g. meandering (Jackson, 1978), some rather more specific vertical sequences may be diagnostic. An example of this comes from the work of Cant and Walker (Cant, 1978) in identifying planar cross-bedded sets in a generally trough-cross-bedded coarse unit, as the result of sedimentation of major flat bars (sand flats) in braided channel situations.

(2) Studies of palaeocurrent directions from bed to bed within a sequence. Data of this sort are not often available in borehole studies and generally require suitable outcrops. Cant and Walker (Cant, 1978, fig. 2) have described palaeocurrent patterns that might be expected in a particular system of bar movement in a braided channel complex.

(3) Studies of the thicknesses of units in any sequence. These may suggest a pattern of accumulation that helps to diagnose the sorts of river system present. An example of this comes from a bed-transport and suspension sequence of Devonian age in Spitsbergen (Fig. 7). Here no two- or three-dimensional channel information is directly visible, but from a vertical sequence log it is possible to examine the thickness of the bed-transport units present, using, in this case, the presence of micro- and mesoform structures as a bed-transport index. The thinnest units form a distinctive type, and are interpreted as the results of short-period sheetfloods. The intermediate units, 1–5 m thick, appear to be accumulations in channels, probably of smaller size than the channel in which the major (16·5 m) unit accumulated. It can be demonstrated that this thicker unit is not simply the result of amalgamation of a number of the intermediate units, because mesoform cross-stratification indicates larger bedforms in this particular case. It is concluded, therefore, that the

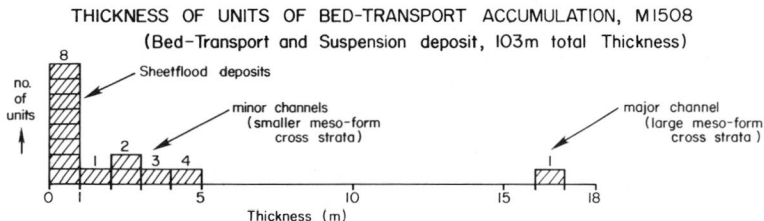

THICKNESS OF UNITS OF BED-TRANSPORT ACCUMULATION, M1508
(Bed-Transport and Suspension deposit, 103m total Thickness)

Fig. 7. Analysis of log from Devonian outcrop in Spitsbergen (for location, etc. see Friend & Moody-Stuart, 1972). Thicknesses of all the various units of continuous bed-load accumulation are plotted, using the presence of meso- and microform bed-load structures as a criterion. This plot allows some suggestions to be made about the channels originally present, see text.

river pattern included a major and some relatively minor channels, and that sheetflood events also occurred periodically.

ACKNOWLEDGMENTS

I am very grateful for the helpful comments of Drs Mike Leeder and Andrew Miall.

REFERENCES

ALLEN, J.R.L. (1963) The classification of cross-stratified units, with notes on their origin. *Sedimentology*, **2**, 93–114.

ALLEN, J.R.L. (1965) A review of the origin and characteristics of Recent alluvial sediments. *Sedimentology*, **5**, 98–191.

ALLEN, J.R.L. (1966) On bed forms and palaeocurrents. *Sedimentology*, **6**, 153–190.

ALLEN, J.R.L. (1970) Studies in fluviatile sedimentation: a comparison of fining-upward cyclothems, with special reference to coarse member composition and growth. *J. sedim. Petrol.* **40**, 298–323.

ALLEN, J.R.L. (1978) Studies in fluviatile sedimentation: an exploratory quantitative model for the architecture of avulsion controlled alluvial suites. *Sedim. Geol.* **21**, 129–147.

BEERBOWER, J.R. (1964) Cyclothems and cyclic depositional mechanisms in alluvial plain sedimentation. *Bull. Kansas geol. Surv.* **169**, 31–42.

BRIDGE, J.S. & LEEDER, M.R. (1979) A simulation model of alluvial stratigraphy. *Sedimentology*, **26**, 617–644.

CANT, D.J. (1978) Development of a facies model for sandy braided river sedimentation: comparison of the South Saskatchewan River and the Battery Point Formation. In: *Fluvial Sedimentology* (Ed. by A.D. Miall). *Mem. Can. Soc. Petrol. Geol., Calgary*, **5**, 627–640.

COLLINSON, J.D. (1978a) Alluvial sediments. In: *Sedimentary Environments and Facies* (Ed. by H.G. Reading), pp. 15–60. Blackwell Scientific Publications, Oxford.

COLLINSON, J.D. (1978b) Vertical sequence and sand body shape in alluvial sequences. In: *Fluvial Sedimentology* (Ed. by A.D. Miall). *Can. Soc. Petrol. Geol., Calgary*, **5**, 577–586.

COTTER, E. (1978) The evolution of fluvial style, with special reference to the central Appalachian Palaeozoic. In: *Fluvial Sedimentology* (Ed. by A.D. Miall). *Mem. Can. Soc. Petrol. Geol., Calgary*, **5**, 361–384.

ERIKSSON, K.A. (1978) Alluvial and destructive beach facies from the Archaean Moodies Group, Barberton Mountain Land, South Africa and Swaziland. In: *Fluvial Sedimentology* (Ed. by A.D. Miall). *Mem. Can. Soc. Petrol. Geol., Calgary*, **5**, 287–312.

FRIEND, P.F. (1978) Distinctive features of some ancient river systems. In: *Fluvial Sedimentology* (Ed. by A.D. Miall). *Mem. Can. Soc. Petrol. Geol., Calgary*, **5**, 531–543.

FRIEND, P.F. & MOODY-STUART, M. (1972) Sedimentation of the Wood Bay Formation (Devonian) of Spitsbergen: regional analysis of a late orogenic basin. *Norsk Polarinstitutt skrifter*, **157**, 1–77.

FRIEND, P.F., SLATER, M.J. & WILLIAMS, R.C. (1979) Vertical and lateral buildings of river sandstone bodies, Ebro Basin, Spain. *J. geol. Soc. London*, **136**, 39–46.

HASZELDINE, R.S. & ANDERTON, R. (1980) A braidplain facies model for the Westphalian B Coal Measures of north-east England. *Nature*, **284**, 51–53.

HOBDAY, D.K. (1978) Fluvial deposits of the Ecca and Beaufort Groups in the eastern Karoo Basin, Southern Africa. In: *Fluvial Sedimentology* (Ed. by A.D. Miall). *Mem. Can. Soc. Petrol. Geol., Calgary*, **5**, 413–430.

JACKSON, R.G., II (1975) Hierarchial attributes and a unifying model of bed forms composed of cohesionless material and produced by shearing flow. *Bull. geol. Soc. Am.* **86**, 1523–1533.

JACKSON, R.G., II (1978) Preliminary evaluation of lithofacies models for meandering alluvial streams. In: *Fluvial Sedimentology* (Ed. by A.D. Miall). *Mem. Can. Soc. Petrol. Geol., Calgary*, **5**, 543–576.

LEOPOLD, L.B. & WOLMAN, M.G. (1957) River channel patterns: braided, meandering and straight. *Prof. Pap. U.S. geol. Surv.* **282B**, 39–85.

LONG, D.G.F. (1978) Proterozoic stream deposits: some problems of recognition and interpretation of ancient sandy fluvial systems. In: *Fluvial Sedimentology* (Ed. by A.D. Miall). *Mem. Can. Soc. Petrol. Geol., Calgary*, **5**, 313–342.

MCLEAN, J.R. & JERZYKIEWICZ, T. (1978) Cyclicity, tectonics and coal: some aspects of fluvial sedimentology in the Brazeau-Paskapoo Formations, Coal Valley area, Alberta, Canada. In: *Fluvial Sedimentology* (Ed. by A.D. Miall). *Mem. Can. Soc. Petrol. Geol., Calgary*, **5**, 441–468.

MIALL, A.D. (Ed.) (1978) *Fluvial Sedimentology. Mem. Can. Soc. Petrol. Geol., Calgary*, **5**, 859 pp.

MIALL, A.D. (1980) Cyclicity and the facies model concept in fluvial deposits. *Bull. Can. Petrol. Geol.* **28**, 59–80.

NAMI, M. & LEEDER, M.R. (1978) Changing channel morphology and magnitude in the Scalby Formation (M. Jurassic) of Yorkshire, England. In: *Fluvial Sedimentology* (Ed. by A.D. Miall). *Mem. Can. Soc. Petrol. Geol., Calgary*, **5**, 431–440.

NIJMAN, W. & PUIGDEFABREGAS, C. (1978) Coarse-grained point-bar structure in a molasse-type fluvial system, Eocene Castisent Sandstone Formation, South Pyrenean Basin. In: *Fluvial Sedimentology* (Ed. by A.D. Miall). *Mem. Can. Soc. Petrol. Geol., Calgary*, **5**, 487–510.

PUIGDEFABREGAS, C. & VAN VLIET, A. (1978) Meandering stream deposits from the Tertiary of the southern Pyrenees. In: *Fluvial Sedimentology* (Ed. by A.D. Miall). *Mem. Can. Soc. Petrol. Geol., Calgary*, **5**, 469–486.

RUST, B.R. (1978) A classification of alluvial channel systems. In: *Fluvial Sedimentology* (Ed. by A.D. Miall). *Mem. Can. Soc. Petrol. Geol., Calgary*, **5**, 187–198.

SCHUMM, S.A. (1960) The shape of alluvial channels in relation to sediment type. *Prof. Pap. U.S. geol. Surv.* **352B**, 30 pp.

SCHUMM, S.A. (1963) A tentative classification of alluvial river channels. *Circ. U.S. geol. Surv.* **477**, 10 pp.

SCHUMM, S.A. (1968) Speculations concerning paleohydrologic controls of terrestrial sedimentation. *Bull. geol. Soc. Am.* **79**, 1573–1588.

SMITH, D.G. (1976) Effect of vegetation on lateral migration of anastomosed channels of a glacial meltwater river. *Bull. geol. Soc. Am.* **87**, 857–860.

STEEL, R. & AASHEIM, S.M. (1978) Alluvial sand deposition in a rapidly subsiding basin (Devonian, Norway). In: *Fluvial Sedimentology* (Ed. by A.D. Miall). *Mem. Can. Soc. Petrol. Geol., Calgary*, **5**, 385–412.

VONDRA, C.F. & BURGGRAF, D.R. JR (1978) Fluvial facies of the Plio-Pleistocene Koobi Fora Formation, Karari Ridge, East Lake Turkana, Kenya. In: *Fluvial Sedimentology* (Ed. by A.D. Miall). *Mem. Can. Soc. Petrol. Geol., Calgary*, **5**, 511–530.

WINSTON, D. (1978) Fluvial systems of the Precambrian Belt Supergroup, Montana and Idaho, U.S.A. In: *Fluvial Sedimentology* (Ed. by A.D. Miall). *Mem. Can. Soc. Petrol. Geol., Calgary*, **5**, 343–360.

Spec. Publs int. Ass. Sediment. (1983) **6**, 355–368

Sandy fluvial point-bar sediments from the Middle Eocene of Dorset, England

A. G. PLINT*

Department of Geology and Mineralogy, Parks Road, Oxford OX1 3PR, U.K.

ABSTRACT

Middle Eocene alluvial sediments occupy the western portion of the Hampshire Basin and are laterally equivalent to estuarine, lagoonal and shallow marine sediments to the east. The alluvial sediments are up to 250 m thick and comprise point-bar sands and laminated interchannel muds. Point-bar sands form erosive-based and upward-fining sequences, 3–15 m thick, which are laterally traceable for hundreds of metres. Basal erosion surfaces are usually overlain by intraformational mudclast conglomerates. Metre-scale cross-bedding of transverse bars sometimes occurs low in the sequence. The bulk of the point-bar is composed of decimetre-scale trough and tabular cross-bedded coarse to fine sand. This may be interbedded with low-angle or horizontally laminated sand, or with convolute bedding. Rarely, the upper part of the point-bar is preserved. This consists mainly of fine, rippled or plane-laminated sand passing up into thinly interbedded fine sand and mud of the levée. Usually the sequence is truncated and locally capped by muds deposited in abandoned chute channels. The point-bar sands enclose large, mud-filled abandoned channels which average 130 m in width and 11 m in depth. Within a single point-bar unit palaeocurrents are fairly consistent, but between units there is a wide divergence. The sedimentary characteristics as a whole suggest deposition on the point-bars of relatively sinuous rivers with a mixed load and variable discharge.

INTRODUCTION

The Hampshire Basin of southern England (Fig. 1) is filled with Palaeocene to Lower Oligocene sediments, up to 720 m thick. The Hampshire Basin is strongly asymmetrical with a near-vertical southern limb, comprising the Isle of Wight and Purbeck Monoclines, and a northern limb which is almost horizontal. The basin has been subject to marginal warping since middle Eocene times (Plint, 1982).

This paper describes middle Eocene alluvial sediments, from east Dorset, which are up to 250 m thick. These sediments have been called the Bournemouth Freshwater Series (Gardner, 1882), the Bagshot Beds (Reid, 1896) or the Bournemouth Formation, by the Institute of Geological Sciences.

They are exposed at Alum Bay (SZ 306584) in the Isle of Wight, on the coast of Poole Bay from Bournemouth Pier (SZ 089907) westwards to Poole Harbour, in Studland Bay (SZ 038828) and in numerous pits in the Wareham area (Fig. 1). These alluvial sediments pass laterally eastwards within a few kilometres into estuarine, lagoonal and shallow marine sediments of the Bracklesham Formation (Cuisian–Lutetian) exposed in the Isle of Wight, Hampshire and Sussex. The Bracklesham Formation is underlain by the London Clay Formation (Lower Eocene) and overlain by the Barton Formation (Upper Eocene); it varies from 150 to 250 m in thickness.

THE SEDIMENTS

General characteristics

The alluvial sediments of Dorset fall into two major categories: (1) coarse- to fine-grained point-bar sands,

* Present address: Department of Geology, University of New Brunswick, P.O. Box 4400, Fredericton, New Brunswick E3B 5A3, Canada.

0141-3600/83/0106-0355 $02.00

Fig. 5. Convolute bedding (facies 8). Note multi-phase deformation with each unit separated by an erosion surface (arrowed). Convolute bedding passes laterally into undeformed decimetre-scale cross-bedding.

Couplets are commonly graded, have a sharp base and range from 2 to 50 mm thick. Occasionally, ripple cross-lamination is seen, sometimes forming isolated sand lenses ('starved ripples'). Fragments of plant debris are common. The facies forms units tens of metres in lateral extent and up to 5 m in thickness. Pronounced lateral variation in thickness may occur, which does not appear to be due to erosion at the base of the unit and may therefore reflect an original topography. The facies invariably occurs at the top of point-bar sequences, and may be overlain either by another point-bar sequence or by interchannel muds.

Thinly bedded fine sand and mud are the result of floods and are typical of levées and flood basins (Fisk, 1944, 1947; Jahns, 1947; Shepard, 1956; Schumm &

Lichty, 1963; Jackson, 1976b; Ray, 1976). Facies 7 is therefore interpreted, both in terms of its sedimentary characteristics and its position in facies sequences, as the deposits of levée or proximal flood-plain environments.

Facies 8: convolute bedded sandstone

Convolute bedding is widely developed in point-bar sands, and forms units metres thick and tens or even hundreds of metres in lateral extent. Convolution is present in coarse to fine sands, with a prevalence in the coarser sands. All gradations from oversteepened foresets (Fig. 4) to intense deformation (Fig. 5) are seen. In several exposures, folds are overturned in a

Fig. 6. Rare dewatering pipe in semi-convoluted transverse bar cross-bedding. Note small-scale normal faulting where the sides of the pipe have collapsed into the fluidized core, now represented by structureless sand in the middle of the pipe. (Compare this structure with plate 16:14 of Curran & Frey, 1975, where an almost identical structure is attributed to a giant burrowing anemone.)

downcurrent direction, as deduced from adjacent cross-bedding. Convolution is most commonly found in association with, and passing into, dune cross-bedding of facies 3. Multiple deformation is common (Fig. 5) and a general repetitive sequence can be recognized:

Deposition of cross-bedded sand

Erosion of top surface ←——— Deformation

It is generally recognized that convolute bedding forms through the partial liquefaction and deformation of water-saturated sediment under the influence of lateral shearing, probably resulting from the movement of large bedforms over an unconsolidated substrate during high-flow stages (Rettger, 1935; McKee, Reynolds & Baker, 1962a, b). The common existence of multiple deformation shows that deposition and deformation took place almost simultaneously, perhaps in a few minutes, before the deposition and subsequent deformation of the next cross-set. Liquefaction by earthquakes (Allen & Banks, 1972) does not seem necessary in this instance, and indeed would be very difficult to reconcile with the features observed. Large-scale dewatering pipes, some showing signs of fluidization (Fig. 6) occasionally accompany convolute bedding.

24

Rhondda Beds of South Wales. *Bull. Am. Ass. Petrol. Geol.* **52**, 2369–2386.

LEEDER, M.R. (1973) Sedimentology and palaeogeography of the Upper Old Red Sandstone in the Scottish Border Basin. *Scott. J. Geol.* **9**, 117–144.

LEOPOLD, L.B. & WOLMAN, M.G. (1960) River meanders. *Bull. geol. Soc. Am.* **71**, 769–794.

LEVEY, R.A. (1978) Bed-form distribution and internal stratification of coarse-grained point-bars, Upper Congaree River, South Carolina. In: *Fluvial Sedimentology* (Ed. by A.D. Miall). *Mem. Can. Soc. Petrol. Geol., Calgary*, **5**, 105–127.

McGOWEN, J.H. & GARNER, L.E. (1970) Physiographic features and stratification types of coarse-grained point-bars: modern and ancient examples. *Sedimentology*, **14**, 77–111.

McKEE, E.D., CROSBY, E.J. & BERRYHILL, H.L. (1967). Flood deposits, Bijou Creek, Colorado, June 1965. *J. sedim. Petrol.* **37**, 829–851.

McKEE, E.D., REYNOLDS, M.A. & BAKER, C.H. (1962a) Laboratory studies on deformation in unconsolidated sediment. *Prof. Pap. U.S. geol. Surv.* **450-D**, 151–155.

McKEE, E.D., REYNOLDS, M.A. & BAKER, C.H. (1962b) Experiments on intraformational recumbent folds in crossbedded sand. *Prof. Pap. U.S. geol. Surv.* **450-D**, 155–160.

MOODY-STUART, M. (1966) High and low sinuosity deposits with examples from the Devonian of Spitzbergen. *J. sedim. Petrol.* **36**, 1102–1117.

NAMI, M. (1976) An exhumed Jurassic meander-belt from Yorkshire, England. *Geol. Mag.* **113**, 47–52.

NIJMAN, W. & PUIGDEFABREGAS, C. (1978) Coarse-grained point-bar structure in a molasse-type fluvial system, Eocene Castisent Sandstone Formation, south Pyrenean Basin. In: *Fluvial Sedimentology* (Ed. by A.D. Miall) *Mem. Can. Soc. Petrol. Geol., Calgary*, **5**, 487–510.

PLINT, A.G. (1980) *Sedimentary studies in the Middle Eocene of the Hampshire Basin.* Unpublished D.Phil. thesis, University of Oxford. 3 volumes.

PLINT, A.G. (1982) Eocene sedimentation and tectonics in the Hampshire Basin. *J. geol. Soc. London*, **139**, 249–254.

RAY, P.K. (1976) Structure and sedimentological history of the overbank deposits of a Mississippi River point-bar. *J. sedim. Petrol.* **46**, 788–801.

REID, C. (1896) The Eocene deposits of Dorset. *Q. Jl geol. Soc. Lond.* **52**, 490–495.

RETTGER, R.E. (1935) Experiments in soft-rock deformation. *Bull. Am. Ass. Petrol. Geol.* **19**, 271–292.

SCHUMM, S.A. (1968) Speculations concerning palaeohydrologic control of terrestrial sedimentation. *Bull. geol. Soc. Am.* **79**, 1573–1588.

SCHUMM, S.A. & LICHTY, R.W. (1963) Channel widening and flood plain construction along the Cimarron River in S.W. Kansas. *Prof. Pap. U.S. geol. Surv.* **352-D**, 71–88.

SCHWARTZ, D.E. (1978) Sedimentary facies, structures and grain-size distribution: the Red River in Oklahoma and Texas. *Trans. Gulf-Cst Ass. geol. Socs* **28**, 473–492.

SHEPARD, F.P. (1956) Marginal sediments of the Mississippi Delta. *Bull. Am. Ass. Petrol. Geol.* **40**, 2537–2623.

SMITH, A.G., BRIDEN, J.C. & DREWRY, G.E. (1973) Phanerozoic World Maps, In: *Organisms and Continents Through Time. Spec. Pap. Palaeont.* **12**, 1–42.

SMITH, N.D. (1972) Flume experiments on the durability of mud clasts. *J. sedim. Petrol.* **42**, 378–384.

STEWART, D.J. (1981) A meander-belt sandstone of the Lower Cretaceous of southern England. *Sedimentology*, **28**, 1–20.

Spec. Publs int. Ass. Sediment. (1983) **6**, 369–384

Possible suspended-load channel deposits from the Wealden Group (Lower Cretaceous) of Southern England

D. J. STEWART

Shell U.K. Exploration and Production Ltd, Shell Mex House, Strand, London, U.K.

ABSTRACT

The Lower Cretaceous Wealden Group of Southern England contains heterolithic sedimentary units ranging from 2 to 10 m thick which cut into varicoloured rootlet- and mudcrack-bearing mudstones similar to those of some recent tropical flood plains. The units display inclined bedding resulting from lateral accretion and have, therefore, been interpreted as river point bars or benches. In addition the units display the following features: (i) sequences generally fine upwards, (ii) they contain considerable amounts of claystone and siltstone either interbedded with very fine-grained quartz sand or beds of sand-grade mudclasts, (iii) the lateral accretion units dip at steep angles, (iv) cross-bedding within the units is mainly of a small scale, and (v) coarse sand- to pebble-sized material is almost entirely intraformational.

These features differ in some respects from the point bars of mixed-load channel systems, which typically form sand-dominated vertical sequences. The Wealden point bars/benches more closely resemble those described from channel systems carrying dominantly suspended loads, found both on mature low-gradient flood plains and tidal flats. This type of point bar is typically very silt/clay rich and the slope (accretion surface) of the point bar or bench is steep owing to the low width/depth ratio of the channel. The thickness of sediment accreted on to the point bar during any one flood cycle appears to have been small, but the rate of point bar migration could have been fairly rapid if floods were frequent.

Variations in the sand content of these point-bar/bench units, and their association with sand-dominated fluvial point-bar units, suggest that Wealden rivers varied in space and time from suspended-load systems to mixed-load systems. The main controls on these variations were probably climate and tectonics, and the consequential alteration of river gradients and sediment supply.

This study provides a rare example of suspended-load point bars or benches in the fossil record. It also highlights the difficulty of distinguishing muddy point-bar deposits from the adjacent flood-plain facies and, in particular, from crevasse splay units which are commonly interbedded with such facies. One must, therefore, consider the overall association of sedimentary facies when interpreting such units.

INTRODUCTION

Schumm (1960, 1968) recognized that streams carrying appreciable suspended loads formed morphologically distinct channels with low width-to-depth ratios and very high sinuosities. However, he did not describe the deposits of suspended-load channels in any detail. Detailed descriptions of such deposits have only recently been published; point benches of Holocene suspended-load channels in eastern Australia (Taylor & Woodyer, 1978; Woodyer, Taylor &

Crook, 1979; Nanson & Page, 1981), developed on low flood-plain gradients.

Other muddy, fine-grained point bars have been described by Jackson (1981), but these appear to be intermediate between the purely suspended-load and the truly mixed-load type channel deposits (i.e. Jackson, 1976).

These recent publications emphasize the variation that is to be found in recent point-bar/bench deposits, variations which should be reflected in the facies models of ancient meandering stream deposits. In particular, differences in hydrodynamic regimes and

0141-3600/83/0106-0369 $02.00

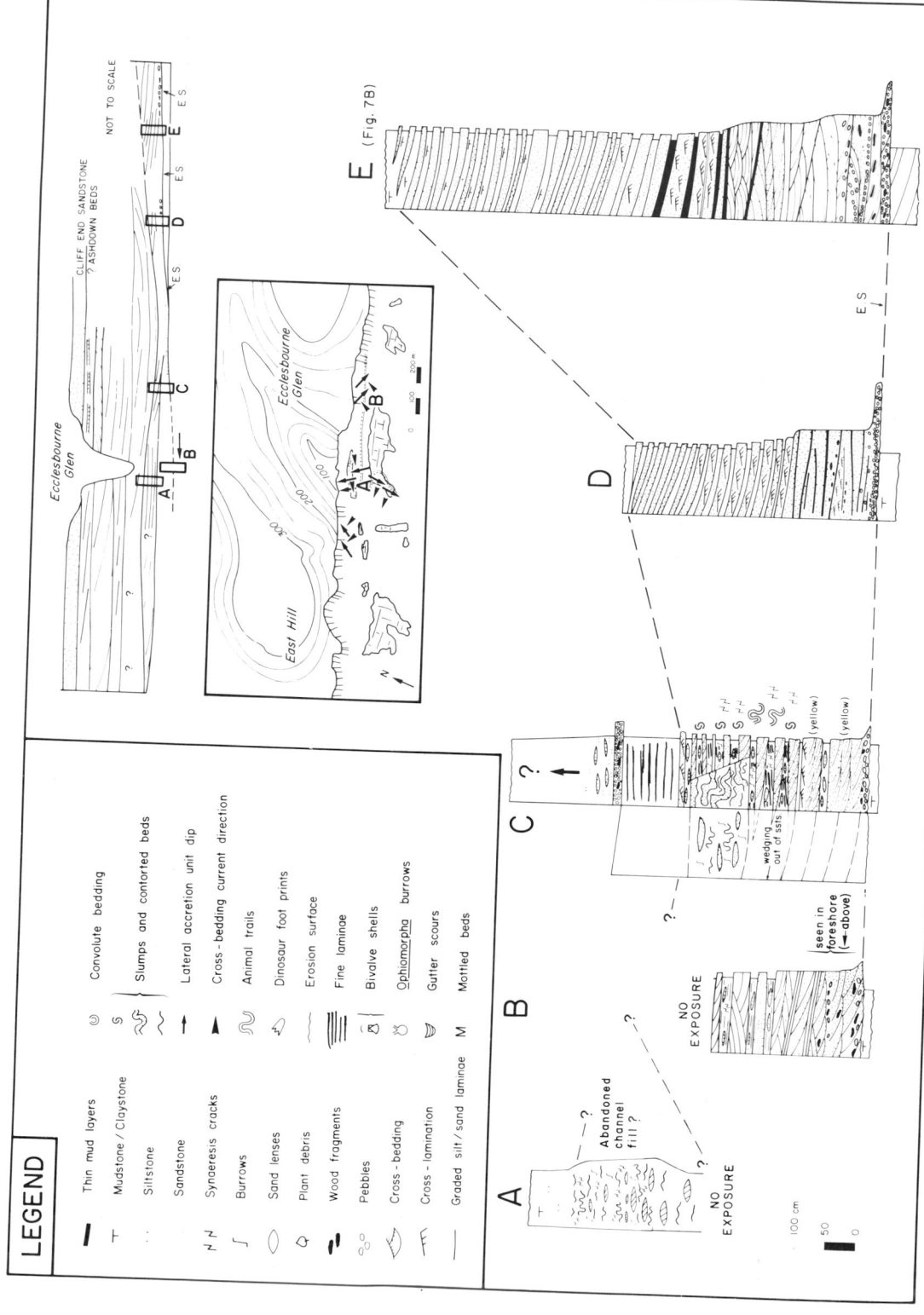

Fig. 5. Lateral variations in vertical profiles through point-bar units at Ecclesbourne Glen (TQ 837099), near Hastings modified from Stewart (1981b) following acquisition of new field data. Four point bars can be mapped out overlying the same erosion surface, which are cut by an abandoned channel (A).

cross-beds that are perpendicular to the accretion direction, there are some that point up the slope of the accretion surface. Also, the small-scale cross-laminations at the top of the units often exhibit random directions.

DISCUSSION

The structures described above seem to record lateral accretion on inclined depositional surfaces with current directions, in general, at right angles to, or climbing up the dip of the inclined surfaces. The structures, therefore, possibly represent lateral accretion of point bars or benches. In many respect the units resemble the interbedded muds and rippled sands of tidal point bars (Reineck, 1958; Bridges & Leeder, 1976) where intraformational lags accumulate on the channel floor and finer-grained sediments settle out on the upper parts of the point bars. In the Wessex Formation, deposition as point bars is supported by the fact that the structures occur in a mud-dominated sequence (the units are often over- and underlain by massive mudstones many metres thick) indicating that the stream banks were muddy and thus, owing to their cohesiveness, favoured meandering (Schumm, 1968). The muddy nature of the point bars would suggest that the streams carried mainly suspended loads, whilst the presence of sands and pebbles composed of intraformational mudstone suggests the periodic occurrence of strong currents that caused the erosion of local bank material (cf. Jackson, 1981). The three-dimensional evidence from Ecclesbourne Glen (Fig. 5) would seem to support this conclusion.

Although the overall environment in which these sediments were deposited was probably fluvial, deposition under tidal influence, as hinted by the resemblance to tidal point bars, cannot be totally ruled out. Many of the Ashdown point bars/benches contain abundant synaeresis cracks, suggesting that the waters in which the sediments were deposited could have been subject to changes in salinity (Donovan & Foster, 1972). Brackish trace fossils (*Ophiomorpha*) (Stewart, 1978a) are present in the basal Wadhurst Clay overlying the Ashdown Sands, and microplankton (*Muderongia*), of possible brackish origin (Batten & Eaton, 1980), occur in the Haddocks Rough 'point bar' unit near the top of the Ashdown Sands. Similar suggestions of brackish influences have also been observed in the Vectis Formation of the Isle of Wight (Stewart, 1978b) and in the top of the Wessex Formation (Stewart, 1978a). It therefore seems

possible that at least some of the point bars/benches at both Hastings and the Isle of Wight could have been deposited under periodically brackish conditions. The source of brackish water was perhaps a distant connection to the open sea (Allen, 1975). Variable salinities may have resulted from tidal influences or dilution of brackish waters by unusually high river-borne discharge of fresh water. The former explanation has some support since sandstones in the Vectis Formation (Stewart, 1978b) and the Top Ashdown Sands (Allen, 1962) have features reminiscent of tidal deposition (e.g. abundant ripple marks, rhythmic bedding, flaser bedding and wavy bedding). However, the overall distribution of fauna and flora, and in particular of spores and microplankton, indicate a dominantly terrestrial influence at most horizons.

In the fluvial setting, the best available recent analogues to these structures appear to be the point benches described from the Barwon River, New South Wales, Australia (Taylor & Woodyer, 1978; Woodyer *et al.*, 1979) and the point bars of the American middle west, recently described by Jackson (1981). Both the Barwon River point benches and the American middle west point bars are composed of finely interbedded parallel/wavy laminated (laminae often result from graded silt/sand layers), cross-laminated sands and graded laminae. Cross-bedded sands commonly form the basal part of reconstructed vertical sequences, and mudlayers become thicker and more important upwards. Other points of similarity include fairly abundant mudclasts, local mudcracks, widespread pedogenic modification and occasional slumping. On the Barwon River a major characteristic of the internal bedding as a whole is the inclined bedding, dipping at angles of up to 23°, which is possibly comparable to the dipping beds of the structures described here.

These points of comparison suggest similar depositional regimes, although one drawback with using the Barwon River as an analogue is the narrow width (perpendicular to the channel axis) of the point benches (11 m, but up to 1·6 km long), owing to the lack of significant channel migration (Taylor & Woodyer, 1978; Woodyer *et al.*, 1979). This is due to minimal erosion on the concave bank of the stream on account of the binding of already cohesive sediment by tree roots, and the very low water velocities encountered.

The relatively wide lateral extent of these Wealden units suggests relatively extended channel migration. The Wessex Formation flood plains were probably gradual migration, but must do so by avulsion. Crevasse splays are not as conspicuous as those noted by Smith (1981) in some temperate anastomosing rivers. They take the form of intricate channel systems radiating from approximate point sources on the original channel (Fig. 3c).

Overbank flooding from active anastomosing channels deposits sandy mud over the floodplain, which includes abandoned anastomosing channels and the relict braids of the underlying sand sheet. The mud is traversed by a network of desiccation cracks, which commonly reach to depths of over a metre and have surface openings up to more than 10 cm wide. On a large scale, the cracks tend to develop a reticulate network with orthogonal junctions, but on outcrop scale they are irregular (Fig. 5). With a few exceptions, the overbank deposits are structureless, due mainly to desiccation shrinkage and, to a lesser extent, root growth. Trampling by cattle is a significant mode of sediment deformation around channels, but is less important on the floodplain, where the mud is baked hard most of the time.

Augering by Veevers & Rundle (1979) and shallow water wells show that the surficial mud is commonly less than 5 m thick, underlain by a sand sheet deposited by the relict braided system. Deeper wells at Canterbury and Barrolka (Fig. 2) show an alternation of sandy mud and sand similar to the surficial alluvium down to the maximum depths penetrated (more than 100 m, Senior, 1970). Other lithotypes present include carbonaceous, evaporite and duricrust units. The duricrusts include silcrete and calcrete, similar to those present in the surficial

alluvium of Cooper's Creek. The carbonaceous units are minor (one thin layer in each borehole), showing that plant material is rarely preserved. This is in marked contrast to temperate anastomosing–fluvial systems, which accumulate thick peat deposits in wetlands (Smith & Smith, 1980).

ANCIENT ANASTOMOSING–FLUVIAL DEPOSITS IN THE CLIFTON FORMATION OF NEW BRUNSWICK, CANADA

Introduction

The Clifton Formation is an Upper Carboniferous (Westphalian C–D) deposit that crops out continuously along the New Brunswick coast west of Bathurst (Fig. 6), as described in outline by Alcock (1935) and Van de Poll (1973). Ball, Sullivan & Peach (1981) divided the Clifton Formation into three members, of which the middle member (B) is over 700 m thick, and contains the strata discussed here. These rocks comprise two distinctive lithostratigraphic units, here informally designated the upper and lower successions (Fig. 7).

The lower succession

The lower succession is predominantly red mudstone with very fine sandstone, and also contains isolated bodies of medium-grained sandstone (Fig. 7). Markov chain analysis of the fine-grained rocks (Legun, 1980) reveals repetition of a coarsening-up,

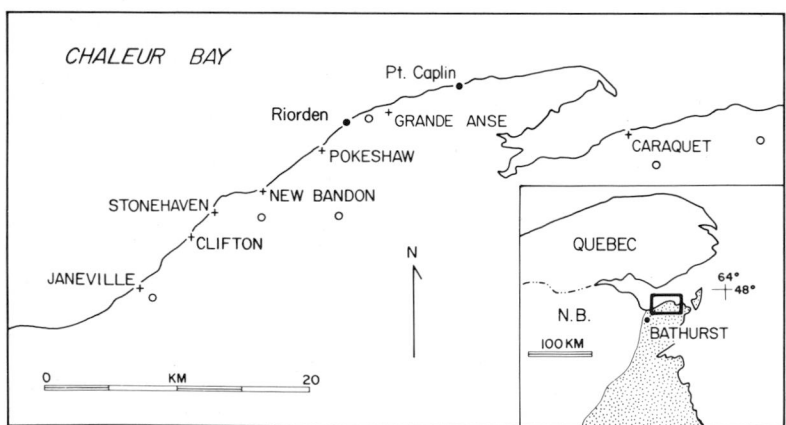

Fig. 6. Location of Clifton–Caraquet section of northern New Brunswick coast. Dotted area on insert indicates Carboniferous Basin of central New Brunswick. Open and closed circles are borehole sites of Carboniferous Drilling Project (Ball *et al.*, 1981) and New Brunswick Power Corporation, respectively.

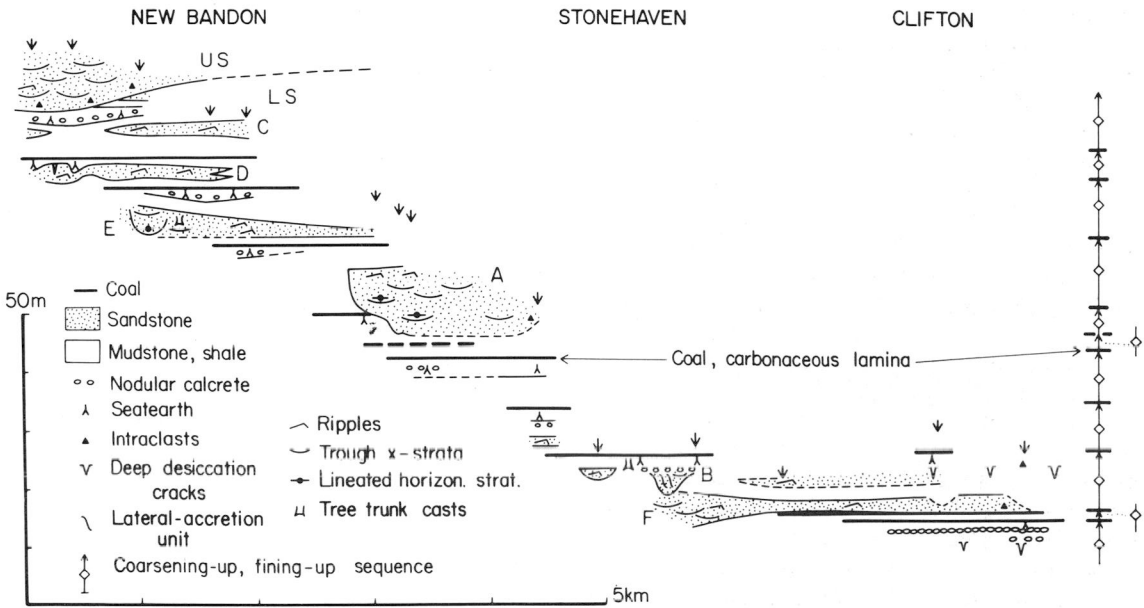

Fig. 7. Composite section of coastal exposure of Member B of Clifton Formation between Clifton and New Bandon, N.B. Arrows indicate locations of measured sections. The spacing of the symbol for calcrete nodules reflects actual spacing, which varies from nodules scattered along a horizon to a continuous calcrete hardpan. US, upper succession; LS, lower succession; A–F, sandstone bodies referred to in text.

fining-up sequence capped by rooted seatearth and a thin coal (Fig. 8). Nodular calcrete is common in the seatearth, but occurs sporadically elsewhere, particularly as root replacements or rhizoliths (Klappa, 1980). Rhizoliths are abundant on vertical surfaces that commonly penetrate about 1 m of strata (Fig. 9), and may also contain packed carbonate nodules. The vertical surfaces are interpreted as deep desiccation cracks with coatings or infillings of calcrete. The repetitive sequence is attributed to filling of floodbasins (slight local depressions on the floodplain) by progradation of depositional lobes (upward coarsening), followed by abandonment (upward-fining, desiccation and pedogenesis). The close association of carbonaceous and calcrete horizons is attributed to fluctuation of the water table in a basically dry climate, capable of producing deep desiccation cracks and persistent calcretes.

In addition to the fine-grained sandstones described above, several discrete medium-grained sandstone bodies are exposed in the cliffs and on the foreshore. They are designated A to F in Fig. 7, and are interpreted as channel sandstones, with associated spillover sandstone units.

Sandstone bodies C to F (Fig. 7) are essentially

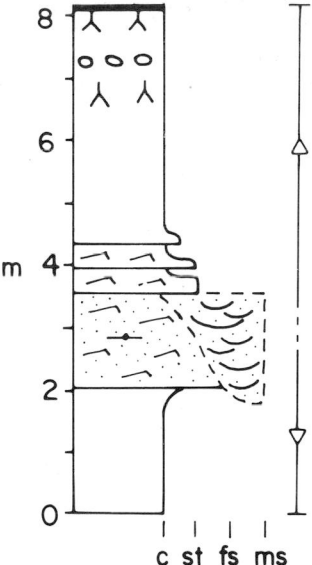

Fig. 8. Repetitive sequence in the lower succession of Member B, with a larger sandstone body included in dotted outline, because it is not a standard part of the repetitive sequence. Triangle with apex up indicates upward–fining sequence; base up is the reverse; c, claystone; st, siltstone; fs, fine sandstone; ms, medium sandstone. See Fig. 7 for explanation of other symbols.

the upper sand-dominated succession of Clifton Member B. The increased grain size could be explained alternatively by tectonic rejuvenation in the form of source uplift or base-level subsidence. The interpretation of sandstone body A as a downcut braided–fluvial complex suggests that an episode of rejuvenation preceded the deposition of the upper succession. However, the association of braided sheet sandstones with abundant plant material and the absence of *in situ* calcrete is hard to explain without invoking a wetter climate. Probably, a combination of increased rainfall and tectonic rejuvenation was responsible for the observed lithological change.

CONCLUSIONS

The Clifton Formation exposed on the northern New Brunswick coast closely resembles recent sediments of Cooper's Creek in Central Australia. The modern sediments of Cooper's Creek are mud-dominated but contain sand in active anastomosing channels. The nature of the channels implies limited lateral or vertical accretion, with relocation occurring by channel avulsion. This is reflected in the ancient record by channel sandstones of limited vertical and lateral extent that are isolated within a mudstone-dominated succession. Features indicative of an arid to semi-arid climatic environment are a high ratio of overbank mud: channel sand, evaporites, duricrusts, deep desiccation cracks and limited organic accumulation.

The shallow-buried braided sand sheet that is relict beneath the floodplain of Cooper's Creek is thought to be analogous to the upper succession of the Clifton coastal exposure. Characteristic features of the recent and ancient examples are: lateral extent (tens of kilometres), abundant erosion surfaces, intraclasts and plant debris. In both cases the braided sand sheet may be attributable to greater rainfall, but this can be demonstrated better for the recent example than for the Clifton Formation, in which tectonic rejuvenation may also have been a factor.

ACKNOWLEDGMENTS

The field work on Cooper's Creek was greatly facilitated by logistic support from Delhi International Oil Corporation and SANTOS Ltd of Adelaide, Australia, with able field assistance from Norrie Hamilton. The New Brunswick study was supported by contract 14SU . 2394-0-0555 from the Department of Energy, Mines and Resources, and both aspects of the project received additional support from the Natural Sciences and Engineering Research Council of Canada (Grant A-2672). We would also like to thank S. Meunier for typing, Edward Hearn for drafting and Andrew Miall and John Collinson for comments on the manuscript.

REFERENCES

ALCOCK, F.J. (1935) Geology of Chaleur Bay Region. *Mem. geol. Surv. Can.* **183**, 146 pp.

BALL, F.D., SULLIVAN, R.M. & PEACH, A.R. (1981) Carboniferous Drilling Project. *Rept Invt.* **18**, *New Brunswick Dept Nat. Resources*, 109 pp.

KLAPPA, C.H. (1980) Rhizoliths in terrestrial carbonates: classification, recognition, genesis and significance. *Sedimentology*, **27**, 613–629.

LEGUN, A.S. (1980) *Sedimentology of the Clifton Formation (Upper Carboniferous) of northern New Brunswick: a semi-arid depositional setting for coal.* Unpublished M.Sc. thesis, University of Ottawa. 97 pp.

MIALL, A.D. (1978) Lithofacies types and vertical profile models in braided river deposits: a summary. In: *Fluvial Sedimentology* (Ed. by A.D. Miall). *Mem. Can. Soc. Petrol. Geol., Calgary*, **5**, 597–604.

RUST, B.R. (1978a) A classification of alluvial channel systems. In: *Fluvial Sedimentology* (Ed. by A.D. Miall). *Mem. Can. Soc. Petrol. Geol., Calgary*, **5**, 187–198.

RUST, B.R. (1978b) Depositional models for braided alluvium. In: *Fluvial Sedimentology* (Ed. by A.D. Miall). *Mem. Can. Soc. Petrol. Geol., Calgary*, **5**, 605–625.

RUST, B.R. (1981) Sedimentation in an arid-zone anasto-mosing fluvial system: Cooper's Creek, Central Australia. *J. sedim. Petrol.* **51**, 745–755.

SENIOR, B.R. (1970) Barrolka, Queensland. *Bureau Min. Res., Geol. Geophys. Expl. Notes*, SG/54/11.

SMITH, D.G. (1981) Peat and coal in modern and ancient anastomosed river deposits. *Abstr. Geol. Ass. Can.* **6**, A-52.

SMITH, D.G. & PUTNAM, P.E. (1980) Anastomosed river deposits: modern and ancient examples in Alberta, Canada. *Can. J. Earth Sci.* **10**, 1396–1406.

SMITH, D.G. & SMITH, N.D. (1980) Sedimentation in anastomosed river systems: examples from alluvial valleys near Banff, Alberta. *J. sedim. Petrol.* **50**, 157–164.

VAN DE POLL, H.W. (1973) Stratigraphy, sediment dispersal and facies analysis of the Pennsylvanian Pictou Group in New Brunswick. *Marit. Sed.* **9**, 72–77.

VEEVERS, J.J. & RUNDLE, A.S. (1979) Channel Country fluvial sands and associated facies of central eastern Australia: modern analogues of Mesozoic desert sands of South America. *Palaeogeogr. Palaeoclim. Palaeoecol.* **26**, 1–16.

Spec. Publs int. Ass. Sediment. (1983) **6**, 393–403

Transient streams in sand-poor redbeds: early–Middle Eocene Kuldana Formation of northern Pakistan

NEIL A. WELLS

Museum of Paleontology/Department of Geological Sciences, The University of Michigan, Ann Arbor, Michigan 48109, *U.S.A.*

ABSTRACT

The Kuldana Formation is a sand-poor redbed complex that prograded into the Persian Gulf-like Kohat depositional basin of northern Pakistan at the end of an early Eocene marine regression. It was formed by a poorly developed system of small, episodic, briefly active, shallowly incised and readily abandoned streams that wandered into a very flat, aggrading, semi-arid basin. Their deposits are about 95% red mud and 5% predominantly trough-bedded and fining-upward channel sand. The streams did not meander, however, because avulsion was too frequent. During low flow they became braided but their bedforms underwent little alteration. Partly because of the basin's flatness, the only available coarse clastics in eastern areas were calcareous soil nodules and, locally, fish and mammal bone fragments. The primary determinants of this stream system seem to have been a below-grade basin floor and a semi-arid climate.

INTRODUCTION

The purpose of this paper is to discuss the unusual characteristics of a redbed complex that temporarily dominated the Kohat basin of north-central Pakistan (Fig. 1A) in early–Middle Eocene times. As will be shown below, the few small sandstones present are composed, in large part, of reworked soil nodules and represent single-storey channels filled with simple, commonly curtailed, fining-upward sequences. Because the stream deposits do not resemble those of either braided or meandering streams (Miall, 1977; Collinson, 1978; Reineck & Singh, 1980), I shall suggest below that they were formed by small and flashy streams, governed primarily by high-flow conditions and shifted by avulsion before meandering could properly develop. First, however, an overview of the geological setting and stratigraphy of the redbeds will be given (both topics are discussed fully in Wells, 1982).

The Kohat area first became a depocentre in the Palaeocene and evolved into an inland sea in the Eocene. In the Palaeozoic and Mesozoic it was part of the continental shelf of the Indian subcontinent, being usually below and occasionally above the Tethyan shoreline. The north-western edge of the subcontinent then lay just west of the study area (see the index map, Fig. 1A). Isopachs of Palaeocene strata show a considerable thickening in the Kohat area (Meissner & Rahman, 1973), demonstrating the formation of a depocentre. Work in progress by the author indicates that most of the region was then a shallowly submerged shelf with reefs and carbonate platforms, whereas the thicker sediments in the Kohat area were deposited in much deeper water. During the early Eocene, the periphery of the Kohat basin became emergent, shallower-water facies prograded into the basin and pelagic forams, corals, non-boring sponges, coralline algae and bryozoa completely or almost completely disappeared from the fauna. This indicates that the basin changed from a submerged depocentre,

0141-3600/83/0106-0393 $02.00

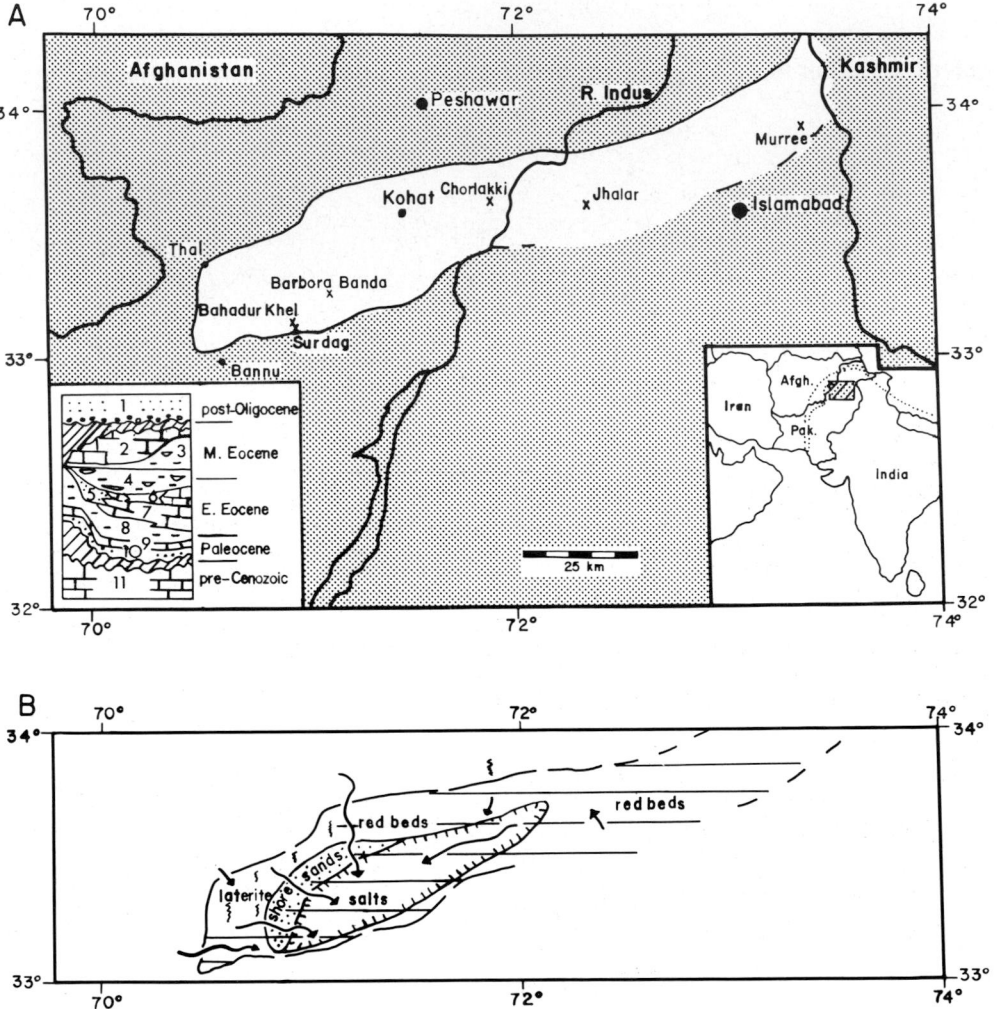

Fig. 1. (A) Present (post-compression) configuration of the mid-late early Eocene Kohat basin. Dotted lines on the index map represent previous edges of the Indian subcontinent. The stratigraphic column shows Palaeogene facies across the northern half of the Kohat basin (Thal to Jhalar). 1 = post-Oligocene molasse, 2 = Kohat Ls. (inland sea), 3 = Upper Kuldana beds (transitional-marine), 4 = Lower Kuldana redbeds (fluvial), 5 = Gurguri Ss. (beach), 6 = Jatta Gypsum (sabkha), 7 = Shekhan–Chorgali–Margala Hill Ls. (carbonate platform), 8 = Panoba–Patala Sh. (slope muds), 9 = Lockhart Ls. (platform and slump), 10 = Hangu Ss. (shoreline clastics), and 11 = pre-Cenozoic shelf sediments.

(B) Present-day distribution of Kuldana redbeds, laterally equivalent laterites, subjacent evaporites, and subjacent/equivalent shoreline sands, with superimposed highly generalized drainage patterns. Laterization began in the NW and redbed progradation in the NE and E when the sea retreated into the southern deep, which then became filled with salt and was later covered by redbeds. The shoreline sands are small deltas and stream-fed cobble beaches that formed on the rim of the remnant evaporite basin.

or offshore basin, to a 200 × *ca.* 75 km Persian Gulf-like inland sea. Relatively falling sea-level at the end of the early-Eocene exposed shallow marine sediments and caused the redbeds to prograde over them into the basin. The regression stranded a gypsum sabkha and fringing carbonates in the NE quarter of the basin and shallow marine clays in the NW, and finally caused the filling of a remnant deep basin in the central south with halite and gypsum (Fig. 1B). The redbed complex was drowned by a short-lived return of the sea in the Middle Eocene. This Middle Eocene marine phase was to be the basin's last, for it was completely dry before

Fig. 2. (A) Part of the Eocene section at Surdag. The horizontal lines delimit the Lower Kuldana redbeds (red clay with two fluvial sandstones) at bottom, transitional-marine marls and variegated clays (Upper Kuldana beds) in the middle and marine limestones and shales of the Kohat Limestone above. X marks a man for scale. Arrows in this and following photos point up-section.

(B) The younger Lower Kuldana sandstone near Barbora Banda (base to right, flow towards camera). It is essentially one dune thick. Note slight erosion by individual scour pools and sudden succession of coarse sediments by red clay, except for thin fine sandstone lens in clay left of penknife.

(C) Planar cross-beds, 90 cm high, at Bahadur Khel. They represent the low-angle straight-crested front of a bar that moved from left to right. Note same-sense ripples at base.

(D) Typical lower Kuldana sandstone, showing an oblique section of a 35 cm thick channel-fill. Note thinness of bed, basal calcareous granulestone and succeeding large-scale trough cross-bedding (flow to left). Cross-beds are overlain by red clay. This is essentially a vertically accreted fill in a low sinuosity channel, as described by Moody-Stuart (1966).

(E) Interbedded clay and mammal-bearing granulestone (marked by the smaller arrow), 50 cm thick, above two well-exposed purple-stained unfossiliferous intraclastic marls (marked x) The sandstones are thought to be fluvial, but occur in a coastal-plain setting.

the late Eocene and was covered by Himalayan molasse in the late Oligocene or early Miocene. The basin has since been considerably deformed.

The redbed complex, which is here informally referred to as the 'Lower Kuldana beds' and which is at present unsatisfactorily grouped with the overlying transitional-marine 'upper Kuldana beds' as the Kuldana Formation, is of interest not only sedimentologically but also palaeontologically and tectonically. Kuldana sandstones contain some of the

Fig. 3. (A) Base of older sandstone at Barbora Banda, showing calcareous granules and two small slightly damaged jaws of primitive artiodactyls, drowned when the stream abruptly changed course.

   (B) Younger sandstone at Barbora Banda, upstream of section shewed in Fig. 2(B). Although this segment is in the centre of the sandstone it is thin, very flat, and non-erosive at base, and it is succeeded abruptly be red clay.

   (C) Mudcracked shale preserved under Kuldana sandstone, showing locally non-erosive establishment of stream course.

   (D) Typical granule lithology: calcareous granules displaying varied patterns of haematite stains. These are believed to be reworked soil nodules. Rectangular white grains are fishbone fragments.

earliest mammalian fossils known from the Indian subcontinent (Gingerich *et al.*, 1979a, b; Russell & Gingerich, 1980; West, 1980). Also, both the biogeography of this early Indian fauna and the question of the formation of the enclosed Kohat basin on India's former continental shelf have implications for the history of the Indian/Asian collision.

THE LOWER KULDANA REDBEDS

   The Lower Kuldana beds are about 95% uniform red shale and 5% channel sandstone. The shale has been locally thinned or thickened and sheared by folding, but the Lower Kuldana beds appear to be as thick as 100–200 m in the southern and eastern areas of the basin. They were never deposited in the north-west. The sandstones, which occur at all levels in the

redbeds, have been largely protected from deformation during folding by movement of the surrounding shale, and can be treated as unrotated tilted blocks.

   The majority of the sandstones are small, thin, simple and single channel fills, less than 6 m wide and 10 times wider than deep. They are isolated in large amounts of red clay and are dominated by basal calcareous granulestone and low-angle tangential trough cross-bedding, with additional small-scale trough cross-bedding (Fig. 2).

Petrography of the granulestones

   The granulestones and coarse sandstones in the eastern half of the basin are composed almost entirely of distinctive, iron-stained, calcareous granules and, locally, bone (Fig. 3A). Quartz and feldspar grains are present only as silt and fine sand in this area, but they, and lithic grains (quartzites, chert, sandstone and limestone) are present in coarser sizes in the western half of the basin. In some beds in the east, rodent teeth are more abundant than epiclastics of the same size. Bone, mostly fish, may locally constitute as much as 40% of the granulestone.

Most calcareous granules are either micritic or have matrix of small calcite rhombs, within or around which may be developed patches of randomly oriented, asymmetrical, intergrown sheaves of coarse and prismatic calcite crystals. A few granules have thin rims of radially elongated calcite. Radial fabric is well enough developed to resemble an oolite in no more than 1 or 2% of the granules. The granules are clearly recrystallized; the effects of degrading recrystallization can be seen in the irregular long crystal edges and the obliteration of fabric. Most granules have concentric rings of haematite and adjacent grains rarely show the same histories of staining (Fig. 3D). A large minority of the granules have internal cracks and look like minuscule septarian nodules.

These features suggest that the granules are reworked soil nodules that grew under arid or semi-arid conditions in calcareous soils. Radial fabrics and rims have been described from calcareous soil nodules by Assereto & Folk (1976). Cracking and episodic haematite staining and coating of growing nodules could occur in an alternating wet and dry soil environment (Siesser, 1973). Pisoliths with similar radial fabrics have also been reported from spring-fed playa pools (Risacher & Eugster, 1979). One microfabric seen with the SEM is a little like those of calcrete nodules with *Microdium* colonies, as described by Esteban (1974, 1976; Chafetz & Butler, 1980). Soil nodules would be common in flat, arid, calcareous mud flats, and minor erosion and winnowing of the clay during floods and stream channel incision could easily have produced significant lag concentrates, particularly in the absence of any other coarse material. Concentration of relatively erodible bone also suggests only a small influx of epiclastics.

Sedimentary structures

The succession of sedimentary structures in the Lower Kuldana sandstones is best described by Markov chain analysis (Gingerich, 1969; Miall, 1973; Cant & Walker, 1976; Reading, 1978) (Table 1 and Fig. 4). Before summarizing the results, however, the categories or microfacies chosen should be discussed.

The category 'channel', which refers solely to the eroded channel floor, was included to show that nearly all channels are isolated in red mud (i.e. to distinguish and show the rarity of multi-storey sandstones). A channel floor refers to a major cross-cutting scour, although its distinction from clay-draped set bound-

Table 1. Sequence in Lower Kuldana Sandstones

	1	2	3	4	5	6	7	8	9	10
				A. Unreduced data						
1 Pebbles		4	8	2	—	—	1	—	2	—
2 Grit	—		30	5	5	4	3	8	28	—
3 Festoons	1	8		7	22	17	4	3	20	3
4 Tabular dunes	2	1	1		4	3	2	—	5	1
5 Plane beds	—	1	6	—		12	6	5	19	3
6 Ripple troughs	—	1	2	—	5		4	—	29	2
7 Other ripples	—	2	—	3	—	—		2	14	—
8 Variegated mud	—	2	—	—	1	2	—		20	1
9 Red mud	—	12	4	—	5	6	2	12		104
10 Channel	14	51	27	4	5	8	2	—	—	
Total										602
				B. Observed–expected						
1 Pebbles		+0·10	+0·34	+0·08	−0·08	−0·09	+0·02	−0·05	−0·12	−0·19
2 Grit	−0·03		+0·21	+0·02	−0·03	−0·05	−0·01	+0·04	+0·07	−0·22
3 Festoons	−0·02	−0·06		+0·04	+0·17	+0·10	0·00	+0·02	−0·03	−0·19
4 Tablular dunes	+0·08	−0·09	−0·08		+0·13	+0·07	+0·06	−0·05	+0·03	−0·14
5 Plane beds	−0·03	−0·13	−0·03	−0·04		+0·14	+0·07	+0·04	+0·12	−0·15
6 Ripple troughs	−0·03	−0·12	−0·09	−0·04	+0·03		+0·05	−0·05	+0·43	−0·16
7 Other ripples	−0·03	−0·05	−0·13	+0·11	−0·08	−0·09		+0·04	+0·43	−0·20
8 Variegated mud	−0·03	−0·05	−0·14	−0·04	−0·04	−0·02	−0·04		+0·53	−0·16
9 Red mud	−0·04	−0·10	−0·14	−0·05	−0·07	−0·07	−0·04	+0·06		+0·47
10 Channel	+0·09	+0·29	+0·08	0·00	−0·05	−0·03	−0·03	−0·06	−0·28	

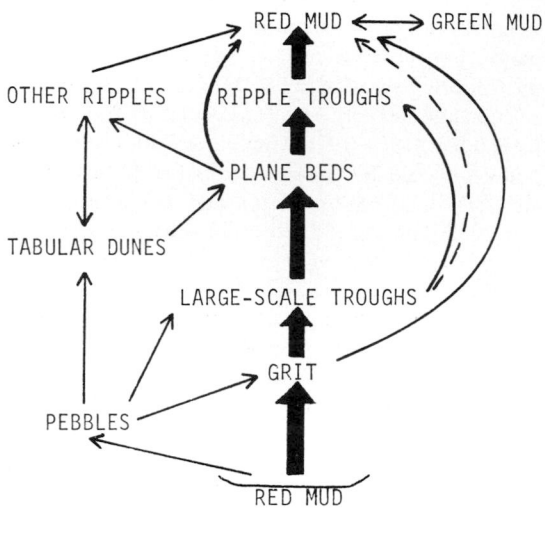

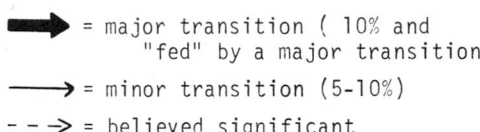

➡ = major transition (10% and
        "fed" by a major transition

⟶ = minor transition (5-10%)

- - -⟶ = believed significant

Fig. 4. Association of facies in Kuldana sandstones: graphic summary of Markov analysis in Table 1.

aries and small in-channel scours is not always easy. Only 10 channels of 114 recorded in detail are clearly multi-storey. In addition, very few channels are deeply incised, suggesting generally slight erosion, and a significant number of sandstones are entirely non-erosive at base (some basal grits preserve underlying mudcracked shale—Fig. 3C). Of 145 transitions from red mud into sandstone, 104 are erosive, but 12 pass to grit, four to festoons, five to plane beds, eight to ripples and 12 to variegated mud, all without showing appreciable erosion. Most channels are single, small and simple (Fig. 2D), less than 10 m wide, with width/depth ratios of between 20 and 6. Very few are thicker than 2·5 m and/or wider than 30 m. Channel bases are generally concave upwards and show no evidence of significant lateral migration. At a few localities the tops of adjacent channel fills are laterally connected, but their bases were separately incised.

'Pebbles' (conglomerates and pebbly sandstones) are present only in the western half of the basin. They are always basal except for three instances where winnowed lags cap large bar deposits. They more commonly pass up into sandstone than grit; grit refers mostly to calcareous granules, and these are less abundant where pebbles are common, probably due to increased abrasion.

'Grit', or calcareous granulestone, and 'festoon' beds are by far the most common facies in the Kuldana sandstones, probably comprising two-thirds to three-quarters of their total volume. 'Festoons' ('large-scale troughs' in Fig. 4) comprise mostly low-angle and tangential, large-scale, trough cross-bedding, but the category also includes a minor yet not precisely determined amount of scour-and-fill and conformable channel fill (Reineck & Singh, 1980, pp. 71–72). 'Grit' and 'festoons' are not mutually exclusive, for the granulestones may be massive, graded, plane-bedded or trough-bedded. 'Grit' was given priority in assigning units to categories, partly because trough-bedded granulestone usually grades up into trough-bedded, coarse-grained sandstone anyway. Only thin, planed-off, basal or lateral sections of large-scale, trough cross-beds are preserved in some channel units (Fig. 2D).

A large number of small, channel-fill sandstones are only one, two or three sets thick. A few well-exposed, one-set-thick beds show preserved dune tops; in others the arrangement of scour pools can be traced. Dune forms may be separated laterally or vertically by thin mud drapes, indicating passage of more than one flood or flood pulse, but nevertheless many sandstones can be interpreted as being the result of only a very few floods. Dunes in thin sandstones seem to have commonly moved around older dunes, locally reworking their toes and sides rather than significantly eroding them or burying them. Preservation of dune tops and some lateral mud drapes therefore suggests that later flow was braided around early high-stage bedforms without necessarily reworking them. Deflection of flow seems to have caused divergence of adjacent dunes by as much as 60°, as measured by foreset dip directions (in different sandstone bodies, three adjacent sets diverge 10°, three at 38°, six at 45°, five at 55° and four at 60°). Preserved bedforms, though possibly biased by their preservation, suggest a predominance of low-lunate to catenary dunes. Maximum preserved dips range from 7° to 25°, but most are between 15° and 21°.

In half of the channel fills, the succession of facies is significantly abridged or curtailed. In 55 of 114, channel floor deposits (coarse lags and/or large dunes) pass directly into red clay without passing through the supposedly ideal sequence of dunes, plane beds, ripples, clay. Eleven more pass to red clay via green or variegated ('variegated') clay. This indicates that avulsion and abandonment of channels was commonplace.

'Tabular dunes' include generally tabular large-scale sets with planar cross-beds, presumably from sand waves/straightcrested megaripples and/or bars with long sections of straight front. Planar cross-bedding seems to be restricted to a few unusually large sandstones and therefore probably indicates large bars in relatively deep streams.

The term 'plane beds' is intended to represent upper flow regime beds, but in practice it probably includes some lower flow regime, coarse-grained plane beds (although plane-laminated granulestone was included in 'grit'), some complete but very low-angle cross-beds and planed-off bases of tangential foresets; it may also exclude some unrecognized fine-grained upper flow regime plane beds.

Small-scale ripples are not well developed in the Kuldana. As a rule they are planed off and rippleform bed surfaces are not preserved. The great majority appear to be trough cross-laminated, suggesting lunate, linguoid and sinuous crested ripples ('ripple [-sized] troughs'). 'Other ripples' comprise mostly straight-crested ripples, plus some climbing ripples.

Variegated mud (green, brown, purplish and/or olive) appears to represent abandoned-channel fill and other low-energy deposits of waterlogged depressions. A thin layer of non-red clay is not uncommon in red clay up to half a metre above a channel deposit, suggesting that minor relief can in some cases be maintained over a channel site for some time after it stops carrying moving water. Despite the large amount of mud in the system, there are very few instances of significant mud deposition within active channels.

As summarized in Fig. 4, my interpretation of a Markov chain analysis indicates that channels are cut into red clay and are floored with conglomerates ('pebbles') where available, channel-lag granulestones ('grit') or even large-scale trough cross-bedded coarse sandstone. Regardless, the basal sequence nearly always fines up into trough-bedded sandstone and ideally passes up through a plane-bedded sequence into trough cross-laminated fine sand ('ripple troughs'). Ripples are usually succeeded by red mud. The succession can be curtailed by a transition to mud at any stage. Deletion of data from atypical sandstones does not change this analysis. Many of these thinning-upward, fining-upward, single-storey sandstones can be explained as the result of only one or two floods. Common curtailment suggests frequent 'premature' abandonment of channels. One would expect a fining-upward sandstone in a clay-dominated deposit to represent a meandering stream but, with minor exceptions, evidence for lateral migration of streams is lacking. In some deposits, early high-flow bedforms seem to have diverted later flow patterns, suggesting in-channel braiding during periods of low flow.

Specific examples of sandstones

Markov analysis provides the description of an 'ideal' Kuldana sandstone. Nevertheless, full description of the Lower Kuldana sandstones is best served by additionally considering some specific exceptions that test the rule and some of the ideal examples that support it.

Exemplary sandstones

Two sandstones south of Barbora Banda epitomise Lower Kuldana sandstones. The upper one is 35 cm thick and occurs 26 m below the top of the Lower Kuldana beds. The lower one, 40 cm–1·5 m thick occurs 27 m further below. Both show flow to the ENE and are exposed in longitudinal section for several hundred metres. The upper sandstone is 'one dune thick' with, essentially, a single layer of 2·5–3 m long and 25–35 cm thick sets of lunate or catenary dune foresets (Fig. 2B). Some dunes are separated laterally by clay drapes, indicating the passage of more than one flood or flood–pulse. Adjacent dunes in the centre of the channel show flow to N 70° E, N 25° E, N 60° E, N 70° E, N 35° E and N 63° E. The locally preserved drapes and moderately divergent flow patterns suggest that renewed flow diverged around old dunes rather than eroding them. Scour pools were locally eroded into the shale, but overall the stream was not erosional, for most basal laminae do not cross-cut underlying mud-cracked shale. The cross-bedding is flat-topped (planed by waning flow) and is succeeded by red mud, which at one point contains a small, 2 cm high lens of laminated fine sandstone (Figs 2B and 3B).

The lower sandstone at Barbora Banda is similar but thicker. The underlying clay is shallowly and unevenly incised by scour pools, and clay coatings are very well developed. The clay coating on top of the basal layer of dunes contains some impressions of sticks that were also stranded by the waning flood. The basal layer itself contains a large number of skulls, jaws and bones of small artiodactyl mammals (even-toed ungulates like deer, goats, etc.) (Fig. 3A). The skeletal material is found only along about 150 m of stream bed, from a point where the material is broken, ground down, toothless and otherwise damaged, to a point upstream where skulls are whole and jaws have almost complete dentitions, and where masses of skeletal material, including an articulated foot, are preserved. The articulated foot implies rapid burial prior to decomposition of ligaments. Farther upstream no fossil material has been found.

My interpretation is that, while in flood, the stream changed course rapidly into a depression that was wooded or scrub-filled, where it caught and drowned a herd of primitive artiodactyls. The inference of a herd comes from the apparent common point of origin of the fossils. (Possibly the animals were browsing and sheltering from the bad weather that caused the flood.) The sudden establishment and abandonment of the streams and their relatively short active lives, and also the flatness of the landscape (inferred in part from the shallow incision and ready abandonment and rearrangement of stream beds), are considered characteristic of Lower Kuldana sandstones. The flatness of the landscape is probably related to (1) the underlying extremely flat gypsum sabkha, (2) the amount of clay in the system, and (3) the aggradation of the system.

A sandstone, found 7·5 km WNW of Barbora Banda, that exemplifies the lack of stream banks is a 5 m wide by 40 m thick lens of entirely flat-bedded granulestone and sandstone with a few small and shallow cross-cutting scours. The bedding indicates largely vertical accretion. The edges of the lens interfinger considerably with the enclosing clay, which indicates that the clay surface was built up as the sandstone accumulated and that the channel was always shallow.

Atypical sandstones

Sandstones are most poorly developed at the base of the formation. Where present, they are little more than small sheet deposits localized in shallow depressions. Most are unchannelled or only slightly channelled. One at Bahadur Khel is 2 m wide and 10 cm thick, has a flat bottom and a convex top, shows planar medium to fine sand passing up into ripples, and changes laterally into red clay through a 3 m belt of green shale with mudcracked gypsum layers. The green shale and gypsum apparently resulted from standing water and waterlogged soils beside the channel.

The most complex Lower Kuldana sandstone is

exposed for over 130 m along strike in the upper fifth of the section at Bahadur Khel. It is a 1–3 m two-storey unit, and it is capped and dominated by a single, 80 m long by 0·7–1·0 m thick, sandstone set of large (up to 1 m high and 7 cm thick), 15°, planar foresets that dip eastward (N 75° E), which is approximately along strike (Fig. 2C). The foresets coarsen toeward, and towards their basal surfaces. The top of the set is locally reworked into a thin pebble lag and it is capped by red clay. Tangential and planar cross-bedding underlie the main unit and replace it laterally. The channel's eastern edge is a wide section of plane-bedded, rippled, very low-angle cross-bedded and shallowly scoured and filled fine to medium sandstone whose beds thin, split, and pass eastward into shale. Cross-beds there indicate flow to between N 70° E and S 65° E. Even farther east on the same level is a separate, oval, 10–20 m wide by 2·7 m thick body of similar, tabular to wedge-shaped, plane-bedded and planar cross-bedded sandstone whose orientations indicate easterly flow.

The major, 1 m high, planar cross-beds are interpreted as foresets formed by a single, large, mid-channel bar, rather than as surfaces of lateral accretion of the sort described by Puigdefabregas & Van Vliet (1978) and Williams (1966), which they greatly resemble. First, measurements of 44 surrounding sets of large-scale cross-beds suggest a mean flow to N 89° E (standard deviation = 23°, total range = N 48° E–S 2° E), which indicates flow parallel to the main cross-beds and not perpendicular as would be the case with epsilon cross-bedding. Poor data on the orientations of basal pebbles and small-scale cross-beds also indicate eastward flow. Secondly, mud is more common between sets under the major set than between its 15° cross-beds, which is the reverse of the expectation for a point-bar sequence. The bar has been only locally altered on top, as shown by a few small channels and pebble lags, and by flow diverted around its sides, as in the north-west corner where several east-moving, pebble-veneered, eastward-rising lower bounding surfaces cut into the planar cross-bedding. The crevasse-splay-like wedge of eastward-thinning sandstones east of the main deposit must be considered as suspended sediments or low velocity (side-channel) bed-load carried over or outside the east bank of the main stream.

In short, the river at Bahadur Khel flowed east, was of moderate to low sinuosity and was much wider and deeper than other lower Kuldana streams. The bar must have formed during bankfull discharge, and flow was surely divided around it at low water. There is a distinct tendency for sandstones in the Bahadur Khel – Surdag area to be larger and more complex than elsewhere, but nevertheless evidence of meandering is slight.

Lastly, the thin, tabular and massive to plane-bedded granulestones and sandstones near Chorlakki are unusual in that (1) they contain a rich and very diverse fossil fauna and (2) they are interbedded with purple shales and other typically Upper Kuldana lithologies (Fig. 2E). One fining-upward granulestone is exposed for 79 m and varies between 15 and 40 cm in thickness. Some are interbedded with clay. As will be discussed elsewhere, the Upper Kuldana beds represent diverse coastal-plain and restricted-marine environments at the front of the transgression, including coastal grasslands, paralic marshes, temporary freshwater lakes and ponds, saline–alkaline lakes, evaporitic lagoons, exposed mudflats, beaches and restricted bays. A description of the Jhalar and Murree sections by Markov chain analysis (unpublished) suggests that the change from red clay to purple seems to be the first effect of transgressive waterlogging. The thinness of the upper Kuldana facies and their highly irregular succession suggests that the environments shifted frequently and almost randomly, as would be expected during the drowning of an exceedingly flat terrain. Environmental diversity and the abundance of water can explain the fauna, which includes artiodactyls, perissodactyls, elephant-sea cow ancestors, primates, carnivores, bats, rodents, insectivores, primitive whales, crocodiles, turtles and fish (particularly catfish), plus freshwater snails (West, 1980; Gingerich *et al.*, 1979a; Buffetaut, 1977, 1978). The rich fluvial fauna implies perhaps that the Chorlakki streams were perennial, whereas non-fossiliferous ones were not. But if the streams were deeper and rarely dry, why are the sandstones so thin? The favoured but unproven explanation is that the granulestones were winnowed veneers or thin channel-floor bars in large, slow, deep and muddy streams that carried a bimodal mud and reworked-nodule load and whose banks were marked by clay–clay contacts that have since been obliterated by the unusually intense deformation of the shales at Chorlakki. Rising sea-level in an apparently tideless basin could easily have produced almost stagnant estuaries and coastal plain rivers that only moved sediment during floods and filled with mud at other times.

DISCUSSION AND SUMMARY

The major component of the Kuldana Formation is red clay. Given that it is continental and fluvial (the formation lies over a sabkha and under a shoreline deposit and contains terrestrial fossils and channel-form sandstones), one would expect the clay to result from overbank floods in a system controlled by meandering streams. However, the sandstones do not appear to represent meandering streams.

The typical lower Kuldana sandstone lies in a single, small, simple, shallowly incised channel that never migrated laterally and that was filled by vertical accretion of reworked calcareous soil nodules and large-scale trough cross-bedded sandstone. It is capped by plane-bedded sandstone and/or trough-bedded ripples unless the succession was curtailed, as frequently happened.

Most sandstones are thin and predominantly high-flow deposits that could have been laid down by only one or two floods. Streams may have held or carried water during low flow (Jackson, 1978, mentions meandering streams that become braided at low flow), but off-peak discharge seems to have been generally insufficient to rework major bedforms. Even the apparently more perennial streams entering the Upper Kuldana coast may have moved most sediment during floods. Generally, then, stream geometry resulted from rare high-flow events. Common curtailed channel-fill sequences (indicating premature abandonment of channels, presumably by avulsion) and one-dune-thick sandstones suggest that the streams were short-lived as well as briefly active. Dominance of high-flow effects results from flashy discharge, which is ultimately controlled by climate. Short stream-lives could preclude the development of meandering, which one would otherwise expect to form on flat and muddy aggrading plains.

Flow was generally inward from the margins of the basin (Fig. 5), as defined by the shorelines of the preceding and following marine incursions (Wells, 1982), but the high variability of palaeocurrent directions between adjacent localities nevertheless suggests that on a small scale the streams wandered all over the landscape, thereby indicating its flatness. Both low to moderate variation of directions of adjacent cross-bedded sets and the concave-upward channel bases indicate moderately sinuous to non-sinuous flow without lateral migration (meandering) (see Moody-Stuart, 1966).

Drainage patterns can help explain atypical streams. Complexity and large stream size seem only

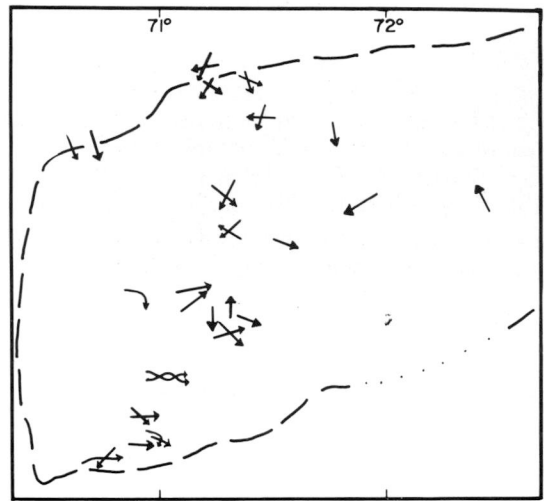

Fig. 5. First-order interpretation of palaeocurrent data, arrayed on a map that has been expanded 2½ times to compensate for north–south compression (use longitudes and N and S basin edges to compare with Fig. 1). Shortening may be greater. Each arrow represents a stream. Sinuous arrows indicate sinuous stream courses and crossed arrows show range of two or more streams crossing the same point.

to have developed high in the formation and in the south-central part of the basin, where both a pre-existing depression and the Kuldana drainage system (Figs 1B & 5) suggest that the highest-order streams should be found. Even these, however, show little evidence of the development of meanders. Similarly, being at the end of a complex network of tributaries may have led to more perennial flow and possibly a longer life and a larger aquatic fauna for streams approaching the Upper Kuldana shoreline. Therefore, the typical Kuldana stream (small, simple, short-lived and with few, if any, fossils) appears to have been of low order.

The impressive aggradation of redbeds suggests that the evaporation/draining of the sea left the basin floor below grade. Aggradation, rather than the basin's flatness, could have kept the streams wide and shallow by limiting their incision and thus controlled avulsion, although the extreme flatness of the underlying sabkha surely contributed to the immaturity of the early drainage system. Overbank deposition could have deposited much of the mud, for frequent avulsion implies common overbank flooding, and channel-form sandstones occur at all levels. However, the ratio of 18·6 m of clay to each metre of channel deposit suggests additional input of clay by other processes, perhaps by sheetwash and wind.

The types and proportions of bedforms in the Kuldana sandstones seem similar to modern deposits described from central and southern Australia by Williams (1968, 1971), and indeed these broad, dry and aggrading plains may represent reasonable modern analogues. Karcz (1972) describes how flash floods in southern Israel change their paths from flood to flood and leave thin deposits that lack pronounced bedform hierarchies, but his deposits seem to be otherwise dissimilar to the Kuldana.

Lastly, palaeocurrent evidence for flow into the basin from the west and north is surprising in that the basin developed on what was India's north-western continental shelf. Flow from the edge of the shelf strongly suggests that orogeny on the continental margin had reversed the palaeoslope. A possible implication is therefore that India had by this time begun to collide with Eurasia, or at least with an Afghan microplate (Wells, 1982).

ACKNOWLEDGMENTS

This research was carried out under the direction of Dr Philip D. Gingerich and was primarily funded by grants to him from the Smithsonian Foreign Currency Program. Additional monies were granted to the author by Sigma Xi, the American Association of Petroleum Geologists, and the Department of Geological Sciences at the University of Michigan. The research was done with the cooperation and support of the Geological Survey of Pakistan. The assistance of Dr S. M. I. Sha, Asif Jah, Mahmoodul Hassan, S. Habib Abbas, and Hassan Shaheed has been particularly valuable. I would also like to thank Drs J. L. Wilson, B. H. Wilkinson and R. G. Jackson for discussions during the research, and Drs J. D. Collinson, P. Friend and C. Puigdefabregas for reviewing the manuscript. George Junne helped with the photography.

REFERENCES

ASSERETO, R. & FOLK, R.L. (1976) Brick-like texture and radial rays in Triassic pisolites of Lombardy, Italy: a clue to distinguish ancient aragonitic pisolites. *Sedim. Geol.* **16**, 205–222.

BUFFETAUT, E. (1977) Données nouvelles sur les crocodiliens paléogènes du Pakistan et de Birmanie. *C. r. hebd. Séanc. Acad. Sci., Paris*, **285**, 869–872.

BUFFETAUT, E. (1978) Crocodilian remains from the Eocene of Pakistan. *Neues Jb. Geol. Paläont. Abh.* **156**, 262–283.

CANT, D.J. & WALKER, R.G. (1976) Development of a braided-fluvial facies model for the Devonian Battery Point Sandstone, Quebec. *Can. J. Earth Sci.* **13**, 102–119.

CHAFETZ, H.W. & BUTLER, J.C. (1980) Petrology of recent caliche pisolite spherulites and speleothem deposits from central Texas. *Sedimentology*, **27**, 497–518.

COLLINSON, J.D. (1978) Alluvial sediments. In: *Sedimentary Environments and Facies* (Ed. by H.G. Reading), pp. 15–60. Elsevier, New York.

ESTEBAN, M. (1974) Caliche textures and *Microcodium*. *Boll. Soc. geol. Ital.* **92**, 105–125.

ESTEBAN, M. (1976) Vadose pisolites and caliche. *Bull. Am. Ass. Petrol. Geol.* **60**, 2048–2057.

GINGERICH, P.D. (1969) Markov analysis of cyclic alluvial sediments. *J. sedim. Petrol.* **39**, 330–332.

GINGERICH, P.D., RUSSELL, D.E., SIGOGNEAU-RUSSELL, S. & HARTENBERGER, J.-L. (1979a) *Chorlakkia hassani*, a new Middle Eocene dichobunid (Mammalia, Artiodactyla) from the Kuldana Formation of Kohat (Pakistan). *Contr. Mus. Paleont. Univ Mich.* **25(6)**, 117–124.

GINGERICH, P.D., RUSSELL, D.E., SIGOGNEAU-RUSSELL, S., HARTENBERGER, J.-L., SHAH, S.M.I., HASSAN, M., ROSE, K.D. & ARDREY, R.H. (1979b) Reconnaissance survey and vertebrate paleontology of some Palaeocene and Eocene formations in Pakistan. *Contr. Mus. Paleont. Univ. Mich.* **25(5)**, 105–116.

JACKSON, R.G. (1978) Preliminary evaluation of lithofacies models for meandering alluvial streams. In: *Fluvial Sedimentology* (Ed. by A.D. Miall). *Mem. Can. Soc. Petrol Geol., Calgary*, **5**, 543–576.

KARCZ, I. (1972) Sedimentary structures formed by flash floods in southern Israel. *Sedim. Geol.* **7**, 161–182.

MEISSNER, C.R. & RAHMAN, H. (1973) Distribution, thickness, and lithology of Paleocene rocks in Pakistan. *Prof. Pap. U.S. geol. Surv.* **716–E**, 6 pp.

MIALL, A.D. (1973) Markov chain analysis applied to an ancient alluvial plain succession. *Sedimentology*, **20**, 347–364.

MIALL, A.D. (1977) A review of the braided stream depositional environment. *Earth Sci. Rev.* **13**, 1–62.

MOODY-STUART, M. (1966) High and low sinuosity stream deposits with examples from the Devonian of Spitsbergen. *J. sedim. Petrol.* **36**, 1102–1117.

PUIGDEFABREGAS, C. & VAN VLIET, A. (1978) Meandering stream deposits from the Tertiary of the southern Pyrenees. In: *Fluvial Sedimentology* (Ed. by A.D. Miall). *Mem. Can. Soc. Petrol. Geol., Calgary*, **5**, 469–485.

READING, H.G. (1978) Facies. In: *Sedimentary Environments and Facies* (Ed. by H.G. Reading), pp. 4–14. Elsevier, New York.

REINECK, H.-E. & SINGH, I.B. (1980) *Depositional Sedimentary Environments*. Springer-Verlag, Berlin. 439 pp.

RISACHER, F. & EUGSTER, H.P. (1979) Holocene pisoliths and encrustations associated with spring-fed surface pools, Pastos Grandes, Bolivia. *Sedimentology*, **26**, 253–270.

RUSSELL, D.E. & GINGERICH, P.D. (1980) Un nouveau primate omomyide dans l'Éocène du Pakistan. *C. r. hebd. Séanc. Acad. Sci., Paris*, **291**, 621–624.

SIESSER, W.G. (1973) Diagenetically formed ooids and intraclasts in South African calcretes. *Sedimentology*, **20**, 539–551.

WELLS, N.A. (1982) Early Eocene paleogeography of north-central Pakistan. *Mem. Geol. Surv. Pakistan* (in press).

WEST, R.M. (1980) Middle Eocene large mammal assemblage with Tethyan affinities, Ganda Kas region, Pakistan. *J. Paleont.* **54**, 508–536.

WILLIAMS, G.E. (1966) Planar cross-stratification formed by the lateral migration of shallow streams. *J. sedim. Petrol.* **36**, 742–746.

WILLIAMS, G.E. (1968) Formation of large-scale trough cross-stratification in a fluvial environment. *J. sedim. Petrol.* **38**, 136–140.

WILLIAMS, G.E. (1971) Flood deposits of the sand bed ephemeral streams of central Australia. *Sedimentology*, **17**, 1–40.

Spec. Publs int. Ass. Sediment. (1983) **6**, 405–420

Morphological characteristics of ephemeral stream channel and overbank splay sandstone bodies in the Permian Lower Beaufort Group, Karoo Basin, South Africa

WILLO M. STEAR

Rand Mines Ltd, P.O. Box 62370, Marshalltown 2107, South Africa

ABSTRACT

Fining-upward cyclic sequences in the lower part of the Beaufort Group of Permian age represent the deposits of an ancient ephemeral stream–playa lake complex analogous with some of those forming in modern semi-arid inland basins. Sheet-like and lenticular sandstone bodies occur as superimposed systems of fluvial channels and overbank splays formed by multiple lateral and vertical accretion. Excellent outcrop conditions provide sufficient transverse cross-sectional exposures of numerous palaeochannels to permit a general two-fold morphological subdivision of the sandstone bodies into sheet and ribbon types. Most of the channel sandstone bodies have a multi-storey composition and many display prominent sets of 'wings' which taper away from the channel-fill deposits as thinner overbank sandstone sheets.

The bulk of the channel-fill deposits consists of high flow regime bedforms and most of the palaeochannels contain features indicative of periodic stage fluctuations. Well-preserved sandstone palaeosurfaces in channel and splay sequences contain a wide variety of sedimentary structures that are attributable to fluctuating hydrodynamic conditions in the falling water stages of floods and during subaerial emergence.

The morphology and internal geometry of the sandstone bodies are adopted as prime criteria for determining the original channel patterns and flow characteristics of the river systems.

The ephemeral streams of the Lower Beaufort were typified by channel patterns that were both straight and highly sinuous and probably displayed various transitions between these two types along their drainage courses.

INTRODUCTION

During the Permian, the Karoo Basin (South Africa) comprised an east–west elongate foreland trough or retroarc basin in which a thick sequence of molasse-type sediments accumulated under the controlling influence of the intensive Gondwanide orogeny (Lock, 1978). The initial rapid downwarping of the basin in Permian times was caused by isostatic compensation of crustal loading in the distant southern orogenic belt. Progressive but sporadic northward encroachment of the rising fold belt on the foreland basin led to renewed thrust-sheet loading and the introduction of large volumes of alluvial detritus to the basin (Rust, 1959; Stear, 1980b). Orogenesis

had reached a peak during the Permian (Visser, 1978), and basin subsidence and sedimentation were accompanied by intermittent, weak volcanism (Martini, 1974; Elliot & Watts, 1974; Lock & Johnson, 1975). By the late Permian, the land-locked basin had probably been segmented into a series of large inland lakes that were gradually filled by prograding and coalescing fluvio-deltaic sequences (Hobday, 1978). Lakes in the southern parts of the basin were shallow and ephemeral (Stear, 1980b). The balance between crustal downwarping in the depositional area and uplift of granitic and metamorphic core rocks in the source area is reflected in the thick pile of fine-grained Lower Beaufort Group sediments that signify no major vertical change in palaeo-environment.

0141-3600/83/0106-0405 $02.00

Fig. 3. Cyclicity in the Lower Beaufort. First-order fining-upward megacycles (1) are made up of numerous second-order fining-upward cycles (2). These, in turn, consist of laterally continuous third-order cycles in the interchannel facies. Prominent channels are illustrated by arrows.

order cycles, whereas the interchannel facies association represents the siltstone- and mudstone-dominant assemblages.

Identification of facies associations in vertical profile is sometimes complicated by poor development of one or other of the two environments. The channel deposits consist mainly of bedload material (less than 5% of siltstone and mudstone) but some channels contain mixed-load deposits (between 5 and 20% of siltstone and mudstone). The inter-channel facies is characterized by suspended-load deposits (more than 20% siltstone and mudstone).

Most descriptions of fluvial sequences emphasize the coarse-grained member deposits although, as in the case of the Lower Beaufort, the deposits of the fine-grained members are frequently proportionally much greater. Analyses of the channel facies association provide an estimate of the river type, whereas the interchannel facies association gives an indication of the palaeoclimate and frequency of channel shifting. The overall sedimentology and taphonomy of the

fossil-bearing inter-channel deposits (Smith, 1979, 1980) indicate that the reptiles of the Lower Beaufort thrived on vast alluvial mudflats (Boonstra, 1969). Seasonal rainfall in a hot, semi-arid climate (Keyser & Smith, 1979; McPherson & Germs, 1979) gave rise to episodic flooding of the mudflats and the formation of ephemeral streams, temporary pools and shallow lakes (Smith, 1979; Stear, 1978, 1980b) which supported a complex assemblage of terrestrial, aquatic and semi-aquatic fauna (Smith, 1979). Calcrete (Keyser & Smith, 1979), silica-replaced gypsum rosettes (Keyser, 1966; Stear, 1978) and various desiccation structures formed during prolonged periods of illuviation (McPherson & Germs, 1979) in low-relief floodbasins. The argillaceous sequences of the Lower Beaufort are, therefore, interpreted as having been deposited in a playa-lake complex characterized by a continuously shifting pattern of ephemeral saline lakes fringed by vast subaerially exposed mudflats (Stear 1980b). During surface runoff the existing playa lakes expanded and large areas

Fig. 4. (A) Sheet sandstone bodies. (B) Ribbon sandstone bodies. Note the sandstone 'wings' that taper away from the channel-fill deposits into the overbank deposits.

previously occupied by dried mudflats were flooded, becoming new sites of shallow ponded water.

CHANNEL SANDSTONE BODIES

Due to the complexity of variables involved in fluvial systems, it is doubtful whether any single parameter can be used on its own as an incontrovertible criterion for recognizing ancient channel types (Jackson, 1978; Collinson, 1978). Instead of categorizing alluvial deposits on the basis of single distilled sequences it would be far more instructive to develop two- and, if possible, three-dimensional fluvial facies models from outcrops where lateral availability of exposure permits a spatial interpretation of the internal organization of channel sandstone bodies in lateral as well as vertical context (Collinson, 1978). The accessibility and excellence of laterally extensive exposures in the Beaufort West district allow for examination of sandstone bodies in transverse sections (perpendicular to palaeochannel flow direction) as well as longitudinal sections (parallel to palaeochannel flow direction). The morphology and internal geometry of channel sandstone bodies, in lateral and vertical profile, are the main criteria adopted in this paper for determining the overall

channel patterns and flow characteristics of the Lower Beaufort ephemeral river systems.

Sandstone body shape

A fluvial sandstone body may represent the deposits of one palaeochannel or it may comprise the total sedimentation of many palaeochannels superimposed one upon the other in such a way that a composite body of sandstone results (Potter, 1967). The term 'palaeochannel' is reserved for the deposits of a single channel in which a single fluvial event took place.

Individual sandstone bodies that define the upper and lower boundaries of second-order cycles in the Lower Beaufort comprise two main morphological types. Tabular or sheet sandstone bodies (Potter, 1963) are the common type (Fig. 4A) although elongate bodies also occur (Fig. 4B). Elongate sandstone deposits are subdivided in the terms of Bersier (1958) into a main or central body ('corps central') which passes laterally into one or more pairs of wings ('ailes d'étalement'). The term 'wings' is used to describe the thinner sandstone sheets which extend outwards from the main channel-fill material into the overbank deposits. There is usually no division between the central body and the wings except that the latter become finer grained away from the channel

27

Fig. 5. A multi-storey palaeochannel sandstone sheet. Note the prominent cutbanks and almost flat erosional base. The individual storeys are marked 1–4.

deposits and eventually peter out into the overbank material (Fig. 4B). The central body represents the channel facies deposit of in-channel flow and the wings the proximal overbank facies deposits of supra-channel flow (Williams, 1975). Friend, Slater & Williams (1979) describe elongate fluvial sandstone bodies as ribbons, and these authors adopt a cross-sectional width (w) to depth (h) ratio of 15:1 in the central body only, to distinguish sheet sandstone bodies (w:h greater than 15) from ribbon sandstone bodies (w:h less than 15). Nami & Leeder (1978) also differentiate between ribbon and sheet channel sandstone bodies in the Jurassic Scalby Formation in England, and use these parameters to interpret the style and magnitude of the palaeochannels.

Palaeochannels in the Lower Beaufort commonly display prominent cutbanks and low relief erosional bases (Fig. 5). Portions of sandstone bodies which constitute the central, channel-fill material are commonly bounded by scoured surfaces. Units separated by prominent scour surfaces are called storeys (Pettijohn, Potter & Siever, 1972). Multi-storeying describes the vertical superposition of storeys in a sandstone sequence (Potter, 1963). Williams (1975) interprets each storey as the remnant of a single palaeochannel. The grouping of such remnants in a multi-storey sandstone body represents the complex sequence of fluvial events in which various palaeochannels shared the same general course. There is no limit to the scale of individual storeys in sandstone bodies of the Lower Beaufort. They range from relatively small units tens of centimetres thick and a few metres wide to large-scale structures hundreds of centimetres thick and many tens of metres wide. Large-scale storeys are the prime building blocks of composite sheet sandstone bodies. Sheet sandstone bodies are usually of the composite

type that, in many places, attains widths of more than 1 km (Turner, 1978) but seldom exceeds 20 m in thickness. Ribbon sandstones consist of isolated bodies tens of metres wide which vary in thickness from 3 to 12 m. Ribbons often have very well-developed sets of wings that usually extend laterally for a few tens of metres away from the central body (Figs 4B and 6). Local thickening of sandstone, particularly in arenaceous zones of first-order cycles, occurs when sheet and/or ribbon sandstone bodies are superimposed one upon the other in stacked fashion. As a result, composite sandstone sheets can attain a local thickness of 50 m or more. Unlike the multiple-channel origin of composite sandstone sheets, some of the isolated sandstone bodies constitute single (uni-storey) channel sequences which contain no internal scour surfaces. Multi-storey sandstone bodies usually display no marked change in lithology from one storey to the next.

The geometry of a few ribbon and sheet sandstone bodies is illustrated by means of line diagrams (Fig. 6) constructed from panoramic photographic mosaics and measured sections of selected outcrops. In each case the exposed face provides a cross-sectional view of the sandstone body perpendicular or slightly oblique to its mean palaeochannel flow direction. Most of the sandstone bodies have a multi-storey composition.

Ribbon sandstone bodies

Friend *et al.* (1979) subdivide ribbons into simple and complex bodies. Simple or uni-storey ribbon sandstone bodies are devoid of internal scour surfaces, whereas complex ribbon sandstone bodies have a multi-storey geometry. The ribbon bodies shown in Fig. 6 (A and B) both contain prominent sets of wings

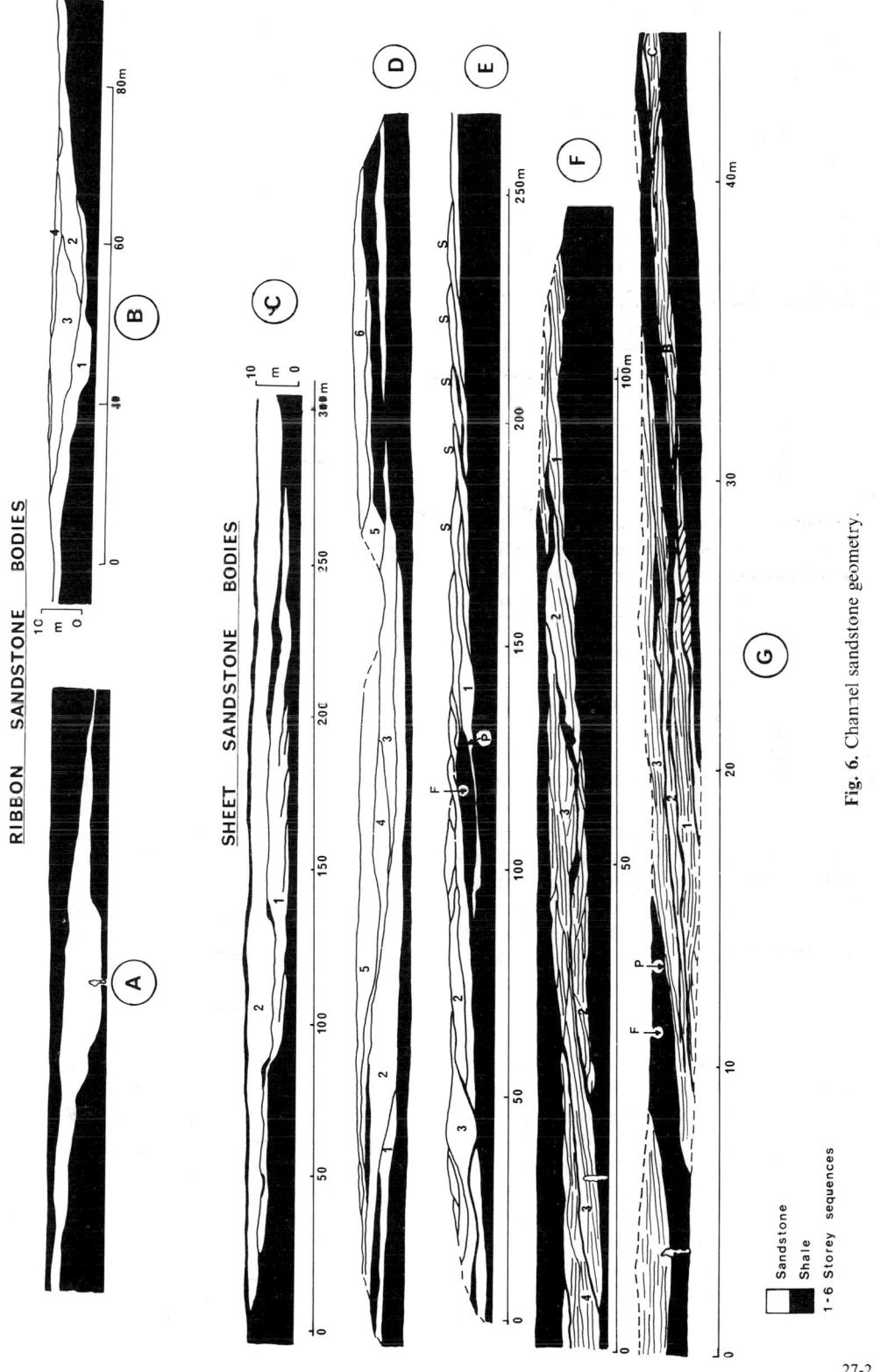

Fig. 6. Channel sandstone geometry.

RIBBON SANDSTONE BODIES

SHEET SANDSTONE BODIES

Sandstone
Shale
1 – 6 Storey sequences

Fig. 10. Levée deposits consisting of interbedded sandstone and shale lenses.

likely to occur in low-sinuosity, upstream reaches of the river systems where flow velocities were high. Sandstone sequences containing a large proportion of ripple lamination and interbedded mudstone are common in laterally accreted units deposited in the more distal, downstream areas on point bars where channel sinuosities were higher. These sequences reflect the more uniform and less vigorous discharge conditions during the waning stage of floods on point bars and in proximal overbank areas. Palaeosurfaces in these units testify to the greater potential for preservation of deposits formed in low-energy, ephemeral stream subenvironments than those formed in high-energy areas within the channels.

OVERBANK SPLAY SANDSTONE BODIES

Crevasse and sheet-splay bodies are usually identified as relatively thin sandstone sheets that taper away from the main river channel sandstone body. Where a definite breach of the river bank took place a crevasse channel results. Well-defined crevasse channel sandstone bodies are not readily developed in the Lower Beaufort. A channel sandstone body is interpreted as being of crevasse channel origin only if it is isolated within levée deposits. Crevasse channel sandstone bodies normally have ribbon shapes. Sandstone units in proximal overbank deposits are generally erosively based and cut discordantly across the underlying strata. Scour surfaces, tool marks, trace fossils and structures formed during runoff and subaerial exposures are common (Smith, 1980; Stear, 1980b).

Prominent levée deposits are generally lacking in the Lower Beaufort. However, in some instances, particularly when a meandering channel origin is implied, very broad wedges composed of alternating thin lenses of sandstone or siltstone capped by mudstone flank the channel sandstone bodies. These are interpreted as low-relief levées (Fig. 10). Due to the poor development of natural levées in most of the fluvial systems of the Lower Beaufort, overbank flooding spread outwards from the channels mainly as sheet-flow over the interchannel areas, depositing very fine sand and silt in broad wings which sometimes coalesced to form extensive sheet-splays. Occasionally, siltstone- or sandstone-filled crevasse channels were cut into finer-grained, wedge-shaped levée deposits (Fig. 11). These proximal overbank crevasse deposits are generally thicker than their distal counterparts. Crevasse channels are commonly filled with siltstone and minor amounts of sandstone, but in many places the upper sections of the fine-grained crevasse channel-fill have been subsequently scoured and plugged by sandstone.

Distal crevasse splay deposits occur either as discrete sandstone lenses or as multiple, laterally connected lenses indicating the anastomosing nature of the small distributary channel systems. The undulatory erosional upper surface displayed by most splay sandstone bodies (Fig. 12) is a product of scouring and reworking by small shallow anastomosing streams that flowed across the tops of the sand sheets. Internal scour-and-fill structures, and siltstone drapes between accretionary units, indicate the multiple episodes of erosion and sedimentation that occurred during formation of the sandstone splays.

Fig. 11. A sandstone crevasse channel cutting through low-angled levée deposits.

Fig. 12. Sheet-splay sandstone bodies in the interchannel facies display undulatory erosional upper surfaces produced by anastomosing rivulets that flowed across the splays.

Recurrent flooding events are reflected by the progradational nature and lateral persistence of the thin, sheet-splay sandstone bodies. Proximal crevasse splays cut obliquely through levée and proximal floodplain deposits but maintain a fairly even base as they thin towards the more distal portions of the playa lake complex. The distal terminations of crevasse- and sheet-splay deposits occur as thin sandstone wedges interfingering with lake deposits.

Sedimentary structures in the overbank sandstone bodies of the Lower Beaufort are similar in most respects to the sedimentary structures in the uppermost units of the channel sandstone sequences. Channel bars, levées and splays in ephemeral fluvial systems are subenvironments that are all characterized by repeated wetting and drying (Picard & High, 1973; Karcz,

1972). Sandstone bodies formed in these subenvironments are sometimes characterized, in the Lower Beaufort, by well-preserved palaeosurfaces exhibiting a wide variety of sedimentary and biogenic structures that reflect the waxing and waning currents and fluctuating hydrodynamic conditions during episodic flow. Sandstone palaeosurfaces provide a detailed sedimentological record of the hydrodynamics of bar formation and also offer valuable evidence of the morphology and behaviour patterns of non-marine organisms which lived during Lower Beaufort times. Stear (1978, 1980b) describes and illustrates some of these palaeosurfaces and draws attention to the similarity in sedimentary structures produced during waning ephemeral sheet-flow in shallow water and structures that form as a result of tidal processes on low-energy beaches. Remnants of these features are quite common in the Lower Beaufort although laterally extensive palaeosurfaces are rare.

SUMMARY AND CONCLUSION

The Lower Beaufort was located in a geographic zone conducive to the development of semi-arid climatic conditions. Large volumes of water that periodically entered the basin probably emanated from heavy seasonal precipitation or glacial melt in the highly mountainous source area. Arid areas are normally characterized by one or more interior drainage systems flowing into a closed basin. These may be temporary or permanent, each with its own centripetal drainage pattern (King, 1963). The lowest portion of an interior drainage system is commonly occupied by a shallow lake or playa. Stream action is episodic and flooding is violent and short-lived, resulting in rapid influx of sediment mainly as sheet-flow. The lower reaches of many internal drainage systems peter out into the alluvium due to loss of water by seepage and evaporation. Characteristic of arid climates is the large variation in rainfall throughout the year and from one year to the next. The great variability often results in freak and, sometimes catastrophic, flash-floods.

Although no proximal, coarse-grained deposit of the Lower Beaufort drainage systems is anywhere preserved, these might have been eroded during uplift following encroachment of the orogenic belt on to the basin, with the result that only the finer-grained detritus, that accumulated in the distal distributary areas, remained. The inland drainage networks preserved in the south-western Karoo are likened to the drainage system of the Indo-Gangetic Plain of northern India, as evidenced by palaeocurrent data indicating a main dispersal trend parallel to the tectonic strike (Gansser, 1964). Seasonally controlled flash-floods periodically introduced large volumes of water and sediment into the interior network and gave rise to the formation of thick wedges of rudaceous material. In this respect the drainage pattern of the Lower Beaufort Basin is analogous to portions of the intracratonic Lake Eyre Basin of central Australia (Bonython & Mason, 1953). Catastrophic flash-floods in the Lake Eyre region transport and deposit masses of sediment over normally desiccated floodplains by means of ephemeral braided streams. Extensive mudflat areas are periodically inundated by floodwaters, forming large playa lakes which gradually evaporate during the dry periods.

Floodwaters that entered the Lower Beaufort Basin were confined, where possible, to incised channels, but due to the excessive volume of the discharge most of the material was deposited by high regime sheet-flow in the proximal overbank areas. This resulted in the deposition of large amounts of sand in wide, elongate, sheet-like bodies. As the floodwaters subsided, a complex of sandy braided channels and shallow playa lakes remained. Repeated ephemeral flood events partially scoured the pre-existing deposits, producing multi-storey sand sequences in which high regime flood deposits predominated. Due to the intensity of the flooding, and the abundance of overbank flow, new channels were rapidly cut and older channels plugged or abandoned. Ribbon sandstone bodies represent straight to slightly sinuous, entrenched streams whereas sheet sandstone bodies are the products of lateral channel migration and sheet-flooding in streams that varied from 25 m to at least 300 m in width and from 1 to 10 m deep. The deeper channels occurred mainly in the thick arenaceous zones in the west of the study area, and the broad but shallow channels mainly in the east, where the overall sandstone:shale ratio in the succession is much less. The formation of laterally persistent, composite sandstone sheets was favoured by a lack of restricting valley walls and prominent levée deposits, resulting in the ability of the channel complexes to migrate randomly across their broad floodplains. An abundance of abandoned channels and cut-offs testifies to the frequency of channel switching and avulsion. The multi-channel sequences probably reflect large fluctuations in channel location caused by repeated avulsion.

The sedimentary environment of the Lower Beaufort uranium province consisted of an ephemeral stream–playa lake complex analogous with some of those formed in modern, semi-arid inland basins. The entire Lower Beaufort succession is, therefore, proposed as a local model for illustrating the sedimentological characteristics of an ephemeral fluvio-lacustrine assemblage. The morphology of channel and overbank splay sandstone bodies illustrates some of the diagnostic features of ancient fine-grained fluvial deposits in the ephemeral stream facies model.

ACKNOWLEDGMENTS

The contents of this paper are an excerpt from the author's original unpublished Ph.D. thesis compiled under the supervision of Professor I. C. Rust of the University of Port Elizabeth.

The author is grateful to Newmont S.A. Ltd, and especially to Mr Vivian Vellet, whose encouragement and keen interest made this study possible.

Appreciation is also extended to Mr A. B. Cadle and Dr I. Stannistreet of the University of the Witwatersrand for critically reviewing the original manuscript, and to Mrs E. Reid-Robertson for typing.

REFERENCES

ALLEN, J.R.L. (1970) Studies in fluviatile sedimentation: a comparison of fining-upwards cyclothems with special reference to coarse-member composition and interpretation. *J. sedim. Petrol.* **40**, 298–323.

BERSIER, A. (1958) Séquences détritiques et divagations fluviales. *Eclog. géol. Helv.* **51**, 854–893.

BONYTHON, C.W. & MASON, B. (1953) The filling and drying of Lake Eyre. *Geogrl J.* **119**, 321–330.

BOONSTRA, L.D. (1969) The fauna of the Tapinocephalus zone (Beaufort beds of the Karoo). *Ann. S. Afr. Mus.* **56**(1), 1–75.

CANT, D.J. (1978) Development of a facies model for sandy braided river sedimentation: comparison of the South Saskatchewan River and the Battery Point Formation. In: *Fluvial Sedimentology* (Ed. by A.D. Miall). *Mem. Can. Soc. Petrol. Geol., Calgary,* **5**, 627–640.

COLE, D. (1980) Aspects of the sedimentology of some uranium-bearing sandstones in the Beaufort West area, Cape Province. *Trans. geol. Soc. S. Afr.* **83**, 375–390.

COLLINSON, J.D. (1978) Vertical sequence and sand body shape in alluvial sequences. In: *Fluvial Sedimentology* (Ed. by A.D. Miall). *Mem. Can. Soc. Petrol. Geol., Calgary,* **5**, 577–586.

ELLIOT, D.H. & WATTS, D.R. (1974) The nature and origin of volcaniclastic material in some Karoo and Beacon rocks. *Trans. geol. Soc. S. Afr.* **77**, 109–111.

FRIEND, P.F., SLATER, M.J. & WILLIAMS, R.C. (1979) Vertical and lateral building of river sandstone bodies, Ebro Basin, Spain. *J. Geol. Soc. Lond,* **106**, 36–46.

GANSSER, A. (1964) *Geology of the Himalayas.* Interscience, London. 289 pp.

HOBDAY, D.K. (1978) Fluvial deposits of the Ecca and Beaufort Groups in the eastern Karoo Basin, Southern Africa. In: *Fluvial Sedimentology* (Ed. by A.D. Miall). *Mem. Can. Soc. Petrol. Geol., Calgary,* **5**, 413–429.

JACKSON, R.G. (1978) Preliminary evaluation of lithofacies models for meandering alluvial streams. In: *Fluvial Sedimentology* (Ed. by A.D. Miall). *Mem. Can. Soc. Petrol. Geol., Calgary,* **5**, 543–576.

JOHNSON, M.R. (1966) *Stratigraphy of the Cape and Karoo Systems in the eastern Cape Province.* Unpublished M.Sc. Thesis. Rhodes University.

KARCZ, I. (1972) Sedimentary structures formed by flash floods in Southern Israel. *Sedim. Geol.* **7**, 161–182.

KEYSER, A.W. (1966) Some indication of an arid climate during deposition of the Beaufort Series. *Ann. geol. Surv. S. Afr.* **5**, 77–80.

KEYSER, A.W. & SMITH, R.M.H. (1979) Vertebrate biozonation of the Beaufort Group with special reference to the western Karoo Basin. *Ann. geol. Surv. S. Afr.* **12**, 1–35.

KING, L.C. (1963) *South African Scenery.* Oliver & Boyd, Edinburgh. 308 pp.

KÜBLER, M. (1977) *The sedimentology and uranium mineralisation of the Beaufort Group in the Beaufort West–Fraserburg–Merweville area, Cape Province.* Unpublished M.Sc. Thesis. University of Witwatersrand.

LOCK, B.E. (1978) The Cape Fold Belt of South Africa: tectonic control of sedimentation. *Proc. geol. Ass.* **89**, 264–281.

LOCK, B.E. & JOHNSON, M.R. (1975) Discussion on "The nature and origin of volcaniclastic material in some Karoo and Beacon rocks" by Elliot & Watts (1974). *Trans. geol. Soc. S. Afr.* **78**, 171.

MARTINI, J.E.J. (1974) On the presence of ash beds and volcanic fragments in the graywackes of the Karoo System in the southern Cape Province (South Africa). *Trans. geol. Soc. S. Afr.* **77**, 113–116.

McPHERSON, J.G. & GERMS, C.J.B. (1979) Calcrete (caliche) in the Beaufort Group of the southern Karoo Basin and its palaeoclimatic significance. *Abstr. Geokongres* 79, Pt 2, *Geol. Soc. S. Afr.* pp. 145–147. Port Elizabeth.

NAMI, M. & LEEDER, M.R. (1978) Changing channel morphology and magnitude in the Scalby Formation (M. Jurassic) of Yorkshire, England. In: *Fluvial Sedimentology* (Ed. by A.D. Miall). *Mem. Can. Soc. Petrol. Geol., Calgary,* **5**, 431–440.

PETTIJOHN, F.J., POTTER, P.E. & SIEVER, R. (1972) *Sand and Sandstone.* Springer-Verlag, New York. 618 pp.

PICARD, M.D. & HIGH, L.R. (1973) Sedimentary structures of ephemeral streams. *Developments in Sedimentology,* **17**, Elsevier, Amsterdam. 223 pp.

POTTER, P.E. (1963) Late Palaeozoic sandstones of the Illinois basin. *Rep. Invest. Ill. St. geol. Surv.* **217**, 92 pp.

POTTER, P.E. (1967) Sand bodies and sedimentary environments—a review. *Bull. Am. Ass. Petrol. Geol.* **51**, 337–365.

PUIGDEFABREGAS, C. (1973) Miocene point bar deposits in the Ebro Basin, Northern Spain. *Sedimentology,* **20**, 133–144.

RUST, B.R. (1978) Depositional models for braided alluvium. In: *Fluvial Sedimentology* (Ed. by A.D. Miall). *Mem. Can. Soc. Petrol. Geol., Calgary,* **5**, 605–625.

RUST, I.C. (1959) On the sedimentation of the Molteno sandstone in the vicinity of Molteno, C.P. *Annale Univ. Stellenbosch,* **37** (A2-10), 165–236.

SCHUMM, S.A. (1961) Effect of sediment characteristics on erosion and deposition in ephemeral stream channels. *Prof. Pap. U.S. geol. Surv.* **353-c**, 31–70.

SLATER, M.J. (1977) *The Oligo-Miocene fluvial molasse sediments of the northern Ebro Basin, Spain.* Unpublished Ph.D. Thesis. University of Cambridge.

SMITH, R.M.H. (1979) The sedimentology and taphonomy of flood-plain deposits of the Lower Beaufort (Adelaide Subgroup) strata near Beaufort West, Cape Province. *Ann. geol. Surv. S. Afr.* **12**, 37–68.

SMITH, R.M.H. (1980) The lithology, sedimentology and taphonomy of flood-plain deposits of the Lower Beaufort (Adelaide Subgroup) strata near Beaufort West. *Trans. geol. Soc. S. Afr.* **83**, 399–413.

STEAR, W.M. (1978) Sedimentary structures related to fluctuating hydrodynamic conditions in flood plain deposits of the Beaufort Group near Beaufort West, Cape. *Trans. geol. Soc. S. Afr.* **81**, 393–399.

STEAR, W.M. (1980a) Channel sandstone and bar morphology of the Beaufort Group uranium district near Beaufort West. *Trans. geol. Soc. S. Afr.* **83**, 391–398.

STEAR, W.M. (1980b) *The sedimentary environment of the Beaufort Group uranium province in the vicinity of Beaufort*

West, South Africa. Unpublished Ph.D. Thesis. University of Port Elizabeth.

TURNER, B.R. (1978) Sedimentary patterns of uranium mineralization in the Beaufort Group of the southern Karoo (Gondwana) Basin, South Africa. In: *Fluvial Sedimentology* (Ed. by A.D. Miall). *Mem. Can. Soc. Petrol. Geol., Calgary*, **5**, 831–848.

VISSER, J.N.J. (1978) Die morfologie van die Karookom (Karboon-Trias) en die implikasie op die rekonstruksie van Gondwanaland. *Tydskr. Natuurwet.* **18**, No. 2, 77–97.

VISSER, J.N.J. & DUKAS, B.A. (1979) Upward-fining fluviatile megacycles in the Beaufort Group, north of Graaff-Reinet, Cape Province. *Trans. geol. Soc. S. Afr.* **82**, 149–154.

WILLIAMS, G.E. (1971) Flood deposits of the sand-bed ephemeral streams of central Australia. *Sedimentology*, **17**, 1–40.

WILLIAMS, R.C. (1975) *Fluvial deposits of Oligo-Miocene age in the southern Ebro Basin, Spain.* Unpublished Ph.D. Thesis. University of Cambridge.

WINTER, H. DE LA R. & VENTER, J.J. (1970) Lithostratigraphic correlation of recent deep boreholes in the Karoo–Cape sequence. *Proc. Pap. 2nd Gondwana Symp.* (Ed. by S.H. Haughton), pp. 395–408. CSIR, Pretoria, South Africa.

Spec. Publs int. Ass. Sediment. (1983) **6**, 421–433

Fluvial distributary channels in the Fletcher Bank Grit (Namurian R2b), at Ramsbottom, Lancashire, England

STEVE A. OKOLO

Department of Geology, The University, Keele, Staffordshire ST5 5BG, U.K.

ABSTRACT

A working quarry in the Fletcher Bank Grit (Namurian R2b) provides excellent cliff-face exposures over a horizontal distance of some 800 m. Vertical and lateral relationships between five major delta distributary channels and their associated bay-fill sequences can be studied.

All five channels are thought to be of low sinuosity, and the lowest three cross-cut one another. The bed forms that produced the channel fills were sandwaves, dunes, ripples and flat beds which participated in variable proportions in four of the five channels. The other channel was probably filled by accretion on alternate bars.

Vertical sedimentation is considered the dominant channel filling mechanism for channels 1, 4 and 5 while side-filling appears to have dominated during the infilling of channels 2 and 3. There is probably a greater channel density in the lower part of the sequence than in the top, suggesting that channel switching may have been more frequent during the deposition of the sediments in the lower part.

INTRODUCTION

Deposits of the Marsdenian Stage (R2) in the south Pennine region of Northern England contain a high proportion of delta top lithofacies. However, in normal exposure, it is usually difficult to see their lateral and vertical relationships.

Fletcher Bank Quarry, a working quarry in Ramsbottom, south-east Lancashire (Figs 1A and 6B), is an exception; with both surface and subsurface data being available (Figs 1C and 6B) it provides a good opportunity for the study of the different elements of the delta top environment in their stratigraphical-sedimentological setting. The lowermost 122 m of a 245 m subsurface section is known from borehole cores, and the section overlying the cores is known from well cuttings. There is an excellent 60 m cliff-face section in the quarry above this. The most prominent quarry face is over 800 m long, and is oriented north–south (Fig. 6B) while the overall palaeocurrent direction is predominantly to the south-

east and in places south-west. Three-dimensional exposure occurs locally through differently oriented faces (Fig. 6B).

Vertical and lateral relationships between various channels are well displayed and the mutual relationships between the channels and their associated interdistributary bay sequences are well exposed.

Five distinct channels are recognizable in the quarry. They are numbered 1–5, where channel 1 is the oldest (Figs 2 and 6A). All the channels except no. 2 are simple. Channel 2 consists of upper and lower divisions separated by a 10 cm thick mudstone unit. For ease of reference, and also because there is no evidence of erosional activity at the base of the upper division, the lower and upper divisions of channel 2 will be referred to as 2a and 2b respectively.

STRATIGRAPHIC-SEDIMENTO-LOGICAL SETTING OF THE PALAEOCHANNELS

The rocks exposed in the Fletcher Bank Quarry belong to the uppermost part of the Fletcher Bank

0141-3600/83/0106-0421 $02.00

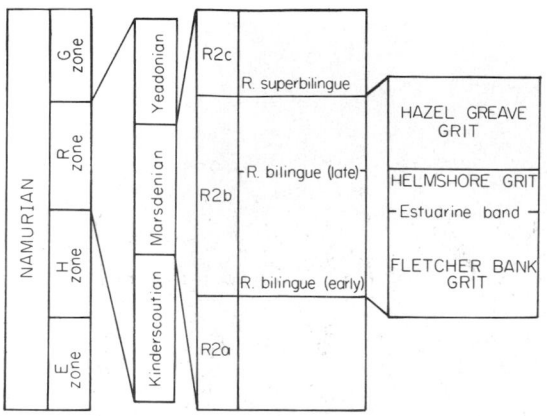

Fig. 3. Stratigraphic location of Fletcher Bank Grit. E = Eumorphoceras, H = Homoceras, R = Reticuloceras, G = Gastrioceras.

locally appear undulatory, though they are generally horizontal over distances of up to 30 m.

Trough cross-bedding described here was probably formed during the migration of trains of dunes (Allen, 1963).

Large-scale, trough-shape bedded sandstone

This lithofacies is very rare. It occurs, as a solitary set 5 m thick, only in channel 4, being well exposed only at the northern part of the quarry (Fig. 6B). It is lithologically similar to the rest of the fill of channel 4. Its base is erosional and trough-shaped. The foresets within the units are discordant with the base and margin of the trough. The upper set boundary is undulatory. The lateral extent of the set, in a direction parallel to the foreset dip, is 30 m. Smaller trough cross-bedded sets occur within the larger foresets of the unit. These trough cross-beds within the foresets are regarded as 'intrasets' (cf. Collinson, 1968). Morphologically, these intrasets resemble the 'scoop-shaped intrasets' of McCabe (1975).

The large-scale trough-shaped bed is regarded as a small, local channel scoured out within channel 4. Its solitary tangential cross-beds are regarded as foresets of an alternate bar which are deposited at the flanks of the bar by the components of the fluviatile currents moving at oblique angle to the true downcurrent trend. In other words, the foresets resulted from a side-filling mechanism in the channel. Side filling may be responsible for the banking of the foresets to the channel margin.

Comparable small channels eroded within fluvial

deposits were shown by Reineck & Singh (1975). Also the genesis of zeta cross-stratification of Allen (1963) is roughly comparable to the formation of the small channel envisaged here.

The intrasets may have been formed by the infilling of scours made by short-lived eddies, associated with fluviatile current in the downcurrent surface of the alternate bar which generated the tangential cross-beds (McCabe, 1975).

DESCRIPTION OF INDIVIDUAL PALAEOCHANNELS

Dimensions of palaeochannels

The length of none of the channels is known. Except for the northern margin of channel 2 and both sides of channel 3 (Figs 2, 6 and 8), no major channel shows its sides. The orientation of channel 3, based on the trend of the groove marks on its base is known. Its calculated width is 62 m. Widths of channels 1, 2, 4 and 5 are difficult to calculate because their true orientations are unknown. Their widths, cited here, are based on the lateral extent of their fills as observed in the main quarry face and may not reflect true width, due to the possible oblique divergence between the orientation of the quarry face and the trend of the channels.

The width of channel 1 is over 800 m, its infill extending beyond the limits of the main quarry face. Channel 2 is about 240 m wide, assuming from its morphology that the exposed width (Fig. 2) is about half the total width (Fig. 8). Based on the same evidence as in channel 1, channel 4 is over 800 m wide. Channel 5 is exposed only at the northern part of the quarry with an exposure width of over 90 m.

The depths of the channels are based on the thickness of their fills. For channels 1, 3 and 4, where there are distinct scours, the depths include the maximum measured relief of the scour. Minimal observed depths of channels 1, 2, 3, 4 and 5 are 29, 10, 5, 11 and 3·6 m respectively.

Erosion surfaces, channel fill sequences and palaeocurrents

Vertical and lateral facies relationships within the five channel fills have been plotted in Fig. 6(A) in an effort to understand their hydrodynamic history. Each vertical section in Fig. 6(A) represents a composite log embracing a substantial lateral extent of the main quarry face. The lateral extents of the northern, central and southern vertical sections are shown in

Fig. 4. Prominent scour of channel 1 base. (A) Scour into mouth bar deposit. (B) Scour into bay fill sequence. Note truncation of sandstone units.

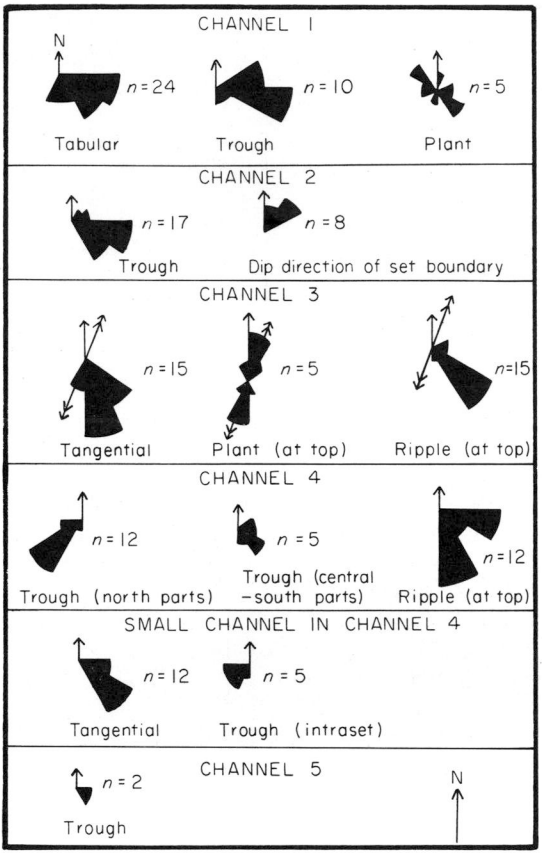

Fig. 7. Palaeocurrent rose diagrams from Fletcher Bank Quarry, with 30° classes. Double arrow indicates channel trend.

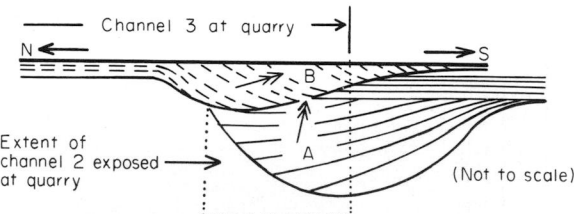

Fig. 8. Sketch diagram illustrating channels 2 and 3 fill geometry, and their cross-cutting relationship. Palaeoflow is indicated by double arrow. (A) Channel 2, filled asymmetrically by inclined strata generated by currents moving in a direction oblique to channel 2 axis (after McKee, 1957). (B) Channel 3, filled asymmetrically by tangentially based foresets deposited at the flank of alternate bar. Note the northwards gradation of tangential cross-beds to horizontal beds.

Channel 3

The erosional surface is trough shaped, though asymmetrical. Its northern margin is steeper than the southern (Figs 2, 6A and 8).

A solitary set of tangential cross-bedding forms virtually the whole in-channel deposit. Traced northwards from the thickest zone of the channel, the upper parts of the set laterally grade to horizontal bedded sandstone (Fig. 2). Vertically both the cross-beds and the horizontal beds grade to a thin unit of trough cross-laminated sandstone very rich in rootlets, logs and *in situ* stigmaria which is particularly well exposed at the northern end. All the sediments of this channel are distinctively brown coloured.

The palaeocurrent directions indicated by the foreset dips within the channel, and the plants and ripples on top of the channel, are fairly consistent though slightly oblique to the channel trend (Fig. 7). This directional anomaly is explicable in the context of side-filling.

Channel 4

At the extreme southern end of the quarry, the base appears irregular and rich in mud clasts. Moving northwards, there is a sharp, horizontal and prominently flat base for some 180 m before it scours a depth of 3·7 m steeply into underlying bay sediments (Figs 2 and 6A). From this locality to the northern parts of the quarry this erosive base undulates irregularly, though never rising up to its former (southern) level.

There is a textural fining upwards and a vertical sequence of sedimentary structures as follows: trough

of 5° to the south. It cross-cuts channel 1 at the southern part of the quarry, eroding its upper lithofacies.

The trough cross-bedding of channel 2A which immediately overlies the erosion surface decreases in size upwards. A 10 cm thick mudstone unit, marking the boundary between 2A and 2B, overlies the trough cross-beds. Parallel laminated sandstone representing the basal unit of 2B grades upwards to trough cross-beds. Striped siltstone caps the sequence.

Grouping the trough axes bearing from both upper and lower parts results in a unimodal palaeocurrent direction to the south-east, and the direction of dip of the set boundaries is roughly perpendicular to the dispersal direction defined by trough axes (Fig. 7), indicating that each set boundary may represent a side-fill surface.

cross-bedded sandstone, horizontal bedded sandstone and trough cross-laminated sandstone. Seatearth and coal cap the sequence.

The trough axis direction of the trough cross-beds, best exposed in the northern part of the quarry, is south-westerly whereas the foreset dip direction from the central and southern parts is south-easterly (Fig. 7). This large directional spread may be indicative of the high angle to the true downcurrent trend at which the dunes migrated. The flow direction of the superimposed ripples is dominantly south-easterly though rather more widely spread (Fig. 7). The foreset dip direction within the small channel is south-easterly, and the trough axis direction of the intrasets also within the small channel is south-westerly (Fig. 7).

Channel 5

The base is sharp, undulatory and cuts into underlying, 2 m thick bay sediments. Relief of scour is up to 1 m. Trough cross-bedded sandstone constitutes the overall infill sediments, while coal on seatearth caps the sequence.

Based on the trough axis direction from the trough cross-bedded sandstone, channel 5 probably flowed south-eastwards (Fig. 7).

Comments

There is no obvious indication that the nature of the substrate has any direct relationship with, or influence on, the relief of the erosion surfaces. Both channels 1 and 4 appear to have excavated substantial scours irrespective of the different nature of their substrates. Indeed, the difference in the observed maximum relief of scouring in channels 1 (6 m) and 4 (3·7 m) appears to relate more to the overall depth of the channel than to the substrate nature.

Channels 1–5 resemble the sand-choked high gradient channels of Campbell (1976), because their sequences generally lack any significant proportion of fine material. Thus although each sequence starts with an erosional base and grades upward through various stages of gradually waning flow, the lithofacies involved are sand dominated.

MECHANISMS OF EROSION, CHANNEL FILLING AND ABANDONMENT

Erosion

Traction currents are regarded as the cutting mechanisms of the five channels, based particularly on

the sedimentological context of the channels. The significant relief of scour on the bases of channels 1, 3 and 4 is indicative of the strong erosive power of these currents. The absence of a mudstone unit between the erosive base and the infill sediments suggests that the same currents may have cut and filled the channels either contemporaneously or shortly afterwards.

Channel filling and abandonment

Two methods of channel filling appear dominant. These are side- and vertical filling, using the terminology of Bluck & Kelling (1963). The method of each channel filling was deduced particularly from the arrangement of the sediments within the channels, and also from the geometry of the channel margin. Where the shapes of the channel margins and fills are conspicuous as in channels 2, 3 and the small channel within channel 4, the interpretation of the method of fill is much easier. Difficulties usually arise because of insufficient exposure, lack of specific evidence and the complexity of the fill, especially when it comprises a mixture of materials deposited by more than one method. A similar problem was reported by Bluck & Kelling (1963).

The mechanisms of channel fill appear to be related to different modes of channel abandonment.

Channel 1 (a vertical fill)

While the transition from the erosive base through horizontal bedded sandstone to tabular cross-bedded sandstone is probably indicative of upward waning flow power, the occurrence of trough cross-bedded sandstone on top of tabular cross-bedded sandstone may be reflecting shallowing water (cf. Harms et al., 1975), as is probably verified by the upward decrease in set thickness of the tabular cross-bedded sandstone and the trough cross-bedded sandstone, and also the progressive vertical transition to other sedimentary structures of lower energy level. Upward reductions of this type usually occur in vertical accretion, even though the contribution by lateral migration is not ruled out. The fairly concave upward base of the channel and the sub-horizontal geometry of the set boundaries within it (Fig. 2) are usually attributed to vertical accretion (Kelling, 1968; Bluck & Kelling, 1963). Therefore, even though conclusive evidence is lacking due to the absence of definitive channel margins, the evidence provided above by channel-fill lithology and geometry, and also the geometry of the channel base, argues in favour of a vertical fill.

distributed across a meander belt (Fisk, 1947) and the area of the quarry (about 400,000 m²) is probably large enough to show them if they are present. Despite a thorough but fruitless search, no epsilon cross-bedding was found, although its absence does not conclusively rule out the involvement of point bars (Collinson, 1978b).

Interpretation

Channel 1 may have constituted a part of a significantly pro-grading delta lobe, based on its ability to build over, and cut into its own mouth bar (Fig. 4), as in the modern Mississippi elongate delta (Brown, 1973).

There is probably a greater channel density in the lower part of the sequence than in the top, based on the mutual erosive relationship of the three lowest channels. Assuming that the overall sequence reflects progradation, channel switching may have been more frequent during the deposition of the sediments in the lower part than in the upper part. The frequency of this avulsion may have been occasioned by the creation of a shorter and steeper course as the delta prograded into the basin. The necessity to create such an alternative course might have arisen due to the over-extension of the parent distributary channel system, resulting in its loss of gradient-advantage to a younger channel, as in the case of the present-day Achafalaya (Gould, 1970). More frequent avulsion can also indicate a rapid rate of progradation in fluvial dominated deltas, because such progradation produces significant shoreline protuberances resulting in the occurrence of gradient advantage (Elliott, 1978).

Seatearths and coals, which are prominent in channels 3, 4 and 5, are not observed in channels 1 and 2. Their absence may be due to post-depositional erosional activity or to non-deposition. Coal on seatearth usually requires a long time to develop (Elliott, 1978). Channels that switch frequently may not have allowed time for coal growth. However, even if coal develops, the chances of its preservation are slim due to post-growth erosional activity which channels 2 and 3 display.

Abandonment of channel 1 may have been followed by a comparatively greater degree of subsidence to account for its relatively greater preserved thickness.

The slow rates of sedimentation that accompanied the abandonment of channel 3 may have generated the persistent marker coal.

While channel 5 changes abruptly from dune to abandonment stage lithofacies, the filling of channel 4 was relatively more gradual, as it grades from trough cross-beds through horizontal beds to ripples before ultimate abandonment.

ACKNOWLEDGMENTS

This study, which forms part of the author's Ph.D. thesis at the University of Keele, England, was carried out during the tenure of a postgraduate scholarship of the Anambra State Government of Nigeria which is gratefully acknowledged. I wish to thank Dr J. D. Collinson for his very inspiring supervision throughout this work, for his availability for discussions at all times, for his very helpful suggestions and for his critical reading of the manuscript. I am also grateful to Professor G. Kelling, Dr T. Elliott and Mr D. Macdonald for reading the manuscript critically, for discussions and for offering very useful suggestions. I thank Mr David Kelsall and Miss Pat Douglass for taking some of the pictures and Mrs Joan Cliff for the typing of this paper.

REFERENCES

ALLEN, J.R.L. (1963) The classification of cross stratified units, with notes on their origin. *Sedimentology*, **2**, 93–114.

BLUCK, B.J. & KELLING, G. (1963) Channels from the Upper Carboniferous Coal Measures of South Wales. *Sedimentology*, **2**, 29–53.

BROWN, L.F. JR (1973) Cratonic basins: Terrigenous clastic models. In: *Pennsylvanian Depositional Systems in north-central Texas. A Guide for interpreting Terrigenous Clastic Facies in a Cratonic basin* (Ed. by L.F. Brown, Jr, A.W. Cleaves and A.W. Erxleben), pp. 10–30. *Gdbk No. 14, Bur. econ. Geol. Univ. Texas, Austin*.

CAMPBELL, C.V. (1976) Reservoir geometry of a fluvial sheet sandstone. *Bull. Am. Ass. Petrol. Geol.* **60**, 1009–1020.

CASEY, J.M. (1980). *Depositional Systems and Basin evolution of the late Palaeozoic Taos Trough, northern New Mexico*. Ph.D. Dissertation, University of Texas, Austin. *Rep. No. UT* 80–1, 236 pp.

COLLINSON, J.D. (1968) Deltaic sedimentation units in the Namurian of Northern England. *Sedimentology*, **10**, 233–254.

COLLINSON, J.D. (1969) The sedimentology of the Grindslow Shales and the Kinderscout Grit: a deltaic complex in the Namurian of Northern England. *J. sedim. Petrol.* **39**, 194–221.

COLLINSON, J.D. (1978a) Alluvial Sediments. In: *Sedimentary Environments and Facies* (Ed. by H.G. Reading), pp. 15–60. Blackwell Scientific Publications, Oxford.

COLLINSON J.D. (1978b) Vertical sequence and sand body shape in alluvial sequences. In: *Fluvial Sedimentology* (Ed. by A.D. Miall). *Mem. Can. Soc. Petrol. Geol. Calgary*, **5**, 577–586.

COLLINSON, J.D. & BANKS, N.L. (1975). The Haslingden

Flags (Namurian G_1) of south-east Lancashire: bar finger sands in the Pennine Basin. *Proc. Yorks. geol. Soc.* **40**, 431–458.

ELLIOTT, T. (1976) The morphology, magnitude and regime of a Carboniferous fluvial-distributary channel. *J. sedim. Petrol.* **46**, 70–76.

ELLIOTT, T. (1978) Deltas. In: *Sedimentary Environments and Facies* (Ed. by H.G. Reading), pp. 97–142. Blackwell Scientific Publications, Oxford.

FISK, H.N. (1947) *Fine Grained Alluvial Deposits and their Effect on Mississippi River Activity.* Mississippi River Comm., Vicksburg. 78 pp.

GOULD, H.R. (1970) The Mississippi Delta Complex. In: *Deltaic Sedimentation Modern and Ancient* (Ed. by J.P. Morgan). *Spec. Publs Soc. econ. Paleont. Miner.,* Tulsa, **15**, 3–30.

HARMS, J.C. & FAHNESTOCK, R.K. (1965) Stratification, bed forms and flow phenomena (with an example from the Rio Grande). In: *Primary Sedimentary Structures and their Hydrodynamic Interpretation* (Ed. by G.V. Middleton). *Spec. Publs Soc. econ. Paleont. Miner.,* Tulsa, **12**, 84–115.

HARMS, J.C., SOUTHARD, J., SPEARING, D.R. & WALKER, R.G. (1975) Depositional environments as interpreted from primary sedimentary structures and stratification sequences. *Soc. econ. Paleont. Miner., Short Course Lecture Notes* **2**, Dallas, Texas. 161 pp.

KELLING, G. (1968) Patterns of sedimentation in Rhondda Beds of South Wales. *Bull. Am. Ass. Petrol. Geol.* **52**, 2369–2386.

MCCABE, P.J. (1975) *The sedimentology and stratigraphy of the Kinderscout Grit Group (Namurian R_1) between Wharfedale and Longdendale.* Unpublished Ph.D. Thesis, University of Keele. 172 pp.

MCKEE, E.D. (1957) Flume experiments on the production of stratification and cross-stratification. *J. sedim. Petrol.* **27**, 129–134.

MCKEE, E.D., CROSBY, E.J. & BERRYHILL, H.L., JR (1967) Flood deposits, Bijou Creek, Colorado, June 1965. *J. sedim. Petrol.* **37**, 829–851.

MOODY-STUART, M. (1966) High- and low-sinuosity stream deposits, with examples from the Devonian of Spitsbergen. *J. sedim. Petrol.* **36**, 1102–1117.

REINECK, H.E. & SINGH, I.B. (1975) *Depositional Sedimentary Environments with Reference to Terrigenous Clastics.* Springer-Verlag, Berlin. 439 pp.

WRIGHT, W.B., SHERLOCK, R.L., WRAY, D.A., LLOYD, W. & TONKS, L.H. (1927) The geology of the Rossendale Anticline. *Mem. geol. Surv. U.K.* H.M.S.O., London. 182 pp.

Spec. Publs int. Ass. Sediment. (1983) **6**, 435–448

Different depositional settings of the Nubian lithofacies in Libya and southern Egypt

DEBA P. BHATTACHARYYA* *and* JOHN C. LORENZ†

Princeton University, Princeton, New Jersey 08540, *U.S.A.*

ABSTRACT

Thick sequences of tabular-planar cross-bedded sands, commonly known as the 'Nubian Sandstone', cover wide areas of north-east Africa, and include two major regressive facies: (a) a widespread pre-Cenomanian fluvial facies locally known as the Messak Sandstone (Libya) or the Basal Clastic and the Desert Rose Sandstones (south-west Egypt) and (b) the Taref Sandstone in central and south-east Egypt that marks a brief period of active northward progradation following the Cenomanian transgressive episode over the north-east African platform.

Although the generalized sandbody morphology and internal sedimentary structures of these closely related sandy sequences appear similar, in detail there are subtle but discernible differences between them which are helpful in deciphering their depositional environments.

The sandbodies in the Messak Sandstone are broadly lenticular with interbedded claystone. The sandstones are texturally immature but mineralogically mature, with tabular-planar cross-beds similar to those found in present-day braided stream deposits. Many of the cross-beds are stacked in a series of sets, up to 4 m thick. Local, tectonically controlled ponding produced thick accumulations of clay. The Basal Clastics and the Desert Rose Sandstone, on the other hand, are characterized by dominantly trough to tabular cross-bedded, immature, lenticular sandbodies and well developed interfluve palaeosols, interpreted together as the deposits of a low-sinuosity stream system.

The spectacularly tabular-planar cross-bedded, texturally mature and broadly lenticular sandbodies of the Taref Sandstone in south-eastern Egypt, in contrast, were deposited as distributary channel bars and migrating sandwaves, with intercalations of interdistributary bay mud deposits on a delta plain.

INTRODUCTION

In 1837 Joseph Rüssegger mapped and described an extensive body of sandstone in the upper Nile valley and adjacent regions of Egypt and Arabia. Subsequently he named this unit the 'Nubischer Sandstein' (Rüssegger, 1847) from its widespread outcrops in the Nubia region of the upper Nile valley, and assigned an early Cretaceous age to it.

Since then the name, 'Nubian Sandstone', has been applied to similar looking sandstones in north-east Africa and adjacent Levant, ranging in age from Cambrian to Cretaceous. As a result their stratigraphic correlation became riddled with confusion (Pomeyrol, 1968).

Much of this confusion is casued by lack of fossils in these non-marine deposits and poorly developed interpretation prior to recent advances in the understanding of fluvial sediments. This resulted in an inability to distinguish between two major regressive, sandy successions: (a) a latest Jurassic–early Cretaceous fluvial sequence and (b) the deposits of a late Cretaceous fluvial and coastal plain complex closely related to the initial stage of the extensive late Cretaceous flooding of Tethys on to the African and

* Present address: Department of Earth and Planetary Sciences, Washington University St Louis, Missouri 63130, U.S.A. and † Present address: Div. 4753 Sandia Laboratories, Albuquerque, New Mexico 87185, U.S.A.

0141-3600/83/0106-0435 $02.00

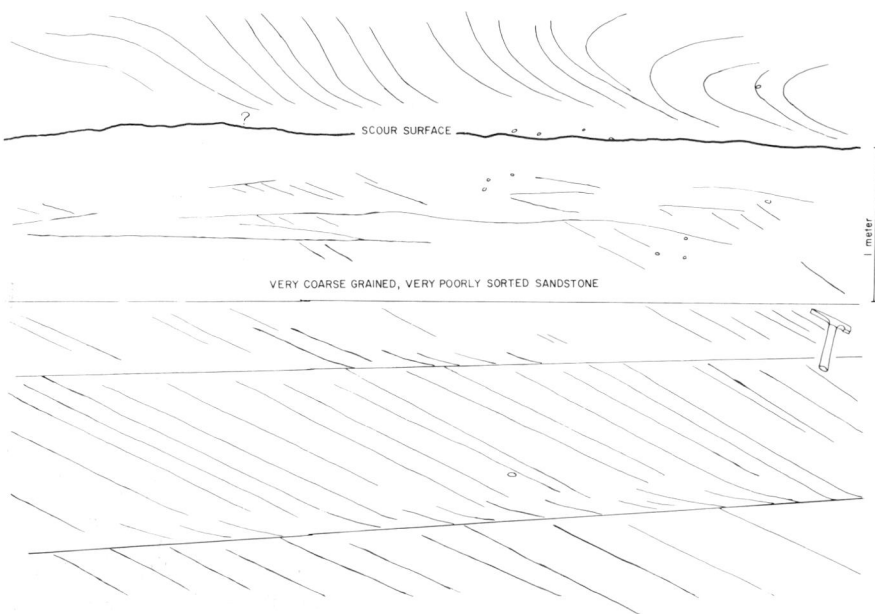

Fig. 4. Large-scale tabular cross-bed cosets in the Messak Sandstone grading up into more erratic small-scale tabular cross-beds. Sequence is overlain by a scour surface and an overturned set of cross-beds.

systems traversing the broad Murzuk Basin. Most of the poorly sorted, tabular-planar sandstone deposits are of high water stage origin in a system of erratic discharge (Lorenz, 1980). Erratic discharge was caused by poorly developed vegetation, which also allowed extensive lateral migration of the system. Although the Messak Sandstones are not perfectly analogous to the deposits of modern braided rivers, at present the Platte River model (Miall, 1977) is most applicable to these sediments.

THE NUBIAN FACIES OF THE SOUTHERN BASIN, EGYPT

In the Southern Basin of Egypt (Fig. 1) the Nubian facies is between 500 and 1200 m thick, and ranges in age from late Jurassic to early late Cretaceous (Barthel & Boettcher, 1978; Klitzsch, 1978; Klitzsch *et al.*, 1979; Ward & McDonald, 1979; Van Houten & Bhattacharyya, 1979; Bhattacharyya, 1980), although the lower age limit is not precisely defined and may extend well into the Jurassic. In the west (Fig. 1) the Nubian sequence above the Precambrian and Palae-ozoic basement is at its thickest, and comprises two major regressive sandy successions, separated by the widespread marginal marine claystone, sandstone and

ironstone deposits of the Cenomanian transgressive episode (Table 1).

In the south-eastern desert, however, the lower regressive fluvial sequence is represented only by a thin mantle of poorly sorted, coarse-grained sandstones and conglomerates over the Precambrian basement. The major part of the 'Nubian' deposits is composed of the upper regressive facies (Table 1 and Fig. 5). Consequently, the total thickness of the Nubian facies there is about 500 m. In both areas, however, the entire sequence is conformably overlain by the marginal marine, variegated claystones of the Senonian (?) Quseir Formation. The Lingula Shale Unit, a thin marine intercalation containing latest Jurassic fossils (Barthel & Boettcher, 1978; Klitzsch, 1978; Klitzsch *et al.*, 1979), further subdivides the lower regressive unit into two subunits, i.e. the Basal Clastic and the Desert Rose Units (Klitzsch, 1978; Klitzsch *et al.*, 1979), in the Western Desert. A tentative correlation of the different Nubian lithofacies units in the Murzuk Basin and the Southern Basin is shown in Table 1.

Eastern Desert

The best developed section of the Nubian lithofacies in the Eastern Desert is between Aswan and Idfu, along the Nile Valley, where as many as seven

Descending ta
th

Dep

Table 1.

| | Murzuk Basin Libya | Southern Basin, Egypt | |
		Western Desert	Eastern Desert
Late		Cretaceous	
Se		Variegated Claystone	Variegated Claystone
		Taref Sandstone Unit, 120 m	Facies 5, 300 m
			Facies 4, 120 m
Tu	Marine carbonate	Plant Bed Unit 100 m	Facies 3b, 30 m
			Facies 2a, 2b and 3a, 30 m
Ce			
Early			
Be		Desert Rose Unit, 300 m	Facies 1, 10–100 m
	Messak Sandstone, 500 m	Lingula Shale Unit, 100 m	
		Late Jurassic	
	Tilemsin Formation	Basal Clastic Unit, 500 m +	
		PC basement	PC basement

The inter
of tabular
downcurrer
were comp
low-angled
coset types
part of one
fluvial, tid
A spectr
sandsheets
probably ir

Studies of ancie
sequences are oft
position within th
tures seen at an
position of a who
palaeochannel is u
several episodes o
have occurred.

A great deal of
available for the
Britain. This oft
thicknesses to be e
individual sandst
body can then be
Subsurface info
north-east Engla
sinuosity sandsto

* Present address
G2 5LJ, U.K.

0141-3600/83/0106-
© 1983 Internatior

distinctive lithofacies have been identified (Bhattach-aryya, 1980, see also Table 1 and Fig. 5).

The pre-Cenomanian part of the sequence is about 10 m thick, composed of lenticular, brownish-yellow, dirty white and grey, coarse-grained kaolinitic sandstones grading upward into a darker grey, structureless sandy mudstone (Facies 1) with vertical, tubular ferruginous concretions (Fig. 6) vaguely resembling root traces. Each of these fining-upward fluvial sequences is commonly between 1 and 2·5 m thick along the axial part, with a maximum thickness of about 6 m, and is conglomeratic at the base (Fig. 6). The lower part of each is generally trough cross-bedded, showing consistent southward dispersal.

Overlying the fluvial sequence of Facies 1 is a pair of sheet sandstones (Facies 2a and 3a, Fig. 5), each up to 10 m thick. Sandwiched between these sheet sandstones is a 10 m thick unit (Facies 2b) of grey to brownish-grey, massive, bioturbated silty mudstone. Facies 2a has a sharp, disconformable basal contact, and is more quartzose and better indurated than Facies 1. The lower part of this unit shows 10–50 cm sets of tabular-planar cross-beds, but the upper part is thoroughly burrowed by *Skolithos*, *Ophiomorpha*, *Diplocraterion* and *Monocraterion*. This, in fact, is the first appearance of such trace fossils in the Nubian sequence of the South-eastern Desert.

Facies 3a has similar characters except that the sandstone is finer grained, better sorted and less burrowed than Facies 2a.

The intervening massive mudstone beds are also

sheet-like, of laterally variable thickness, and contain local palaeosol horizons.

These three sub-facies have been interpreted as the deposits of strand plains and coastal marshes that marked the advent of the Cenomanian transgression over southern Egypt (Bhattacharyya, 1980).

Facies 3b, between Aswan and Idfu, consists of claystone, sandstone and associated oolitic ironstone in at least three shoaling-upward sequences with an aggregate thickness of about 30 m. Each shoaling-upward unit starts with a laminated kaolinitic claystone at the base with a sharp lower contact, and gradually passes upward into a sandstone or oolitic ironstone or both. The clay beds contain minute, carbonized plant debris, mostly oxidized brown on exposure. The sandstones are medium- to fine-grained, well sorted, and show well developed tabular-planar cross-beds in 10–30 cm sets in the lower part. The upper part is thin-bedded (5–10 cm) and current-rippled. The top 10–50 cm of the sandstones/iron-stones are profusely burrowed by *Skolithos*, *Diplocraterion*, *Teichichnus*, *Chondrites* and *Thalassin-oids*. Muddy, oolitic ironstone commonly occurs at the top of each shoaling sequence, whereas well-sorted ferruginous oolites generally interfinger with cross-bedded and ripple-bedded sand. Ferruginized moulds of marine bivalves *Inoceramus* and *Isocardia*, though rare, only occur within this part of the sequence. Facies 3b has been interpreted as the marginal marine deposits of a shoaling mudflat and breaker-bar association.

A 50–120 m thick sequence of Facies 4 comprises

distinctive c
sandstones

Fieldworl
Oil Compai
supported l
Sciences R
South Car
reviewed or
many helpf
benefit of
unknown r

N

Recent 1
Unit' in th
Bisewski, 1
this unit u
limit of tl
therefore,
Jurassic as
of the 'Do
Aptian an
These, h
the lithofa
of events (
basins. In:
there, the
well with 1
subsurfac

ATTIA, M
deposits
262 pp.
BARTHEL,
Formati
Egypt) ;
Nubian
BHATTACH
Formati
ironston
sity, Pri
BOREAU,
jurassiq
somm. !
BUROLLET
Tectoni(
CONANT,
tectonic
51, 719

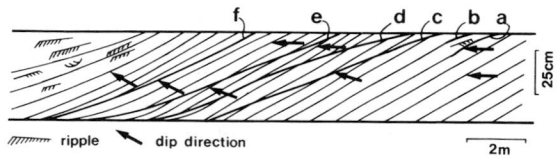

Fig. 9. Idealized west–east section through a coset parallel to palaeocurrent showing types of bounding surfaces. Horizontal scale is compressed. The sequence of bounding surfaces a–f from right to left would be produced by a flow separation which became increasingly three-dimensional as the sandwave (i.e. coset) crestline became more oblique to palaeocurrent (see text). Ripples on left result from a decrease in flow stage. Arrows within the coset show dip direction at their point, oriented on a circular 360° compass scale.

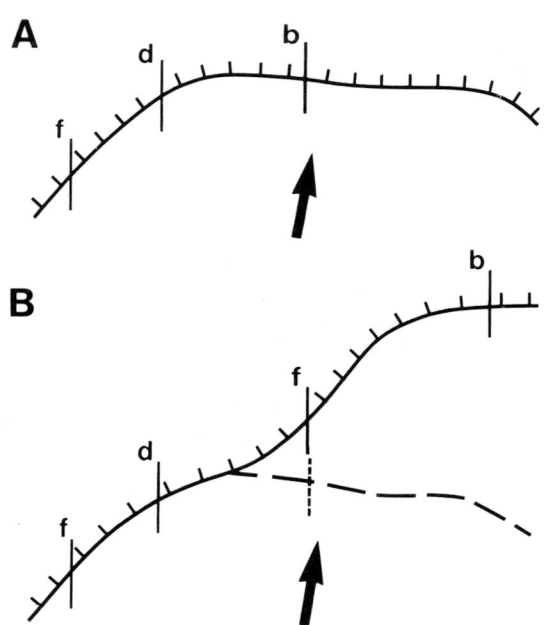

Fig. 10. Hypothetical plan view of a sandsheet crestline which deposited a coset. Large arrows show palaeocurrent direction. Sections b, d, f refer to type of bounding surface present at that point (cf. Fig. 9). The type of surface was influenced by the position of the crestline, thus a downcurrent change from b to f was produced under a steady flow as the crestline shape changed.

over the leeside of a large bar. The whole 10 m thick sandstone was deposited as part of one bar within a 1·9 km wide low-sinuosity river. Analogous arrangements of cross-bed sets occur on the leesides of other modern and ancient large fluvial, tidal shelf and aeolian bedforms.

Bounding surfaces separate foresets deposited by different bedforms. The shapes of these bounding surfaces reflect the orientation of a larger sandwave crestline relative to the palaeoflow which passed over it.

ACKNOWLEDGMENTS

I would like to thank Dr R. Anderton for his enthusiastic guidance and supervision and an early reading of the manuscript. Dr B. J. Bluck is thanked for several patient and stimulating discussions. The manuscript was much improved after comments by N. H. Allen, G. A. Blackbourn and two referees. This work was supported by a N.E.R.C. studentship and completed using B.N.O.C. facilities; both are gratefully acknowledged.

REFERENCES

ALLEN, J.R.L. (1968) *Current Ripples.* North-Holland, Amsterdam. 433 pp.

BANKS, N.L. (1973) The origin and significance of some downcurrent dipping cross-stratified sets. *J. sedim. Petrol.* **43**, 423–427.

BLUCK, B.J. (1971) Sedimentation in the meandering River Endrick. *Scott. J. Geol.* **7**, 93–138.

BLUCK, B.J. (1979) Structure of coarse grained braided stream alluvium. *Trans. Roy. Soc. Edin.* **70**, 181–221.

BLUCK, B.J. (1980) Structure, generation and preservation of upward fining braided stream cycles in the Old Red Sandstone of Scotland. *Trans. Roy. Soc. Edin.: Earth Sci.* **71**, 29–46.

BLUCK, B.J. (1981). Upper Old Red Sandstone, Firth of Clyde. In: *Field Guides to Modern and Ancient Fluvial Systems in Britain and Spain* (Ed. by T. Elliott), pp. 5.14–5.20. International Fluvial Conference, Keele University, U.K.

BROOKFIELD, M.E. (1977) The origin of bounding surfaces in ancient aeolian sandstones. *Sedimentology,* **24**, 303–332.

CANT, D.J. & WALKER, R.G. (1978) Fluvial processes and facies sequences in the sandy braided S. Saskatchewan River, Canada. *Sedimentology,* **25**, 625–648.

COLLINSON, J.D. (1970) Bedforms of the Tana River, Norway *Geogr. Annlr,* **52 A**, 31–56.

HARMS, J.C., SOUTHARD, J.B., SPEARING, D.R. & WALKER, R.G. (1975) Depositional environments as interpreted from primary sedimentary structures and stratification sequences. *Soc. econ. Paleont. Miner., Tulsa, Short Course Notes,* **2**. Dallas. 161 pp.

HASZELDINE, R.S. (1981) *Westphalian B coalfield sedimentology in NE England and its regional setting.* Unpublished Ph.D. Thesis. University of Strathclyde, U.K. 229 pp.

JOHNSON, H.D. (1977) Shallow marine sand bar sequences, an example from the Late Precambrian of N. Norway. *Sedimentology,* **24**, 245–270.

JONES, C.M. (1979) Tabular cross-bedding in Upper Carboniferous fluvial channel sediments in the S. Pennines, England. *Sedim. Geol.* **24**, 85–104.

JONES, C.M. & MCCABE, P.J. (1980). Erosion surfaces within giant fluvial cross-beds of the Carboniferous in N. England. *J. sedim. Petrol.* **50**, 613–620.

LAND, D.H. (1974) Geology of the Tynemouth district. *Mem. geol. Surv. U.K.* **15**, HMSO, London.

LEVELL, B.K. (1980) A late Precambrian tidal shelf deposit, the Lower Sandfjord Formation, Finnmark, N. Norway. *Sedimentology,* **27**, 539–557.

Spec. Publs int. Ass. Sediment. (1983) **6**, 457–471

Structural and sedimentological controls of coal deposition in the Nongoma graben, northern Zululand, South Africa

BRIAN R. TURNER* *and* MICHAEL K. G. WHATELEY†

*Department of Geology, The University, Newcastle upon Tyne NE1 7RU, U.K. and †Southern Sphere Mining and Development Company (Pty) Ltd, Box 50065, Randburg 2125, South Africa

ABSTRACT

The Nongoma graben in northern Zululand developed in response to crustal thinning and the first phase of extensional tectonics (rifting) prior to continental break-up and the separation of east and west Gondwanaland. Sedimentation contemporaneous with graben formation led to the deposition of a thick sequence of coal-bearing fluvio-deltaic Ecca Group (Permian) sediments, controlled mainly by episodes of tensional stress build-up and release in the crust, and not to factors inherent in the depositional system, as on the flanking craton to the north.

A lower progradational deltaic phase of deposition is succeeded by a fluvial depositional phase characterized by fining-upward sequences which themselves show a gross fining-upward trend throughout the succession. Early fluvial deposition was dominated by high-gradient, low-sinuosity (non-braided) bedload channels. During later fluvial deposition, tectonic events controlling deposition were spaced increasingly further apart. Although low-sinuosity channels developed at first, stabilization of source and depositional site promoted lower gradients, increased production of fines and produced high-sinuosity channels due to denudation and possible drifting of the source area. Thus, late fluvial deposition is dominated by fining-upward depositional couplets, comprising low-sinuosity channel deposits overlain by high-sinuosity channel deposits. Fluvial deposition was terminated by a transgressive deltaic depositional phase.

Economically important coals occur at or near the top of fining-upward sequences associated with the high-sinuosity channel facies. Within this facies thick, laterally persistent coal seams formed in extensive floodbasin peat swamps. Overbank flooding was infrequent, with the result that the coals are low ash, low volatile, bright coals containing few shale partings. Depositional modelling indicates that: coal trends are seldom influenced by the thick, laterally extensive sand sheets at the base of the low–high-sinuosity channel depositional couplets; the better-quality coals are of distal floodbasin origin; roof conditions are generally good and the effects of differential compaction minimal. Because of the strong structural imprint on sedimentation, the coals differ in depositional setting and physico-chemical properties from coals in the nearby northern Natal Coalfield where more stable conditions prevailed. This has wider implications for the structural evolution of this part of Africa and the location of coal elsewhere in the southern hemisphere.

INTRODUCTION

The Ecca Group within the main Karoo Basin of South Africa is a thick, laterally extensive sequence of clastic sediments containing economic deposits of coal. Depositional models for coal exploration within the basin have been mainly concerned with the

regional picture in which fluvio-deltaic complexes of Permian age prograded across a shallow shelf or stable platform that deepened southwards into an unstable slope and basin setting. The location and character of the coal within the framework of this regional model were largely controlled by factors inherent in the depositional system (Hobday, 1973).

The Nongoma graben in northern Zululand lies on

0141-3600/83/0106-0457 $02.00

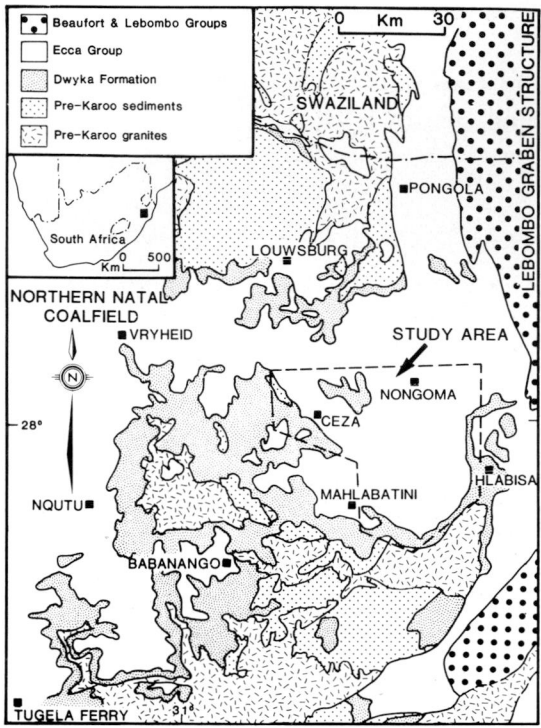

Fig. 1. Generalized geological map of northern Zululand and location of study area.

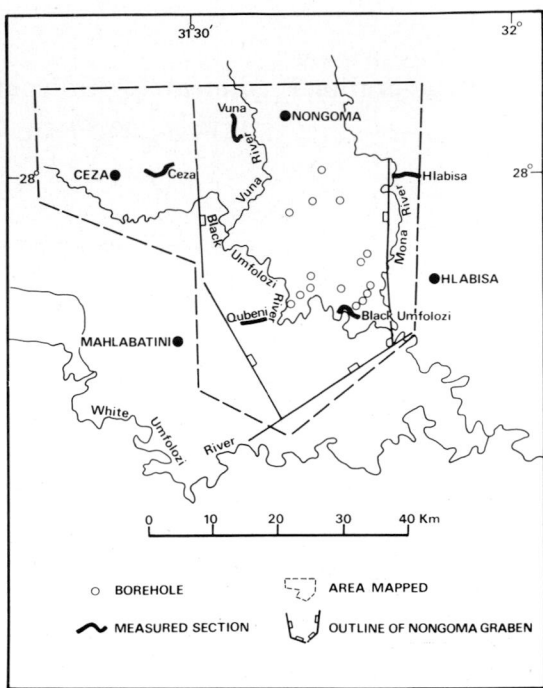

Fig. 2. Location of measured stratigraphic sections and boreholes in the study area.

the eastern margin of the north-east Karoo Basin between the Lebombo graben structure and the northern Natal Coalfield (Fig. 1). Although coal has been known in northern Zululand for some time it was thought to be of little economic significance (Petrick, 1975; De Jager, 1976). Renewed interest in the coal, and recent exploration, have now established that the coal is generally a high-grade anthracite of considerable economic importance, and that it differs from the coal in the nearby northern Natal Coalfield in a number of ways. Zululand was located closer to the edge of the Kaapvaal craton, with the result that the major controls on coal accumulation and its conversion to anthracite are related to the development of half-graben basins, such as the recently recognized Nongoma graben (Whateley, 1980), pre-dating continental rifting. Preferential sites for rifting can be traced back into the Archaean and Proterozoic basement rocks, with later structural development favouring the preservation of unexpectedly thick coal-bearing sequences within the Nongoma graben. Similar coal measures are to be anticipated in adjacent basins which evolved along similar lines.

Because of the poor exposure, complete sections are not known and outcrop information is based on five major sections. Elsewhere considerable use has been made of continuously cored drill-holes and wireline logs, especially the gamma log. Correlations and environmental interpretations are based on detailed descriptions of the measured sections and logged cores, augmented by wireline log patterns. The location of the major sections and logged drill-holes is shown in Fig. 2. Additional information from photogeological mapping and detailed field mapping of selected localities, together with data from logged cores, provided evidence of the structural setting of the area, and the fault block nature of the terrain (Whateley, 1980).

The main purpose of this paper is to show the way in which the structural evolution of northern Zululand, and particularly the Nongoma graben, has influenced patterns of sedimentation and the character and location of the associated coals, and why the coals differ from those in the nearby Natal Coalfield.

STRUCTURAL SETTING

Southern Africa comprises stable cratons separated by circumcratonic metamorphic zones or mobile belts

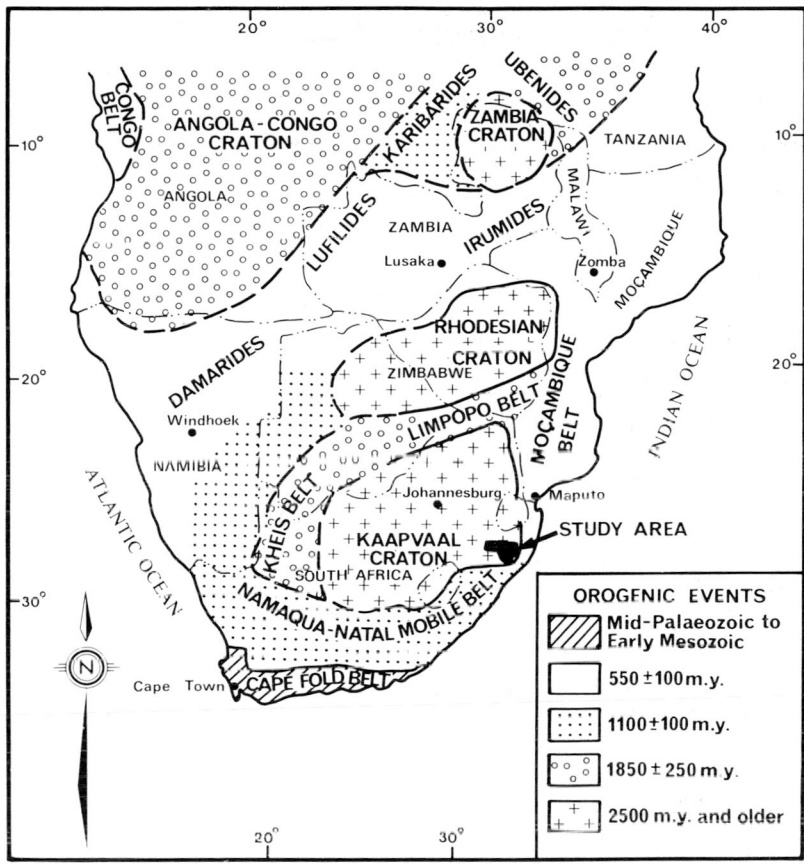

Fig. 3. Structural setting of study area.

(Fig. 3) (Clifford, 1971). The cratons evolved about 3000 my ago and continued to grow throughout the Proterozoic, culminating in late Phanerozoic continental drift (Pretorius, 1974). Because of the stability of the cratons, a number of intracratonic sedimentary basins developed including the Pongola basin, which contains a relatively undeformed sequence of Proterozoic rocks that form most of the pre-Karoo basement in northern Zululand.

The study area lies on the south-east edge of the Kaapvaal craton at the junction of two major linear mobile belts (Fig. 3): the eastern sector of the Namaqua–Natal mobile belt (Proterozoic) and the southerly extension of the Mozambique mobile belt which manifests itself as the north–south-trending Lebombo graben structure (Phanerozoic). Within the relatively stable craton area two main structural trends predominate: a north-west to south-east trend and a north-east to south-west trend. Both trends were inherited from Archaean times (Pretorius, 1974),

but they reacted to later superimposed Proterozoic and Palaeozoic stress fields to produce and modify a simple cross-folded and cross-faulted structural system of domes and basins which determined the location and development of late Palaeozoic and Mesozoic sedimentary basins.

The Namaqua–Natal and Mozambique mobile belts represent zones of weakness which were more susceptible to faulting than the relatively stable cratons. As a result, Karoo sediments deposited over such zones occur in half-grabens, for example at Subi/Bubye, Soutspanberg, Waterberg and Nongoma. These structures can be related to incipient rifting which developed parallel to basement structures in response to crustal thinning and tensional stresses prior to continental break-up. The existence of an east–west tensional stress in this area has been documented by Du Toit (1929).

Although the earliest rift between east and west Gondwanaland occurred about 180 my ago (Dingle &

Scrutton, 1974), fragmentation is thought to have been initiated some 300 my prior to actual break-up (Pretorius, 1974). Thus the forces responsible for the fragmentation of Gondwanaland were initiated approximately 450 my ago, which coincides with the date of the last tectono-thermal event of the Mozambique mobile belt (Holmes, 1951). This suggests that incipient fracturing and faulting, including narrow, linear pre-split grabens, were active in pre-Karoo times and that these structures exerted a significant influence on sedimentation in this area.

Narrow, linear, graben-like sedimentary basins occur on passive continental margins prior to initial split (Bott, 1976). One mechanism for this graben-like subsidence is thought to be crustal stretching due to tension caused by thickness variations in the ductile mantle beneath continental crust. Faulting takes up some of the extension caused by tensional stress. This reduces pressure on the crust and enables more dense material to rise, producing a high geothermal gradient. Many mobile belts are characterized by high geothermal gradients. However, the Karoo sediments in northern Natal were not affected by heat because the time interval between the development of the mid-Proterozoic Namaqua–Natal mobile belt and the formation of the sediments was considerable (850 my), resulting in a very much diminished heat flow. In contrast the Mozambique mobile belt is characterized by a high geothermal gradient which caused extensive devolatilization of the coals in northern Zululand, Swaziland and Empangeni in the south producing a regional conversion of the coal to anthracite prior to late Triassic magmatism. Since the date of the last geothermal event in the Mozambique mobile belt was about 450 my ago (Holmes, 1951) the forces responsible for crustal thinning and graben development probably operated from this time onwards. Also, the mobile belt is a zone of thinner, weaker crust, more susceptible to splitting. Thus, incipient rifting probably began at about this time, some 150 my before that suggested by Pretorius (1974). The site of fragmentation may have been the Mozambique mobile belt (Lebombo structure) between the mainland and Malagasy, with crustal extension and thinning having a greater effect on the weaker mobile belt than the more stable craton edge (study area) (Fig. 4). Evidence for crustal thinning over the Lebombo structure is a large, positive gravity anomaly attributed partly to Karoo basalts and partly to a crustal effect (thinning) (Darracott, 1974).

The final stages of disruption, beginning about 180 my ago, superimposed a post-Karoo tectonic

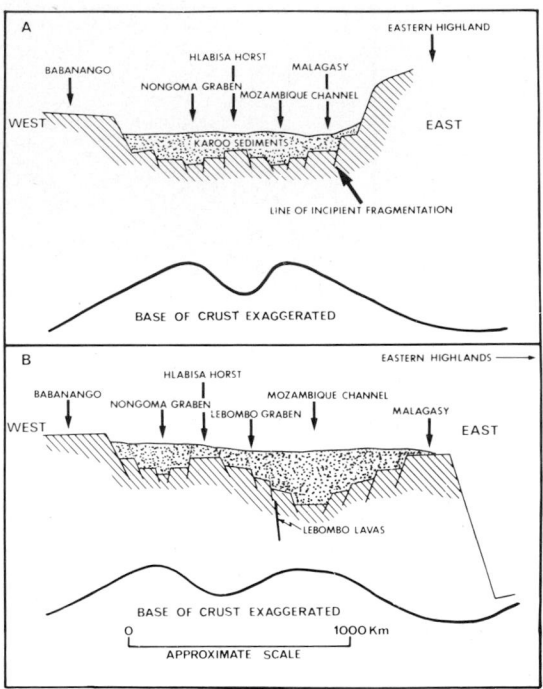

Fig. 4. Schematic east–west cross-sections showing the suggested development of Nongoma and Lebombo grabens by crustal thinning and the site of incipient fragmentation (A). Note that the Lebombo graben or 'Natal Monocline' is in fact a tilted half-graben (B).

imprint on the area. The fractures active during basin subsidence, and probably responsible for the over-thickening of the sediments in the area, were reactivated in post-Karoo times by north-east to south-west tension as the Falkland Plateau broke away from the African continent (Scrutton & Dingle, 1976). This resulted in block faulting and the present configuration of the Nongoma graben which is resonsible for the preservation of the coal measures in this area.

STRUCTURAL CONTROLS ON SEDIMENTATION

General characteristics of the sediments

The coal-bearing Ecca Group sediments (late Carboniferous-Permian) in the north-eastern Karoo Basin have been divided into a lower shale unit, a middle sandy unit containing coal and an upper shale unit. These units have been interpreted in terms of a series of deltaic and fluvial complexes that prograded mainly southwards across a shallow open-shelf facies,

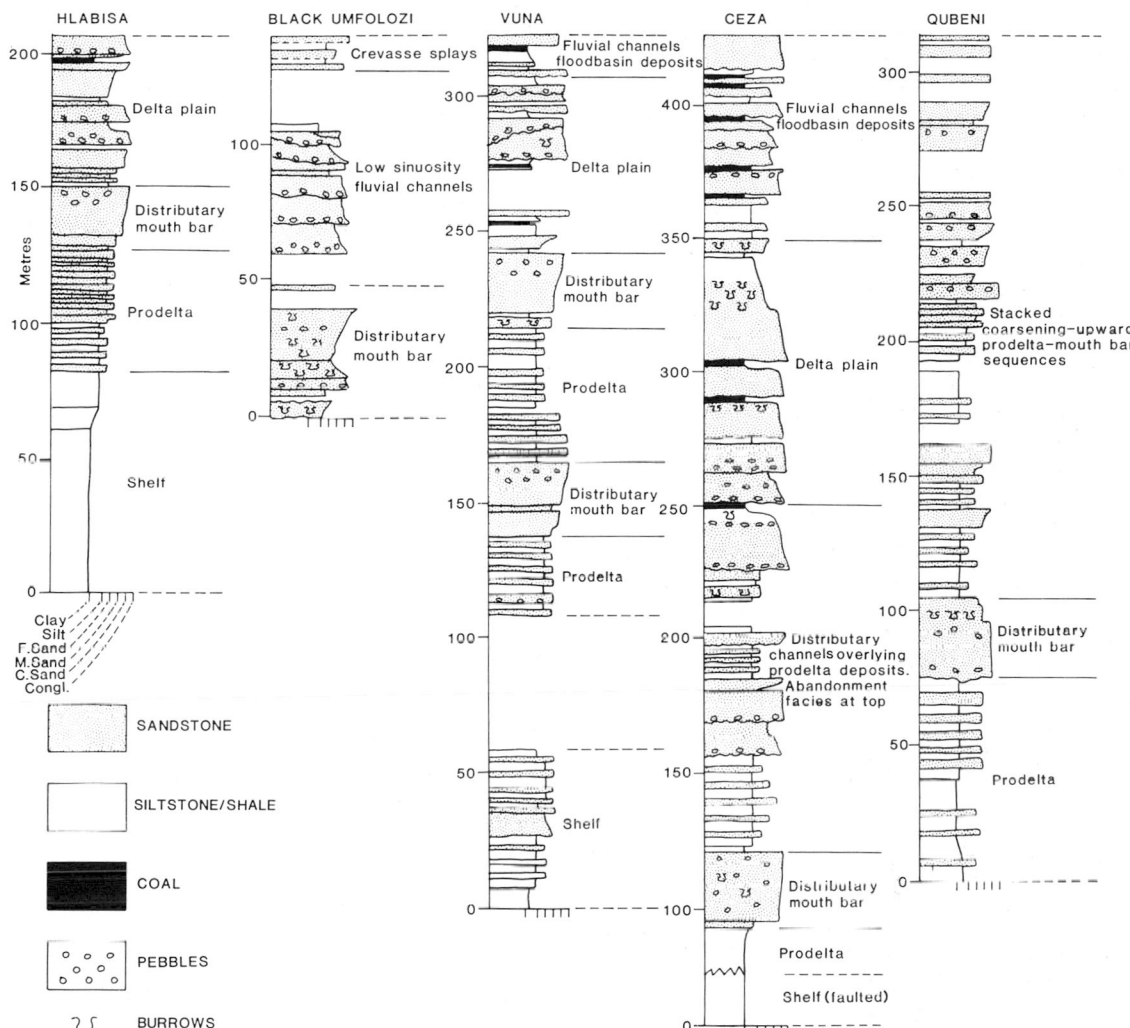

Fig. 5. Measured stratigraphic sections showing coarsening-upward and fining-upward depositional trends and interpreted depositional environments. Location of sections shown in Fig. 2.

with sedimentation being terminated by a transgressive deltaic phase leading to deposition of the upper shelf shales (Hobday, 1973; Cadle, 1974; Mathew, 1974; Whateley, 1980).

A similar sequence of sediments was deposited in the Nongoma graben where the Ecca Group comprises a lower progradational deltaic phase, a middle fluviatile phase and an upper transgressive deltaic phase (Fig. 5). However, within the graben the sequence is much thicker and more sandy. We infer that the repetition of deltaic and fluvial depositional sequences within the succession was largely controlled by tectonism and not by factors inherent in the depositional system (shifting) as on the flanking craton to the west and north-west (cf. Hobday, 1973, 1978).

Deltaic depositional sequences

Deposition was initiated following isostatic uplift of deglaciated highlands to the north and east. High-gradient, coarse-bedload streams carried arkosic detritus basinward. Because of the low energy of the water body the development of beaches, barriers and other features indicative of active coastal and

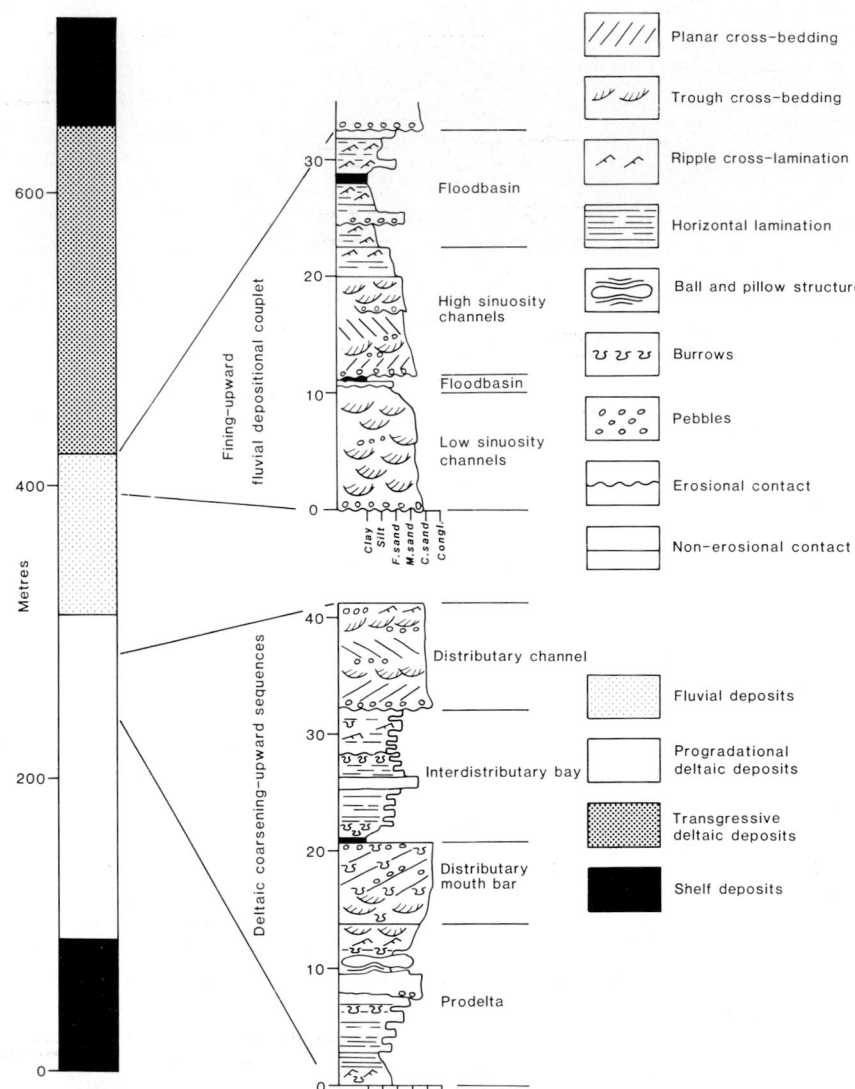

Fig. 6. Generalized stratigraphic column of the Ecca Group in the Nongoma graben, northern Zululand, showing coarsening-upward deltaic sequence and fining-upward fluvial depositional couplet of low-sinuosity and high-sinuosity channel deposits.

shallow-water marine environments was inhibited. As a result, fine-grained detritus was carried basinward, acting as a base for fluvially induced progradation and the development of a series of coarsening-upward daltaic sequences (Figs 5 and 6). These sequences comprise alternating thin beds of siltstone and sandstone of the prodelta environment gradationally overlain by coarse-grained, cross-bedded and locally bioturbated sandstones of the distributary mouth bar. The poor sorting of the delta-front sands, lateral

continuity of the mouth bar deposits and vertical stacking of progradational sequences suggest frequent delta shifting and the development of numerous coalesced, sand-rich, fluvially dominated delta lobes prograding into a shallow (105 m deep) water body characterized by low wave energy and weak tidal currents with little or no sediment reworking (Whateley, 1980).

Because of the rapid rate of progradation under conditions of rapid basin subsidence, the lower deltaic depositional phase comprises relatively thick (up to 105 m) coarsening-upwards sequences. Similar, but thinner sequences were deposited during the upper transgressive deltaic phase, although the transition from prodelta to shelf is less distinct. This is attributed to tectonic events of lesser magnitude and reduced relief between the source and depositional basin, with the result that the shelf was shallower and flatter compared with the deeper shelf environment across which the earlier deltas prograded. The delta plain is dominated by interdistributary bay deposits (Fig. 6) which include: (1) small-scale coarsening-upward bay-fill and minor mouth bar sequences of locally bioturbated and rippled siltstone overlain by coarse trough cross-bedded sandstone; (2) erosively based, often lenticular, coarse crevasse-splay sandstone and (3) carbonaceous siltstone and shale with rare coal of the restricted bay and marsh environments accompanying delta abandonment and decay. Cut into the interdistributary bay deposits are erosively based, rippled and cross-bedded coarse distributary channel sandstones which grade upstream into channels of the alluvial plain.

Fluvial depositional sequences

The alluvial plain deposits are characterized by two types of fining-upward sequences (Figs 5 and 6) which themselves show a gross fining-upward trend throughout the succession. The first type dominates the lower part of the fluvial depositional phase, but higher up it forms the basal part of a more complex, tectonically controlled depositional couplet, together with the second type of fining-upward sequence. Each fining-upward sequence can be identified with a particular channel type, the lower one with low-sinuosity channels and the upper one with high-sinuosity channels (Fig. 7).

Fining-upward sequences assigned to the low-sinuosity channel deposits consist of a lower scoured surface overlain by conglomerate, coarse to medium-grained sandstone and locally developed, subordinate amounts of fine sandstone, siltstone, shale and coal (Fig. 7). The conglomerate contains extrabasinal clasts of quartz and feldspar and intrabasinal clasts of siltstone and shale. The sandstones are tabular, sheet-like deposits up to 22 m thick which can be traced laterally in boreholes for more than 10 km in a direction normal to the palaeoslope. Internal sedimentary structures consist of large-scale trough cross-beds from 1 to 2 m thick arranged in cosets. Foreset azimuths show a unimodal distribution to the west and north-west during the early phase of fluvial deposition, but during later fluvial deposition there was a change in trend to the south-west coincident with the high-sinuosity channel deposits forming the upper part of the depositional couplets. Where present, the upper part of the sequence consists of alternating beds of laminated and bioturbated fine sandstone and siltstone with rippled tops, capped by shale and thin coal. Sandstone geometry and the dominance of sand and traction current structures suggest deposition within high-gradient, bedload-dominated, low-sinuosity channels (Schumm, 1972) (Fig. 7). In view of the absence of channel bars the channels were probably non-braided types. Rapid channel shifting produced stacked, tabular, sheet-like sandstones and effectively limited the growth of stabilizing vegetation and the development and preservation of fine overbank deposits. The dominance of large-scale troughs is interpreted in terms of a highly seasonal discharge favouring the preservation of high-stage, sinuous-crested dune bedforms (Cant, 1978).

The high-sinuosity channel deposits have a much lower proportion of sandstones to finer-grained sediments arranged in fining-upward sequences (Figs 6 and 7). Each sequence is erosively based and comprises a basal conglomerate of extrabasinal and intrabasinal clasts, overlain by lenticular, laterally restricted stacked channel sandstones. Individual sandstones range from 1 to 6 m thick, and in sections normal to the channel they extend laterally for a few tens of metres. Many channels have asymmetric, triangular profiles and well-developed lateral accretion surfaces dipping in opposite directions within superimposed sand bodies. The lower part of the these stacked sequences comprises coarse-grained, planar and trough cross-bedded sandstone graded-up into finer-grained, flat-bedded and ripple cross-laminated sandstone. Azimuths of planar and trough foresets reveal a bimodal distribution directed towards the south-west. The upper part of the sequence consists of thin, coarse-grained, massive sandstone with alternating layers and lenses of laminated and rippled dark grey carbonaceous siltstone and shale. Where coal is present it generally terminates the sequence, resting directly on siltstone or shale with no intervening seatearth. Thicker (20–60 cm), erosively based, trough cross-bedded sandstone, non-erosive, massive sandstone, and trough cross-bedded sandstone, forming part of the small coarsening-upward sequences with the underlying sediments, occur in the upper part of the fining-upward sequence.

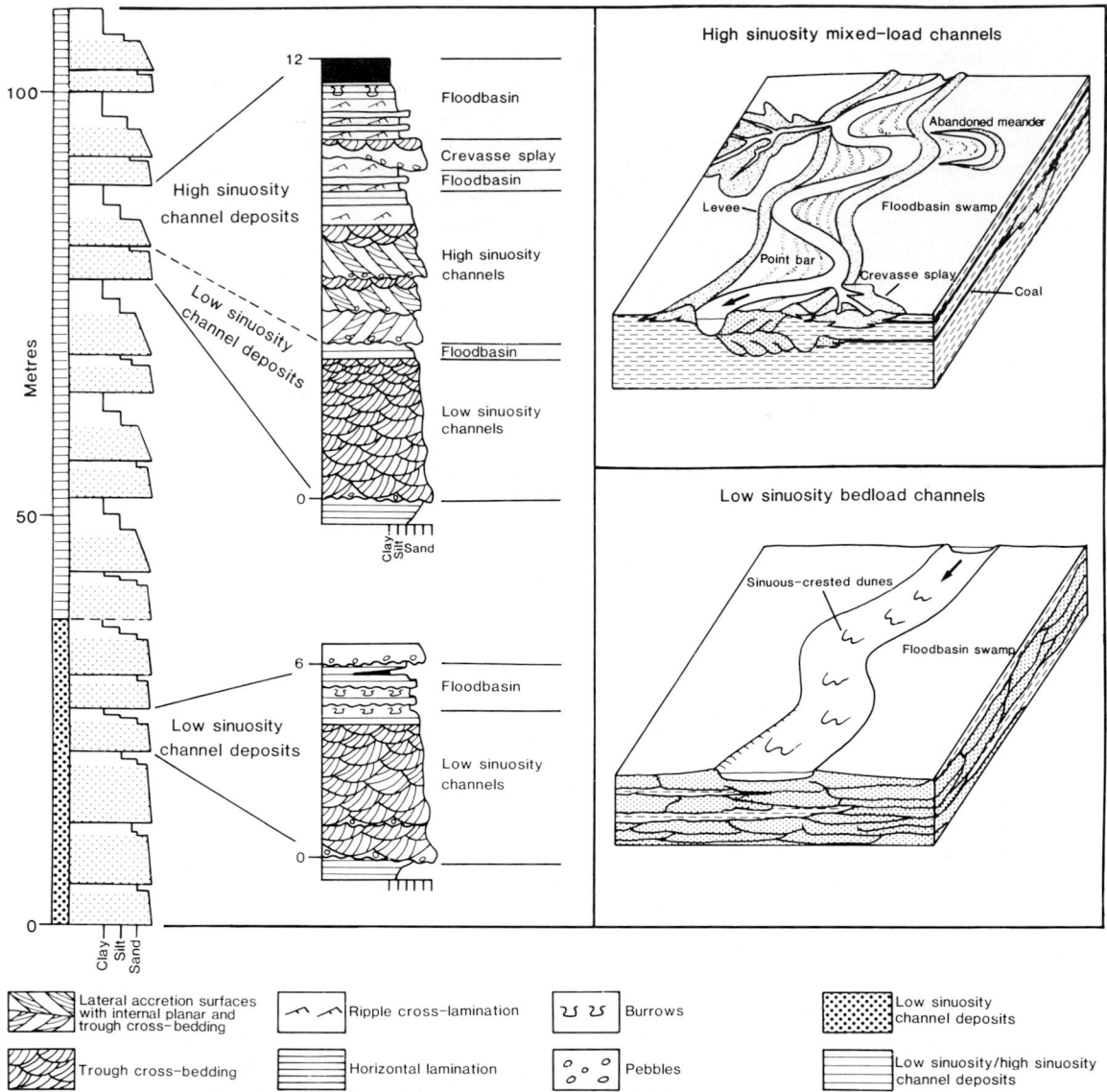

Fig. 7. Generalized stratigraphic column of Ecca Group fluvial deposits in the Nongoma graben showing: (1) distribution of low-sinuosity channel deposits and low/high-sinuosity channel deposits (fluvial depositional couplets), (2) details of typical fining-upward sequences generated by these two types of deposit and (3) generalized interpretive depositional models.

The fully developed fining-upward sequence corresponds with the high-sinuosity, meandering-stream floodplain model of Allen (1965) and others. The sandstones were deposited within small meandering channels as laterally accreting point-bar sands (Fig. 7). Their geometry and stacked nature suggests meander cut-off and channel shifting within the confines of a relatively narrow, stable meander belt on the flood-plain. Because of slow rates of floodplain aggradation point bars were eroded by the younger migrating channels, thereby preventing their complete preservation, except for the younger sandstone at the top of the sequence which records avulsion and abandonment of the complete meander belt. Flood discharge and overbank flow carried finer clastics into the flood-basin where relatively thick deposits, including peat swamps, accumulated. The thicker sandstones in the upper part of the sequence are interpreted as proximal and distal crevasse splay sands (Fig. 7) which entered the floodbasin via ephemeral crevasse channels. Long-term occupation of crevasse channels generated progradational coarsening-upward bay-fill type sequences as described by Coleman (1976).

Sedimentary response to contemporaneous tectonism

The lower Karoo sediments, especially the Dwyka and lower Ecca, thicken towards the south-east into the Natal trough (Haughton, 1969). A similar pattern of thickening is discernible during middle Ecca times except that the depocentre shifted northwards to lie between Tugela Ferry and the Nongoma graben (Fig. 1). The cause of this southerly thickening of the sediment pile, however, is not the same in the two areas. At Tugela Ferry, on the more stable craton, it is attributed to rapid basinward progradation and repeated delta shifting and stacking in response to a high rate of fluvial input and relatively slow rate of basin subsidence. Because of lateral facies changes on the craton, Hobday (1973) correlated the deltaic sequences of Tugela Ferry with the entire middle Ecca of northern Natal, including the fluvial coal-bearing strata.

In northern Zululand, the middle Ecca fluvio-deltaic sequences thicken dramatically when traced to the south-east across the craton edge into the Nongoma graben, suggesting significant vertical movement during deposition. The basal coarsening-upward sequence increases in thickness, whilst the ones above increase in both thickness and number. For example,

at Hlobane on the craton, three coarsening-upward sequences occur within 82 m of sediment, compared to at least fifteen similar sequences within some 240 m of sediment at Qubeni in the graben (Fig. 8). A similar effect can be seen when comparing the alluvial plain/delta plain sequences, but further comparisons are limited because of the erosion and removal of much of the upper transgressive deltaic depositional phase (Whateley, 1980).

These increases in thickness and repetitive pattern of sedimentation are related to the structural development of the graben and the influence of growth faulting, which was most pronounced during deposition of the lower progradational deltaic phase and the early coal measures. However, the absence of features such as mud lumps associated with differential loading and subaqueous mass movement at the rapidly prograding delta margin suggests that growth faulting may be mainly structural in origin as a result of repeated movement of an unstable rifted basement. The growth faults enhanced by sediment density contrasts were buried once equilibrium had been attained, only to be reactivated again during post-depositional structural block settling and block faulting associated with continental break-up.

Some additional evidence of tectonic influence during this phase of deposition is provided by ball and pillow structures and slump folds in the prodelta sediments. There are no elongate pillow forms or thrust planes between pillows, and no slip faults associated with slump folds indicative of downslope movement under gravity. This suggests a tectonic origin for these structures, possibly related to seismically induced liquefaction.

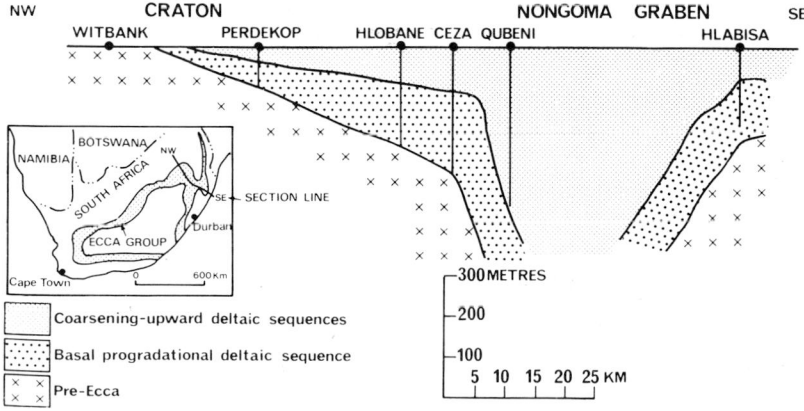

Fig. 8. Thickening of middle Ecca lower deltaic sequences in northern Zululand when traced south-east across the craton edge into the Nongoma graben.

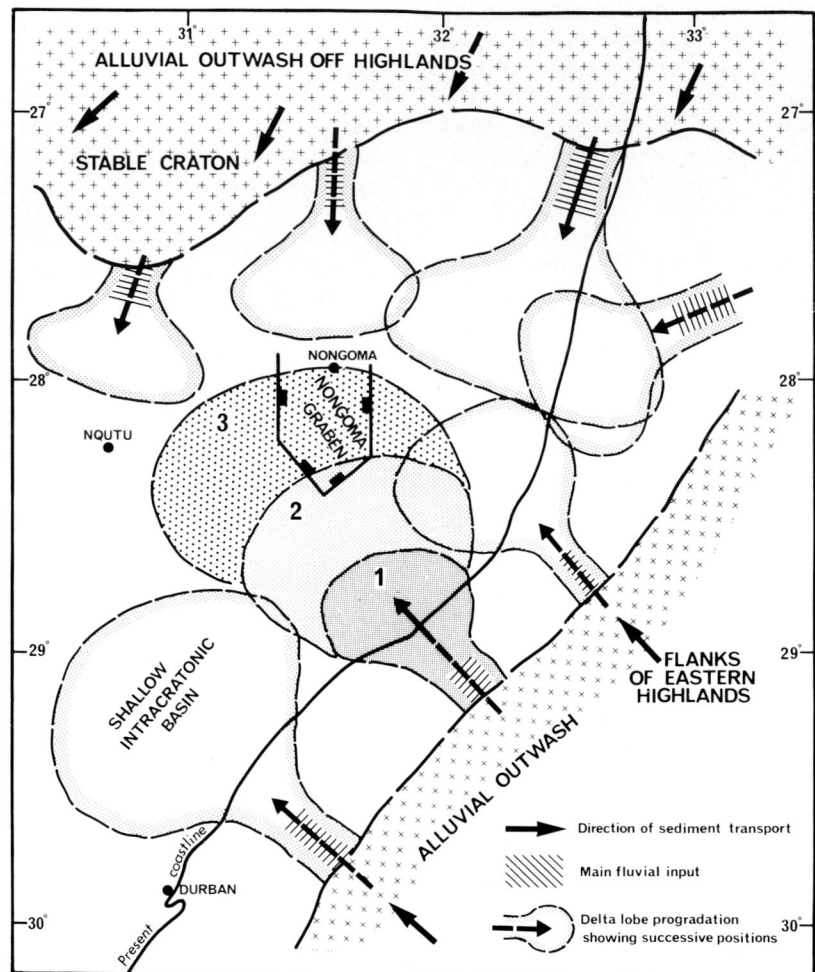

Fig. 9. Schematic reconstruction of early Ecca Group palaeogeography and location of Nongoma graben. Numbers 1, 2 and 3 denote successive prograding delta lobes.

The early stages of extensional disruption and crustal thinning associated with the development of pre-Gondwanaland split graben structures were dominated by vertical displacement contemporaneous with sedimentation. Repetition of depositional sequences is believed to represent a response to major episodes of vertical movement of the basin floor relative to the source areas due to: (1) tensional stress build-up in the crust accompanied by erosion of the flanking craton and (2) tensional stress release causing basin subsidence. Because of the rapid rate of basin subsidence and abundant availability of sediment the depositional system exhibited strong progradational tendencies.

The regional palaeoslope during deposition of the lowermost coarsening-upward deltaic sequence and the underlying Dwyka lies towards the south-west (Matthews, 1970; Whateley, 1980). Above this the deltaic sequences show a marked change in trend as the structures associated with build-up and release of crustal tension (vertical movement) became more effective and sediments derived from a tectonically active 'Eastern Highland', prograded rapidly west and north-west across the deepening graben as a series of inferred delta lobes (Fig. 9). This same pattern is evident during the initial phase of fluvial deposition when build-up and release of crustal tension promoted the development of high-gradient, low-sinuosity bedload channels. Because of the rapid rate of basin subsidence and lateral shift of the channels, plant

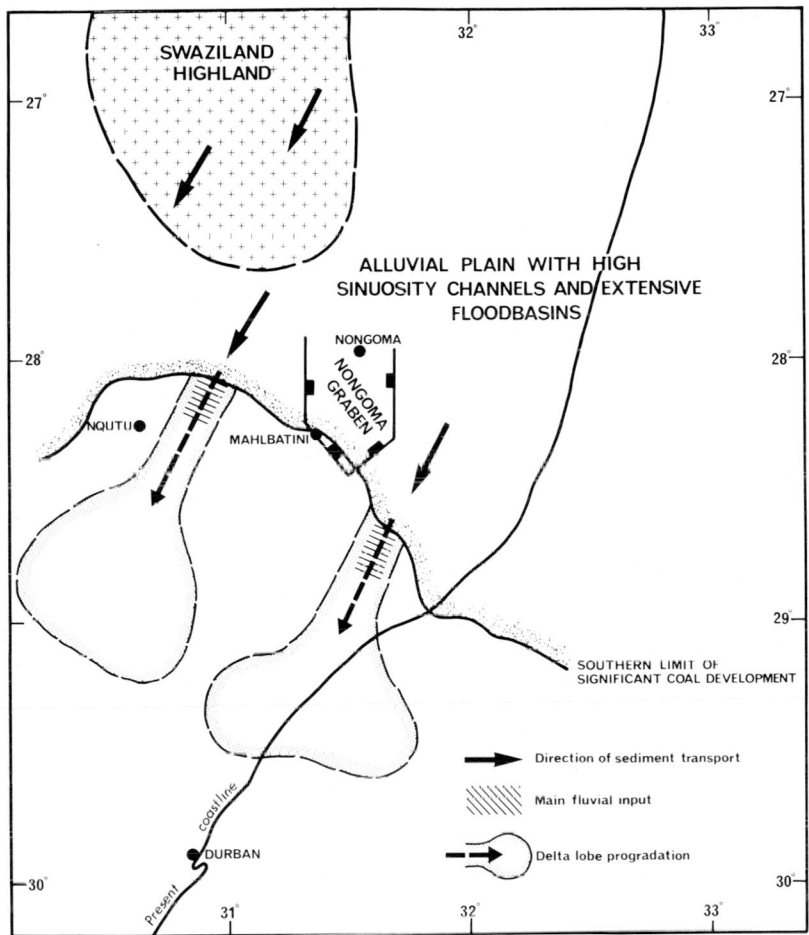

Fig. 10. Schematic reconstruction of Ecca Group palaeogeography during deposition of upper fluviatile coal measures and location of Nongoma graben.

growth could not be maintained for sufficiently long periods of time to form substantial peat swamps and thick coals.

Following this, the 'Eastern Highlands' declined in importance as a result of relative tectonic stability, source area denudation and possibly early drifting, and the south-westerly trend reaffirmed itself. The inferred delta lobes and alluvial plain during this phase of graben development are shown in Fig. 10. Fluvial depositional sequences suggest that episodes of build-up and release of crustal tension at this time became spaced increasingly further apart.

Thus, although low-sinuosity channels developed at first, stabilization of source and depositional site promoted lower gradients, increased production of fines and high-sinuosity stable channels, forming a complex fluvial depositional couplet (Fig. 7). An overall increase in stabilization of the source and depositional site during fluvial deposition is reflected in the overall fining-upward trend (decreasing sand/shale ratio) and the remarkably uniform thickness of the upper, coal-bearing part of the fluvial depositional, phase over an area of about 130 km².

Fluvial deposition was terminated by renewed tectonic activity and basin subsidence leading to the deposition of a thin (5–15 m), reworked, wave-rippled transgressive sand sheet (destructional unit) and a further phase of deltaic sedimentation. However, tectonism and relief differences were probably more subdued than in early Ecca times, with the result that coarsening-upward sequences are less well defined, and as transgression proceeded (basin subsidence greater than sediment supply) they became progressively overlain by finer-grained deposits of the prodelta and shelf environments.

COAL

Coal seams are associated with the fluvial and deltaic phases of deposition where they occur at or near the top of fining-upward and, less commonly, coarsening-upward sequences. The seams consist of low-ash, low-volatile, bright coal with minor dull coal and occasional shale partings. However, during deposition of the lower progradational deltaic phase and early fluvial phase of deposition (low-sinuosity channel deposits) when tectonic activity exerted the greatest influence on sedimentation, subsidence outstripped the rate of plant growth and peat accumulation with the result that the peat swamps foundered and the coals are thin and localized. Thicker, laterally more persistent, economic coals occur at or near the top of fining-upward sequences within the high-sinuosity channel deposits which were laid down under increasingly stable tectonic conditions favouring the development of more confined, stable meander belts flanked by extensive floodbasin peat swamps. For example, the most important coal zone towards the top of the succession between the main seam and upper anthracite marker maintains a constant thickness of between 35 and 40 m over an area of some 130 km².

The better-quality, thicker coals (> 2 m) in this area are confined to the graben. This relationship suggests that, within the graben, channel activity and sandstone deposition may have been preferentially concentrated along the active fault margins. A comparable relationship between depositional facies and faulting was noted by Horne *et al.* (1978) in the Carboniferous of West Virginia and Pennsylvania. If this interpretation is correct, then the better-quality, thicker coals are located between the faulted margins and, because of the high water table in response to slow basin subsidence, they are predominantly low ash, bright coals. In contrast, time-equivalent coals on the flanks of the graben are thin dull and shaly. Thus the ultimate control on coal distribution was the shape and extent of the graben. On the flanking craton in comparison, there were no such restraints, with the result that contemporaneous topography exerted the greatest influence on coal type and distribution.

The effects of differential compaction are minimal, although in some seams the thicker coals tend to be located above the more highly compactible argillaceous floodbasin deposits, and the thinner coals above less compactible sandy horizons. Thus the thickness and quality of the coal can be predicted to some extent by its location within a proximal or distal floodbasin setting relative to the proximity of contemporaneous active channels. In proximal settings the coals tend to be thinner and the ash content, frequency and size of coal splits and adverse mining gradients increase towards the channel loci. The fluvial depositional model erected for the coal measures futher predicts that the coal seams will generally be overlain by laterally extensive, thick (> 2 m) sheet sandstones at the base of the overlying depositional couplet, and that these will tend to be thicker over the more argillaceous floodbasin facies because of the effects of differential rates of sand–mud compaction. Thus roof conditions are generally good, differential compaction effects are minimized and, where the roof-rock is of fresh water rather than marine or brackish water in origin, the sulphur content of the coal is low. A possible disadvantage is the more widespread effect of erosion of the seams, although this will be offset to some extent by the relatively short occupancy of the low-sinuosity channels. The nature of the channels and pattern of sedimentation tends to smooth the palaeotopography prior to peat accumulation. Hence the influence of palaeotopography on depositional trends was limited and the seams more widely spread and of uniform thickness.

Limited trenching operations and studies of well logs show that some of the coal seams are overlain by coarsening-upward sequences of silty-shale, siltstone and fine sandstone which resemble modern bay-fills (Saxena, 1976). Coarsening-upward sequences of this type generally provide good roof conditions because the strength of the rock increases upwards (Horne *et al.*, 1978).

The preservation of peat requires a water table which is high enough to cover the decomposing vegetation without drowning it; this situation can be explained in terms of a balance between compactional subsidence and swamp aggradation, or a rising water table in response to compactional subsidence. Continuation of the process of coal formation is accompanied by changes in physical, chemical and petrographic properties which take place in stages, each stage producing coal of different rank. Many factors influence rank but temperature is generally regarded as the most decisive, in that coal rank increases as the temperature or geothermal gradient rises. In the Nongoma graben the high geothermal gradient responsible for the conversion of the coal to anthracite, and the initially high rate of basin subsidence, are a direct result of the tectonic setting. In spite of this, thick localized coals with numerous shale partings typical of rapidly subsiding basins punctuated by numerous vertical displacements are lacking because the important coals formed during a later, more stable tectonic phase. When traced away from the graben on to the relatively stable craton where the geothermal gradient was lower, the coals are generally lower in rank, but still contain some anthracite attributed to the local thermal effects of post-Karoo intrusions (De Jager, 1976).

The Permian coals of the southern hemisphere developed after a major ice age, in a humid, cool temperate climate with seasonal rainfall, as evidenced by the fresh feldspars in the sandstones, the absence of leaching of the floor of the coal seams, and the marked seasonal growth of the coal measure plants (Plumstead, 1961). The vegetation generally grew in shallow water or waterlogged conditions. Pollen analysis by Hart (1966) suggests that the vegetation was probably of the shallow-rooted Arctic or Tundra type, which may account for the lack of rootlet horizons and seatearths. Accumulation of the decayed shallow-rooted, *in situ* vegetation then provided the substrate for the growth of later, better-rooted plants.

The low-ash, bright coal in northern Zululand implies the maintenance of a consistently high water table in order to prevent extensive oxidation of vegetation. Additional evidence in support of a relatively high water table is the relatively high pyritic sulphur content of the coal, which is twice that of organic sulphur. Although factors such as rainfall, temperature and humidity influence the level of the water table, these will affect the entire area and not just the graben. Thus the level of the water table is considered to be primarily a function of tectonism and to a lesser extent of the rate of growth of plant material. Most coal seams in northern Zululand consist of bright coal which near the top of the seam grades into dull coal; shale partings are seldom developed, particularly during deposition of the upper coal measures. The nature of the seam profile is related to the growth of the peat swamps relative to the drop in ground water level. For example, repeated basin subsidence results in bright coal high in vitrinite interlayered with shale partings as in the Waterberg Coalfield (De Jager, 1976). At the opposite extreme, in a more stable environment, dull coal forms the major component of the seam profile. In northern Zululand the coals formed during a relatively stable

phase of graben development, when the rate of subsidence was sufficiently high to maintain a high water table, yet not high enough to promote frequent overbank flooding and the introduction of large amounts of sediment into the peat swamp. Overbank flooding may also have been limited by any well-developed levées along major channels and by abandoned meander bends of channels (Flores, 1981). This in turn allowed for a dense growth of vegetation which would further impede sediment influx.

Because the ratio of reactive (vitrinite and exinite) to non-reactive (inertinite) macerals is higher than in most other South African coals (1:6:1) (De Jager, 1976) they are suitable for use as metallurgical coal (high fixed carbon content) without blending. However, unlike the coals in the nearby Northern Natal Coalfield they do not possess good coking properties owing to the drastic loss of volatiles on regional conversion to anthracite.

SUMMARY AND CONCLUSIONS

During the Permian northern Zululand was subjected to crustal thinning and tensional stress; this resulted in the formation of half-graben basins, such as the Nongoma graben, associated with continental rifting and the break-up of east and west Gondwanaland. Sedimentation, contemporaneous with graben development, led to the deposition of a thick sequence coal-bearing fluvio-deltaic Ecca Group sediments. This was controlled mainly by episodes of build-up and release of tensional stress in the crust and not by factors inherent in the depositional system (shifting) as on the flanking craton to the west and north-west where similar coal-bearing sediments occur. A lower progradational deltaic phase of deposition was followed by a fluvial phase of deposition characterized by fining-upward sequences. Early fluvial deposition was dominated by high-gradient, low-sinuosity bedload channels, prograding to the west and north-west. During later fluvial deposition, episodes of build-up and release of tensional stress in the crust became spaced increasingly further apart and, although low-sinuosity channels developed at first, stabilization of source and depositional site promoted lower gradients, increased production of fines and high-sinuosity channels. Thus the upper part of the fluvial depositional phase is dominated by complex depositional couplets of low- and high-sinuosity channel deposits, with palaeocurrent trends indicating a source area to the north and east. Renewed basin subsidence and transgression terminated fluvial deposition.

Economically important coals occur at or near the top of fining-upward sequences within the high-sinuosity channel deposits, when tectonic conditions were more stable and favoured the development of extensive floodbasin peat swamps. Basin subsidence was able to maintain a water table sufficiently high for the preservation of plant material but not high enough to promote frequent overbank flooding and the introduction of numerous shale partings into the coal. As a result the coals are low-ash, low-volatile bright coals with minor dull coals and some shale partings; but because of crustal attenuation and high geothermal gradient they have regionally formed anthracite. The depositional model for the coal measures indicates that: (1) the thicker, better-quality coals are of distal floodbasin origin, (2) the roof conditions are generally good, (3) the effects of differential compaction are minimal, and (4) depositional trends are seldom influenced by thick, laterally extensive sheet sands at the base of fluvial depositional couplets.

ACKNOWLEDGMENTS

This paper is based largely on a thesis completed by the junior author at the University of the Witwatersrand, Johannesburg, under the supervision of the senior author. We are indebted to the management of Southern Sphere Mining and Development Company (Pty) Limited for financial support and permission to publish the results of the work. The manuscript benefited considerably from the constructive comments of Mary Monro and Maurice Tucker. Elizabeth Walton is thanked for typing the manuscript.

REFERENCES

ALLEN, J.R.L. (1965) A review of the origin and characteristics of recent alluvial sediments. *Sedimentology*, **5**, 89–191.

BOTT, M.H.P. (1976) Formation of sedimentary basins of graben type by extension of the continental crust. *Tectonophys.* **36**, 77–86.

CADLE, A.B. (1974) *A subsurface sedimentological investigation of parts of the Ecca and Beaufort Groups in the northeastern Karroo Basin.* Unpublished M.Sc. Thesis. University of Natal, Pietermaritzburg. 106 pp.

CANT, D.J. (1978) Development of a facies model for sandy braided river sedimentation: comparison of the South Saskatchewan River and the Battery Formation. In: *Fluvial Sedimentology* (Ed. by A.D. Miall). *Mem. Can. Soc. Petrol. Geol., Calgary*, **5**, 627–639.

CLIFFORD, T.N. (1971) Location of mineral deposits. In: *Understanding the Earth* (Ed. by I.G. Gass, P.J. Smith and R.C.L. Wilson), pp. 315–326. Artemis Press, Sussex.

COLEMAN, J.M. (1976) *Deltas: Processes of Deposition and Models for Exploration.* Continuing Education Publication Company, Inc., Champaign, Illinois. 102 pp.

DARRACOTT, B.W. (1974) On the crustal structure and evolution of southeastern Africa and the adjacent Indian Ocean. *Earth planet. Sci. Lett.* **24**, 282–290.

DE JAGER, F.S.J. (1976) Coal. In: *Mineral Resources of South Africa*, 5th ed. (Ed. by C.B. Coetzee). *Handbk geol. Surv. S. Afr.* **7**, 322–323. Government Printer, Pretoria.

DINGLE, R.V. & SCRUTTON, R.A. (1974) Continental breakup and the development of post-Palaeozoic sedimentary basins around southern Africa. *Bull. geol. Soc. Am.* **85**, 1467–1474.

DU TOIT, A.L. (1929) The volcanic belt of the Lebombo—a region of tension. *Trans. R. Soc. S. Afr.* **18**, 189–217.

FLORES, R.M. (1981) Coal deposition in fluvial paleoenvironments of the Paleocene Tongue River Member of the Fort Union Formation, Powder River area, Powder River Basin, Wyoming and Montana. In: *Recent and Ancient Nonmarine Depositional Environments: Models for Exploration* (Ed. by F.G. Ethridge and R.M. Flores). *Spec. Publs Soc. econ. Paleont. Miner., Tulsa*, **31**, 169–190.

HART, G.F. (1966) *Lower Karroo biostratigraphy of parts of southern Africa.* Unpublished Report. Bernard Price Institute of Palaeontological Research, University of the Witwatersrand, Johannesburg. 305 pp.

HAUGHTON, S.H. (1969) *Geological History of Southern Africa.* Geological Society of South Africa, Capetown. 535 pp.

HOBDAY, D.K. (1973) Middle Ecca deltaic deposits in the Muden–Tugela Ferry area of Natal. *Trans. geol. Soc. S. Afr.* **76**, 309–318.

HOBDAY, D.K. (1978) Fluvial deposits of the Ecca and Beaufort Groups in the eastern Karoo Basin, southern Africa. In: *Fluvial Sedimentology* (Ed. by A.D. Miall). *Mem. Can. Soc. Petrol. Geol., Calgary*, **5**, 413–430.

HOLMES, A. (1951) The sequence of pre-Cambrian orogenic belts in south and central Africa *Rep. 18th int. geol. Congr. London*, 1948, part 14, 254–269.

HORNE, J.C., FERM, J.C., CARUCCIO, F.T. & BAGANZ, B.P. (1978) Depositional models in coal exploration and mine planning in Appalachian region. *Bull. Am. Ass. Petrol. Geol.* **62**, 2379–2411.

MATHEW, D. (1974) *A statistical and palaeoenvironmental analysis of the Ecca Group of northern Zululand.* Unpublished M.Sc. Thesis. University of Natal, Pietermaritzburg. 158 pp.

MATTHEWS, P.E. (1970) Palaeorelief and the Dwyka glaciation in the eastern region of South Africa. *Proc. Papers, 2nd Gondwana Symposium, South Africa*, pp. 491–499.

PETRICK, A.J. (1975) *Commission of Enquiry into the Coal Resources of the Republic of South Africa.* Government Printer, Pretoria. 153 pp.

PLUMSTEAD, E.P. (1961) The Permo-Carboniferous Coal Measures of the Transvaal, South Africa—an example of the contrasting stratigraphy in the southern and northern hemispheres. *4e Congrés de Stratigraphie et de Géologie du Carbonifère, Heerlen*, 1958, **2**, 545–550.

PRETORIUS, D.A. (1974) The structural boundary between the

Kaapvaal and Sonama crustal provinces. *Inf. Circ. 88, Economic Geology Research Unit, University of the Witwatersrand, Johannesburg,* 35–38.

SAXENA, R.S. (1976) Modern Mississippi delta–depositional environments and processes. *Am. Ass. Petrol. Geol. Soc. econ. Paleont. Miner. Field Trip Guidebk, New Orleans, May 23–26.* 125 pp.

SCHUMM, S.A. (1972) Fluvial paleochannels. In: *Recognition of Ancient Sedimentary Environments* (Ed. by J.K. Rigby and W.K. Hamblin). *Spec. Publs Soc. econ. Paleont. Miner., Tulsa,* **16**, 98–107.

SCRUTTON, R.A. & DINGLE, R.V. (1976) Observations on the processes of sedimentary basin formation at the margins of southern Africa. *Tectonophys.* **36**, 143–156.

WHATELEY, M.K.G. (1980) Deltaic and fluvial deposits of the Ecca Group, Nongoma graben, northern Zululand. *Trans. geol. Soc. S. Afr.* **83**, 345–352.

Spec. Publs int. Ass. Sediment. (1983) **6**, 473–483

Analysis of the Upper Devonian Munster Basin, an example of a fluvial distributary system

JOHN R. GRAHAM

Department of Geology, Trinity College, Dublin 2, Ireland

ABSTRACT

The Munster Basin, covering an area of more than 12,500 km², was the site of fluvial sedimentation during the Upper Devonian. The basin was of half-graben style, with a faulted northern margin and a maximum thickness of 6–7 km just to the south of this margin. There is evidence for an eastern basin margin separating this area from the Anglo-Welsh basin, and some indication from palaeocurrents of a western margin. The southern margin cannot presently be determined and an original southward connection to marine strata is uncertain.

The basin fill is described in terms of four major facies: (a) a localized coarse marginal facies with alluvial fans and, more rarely, aeolian sediments, (b) a coarse-grained fluvial facies dominated by 'in channel' deposition, (c) a fine-grained fluvial facies with rare or poorly defined channels, (d) a fluvial coastal plain facies. Facies (d) occurs only at the top of the succession and represents time-dependent changes associated with the major Tournaisian transgression. For most of the Upper Devonian facies (b) and (c) predominate, with facies (b) occupying the more marginal and facies (c) more central parts of the basin. The facies pattern is best explained by means of a major fluvial distributary system. Although there was a local input of sediment from the basin margins the petrography generally indicates a distant source.

INTRODUCTION

The Munster Basin covers an area of over 12,500 km² in the southern part of Ireland (Fig. 1). This basin records the later stages of denudation of the Caledonides which were being uplifted from late Silurian to at least mid-Devonian. The strata were all deformed during the Hercynian orogeny and most mudrocks show a penetrative cleavage. Biostratigraphic data for this basin are limited and provided mainly by spores and fish. They suggest that most of the fill is of Upper Devonian age with the basal parts probably extending down into the Middle Devonian. There is everywhere an upward transition into rocks of Lower Carboniferous age.

The limits of the Munster Basin can be defined on the basis of the thickness of the sedimentary pile (Naylor & Jones, 1967; Clayton *et al.*, 1980; Naylor

et al., 1981) taken where the Old Red Sandstone facies exceeds 1 km. The generalized isopachs (Fig. 1) are only minimum values away from the exposed basin margins as pre-Old Red Sandstone strata are not exposed in the central parts of the basin. Despite this there are two reasons for suggesting that they approximately demonstrate the true thickness variation. First, the age of the oldest exposed strata near the northern margin on the Iveragh Peninsula (Russell, 1978) is approximately the same as that for the oldest exposed strata on Clear Island in the extreme south (Clayton & Graham, 1974). Secondly, the limited gravity data of Murphy (1960) show good agreement with the isopach data derived from measured sections and have been used by Naylor & Jones (1967), Clayton *et al.* (1980) and Naylor *et al.* (1981) to construct an asymmetrical basin profile and to suggest the presence of a half-graben basin. Along the northern margin, the north–south changes in thickness

0141-3600/83/0106-0473 $02.00

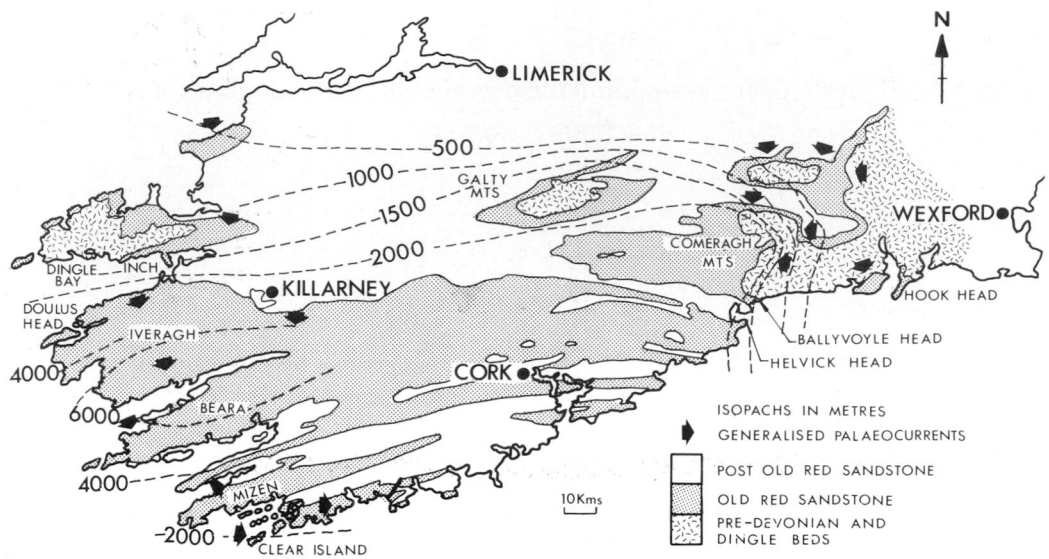

Fig. 1. Location map of the Munster Basin with generalized isopachs and palaeocurrents.

of the Old Red Sandstone succession are very marked and occur over about 10 km. This must imply some faulting at depth and it is probable that accommodation along these faults was contemporaneous with sedimentation (Naylor & Jones, 1967; Clayton *et al.*, 1980). A north–south trending eastern margin can be clearly seen from isopachs and palaeocurrents (Fig. 1; Penney, 1980), whereas a western margin can only be inferred from facies patterns and palaeocurrents (Figs 1 and 7). Although the basin floor is shallowing to the south, there is no evidence for the location of a southern margin. There is however little positive evidence for connection of the Munster and Anglo-Welsh basins throughout most of the Upper Devonian as suggested by Allen (1979).

MAJOR FACIES OF THE MUNSTER BASIN

The fill of the Munster Basin can be described in terms of four major facies: (a) a coarse marginal facies often with alluvial fans and more rarely with aeolian sediments; (b) a coarse-grained fluvial facies dominated by 'in channel' deposition; (c) a fine-grained fluvial facies with rare or poorly defined channels; (d) a fluvial coastal plain facies.

Facies (a): coarse marginal facies

This facies crops out near the northern and eastern margins of the basin where the Old Red Sandstone can

be seen to rest unconformably on older rocks. It consists predominantly of conglomerates and pebbly sandstones. Best known is the unit in the north-west of the basin termed the Inch Conglomerate Formation (Horne, 1971, 1974, 1975; Capewell, 1951, 1965). The facies also constitutes most of the Coumshingaun Conglomerate Formation in the north-east of the basin (Capewell, 1957b; Penney, 1980), the Pigeon Rock Formation in the Galty Mountains (Doran, Holland & Jackson, 1973) and the lower part of the Old Red Sandstone in coastal sections at the eastern end of the basin at Ballyvoyle Head and Helvick Head (Fig. 1) (MacCarthy, Gardiner & Horne, 1978). Several of the coarser conglomerates are unstratified, show graded tops, inverse coarse tail grading in their lower parts and some imbrication. This combination of features suggests a debris-flow origin. However, most of the conglomerates are clast supported (Fig. 2), show clear stratification and indicate bedload transport. These various formations have been interpreted as alluvial fan (Horne, 1975) and proximal braided fluvial deposits (Penney, 1980). Locally aeolian deposits are interbedded with these conglomerates (Horne, 1971, 1975).

The age and lateral extent of these deposits are still somewhat problematical. It is possible that they are of markedly different ages. It is not generally possible to trace individual conglomeratic levels into the more central parts of the basin. However in the north-west part of the basin south of Dingle Bay, coarse conglomerates with some debris-flow deposits occur at Doulus Head (Fig. 1), and these can be seen to thin

Fig. 2. Coarse marginal facies—Inch Conglomerate Formation. (A) Typical poorly stratified boulder conglomerates, Inch shore (Q 631006). Beds dip and young towards bottom right. (B) Finer-grained stratified conglomerates, Red Cliff near Inch (Q 611004).

and fine rapidly southwards and to interfinger with fine-grained fluvial facies (facies (c) below). These conglomerates have a similar petrography to the distinctive Inch Conglomerate Formation seen north of Dingle Bay and probably share the same source area. It is also probable that these conglomerates, which contain clasts of a varied metamorphic suite, represent locally complex distribution of sediment from fault-bounded basement slivers (Horne, 1970, 1975; Naylor *et al.*, 1981). In all cases except the Inch Conglomerate Formation the clasts are clearly derived from the local sub-Old Red Sandstone.

This facies is thought to have accumulated at the margins of the main depositional area and also in places to record the retreat of the basin margins, probably by pediplanation as discussed by Capewell (1965) and Doran *et al.* (1973). Accumulation was due to favourable local factors such as banking against scarp features as in the Galty Mountains (Doran *et al.*, 1973) or proximity to areas of sub-Old Red Sandstone surface with appreciable relief and local sediment supply.

Facies (b): coarse-grained fluvial facies

This facies occupies a slightly more basinal position than facies (a) although there is often insufficient control to demonstrate clear lateral gradations between the two. The facies is best seen in the Chloritic Sandstone Formation (Capewell, 1957a; Husain, 1957), the Grey and Green Sandstone Formations (Walsh, 1968), the Sherkin Formation (Graham & Reilly, 1972) and parts of the Comeragh Conglomerate and Nier Sandstone Formations (Penney, 1980; Capewell, 1957b; Boldy, personal communication). Successions are dominated by sandstones and pebbly

sandstones showing large-scale cross-stratification and flat bedding as their main internal structures (Figs 3 and 4). In the northern part of the basin exposures of the Chloritic Sandstone Formation south and east of Killarney (Figs 1 and 7) comprise coarse-grained, often pebbly sandstones with very minor amounts of mudrocks. Erosion surfaces are common but the amount of downcutting is characteristically less than 1 m. Trough cross-bedding produced by migrating dunes is the commonest structure, with flat and low-angle cross-bedding and planar tabular cross-bedding less common. In general terms these rocks can be interpreted as the stacked deposits of fluvial channels. When traced to the south-west (Capewell, 1957a; Russell, personal communication) and to the south (Husain, 1957) there is a clear fining of grain size, increase in interbedded mudrocks, decrease in sandstone bed thickness and decrease in the amount of large-scale cross-stratification and flat bedding. There is also an increase in the importance of flat-bedded sandstones relative to cross-bedded sandstones (compare Fig. 3A, B). Individual beds of flat bedded and cross-bedded sandstone are frequently separated by muddy drapes, each probably representing the final deposits of a flood. Although the north–south changes involve tracing this unit across fold structures, there is excellent mapping control; changes along the Iveragh Peninsula to the west and south-west can also be easily traced through well-exposed ground. At its southern and western limits the 'Chloritic Sandstone Formation' ceases to be a mappable unit. Thus the field evidence clearly indicates that facies (b) is passing downstream into facies (c) described below. The large amounts of chlorite which give rise to the name of this formation are both detrital and authigenic. Although authigenic

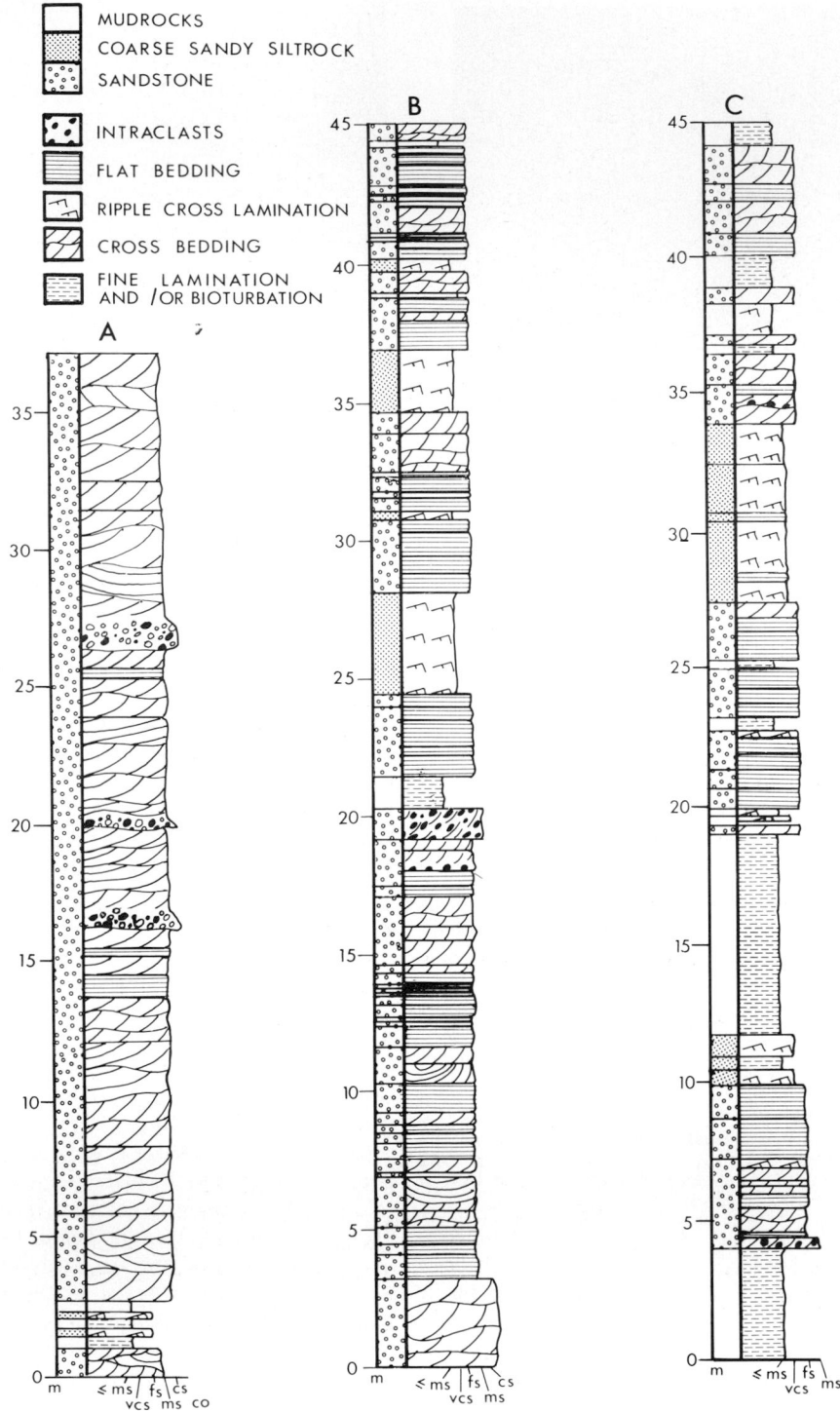

Fig. 3. Representative logs of the coarse-grained fluvial facies showing variation between proximal (A) and more distal (B, C) types. (A) Chloritic Sandstone Formation, 9 km south-west of Killarney (V 928825). (B) Chloritic Sandstone Formation, 18 km south-west of Killarney (V 863768). (C) Sherkin Formation, 6 km east-north-east of Clear Island (W 03424).

Fig. 4. Coarse-grained fluvial facies. Typical exposure of the Chloritic Sandstone Formation 14 km south-east of Killarney (W 076824).

chlorite is predominant, allogenic chlorite is present throughout and is locally abundant (Capewell, 1957a; Walsh, 1968). This formation also contains fragments of chert, jasper and devitrified volcanic rocks (Husain, 1957).

The Sherkin Formation (Graham & Reilly, 1972), which crops out only in the south-west of the basin in the core of a major anticline, is also typical of this facies. The formation appears to be interfingering with facies (c) near the eastern extremity of its outcrop, but investigation is hampered by an easterly fold plunge. A northerly increase in thickness of the Old Red Sandstone indicated by the regional isopach pattern (Fig. 1) prevents definitive statements concerning its extent in that direction. Thus, in this part of the basin, it is more difficult to separate spatial and temporal changes. However, the sediments resemble very closely those of the finer, more distal parts of the Chloritic Sandstone Formation (Fig. 3C). There is little sediment coarser than medium sand size, and channelling at the base of and within sandstone bodies, with one notable exception described by Graham & Reilly (1972), is generally limited, although erosion surfaces are common. There is a common stoss-side preservation of dunes, a lack of obvious cyclicity and an apparent absence of lateral

accretion bedding. Rapid deposition in channels of low sinuosity is suggested (Graham & Reilly, 1972). There is a strong westerly component to the palaeocurrents for this formation (Graham & Reilly, 1972) and also for more isolated sand bodies higher in the succession in this area (Graham, unpublished). These sandstones show little evidence of detrital chlorite or any of the other fragments typical of the Chloritic Sandstone Formation and thus probably represent a different source area.

In the north-eastern part of the basin this facies is extensively represented in the Comeragh Mountains where it has been described by Penney (1980). There are again suggestions of a passage south and west into finer-grained facies. However, there is a lack of detailed work on most of the succession towards Cork city so that at present these suggestions remain tentative.

Facies (c): fine-grained fluvial facies

This is the most voluminous facies in the Munster Basin. It constitutes most of the succession in all the major west coast peninsulas, where up to 6 km of succession is seen. The multiplicity of lithostratigraphic terms applied to thick successions of this facies, e.g. Valentia Slate Formation, St Finan's Sandstone Formation, Ballinskelligs Sandstone Formation (Capewell, 1975); Caha Mountain Formation, West Cork Sandstone Formation (Coe & Selwood, 1968) and Castlehaven Formation (Reilly & Graham, 1976), serve only to conceal an overall lithological similarity. The main features of this facies are the absence or very limited development of large-scale bedded (flat-bedded or cross-bedded) sandstone bodies, the predominance of siltrock and fine sandstone displaying lamination or small-scale cross-lamination produced by ripples— often as climbing sets, common bioturbation and desiccation cracks (Figs 5 and 6).

With the exception of a few sandstone bodies, sediment size does not exceed fine sand and is most commonly very fine sand and silt. Beds dominantly composed of clay-size material are also rare. Most of the sediment which is around the sand–silt grain boundary occurs as beds with sharp bases and ripple cross-lamination which shows common vertical and lateral gradations to flat lamination (Fig. 6B). Some of the flat laminated bedding planes show primary current lineation but many do not. Climbing ripple drift is common, angles of climb being steeper in the finer grain sizes. Beds which contain sand are usually 0·05–0·5 m thick and frequently show an upward

fining of grain size, an upward decrease in flat lamination and an increase in climbing ripple drift. Bases of beds are commonly erosive but rarely cut down more than a few centimetres. The overall aspect of these beds is sheet-like, although they have been observed to pass laterally into large-scale bedded sandstones and also into finer siltrocks showing mainly fine lamination and an alternation of coarser and finer grain sizes on the scale of a few centimetres. In the latter case, individual coarser silt beds (usually 5–50 mm) commonly have sharp bases and gradational tops, and fine upwards. It is in these finer beds that partial to complete destruction of primary structures by bioturbation and polygonal desiccation cracks is most common.

Sandstone bodies which occur within this facies usually show very limited basal erosion (less than 0·3 m). Major channel features and lateral accretion surfaces are absent. Sandstone bodies are commonly 2–3 m thick but are invariably composed of individual sandstone beds 0·1–0·4 m in thickness separated by mud drapes and/or erosion surfaces. Identical sandstone beds also occur singly. These sandstone beds are broadly lenticular over about 10 m and show clear fining upwards. In contrast, there is no clear overall trend observable in the sandstone bodies as a whole. Internal structures are flat bedding with primary current lineation, trough cross-bedding and, more rarely, tabular cross-bedding. Small-scale cross-lamination produced by ripples occurs in the top few centimetres of many individual sandstone beds.

Facies (c) sediments can be readily interpreted as a series of vertical accretion deposits produced during successive floods. This explains the fining-upwards motif visible in all parts of the facies, with the thicker (up to 0·5 m) beds representing relatively proximal or more powerful floods and the thinner (5–50 mm) beds representing relatively distal or weaker floods. This distinction in terms of grain size and thickness is largely for ease of description, and there is probably a complete continuum of these beds. The general sheet-like nature of the deposits and the lack of erosion suggests accretion on broad, flat, non-channelized alluvial plains. This facies has much in common with the 'fine horizontally laminated rock type' and 'red argillite rock type' of Winston (1978) and the 'floodbasin' deposits of Steel & Aasheim (1978).

Facies (d): fluvial coastal plain facies

This facies is characterized by the presence of numerous, relatively thick, often multi-storey sandstone bodies and by the common presence of interbedded sandstones and mudrocks interpreted as levée and crevasse splay sediments (Graham, 1975). This facies typically is green or grey in colour rather than red, probably reflecting a higher water table. Several of the mudrocks are massive but others are finely laminated and may represent localized lacustrine deposits (Graham, 1972; Pilling, personal communication). Lateral accretion surfaces have been recognized and plant fragments are widespread. Relatively stable channel systems, some at least of which were sinuous, are indicated. The presence of fining-upward cycles within this facies has been demonstrated (Graham, 1975).

FACIES RELATIONSHIPS

There is a clear organization of the four facies described above within the Munster Basin. Facies (d) occurs almost everywhere at the top of the local Old Red Sandstone succession (Fig. 7). The occurrence of this facies is due to temporal changes in the regional tectono-sedimentary framework. The presence of major channel belts, levées, floodplains and temporary lakes together with the generally higher water table are interpreted as due to climatic changes associated with the subsequent marine transgression. This facies also occurs outside the basin margins (e.g. Colthurst, 1978) which appear to have been overstepped in latest Devonian times (Clayton et al., 1980). This overstepping of the margins is probably also reflected in the fining upwards which occurs in the upper part of many of the Old Red Sandstone successions within the basin, although gradual reduction of relief with time may also have been a contributory factor.

However, it can be clearly seen from Fig. 7, and from the discussion above, that these vertical changes are superimposed on a large-scale lateral variation which existed throughout much of the basin's history. Thus there is a predominance of facies (c) in the central parts of the basin in a downstream position from coarser-grained, more channelized flows represented by facies (b). These downstream changes are seen as a decrease in maximum and mean grain size, an increase in the proportion of mudrocks and a decrease in thickness of sandstone bodies. There is also a downstream increase in the ratio of flat bedding to cross-bedding within the sandstones and a decrease in set thickness, both probably related to channel depth. A consistent palaeocurrent pattern suggests that these trends are meaningful even where there is a lack of precise correlation.

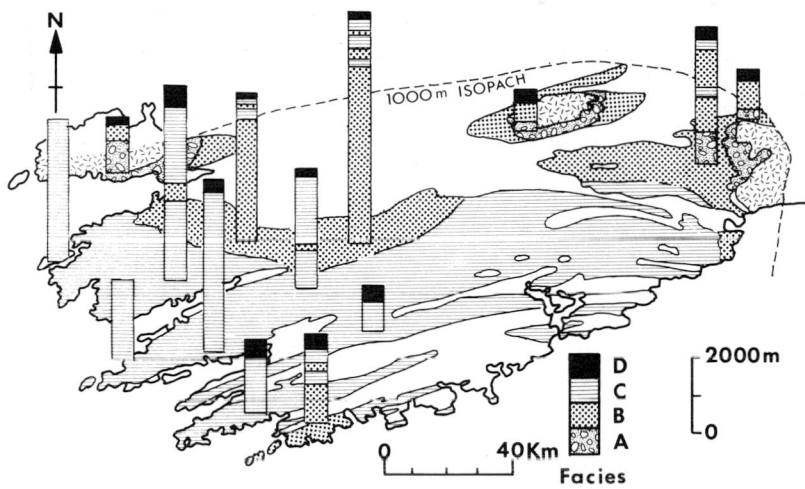

Fig. 7. Lateral and vertical distribution of facies within the Munster Basin. Columns are composites for small areas with location indicated by the base of the column. Background map shows areas where facies (a)–(c) are dominant. In addition to the author, data come from Russell (1978), Walsh (1968), Husain (1957), Penney (1980), Reilly & Graham (1976), Gardiner & Horne (1972) and Khuda (1953).

DEPOSITIONAL MODEL

Friend (1978) discussed four major fluvial basin fills, three of them Old Red Sandstone successions, and interpreted them in terms of a terminal fan model. This model applies remarkably well to the Munster Basin in almost all its features. The facies pattern described above suggests that for most of the Upper Devonian the river systems were depositing their loads within the Munster Basin in terminal fans. The Munster Basin is of similar size to those described by Friend (1978) and a situation near uplifting source mountains can also be inferred. Sedimentation rate can be estimated using the available biostratigraphical data discussed above to suggest sedimentation throughout the Frasnian and Famennian. Using isotopic age data from Gale, Beckinsale & Wadge (1979) and data for the relative lengths of stages from Ziegler (1978), a maximum sedimentation rate of about 6 km in 15 My or 0.4 m $(10^3 \text{ yr})^{-1}$, very similar to the rates estimated by Friend (1978), has been calculated.

I concur with Friend that orogenic factors and the nature of Devonian vegetation were important in the generation of these terminal fans and suggest that climate had a much smaller effect than these two. Microfloras from the Munster Basin are rare mainly because suitable preservation conditions were rare, i.e. most of the sediment was oxidized. Where they do exist, very diverse floras are indicated and there is no reason to suggest that their preservation implies markedly different or specialized environments sedi-

mentologically. Similarly bioturbation, often intense, is present throughout and there was clearly an abundant indigenous biota. Thus there was probably a lack of sediment-binding vegetation rather than a lack of terrestrial flora *per se* which was important in decreasing channel-bank cohesion, allowing rapid widening of channels during floods. The limited development of pedogenic carbonate—which is usually present only as isolated nodules—lack of evaporites and rare indications of aeolian activity suggest that the climate was at least seasonally humid. Comparison with published data suggests that sedimentation rate is unlikely to have been sufficient in itself to explain these observations. Conversely, the abundance of desiccation cracks and the absence of laminites or other typical lacustrine sediments for most of the succession suggests that standing bodies of water existed, if at all, only for short periods following major floods.

A crude calculation of the amount of Upper Devonian sediment in the fluvial Munster Basin is 38,000 km³. It is obvious that the present marginal areas, where alluvial fans of facies (a) are seen, made only a small contribution to this. It is suggested that the major sediment source lay well to the north of the Munster Basin. This area was predominantly one of erosion, with a well-developed river system which fed a major inland distributary system in the south.

This model is consistent with several aspects of Devonian palaeogeography in Ireland. The Munster Basin sediment is generally rather fine grained and contains a considerable amount of detrital chlorite even though its margins show rocks virtually devoid of chlorite. Chlorite schists are present in much of the Dalradian succession to the north and probably were much more common at higher levels of the tectonic pile. It is possible that these schists are now present in dismembered form in the Munster Basin although distance of transport seems large for the preservation of chlorites. In addition the heavy mineral suites indicate a predominantly metamorphic source area (Capewell, 1957a; Husain, 1957). There is very little indication of any Upper Devonian sediment in the north of Ireland, suggesting that the area was being eroded at this time. However, more petrographic work is necessary before source area, and thus the length of the river systems, can be specified with any certainty.

On a regional scale the application of a terminal fan model to the Munster Basin raises the question of how much sediment reached the sea which lay at an unknown distance to the south. Intuitively the chances of bypassing such a fine grained terminal fan system would seem to be small.

ACKNOWLEDGMENTS

My thanks to Ken Russell and to the staff and research students at Trinity College for many helpful discussions, and to Elaine Cullen for diagrams. The manuscript has benefited from comments by John Bridge, Andrew Miall and David Macdonald.

REFERENCES

ALLEN, J.R.L. (1979) Old Red Sandstone facies in external basins, with particular reference to southern Britain. In: *The Devonian System* (Ed. by M.R. House, C.T. Scruton and M.G. Bassett). *Spec. Pap. Palaeont. palaeont. Ass., London*, **23**, 65–80.

CAPEWELL, J.G. (1951) The Old Red Sandstone of the Inch and Annascaul district, Co. Kerry. *Proc. R. Ir. Acad.* **54B**, 141–168.

CAPEWELL, J.G. (1957a) The stratigraphy and structure of the country around Sneem, Co. Kerry. *Proc. R. Ir. Acad.* **58B**, 167–183.

CAPEWELL, J.G. (1957b) The stratigraphy, structure and sedimentation of the Old Red Sandstone of the Comeragh Mountains and adjacent areas, County Waterford, Ireland. *Q. Jl geol. Soc. Lond.* **112**, 393–410.

CAPEWELL, J.G. (1965) The Old Red Sandstone of the Slieve Mish, Co. Kerry. *Proc. R. Ir. Acad.* **64B**, 165–174.

CAPEWELL, J.G. (1975) The Old Red Sandstone Group of Iveragh, Co. Kerry. *Proc. R. Ir. Acad.* **75B**, 155–171.

CLAYTON, G. & GRAHAM, J.R. (1974) Miospore assemblages from the Devonian Sherkin Formation of south-west County Cork, Republic of Ireland. *Pollen Spores*, **16**, 565–588.

CLAYTON, G., GRAHAM, J.R., HIGGS, K., HOLLAND, C.H. & NAYLOR, D. (1980) Devonian rocks in Ireland: a review. *J. Earth Sci. R. Dubl. Soc.* **2**, 161–183.

COE, K. & SELWOOD, E.B. (1968) The Upper Palaeozoic stratigraphy of West Cork and parts of South Kerry. *Proc. R. Ir. Acad.* **66B**, 113–131.

COLTHURST, J.R.J. (1978) Old Red Sandstone rocks surrounding the Slievenamon inlier, Counties Tipperary and Kilkenny. *J. Earth Sci. R. Dubl. Soc.* **1**, 77–103.

DORAN, R.J.P., HOLLAND, C.H. & JACKSON, A.A. (1973) The sub-Old Red Sandstone surface in southern Ireland. *Proc. R. Ir. Acad.* **73B**, 109–128.

FRIEND, P.F. (1978) Distinctive features of some ancient river systems. In: *Fluvial Sedimentology* (Ed. by A.D. Miall). *Mem. Can. Soc. Petrol. Geol., Calgary*, **5**, 531–542.

GALE, N.H., BECKINSALE, R.D. & WADGE, A.J. (1979) A Rb–Sr whole rock isochron for the Stockdale Rhyolite of the English Lake District and a revised mid-Palaeozoic time scale. *J. geol. Soc. London*, **136**, 235–242.

GARDINER, P.R.R. & HORNE, R.R. (1972) The Devonian and Lower Carboniferous clastic correlatives of Southern Ireland. *Bull. geol. Surv. Ireland*, **1**, 335–366.

GRAHAM, J.R. (1972) *The sedimentation of Devonian and Carboniferous rocks in south-west County Cork.* Unpublished Ph.D. Thesis. University of Exeter. 169 pp.

GRAHAM, J.R. (1975) Deposits of a near-coastal fluvial plain—the Toe Head Formation (Upper Devonian) of south-west Cork, Eire. *Sedim. Geol.* **14**, 45–61.

GRAHAM, J.R. & REILLY, T.A. (1972) The Sherkin Formation (Devonian) of south-west County Cork. *Bull. geol. Surv. Ireland*, **1**, 281–300.

HORNE, R.R. (1970) A preliminary reinterpretation of the Devonian palaeogeography of western County Kerry, Ireland. *Bull. geol. Surv. Ireland*, **1**, 53–60.

HORNE, R.R. (1971) Aeolian cross stratification in the Devonian of the Dingle Peninsula, County Kerry, Ireland. *Geol. Mag.* **108**, 151–158.

HORNE, R.R. (1974) The lithostratigraphy of the late Silurian to early Carboniferous of the Dingle Peninsula, County Kerry. *Bull. geol. Surv. Ireland*, **1**, 395–428.

HORNE, R.R. (1975) The association of alluvial fan, aeolian and fluviatile facies in the Caherbla Group (Devonian), Dingle Peninsula, Ireland. *J. sedim. Petrol.* **45**, 535–540.

HUSAIN, S.M. (1957) *The geology of the Kenmare Syncline, Co. Kerry, Ireland.* Unpublished. Ph.D. Thesis. University of London. 387 pp.

KHUDA, M.M. (1953) *The stratigraphy and structure of the south-western part of County Tipperary and the petrography of the Old Red Sandstone.* Unpublished Ph.D. Thesis. University of London.

MACCARTHY, I.A.J., GARDINER, P.R.R. & HORNE, R.R. (1978) The lithostratigraphy of the Devonian—Early Carboniferous succession in parts of Counties Cork and Waterford, Ireland. *Bull. geol. Surv. Ireland*, **2**, 265–305.

MURPHY, T. (1960) Gravity anomaly map of Ireland (sheet 5—South-west). *Geophys. Bull., Dubl.* **18**.

NAYLOR, D. & JONES, P.C. (1967) Sedimentation and tectonic setting of the Old Red Sandstone of Southwest Ireland. In: *International Symposium on the Devonian System* (Ed. by D.H. Oswald). *Alberta Soc. Petrol. Geol., Calgary*, **11**, 1089–1099.

NAYLOR, D., SEVASTOPULO, G.D., SLEEMAN, A.G. & REILLY, T.A. (1981) The Variscan fold belt in Ireland. *Geol. Mijnb.* **60**, 49–66.

PENNEY, S.R. (1980) A new look at the Old Red Sandstone succession of the Comeragh Mountains, County Waterford. *J. Earth Sci. R. Dubl. Soc.* **3**, 155–178.

REILLY, T.A. & GRAHAM, J.R. (1976) The stratigraphy of the Roaringwater Bay area of South-west County Cork. *Bull. geol. Surv. Ireland*, **2**, 1–13.

RUSSELL, K.J. (1978) Vertebrate fossils from the Iveragh Peninsula and the age of the Old Red Sandstone. *J. Earth Sci. R. Dubl. Soc.* **1**, 151–162.

STEEL, R.J. & AASHEIM, S.M. (1978) Alluvial sand deposition in a rapidly subsiding basin (Devonian, Norway). In: *Fluvial Sedimentology* (Ed. by A.D. Miall). *Mem. Can. Soc. Petrol. Geol., Calgary*, **5**, 385–412.

WALSH, P.T. (1968) The Old Red Sandstone west of Killarney, Co. Kerry, Ireland. *Proc. R. Ir. Acad.* **66B**, 9–26.

WINSTON, D. (1978) Fluvial systems of the Precambrian Belt Supergroup, Montana and Idaho, U.S.A. In: *Fluvial Sedimentology* (Ed. by A.D. Miall). *Mem. Can. Soc. Petrol. Geol., Calgary*, **5**, 343–360.

ZIEGLER, W. (1978) Devonian. In: *Contributions to the Geological Time Scale* (Ed. by G.V. Cohee, M.F. Glaessner and H.D. Hedberg). *Am. Ass. Petrol. Geol. Studies Geol.* **6**, 337–339.

Spec. Publs int. Ass. Sediment. (1983) **6**, 485–497

Recent and Tertiary fluvial carbonates in Central Spain

S. ORDÓÑEZ* *and* M. A. GARCÍA DEL CURA†

* *Departamento de Petrología, Facultad de Geología, Universidad Complutense, Ciudad Universitaria, Madrid 3, and* † *Instituto de Geología Económica CSIC, Madrid, Spain*

ABSTRACT

Present-day and Neogene fluvial carbonates of Central Spain allow us to describe several petrological types: (a) *spring and waterfall carbonates* in wedge-shaped tufa buildups with moss tufa and karstification features, (b) *water flow carbonates*: encrustation tufa on plants (vertical tube facies and crossed tubes facies) and related detrital forms are shown. The stromatolitic buildups genetically associated with blue-green algae colonization on hard- or coarse-grain bottoms are also described. The blue-green algae colonization of erratic objects leads to the formation of oncolites (high-density colonization) or oolites (low-density colonization?). The biological data and the textural and structural properties of these carbonates are compared with other similar Tertiary and present-day carbonates. The Upper Neogene fluvial carbonates of the Madrid basin allow us to develop a distribution model of water flow carbonates.

Finally, a conceptual genetical model of fluvial carbonates is proposed. The main factors that influence carbonate features are: environmental energy, colonization morphology by blue-green algae and water properties.

INTRODUCTION

In this paper petrological, mineralogical and biological data from Tertiary and present-day carbonates of Central Spain are presented.

Theoretical support of this paper, on the genetic aspects of fluvial carbonates, is given by Golubic (1973). This author gives a classification of carbonates, supported in their position in the fluvial longitudinal profile and shows, in an ideal fluvial system, two different sections: a proximal section with oncolitic features related to algal activity, and a distal section with stromatolitic crusts, also related to algal activity.

Thrailkill's (1968) model of the physicochemical precipitation and dissolution of calcium carbonate (calcite), is the most important contribution used in the sketch of the genetic models.

Our observations and petrographic studies on present-day carbonates, oncolites and stromatolites, situated in the Central Spain rivers, as well as on the water composition, associated floral and meso-

structural aspects, allow us to see the state of the palaeoenvironmental problem (Ordóñez, García del Cura & Carballal, 1978; Ordóñez & Gonzalez, 1979; Ordóñez, Gonzalez & García del Cura, 1979, 1981; Ordóñez, Carballal & García del Cura, 1980).

In the Tertiary basins of Central Spain (Tajo and Duero Rivers) it has been demonstrated by Ordóñez & García del Cura (1977a) and Ordóñez, Gonzalez & García del Cura (1982) that a variety of carbonate deposits can occur in fluvial systems.

Low magnesium calcite is the main mineralogical phase of these carbonates, and the old fluvial carbonates have an isotopic composition close to that of present-day carbonates. The averages of all the data on isotopic composition are δC^{13}_{PDB} $-8 \cdot 06 \pm 0 \cdot 35$ and δO^{18}_{PDB} $-7 \cdot 18 \pm 0 \cdot 19$.

SPRING AND WATERFALL CARBONATES

Tufa buildups, related to waterfalls, have an approximate volume of 10^5 m^3. Their morphology is wedge-shaped and their maximum thickness is close

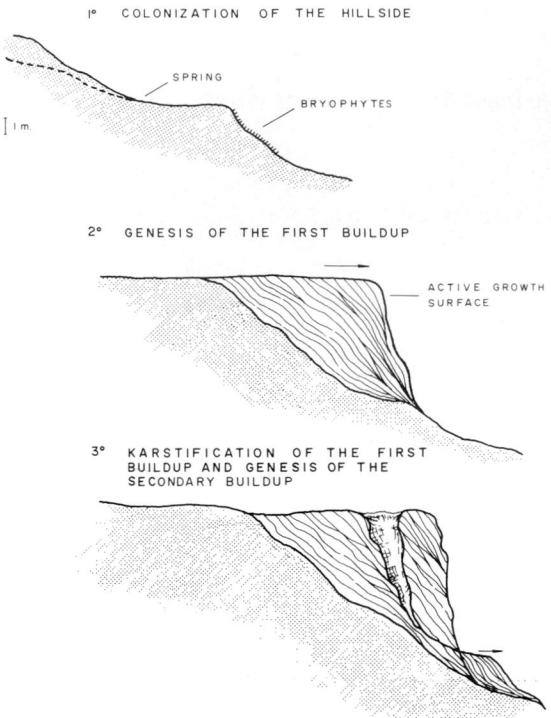

1° COLONIZATION OF THE HILLSIDE

SPRING

BRYOPHYTES

1 m.

2° GENESIS OF THE FIRST BUILDUP

ACTIVE GROWTH
SURFACE

3° KARSTIFICATION OF THE FIRST
BUILDUP AND GENESIS OF THE
SECONDARY BUILDUP

Fig. 1. Sedimentological model of waterfall carbonates.

to 20 m. Their growth in the horizontal dimension is related to the hillside slope (Fig. 1). The internal structure of these buildups is laminated with vertical dips in the frontal part of the buildup. Every lamina represents a regular annual growth, with a thickness of the order of 8–10 cm. We have identified $CaCO_3$-encrusted bryophytes, of the genus *Cratoneuron* and *Eucladium*.

The growth of the buildup prevents the water from reaching the active growth surface. Water can percolate through the buildup and dissolve the calcium carbonate, giving pseudostalactitic structures in the solution cavity. The morphology of the structures is similar to stalactites, but they are formed of a non-dissolved residual tufa with a calcium carbonate rim. Springs located in the lower part of the buildup can lead to a new buildup (Fig. 1, part 3).

In the River Tajuña, in the Brihuega–Masegoso Section, we have found a working example of these buildups (Ordóñez & Gonzalez, 1979). The frontal part of the buildup is colonized by *Cratoneuron commutatum*. This moss is encrusted by low magnesium calcite, the water composition is 0·70 mm/l Ca^{2+},

0·009 mm/l Mg^{2+}, 0·55 mm/l CO_2, and the pH reaches a value of 7·3 (data on Prado Spring).

Petrological obervations on carbonates of these buildups reveal a typical moss tufa texture (Irion & Muller, 1968). The internal structure of these buildups is closely related to the hillside slope, but on the more gentle slopes higher plants grow, and are also encrusted by carbonates. Gastropods which are commonly associated with the higher plants are also present. The carbonate precipitation is not related to biogenic activity, but acts indiscriminately on both living and dead plants.

In the Mundo River valley (S.E. Spain) buildups are present which have been studied by Calvo, García del Cura & Ordóñez (1979). Their internal structures are similar to the waterfall carbonates of the River Tajuña, but the microtextural features are more complex than the algal tufa of Irion & Muller (1968), since complex microstromatolitic textures have been identified. The living plants on the frontal part of these buildups, at the present time, are *Cratoneuron commutatum* and *Eucladium verticillatum*, species which are present in the present-day tufa of Germany (Irion & Muller, 1968). Karstification features are also present in some buildups.

The climatic significance of the waterfall carbonates in the stratigraphic record is non-specific, and their main significance is as palaeospring and palaeohydrological indicators.

WATER FLOW CARBONATES

Two recent types can be distinguished:

(a) encrustation tufa on plants and related detrital forms;

(b) colonization by blue-green algae on the channel bottom or on erratic objects.

Encrustation tufa on plants and related detrital forms

In the wider part of many valleys of the upper part of rivers in central Spain we were able to study subrecent and recent deposits of encrusting plants displaying three main facies, whose relative situation is shown in Fig. 2.

Vertical tubes facies (VTF), are associated with higher plants located in the flood plain. Their most important petrographical aspects are the presence of stromatolitic facies (spongiostromata and porostromata), located around the stalks of rushes and other plants. The carbonate coat has a thickness of 1 cm or more, and the porostromata structure is the main

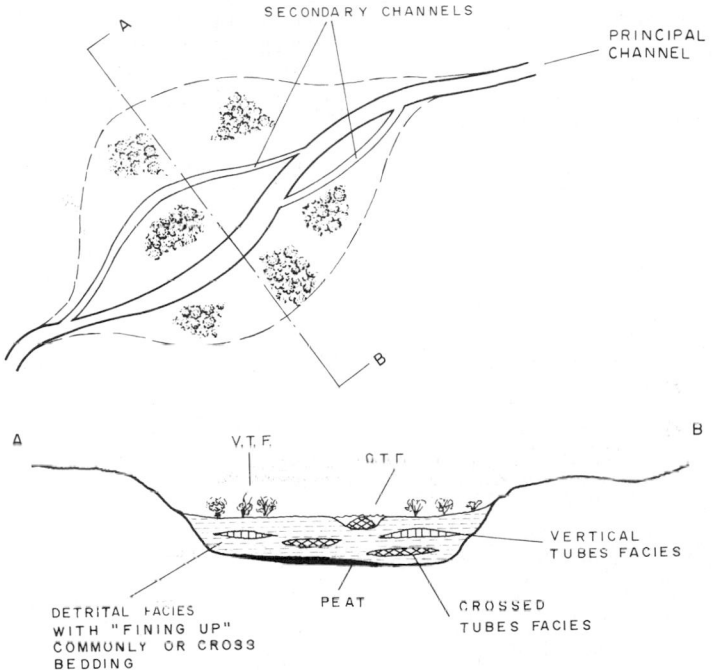

Fig. 2. Sedimentological and stratigraphic model of encrusted tufa carbonates.

feature. Trapping processes of detrital algal crusts (1–1·5 mm), and quartz grains (0·2 mm) have been identified.

Crossed tubes facies (CTF), are located in the principal active channels. They show a higher porosity and can be used as palaeodirectional indicators (Fig. 3C). The carbonate coat has a thickness of several millimetres, and spongiostromata structure is the main feature with an homogenous micritic carbonate.

Detrital tufa facies, the erosional processes, related to episodic floods, partially break and crumble the delicate tufa structures. The sedimentary structures in the detrital materials are very variable (cross-bedding, fining-up sequences, rhythmites).

In the lower part of some deposits, peat crops out with a thickness of about 1 m.

In the River Tajuña (Abadanes–Masegoso section), recent fluvial terraces display a sequential fluvial deposit. The basic sequential unit is a detrital sand or silt tufa with edaphic features, above this a sporadic layer of vertical or crossed tube tufa and above this fining-upward detrital deposits (Fig. 4). Clasts are mainly fragments of tufa (Fig. 5). This deposit displays laminations and small migration ripples. Locally, cut-and-fill structures with channel-lag terrigenous deposits (non-tufa clasts) are present.

'Crossed tubes facies' are located on a subrecent terrace of the River Tajuña, are genetically related with an encrustation process on charophytes, and the flow direction is fossilized. The main microscopic feature is a stromatolitic structure (Fig. 6). Ostracoda and terrigenous grains with a size near 0·03 mm are also identified. The bulk porosity is higher than the 'charophyta tufa' of Lang & Lucas (1970), described from Afghanistan. 'Vertical tubes facies' (VTF) are also present and these have a genetic significance different from that of crossed tubes facies (CTF), because they (VTF) are located on the flood plain, while CTF occur in active channels. A conceptual model of fluvial carbonates related to natural dams can be proposed: peat is the main sediment in the initial stage of refilling of the swamp located behind the natural dam. When the depths of water in the swamp have decreased, and also at the shoreline of the swamp, plants begin to grow, and become encrusted by carbonate precipitation. Episodic floods destroy tufa buildups and the clasts are transported into the swamp as a prograding microdelta. However, the natural dam is formed by the hard bottom in this example, whilst in Lang & Lucas' (1970) example it is related to biological buildups (tufa dam).

Locally, soft bio-physicochemical carbonates, with some idiomorphic crystals of calcite, are developed. At High Henares River valley (central Spain), in a terrace

Fig. 3. (A) Waterfall tufa with internal structures (Tajuña River). (B) Pseudostalactite facies related to infiltration processes in waterfall tufa buildups (Tajuña River). (C) Crossed tubes facies as palaeogeographic indicator (Tajuña River). (D) Stromatolitic facies around stems in the encrusted tufa facies (Tajuña River).

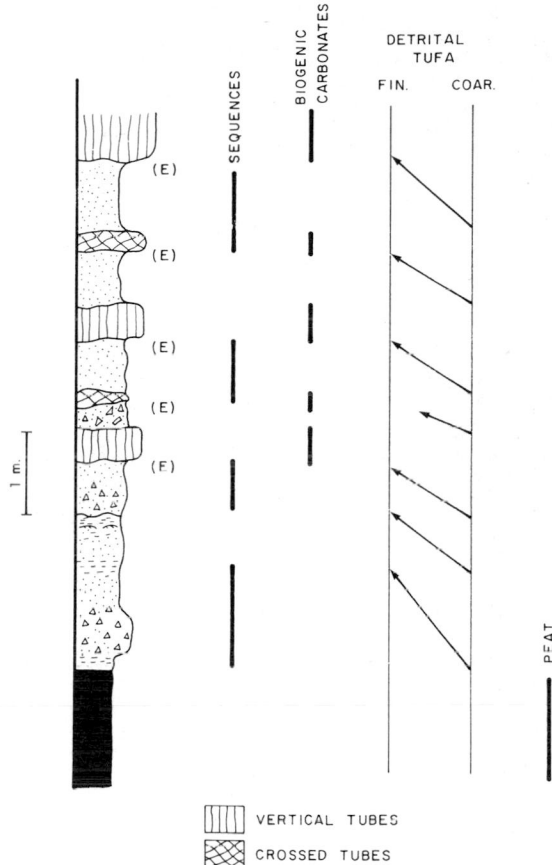

Fig. 1. Sequence of fluvial deposits in the River Tajuña (Abadanes–Masegoso section).

(Campiña Terrace) of 1257 ± 190 yr age, we found a deposit of very homogeneous and incoherent carbonate with many gastropod shells and peat in thin layers of several millimetres thickness.

Colonization by blue-green algae on the channel bottom or on erratic objects

Our data on present-day carbonates can help in recognition of the carbonate precipitation conditions associated with colonization by blue-green algae.

Stromatolitic and oncolitic buildups are developed at Dulce River valley (Central Spain). The main mineralogical component of these buildups is low magnesium calcite (Ordóñez *et al.*, 1978, 1982).

The chemical composition of the water is shown in Table 1. These analytical data support our opinion that photosynthetic processes play an important role in the carbonate genesis because the water is undersaturated in calcite.

Stromatolites

The stromatolitic deposits are located on the hard bottom of the artificial channel and their maximum growth is at depths between 10 and 20 cm. The water flow velocity is about 50 cm sec^{-1}.

The stromatolites may be classified as SH-V with the infrastructure LLH-C or more commonly homogeneous, according to Logan, Rezak & Ginsburg (1964).

The average thickness of these stromatolites is 2 cm and they are formed by a pair of dark (0·07 mm) and light (0·11 mm) laminae (Fig. 7C). Light laminae have prismatic crystals (10–35 μm) with a radial fibrous arrangement and interstitial rim cement. Remains of blue-green algae sheaths are observed in intercrystalline positions. Dark laminae have a microcrystalline texture.

The stromatolites' growth surface is defined by an algal mat, with a stratification 'in space' or 'instantaneous biological stratification' (see Monty, 1976). The external part of the algal mat is formed by *Pleurocapsa minor* Hansg. Geit., which has the maximum growth in the autumn and spring (Fig. 7E). The internal part of the algal mat is formed by *Phormidium incrustatum* (Nägeli) Gomont, which shows maximum growth in the autumn. *Rivularia* sp. is also found in this structures (Fig. 7F).

Filaments are normal to the surface of stromatolites (Fig. 7A). *Nostoc sphaericum* Vauch replaces the blue-green algae during the winter.

Oncolites

There are no true oncolites, although oncoids are the main structures. The nuclei of these oncoids are fragments of sedimentary rocks, bricks, slag or other wastes of human activity (Fig. 7B). They are located on soft channel bottoms, less deep than those colonized by stromatolites. The flow regime is subcritical and laminar. Their size is about 1–4 cm, with an average value of 2 cm. The number of laminae around the nucleus varies between 4 and 30. The laminar fabric is given by a couplet of dark and light laminae. Dark laminae have an average thickness of 49 μm with a crystal size of 1–2 μm. Light laminae have a thickness greater than 400 μm. They have an arborescent texture, which displays a trapping effect of quartz grains (0·3–0·5 mm) and diatoms. Often, the most external arborescent lamina (porostromata) is thicker than other arborescent laminae.

These oncoids can be classified as SS-C/LLH-C after Logan *et al.* (1964). However, the oncoids have

Fig. 5. Detrital tufa facies show a fining-up sequence (Tajuña River).

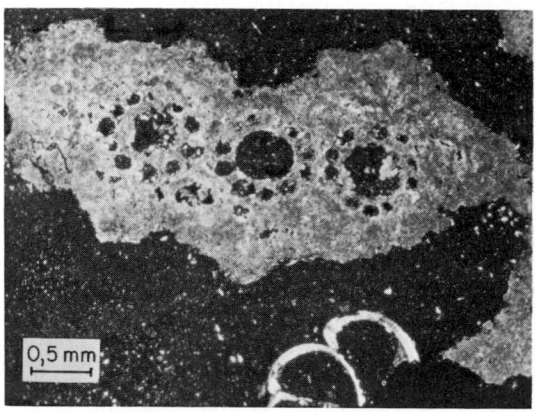

Fig. 6. Crossed tubes facies thin section (Charophyta tufa) (Tajuña River).

an asymmetrical fabric, because the arborescent structure is better developed on the upper side. Sometimes the bottom side is covered by compact dark laminae (spongiostromatae). The symmetry modifications caused by differential growth alter the movement threshold by shifting the centre of gravity of the oncolite.

Algal mats cover all oncolite surfaces and are partially fossilized by new carbonate coats. The main algal communities of algal mats are *Schizotrix*, in the early spring, and *Lyngbia* genus, *Phormidium* subgenus (*Ph. cf. incrustatum* (Nägeli) Gomont) in the later spring. These coexist with *Schizotrix* in the lower parts of the algal mat. This algal association is related to the development of arborescent structures during the summer, but in the autumn the most abundant alga is *Phormidium* sect. *Moniliformia*.

Table 1. Water composition of oncolite and stromatolite environment

	February		June		September		November	
	s.w.	o.w.	s.w.	o.w.	s.w.	o.w.	s.w.	o.w.
pH	7	7	7	7	7·1	7·2	6·9	7·1
v (m s g^{-1})	0·5	0·01	0·55	—	0·51	0·05	0·52	—
CO_3^{2-} mm	0·82	1·00	0·27	0·55	0·91	0·91	0·91	1·00
SO_4^{2-} mm	1·40	1·27	0·72	0·64	0·52	0·47	1·00	0·80
Cl^- mm	0·42	0·71	0·56	0·71	0·42	0·28	—	—
Mg^{2+} mm	0·40	1·10	0·90	0·30	0·80	0·80	0·35	0·43
Ca^{2+} mm	2·50	3·05	1·50	1·50	1·80	1·30	0·95	0·85
Log PAI	−5·69	−5·52	−6·39	−6·08	−5·79	−5·93	−6·06	−6·07

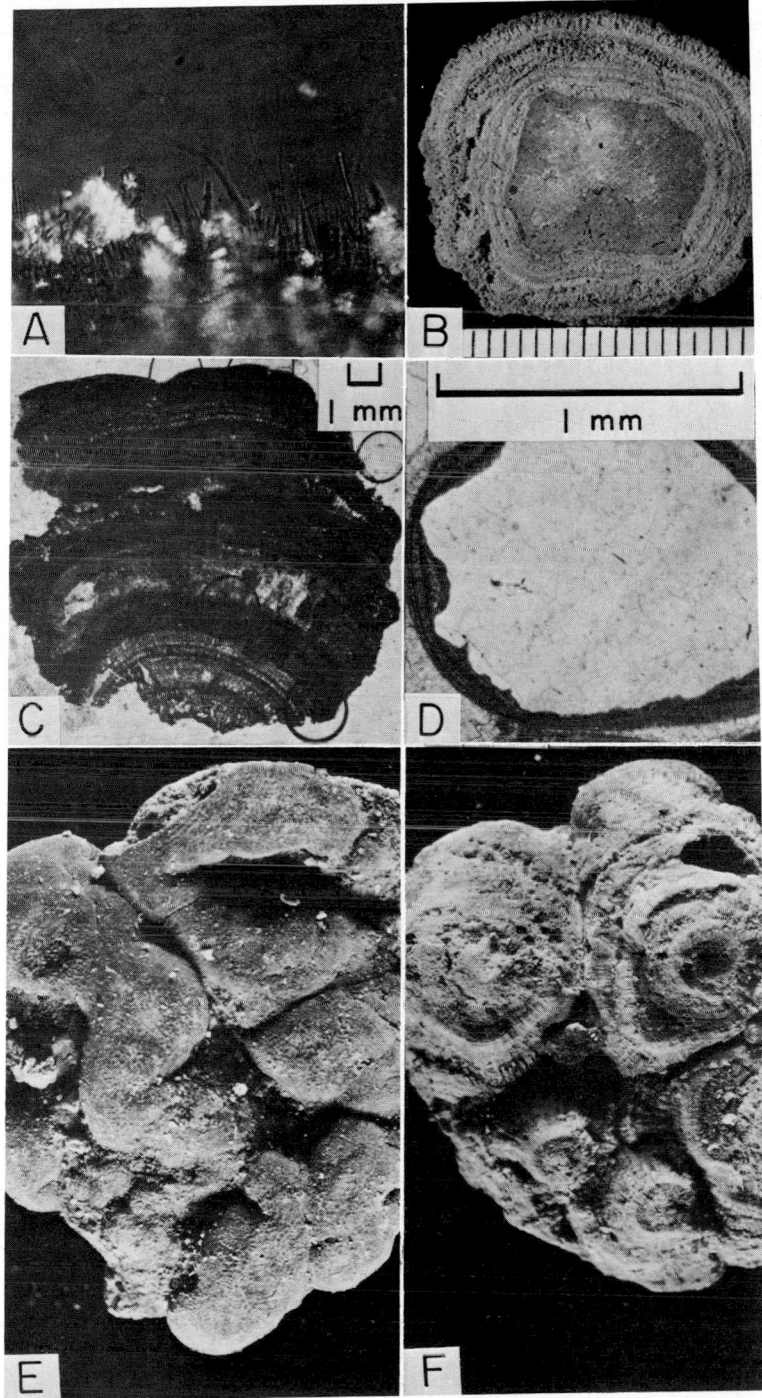

Fig. 7. (A) Blue-green algae thallus on present-day stromatolitic surface (Dulce River valley). (B) Section of present-day oncolites with detritic nucleus (Dulce River valley). (C) Section of present-day stomatolites (Dulce River valley). (D) Ooid in fluvial tertiary carbonates (Madrid Basin). Present-day stromatolites (Dulce River valley). (E) External surface. (F) Internal view.

The oncolites show algal filaments with an interlaced arrangement. This arrangement is different from the vertical arrangement of the stromatolites. The biozonation of the algal mats is 'in time' or 'historical biomineralogical stratification'.

Comparison with similar documented structures

Jones & Wilkinson (1978) have shown Michigan Lake oncolites with an individual lamina thickness similar to our fluvial oncolites. However their arborescent structure is developed on all the surfaces of the oncolites though these oncolites also displayed a larger thickness in the upper side. The dark lamina is genetically related, according to these authors, to an inorganic precipitation coincident with a reduced growth of algae during autumn and winter.

Monty & Mas (1981) in centimetre-sized oncolites of the Hoyoux creek (Belgium) found *Phormidium foveolarum* (Mont) Gomont associated with imbricated fibrous-like calcite units. The fibrous appearance results from the fact that each component crystal grew around filamentous bundles of dense flabellate colonies of these algae, the filament widths of which lie around 1·5 μm. These authors also found this alga associated with *Phormidium incrustatum* (Nägeli) Gomont in freshwater stromatolites from the Bois d'Hautmont Creek (Belgium). Geurts (1975) has identified, in Belgium, tufas of the encrustant blue-green alga *Phormidium incrustatum* (Nägeli) Gomont. According to Geurts *Phormidium* grows very fast in spring with a hair-like morphology; its growth is slower during summer and is retarded during autumn and winter.

The oncolites of Lake Constance (Schafer & Stapf, 1978) are similar to our fluvial oncoids in terms of both textural and algal association data. Dark laminae are related to *Schizothrix* activity. Spongy microstructure laminae are due to *Phormidium–Calothrix/Dichothrix*. *Rivularia* colonies occur singly or randomly throughout the entire oncolite. Lake Constance oncolites have two main structural types: the first is hard and smooth with more numerous *Schizothrix* layers. The other is spongy with a dominant activity of *Phormidium–Calothrix/Dichothrix*. However, we have only one structural type in the fluvial oncolites of the River Dulce. This consists of regularly alternating dark or dense micrite layers and light or spongy microstructure layers.

We have mainly found, in the Neogene fluvial oncolites (Duero Basin, Central Spain), structures with a predominantly spongy microstructure but also others with alternating dark and light laminae (Ordóñez & García del Cura, 1977b). We have obtained sheaths by acid attack but it was not possible to classify the blue-green algae species. These fluvial oncolites have a nucleus which is formed by calcareous sand and they reach sizes of 32 cm. Size distributions of neogene oncolites are poorly sorted and display a close packing. Detrital layers are common and there are also detrital grains trapped in the arborescent structures.

Other Tertiary fluvial oncolites are those of the Palaeogene of the Ebro Basin (N.W. Spain) (Anadon & Zamarreño, 1981). These show pairs of light-coloured, generally thicker (up to 400 μm) somewhat grumelous laminae with generally thinner, dark-coloured micritic laminae formed by the rapid addition of elementary micritic films. These are similar to these formed by present-day *Schizothrix* activity. The light laminae with filamentous structures compare with those related to present-day activity of *Phormidium incrustatum* or *Dichothrix/Calothrix*. The filamentous laminae are often invaded by horizontal, thin, micritic films.

Glazek (1965) studied present-day oncolites in the streams of the Red River Basin (North Vietnam); their diameters ranged from 0·05 to 12·8 cm, and their surfaces were usually smooth, but sometimes with small protuberances. They were formed by layers whose thickness ranged from 1 to 8 mm and they had numerous undulating laminae of 0·05–0·5 mm. These microlaminae are distinctly undulating and their boundaries are indistinct. Glazek (1965) also found diatoms (*Navicula*) and cellular membranes of indeterminate green algae or blue-green algae associated with the oncolites. The tropical climate may be responsible for putrefaction of algal material (development of saprophytic bacteria (Actinomycetales)). In North Vietnam, the dry season (October–May) favours the development of algae because the suspended load in the rivers is less than during the wet season. Glazek (1965) attributed the microcrystalline layers (dark laminae) to physicochemical processes and the porous layers (light laminae) to physiological or/and biochemical processes related to algal activity.

Glazek (1965) also studied the recent calcitic oncolites from streams of the Lejowa Valley (Tatra Mountains, Poland). These oncolites are associated with tufa and are located in the vicinity of three small springs. They are irregular in shape and flattened. The longest diameter ranges from 2 to 10 cm. The surface is frosted and porous with small protuberances. Their nuclei are rock fragments and their carbonate coating is micritic, locally forming irregular layers up to 1 mm thick. After digestion of these oncolites with acetic acid, Glazek identified green algae, diatoms and blue-green algae (*Plectonema tomasinianum* (Kuetz) Born) in the insoluble residues.

Braithwaite (1979) described recent fluvial pisolites and laminated crystalline crusts in Dyfed, South Wales. These are formed by magnesium calcite and are generated by physicochemical processes in supersaturated water with respect to calcite. The slope is an important element in the distribution of deposits. The size of accretionary bodies ranges from 5 to 30 mm of diameter. These are usually highly spherical, but larger bodies become flattened. The accretionary bodies also show re-entrants. They are more irregular in shape than cave pearls. This is the most distinguishing feature between our fluvial oncolites and those described by Braithwaite (1979).

Laminae are of two kinds: layers of micrite which occasionally show a distinct clotted texture; these form the substrate for growths of coarse, radial, bladed crystals that show the characteristic features of decrease in numbers and progressive grain enlargement which would be expected in a cement deposited on a free surface. Micrite branching filaments (algae?) have been observed within such crystals. Braithwaite (1979) interprets the clotted texture as accretion surfaces of fine sediments, but the presence of probable algal filaments does not rule out a biological influence on their genesis.

Nuclei in the smaller bodies may be spherical crystal aggregates, micritic pellets or, occasionally, quartz grains. Some are composite and the larger bodies commonly include micritic pellets and first-order pisolites. A suggestion of Braithwaite's paper is that the pisolites do not require continuous agitation.

In our opinion, Holocene pisolites associated with spring-fed surface pools (Pastos Grandes, Bolivia) (Risacher & Eugster, 1979) are similar to cave pearls. These pisolites have concentric bands made up by couplets of clear and dark laminae. The clear laminae show crystals of calcite with radial arrangement; they are composed of radial fibrous calcite fans, in section, or calcite cones in three dimensions.

Calcitic fluvial oolites of late Pleistocene age are described by McGannon (1975) from south central, Texas. Their size ranges from 1 to 2 mm diameter. There is no direct evidence in thin sections and in insoluble residues for the biological origin of the carbonate coats, but algae may be responsible, in part, for their origin. Algae may provide a lot of material for nuclei: calcified sheaths have been identified in

some nuclei. The dark, mottled, microcrystalline calcite which constitutes most oolite nuclei may well be the result of direct algal precipitation and algae probably also influence precipitation indirectly by extracting CO_2 from the water and thus facilitating the precipitation of calcite to form the oolite laminae. The depositional environment of these oolites must be interpreted as a turbulent one, at least in part.

AN EXAMPLE OF FLUVIAL CARBONATE DISTRIBUTION

Fluvial facies have been studied in the upper part of Madrid basin refillings (Ocaña-Tarancón, central Spain) (U.T.S. 4 of Megias, Ordóñez & Calvo, 1982).

The fluvial facies consists of flood-plain deposits with reddish silts with some calcimorphic soil-layers locally superimposed. Tufa and other biogenic or physicochemical carbonates are also present. In the channel deposits, besides detrital materials with longitudinal cross-bedding, stromatolite buildups are present. This fluvial facies is about 30 m thick and it extends as a band for about 40 km and is 20 km wide. Some outcrops are very favourable for trying to outline a distribution model for fluvial carbonates.

This model is shown in Fig. 8. Its dimensions are of the order of tens of metres. It is located in a meandering-type fluvial network, with a low density of channels on the flood plain. A luxuriant vegetation covered the flood plain and is responsible for the lack of bedload and suspended load. The network is also close to an equilibrium profile.

The steep side of the channel is often covered by bryophytes, which are encrusted by carbonates. The gently sloping side of the channel is commonly colonized by higher plants, and develops a 'vertical tubes facies'. On both sides of a channel these tufa carbonates would be interfingering with stromatolitic buildups. These buildups show a mamillary morphology in the less deep water, and change to a more planar morphology in the direction of the channel axis. The axis of the channel is often filled by conglomerate–sand bodies, with longitudinal cross-bedding. These bodies are composed mainly of oncolites, tufas and stromatolitic buildup fragments, other detrital components (quartz, quartzite and limestone), bivalve shells (*Unio* s.p.) colonized and encrusted by stromatolitic buildups and often by oolites.

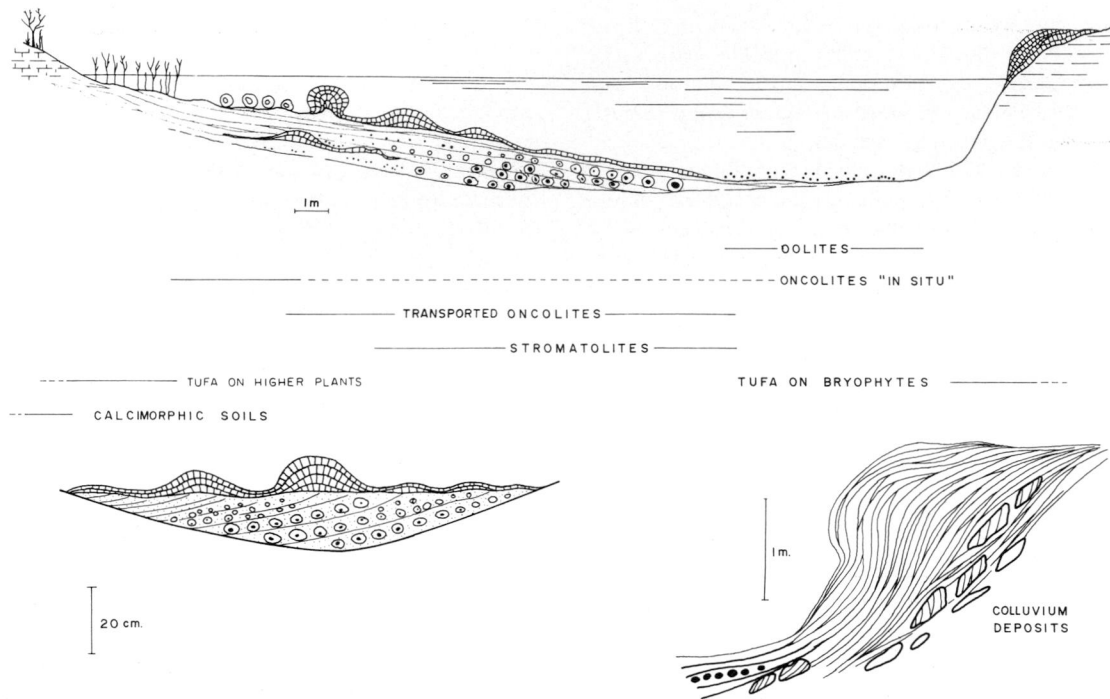

Fig. 8. Idealized sedimentological model of fluvial carbonates in the Madrid Basin.

CONCEPTUAL MODEL OF FLUVIAL CARBONATES

It is difficult to establish the conditions under which the different biogenic fluvial carbonates are formed. Our present-day data allow us to establish the main factors that influence genetic aspects, which are: environmental energy, degassing mechanisms, colonization morphology (by blue-green algae) and water properties.

The environmental energy is related to degassing mechanisms. Agitated waters quickly lose dissolved CO_2 and the water reaches a supersaturated state. However, in the undersaturated water with laminar flow only, biological activity can depress CO_2 pressure and reach a microenvironmental supersaturated state. The environmental energy is also related to biological colonization density and it has a stabilizing effect on clastic sediments. The colonization support can be living plants, channel bottom or erratic objects. Living plants are colonized by epiphytic blue-green algae and these are responsible for carbonate precipitation with microstromatolitic structures around the stalks of higher plants. When biological activity stops during the winter or when the environment energy prevents the biological colonization, only physicochemical carbonates may be generated. Crystal morphology of these carbonates is only related to nucleation rate and/or supersaturated water state. Channel bottom colonization is possible when the size of channel lag deposits ranges from coarse sand to gravel or when the bottom is a hard one. A soft bottom is not easily colonized because the bottom erodibility prevents the growth of benthonic blue-green algae. Biological colonization on the bottom can be total or patchy.

Patchy colonization is developed under either shallow or deep conditions and both wet and dry ones. Total colonization is related to planar stromatolites and patchy colonization is related to mamillary ones.

The colonization of erratic objects can be developed on all bottom types. The erratic objects have a hard surface and can be stalks of dead plants, tufa and stromatolitic fragments, gastropod or bivalve shells, rock fragments, brick or other fragments. The oncolite morphology is related to that of the colonized object. The cyclical annual growth of blue-green algae is responsible for successive carbonate coats.

An important additional problem is the origin of oolites. These are situated at the deepest part of the channel, and their size suggests that they are usually

Table 2. Conceptual model of fluvial carbonates

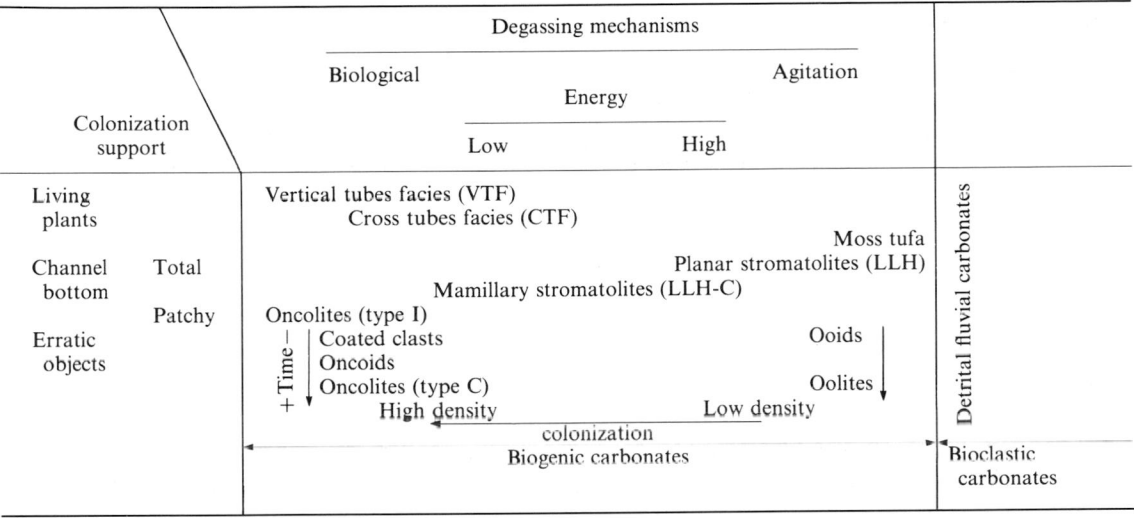

Colonization support		Degassing mechanisms			Detrital fluvial carbonates
		Biological		Agitation	
			Energy		
		Low	High		
Living plants		Vertical tubes facies (VTF)			
		Cross tubes facies (CTF)			
				Moss tufa	
Channel bottom	Total		Planar stromatolites (LLH)		
		Mamillary stromatolites (LLH-C)			
Erratic objects	Patchy	Oncolites (type I)			
		Coated clasts	Ooids		
		Oncoids			
		Oncolites (type C)	Oolites		
		High density	Low density		
		colonization			
		Biogenic carbonates			Bioclastic carbonates

(Time axis labelled −Time / +Time on the left of the oncolite block)

transported by saltation. It is not possible for high-density colonization to occur under these conditions, but precipitation can occur on the nucleus if the water is supersaturated with respect to calcite. Mechanical friction during transport partially erodes the carbonate coat. A situation that alternates between erosion and precipitation on a nucleus can be responsible for the successive laminae. A schematic diagram of this discussion is shown in Table 2.

Daylight and oxygenated waters are necessary for the growth of blue-green algae. Suspended solids (turbidity) in water decrease light penetration and hence photosynthetic activity. The dissolved oxygen is consumed by respiration and it is replenished by photosynthesis and by aeration through the surface of the water. The aeration processes are negligible in water with laminar flow. Only photosynthetic activity is an important source of oxygen in these waters. Therefore, the absence of suspended solids in water is necessary to ensure the necessary daylight and oxygen level. It can be related to geomorphological aspects: stability of the drainage network and the abundance of vegetation. We have been able to see, as a result of an artificial small addition of nutrient-rich water, with consequent oxygen depletion, the destruction of blue-green algal mats and therefore of the biogenic buildups.

The water composition reflects a closed origin of karstic springs and small streams, but we have also actually found stromatolitic structures in hard-bottom artificial channels of the river Tajo. However, the river Tajo source and tributary rivers are commonly located on Mesozoic and Tertiary carbonate materials.

All biogenic fluvial carbonates can be broken and transported by episodic flood but this fact is only clearly proved when the colonization support is living plants on channel bottoms. However, the colonization of erratic objects and their related structures makes it very difficult to interpret whether they are *in situ* or transported. Only a strong mixing with non-coated detrital clasts allows us to identify a detrital origin.

CONCLUDING REMARKS

All freshwater (lacustrine and fluvial) carbonates directly or indirectly concerned with benthonic blue-green algal activity present similar structural aspects. These structural aspects are influenced by substrate morphology. Microstructural features (arborescent and laminated) are conditioned by the association of colonizing blue-green algae and ecological factors. The geographical or palaeogeographical aspects are less important in determining structural and microstructural (and/or textural) aspects. The previous considerations may be explained by the fact that small temporary (Tertiary and Quaternary) variations of blue-green associations and their geographical distribution have not affected morphology. This is supported by the constancy of blue-green algae associations in all present-day freshwater environments.

Sequential analysis and sedimentological and struc-

Table 3. Conceptual genetic model of fluvial carbonates

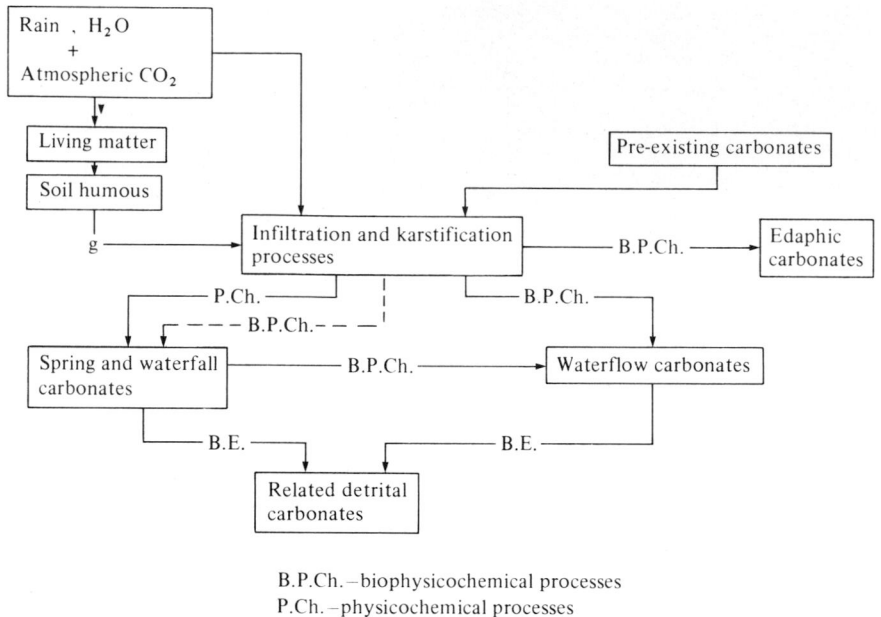

B.P.Ch.–biophysicochemical processes
P.Ch. –physicochemical processes
B.E. –breaking and erosional processes

tural aspects provided a geological tool for palaeo-geographical interpretations. Thus biogenic lacustrine sediments are related to mud carbonate flats; water-flow carbonates are related to channel morphology and are located in fluvial sequences, and waterfall carbonates present a characteristic slope lamination. Biogenic carbonates in the stratigraphical record are thus good palaeoenvironmental indicators.

The dissolution of pre-existing carbonates is related to high CO_2 pressure in soils (microbiological activity). The water percolating across the carbonate bodies can reach the surface as a spring on a hillside. The water is saturated with respect to carbonates by the high CO_2 pressure, and the fast degassing processes lead to supersaturation of the water. These degassing processes may be related with photosynthetic activity and/or with waterfall agitation. The role of plants in calcium carbonate precipitation is not resolved. We think that the photosynthetic processes might aid the degasification but their contribution is not necessary. An attempt at a conceptual genetic model is shown in Table 3.

The genetic conditions for the precipitation of flow carbonates are the following: (a) waters with a $p_{CO_2} > 10^{-3}$; (b) carbonate- and bicarbonate-rich waters (these are related to a fluvial network

superimposed on old carbonate sediments); (c) light requirement conditions necessitate shallow water with a depth less than 1 m; and (d) the absence of suspended load is also a very important condition because turbidity has an inhibiting effect on growth of blue-green algae. Therefore a near-equilibrium fluvial network is a geomorphological condition for the precipitation of fluvial carbonates.

Bioclastic carbonate deposits may be formed if biogenic carbonates are destroyed and transported by episodic floods or flows.

ACKNOWLEDGMENTS

The authors wish to acknowledge and thank Dr J. D. Collinson for his valuable suggestions as well as for his assistance in the English version of the text.

We are indebted to Drs Carballal and Gonzalez Martin for their collaboration.

REFERENCES

ANADON, P. & ZAMARREÑO, I. (1981) Paleogene nonmarine algal deposits of the Ebro Basin, northeastern Spain. In: *Phanerozoic Stromatolites* (Ed. by C.L.V. Monty), pp. 140–154. Springer-Verlag, Berlin.

BRAITHWAITE, C.J.R. (1979) Crystal textures of recent fluvial pisolites and laminated crystalline crusts in Dyfed, South Wales. *J. sedim. Petrol.* **49**, 181–194.

CALVO, J.P., GARCÍA DEL CURA, M.A. & ORDÓÑEZ, S. (1979) Edificios tobaceos en el Valle del Rio Mundo (Prov. de Albacete). *Actas de la IV Reunión del Grupo de Trabajo del Cuaternario. Banyoles*, pp. 23–32.

GEURTS, M.A. (1975) Formation de travertins postglaciaires en Belgique. In: *Colloque" Types de croutes calcaires et leur repartition régionale"* (Ed. by T. Vogt), pp. 76–79. Strasbourg.

GLAZEK, J. (1965) Recent oncolites in streams of North Vietnam and of the Polish Tatra Mts. *Roczn. pol. Tow. dendrol.* **XXXV**, 221–242.

GOLUBIC, S. (1973) The relationship between blue-green algae and carbonate deposits. In: *The Biology of Blue-green Algae* (Ed. by N.G. Carr and B.A. Whitton), pp. 434–472. Blackwell Scientific Publications, Oxford.

IRION, G. & MULLER, G. (1968) Mineralogy, petrology and chemical composition of some calcareous tufa from the Schwäbische Alb, Germany. In: *Recent Developments in Carbonate Sedimentology in Central Europe* (Ed. by G. Müller and G.M. Friedman), pp. 157–171. Springer-Verlag, Berlin.

JONES, F.G. & WILKINSON, B.H. (1978) Structure and growth lacustrine pisoliths from recent Michigan marl lakes. *J. sedim. Petrol.* **48**, 1103–1110.

LANG, J. & LUCAS, G. (1970) Contribution à l'étude de biohermes continentaux: barrages des lacs de Band-e-Amir (Afghanistan central). *Bull. Soc. géol. Fr.* (Ser. 7), **12**, 834–842.

LOGAN, B.W., REZAK, R. & GINSBURG, R.N. (1964) Classification and environment significance of algal stromatolites. *J. Geol.* **72**, 68–83.

McGANNON, D.E. (JR) (1975) Primary fluvial oolites. *J. sedim. Petrol.* **45**, 719–727.

MEGIAS, A.G., ORDÓÑEZ, S. & CALVO, J.P. (1982) Un essai de synthèse lithostratigraphique du Bassin de Madrid. *Sci. Terre* (in press).

MONTY, C.L.V. (1976) The origin and development of cryptalgal fabrics. In: *Stromatolites* (Ed. by M.R. Walter), pp. 193–249. Elsevier, Amsterdam.

MONTY, C.L.V. & MAS, J.R. (1981) Lower Cretaceous (Wealdian) blue-green algal deposits of the province of Valencia, Eastern Spain. In: *Phanerozoic Stromatolites* (Ed. by C.L.V. Monty), pp. 85–120. Springer-Verlag, Berlin.

ORDÓÑEZ, S., CARBALLAL, R. & GARCÍA DEL CURA, M.A. (1980) Carbonatos biogénicos actuales en la Cuenca del Rio Dulce (Prov. de Guadalajara). *Boll. R. Soc. Esp. Hist. Natural (Geol.)* **78**, 303–315.

ORDÓÑEZ, S. & GARCÍA DEL CURA, M.A. (1977a) Calcareous tufas associated to the Middle Terrace and Campiña Terrace of High Henares Basin (Central Spain) and their climatologic and geomorphological significance. *10th int. Congr. INQUA*, **gI**. Sec. A, 337.

ORDÓÑEZ, S. & GARCÍA DEL CURA, M.A. (1977b) Facies oncolíticas en medio continental. Aplicación al sector S.E. de la Cuenca del Duero. *Estudios geol. Inst. Invest. geol. Lucas Mallada*, **33**, 459–466.

ORDÓÑEZ, S., GARCÍA DEL CURA, M.A. & CARBALLAL, R. (1978) Stromatolites and oncolites in the modern fluvial environment. *Abstr. 10th int. Congr. Sedimentology. Jerusalem*, **2**, 489–490.

ORDÓÑEZ, S. & GONZALEZ, J.A. (1979) Formaciones tobáceas del valle del rio Tajuña entre Brihuega y Masegoso (Provincia de Guadalajara). *Estudios geol. Inst. Invest. geol. Lucas Mallada*, **35**, 205–212.

ORDÓÑEZ, S., GONZALEZ, J.A. & GARCÍA DEL CURA, M.A. (1979) Génesis y significado de las tobas de cascada de briofitas. *Actas de la IV Reunión del Grupo de Trabajo del Cuaternario, Banyoles*, pp. 171–178.

ORDÓÑEZ, S., GONZALEZ, J.A. & GARCÍA DEL CURA, M.A. (1982a) Carbonatos fluviales en la Mesa Ocaña-Tarancón. *Actas del IX Congreso Nacional de Sedimentología, Salamanca* (in press).

ORDÓÑEZ, S., GONZALEZ, J.A. & GARCÍA DEL CURA, M.A. (1981) Carbonatos fluviales paraactuales en el valle del Rio Tajuña. *Actas de la V Reunión del Grupo de Trabajo del Cuaternario, Sevilla*, pp. 280–293.

RISACHER, F. & EUGSTER, H.P. (1979) Holocene pisoliths and encrustations associated with spring-fed surface pools, Pastos Grandes, Bolivia. *Sedimentology*, **26**, 253–270.

SCHAFER, A. & STAPF, K.R.G. (1978) Permian Saar-Nahe Basin and Recent Lake Constance (Germany): two environments of lacustrine algal carbonates. In: *Modern and Ancient Lake Sediments* (Ed. by A. Matter and M.E. Tucker). *Spec. Publs int. Ass. Sediment.* **2**, 83–107. Blackwell Scientific Publications, Oxford.

THRAILKILL, J. (1968) Chemical and hydrologic factors in the excavation of limestone caves. *Bull. geol. Soc. Am.* **79**, 19–46.

Economic aspects

Spec. Publs int. Ass. Sediment. (1983) **6**, 501–515

Basin facies analysis of coal-rich Tertiary fluvial deposits, northern Powder River Basin, Montana and Wyoming

ROMEO M. FLORES

U.S. Geological Survey, Denver, Colorado 80225, U.S.A.

ABSTRACT

Facies analysis of the Palaeocene Tongue River Member of the Fort Union Formation in the northern Powder River Basin provided an understanding of the origin of associated thick coal deposits. Lithostratigraphic synthesis of the lower part of the Tongue River Member recognized coals associated with meander belt and lacustrine–floodplain lithofacies. The meander belt lithofacies is characterized by a high density of closely spaced channel sandstones and abandoned channel deposits, interspersed with subordinate levée deposits. This lithofacies grades laterally and alternates vertically with the lacustrine–floodplain lithofacies that consists of crevasse-splay and crevasse delta–lake deposits. Associated with both of these lithofacies is the backswamp lithofacies that consists of coal beds (as much as 20 m in thickness) and carbonaceous shales. The accumulation of thick coals was controlled by the following interrelated factors: localized aggradation of fluvial channels, subsidence due to basement tectonic control and differential compaction of sediments, length of time of peat accumulation, nature of the backswamp's palaeoflora, and palaeoclimate. These factors were present in an alluvial plain that was situated in an intermontane basin. The alluvial plain contained a north-eastward flowing trunk–tributary system that drained into the Cannonball Sea.

INTRODUCTION

Although the Tertiary coal deposits of the Powder River Basin in Montana and Wyoming (Fig. 1) have been mined since the 1880s, their characteristics have only recently been directly related to their depositional environments (Law, 1976; Obernyer, 1978; Flores, 1979a, b, 1981; Ethridge, Jackson & Youngberg, 1981). An attempt by Flores (1981) to characterize in detail the lithofacies of the Palaeocene Tongue River Member of the Fort Union Formation has shed some light on the fluvial settings of its coal deposits. Although specific lithofacies were described in Flores (1981), their basin-wide relationships and associations were not established. Thus this paper reconstructs the depositional history of economically important coals of the lower part of the Tongue River Member in the northern Powder River Basin. The accumulation of

these coal deposits and their relationship to fluvial environments are demonstrated by palaeogeographic reconstructions of temporally equivalent lithofacies associations. The reconstructions are performed in the northern Powder River Basin where rocks are extensively exposed along highly dissected valleys of drainage systems. Here, the rate of lateral variation among lithic types of the Tongue River Member is gradual and observable within the spacing of available outcrop.

The basin-wide relationships and associations of the lithofacies were determined from 500 closely spaced measured sections and 144 drill-holes (Culbertson, 1980; Culbertson, Gaffke & Correia, 1980; Kent, Berlage & Boucher, 1980). A network of cross-sections, constructed from 95% of the total measured sections which average 0·4 km apart, spans about 300 km. Additional cross-sections were constructed from 75%

0141-3600/83/0106-0501 $02.00

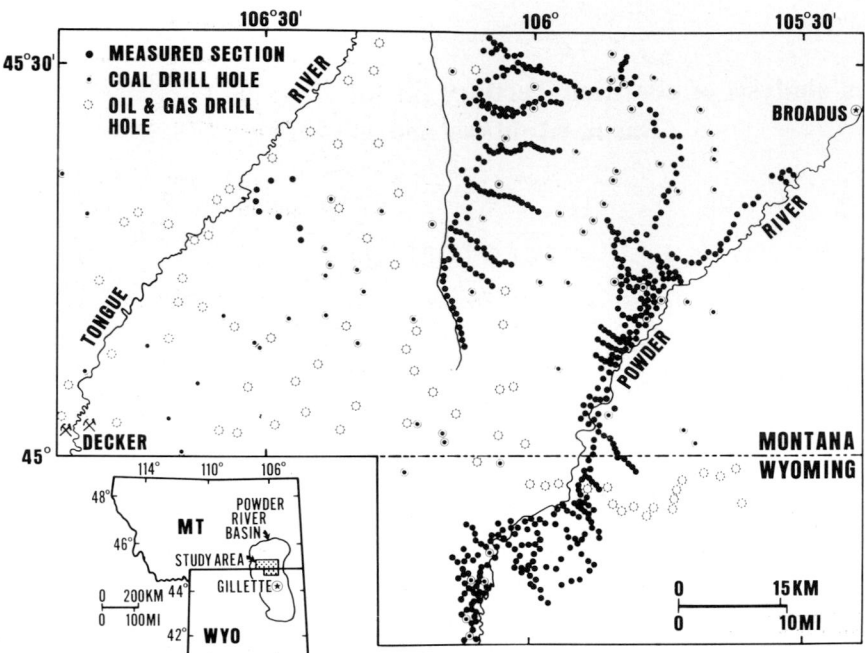

Fig. 1. Index map and distribution of measured sections and drill-holes in the study area in the northern Powder River Basin (see inset map), Montana and Wyoming.

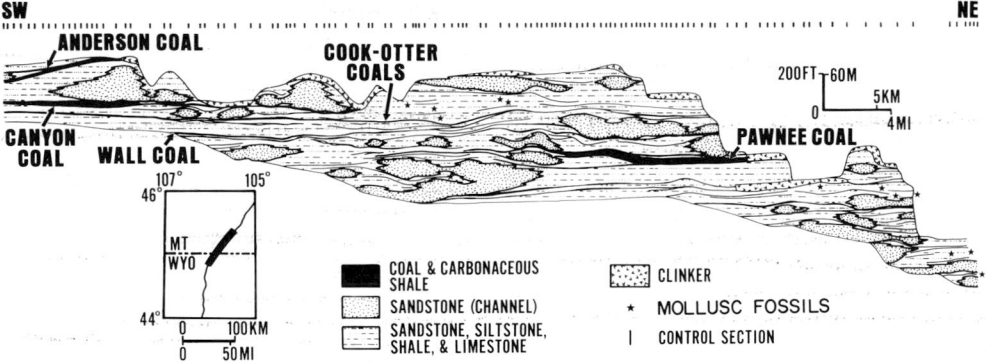

Fig. 2. Cross-section showing environmental and stratigraphic relationships of the fluvial channel sandstone, lacustrine–floodplain sandstone, siltstone, shale and limestone, and backswamp coal and carbonaceous shale beds along the Powder River (see inset map for location).

of the total drill-holes which average 3·2 km apart, spanning about 250 km. Each section and drill-hole contained key marker beds of sandstone, limestone and coal that were used for correlation. The cross-sections, portions of which are shown in Figs 2, 3 and 4, were utilized to establish regional lithofacies relationships and associations.

LITHOFACIES AND DEPOSITIONAL ENVIRONMENTS

Lithostratigraphy of the Tongue River Member has been described in detail by Canavello (1980), Lynn (1980), and Flores (1981). On the basis of the spatial arrangement of the lithic types (Figs 2, 3 and 4), two laterally coexisting major lithofacies are recognized: a

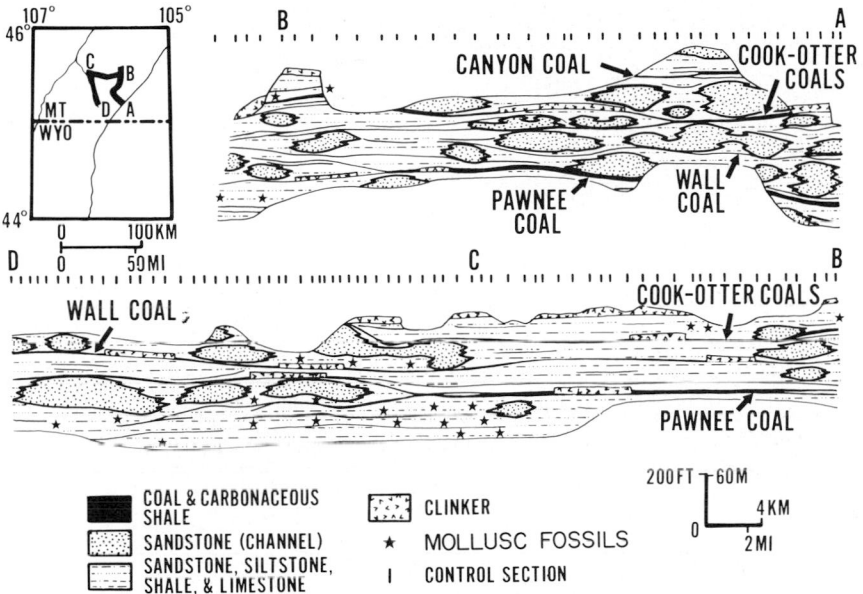

Fig. 3. Cross-section showing environmental and stratigraphic relationships of the fluvial channel sandstone, lacustrine–floodplain sandstone, siltstone, shale and limestone, and backswamp coal and carbonaceous shale beds along Bloom Creek, Pumpkin Creek and Otter Creek (see inset map for location). Modified from Lynn (1980).

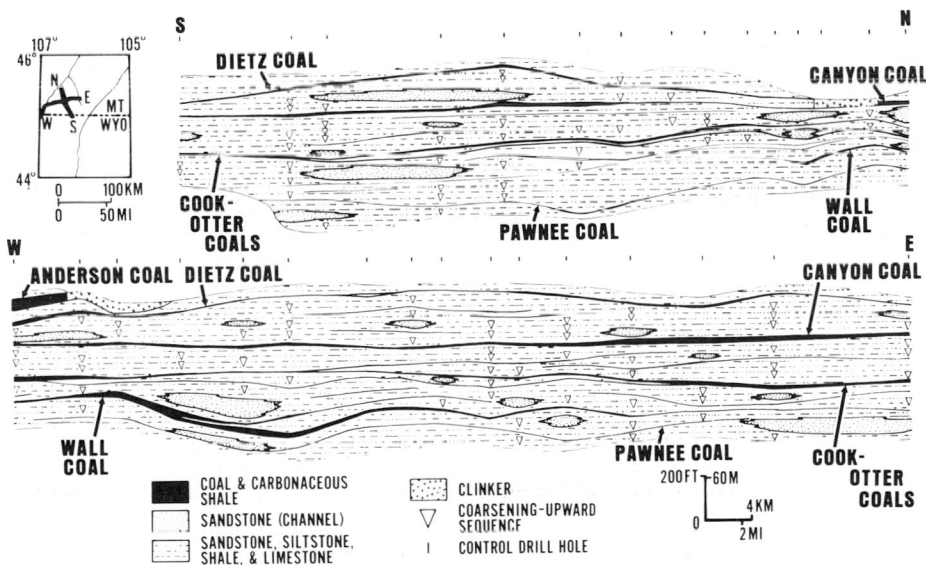

Fig. 4. Cross-section showing environmental and stratigraphic relationships of the fluvial channel sandstone, lacustrine–floodplain sandstone, siltstone, shale and limestone, and backswamp coal and carbonaceous shale beds between the Otter Creek and Tongue River (see inset map for location). Coal correlations from Culbertson (1980).

Fig. 5. Point-bar deposits separated by lateral accretion surfaces that dip from left to right in a meander channel sandstone.

meander belt and a lacustrine–floodplain lithofacies. A subordinate backswamp lithofacies, which consists of coal and carbonaceous shale, is interbedded with both of these lithofacies. The meander belt lithofacies is characterized by high concentrations of channel sandstones. The lacustrine–floodplain lithofacies is dominated by finer-grained sandstone, siltstone and shale, as well as abundant limestone and ironstone.

Meander belt lithofacies

The meander belt lithofacies (Fig. 5) consists of a series of abundant sandstone and subordinate fine-grained detrital bodies separated by erosion surfaces as well as marginal fine-grained deposits. The sandstone bodies are basally scoured, fine to coarse grained, and form fining-upward sequences. They are as much as 38·1 m thick and up to 14·5 km wide. The sandstones may consist of more than one body that include lateral accretion deposits separated by lateral accretion surfaces which dip at an average of 15°. The lower part of the sandstone bodies contain abundant large-scale festoon and subordinate tabular cross-beds locally interrupted by large convolute lamination; the upper part contains small-scale festoon cross-beds,

convolute lamination and ripple lamination. The sandstones contain lag deposits of rounded sandstone, siltstone, shale, limestone, ironstone, carbonaceous shale, coal spar and freshwater mollusc fragments. These characteristics of the sandstone bodies suggest deposition in sinuous channels that were filled with point-bar sands and whose floors were also filled with reworked deposits of lacustrine–floodplain and backswamp lithofacies.

The fine detrital deposits consist of siltstone, shale and carbonaceous shale. As shown in Fig. 6 the interbedded sequence of siltstone, shale and carbonaceous shale assumed a depositional attitude subparallel to an underlying concave erosion surface. The carbonaceous shale lies near the top of the erosional surface and overlies the siltstone and shale. The siltstone shows ripple laminations and the shale contains a few vertical burrows. These features of the deposits indicate accumulation in an abandoned channel.

Thick sequences of siltstone, shale, carbonaceous shale, ironstone and silty sandstone border channel deposits. The sandstone and siltstone display ripple drift and lenticular ripple laminations. This lithologic sequence is commonly rooted and contains tree

Fig. 6. Abandoned channel deposit (AC) consisting of siltstone, shale and carbonaceous shale. Arrows indicate erosion surface.

stumps in growth position an important feature of a levée deposit. The interbedded sequence of variable-size particles reflects the fluctuating flood-water stages.

The meander belt lithofacies, illustrated by Figs 2, 3 and 4, volumetrically comprises as much as 70% of the study interval in places. Where a high density of channel and levée deposits is present, the deposits occur in isolated, offset clusters (Fig. 2) and the deposits in each cluster are arranged either *en echelon* or vertically stacked (Fig. 5). Where the deposits are stacked, they represent deposition in a meander belt localized in the alluvial plain for some length of time prior to avulsion. The avulsion of the meander belt to a new position in the alluvial plain probably resulted from a response to a more favourable gradient created by depositional palaeotopography and differential compaction. That the channel deposits were laid down in freely meandering channels is indicated by their *en echelon* arrangement, resulting from channel migration by avulsion due to neck or chute cutoff.

Lacustrine–floodplain lithofacies

The lacustrine–floodplain lithofacies consists of sequences of interbedded sandstone, siltstone, shale,

limestone and ironstone. The sequences of these rocks, based on vertical grain-size variation, nature of fossil content and rock association, can be grouped into two types. One type is a coarsening-upward sequence consisting of shale and siltstone in the lower part and tabular sandstone that is locally replaced by a lenticular sandstone in the upper part. The shale and siltstone interbedded with a few ironstone layers display abundant vertical and horizontal burrows. The siltstone shows common lenticular ripple lamination. The tabular sandstone is a very fine to medium grained, coarsening-upward deposit and displays a gradational or sharp basal contact. Internal structures of the tabular sandstone are dominated by various combinations of ripple drift lamination, ripple lamination, parallel lamination, small-scale festoon cross-bedding and convolute lamination. The sandstone also contains a few vertical burrows and degassing-like structures. The lenticular sandstones that laterally replace the tabular sandstones show erosional bases, festoon cross-bedding, and ripple lamination. This sandstone passes laterally into small, isolated channel deposits in a few places. The coarsening-upward sequence averages 4·6 m in thickness and is as much as 12·9 km in lateral extent parallel to channel deposits. The sequence occurs either as a single or a stack of coarsening-upward sequences.

The second type of sequence consists of interbedded silty sandstone, siltstone and shale interrupted by limestone. The silty sandstone, which ranges from a few millimetres to 2 m in thickness, is uniformly distributed throughout the sequence. The sandstone exhibits coarsening-upward as well as homogeneous grain sizes from bottom to top. Internal structures of the sandstone consist of parallel lamination marked by concentrations of finely divided plant fragments and ripple lamination. The sandstone may contain vertical escape structures of freshwater bivalves. The siltstone, shale and limestone interbedded with the sandstone contain abundant freshwater bivalve and gastropod fossils. The siltstone, which is parallel laminated and lenticular bedded, commonly contains vertical and horizontal burrows as does the shale. Generally, the limestone is micritic, thin and lenticular. A few limestone beds display coquina-like concentrations of freshwater mollusc fossils. The fossiliferous sequence is up to 10 m in thickness and ranges from a few hundred metres to 20 km in lateral extent. This

sequence grades laterally and vertically into the coarsening-upward sequence.

The characteristics of these lithofacies sequences and their associations suggest that the first type of sequence represents a crevasse splay–channel deposit and the second type of sequence is a crevasse delta–lake deposit. The overall crevasse splay–channel deposit probably reflects deposition during flood conditions where the fluvial channels at bankfull stage breached levées and deposited fan-shaped sediments into the floodplain. The abundant occurrence of ripple laminations in combination with festoon cross-beds suggests rapid sedimentation by high-velocity flow of floodwaters into the floodplain probably generated during flash floods. Overlapping and local merging of crevasse splays produced continuous sheet-like deposits particularly near channels. Continuous progradation of crevasse splays farther into the floodplain led to their debouching into lakes. Pulses of sediment discharge into the lakes from crevasses during flash floods spread sands and silts and disturbed filter-feeding fauna at the botton. Suspended sediments forming shales were deposited in the lakes farthest from crevasse delta point sources. The limestone

Fig. 7. Pawnee coal (11 m thick) and a 1·52 m thick rider coal split by crevasse channel sandstone (CS) and associated crevasse-splay siltstone and shale.

probably represents precipitation of carbonates in the lakes during cessation of detrital influx or non-flood condition, and/or in parts of the lakes not subjected to detrital sedimentation. Progradation of crevasse splays and filling of the lakes by crevasse deltas may have provided another mechanism of avulsion by increased proportion of discharge along one of the crevasse channels. Figs 2, 3 and 4 show localized as well as widespread distribution of the fossiliferous crevasse delta-lake deposit, suggesting that lacustrine processes in the floodplain occurred in various sizes, perhaps controlled by local or regional subsidence and compaction of underlying sediments.

Backswamp lithofacies

The backswamp lithofacies (Fig. 7) occurs as interbeds in both meander belt and lacustrine–floodplain lithofacies. The major deposits of the backswamp lithofacies are coal and carbonaceous shale that consists of clay and silt-size particles mixed with varying proportions of organic material. The coal beds are underlain, overlain and grade laterally into and interbed with the carbonaceous shale beds. A minor amount of organic-rich siltstone and sandy siltstone occurs as partings in the coal beds. The coal beds range from a few centimetres to as much as 20 m in thickness. The major coal beds attain thicknesses of 12·2 m for the Pawnee coal, 20 m for the Wall coal, 7·6 m for the Cook coal and 10·1 m for the Canyon coal.

An investigation of the palynology of a few beds of the Canyon coal by Tschudy (USGS, personal communication) indicates that it is composed of remains of a combination of woody and herbaceous plants. Table 1 shows the varieties of palaeoflora and their relative abundances in the coal beds based on spore and pollen analysis. Although these data may not reflect statistically significant representations of the overall palaeoflora of the backswamps in the Powder River Basin, they nevertheless provide a brief survey of the frequency distribution of the woody plants (75%) versus the herbaceous plants (25%).

As illustrated in Figs 2 and 3, the individual coal beds can be traced laterally along outcrop for as much as 20 km before splitting or merging. The lateral continuity of coal beds breaks down on passage into the meander belt lithofacies. The discontinuity of coal beds associated with this lithofacies resulted from their erosion by channels, splitting by overbank detritus, and thinning over the meander belt deposits. In a few cases, as shown in Fig. 3 (cross-section D–B), thick coal beds, such as the merged Otter and Cook beds, are immediately above meander belt deposits. This situation suggests that abandoned meander belt deposits make good swamp platforms. However, the generally poor peat-forming conditions in swamps marginal to the meander belts simulate those present in well-drained backswamps described by Coleman (1966).

In contrast, coal beds in the study interval dominated by the lacustrine–floodplain lithofacies are more laterally continuous than those associated with the meander belt lithofacies as illustrated in Fig. 4. This cross-section, using subsurface coal-bed correlations by Culbertson (1980), indicates occurrence of laterally extensive coals. The coal beds associated with the lacustrine–floodplain lithofacies are commonly split by crevasse–splay deposits (Fig. 7) as demonstrated by the Pawnee, Otter and Cook coal beds. As illustrated in Figs 2 and 3, thick coal beds such as the Pawnee and Wall occur in association with lacustrine–floodplain lithofacies near and distant from channels, respectively. In both cases, however, the coals are thickest at locations distant from the channel deposits; these sites represent favourable conditions for prolonged peat accumulation. These sites of favourable coal accumulation are similar to the poorly drained backswamps described by Coleman (1966).

PALAEOGEOGRAPHIC RECONSTRUCTIONS

Reconstructions of the palaeogeography of the study area are made possible by the three-dimensional control of the study interval. The palaeogeographic reconstruction is based on the spatial distributions of lithic types among the lithofacies. Palaeogeographic reconstructions are illustrated in Figs 8, 9, 10 and 11. Each palaeogeographic map was constructed from temporally equivalent lithofacies immediately underlain by and/or slightly coincident to the thick Pawnee, Wall, Cook and Canyon coal beds.

Figure 8 illustrates the palaeogeographic characteristics of the lithofacies associations during the time preceding accumulation of the Pawnee coal. At this time the meander belt constructed by the major fluvial drainage system occupied the central part of the study area. The meander belt was built by deposits of active and abandoned courses of the fluvial channels and associated levée deposits which were oriented in a northerly direction. At the western part of the study area, the alluvial plain was aggraded by a subordinate

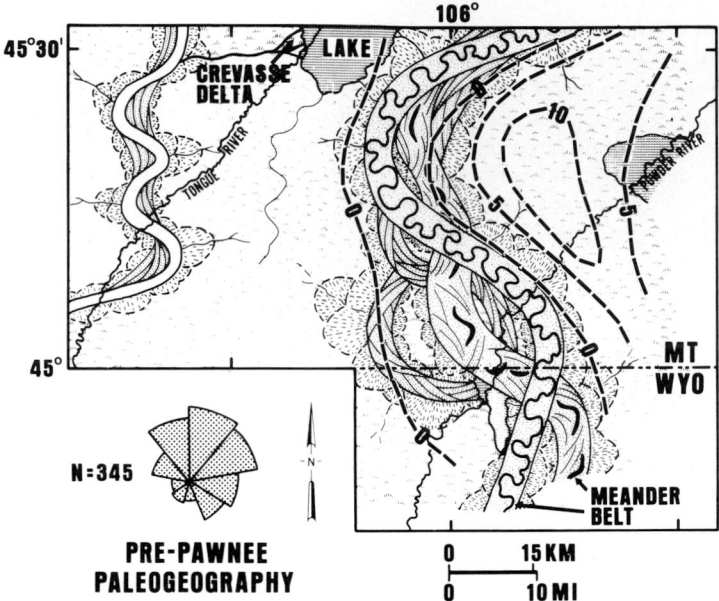

Fig. 8. Pre-Pawnee coal palaeogeographic map. The rose diagram of the cross-beds was constructed from the fluvial channel sandstones deposited at the eastern part of the study area. The isopach of the Pawnee coal bed is displayed in metres by the dashed line. N = total number of observations.

fluvial channel system that served as a tributary of the trunk stream to the east. The floodplain between these two drainage systems probably contained a well-drained backswamp at this time. However, the floodplain east of the trunk stream formed a poorly drained backswamp that served as a setting for the initial accumulation of the Pawnee coal. This floodplain probably remained a topographic low because of the high compactibility of the underlying lacustrine–floodplain sediments accentuated by the low compactibility of adjoining stacked channel deposits. This floodplain maintained through time a poorly drained backswamp that accumulated vegetable debris and partly encroached over the meander belt deposit. The isopach of the Pawnee coal bed, which is superimposed on the palaeogeography in Fig. 8, shows the thickest part of the bed centrally located in the floodplain. The direction of elongation of uniformly thick coal is north-west to south-east, which is subparallel to the length of the meander belt deposits.

Figure 9 shows the palaeogeographic map of the area preceding the accumulation of the Wall coal. At this time, the meander belt of the fluvial channel shifted eastward directly over the Pawnee backswamp. The differential subsidence during Pawnee deposition controlled the site of avulsion of the succeeding channel complex. These channels were oriented in a

northerly direction. The distal part of the lacustrine–floodplain environment to the west, formerly occupied by a tributary channel system, now formed a poorly drained backswamp in which the Wall coal initially accumulated. The abandoned secondary meander belt deposits may have served as a positive platform that supported initial growth of woody vegetation in the backswamp. As the backswamp remained a low-lying area enhanced by differential compaction of the underlying, generally fine-grained sediments and the adjoining stacks of coarser-grained sediments of the trunk stream, poor drainage conditions and accompanying vegetal growth were maintained. The pre-Wall meander belt to the east was eventually partly overrun by the backswamp. The isopach of the Wall coal is superimposed on the palaeogeography. The direction of elongation of uniformly thick coal is in a north–south direction, and is subparallel to the length of the fluvial channel and associated levée deposits of the meander belt to the east.

The palaeogeography previous to deposition of the Cook coal is illustrated in Fig. 10, and shows localization of north–south oriented meander belts of fluvial channels over the same site as in the belt during pre-Wall coal time. The vertical stacking of fluvial channels was probably in part a result of subsidence by differential compaction of sediments and basement

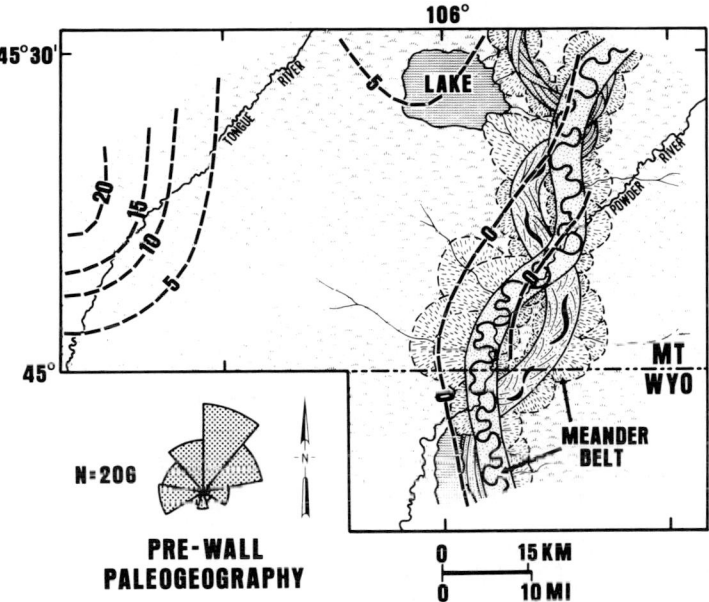

Fig. 9. Pre-Wall coal palaeogeographic map. The rose diagram of the cross-beds was constructed from the fluvial channel sandstones deposited at the eastern part of the study area. The isopach of the Wall coal bed is shown in metres by the dashed line. N — total number of observations.

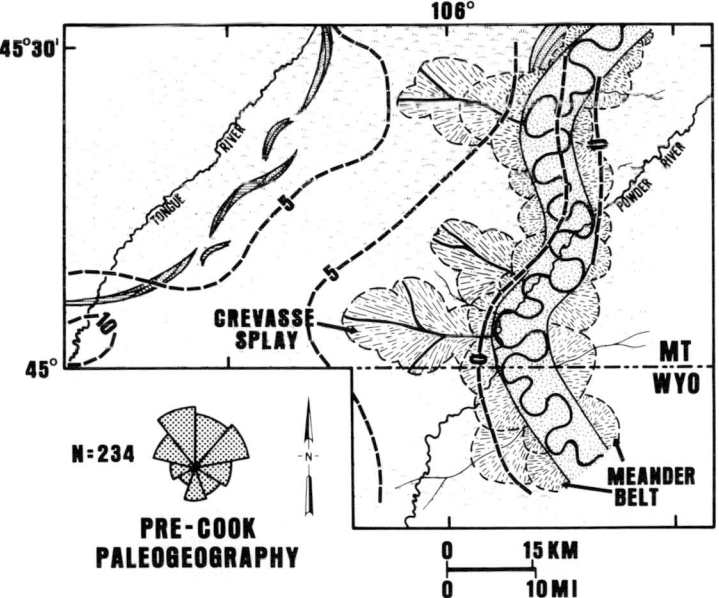

Fig. 10. Pre-Cook coal palaeogeographic map. The rose diagram of the cross-beds was constructed from the fluvial channel sandstones deposited at the eastern part of the study area. The isopach of the Cook coal bed is displayed in metres by the dashed line. N = total number of observations.

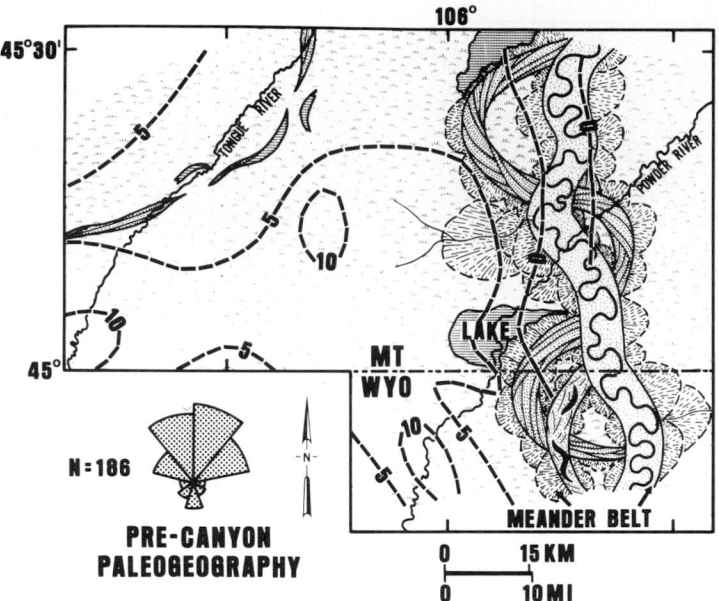

Fig. 11. Pre-Canyon coal palaeogeographic map. The rose diagram of the cross-beds was constructed from the channel sandstones deposited at the eastern part of the study area. The isopach of the Canyon coal bed is shown by the dashed line (in metres). N = total number of observations.

tectonic control along this zone of meander belt deposits. The floodplain to the west of the meander belt, which was previously aggraded by a tributary channel system, formed a poorly drained backswamp in which the Cook coal initially accumulated. The coal is thickest where it merged with the underlying Otter coal at the distal parts of the floodplain. The Cook coal bed is split by crevasse splay-deposits that originated from the trunk stream to the east, which was eventually partly encroached by the backswamp. The isopach of the Cook coal bed is superimposed on the palaeogeography. The direction of elongation of uniformly thick coal is in a north-east to south-west direction, which is subparallel to the length of the slightly contemporaneous trunk stream to the east.

The characteristics of the palaeogeography previous to deposition of the Canyon coal are shown in Fig. 11. Fluvial channel and overbank deposition along the meander belt to the east continued and was partly coeval to a tributary system to the west, which was in the process of abandonment during pre-Canyon time. The fluvial channels along the meander belt to the east are oriented in a northerly direction. The floodplain to the west of this meander belt formed a poorly drained backswamp in which the Canyon coal initially

accumulated. Through time, the backswamp partly prograded over the meander belt deposits as fluvial channel avulsion was directed into a new site on the alluvial plain. The isopach of the Canyon coal (Fig. 11) shows direction of elongation of uniformly thick coal along a north-west to south-east orientation subparallel to the length of the meander belt.

The areas between zero isopachs of the coals on each map represent zones where the coal bed locally grades laterally into carbonaceous shale and/or splits into numerous thin coal beds as a result of merging with the contemporaneous channel complex and the topography effect of an earlier channel complex.

These palaeogeographic relationships are summarized in Fig. 12, which illustrates the sequences of development of meander belts, coeval lacustrine–floodplains and associated backswamps of the alluvial plain. The sequence of events from A to D clearly shows localization of the trunk stream channels along the meander belt at the eastern part of the study area from the pre-Pawnee time to the top of the Canyon coal, which includes an interval of approximately 183 m. This process of localization of fluvial channels enhanced passive sedimentation in the floodplain to the west, which was highlighted by successive accumulations of thick Wall, Otter–Cook and Canyon

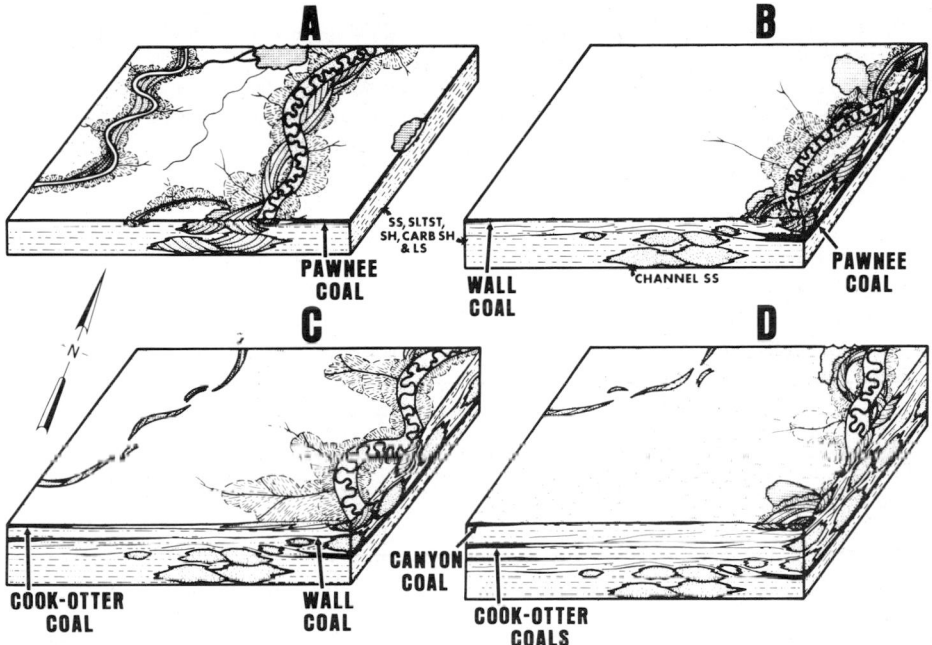

Fig. 12. Schematic block diagrams showing sequences of meander belt, lacustrine–floodplain and backswamp deposition during the pre-Pawnee (A), pre-Wall (B), pre-Cook (C) and pre-Canyon (D) times.

coals. Under these conditions, areas along the floodplain that were not reached by overbank-crevasse sediments probably permitted continuous accumulation of these successive coals to form coalesced beds of the order of magnitude of the Dietz-Anderson coal in the Decker Mines, Montana (Sholes & Cole, 1981), and the Wyodak-Anderson coal in the Gillette coal field, Wyoming (Law, 1976). The general north–south axis of this floodplain partly controlled the patterns of coal distribution and directions of elongation, which are along north-west to south-east, north to south, and north-east to south-west orientations.

The directions of elongation of the thick coals and vertical stacking, clustering and northerly orientation of the meander belt deposits in the study area probably were influenced by basement-controlled fracture zones mapped as lineaments (R. W. Marrs and G. L. Raines, personal communication) shown in Fig. 13. The coal elongations, particularly along the north-west to south-east and north-east to south-west directions, may have been partly controlled by recurrent subsidence along parallel-oriented lineaments throughout the Tongue River time. Abnormally thick peat may have been generated in a slowly subsiding floodplain. The lineament (PR) that trends N. 35° E. from west of Casper, Wyoming, to the west

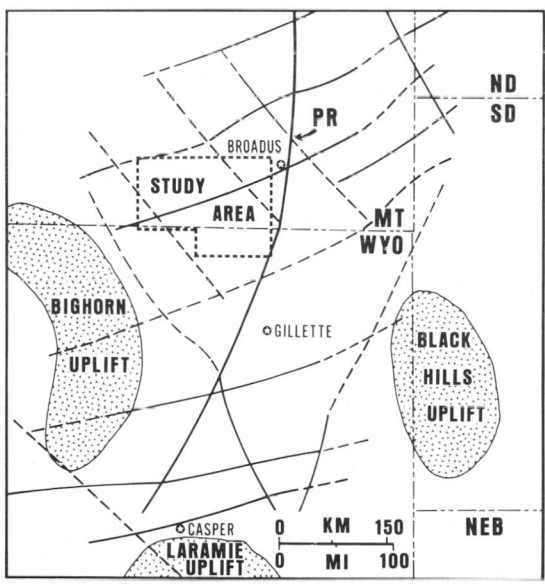

Fig. 13. Structural lineaments in the Powder River Basin as mapped by R.W. Marrs and G.L. Raines (personal communication). Lineament PR, which runs adjacent to the Powder River, represents the major linear zone of high concentration of linear fractures immediately east of the study area.

of Biddle, Montana, and due north from east of Broadus to Terry, Montana probably controlled the clustering and stacking of meander belts during the Tongue River time. During this time the meander belts developed immediately west and followed north-easterly courses parallel to the lineament. These lineaments in the study area are related to zones of weakness along basement-block boundaries that were recognized by Weimer (1961), McKee (1967) and Slack (1981) as contributing to stratigraphic variation in thickness, facies distribution, and alteration in palaeoshoreline orientations in Permian through Cretaceous rocks of the Powder River Basin.

SUMMARY OF ORIGIN OF THE COALS

The extraordinary thickness of coals, as much as 20 m in the study area and up to 61 m in adjacent areas, is commonplace in the Powder River Basin. Although these coals were accumulated in poorly drained backswamps of fluvial systems, their unusual thickness is an enigma. Obernyer (1978) in a study of the Lake-de-Smet coal, which is 61 m thick in the western part of the Powder River Basin, suggested accumulation for an unusually long time. Assuming that the rate of peat accumulation is 0·15 m/century and a compaction ratio of 1·52 m of peat for 0·3 m of coal, Obernyer suggested that at this rate of accumulation and compaction the life span of the Lake-de-Smet backswamp was as much as 200,000 years.

On the whole, the alluvial-plain setting for the accumulation of thick coals such as those in the Tongue River Member seems unlikely, given the unstable conditions inherent to the environment. Frequent interruptions of the backswamps by crevasse splay and overbank sediments, and frequent avulsion of meander belts appear to produce unfavourable conditions for prolonged and continued accumulation of thick peat. Although all these fluvial processes created unstable conditions in the alluvial plain, the fact remains that associated coals are thick in the Tongue River Member. Therefore, the following combination of interrelated factors is proposed to have controlled the accumulation of the thick coals: (1) localized aggradation of fluvial channels, (2) subsidence due to basement tectonic control and differential compaction of sediments, (3) length of time of peat accumulation, (4) nature of the backswamp's palaeoflora and (5) palaeoclimate.

The localized aggradation of fluvial channels in the Tongue River Member is indicated by clustering of meander belt deposits in the interval between the Pawnee and Canyon coal beds in the study area. Prior to this localized aggradation, similar clustering of meander belt deposits to the east occurred in the stratigraphic interval below the Pawnee coal (Fig. 2). The concentration of active deposition within a narrow zone in the alluvial plain created areas in the poorly drained backswamps that were uninterrupted by crevasse and overbank sedimentation for an extended period of time. The concentration of meander belt deposits along the alluvial plain at any one place and time may have been controlled in part by prolonged subsidence due to basement zones of weakness.

The difference in compactibility of the generally coarse-grained meander belt lithofacies and adjoining fine-grained lacustrine–floodplain lithofacies created positive and negative topographic sites in the alluvial plain. The low compactibility of the meander belt lithofacies transformed its site into a relatively higher topography than the adjoining site of highly compactible lacustrine–floodplain lithofacies and associated peat deposits. The low-lying areas, where not subjected to periodic influx of crevasse splay and overbank sediments, formed poorly drained backswamps maintaining stagnant bodies of water.

The accumulation of thick peats in the poorly drained backswamps is a function of time. This condition is illustrated in Fig. 14 in which peat initially accumulated in the poorly drained backswamp (stage 1) that is formed in the floodplain away from the direction of avulsion. Accumulation of thicker peat (stage 2) in the poorly drained backswamp was sustained as active-overbank and crevasse-splay deposition continued to be diverted into the well-drained backswamp at the opposite side of the meander belt. As the crevasse splay prograded, debouched and filled up a lake (stage 3), an increased discharge along one of the conduits of the crevasse delta may have initiated avulsion, and thus diverted active detrital sedimentation farther into the well-drained backswamp. This process permitted generation of additional peat in the poorly drained backswamp. Finally, as more peat was generated in the poorly drained backswamp (stage 4), early-formed peat was compacted by the weight of later-formed peat and continued subsidence of the backswamp permitted establishment of internal drainage.

The generation of thick coals in the poorly drained backswamps is controlled also by the nature of the palaeoflora. In order to sustain, for a long period of time, the generation and regeneration of peat in the

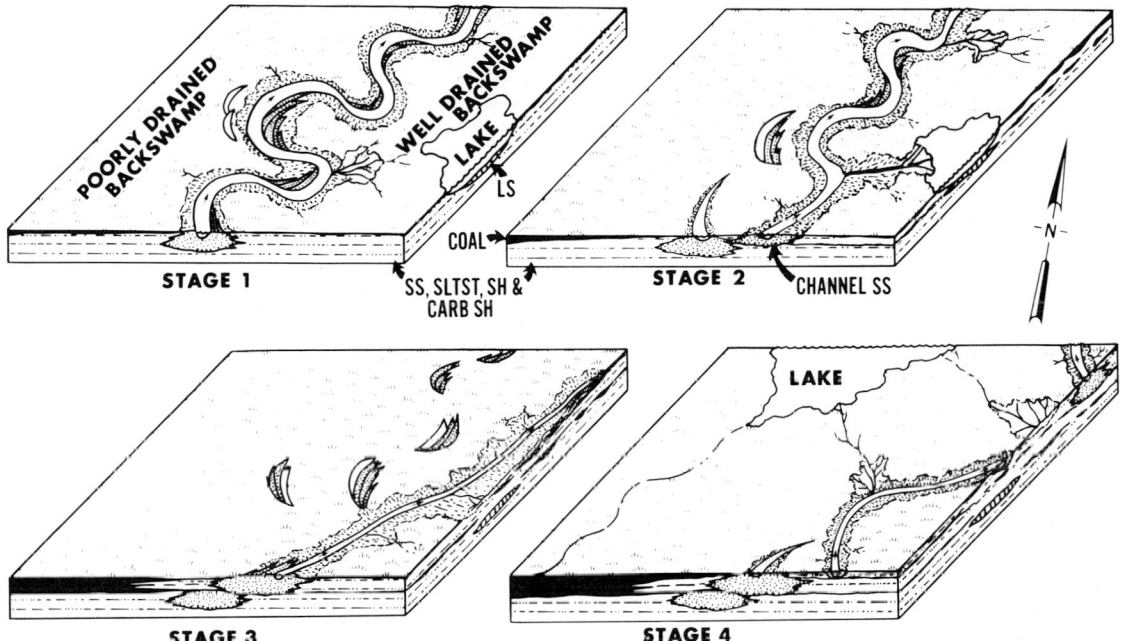

Fig. 14. Schematic block diagrams of sequence of events (stages 1, 2, 3 and 4) showing accumulation of thick peat in a poorly drained backswamp for a long period of time.

poorly drained backswamp, Flores (1981) proposed that the peat bog model of Moore & Bellamy (1976) best describes this type of accumulation in the Powder River Basin. This dome-shaped peat-bog model invoked generation of peat by woody plants in hummocks and by herbaceous plants in hollows. These various palaeofloras (Table 1) were probably distributed alternately within the poorly drained backswamps. It is suggested that the hummocks formed larger palaeofloral sites than those of the hollows. Vegetation, in the form of woody plants,

Table 1. Percentage frequency distribution of palaeoflora based on analysis of pollen and spores of the Canyon coal

Palaeoflora pollen and spores	%
Tree pollen (including shrubs)	39·5
Taxodiaceous pollen	1·5
Bisaccate conifer pollen	22·0
Fern spores and monosulcate pollen	8·5
Moss spores	7·5
Sphagnum	1·5
Other*	19·5

* Other includes angiosperm pollen grains that may represent pollens of trees and herbaceous plants whose affinity to modern plants is not known.

probably became more homogeneous towards the overbank and abandoned meander belt areas. In the model, a layer of peat is generated in both hummocks and hollows that alternated vertically in the next regeneration of peat such that old hummocks were replaced vertically upward by hollows and vice versa. These vertical alternations of palaeoflora continued through time, and regeneration of peat was maintained by perching of ground-water tables through capillarity or by sustaining levels of ground-water tables due to subsidence by compaction and/or tectonic control. The regeneration of peat ceased when the rise of the ground-water table could not keep pace with organic accumulation.

Palaeobotanical investigations by Leo Hickey (1977, 1980) of Palaeocene and Eocene deposits in the Bighorn and Williston Basins immediately west and north-east, respectively, of the Powder River Basin show mean annual temperature of these basins from 10° to 15 °C. Based on this indirect evidence, it is presumed that similar climate conditions existed in the Powder River Basin. The temperate climate of the Powder River Basin and adjacent areas during the Palaeocene may have been an ideal condition for the development of peat bog swamps.

In summary, all the preceding factors that

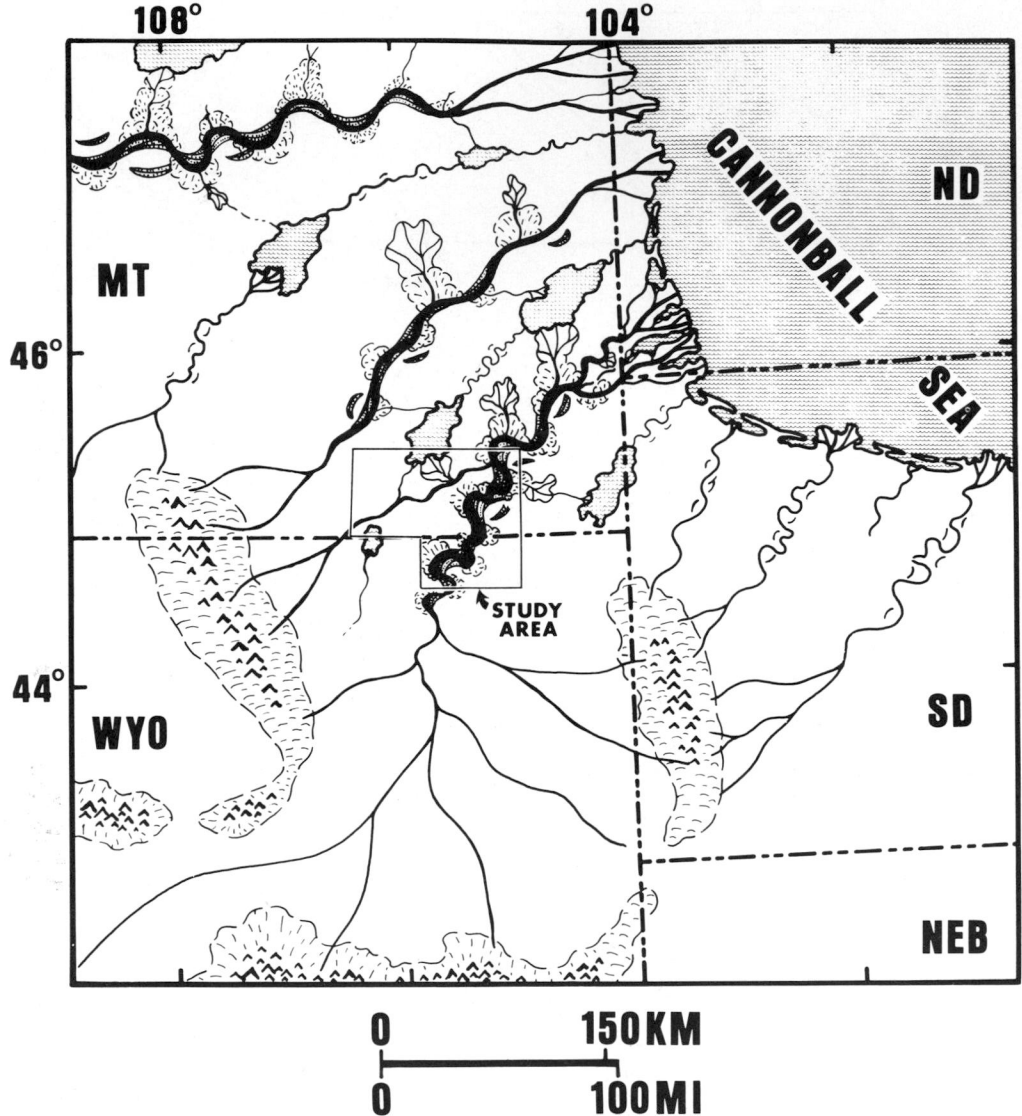

Fig. 15. Generalized palaeoenvironmental reconstruction of the alluvial plain and drainage pattern in the intermontane Powder River Basin and adjoining coastal plain and marine environments in the Williston Basin.

controlled accumulation of thick peats may have been present in an alluvial setting in the Powder River Basin not unlike the palaeoenvironmental reconstruction shown in Fig. 15. The trunk stream and associated tributaries flowed to the north-north-east and eventually drained into the Palaeocene Cannonball Sea (Cvancara, 1976) in North-west South Dakota and western North Dakota. The Tongue River Member in western North Dakota was probably deposited in a lower delta-plain environment (Jacob, 1976).

REFERENCES

CANAVELLO, D.A. (1980) *Geology of some Paleocene coal-bearing strata of the Powder River Basin, Wyoming and Montana.* Unpublished M.S. Thesis. North Carolina State University, Raleigh. 63 pp.

COLEMAN, J.M. (1966) Ecological changes in a massive fresh-water clay sequence. *Trans. Gulf-Cst Ass. geol. Socs* **XVI**, 159–174.

CULBERTSON, W.C. (1980) Diagrams showing correlation of Paleocene coal beds underlying the Birney 1° × 1/2° quadrangle, Big Horn, Rosebud, and Powder River Counties. *Open-File Rep. U.S. geol. Surv.* **80-666**.

CULBERTSON, W.C., GAFFKE, T.M. & CORREIA, G. (1980) Coal exploratory holes drilled in 1978–79 in Birney 1° × 1/2° quadrangle, Big Horn and Rosebud Counties, Montana, for coal beds in the Tongue River Member of the Paleocene Fort Union Formation. *Open-File Rep. U.S. geol. Surv.* **80-55.**

CVANCARA, A.M. (1976) Geology of the Cannonball Formation (Paleocene) in the Williston Basin, with reference to uranium potential. *Rep. Invest. North Dakota geol. Surv.* **57,** 21 pp.

ETHRIDGE, F.G., JACKSON, T.J. & YOUNGBERG, A.D. (1981) Floodbasin sequence of a fine-grained meander belt subsystem—the coal-bearing lower Wasatch and upper Fort Union Formations, southern Powder River Basin, Wyoming. In: *Recent and Ancient Nonmarine Depositional Environments; models for exploration* (Ed. by F.G. Ethridge and R.M. Flores). *Spec. Publs Soc. econ. Paleont. Miner.*, Tulsa, **31,** 191–289.

FLORES, R.M. (1979a) Coal depositional models in some Tertiary and Cretaceous coal fields in the U.S. Western Interior. *Org. Geochem.* **1,** 225–237.

FLORES, R.M. (1979b) Coal variations in fluvial deposition of Paleocene Tongue River Member of the Fort Union Formation, Powder River area, Wyoming and Montana [Abs]. *Bull. Am. Ass. Petrol. Geol.* **63,** 826.

FLORES, R.M. (1981) Coal deposition in fluvial paleoenvironments of the Paleocene Tongue River Member of the Fort Union Formation, Powder River area, Powder River Basin, Wyoming and Montana. In: *Recent and Ancient Nonmarine Depositional Environments: models for exploration* (Ed. by F.G. Ethridge and R.M. Flores). *Spec. Publs Soc. econ. Paleont. Miner.*, Tulsa, **31,** 169–190.

HICKEY, L.J. (1977) Stratigraphy and paleobotany of the Golden Valley Formation (Early Tertiary) of western North Dakota. *Mem. geol. Soc. Am.* **150,** 183 pp.

HICKEY, L.J. (1980) Paleocene stratigraphy and flora of the Clark's Fork Basin. In: *Early Cenozoic Paleontology and Stratigraphy of the Bighorn Basin, Wyoming* (Ed. by P.D. Gingerich). *Univ. Mich. Pap. Paleont.* **24,** 33–49.

JACOB, A.F. (1976) Geology of the upper part of the Fort Union Group (Paleocene), Williston Basin, with reference to uranium. *Rep. Invest. North Dakota geol. Surv.* **58,** 49 pp.

KENT, B.H., BERLAGE, L.J. & BOUCHER, E.M. (1980) Stratigraphic framework of coal beds underlying the western part of the Recluse 1° × 1/2° quadrangle, Campbell County, Wyoming. *U.S. geol. Surv. coal Invest. Map* **C–81C.**

LAW, B.E. (1976) Large scale compaction structures in the coal-bearing Fort Union and Wasatch Formations, northeast Powder River Basin, Wyoming. *28th Ann. Field Conf., 1976, Gdbk, Wyo. geol. Soc.* pp. 221–229.

LYNN, L.R. (Jr) (1980) *Stratigraphic framework of some Tertiary coal-bearing alluvial strata, Powder River Basin, Wyoming and Montana.* Unpublished M.S. Thesis. North Carolina State University, Raleigh. 94 pp.

MCKEE, E.D. (1967) Paleotectonic investigations of the Permian system in the United States. *Prof. Pap. U.S. geol. Surv.* **515,** 271 pp.

MOORE, P.D. & BELLAMY, P.J. (1976) *Peatlands.* Springer-Verlag, New York. 221 pp.

OBERNYER, S. (1978) Basin-margin depositonal environments of the Wasatch Formation in the Buffalo-Lake de Smet area, Johnson County, Wyoming. In: *1977 Proceedings of the Symposium on the Geology of Rocky Mountain Coal* (Ed. by H.E. Hodgson). *Res. Ser. Colo. geol. Surv.* **4,** 49–65.

SHOLES, M.A. & COLE, G.A. (1981) Depositional history and correlation problems of the Anderson-Dietz coal zone, southeastern Montana. *Mount. Geol.* **18,** 35–45.

SLACK, P.B. (1981) Paleotectonics and hydrocarbon accumulation, Powder River Basin, Wyoming. *Bull. Am. Ass. Petrol. Geol.* **65,** 730–743.

WEIMER, R.J. (1961) Uppermost Cretaceous rocks in central and southern Wyoming. *16th Ann. Field Conf. 1961, Gdbk, Wyo. geol. Ass.* pp. 17–28.

Spec. Publs int. Ass. Sediment. (1983) **6**, 517–532

Fluvial deposits and hydrocarbon accumulations: examples from the Lloydminster area, Canada

PETER E. PUTNAM

Husky Oil Operations Ltd, P.O. Box 6525, Postal Station 'D' Calgary, Alberta T 2P 3G7, Canada

ABSTRACT

Thick, multi-storied, narrow, channel sandstones are found over a large area of west-central Saskatchewan and east-central Alberta within the upper part of the lower Cretaceous Mannville Group (Albian). The channel system has known areal dimensions of several thousand km² and is interpreted as representing deposition within an anastomosed channel complex. Between the channel sandstones are found thinner sheet-like sandstones which are interbedded with shales, mudstones, siltstones, and coals.

Both channel and inter-channel sandstones are important heavy oil and gas reservoirs. The spatial distribution of hydrocarbons and oil gravities can be directly related to the depositional facies. When oil-filled, the discontinuous inter-channel sheet sandstones usually contain lighter oils relative to channel sandstone reservoirs. The channel sandstones form good reservoirs only where the channel trend corresponds with localized structural highs or when channel trend is perpendicular to regional dip.

The trapping mechanism for hydrocarbons is mainly stratigraphically controlled. Sandstone pinchouts and shale-filled channels can act as up-dip permeability barriers to hydrocarbon migration. Structural traps caused mainly by differential compaction are also present. Due to the regionally continuous nature of the channel sandstones, they commonly act as aquifers. Consequently, the oil may be flushed away or else it becomes highly degraded. The interchannel sandstones are regionally discontinuous bodies. Consequently, the oil found within them is not so highly degraded.

The recognition of some fluvial lithofacies in the subsurface can be problematic. For example, distinguishing siltstone/shale breccias, a potential hydrocarbon reservoir, from *in situ* siltstone/shales, a potential hydrocarbon trap, on the basis of log responses alone is a difficult exercise.

INTRODUCTION

Clastic sedimentologists throughout the world are commonly informed that their work has relevance to the petroleum industry. However, with respect to fluvial sediments, it is rare to find published examples which demonstrate the direct relationship between depositional facies and hydrocarbon occurrence (Flores & Ethridge, 1981).

This work attempts to outline the relationship between hydrocarbon accumulations and fluvial facies within the subsurface lower Cretaceous upper Mannville Group (Albian) of the Lloydminster area of Alberta and Saskatchewan (Fig. 1). The major aim of

0141-3600/83/0106-0517 $02.00

this work is to show that the hydrocarbon accumulations are strongly controlled by the depositional facies. The primary data base for the present work consists of several thousand electric well logs and over 100 cores. Other data from micropalaeontology, palynology, petrography, organic geochemistry, geophysics, and hydrodynamics were also incorporated into the conclusions.

GENERAL GEOLOGY

The stratigraphic interval under discussion is the upper Mannville Group (Albian) of Vigrass (1977), which is the uppermost unit of the lower Cretaceous Mannville Group of the central Canadian Plains (Fig. 2). Subsurface stratigraphic subdivision is commonly

Fig. 1. Location of the study area.

an inexact science and Mannville Group stratigraphy is most ambiguous. The stratigraphic format of this work is based on the genetic relationship between three units which are in present use by the oil industry (Vigrass, 1977; Putnam, 1980).

The unit immediately underlying the upper Mann-

WICKENDEN (1948)			VIGRASS (1977)		PUTNAM (1982 b)
COLORADO GROUP	Joli Fou Formation		Joli Fou Formation		Joli Fou Formation
MANNVILLE GROUP	UPPER DIVISION	O'Sullivan Member	UPPER	Colony Member	Upper Mannville
				McLaren Mbr.	
				Waseca Member	
	MIDDLE DIVISION	Borradaile Mbr.	MIDDLE	Sparky Member	Middle Mannville
				General Petroleums Member	
		Tovell Member		Rex Member	
		Islay Member		Lloydminster Mbr.	
	BASAL DIV.	Cummings Member	LOWER	Cummings Mbr.	
		Dina Member		Dina Member	Lower Mannville
PALEOZOIC			PALEOZOIC		PALEOZOIC

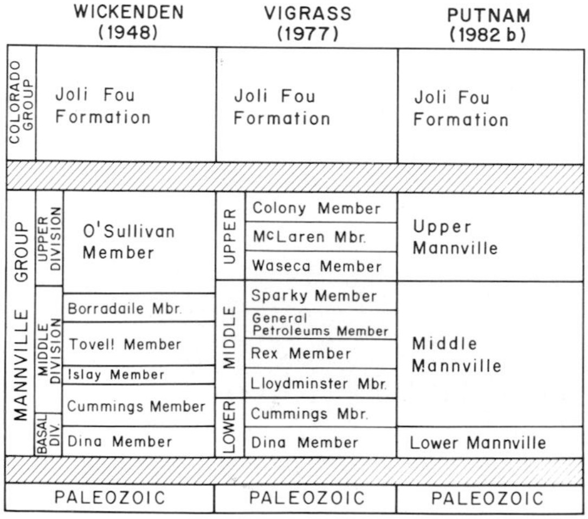

Fig. 2. Stratigraphic nomenclature of the Mannville Group within the Lloydminster area (from Putnam, 1982b).

ville Group, the so-called Sparky Formation, represents deposition within wave-dominated shoreline environments (Putnam, 1982b). The unit overlying the Mannville Group, the Joli Fou Formation, is a dark grey, *Inoceramus*-bearing, transgressive marine shale, The contacts between the upper Mannville Group and the Sparky and Joli Fou Formations are respectively conformable and disconformable.

The Mannville Group dips gently (1.9 m km^{-1}) to the south-west (Orr, Johnston & Manko, 1977) and is, for the most part, undeformed within the study area. However, salt solution collapse features observed within underlying Palaeozoic evaporites have created a complex structure in localized areas (Maycock, 1967; Orr *et al.*, 1977).

SANDSTONE GEOMETRY

Within the upper Mannville Group there are two main sand body types: (i) 'shoestring' sandstones, subsequently referred to as channel sandstones and (ii) sheet-like, inter-channel sandstones. Discussions of the geometry of the upper Mannville sandstones can be found in recent articles by Putnam & Oliver (1980) and Putnam (1982a). Only the general aspects will be repeated here.

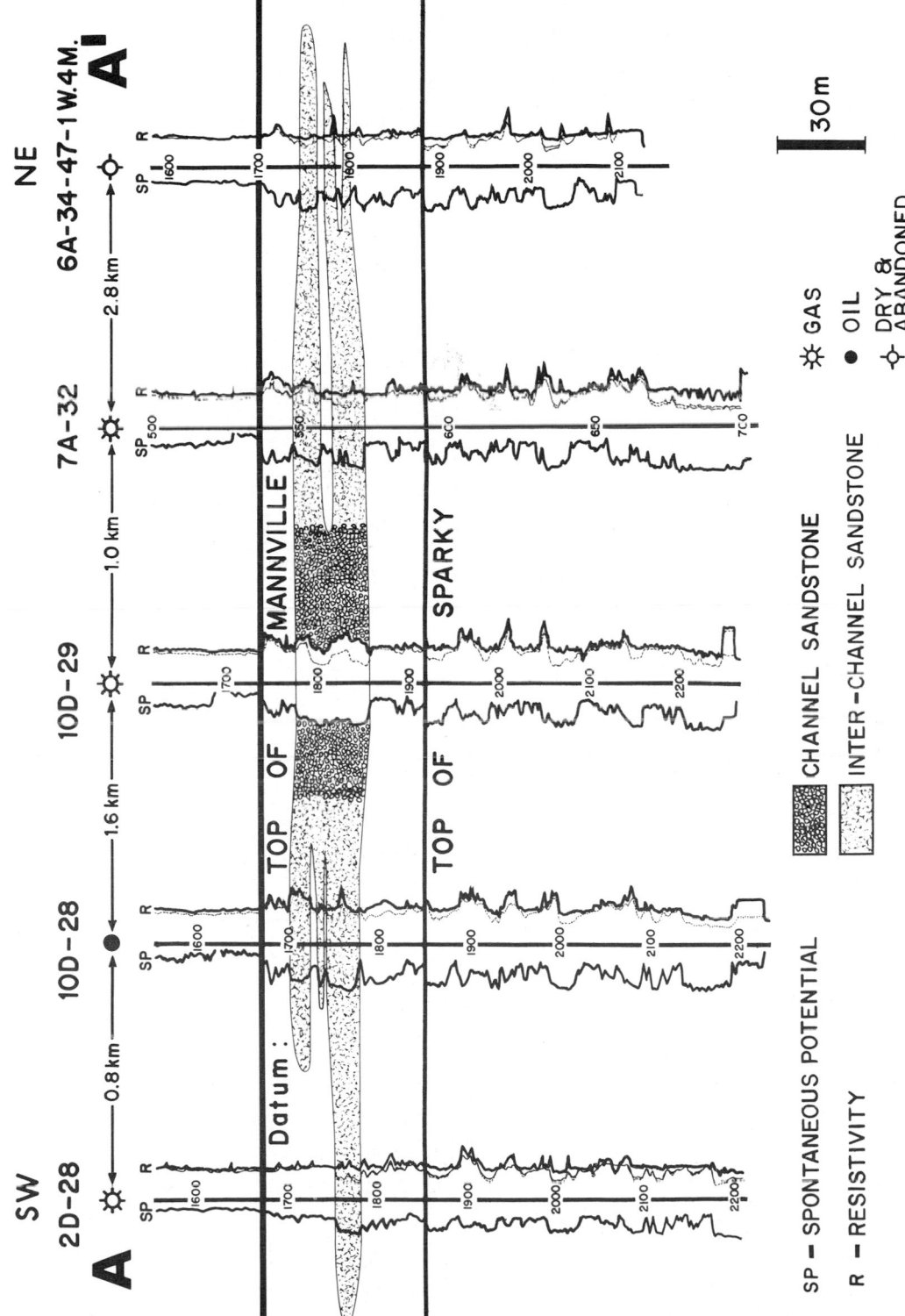

Fig. 3. Representative cross-channel cross-section (modified after Putnam, 1982a).

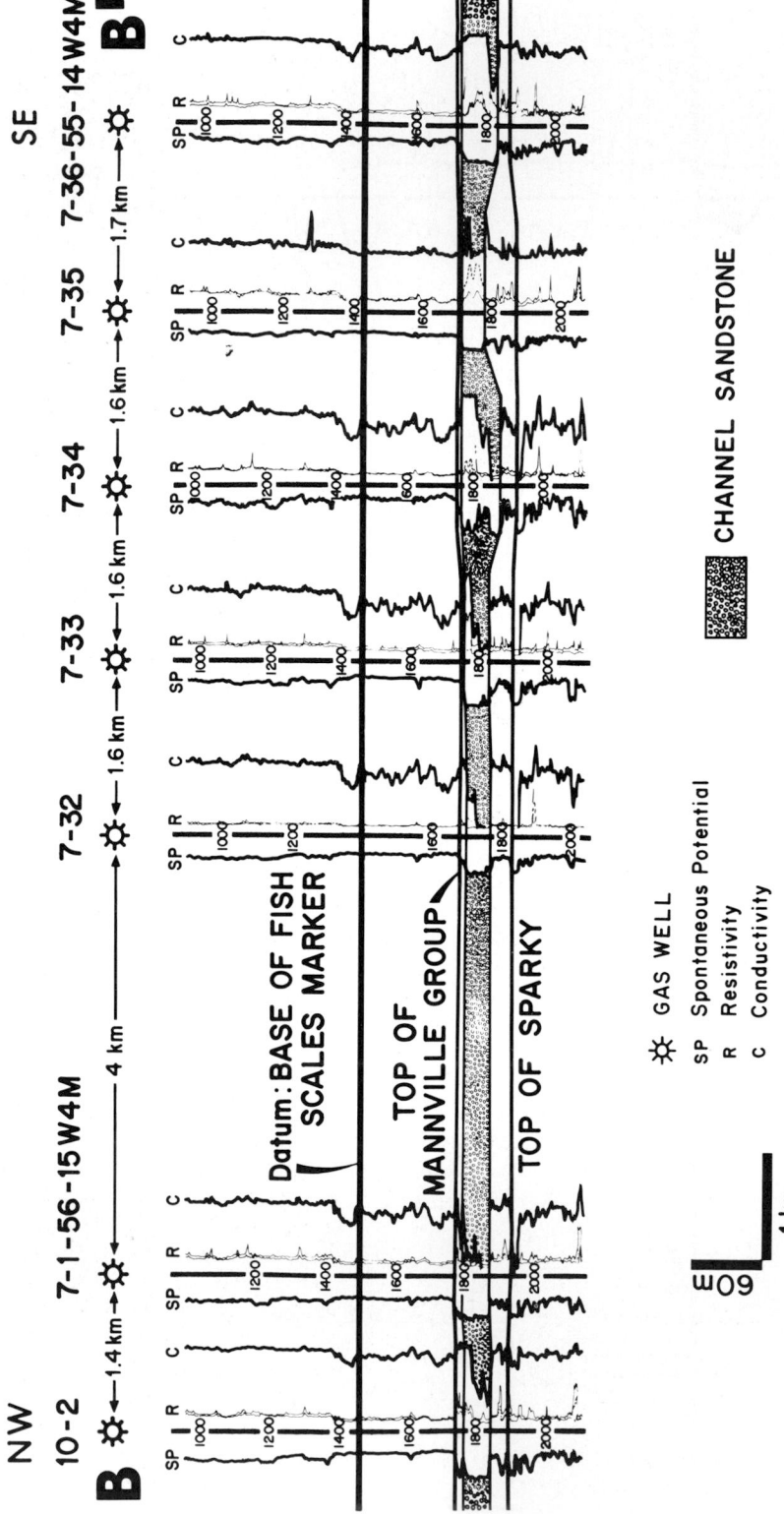

Fig. 4. Representative along-channel cross-section (modified after Putnam & Oliver, 1980).

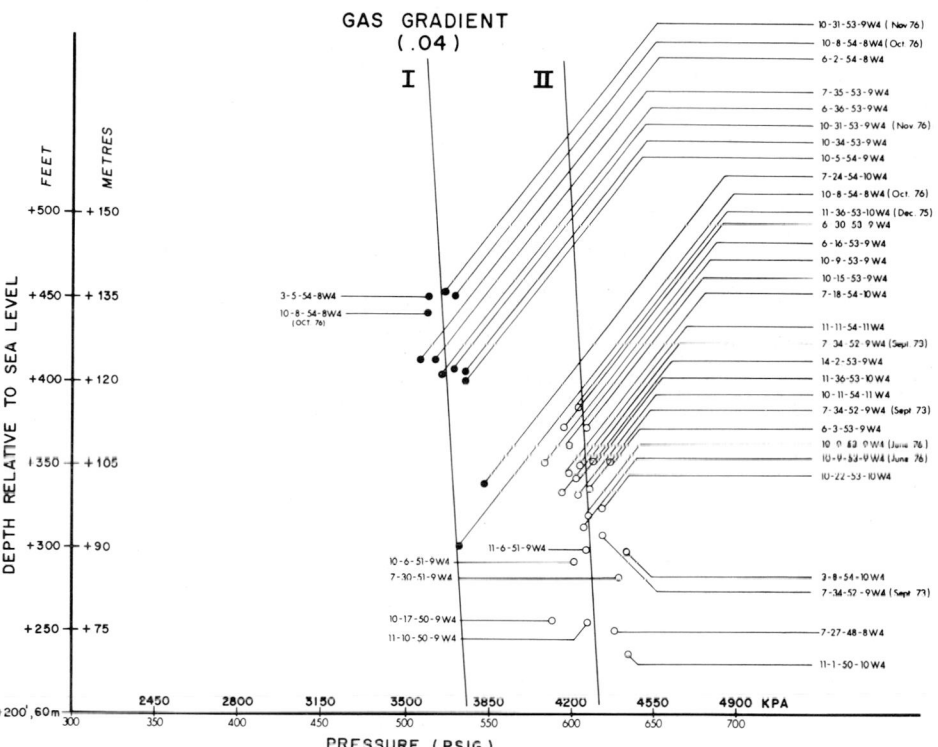

Fig. 5. Pressure–depth graph for the upper Mannville of the Myrnam area, east-central Alberta (modified after Putnam & Oliver, 1980).

The channel sandstones are thick, narrow, elongate bodies. Based on well log and geophysical data, the channel sandstones can be as narrow as 300 m and as thick as 35 m (Putnam, 1980, 1982a; Putnam & Oliver, 1980). Figure 3 is a cross-section across a typical channel sandstone. The 25 m thick channel sandstone in the central well (10D-29-47-1W4, 541–566 m; 1775–1857 feet) splits laterally into several thinner, sheet-like sandstones which get progressively thinner with increasing distance from the channel sandstone.

The elongate nature of the channel sandstones, as shown by a well log cross-section, is documented in Fig. 4. Seismic data associated with subsequent well information have also demonstrated the elongate nature of these deposits (Putnam & Oliver, 1980; Putnam, 1982a).

Formation pressure data can be used to distinguish between separate channel reservoirs. Within Fig. 5 the two separate pressure groupings are inferred to correspond to two separate channel sandstones. Figure 6 is a well log cross-section which shows the spatial relationship of the two channel sandstones. It

is important to note the inter-channel sheet sandstones which plot within one of the two pressure groupings (e.g. 10-31-53-9W4). Such an occurrence is interpreted to imply direct horizontal communication between channel and inter-channel sandstones. The data within Figs 5 and 6 also imply that the inter-channel sheet sandstones are not widespread deposits.

Figures 7 and 8 are maps of the areal distribution of some of the thickest channel sandstones. Only those wells which contain 16 m or more of vertically continuous sandstone were used to produce these maps. It has been shown previously (Putnam, 1980) that channel sandstones less than 16 m thick probably exist and that the distribution of sandstones is far more complicated than that shown in Fig. 7. The more complex channel distribution within the Saskatchewan side of Fig. 7 is due in part to the greater well density there. The Alberta side is more gas-prone, and wells are commonly spaced at one per square mile (2.56 km²). In Saskatchewan, where more oil fields exist, wells can be less than 200 m apart. Figure 7 does not include the locations of shale-filled channels which are

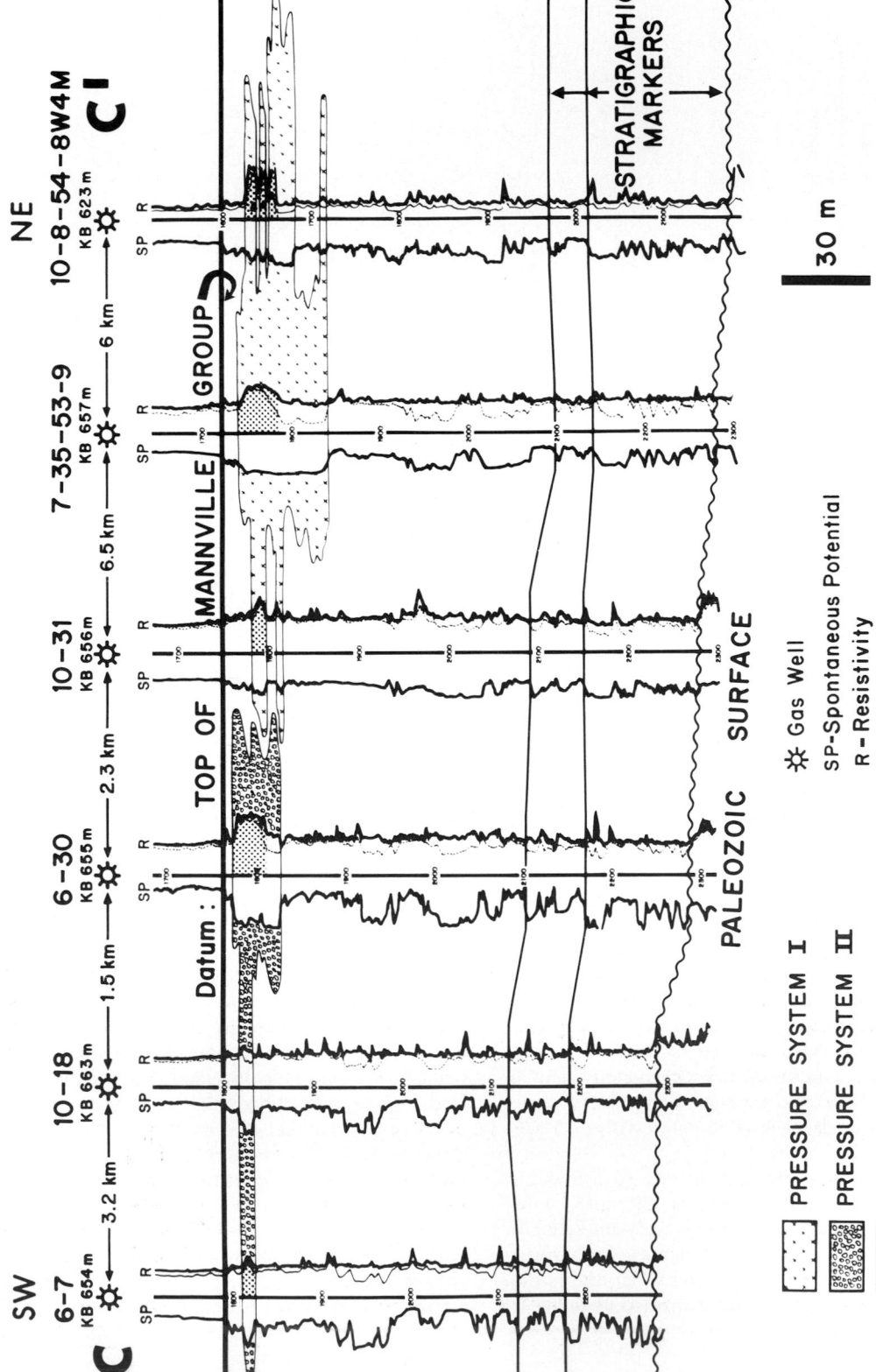

Fig. 6. Cross-section showing spatial and pressure relationships of the two channel sandstones found in the Myrnam area, Alberta.

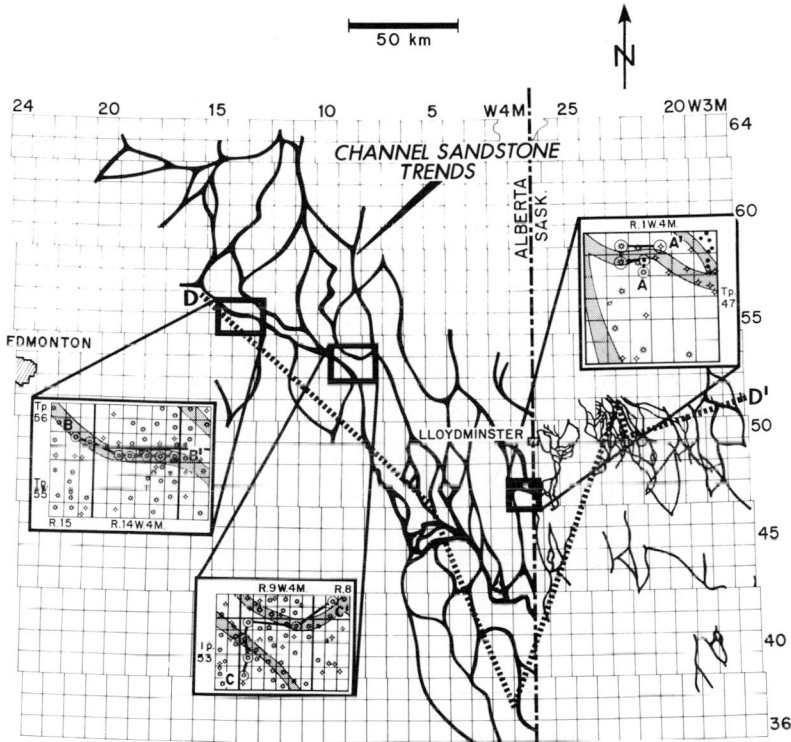

Fig. 7. Areal distribution of the thickest channel sandstones (16 m minimum).

also present in the upper Mannville (Gross, 1980; Putnam, 1981).

The channel sandstones are found in a variety of stratigraphic positions. The map view of channel sandstone distribution shown in Fig. 7 is misleading because it does not account for stratigraphic variability. Figure 8 shows that, in west-central Saskatchewan, more than one stratigraphic horizon within the upper Mannville possesses thick channel sandstones. Figure 9 is a regional cross-section which documents the highly variable stratigraphic position of the channel sandstones across the Lloydminster region. The absence of traceable chronostratigraphic units within the entire Mannville Group makes it impossible to equate, in a temporal sense, channel sandstones located within different areas at the same stratigraphic position.

LITHOLOGICAL DESCRIPTIONS

Coarse channel fill deposits

Coarse channel deposits are composed dominantly of very fine to coarse-grained, well-sorted, subangular

to subrounded sandstones. The sandstones are generally friable. Where lithification has occurred the sandstones are cemented by siderite, calcite, and/or dolomite. Oil stains and cements are present within some sandstones.

The channel sandstones exhibit a variety of sedimentary structures such as angular rip-up clasts (commonly formed of grey siltstones, grey shales, coal or ironstone), high-angle cross-beds, ripples, climbing ripples, plane beds and massive beds. Within some wells shale and siltstone clasts can be found in great abundance within intervals several metres thick (Fig. 10). These clasts are rarely imbricated but are more commonly found in a chaotic arrangement within a sand matrix. These clast-rich zones are commonly found above clean sandstones and not necessarily as a basal lag deposit. The breccia zones are commonly overlain by tilted and slumped (but not brecciated) muddy siltstones which are in turn overlain by flat-lying muddy siltstones. The vertical sequence of sedimentary structures and lithologies found within channel sandstones rarely show distinct patterns. Shale and siltstone drapes or interbeds are extremely rare within the channel sandstones. From the well

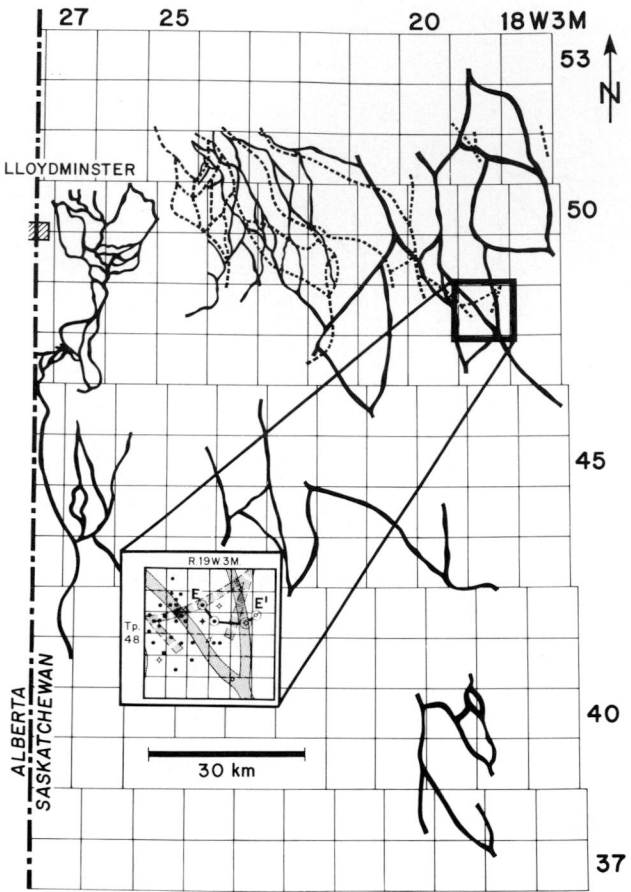

Fig. 8. Areal distribution of the thickest channel sandstones within west-central Saskatchewan showing the stratigraphic variability of some channel sandstones. Dashed lines are channel sandstones found within the interval between top of the Mannville Group and arbitrary plane 23 m (75 ft) below. Solid lines are channel sandstones found within interval between arbitrary plane 23 m below Mannville top to top of the Sparky formation.

logs, the channel sandstones commonly appear to be single, large-scale, fining-upward sequences. However, Tilley & Last (1980) report that the channel sandstones are in fact multi-storied, based on grain size variations. They state that the channel sandstones comprise several fining-upwards cycles, each of which is 2–6 m thick.

The contacts between channel sandstones and underlying rocks (commonly coal) are sharp or erosional. The upper contacts are commonly gradational but can be sharp.

Inter-channel sheet sandstones

The inter-channel sandstones have essentially the same composition as the channel sandstones. How-

ever, the inter-channel sandstones can be more poorly sorted relative to the channel sandstones.

Sedimentary structures found within the inter-channel sandstones are the same as those seen within the channel sandstones except for the additional presence of thin (less than 1 dm) normally graded beds and root structures. The most common sedimentary structures observed are current and climbing current ripples. The inter-channel sandstones are interbedded with siltstones, mudstones, shales and coals. The contacts between lithologies can be sharp or gradational.

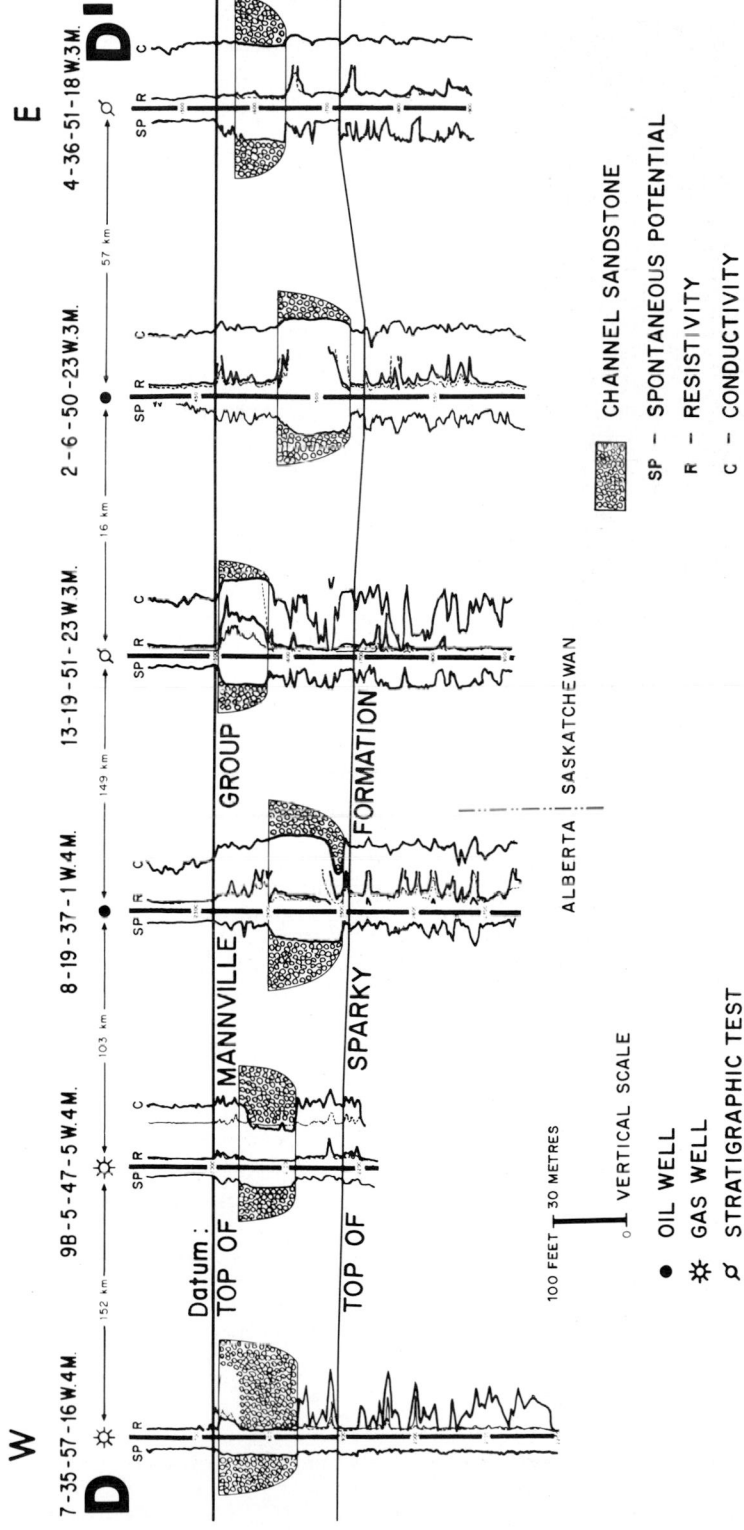

Fig. 9. Regional cross-section showing the stratigraphic variability of some channel sandstones over a distance of 250 km (modified after Putnam, 1982a).

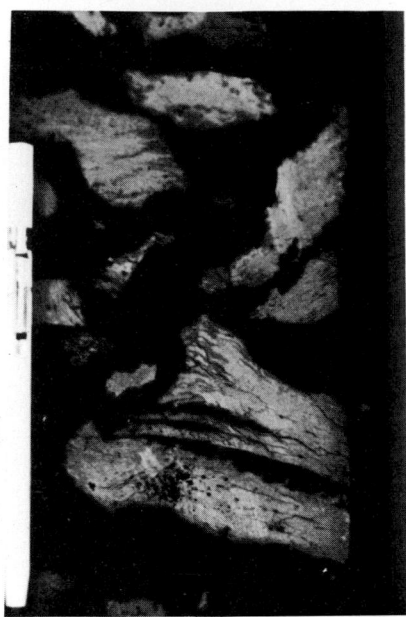

Fig. 10. Siltstone and shale breccia encased within an oil-saturated fine-grained sand matrix. Upper Mannville, Husky Pikes Peak A6-6-50-23W3.

Inter-channel fine-grained rocks

The inter-channel siltstones, mudstones and shales are generally parallel-bedded and commonly exhibit slight to extreme bioturbation. These rocks are commonly carbonaceous and rooted. Well-preserved leaves and wood fragments can be observed along parting planes.

DEPOSITIONAL ENVIRONMENTS

The upper Mannville has been interpreted as representing deposition within an anastomosed fluvial network by Smith & Putnam (1980) and Putnam & Oliver (1980). (For more comprehensive arguments concerning environmental conclusions the reader is referred to these two papers.) Within such a setting the thick 'shoestring' sandstones represent active channel fill deposits. The thinner, sheet-like, inter-channel sandstones are overbank, levee and crevasse-splay deposits. The siltstones, shales, mudstones and coals of the inter-channel areas are lacustrine and wetland deposits. Figure 11 is a schematic block diagram which documents the major depositional facies of the upper Mannville interval.

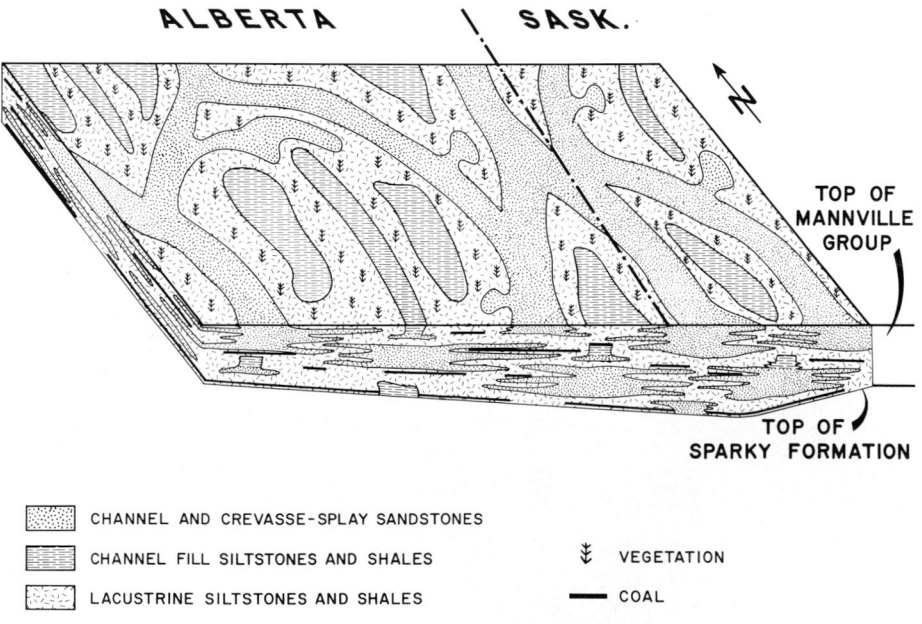

Fig. 11. Schematic block diagram illustrating upper Mannville depositional facies.

Table 1. Comparison of the upper Mannville channel system to the anastomosing upper Columbia River of south-eastern British Columbia. Columbia River data taken from Smith (1983) and Smith (1981, pers. comm.).

	UPPER MANNVILLE	COLUMBIA RIVER
CHANNEL FILLS		
Lithology	Dominantly sandstone, some shale fills	Dominantly sand, some mud fill
Thickness (max.)	approx. 35 m	10 m
Width (min.)	approx. 300 m	40 m
Sedimentary Structures/ Bedforms	High-angle (30°) cross-beds with laminae up to 2 cm thick.	Dominant bedforms are sand waves with main sedimentary structure being tabular planar cross-beds.
Nature of Contacts	Scoured lower contact commonly overlying coal, upper contact can be sharp to gradational.	Scoured lower contact commonly overlying peat or mud, upper contact can be sharp to gradational.
Textural Trends	Multi-storied with individual fining-upward stories 2–6 m in thickness. Channel sandstones commonly exhibit an overall upwards-fining appearance.	Multi-storied with individual fining-upwards stories about 1 m thick.
INTER-CHANNEL SAND DEPOSITS		
Thickness (average) and Area	6–9 m, thicker near major channel sandstones. Several km² up to tens of km².	2–3 m near active channel crevasse and thinning to 0.3 m. Larger splays up to 1 km².
Sedimentary Structures/ Bedforms	Current ripples, climbing current ripples, wave ripples.	Current ripples, parallel laminations, rare cross-beds, bioturbation by rooting. Bedforms are ripples, dunes, transverse bars.
Nature of Contacts	Scoured lower contact proximal to major channel sandstones, gradational lower contact away from major channel sandstones. Upper contacts sharp to gradational.	Scoured lower contact proximal to channel crevasse, gradational lower contact away from crevasse. Upper contacts sharp to gradational.
Vertical Textural Trends	Fining-upwards proximal to major channel sandstones, coarsening-upwards away from major channel sandstones.	Fining-upwards proximal to channel crevasse, coarsening-upwards away from crevasse.
INTER-CHANNEL FINE-GRAINED DEPOSITS		
Lithology	Mudstone, shale, siltstone, coal.	Clay, silt, peat.
Sedimentary Structures	Thin parallel laminae (scale of mm), bioturbation in varying degrees, coarser siltstones commonly current and wave-rippled, soft sediment deformation features.	Thin parallel laminae (scale of mm), rare bioturbation by waterfowl

As a comparative example, the characteristics of the upper Mannville fluvial system are contrasted with those of the upper Columbia River of south-eastern British Columbia (Smith, 1983; Table 1). In the Columbia River the channels are relatively deep (up to 15 m) and narrow (30–50 m). Channel sands comprise stacked fining-upwards flood cycle deposits, and the dominant sedimentary structures are planar cross-beds which are deposited by the downstream migration of sandwaves. The channel sandstones of the upper Mannville are thick (up to 35 m), narrow (300 m) and formed of several fining-upwards cycles 2–6 m thick (Tilley & Last, 1980). The dominant sedimentary structures are large-scale, high-angle (30°) cross-beds.

Upper Mannville channel sandstones commonly sit on coals (e.g. Fig. 16). Peat horizons can act as a control of channel incision during avulsion. Modern anastomosed channels propagate by avulsion processes (Smith & Putnam, 1980). Peat, vegetation and mud are media which tend to inhibit erosion and limit lateral migration (Schumm, 1968; Smith, 1976; Rust, 1981). Anastomosed channels do not migrate laterally

to an appreciable extent (Smith & Putnam, 1980) although some migration does occur (Smith, 1983). Therefore, many upper Mannville channels were probably initiated by avulsion processes. Following the initial scouring event, vertical accretion occurred.

The siltstone and shale breccias found in some channel deposits were incorporated into the channel sediment through a combination of bank slumping synchronous with rapid sand sedimentation. The shale/siltstone breccias associated with channel deposits are very restricted in lateral occurrence. Bank slumps similar to those observed in the upper Mannville can be seen along the margins of modern anastomosed channels. The common presence of shale/siltstone breccias overlying channel sandstones (Fig. 16) in the upper Mannville may indicate coarse clastic deposition during channel abandonment. The appearance of breccia fragments 'frozen' chaotically within a sand matrix attests to rapid deposition. The overlying tilted-to-flat-lying siltstones represent the final attainment of quiescent conditions which led to channel filling by fine-grained clastics.

The crevasse-splay sands of the Columbia River are

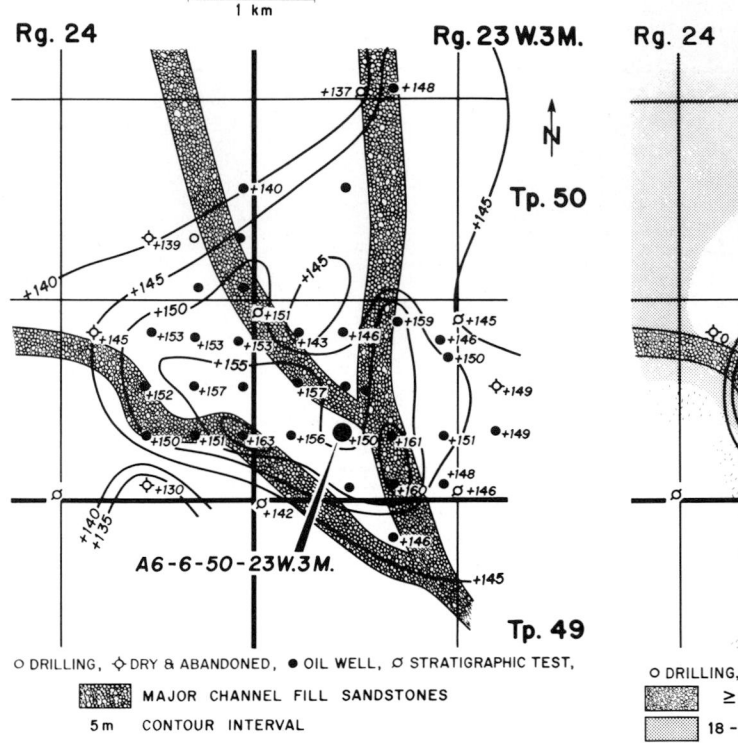

Fig. 12. Structure on top of the Mannville Group relative to the main channel sandstones of the Pikes Peak area of west-central Saskatchewan.

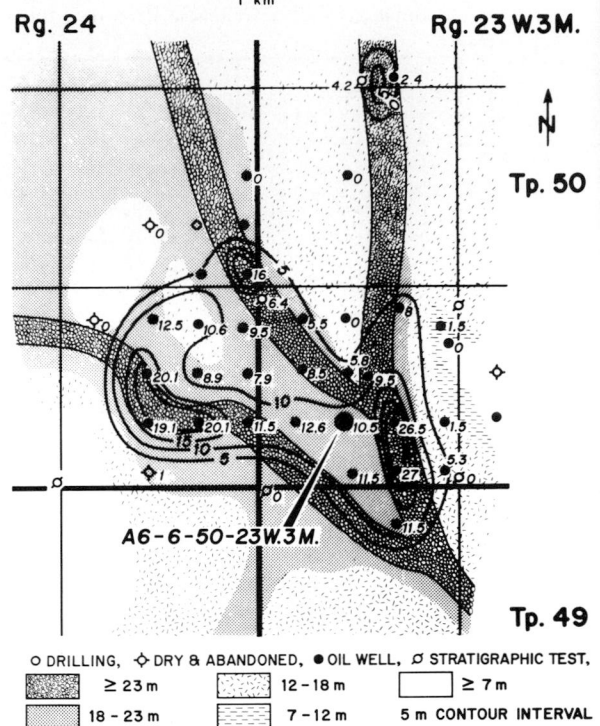

Fig. 13. Net pay in relation to the main channel sandstones of the Pikes Peak area of west-central Saskatchewan. Legend symbols refer to gross sandstone thicknesses found between an arbitrary datum, 23 m below the top of the Mannville Group, and the top of the Sparky interval.

lobate bodies of variable dimensions. They have sharp lower contacts proximal to the main channel, and gradational contacts away from the main channel. The most common sedimentary structures are current ripples. These sands have a high content of wood fragments and leaves which are concentrated along the avalanche faces of bedforms. When buried, these sands become encased within the overbank wetland deposits of silts, clays and peats. The inter-channel sandstones of the upper Mannville are also sheet-like with sharp, lower contacts near channel sandstones and variable vertical profiles further away (Fig. 3). The most common sedimentary structures are current and climbing current ripples. These sandstones are found encased in quiet water siltstones, shales, mudstones and coals.

Micropalaeontological and palynological analyses of upper Mannville shales did not recover any marine indicators such as foraminifera or dinoflagellates (Singh, 1964; Putnam & Oliver, 1980; Husky Oil, unpublished information).

HYDROCARBON DISTRIBUTIONS, FACIES AND TRAP TYPES

Hydrocarbons are found within both channel and inter-channel sandstones. With respect to gas reservoirs, the groupings of different wells within the same pressure systems, as shown in Figs 5 and 6, indicates the presence of direct hydrodynamic connection between channel and inter-channel sandstones. Additional evidence for communication between these facies types is furnished by the consistent fluid contacts across facies boundaries (Fig. 6). The major trapping mechanisms for gas within these rocks are lateral facies changes in an updip direction (Putnam & Oliver, 1980). There is a regional dip to the south-west and the channel sandstones trend dominantly north-west by south-east (Fig. 7). In an updip direction (north-east) the channel sandstones of Fig. 6 feather into interbedded sheet sandstones, siltstones, mudstones and coals. The sheet sandstones in turn pinch out further to the north-east. Therefore, the

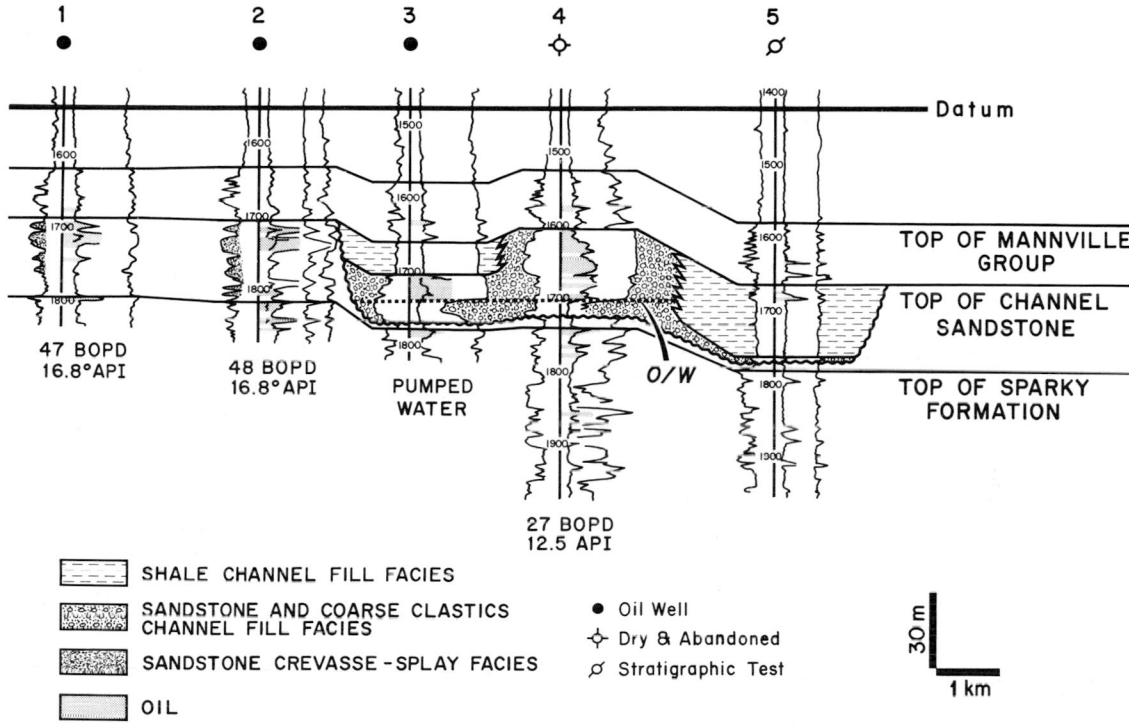

Fig. 14. Structural cross-section through an area of west-central Saskatchewan showing the relationship between facies, fluid distributions, production rates, and oil quality.

updip facies change (from channel sandstones to inter-channel sandstones and finally to siltstones and shales) forms the hydrocarbon trap.

An example of a major oil accumulation within channel sandstones is found within Figs 12 and 13. Within Fig. 12 note the position of structural highs relative to the trend of the thickest channel sandstones. Within Fig. 13 note the relationship between oil leg thickness and the position of the thickest channel sandstones. There is a relationship between facies, structure and hydrocarbon distribution. The thickest sandstones were not as compactible as the adjoining strata, consequently they formed structural highs due to differential compaction. The thickest sandstones (also the best reservoir rocks), because of their high structural position, became the loci of oil accumulation.

Figure 14 demonstrates how reservoir continuity can be broken up by the presence of channel shales. Wells 1 and 2 contain correlatable oil-bearing sheet sandstones within the upper Mannville. Coeval strata within well 3 contain only shale. The shale in well 3 overlies a 20 m (65 ft) thick channel sandstone. Another channel shale can be observed within well 5.

Depositional facies also influence oil quality within the upper Mannville. When oil becomes trapped within channel sandstones it invariably possesses a lower gravity and higher viscosity than stratigraphically equivalent oil found within inter-channel sandstones. In Fig. 14, well 4 has an API oil gravity of 12·5° whereas the oil within wells 1 and 2 both have a gravity of 16·8°. Well 4 produced at a rate of 27 barrels per day, whereas wells 1 and 2 produce at 47 and 48 barrels per day respectively. It is concluded that the channel sandstones, due to their apparently regionally continuous nature, act as regionally dynamic aquifers. As such, they are capable of transporting waters with relatively high oxygen contents. These oxygenated waters can chemically degrade the oil and can also transport aerobic bacteria which biodegrade oil (L. Snowdon, pers. comm). Some drill stem and production tests of the channel sandstones have recovered fresh water which contains less than 3000 ppm of dissolved solids. The inter-channel sheet sandstones are not as regionally continuous as the channel sandstones. Consequently, the oil found within them tends to be lighter relative to the channel sandstones. Thus, primary production rates associated

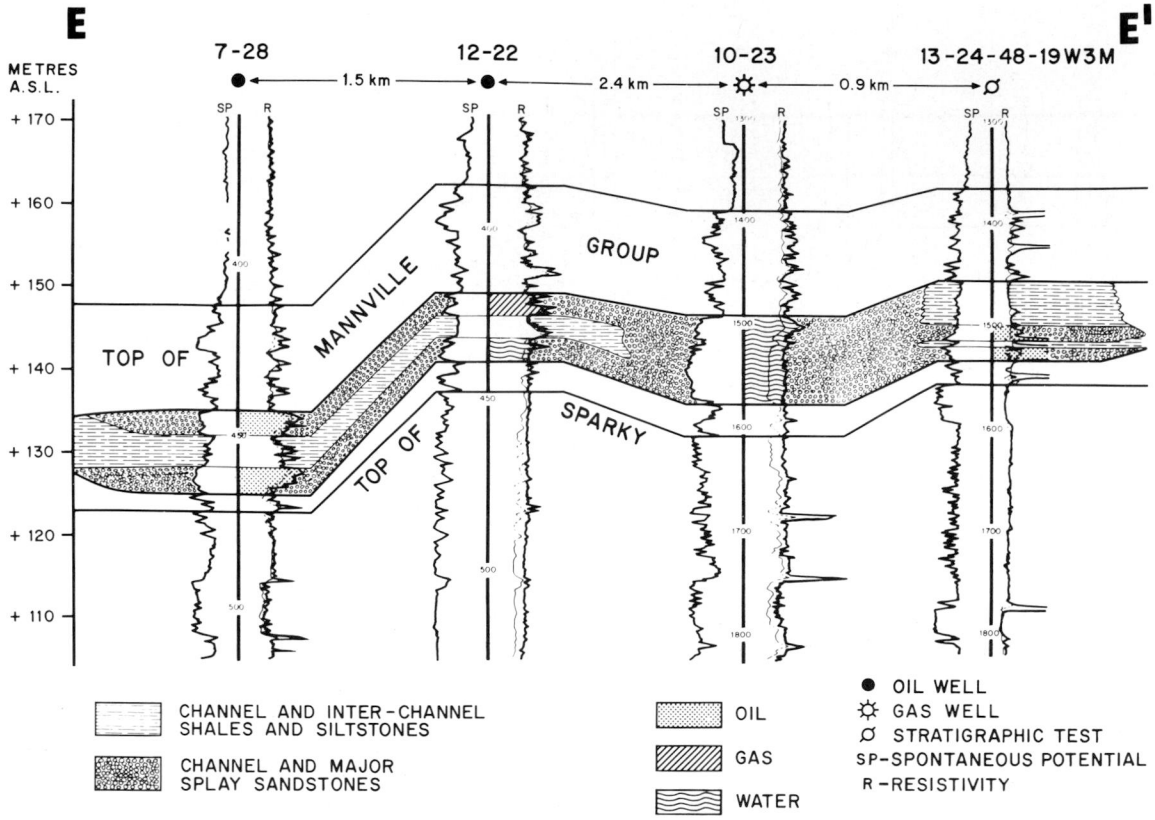

Fig. 15. Structural cross-section through an area of west-central Saskatchewan showing the distribution of fluids in relation to facies.

with inter-channel sandstones are higher than those associated with channel sandstones. Despite the significant amount of oil within the upper Mannville channel sandstones shown in Figs 12 and 13, there is no primary production due to the high viscosity of the oil.

Migrating waters associated with the channel sandstones are also capable of flushing hydrocarbons from these rocks. Within parts of west-central Saskatchewan, water-saturated channel sandstones occur which are laterally equivalent to gas- and oil-saturated inter-channel sheet sandstones as shown in Fig. 15. A core taken from well 10-23-48-19W3 shows that the channel sandstone has a residual oil stain, meaning that some time in the past oil has migrated through these rocks. The regional continuity of the channel sandstones means that these rocks can be flushed of hydrocarbons unless they are found on structural highs (e.g. Fig. 12) or are oriented perpendicular to the regional dip (e.g. Fig. 6). A

consequence of the flushing process is sometimes to isolate petroleum within the less continuous (in a regional sense) inter-channel sheet sandstones as shown in Fig. 15. Therefore, apparently laterally continuous hydrocarbon-bearing sheet sandstones can be broken up by water-flushed channel sandstones.

Within the upper Mannville of the Lloydminster area, an interesting phenomenon exists in the form of 'shaled out' reservoirs. Fig. 16 shows that well A6-6-50-23W3 exhibits a subdued spontaneous potential (SP) log response between 496 and 510 m. Such a response is normally indicative of little permeability. However, a high-resistivity (R) log response is recorded for the same interval. The gamma ray (GR) log for the same interval has a ragged response indicative of interbedded sandstones and shales. Immediately below the 510 m level is found an 11 m thick oil-filled sandstone (510–521 m) which is easily recognized on all three log patterns. In the interval

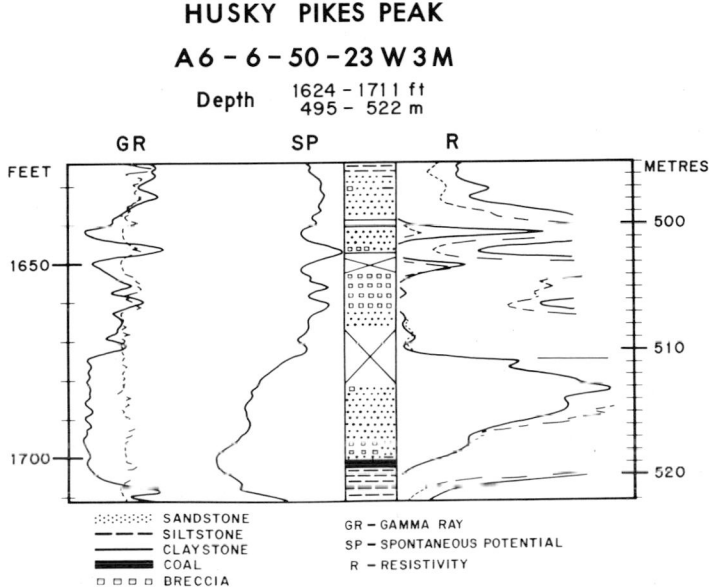

Fig. 16. Gamma ray (GR), spontaneous potential (SP), and resistivity (R) well logs compared to core lithology within part of the upper Mannville of well A6-6-50-23W3.

from 496 to 510 m, the SP and GR log characters would normally lead one to conclude that there was relatively poor reservoir development. The high resistivity value associated with the same interval is problematic. A core taken from well A6-6-50-23W3 shows that this interval is indeed mostly shale and siltstone (Fig. 16). However, the shale/siltstone is in the form of very large clasts and blocks (Fig. 10) which can attain thicknesses of several metres. These shale/siltstone blocks have an oil-saturated sand matrix which explains the high-resistivity log response seen in Fig. 16. To conclude, shale breccias influence log responses in that, if breccia fragments are plentiful, high GR and low SP log responses will result which are indicative of poor reservoir development, although in fact the breccias are reservoir rocks capable of containing hydrocarbons.

From the perspective of subsurface exploration and exploitation, the presence of both shale/siltstone breccias and shale-filled channels poses a perplexing problem. Both are localized deposits within the upper Mannville. In one case, a shale-filled channel would make an excellent updip stratigraphic trap (e.g. Fig. 14). However, if, as in the case of Fig. 13, the 'shale' is in fact a breccia, then it must be considered to be a potential reservoir and not a trapping mechanism. Consequently, an exploration programme must be

geared to look updip in the case of the shale breccia and downdip in the case of a shale-filled channel. With respect to enhanced recovery techniques such as steam flooding, the presence of an updip shale-filled channel will block and deflect the flood front whereas a shale breccia will act as a potential flood sink. Without the presence of cores, an unequivocal interpretation of either shale filled channels or breccias is not always possible within the subsurface. If intervals similar to the upper Mannville within well A6-6-50-23W3 were not cored and did not contain hydrocarbons, the tendency of the geologist may be to interpret the section as possessing a basal channel sandstone which has 'shaled out' in overlying beds. His interpretation would be based on the subdued SP and ragged GR log responses alone. An examination of well cuttings may show the presence of the sand matrix which occurs within the breccia zones.

The above examples help to illustrate the complexity of hydrocarbon distributions found within an extremely complicated subsurface fluvial sequence. The understanding of the geometrical and depositional relationships between channel and inter-channel sandstones is crucial to the understanding of fluid distributions, trap types, and the flood paths of fluids used to stimulate production. An anastomosed fluvial interpretation is consistent in explaining the observed lithofacies and fluid relationships within the upper Mannville of the Lloydminster area.

ACKNOWLEDGMENTS

The author wishes to acknowledge the management of Husky Oil Operations Ltd for granting permission to publish this work. The manuscript was greatly improved by reviews from T. A. Oliver, E. M. Manko, R. J. Knight, D. G. Smith, J. R. McLean and an anonymous reviewer.

REFERENCES

FLORES, R.M. & ETHRIDGE, F.G. (1981) Nonmarine deposits and the search for energy resources and minerals. In: *Recent and Ancient Nonmarine Depositional Environments: Models for Exploration* (Ed. by F.G. Ethridge and R.M. Flores). *Spec. Publs Soc. econ. Paleont. Miner., Tulsa,* **31,** 1–18.

GROSS, A.A. (1980) Mannville channels in east-central Alberta. In: *Lloydminster and Beyond: Geology of Mannville Hydrocarbon Reservoirs* (Ed. by L.S. Beck, J.E. Christopher and D.M. Kent). *Spec. Publs Sask. geol. Soc., Regina,* **5,** 33–63.

MAYCOCK, I.D. (1967) Mannville Group and associated Lower Cretaceous rocks in southern Saskatchewan. *Rep. Sask. Dept. Miner. Resour.* **96,** 108 pp.

ORR, R.D., JOHNSTON, J.R. & MANKO, E.M. (1977) Lower Cretaceous geology and heavy-oil potential of the Lloydminster area. *Bull. Can. Petrol. Geol.* **25,** 1187–1221.

PUTNAM, P.E. (1980). Fluvial deposition within the upper Mannville of west-central Saskatchewan: stratigraphic implications. In: *Lloydminster and Beyond: Geology of Mannville Hydrocarbon Reservoirs* (Ed. by L.S. Beck, J.E. Christopher and D.M. Kent). *Spec. Publs Sask. geol. Soc., Regina,* **5,** 197–216.

PUTNAM, P.E,. (1981) Depositional environments of channel and interchannel sandstones within the Waseca Formation (Lower Cretaceous) of the Lloydminster area with examples from Rivercourse, Alberta. In: *Proc. Can. Soc. Petrol. Geol. Annual Core and Field Sample Conference* (Ed. by F.A. Stoakes), pp. 28–31. Calgary.

PUTNAM, P.E. (1982a) Fluvial channel sandstones within the upper Mannville (Albian) of east-central Alberta: geometry, petrography, and paleogeographic implications. *Bull. Am. Ass. Petrol. Geol.* **66,** 436–459.

PUTNAM, P.E. (1982b) Regional aspects of the petroleum geology of the Lloydminster heavy oil fields, Alberta and Saskatchewan. *Bull. Can. Petrol. Geol.* **30** (in press).

PUTNAM, P.E. & OLIVER, T.A. (1980) Stratigraphic traps in channel sandstones in the upper Mannville (Albian) of east-central Alberta. *Bull. Can. Petrol. Geol.* **28,** 489–508.

RUST, B.R. (1981) Sedimentation in an arid-zone anastomosing fluvial system. *J. sedim. Petrol.* **51,** 745–756.

SCHUMM, S.A. (1968) Speculations concerning paleohydrologic controls of terrestrial sedimentation. *Bull. geol. Soc. Am.* **79,** 1573–1588.

SINGH, C. (1964) Microflora of the Lower Cretaceous Mannville Group, east-central Alberta. *Bull. Res. Coun. Alberta,* **15,** 239 pp.

SMITH, D.G. (1976) Effect of vegetation on lateral migration of anastomosed channels of a glacial meltwater river. *Bull. geol. Soc. Am.* **87,** 757–860.

SMITH, D.G. (1983) Anastomosed fluvial deposits: modern examples from Western Canada. In: *Modern and Ancient Fluvial Systems* (Ed. by J.D. Collinson and J. Lewin). *Spec. Publs int. Ass. Sediment.* **6,** 155–168. Blackwell Scientific Publications, Oxford.

SMITH, D.G. & PUTNAM, P.E. (1980) Anastomosed fluvial deposits: modern and ancient examples in Alberta, Canada. *Can. J. Earth Sci.* **17,** 1396–1406.

TILLEY, B.J. & LAST, W.M. (1980) Upper Mannville fluvial channels in east-central Alberta. In: *Lloydminster and Beyond: Geology of Mannville Hydrocarbon Reservoirs* (Ed. by L.S. Beck, J.E. Christopher and D.M. Kent). *Spec. Publs Sask. geol. Soc., Regina,* **5,** 217.

VIGRASS, L.W. (1977) Trapping of oil at intra-Mannville (Lower Cretaceous) disconformity in Lloydminster area, Alberta and Saskatchewan. *Bull. Am. Ass. Petrol. Geol.* **61,** 1010–1028.

WICKENDEN, R.T.D. (1948) The Lower Cretaceous of the Lloydminster oil and gas area, Alberta and Saskatchewan. *Pap. geol. Surv. Can.* **48–21,** 15 pp.

Spec. Publs int. Ass. Sediment. (1983) **6**, 533–547

Fluvial architecture of Jurassic uranium-bearing sandstones, Colorado Plateau, western United States*

NOEL TYLER† *and* FRANK G. ETHRIDGE

Department of Earth Resources, Colorado State University, Fort Collins, Colorado 80523, U.S.A.

ABSTRACT

Vanadium- and uranium-bearing sandstones of the Salt Wash Member of the Morrison Formation in south-western Colorado were deposited in a fluvial system that contained a range of channel types. Broad and deep low-sinuosity streams, which were the principal drainage elements, deposited dip-elongate sandstone units. Amalgamation of individual sand bodies resulted in multilateral sandbelt geometry of the lower and upper intervals of the Salt Wash. Minor components of the fluvial system were meandering tributaries of the low-sinuosity streams and crevasse channel and associated splays formed during flooding of the trunk and tributary system.

The Slick Rock uranium district is located in the zone of convergence of smaller streams into trunk rivers. The trunk streams are characterized by individual depositional units stacked into two zones of higher sandstone content. The northernmost of these axes strongly influenced the pattern of migration and concentration of uranium. Significant ore deposits in the district are developed within and along the margins of the axis, which is principally composed of low-sinuosity stream deposits. Excellent downdip interconnectedness of these sandstones made them the major conduits of uraniferous ground-water flow. Smaller deposits are contained within meandering stream sediments. Crevasse splay sequences are essentially barren. Vertical interconnectedness between multi-storied depositional units results in local stacking of mineral deposits.

INTRODUCTION

Accumulations of epigenetic uranium in sandstone result from the interaction of mineralized ground water either with reduced host rock (i.e. roll-type uranium deposits of Wyoming and South Texas, U.S.A.) or with stagnant ground water and organic material contained within the host aquifer (i.e. peneconcordant deposits of the Colorado Plateau, U.S.A.). The ore-bearing sandstones are therefore both the plumbing system through which the mineralizing solutions migrate and the hosts of mineralization. The pattern of ground-water migration and consequently the localization of ore minerals is largely controlled by the geometry and connectedness of the framework sandstones. Primary connectedness is a complex attribute dependent on the arrangement (packing) of sand bodies in space, the frequency and extent to which they touch and the degree to which they are in mutual communication (Allen, 1978). Geometry, arrangement of sand bodies in space and connectedness jointly describe the *architecture* (Allen, in Miall, 1978) of a fluvial sequence.

The concept of fluvial architecture (the way in which various elements in a fluvial sequence are stacked, Miall, 1978) is relatively new. Miall (1978) reviewed the origin and evolution of the concept. Notable field analyses that bear further mention are those of Moody-Stuart (1966) and Campbell (1976). Recent attention has focused on the modelling of fluvial architecture (e.g. Allen, 1978; Bridge & Leeder, 1979).

---

* Publication authorized by the Director, Bureau of Economic Geology, The University of Texas at Austin.

† Present address: Bureau of Economic Geology, The University of Texas at Austin, Austin, Texas 78712, U.S.A.

0141-3600/83/0106-0533 $02.00

The Salt Wash Member of the Morrison Formation, which is the subject of this paper, was the object of several pioneering studies of fluvial architecture. Stokes (1954) mapped sedimentary trends and their relation to ore bodies. Craig *et al.* (1955) and Mullens & Freeman (1957) documented the broad geometry and distribution of sandstone within the Member. Shawe, Simmons & Archbold (1968), who undertook a more site-specific study of the Salt Wash, documented the multilateral nature of the sandstones. Recently, Peterson (1980) and Tyler (1981) examined the relation between sedimentology and ore deposits in two of the uranium districts of the Salt Wash Member.

This paper describes the relation between fluvial architecture and mineralization in a part of the Salt Wash Member of the Morrison Formation in the western United States. The Slick Rock vanadium and uranium district of south-western Colorado (Fig. 1) was chosen for detailed study because of the wealth of subsurface information arising from over 80 years of exploration and mining. Special emphasis is placed on the internal arrangement of depositional units, sand-body geometry, spatial arrangement of sand bodies, connectedness between sandstones and overall distribution of sandstone within the Member. This

paper concludes with a discussion of the relationship between fluvial architecture and the location of vanadium–uranium ore bodies.

GEOLOGICAL SETTING

The Upper Jurassic Morrison Formation (Fig. 2) constitutes the uppermost unit of an assemblage of stable-shelf non-marine rocks on the Colorado Plateau. This formation extends throughout the northern Rocky Mountain to the border between the United States and Canada, westward into central Utah, and eastward into Kansas (Campbell, 1976). The Salt Wash Member crops out over much of the stable platform of the Colorado Plateau of the western United States in Arizona, New Mexico, Colorado and Utah. The assemblage has a clearly defined, broad, fan-shaped geometry (Fig. 1). The thickness and grain size of the member decrease towards the north-east. Palaeocurrent azimuths follow a radiating pattern towards the north, east and south-east. On the basis of these criteria it has been suggested that the unit was deposited by an aggrading distributary system of braided channels on a fan-shaped alluvial plain or alluvial fan (Craig *et al.*, 1955; Mullens & Freeman,

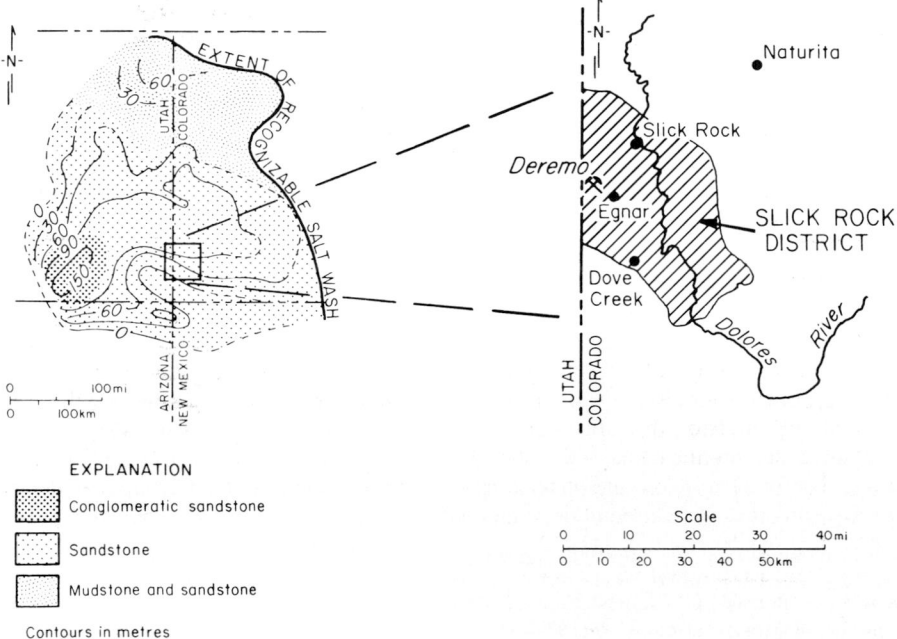

Fig. 1. Geometry of the Salt Wash Member of the Morrison Formation (modified from Craig *et al.*, 1955) and geographic location of the Slick Rock District within this sedimentary sequence.

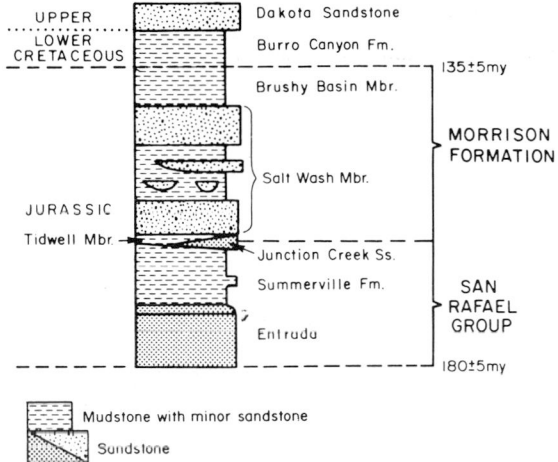

Fig. 2. Stratigraphic setting of the Salt Wash Member of the Morrison Formation.

mixed-load sinuous streams; and an upward-coarsening sandstone facies with variable sedimentary structures deposited as crevasse channels and splays. These sandstone facies have distinctive internal organizations and three-dimensional geometries; these and other characteristics are compared and contrasted in Table 1. The non-framework component of the Salt Wash is collectively assigned to the interbedded mudstone, siltstone and thin sandstone facies.

Poor exposure of the non-framework sediments of the Salt Wash makes mapping of individual depositional units impossible. For this reason the geometry and distribution of these sediments are discussed only in relation to the distribution of the framework sandstones of the Salt Wash. Sub-environments of deposition observed in these floodplain sediments are fully discussed elsewhere (Tyler, 1981).

FACIES AND GEOMETRY OF THE FRAMEWORK SANDSTONES

Low-sinuosity channel sandstones

Internal organization

Low-sinuosity channel sandstones represented by the large-scale cross-bedded sandstone facies are the most common sandstone deposits in the Salt Wash. The facies generally rests on a marked erosion surface (with reliefs of up to 4 m) along which intraformational mudclast conglomerates are abundant. Large-scale trough cross-bedding up to 1 m thick with gently dipping foresets that have tangential bases is the dominant stratification type. These cross-bed sets are characterized by high width-to-depth ratios and an absence of topsets. Foresets are locally accentuated by heavy minerals or mudflakes. The sandstones are predominantly fine grained and poorly to moderately sorted. Fragmental plant material, large logs (without root systems), bones and bone fragments are common, as are pedogenic limestone nodules reworked out of the adjacent floodplain deposits.

Minor components of this sandstone facies are tabular cross-beds, ripple-laminated and horizontally laminated sandstone and siltstone, and thin mud stringers (Table 1).

A striking feature of the large-scale cross-bedded sandstone facies is very large (giant) trough cross-bedding in sets up to 60 m wide and 4 m thick. Foresets on the lateral margins of these giant cross-beds dip steeply (Fig. 3) and are commonly

1957). More recent studies concluded that the Salt Wash fluvial system is more complex, consisting of both low- and high-sinuosity channel deposits that interfinger with well-drained floodplain and lacustrine sediments (Ethridge *et al.*, 1980; Peterson, 1980; Tyler, 1981).

In the Slick Rock district, the Salt Wash Member overlies and erodes the Tidwell Member, a floodplain-dominated fluvial deposit that defines the base of the Morrison Formation (Fig. 2). The Morrison rests upon the aeolian Junction Creek sandstone and lacustrine-evaporative deposits of the Summerville Formation. Conformably overlying the Salt Wash Member is a thick, poorly exposed shale sequence, the Brushy Basin Member, which constitutes the upper member of the Morrison Formation. This member grades into the Lower Cretaceous Burro Canyon Formation, a similar assemblage of mudstones and minor sandstones.

The Salt Wash succession in the study area can be separated into three lithologic units; (1) lower and (2) upper sandstone-dominated intervals separated by (3) an interval of variable mudstone content interbedded with between three and seven sandstone ledges. The Salt Wash is interpreted as having been deposited in a complex fluvial system that contained a range of coexisting channel types. Sandstone facies recognized in outcrop, cores and electric logs are interpreted as follows: large-scale cross-bedded sandstone facies deposited by low-sinuosity streams; fining-upward, medium-to-small-scale, trough cross-bedded sandstone facies deposited during the lateral migration of

Table 1. Depositional characteristics of the three major sandstone facies of the Salt Wash Member

	Large-scale cross-bedded facies	Medium- to small-scale trough cross-bedded facies	Thin, upward-coarsening facies
Nature of contacts	Lower: erosional Upper: abrupt, locally gradational	Lower: erosional Upper: transitional	Lower: planar, only locally erosive Upper: sharp to gradational
Primary sedimentary structures	Giant trough cross-beds at base overlain by gently-dipping, high w/d trough cross-beds; planar cross-beds, ripple and horizontal stratification are accessory	Low w/d, steeply dipping trough cross-beds dominant; grade vertically into ripple and horizontal lamination	Extremely variable; often an upward increase in scale of structures
Vertical profiles	Blocky; thin fining-upward zone locally developed in upper part of facies; average thickness 10 m	Upward-fining; average thickness 5 m	Generally upward-coarsening, locally highly variable; average thickness 1 m
Geometry of depositional units	Strike: lensoid, concave based Dip: elongate Lateral amalgamation results in sheet geometry	Strike: lensoid Dip: lensoid	Strike and dip $\left\{\begin{array}{l}\text{lobate;}\\ \text{planar base,}\\ \text{convex upper}\\ \text{surface}\end{array}\right.$
Isolith pattern	Broad, straight channels eastward oriented	Beaded or ribbon-like sand thicks, oriented obliquely to palaeodip	Not recognized; inferred lobate or fan shape
Electric log pattern	Thick, blocky	Hemi-conical, often stacked	Thin, inverted Christmas tree
Interpretation	Low-sinuosity streams	Meandering streams	Crevasse channel and splay

w/d: ratio of width to depth.

contorted. Towards the axes of the troughs foresets become gently inclined. Foreset intrasets (Collinson, 1968) oriented obliquely to the master set are common. The giant trough sets, which average 10–15 m wide, occur as mutually erosive, multilateral structures.

Detailed mapping and numerous measured sections through the large-scale cross-bedded facies did not reveal any regular vertical arrangement of sedimentary structures or grain size. Rather, vertical and lateral variability in vertical sequences is characteristic of the facies and is well illustrated in the detailed cross-section through the upper sandstone interval of the Salt Wash (Fig. 3). However, it was observed in the field that the largest of the giant cross-beds occurred near or on the erosional surface at the base of this facies (Fig. 3) and that there is an occasional progressive decrease in the scale of sedimentary structures in the upper 50 cm of individual depositional units.

Palaeocurrent analysis revealed that the streams responsible for deposition of this facies flowed towards the east and north-east.

Geometry of sandstone units

The large-scale cross-bedded sandstone facies is characterized by abrupt lower and upper contacts,

which are easily recognizable both in electric logs and in outcrop. The lack of a regular variation in grain size and sharp contacts of the facies result in a dinstinctive cylindrical pattern on electric logs. An erosion surface marks the base of the assemblage. Mudstones rest directly upon ripple- or horizontally laminated sandstones at the upper contact. Facies thickness averages 10 m, but it varies greatly and a maximum thickness of 25 m has been recorded.

Two distinct cross-sectional geometries are recognized in this facies. In the lower and upper sandstone-dominated intervals of the Salt Wash, amalgamation of individual depositional units results in a multilateral, sheet-like geometry. Due to truncation and reworking of individual depositional units, it is difficult to estimate the pristine dimensions of the palaeochannels. However, in one instance a lenticular, asymmetric channel-fill sequence near the top of the lower sandstone unit was measured to be 250 m wide and 15 m deep. Juxtaposition and amalgamation of these individual depositional units created sand belts more than 20 km wide (Fig. 4).

The geometry of the large-scale cross-bedded facies in the floodplain-dominated middle interval differs from that in the sand-dominated intervals. Channel sequences here exhibit a multi-storied geometry.

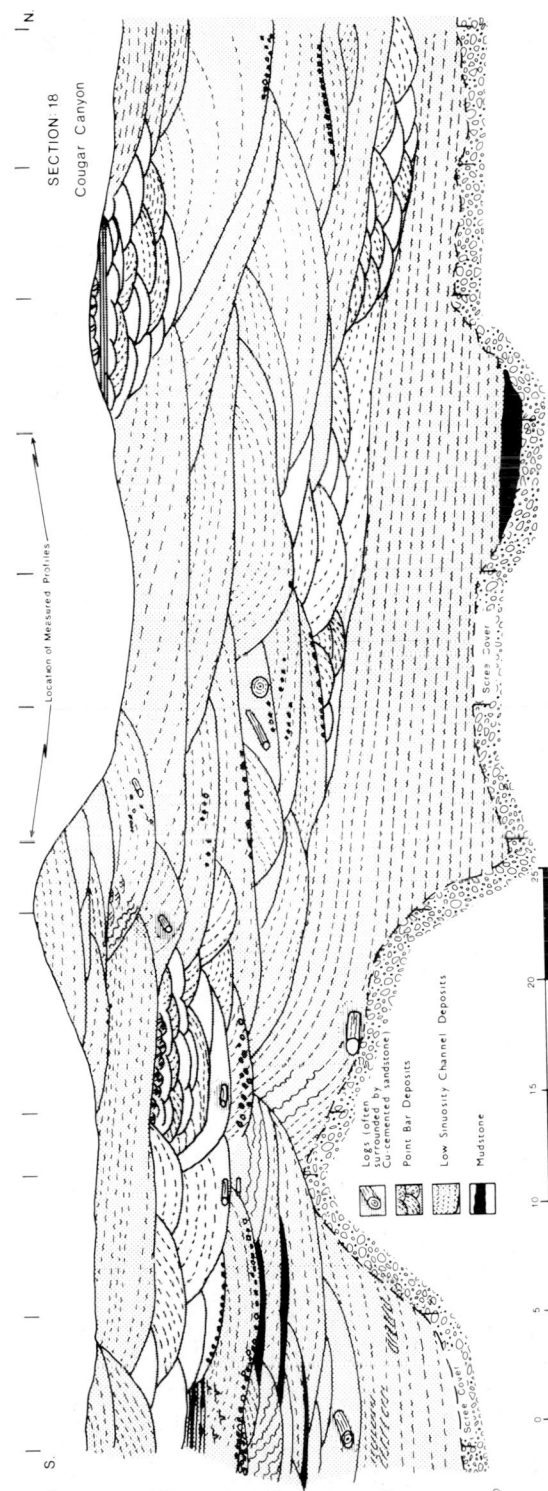

Fig. 3. Detailed cross-section through the upper sandstone interval of the Salt Wash illustrating the lateral variability of the large-scale cross-bedded sandstone facies. Giant trough cross-beds with contorted lateral margins occur at the base of the facies, which is dominated by large-scale trough cross-beds. Thin, asymmetric lensoid point-bar cycles are interbedded within this facies.

35

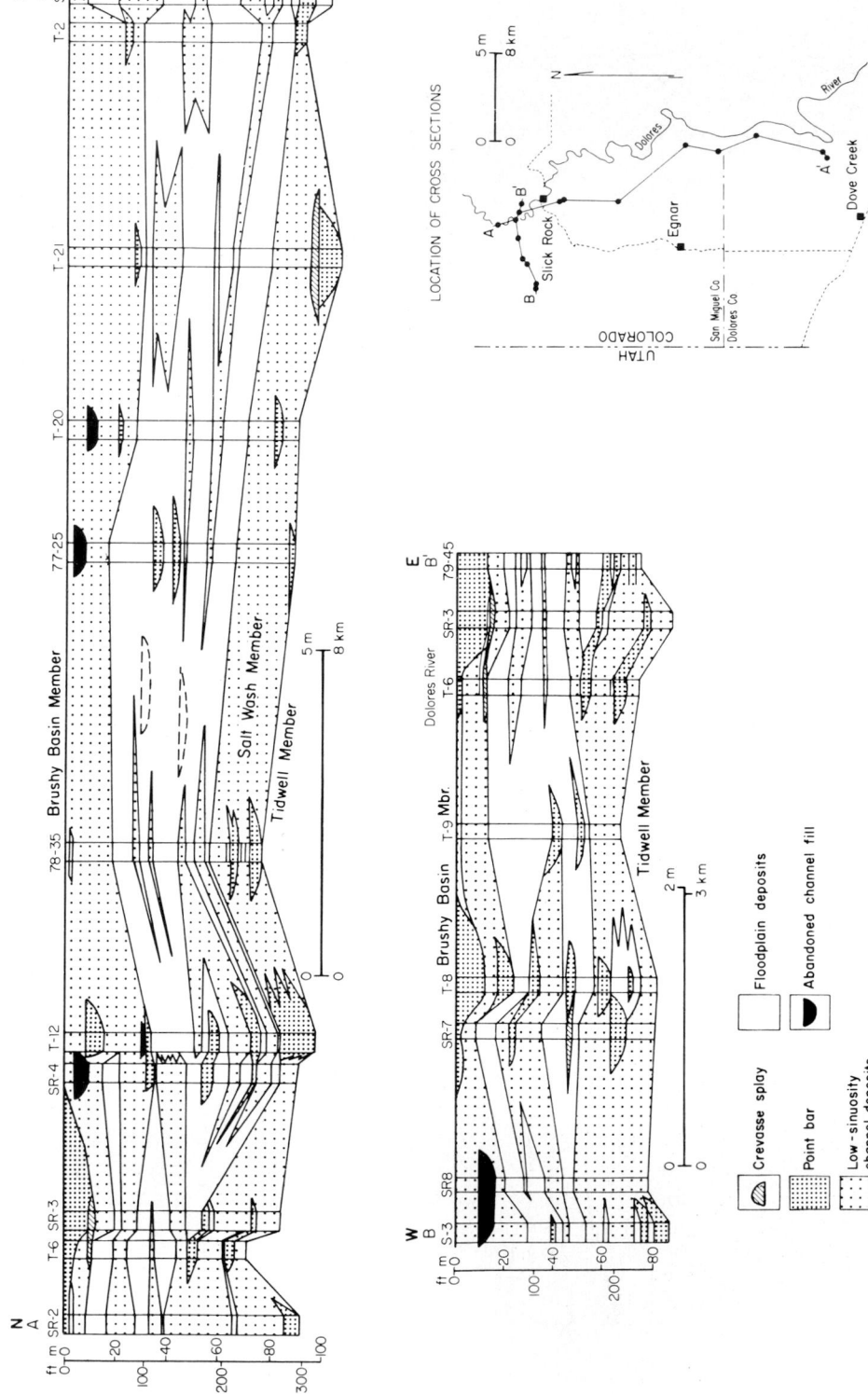

Fig. 4. Strike (A–A′) and dip (B–B′) oriented cross-sections through the Salt Wash Member. Amalgamation of dip-elongate low-sinuosity channel deposits created sand belts in the lower and upper intervals of the Salt Wash. Point-bar and crevasse splay depositional units exhibit lenticular geometries.

Lateral margins of the sand bodies interfinger with adjacent floodplain deposits. Laterally and vertically truncated tabular mudstones within these sandstones suggest there were multiple periods of sedimentation. The widths of these amalgamated sand belts vary between hundreds of metres and several kilometres as compared to the tens of kilometres in the sand-dominated intervals.

The plan geometry of this facies is illustrated in the isopach map of the upper sandstone interval. Two broad, slightly sinuous belts of thickened sand traverse the district from west to east (Fig. 5). These dip-oriented zones of higher sandstone content are considered to represent the plan morphology of the large-scale cross-bedded facies. On dip-oriented cross-sections, the sandstone units comprise laterally persistent layers (Fig. 4) that bifurcate towards the east (Tyler, 1981).

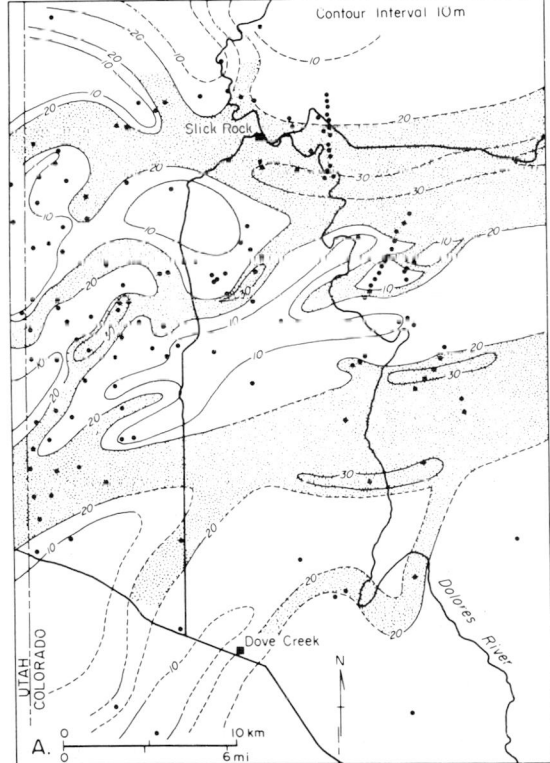

Fig. 5. Isopach map of the upper sandstone-dominated interval of the Salt Wash. Eastward-oriented thickened sand bodies represent the amalgamated deposits of low-sinuosity channel streams, while sinuous ribbons, aligned obliquely to the palaeoslope, are inferred to be the deposits of a meandering tributary system.

Interpretation

The laterally persistent, palaeodip-oriented, large-scale cross-bedded facies is interpreted to represent deposition in deep, slightly sinuous bedload channels that traversed the district from west to east. Modern analogues are parts of the Niger–Benue system (NEDECO, 1959) and possibly parts of the Red River (Schwartz, 1978). Course change was accomplished by avulsion of streams that had aggraded above the adjacent floodplain, rather than by lateral migration.

The mutually erosive, multilateral, giant trough cross-beds that consistently occur at the base of the facies represent the deposits of giant ripples. An alternative interpretation is that the troughs are channel-fill deposits. However, their location in the facies, the laterally truncated boundaries with contiguous giant cross-beds, and their internal structure all indicate deposition by downstream-migrating giant ripples. Primary bedforms of a similar magnitude have been reported from monsoonal belt rivers that experience high seasonal floods, for example the Yamuna River (Singh & Kumar, 1974). The scale of these giant cross-beds and comparison with modern analogues suggests that the Salt Wash fluvial system was subjected to periodic flooding. Avulsion of channels took place during these possibly seasonal flood events. Early erosion was followed by deposition of giant cross-beds upon the basal erosional surface.

The broad but relatively shallow trough cross-beds that comprise the majority of the structures originated under conditions of base ('normal') flow. High width-to-depth ratios, the absence of topsets, the low amplitude of the basal erosion surface, and the low angle of dip of the foreset of most trough crossbeds indicates that these structures are the erosional remnants of larger megaripples that were truncated by subsequent bedforms migrating down the palaeo-channel. Small-scale tabular cross-beds were probably produced by the downstream migration of straight crested dunes. The size of the tabular cross-beds suggests deposition during low flow. The upward thinning of sedimentary structures in the upper parts of individual sandstone units reflects waning current strengths following avulsion and channel abandonment.

High-sinuosity channel sandstones

Internal organization

High-sinuosity channel sandstones, as represented by the medium- to small-scale trough cross-bedded sandstone facies, is characterized by a progressive vertical decrease in grain size (Table 1) and scale of sedimentary structures. The facies overlies an erosion surface along which mud clasts are common. Trough cross-bedding is the dominant sedimentary structure. Locally, tabular cross-beds and horizontal- and ripple-stratification are present. Trough cross-beds of this facies have low width-to-depth ratios and steeply dipping foresets. The erosion surface of individual cross-bed sets has a greater relief than in the aforementioned facies. Grain size varies from medium- to fine-grained sandstone at the base to very fine-grained sandstone and siltstone at the top of the sequence. The upper part of the facies contains planar, ripple- and climbing-ripple-stratification and abundant burrows. This fining- and thinning-upward sequence is occasionally associated with lateral accretion surfaces.

The thickness of the medium- to small-scale trough cross-bedded sandstone facies varies between 4 and 9 m. The sharp basal contact and upward-fining grain size imparts a 'Christmas-tree' pattern to electric logs.

Palaeocurrent vectors of this facies are highly variable. They have low consistency ratios and are oriented towards the north-east or south-east (Tyler, 1981). The streams that deposited these sediments therefore flowed obliquely across the palaeoslope.

Geometry of depositional units

Individual depositional units of this facies are distinctly lensoid (Table 1) and from detailed mapping appear to be asymmetrical in shape (Fig. 3). On a larger scale amalgamated depositional units retain their lensoid geometry both in strike- and dip-oriented sections (Fig. 4).

Isopach maps of the upper sandstone interval of the Salt Wash illustrate narrow sinuous zones of high sandstone that converge with the broad sheet sandstones of the low-sinuosity channel facies (Fig. 5). These narrow 'beaded' or 'ribbon' sands are considered to represent the plan morphology of the lensoid, medium- to small-scale trough cross-bedded sandstone facies.

Interpretation

Lenticular, fining- and thinning-upward sandstone packages are interpreted to represent point-bar deposits. The meandering pattern exhibited by the 'ribbon sands' in Fig. 5 and the presence of epsilon cross-beds confirm this interpretation; other researchers have reached a similar conclusion. In an early study, Stokes (1954) mapped curving sedimentary trends (interpreted as old river bends) and their relation to ore deposits west of our study area. Exploratory drilling has revealed sinuous sandstone ribbons in the subsurface (P. Rubick, personal communication, 1980).

Crevasse channel and splay sandstones

Internal organization

The coarsening-upward sandstone facies, representing crevasse channel and splay sandstones, is the most variable of the three facies. The internal organization of sedimentary structures varies laterally within individual depositional units, but varies even more dramatically between depositional units. Although vertical increases in grain size with height in the column are common, other profiles (blocky, fining-upward) have also been observed. Unifying characteristics that identify a sandstone package as part of this facies are the distinctive geometry of the depositional units, the variability of sedimentary structures, facies thickness and facies association.

The base of the facies is generally planar and is only locally erosive into underlying assemblages. A variety of cross-stratification structures rests on the basal contact; these include stacked cosets of tabular cross-beds of variable thickness (20–50 cm), trough cross-beds with low width-to-depth ratios, single tabular cross-beds (up to 1 m thick) resting upon a thin ripple-laminated zone, and interstratified trough and tabular cross-beds. Upward-coarsening cycles are concomitant with an increase in the scale of sedimentary structures.

Both single and composite depositional units are recognized in the facies. Single units comprise upward-coarsening profiles overlain either by a thin transitional zone or by siltstone or mudstone. These simple profiles attain a maximum thickness of 2 m. Composite profiles contain thin interbedded mudstones (a few centimetres to 30 cm thick) that terminate minor upward-fining cycles within the broad, upward-coarsening sandstone progradational unit. Reactivation and truncation surfaces are common in these units. Composite depositional units are 1–2 m thick. Depositional units of this facies are thinner than the two other sandstone facies recognized in the Salt Wash.

Fig. 6. Distinctive geometry of a crevasse splay deposit (C.S.) viewed obliquely to palaeoslope (towards the north-east). Basal contact is planar, non-erosive. Upper contact is abrupt and convex, Slick Rock Hill.

Geometry of depositional units

Lensoid cross-sectional geometries in both dip and strike section suggest a lobate geometry (Table 1). A sharp planar surface that is only locally erosive marks the base of the facies. The upper contact is convex (Fig. 6) and both abrupt and gradational boundaries have been observed. This distinctive geometry is best preserved when the unit is interbedded within fine-grained floodplain clastics. In the sand-dominated parts of the Salt Wash, geometries are largely destroyed by truncation and amalgamation with adjacent sandstone bodies. Individual depositional units vary from less than ten to a few hundreds of metres wide. Upward coarsening of this sandstone facies imparts a funnel-shaped pattern to electric logs.

Interpretation

Lobate, coarsening-upward sandstone cycles are interpreted to be crevasse splay deposits. Crevasse splays are fan- or tongue-shaped sandstone bodies introduced into the floodplain by crevassing of a levée during flooding (Collinson, 1978). Geometries and the internal organization of crevasse splays deposited on modern alluvial plains have not been well documented.

However, crevasse splays in the interdistributary areas of delta plains have received considerable attention. In the deltaic setting, splays can comprise semi-permanent, prograding and upward-coarsening minor mouth bar/crevasse channel couplets, or they may be deposited by single sudden incursions of sediment-laden waters into the bay producing locally wide levée aprons (Elliott, 1978). Numerous small anastomosing streams comprise the crevasse channels in the latter case and the resultant deposits take the form of isolated small channel lenses each separated by a thin mud drape. Alternatively, rather than being confined to channels, the flow may develop into a density current and deposit an erosive-based sand lobe (Elliott, 1978).

The variability of crevasse splay deposits in the Salt Wash is therefore a function of the diverse processes active at the river–floodplain interface. The composite crevasse splays were probably semi-permanent features that were active during high river stage and were temporarily abandoned at low stage when suspension sedimentation dominated. Subsequent flooding of the river resulted in truncation of the mudstones and the formation of reactivation surfaces in the sandstones. The vertical increase in grain size is the result of successive flood events aggrading increasingly coarser grained sand over the fan-shaped crevasse splay surface. The larger, abruptly based and convex-topped, simple lenses commonly observed in the Salt Wash probably result from deposition by single-pulse, unconfined density currents. Smaller isolated lenses represent the deposits of anastomosing crevasse streams on the river flanks.

The lateral variation in sedimentary structures probably represents proximal–distal relations. In the proximal parts of the splays, stacked cosets of trough cross-beds resting on an erosive base are inferred to represent deposition in crevasse channels that scoured into the underlying floodplain assemblage. In the distal parts of the splay, isolated or stacked sets of tabular cross-beds that rest on a planar, non-erosive base were probably formed by the avalanching of sediment-laden water over the edge of the crevasse fan under conditions of unconfined flow.

INTERCONNECTEDNESS OF SANDSTONE DEPOSITIONAL UNITS

Interconnectedness of depositional units is a complex attribute dependent on their packing and the frequency and extent to which they touch (Allen, 1978). In the Salt Wash, interconnectedness is further dependent on vertical and horizontal location in the assemblage and environments of deposition. In the sandstone-dominated lower and upper intervals of the Salt Wash (Fig. 4) interconnectedness of individual depositional units with adjacent and under- and overlying depositional units is well developed. All three sandstone facies are interbedded in these zones, and contacts between sandstone bodies are non-sealing. Sandstones in the floodplain-dominated middle interval are generally isolated, therefore interconnectedness is poor (Fig. 4). Exceptions occur near Slick Rock and Dove Creek, where stacking of depositional units results in partial vertical interconnectedness of sandstone bodies.

The juxtaposition and amalgamation of depositional units in the lower and upper intervals of the Salt Wash destroyed the original geometry of sandstone facies. This merging of sand bodies reduced the importance of geometry with respect to connectedness of depositional units. However, in the middle, floodplain-dominated interval the initial interconnectedness (or lack of connectedness) of individual depositional units is preserved. In strike section, lensoid point bar and tabular low-sinuosity channel deposits occur as isolated pods in a 'matrix' of floodplain sediments (Fig. 4). Interconnectedness is poor. In palaeo-dip section, point-bar deposits retain their isolated, lenticular geometry. Low-sinuosity channel deposits comprise laterally persistent sand sheets that pinch and swell, bifurcate and merge down the palaeoslope (Tyler, 1981, plate 1c). Interconnectedness between adjacent deposits is high. These highly interconnected, low-sinuosity channel deposits comprise the major conduits of ground-water flow in the Salt Wash.

In general, the fan-shaped geometry and restricted areal extent of the crevasse splay sandstones cause these deposits to have a low interconnectedness.

SANDSTONE DISTRIBUTION

In addition to the preferred vertical location of sandstones in the lower and upper intervals of the Salt Wash, the coarser clastics exhibit a preferential areal distribution. Total sandstone content varies from less than 40% to more than 70% of the Salt Wash column. Areas of thickened sandstone comprise two palaeodip-aligned, eastward-trending linear belts (illustrated by the percentage sand isolith map of the Salt Wash, Fig. 7) separated by intermediate sandstone contents. Increases in the thickness of the sandstone-dominated lower and upper intervals of the Salt Wash (Figs 8A and 5, respectively) and an increase in the sandstone content of the middle interval (Fig. 8B) account for the dip-aligned thickened sand belts reflected in the isolith map of the member.

Comparison of isopach and sand percentage maps of the three intervals (Figs 5 and 8A, B) reveals that the northern (Slick Rock) thick sand belt remained static throughout deposition of the Salt Wash, and was probably localized by the rising Gypsum Valley salt ridge (Tyler, 1981). The Dove Creek belt was unaffected by local salt tectonics and shifted its position and orientation through time (see Figs 5 and 8A, B). For this reason the Slick Rock belt is more sharply defined and contains a higher percentage of sandstone than does the Dove Creek belt.

Isolith and isopach maps illustrate that the Salt Wash in the study area was deposited in a zone of confluence of smaller steams into major or trunk drainages, and the merging of tributaries into these trunk streams. The broad fluvial pattern was one of convergence (Figs 5 and 8) and can be described as dendritic. East-trending, dip-oriented thickened sand belts are inferred to be the deposits of the trunk streams of this dendritic system.

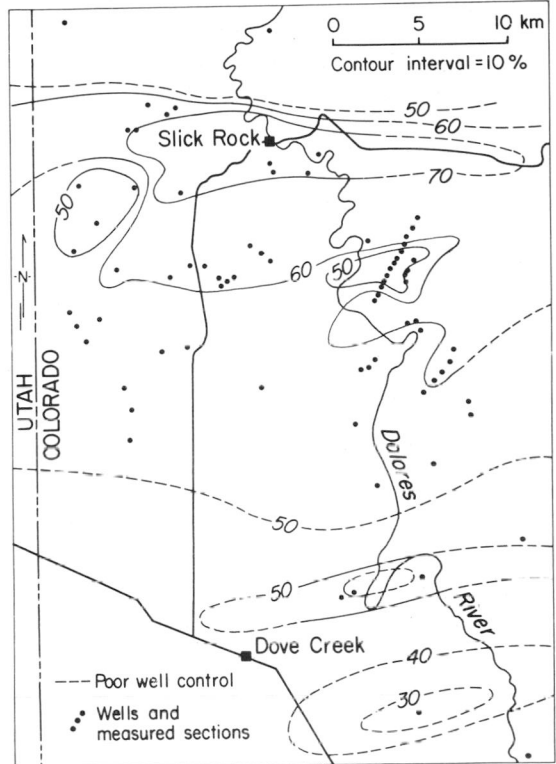

Fig. 7. Sandstone percentage map of the Salt Wash Member illustrating two palaeodip-oriented belts of high sandstone content.

DEPOSITIONAL SETTING OF THE FRAMEWORK SANDSTONES

Framework sandstones of the Salt Wash in the Slick Rock District were deposited by a range of coexisting channel types. Channel arrangement was dendritic. Broad and deep low-sinuosity streams that were seasonally subjected to major floods flowed across the district from west to east. Meandering streams which, according to their palaeocurrent and isopach and isolith patterns, were tributaries of the major low-sinuosity streams, rose on the adjacent well-drained floodplain. This tributary system recycled sediment deposited on the floodplain by crevasse channels and splays during flood events. Sandstones of the Salt Wash were thus deposited by a dynamic system of aggrading channel types that deposited and recycled sediment continually.

Convergence of smaller streams and tributaries into major or trunk streams gave rise to thicker accumu-

lations of sandstone in two east-trending belts. Local salt tectonics in the form of the rising Gypsum Valley anticline localized the northern belt by restricting migration of the trunk stream. The less well-defined southern belt, which was apparently unaffected by local salt tectonics, shifted its orientation and position during the deposition of the Salt Wash.

FLUVIAL ARCHITECTURE AND ORE DEPOSITS

Epigenetic uranium deposits in sandstone host rocks account for 95% of the past production of uranium in the United States. These deposits constitute at least 95% of the reserves and 79% of the potential United States uranium resources (Young, 1978). The importance of the Colorado Plateau as a uranium-producing province in underscored by the fact that it has accounted for 70% of past uranium production in the United States (Young, 1978). Uranium ore is contained in 28 sedimentary formations ranging in age from Pennsylvanian to Pliocene (Wood, 1956). Twenty-four of these formations are terrigenous clastic deposits, three are carbonates, and one is coal (Young, 1978). The most important uranium-producing formations are the Triassic Chinle Formation, the aeolian Entrada Sandstone (Jurassic) and the fluvial Morrison Formation (Jurassic).

In the Slick Rock District most of the ore production is derived from the Salt Wash Member of the Morrison Formation. Ninety per cent of the vanadium–uranium ore mined comes from the upper sand-dominated interval (Chenoweth, 1978). Uranium deposits range in size from a few tons of ore produced from one-man surface workings to cumulative productions exceeding 1,000,000 kg of U_3O_8 produced by large-scale underground operations (e.g. Deremo Mine, Figs 1 and 9). Ore deposits are confined to the northern half of the district, and their distribution is apparently random if the sizes of the ore deposits are not considered. However, a contour map of cumulative uranium production through 1971 shows that the larger ore deposits are aligned along distinct trends (Fig. 9) that are west–east oriented in the northern part of the study area and south-west–north-east oriented in the west central part of the district. This alignment of ore deposits parallels the depositional fabric of the framework sandstones of the Salt Wash Member (Figs 5 and 7).

Comparison of the production isopleth map with the isopach map of the upper sandstone interval (the

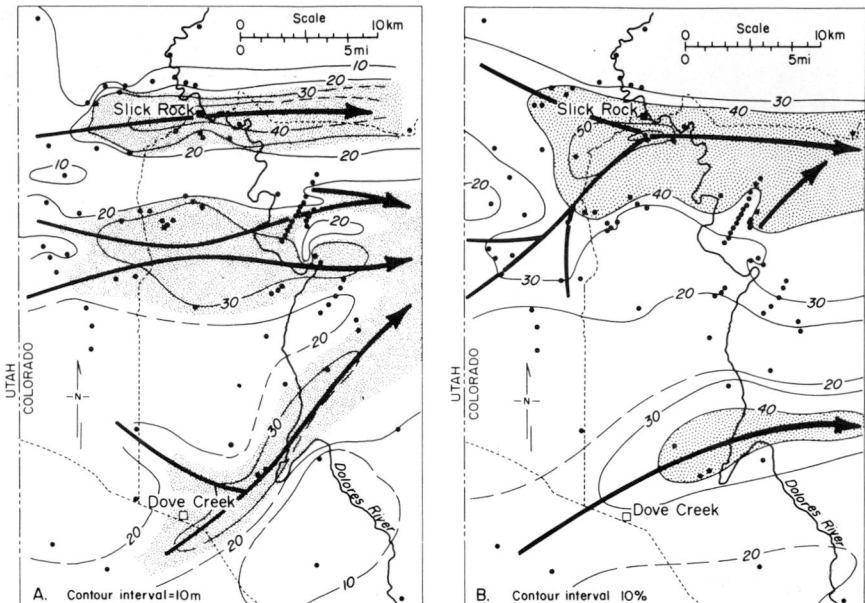

Fig. 8. Isopach map of the lower interval of the Salt Wash Member (A), and percentage sand isolith map of the middle interval (B), illustrating eastward-trending thickened sand bodies that transect the district. Arrows emphasize the directions of flow and the convergence of streams into trunk systems.

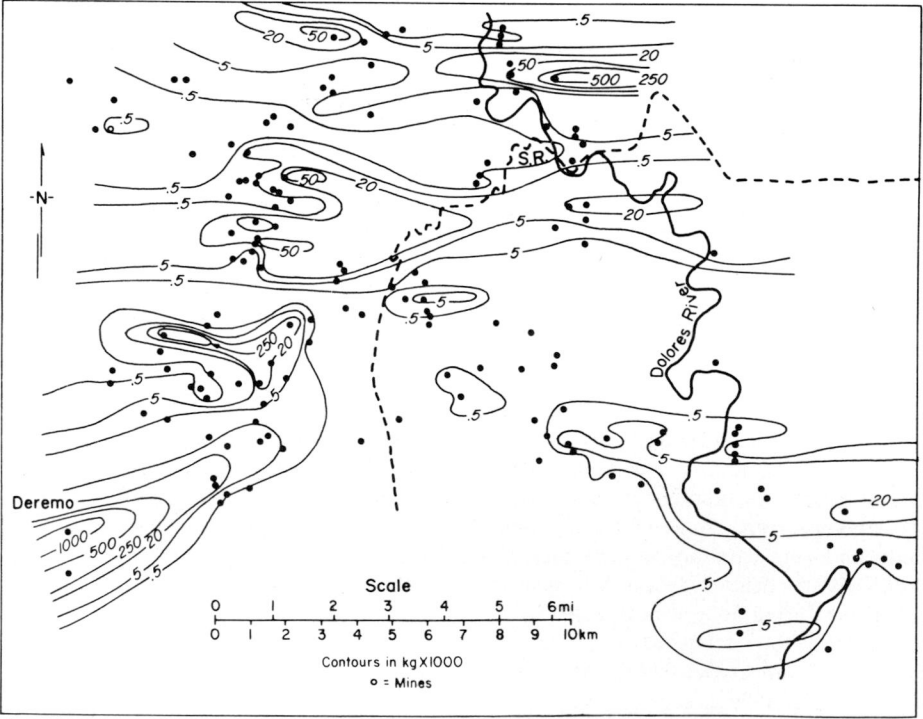

Fig. 9. Isopleth map of cumulative uranium production from the Slick Rock district prior to 1971 (data obtained from Nelson-Moore, Collins & Hornbaker, 1978). The larger ore deposits align west-east and north-east trends parallel to the depositional fabric of the host sandstones.

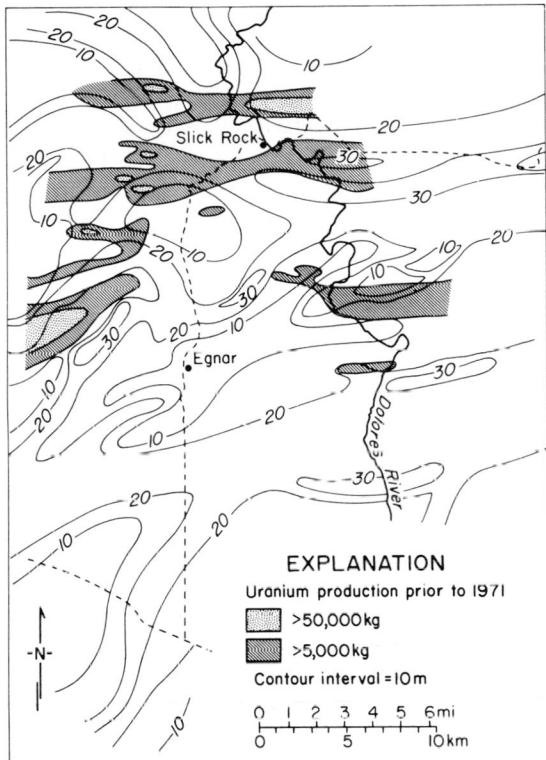

Fig. 10. Combined production isopleth-upper sandstone interval isopach map. Major ore deposits of the district (with cumulative productions greater than 50,000 kg U_3O_8) occur within or along the margins of the Slick Rock trunk stream, or within the tributary system near their confluence with the trunk. Smaller ore deposits have a more widespread distribution.

channel sandstones. The uranium ore occurs as strata-bound tabular bodies that are thin compared to their lengths. Long axes of the ore deposits are parallel or subparallel to the local sedimentary fabric. Boundaries are diffuse and are defined by economic limits rather than by abrupt changes in ore grade. Locally, vertical connectedness between low-sinuosity channel sandstones results in stacked ore bodies.

Point-bar sequences are less commonly mineralized. Less than one-quarter of the 43 ore bodies observed occurred in lower point-bar sequences; these ore bodies were smaller than the vanadium–uranium ore deposits in low-sinuosity channel sandstones. Boundaries of the ore zones in this sandstone facies are more abrupt, occasionally being defined by epsilon cross bedding. The ore often takes the form of a massive replacement of the host sandstone obliterating the primary sedimentary structures, whereas ore zones in low-sinuosity channel sandstones are characterized by mineralization localized along foresets.

Mineralization is rarely observed in upper point-bar, or crevasse channel and splay sandstones. A significant site of minor, patchily developed ore bodies is at permeability barriers associated with abrupt changes in lithology (e.g. at the base and upper contacts of sandstone sequences). Six of the 43 ore bodies examined were associated with permeability barriers; this relationship is independent of facies.

Discussion

The close association between ore deposits and low-sinuosity channel sandstones indicates that these palaeodip-elongate highly interconnected sandstones acted as the major conduits for uranium-bearing ground-water migration. The source of uranium is generally considered to lie in the diagenetic destruction of volcanic detritus contained within the Salt Wash and the overlying Brushy Basin Member. Uranium-bearing ground waters migrated downdip along transmissive conduits until encountering reductants in the form of vanadium(III)-bearing organic complexes contained within stagnant ground water (Granger & Warren, 1981) and, locally, fossil plant debris. Reduction and precipitation of uranium resulted in tabular bodies aligned parallel to the sedimentary fabric. (For a more detailed description of ore-body genesis see Rackley, 1976, Nash, Granger & Adams, 1981, Thamm, Kovschak & Adams, 1981, Tyler, 1981, and others.) Although the ore bodies have a distinct dip orientation, they are also concentrated into an elongate strike-parallel belt (the Uravan Mineral Belt,

source of much of the ore mined before 1971) reveals that the larger ore deposits are preferentially contained within sandstone of intermediate thickness (Fig. 10). Furthermore, the more important ore deposits (with pre-1971 productions of 50,00 kg of U_3O_8) tend to occur within and along the lateral margins of the Slick Rock trunk stream or within the meandering tributaries near their confluence with the trunk (Fig. 10). Less significant deposits (more than 5000 kg U_3O_8 produced) have a more widespread distribution, being located within the Slick Rock trunk, along the margins of both trunks, and within the tributary system (Fig. 10).

The influence of fluvial architecture on the location of uranium deposits on a local scale is illustrated by the fact that ore bodies most commonly occur within the dip-elongate, highly interconnected, low-sinuosity

Fischer & Hilpert, 1952) roughly concentric with the truncated proximal portion of the Salt Wash as well as the outer limits of the Salt Wash Member (Fig. 1). The Slick Rock district is located in the southern part of this belt. It is considered significant, although possibly fortuitous, that this clustering of ore deposits coincides with the zone of convergence of streams (see Figs 5 and 8) midway down the palaeoslope of the member.

Ore bodies within low-sinuosity channel sandstones are larger but are less concentrated (as indicated by alteration and scintillometer readings) than the ore bodies in point-bar sequences. This suggests that the geometry of depositional units influences not only paths of fluid migration but also ore grade. It has not yet been determined whether the higher grade ore in point-bar sequences is a consequence of (1) a higher concentration of reductants in the stagnant ground waters in these lenticular units; or (2) facies boundaries of the sinuous channel sandstones having confined and concentrated the movement of the uranium-bearing mineralizing solutions; or (3) a combination of both these factors.

The close association between the major ore deposits of the Slick Rock uranium district and the Slick Rock trunk stream suggests that this trunk (or fluvial axis) was the principal zone of movement of the mineralized ground waters. It has been shown that the less well-defined Dove Creek trunk was characterized by changes in position and orientation during Salt Wash sedimentation. This instability is reflected by the paucity of mineralization associated with the trunk. The Slick Rock trunk, stabilized by local salt tectonics, experienced only minor changes in position, and therefore comprises an axis of highly interconnected, multi-storied, dip-elongate depositional units that contain a comparative abundance of ore deposits. Recognition of trunk elements, such as the Slick Rock trunk, within a potentially mineralized fluvial sequence may prove to be a useful exploration tool.

ACKNOWLEDGMENTS

Funding for this project was provided by a grant from the Branch of Uranium and Thorium Resources, United States Geological Survey. One of us (N.T.) gratefully acknowledges the financial support of the Trustees of the Jim and Gladys Taylor Educational Trust. We would like to thank the many geologists with whom discussions were held, in particular Drs H. Granger, D. Shawe and F. Peterson. Our thanks are extended to companies who allowed us access to their properties and data files. An early draft of this manuscript was reviewed by Drs B. Turner, W. E. L. Minter and E. S. Belt, and by an anonymous reviewer; their help is greatly appreciated. Several of the figures were redrafted under the supervision of Dan F. Scranton. Preliminary editing was undertaken by Amanda R. Masterson and word processing by Margaret Chastain.

REFERENCES

ALLEN, J.R.L. (1978) Studies in fluviatile sedimentation: an exploratory quantitative model for the architecture of avulsion-controlled alluvial suites. *Sedim. Geol.* **21**, 129–147.

BRIDGE, J.S. & LEEDER, M.R. (1979) A simulation model of alluvial stratigraphy. *Sedimentology*, **26**, 617–644.

CAMPBELL, C.V. (1976) Reservoir geometry of a fluvial sheet sandstone. *Bull. Am. Ass. Petrol. Geol.* **60**, 1009–1020

CHENOWETH, W.L. (1978) Uranium in western Colorado. *Mountain Geol.* **14**, 89–96.

COLLINSON, J.D. (1968) The sedimentology of the Grindslow Shales and the Kinderscout Grit: a deltaic complex in the Namurian of northern England. *J. sedim. Petrol.* **39**, 194–221.

COLLINSON, J.D. (1978) Alluvial sediments. In: *Sedimentary Environments and Facies* (Ed. by H.G. Reading), pp. 15–60. Elsevier, New York.

CRAIG, L.E., HOLMES, C.N., CADIGAN, R.A., FREEMAN, V.L., MULLENS, T.E. & WEIR, G.W. (1955) Stratigraphy of the Morrison and related formations, Colorado Plateau region, a preliminary report. *Bull. U.S. geol. Surv.* **1009E**, 125–166.

ELLIOTT, T. (1978) Deltas. In: *Sedimentary Environments and Facies* (Ed. by H.G. Reading), pp. 97–142. Elsevier, New York.

ETHRIDGE, F.G., ORTIZ, N.V., SUNADA, D.K. & TYLER, N. (1980) Laboratory, field and computer flow study of the origin of Colorado Plateau type uranium deposits, second interim report. *Open file Rep. U.S. geol. Surv.* **80–805**, 81 pp.

FISCHER, R.P. & HILPERT, L.S. (1952) Geology of the Uravan mineral belt. *Bull. U.S. geol. Surv.* **988-a**, 1–13.

GRANGER, H.C. & WARREN, C.G. (1981) Genetic implications of the geochemistry of vanadium-uranium deposits in the Colorado Plateau Region. *Progr. Abstr. Rocky Mountain Section, Am. Ass. Petrol. Geol., 29th Ann. meeting*, p. 24. Albuquerque, New Mexico.

MIALL, A.D. (1978) Fluvial sedimentology: An historical review. In: *Fluvial Sedimentology* (Ed. by A.D. Miall) *Mem. Can. Soc. Petrol. Geol., Calgary*, **5**, 1–48.

MOODY-STUART, M. (1966) High and low sinuosity stream deposits, with examples from the Devonian of Spitzbergen. *J. sedim. Petrol.* **36**, 1102–1117.

MULLENS, T.E. & FREEMAN, V.L. (1957) Lithofacies of the Salt Wash Member of the Morrison Formation, Colorado Plateau. *Bull. geol. Soc. Am.* **68**, 505–526.

NEDECO (1959) *River Studies and Recommendations on Improvement of Niger and Benue*. North-Holland, Amsterdam. 1000 pp.

NASH, J.T., GRANGER, H.C. & ADAMS, S.S. (1981) Geology and concepts of genesis of important types of uranium deposits. *Econ. geol. 75th Anniv. Volume*, pp. 63–116.

NELSON-MOORE, J.L., COLLINS, D.B. & HORNBAKER, A.L. (1978) Radioactive mineral occurrences of Colorado and bibliography. *Bull. Colorado geol. Surv.* **40**.

PETERSON, F. (1980) Sedimentology of the uranium-bearing Salt Wash Member and Tidwell unit of the Morrison Formation in the Henry and Kaiparowits Basins, Utah. *U.G.A.—1980 Henry Mountains Symposium*, pp. 305–322.

RACKLEY, R.I. (1976) Origin of western-states type uranium mineralization. In: *Handbook of Stratabound and Stratiform Ore Deposits II* (Ed. by K.H. Wolf), **7**, pp. 89–156. Elsevier, New York.

SCHWARTZ, D.E. (1978) Sedimentary facies, structures, and grain size distribution: the Red River in Oklahoma and Texas. *Trans. Gulf-Cst Ass. geol. Socs* **28**, 473–491.

SHAWE, D.R., SIMMONS, G.C. & ARCHBOLD, N.L. (1968) Stratigraphy of the Slick Rock District and vicinity, San Miguel and Dolores Counties, Colorado. *Prof. Pap. U.S. geol. Surv.* **576-A**.

SINGH, I.B. & KUMAR, S. (1974) Mega- and giant-ripples in the Ganga, Yamuna, and Son Rivers, Uttar Pradesh, India. *Sedim. geol.* **12**, 53–66.

STOKES, W.L. (1954) Some stratigraphic, sedimentary, and structural relations of uranium deposits in the Salt Wash sandstone, *U.S. Atomic Energy Comm.* **RME-3102**. 49 pp.

THAMM, J.K., KOVSCHAK, A.A. & ADAMS, S.S. (1981) Geology and recognition criteria for sandstone uranium deposits of the Salt Wash type, Colorado Plateau Province. *Final Rep, Bendix Field Eng. Corp.* **GJBX-6(81)**, Grand Junction, Colorado.

TYLER, N. (1981) *Jurassic depositional history and vanadium-uranium deposits, Slick Rock District, Colorado Plateau.* Unpublished Ph.D. Thesis. Colorado State University, Fort Collins. 251 pp.

WOOD, H.B. (1956) Relations of the origin of the host rocks to uranium deposits and ore production in the western United States. *Prof. Pap. U.S. geol. Surv.* **300**, 533–541.

YOUNG, R.G. (1978) Depositional systems and dispersal patterns in uraniferous sandstones of the Colorado Plateau. *Utah Geol.* **5**, 85–102.

Spec. Publs int. Ass. Sediment. (1983) **6**, 549–562

The Saaiplaas Quartzite Member: a braided system of gold- and uranium-bearing channel placers within the Proterozoic Witwatersrand Supergroup of South Africa

S. G. BUCK*

Geology Department, Anglo American Corporation of South Africa, P.O. Box 20, Welkom 9460, South Africa

ABSTRACT

Channel-like bodies of quartz arenite occur at a number of stratigraphic levels within a sequence of diamictic quartz wackes in the Harmony Formation (Witwatersrand Supergroup) of the Welkom Goldfield, South Africa. The lowermost level of channel bodies, referred to as the Saaiplaas Quartzite Member, contain economic concentrations of gold and uranium. The quartz arenites and quartz wackes are the deposits of an extensive alluvial fan system which prograded northwards across the Welkom Goldfield. Quartz arenite channel bodies accumulated from networks of semi-perennial sandy braided streams during periods of frequent rainfall, while the quartz wackes were deposited by ephemeral mud flows during periods of less frequent rainfall.

The precious minerals of the Saaiplaas Quartzite Member were derived locally from the reworking of the underlying Steyn Placer. Gold, uranium and pyrite were segregated from lighter sand grains during bedload transportation and accumulated as discrete deposits upon scour surfaces and within the matrices of conglomerates, the lowermost laminae of horizontally laminated quartz arenites, the upper foreset laminae of trough cross-bedded quartz arenites and the stoss surfaces of rippled quartz arenites. Reworking of the upper parts of the bedforms during the accumulation of the quartz arenites resulted in the preferential preservation of precious minerals as segregations upon scour surfaces and within horizontally laminated quartz arenites and conglomerates.

INTRODUCTION

The gold- and uranium-bearing placers of the Proterozoic Witwatersrand Supergroup of South Africa have been widely interpreted as fluvial sediments which accumulated upon a number of large alluvial fans around the margin of an extensive lacustrine basin (Pretorius, 1976; Vos, 1975; Minter, 1978). The placers vary considerably in character between: (a) regionally extensive, sheet-like placers overlying major intraformational unconformities, e.g. the Vaal Placer; (b) regionally extensive sheet-like placers upon disconfirmities, e.g. the Steyn Placer; (c) multi-storey lenticular placers, e.g. the Eldorado

* Present address: Sedimentology Research Laboratory, Department of Geology, University of Reading, Reading RG6 2AB, U.K.

0141-3600/83/0106-0549 $02.00

Placers and (d) channel placers, e.g. the Saaiplaas Quartzite Member.

The regionally extensive, sheet-like placers provide the major economic horizons of the Witwatersrand Goldfields, and as a result have been extensively described (Minter, 1970, 1976, 1978; Simms, 1969; Antrobus & Whiteside, 1964; De Kok, 1964; McKinney, 1964; Pretorius, 1964). Other types of placers, although richly mineralized, have been economically less important due to their lateral discontinuity. They are therefore poorly documented in the published literature (Minter, 1978; Smith & Minter, 1980), although a vast amount of information has been collected by mining companies during the exploration and exploitation.

This paper documents a network of channel placers in the Welkom Goldfield (Fig. 1) belonging to the Saaiplaas Quartzite Member of the Harmony Form-

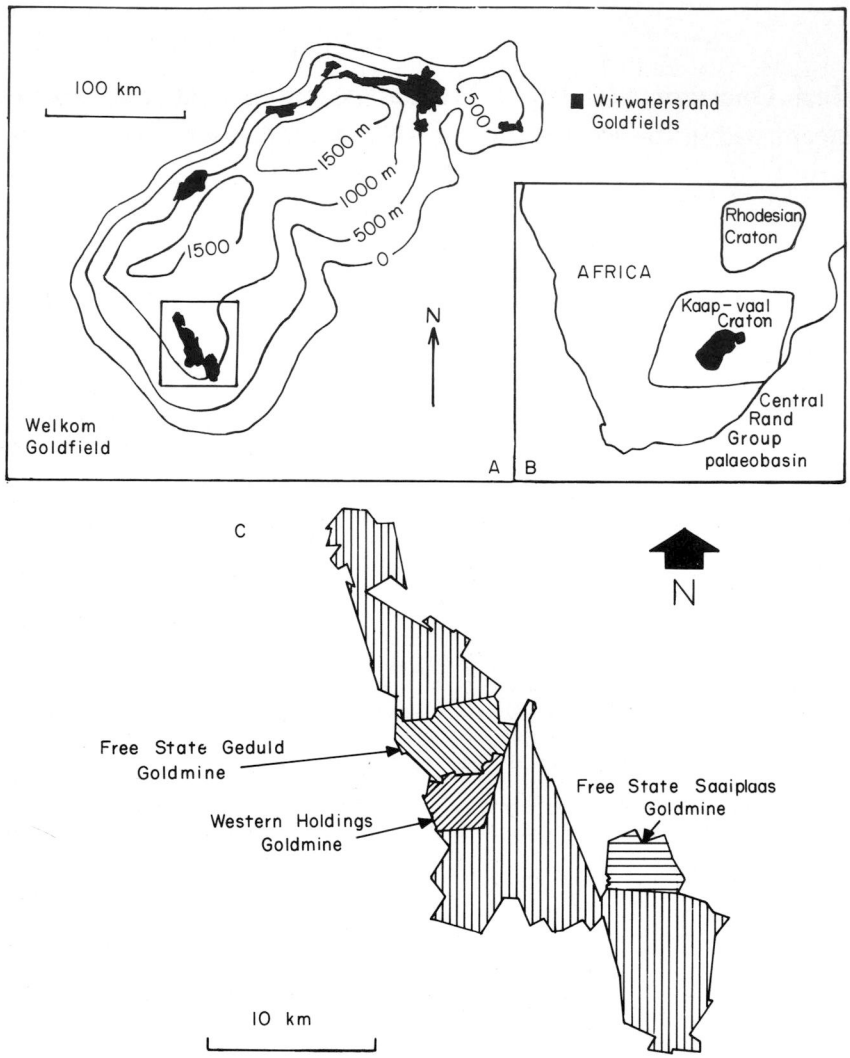

Fig. 1. Location plans. (A) Isopach map of the Central Rand Group palaeobasin showing the location of the Welkom Goldfield. (B) The position of the Central Rand palaeobasin within the Kaap–Vaal Craton. (C) Plan of the mining areas of the Welkom Goldfield showing the position of the three gold mines which exploit the Saaiplaas Quartzite Member.

ation (Fig. 2). They are unusual in that they consist of finer-grained clastic sediments than those normally forming the Witwatersrand placers, and are interbedded with quartz wackes which are texturally diamictites. These placers illustrate how the reworking of an older placer can provide a local source of precious minerals to sediments which normally would be barren.

GEOLOGICAL SETTING

The Harmony Formation occurs stratigraphically towards the centre of the Central Rand Group, where it overlies the gold- and uranium-bearing sediments of the Steyn Placer (Fig. 2). The Steyn Placer is the major economic horizon of the southern part of the Welkom Goldfield, having been exploited across an area greater than 200 km². It comprises a regionally extensive, almost sheet-like horizon of laterally coalescing channel bodies of conglomerate and quartz

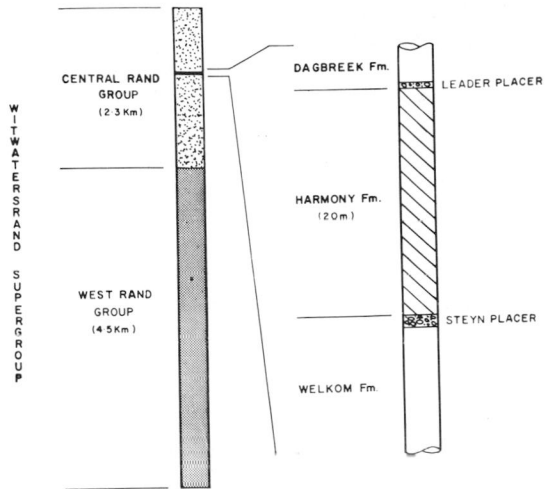

Fig. 2. Schematic stratigraphic column of the Witwatersrand Supergroup in the Welkom Goldfield showing the detailed position of the Harmony Formation.

arenite lying disconformably upon the quartz arenites of the Welkom Formation. Minter (1978) interpreted the Steyn Placer as fluvial deposits resulting from extensive lateral migration of a large network of perennial braided streams flowing towards the north-east.

The lowermost sediments of the Harmony Formation comprise laminated shales, thinly interbedded with laminated and ripple cross-laminated silts. These fine-grained sediments reach a maximum thickness of only 3 m, but occur throughout the Welkom Goldfield over an area in excess of 450 km². They are interpreted as resulting from a lacustrine transgression of the Witwatersrand Basin. The highly winnowed upper deposits of the Steyn Placer and the uniformly planar base to the lacustrine sediments is attributed to the reworking and deflation of the uppermost placer sediments as the transgression advanced.

The remaining 20 m of the Harmony Formation consist predominantly of quartz wackes exhibiting a diamictite texture; channel-like bodies of quartz arenite occur at a number of stratigraphic levels with this sequence (Fig. 3). The lowermost channel bodies are incised into the lacustrine silts and shales and into the sediments of the underlying Steyn Placer. They contain economic concentrations of gold and uranium, and are referred to as the Saaiplaas Quartzite Member. These placers have been extensively mined in the eastern parts of the Welkom Goldfield at Free State Saaiplaas Goldmine, and in the western parts of the goldfield at Free State Geduld and Western Holdings Goldmines (Fig. 1). Recent exploration within the goldfield has revealed the presence of channel placers in the intervening areas; exploitation of these will commence in the near future. Quartz arenite channel bodies at higher levels within the Harmony Formation are unfortunately barren of precious minerals, although they are identical in all other characteristics to the channel placers of the Saaiplaas Quartzite Member.

The Leader Placer overlies the Harmony Formation

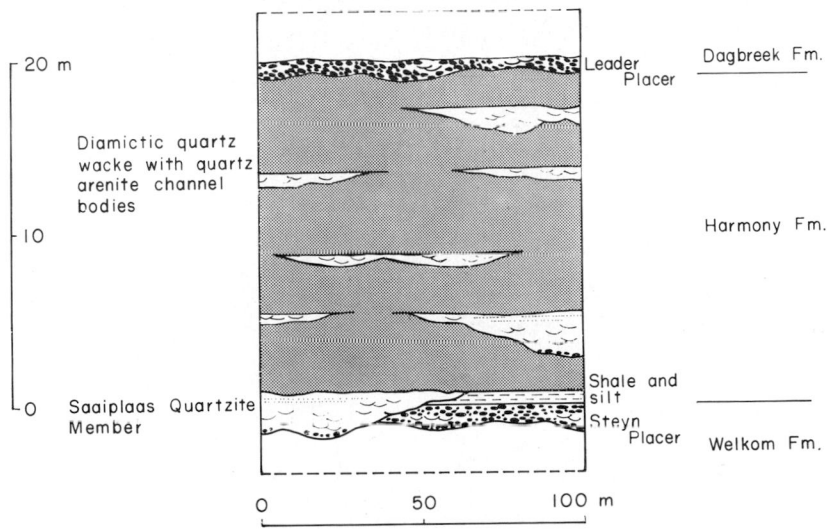

Fig. 3. A schematic section showing the stratigraphic relationship of the Saaiplaas Quartzite Member and other quartz arenite channel bodies within the Harmony Formation.

unconformably, and forms the basal deposits of the Dagbreek Formation (Fig. 2). It comprises a regionally extensive sheet-like horizon of coarse clastic sediments very similar in lithology and structure to those of the Steyn Placer, but differs considerably in assemblage of structures, palaeocurrent direction, and precious mineral content. Smith & Minter (1980) interpreted the Leader Placer as the proximal deposits of perennial braided streams flowing towards the east. The Steyn and Leader Placers accumulated within very similar depositional environments, but their differences imply rather important changes in the provenance and direction of progradation of the fluvial systems within that part of the Witwatersrand Basin.

OBSERVATIONS

The Saaiplaas Quartzite Member

Observations made in the underground workings at Free State Saaiplaas Goldmine over the past twenty years have established, in great detail, the distribution and the variation in thickness of the quartz arenite channel bodies comprising the Saaiplaas Quartzite Member in that part of the Welkom Goldfield (Fig. 4). These channel bodies exhibit a high degree of lateral connection, forming a braided distribution pattern which progressively diverges towards the north. Large areas occur between the channel bodies across which the quartz arenites are absent. A twofold hierarchy of channel body size is evident from the braided network of channel bodies. Large channel bodies comprise the greater proportion of the braided network, and vary between 100 and 250 m in width and 1·2 and 3·0 m in thickness. Smaller channel bodies, up to 50 m wide and less than 1 m thick, traverse the areas between the larger channel bodies, from which they may deviate in direction by as much as 70°.

A single, very large channel body comprises the Saaiplaas Quartzite Member in the western part of the Welkom Goldfield across Free State Geduld and Western Holdings Goldmines (Fig. 1). It is some 500 m wide, up to 6 m thick and exhibits a low-sinuosity pattern orientated in a northerly direction.

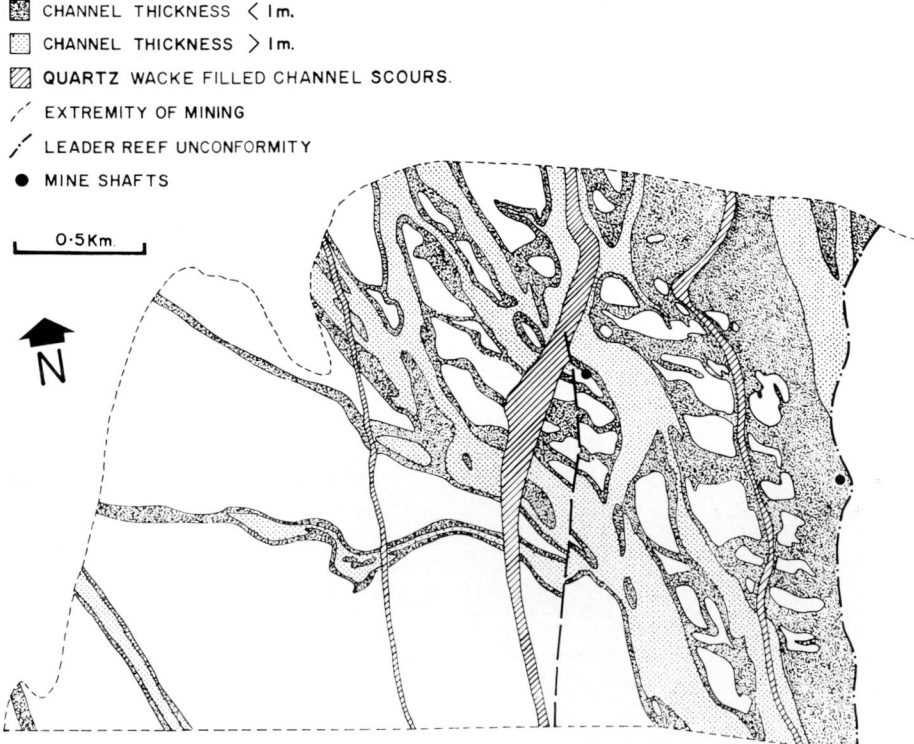

Fig. 4. Distribution and variation in thickness of the channel bodies of the Saaiplaas Quartzite Member and the quartz-wacke-filled channel scours at Free State Saaiplaas Goldmine.

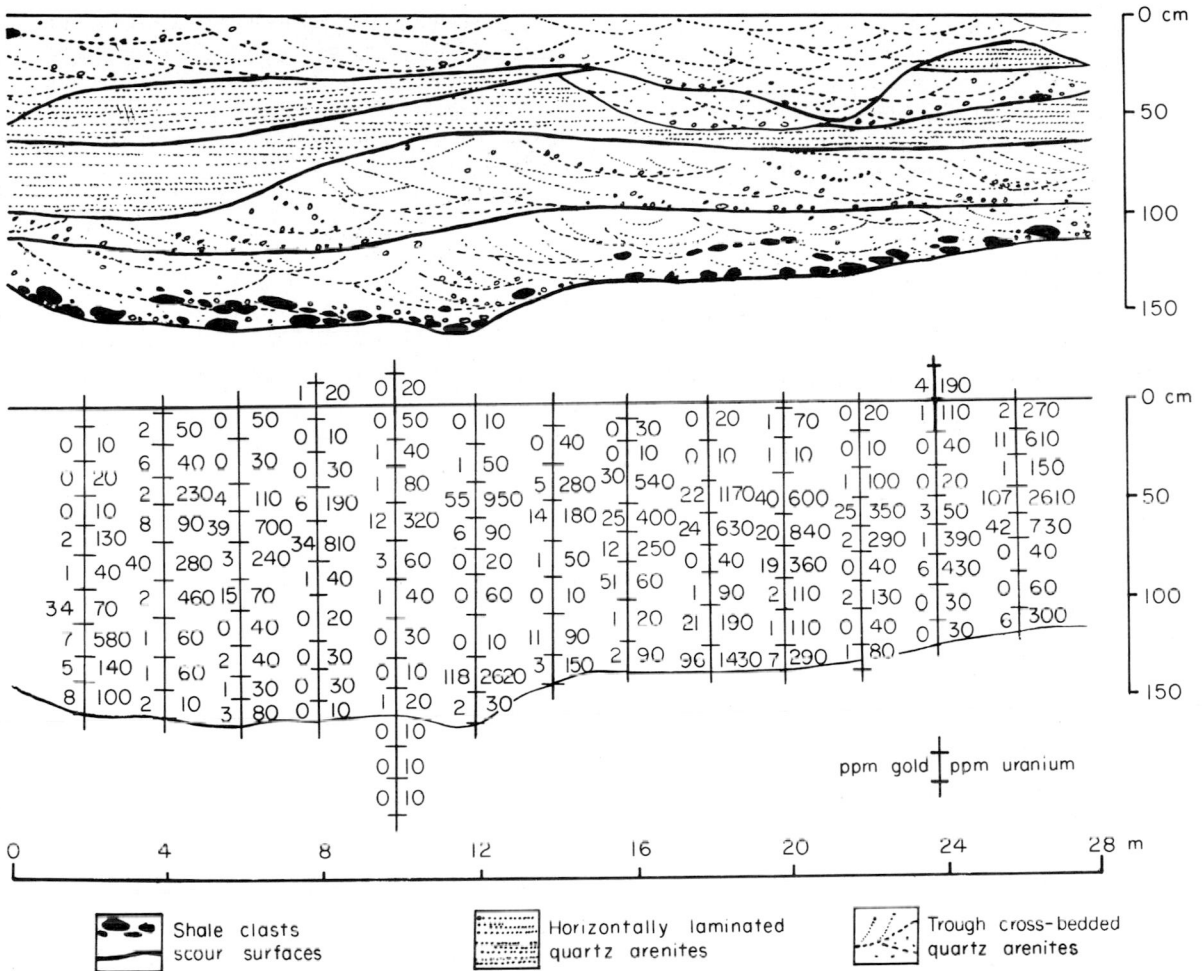

Fig. 5. Channel-fill sequence (A) related to the gold and uranium content (B) within a portion of a channel body of the Saaiplaas Quatzite Member in 9-27 Raise, Free State Saaiplaas Goldmine.

Similar sedimentary features are exhibited by all the channel bodies of the Saaiplaas Quartzite Member presently exposed throughout the Welkom Goldfield. The lithofacies consist predominantly of coarse-grained and pebbly quartz arenites, with subordinate conglomerates and shales. The quartz arenites and conglomerates are well sorted and well packed, and contain negligible amounts of clay other than in the form of shale clasts. Consequently they are white to light grey, and strongly contrast with the enclosing beds of waxy brown quartz wackes. The sand fraction of the quartz arenites comprises mostly quartz, although chert, feldspar and pyrite are important minor constituents. Pebbles are predominantly composed of vein quartz, with subordinate compositions of chert and shale, although pebbles of quartzite,

porphyry lava and quartz wacke are present in minor quantities. The pebbles are well rounded and highly spherical, with the exception of subangular and tabular shale clasts.

Primary sedimentary structures arc abundant and well preserved within the sediments of all channel bodies. The quartz arenites are most commonly either trough cross-bedded or horizontally laminatcd, rarely tabular cross-bedded and ripple cross-laminated. Conglomerates are crudely stratified, while shales exhibit a dclicate horizontal lamination. Channel bodies consist almost entirely of trough cross-bedded quartz arenites in the lower part of the channel fill sequence, and of horizontally laminated and trough cross-bedded quartz arenites in the upper part (Fig. 5).

Trough cross-bedded quartz arenites in the lower

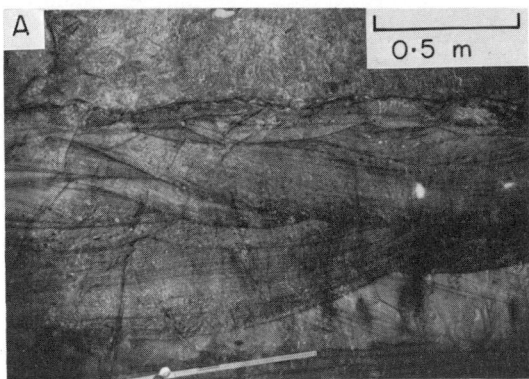
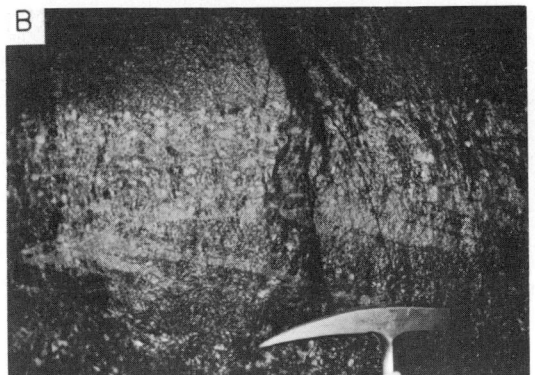

Fig. 6. Trough cross-bedded quartz-bedded quartz arenites of the Saaiplass Quartzite Member. (A) Transverse view of a large coset, containing two large clasts of shale (white areas within the right-hand side of the photograph) (B) Segregations of pyrite upon the upper parts of foreset laminae (scale: hammer head—25 cm).

parts of channel fills occur in cosets between 20 and 150 cm thick (Figs 5 and 6A). The bottom contact of each coset is a prominent scour surface and is commonly overlain by a lag deposit of pebbles and heavy minerals. Lag deposits vary laterally from scattered pebbles to conglomerate beds up to 30 cm thick, the conglomerates forming irregular beds which persist laterally for distances usually much less than 20 m. Individual trough sets range in thickness up to 60 cm, and are separated from neighbouring sets by erosional discordances (Fig. 6A). Individual set size tends to decrease upwards in the thicker cosets (Fig. 6A). Rarely, the tops of cosets are draped by thin shale beds, but are more commonly overlain erosively by the succeeding coset.

Pyrite, although finely dispersed throughout the channel sediments, occurs within the lower part of the channel fill sequence as thick segregations associated with the conglomerates and trough cross-bedded quartz arenites. The matrices of the conglomerates are commonly crudely stratified due to alternations in the composition between horizons consisting essentially of quartz sand and horizons consisting predominantly of pyritic sand. Segregations of pyrite within trough cross-bedded quartz arenites are restricted to the upper parts of the foreset laminae (Fig. 6B). These pyritic foresets decrease in thickness towards the base of a set, and usually die away before reaching the base.

Horizontally laminated quartz arenites occur in beds between 3 and 30 cm thick which extend laterally for several tens of metres. The beds erosively overlie trough cross-bedded quartz arenites, and are themselves either overlain erosively by trough cross-bedded quartz arenites or draped by thin shale beds. They consist of well-sorted, coarse-grained quartz sands in

which mineral segregations of pyrite and chert commonly enhance the horizontal lamination, while sporadic very coarse laminae tend to erosively truncate underlying laminae (Fig. 7A). Pyritic laminae, close to the base of the beds of horizontally laminated quartz arenites, are usually thicker and more laterally persistant than pyritic laminae towards the tops. Less commonly, the laminae consist of black sooty seams of carbon. These carbonaceous laminae are up to 3 mm thick, but are mostly less than 1 mm. Bedding surfaces or carbonaceous laminae commonly exhibit a parting lineation, but are rarely ornamented by a simple pattern of asymmetric ripple marks (Fig. 7B). Carbon occurs as thicker segregations upon the lee side of the ripples, while the stoss sides prove to be richly mineralized with gold and uranium.

The channel bodies of the Saaiplaas Quartzite Member are mostly incised into the lacustrine silts and shales which comprise the basal deposits of the Harmony Formation (Fig. 3). The erosional surfaces of the channel bodies against these fine-grained, cohesive sediments are characterized by abundant sole markings (Fig. 8). The style, size and profusion of these sole marks changes rapidly across the channel bodies, comprising small, isolated flutes along the outer edges of the channel bodies, becoming progressively larger and more numerous towards the centre. Current crescents occur together with flute marks away from the channel edges while, towards the centres of the channels, the sole marks become superimposed, developing into a linear system of furrow-like scours. Clasts derived from the erosion of the lacustrine shales occur scattered within the conglomerates and the trough cross-bedded quartz arenites of the channel bodies. These clasts range up

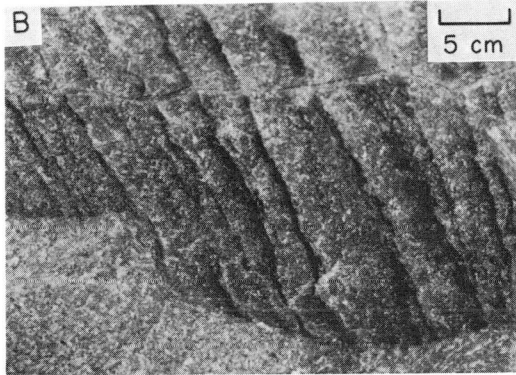

Fig. 7. Horizontally laminated quartz arenites of the Saaiplaas Quartzite Member. (A) Segregations of pyrite within the horizontal laminae (scale: hammer head—25 cm). (B) Ripple-marked under-surface of a carbonaceous lamina. The carbon occurs more thickly upon the lee-side of the ripples.

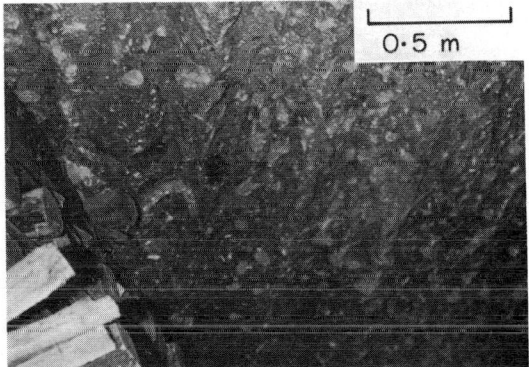

Fig. 8. Sole marks preserved upon the underside of channel bodies scoured into the lacustrine shales.

to 80 cm in diameter, are subangular and plastically deformed, and have obviously been transported only short distances.

Analysis of the foreset orientation of the trough cross-bedding within the Saaiplaas Quartzite Member reveals a unimodal palaeocurrent direction towards the north (Fig. 9B), sub-parallel to the orientation of the channel bodies. The distribution of these palaeocurrent directions is considerably more variable than those obtained for the Steyn Placer (Fig. 9A) or from the scoured surfaces between beds of quartz wacke (Fig. 9C).

The position of greatest concentration of gold and uranium in the channel placers varies from locality to locality depending upon the lithofacies present. Conglomerate beds are commonly richly mineralized, especially when the matrices consist mostly of pyrite. However, the conglomerate beds of the Saaiplaas Quartzite Member rarely attain the thickness or the wide lateral continuity necessary to contribute significantly to the overall economic potential of the channel bodies. Gold and uranium is restricted within the trough cross-bedding to foresets and trough scour surfaces which exhibit visible segregations of pyrite. Concentrations of the precious minerals can be relatively high in such locations. However, the greater proportion of unmineralized foresets results in the trough cross-bedded quartz arenites having subeconomic grades. In contrast, the lower parts of horizontally laminated quartz arenites are commonly characterized by numerous pyritic laminae and sporadic carbonaceous laminae, both of which contain high concentrations of gold and uranium. These sediments are a common and laterally extensive facies within many of the channel placers and therefore provide a major economic horizon. Prominent scour surfaces at any position within the channel-fill sequence of the channel bodies are usually accompanied by high levels of gold and uranium mineralization, and contribute enormously to the overall economic viability of the channel placers. Figure 5 illustrates a sampled portion of the Saaiplaas Quartzite Member at Free State Saaiplaas Goldmine. In this particular example the chip sampling has identified the horizontally laminated quartz arenites within the middle of the channel fill sequence to be the major horizon of mineralization.

The quartz wackes

The quartz wackes of the Harmony Formation are consistently uniform in their composition and appearance throughout the Welkom Goldfield. They consist predominantly of sand with appreciable

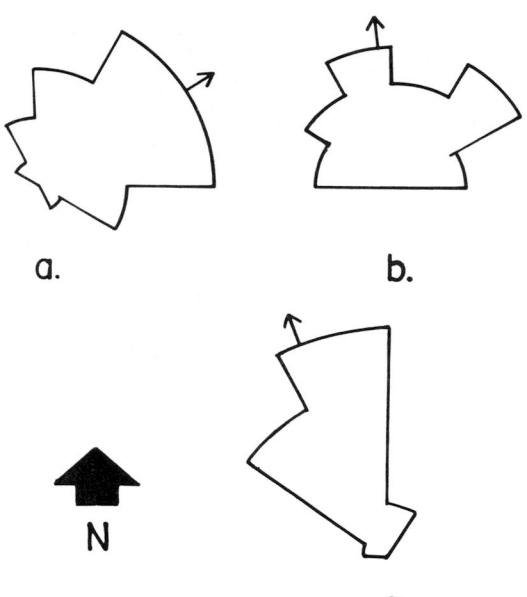

Fig. 9. Rose diagrams showing the distribution of palaeo-currents at Free State Saaiplaas Goldmine for (A) the Steyn Placer, (B) the Saaiplaas Quartzite Member, and (C) scour marks upon the bedding surfaces of the quartz wackes.

amounts of clay and small pebbles. They are poorly sorted, with the pebbles randomly scattered within a clay-rich accumulation of variably sized sand grains. The textural term 'diamictite', introduced by Flint, Sanders & Rodgers (1960a, b) for a non-sorted or poorly sorted sediment that consists of sand and/or larger particles in a muddy matrix, is most appropriate for the quartz wackes. The pebbles are composed predominantly of quartz (62%) and chert (38%), although pebbles of shale, quartzite and quartz wacke are occasionally preserved. The sand fraction consists almost entirely of quartz with scattered grains of chert, shale and quartzite, while the clay fraction is sericitic. The abundance of sericite within the quartz wackes provides a distinctive coloration described as waxy brown by the local geologists. Absence of any fabric organization or sedimentary structures within the quartz wackes is very noticeable.

The quartz wackes are indistinctly bedded in units up to 1·5 m thick. Rarely the bases of beds exhibit trough- and channel-shaped scours, which are occasionally overlain by thin veneers of muddy quartz arenites. Three large channel-scours at the base of quartz wacke beds have been observed and mapped at Free State Saaiplaas Goldmine (Fig. 4). Their courses have been determined for distances of

approximately 2 km, and have proved to be straight to slightly sinuous, trending in a northerly direction. They vary between 20 and 150 m in width and are up to 6 m deep. Bends of the channel-scours exhibit asymmetric cross-sections with steeply sloping banks along the outer portion of the curve, while straighter reaches of the channel-scours have more symmetrical cross-sections. Muddy quartz arenites and small pebble conglomerates occur as thin, impersistent basal deposits upon the flanks of the channel-scours, while the remainder of the channel-fill comprises diamictic quartz wackes. Palaeocurrents, determined from the asymmetry of the trough scours, reveal a unimodal distribution towards the north (Fig. 9C).

Thin beds of quartz arenite, occurring within the quartz wackes, are very obvious due to the strong contrast in their colours. These quartz arenite beds are commonly deformed and appear as horizons of detached, pod-like bodies (Fig. 10) similar to the pillow structures described by Conybeare & Crook (1968).

Fig. 10. A pod-like and deformed body of quartz arenite (light area in the centre of the photograph) within diamictic quartz wacke.

INTERPRETATION

The Saaiplaas Quartzite Member

The channel-like geometry, the abundance of traction deposits, repetition of fining-upward facies sequences and the strongly unimodal palaeocurrent pattern of the Saaiplaas Quartzite Member can be used to interpret its depositional environment. These features are comparable with the deposits of modern fluvial environments (Collinson, 1978; Walker & Cant, 1979). The large numbers of channel bodies comprising the Saaiplaas Quartzite Member are therefore interpreted as the deposits of an extensive fluvial system which traversed the Welkom Goldfield in a northerly direction; each quartz arenite channel body is considered to represent the deposits of a single river. The highly connected pattern of the channel bodies closely resembles the braided patterns of river channels upon many of the larger modern alluvial fans (Bull, 1977). The fluvial system of the Saaiplaas Quartzite Member was probably comparable in size to the Kosi River (Gole & Chitale, 1966).

The absence of quartz arenites, or facies-equivalent deposits, across the areas between the channel bodies indicates that deposition, and presumably the fluvial activity, was confined to the zones defined by the channel bodies, suggesting that the rivers were entrenched into the surface of the alluvial fans (Bull, 1977).

Interpretation of the textures, structures and sequences exhibited by the channel bodies provides an insight into the fluvial processes active during deposition. The basal scoured surfaces and the conglomerates are comparable to channel lag deposits and represent erosion and residual accumulation of the coarser clastic debris within the channels during the more powerful stream flows (Allen, 1965). Ripple cross-lamination, trough cross-bedding and horizontal lamination within the quartz arenites result from the accretion of sand upon ripples, dunes and upper-flow-regime plane beds respectively, under conditions of progressively greater stream power (Allen, 1970). Vertical segregation of the individual sets of trough cross-bedding by erosional discordances is attributed to conditions of slow sedimentation rates associated with relatively high rates of bedform migration, during which erosion between successive dunes resulted in the preservation of only the lower parts of the foresets deposits (Allen, 1971). The thin beds of shale draping across quartz arenite and conglomerate beds resemble suspension deposits which accumulated during periods of slack water (Collinson, 1978).

The sole marks preserved upon the scoured bottoms of channel bodies incised into the lacustrine sediments closely resemble the sole marks illustrated by Conybeare & Crook (1968) and described by Allen (1970) as forming by the differential erosion of cohesive mud through turbulent unidirectional flow. The increase in the size and complexity of the sole marks away from the edges of the channels suggests that the fluvial conditions were progressively more turbulent in the deeper parts of the channels. Support for this is provided by shale clasts resting upon the basal scour surfaces, which increase in size towards the centre of the channel bodies.

The cosets of trough cross-bedding, characterized by a lag conglomerate and decrease upward in set size, are overlain by a thin shale drape, typical of the fining-upward facies sequences of sandy fluvial systems (Collinson, 1978). These sequences result from progressively decreasing bed-shear stresses during the waning phase of a flood episode (Turner, 1978). The shale drapes directly overlying the cosets suggest that the fluvial episodes waned in strength too quickly to allow sufficient reaction time for the bed to adjust to ripple bedforms before losing competence to transport sand (Allen, 1977).

The internal scour surfaces and carbon seams within the horizontally laminated quartz arenites also imply fluctuations in the strength of the streams. Periods of strong flow were responsible for erosion of the plane beds, and periods of slack stream flow allowed carbonaceous matter to settle from suspension (Brady & Jobson, 1973). Similar features have been described for horizontally laminated sands from Bijou Creek (McKee, Crosby & Berryhill, 1967) which have been attributed to deposition from more than one flood episode (Picard & High, 1973). The evidence of sequence repetition suggests that the channel bodies had a rather protracted history of sedimentation from rivers characterized by periodic flash floods (Schumm, 1977).

The predominance of trough cross-bedded quartz arenites in the lower part of the channel bodies, and of horizontally laminated quartz arenites in the upper part, closely resembles the channel-fill sequences of many modern, sand-dominated rivers (Collinson, 1978). In such rivers, the trough cross-bedded and horizontally laminated sand facies develop at the same time from the accretion of dunes in the deeper water of stream channels and upper-flow-regime plane beds in the shallow water across sand bars, respectively (Collinson, 1978). The lateral accretion of bar deposits across channel deposits results in the superimposition of these facies within the channel-fill sequence (Cant & Walker, 1976). This interpretation for the Saaiplaas Quartzite Member suggests that each channel body represents the zone of activity of a laterally migrating and accreting system of channels and bars (Allen, 1965).

The above interpretations of the Saaiplaas Quartzite Member are compatible with sedimentation from a system of sandy braided rivers (Miall, 1977). The South Saskatchewan River (Cant, 1978), the Platte River (Smith, 1970, 1971), and the Bijou Creek

(McKee *et al.*, 1967) have been proposed by Miall (1977, 1978) as modern examples of differing sandy braided rivers. The sediments of the Saaiplaas Quartzite Member show many similarities, as well as important differences to the deposits of all these rivers. The sedimentary facies and facies sequence are, however, almost identical to those described by Rust (1978) from the sandstone units of the Malbaie Formation of Quebec, Canada. Rust (1978) considered these sandstones to be typical of the proximal facies of semi-perennial or perennial sandy braided streams. The deposits of the Bijou Creek were re-interpreted by Rust (1978) as the proximal facies of ephemeral sandy braided streams, while the Platte and South Saskatchewan Rivers were equated with the distal facies of perennial sandy braided streams. A semi-perennial to perennial interpretation for the streams of the Saaiplaas Quartzite Member is compatible with the high degree of reworking and slow rates of aggradation indicated by its sediments.

The occurrence of high concentrations of gold and uranium associated with pyrite suggests that these minerals were concentrated together. Coetzee (1965) has shown, for other Witwatersrand placers, that these minerals are in hydraulic equilibrium with each other and with the sand fraction of the sediment. Visible concentrations of pyrite within the sediments of the Saaiplaas Quartzite Member were observed: (a) in the matrices of conglomerates, (b) upon scour surfaces, (c) within the foreset laminae of trough cross-bedded quartz arenites and (d) within the laminae of horizontally laminated quartz arenites. Concentrations of gold and uranium were also found upon the stoss-sides of ripples associated with carbonaceous laminae in the horizontally laminated quartz arenites. Flume studies of the activity of heavy minerals, within sands (Brady & Jobson, 1973; McQuivey & Keefer, 1969) and within gravels (Minter & Toens, 1970), under alluvial flow conditions have shown that heavy minerals tend to accumulate in identical locations as the segregations of pyrite observed in the Saaiplaas Quartzite Member. Petrological studies by Koen (1961), Saager (1970) and Hallbauer (1972) have revealed that many of the gold, uraninite and pyrite grains within the Witwatersrand placers exhibit shapes and surface textures characteristic of detrital particles. The gold and uranium minerals of the Saaiplaas Quartzite Member are therefore considered to have accumulated, along with pyrite, as detrital heavy mineral segregations during the deposition of sand from the bedload of the braided streams.

The channel bodies of quartz arenite which occur at higher stratigraphic levels within the Harmony Formation consist of identical sedimentary deposits as the channel bodies of the Saaiplaas Quartzite Member, but lack gold and uranium. They are interpreted as having accumulated by similar fluvial processes but with gold and uranium being absent from their detrital assemblage. The fact that the channel bodies of the Saaiplaas Quartzite Member contain gold and uranium, and are incised into the sediments of the Steyn Placer, strongly suggests that these minerals were derived from reworking of the Steyn Placer. Precious minerals were therefore locally released into the streams of the Saaiplaas Quartzite Member, where the slow rates of aggradation and the high degree of reworking caused them to be reconcentrated rather than dispersed. The development of similar fluvial conditions at later times during the accumulation of the Harmony Formation, however, lacked a local supply of gold and uranium and so the quartz arenites deposited within these channels remained barren.

The quartz wackes

An absence of organisms capable of homogenizing stratified sediments during the Proterozoic (Cloud, 1968) leads to the conclusion that the composition, diamictite texture, absence of sedimentary structure and indistinct bedding are primary depositional characteristics of the quartz wackes. These features indicate that clay, sand and pebbles accumulated together without any depositional sorting and that individual beds of quartz wacke accumulated very rapidly, most probably by mass-emplacement. The uniform character of the quartz wackes throughout the Welkom Goldfield suggests that similar depositional processes were active across a large area, and that the depositional processes were not capable of causing any regional sorting of the clastic debris. The pattern, orientation and variability of cross-sectional symmetry shown by the channel-scoured bedding surfaces of the quartz wackes closely resemble low-sinuosity fluvial channels incised into cohesive sediments (Schumm, 1977). This suggests that they result from fluvial erosion and modification of pre-existing quartz wacke beds. These erosion surfaces, together with the quartz arenite channel bodies, suggest that fluvial conditions commonly prevailed in the periods between deposition of quartz wacke beds. The deformation and pillowing of many of the thinner beds of quartz arenite into underlying quartz wacke beds indicates that the quartz wackes were poorly packed, water-saturated sediments, which acted hydroplastically when subjected to loading by overlying deposits (Conybeare & Crook, 1968).

The observed and interpreted characteristics of the quartz wackes are very similar to those of mud-flow deposits on modern, arid-region alluvial fans (Bull, 1977). Prerequisites for modern mud flows are ephemeral rainfall, rapid run-off and an abundance of detritus, especially clay and silt (Bull, 1977). Observations made by Hampton (1972) suggest that mud flows originate from landslides of water-saturated sediment which, through agitation, disintegration and the incorporation of surface waters, become remoulded into the mud flows. A mud flow moves under the influence of gravity, as a viscous mass in which clay and water combine as a fluid phase of finite cohesive strength that buoyantly supports the coarser debris (Middleton & Hampton, 1973). Deceleration of a mud flow is caused by either a reduction of slope or a loss of moisture. Some modern mud flows have been reported to flow for many kilometres on relatively low slopes, where they roll across and bury all other sediments (Fisher, 1971). Mud-flow deposits are typically less than 2 m thick, and comprise ill-sorted mixtures of sand and pebbles in a clay-rich sand matrix (Bull, 1977). They occur as lobate bodies extending down the palaeosurface for several kilometres, but are only a fraction of a kilometre in width.

Modern arid-region alluvial fans commonly consist of deposits of both mud-flow and fluvial processes (Blissenbach, 1954; Bull, 1977). Streams developed during fluvial episodes are commonly incised into mud-flow deposits in the more proximal parts of the fans, and form entrenched networks of channels which become shallower and diverge down fan (Bull, 1977).

The quartz arenites and quartz wackes of the Harmony Formation are interpreted, by analogy with modern alluvial fan deposits, to represent the aggradational products of an ancient system of alluvial fans which advanced northwards across the lacustrine shales into the Witwatersrand Basin. The quartz wackes are deposits of mud flows resulting from periods of ephemeral rainfall. Long periods of non-deposition between mud flows are indicated by the channelled and scoured bedding surfaces between succeeding quartz wacke beds. In contrast, the quartz arenite channel bodies imply periods of prolonged fluvial activity during which diverging networks of sandy braided streams incised into the surface of the alluvial fans. The interstratification of quartz arenites with quartz wackes suggests an alternation between fluvial and mud-flow conditions in response to climatic fluctuations.

DISCUSSION

The gold, uranium and pyrite minerals of the Saaiplaas Quartzite Member have been interpreted earlier as detrital heavy minerals. Their occurrence as discrete segregations associated with sedimentary structures, and the greater abundance of these segregations within certain lithofacies than others, suggests that they were deposited and preserved under specific hydraulic conditions. Consideration of the processes responsible for the sorting of heavy from light minerals, and for the accumulation of the various sedimentary structures, allows deductions to be made concerning the nature of these hydraulic conditions.

Sand transported in suspension within a slowly decelerating stream experiences sorting of the heavy from light minerals according to their relative settling velocities (Brady & Jobson, 1973). Settling velocity is related to the difference in the gravitational forces promoting deposition and the fluid lift and drag forces inhibiting settling (Allen, 1970). Heavy mineral grains settle more rapidly than similar-sized light mineral grains, and have equivalent settling velocities to larger light mineral grains. Sand moves intermittently when transported within the bedload of a stream, and requires entrainment prior to each burst of saltation or rolling. The sand therefore experiences sorting of heavy from light minerals according to differences in entrainment velocities (Brady & Jobson, 1973), which relates to the difference in the fluid lift and drag forces promoting entrainment and the gravitational and frictional forces inhibiting movement (Brady & Jobson, 1973).

The fluid lift forces are produced by the difference in fluid pressure between the lower and upper sides of a particle, and are greater for larger particles upon the bed of a stream (Brady & Jobson, 1973). The frictional forces experienced by a particle are a function of the size of the particle with respect to the size and packing of the grains within the bed (Allen, 1970). Progressively greater frictional forces occur as the size of the grains forming the stream bed increases relative to the particle size, and from increasingly more open packing arrangements of the bed grains (Brady & Jobson, 1973). The frictional and fluid forces acting upon a particle therefore relate directly to particle size. In consequence, bedload transport of sand leads to segregations of minerals of similar density and size, while the settling of sand from suspension produces graded deposits in which small heavy mineral grains occur together with larger light mineral grains of equivalent settling velocity.

McQuivey & Keefer (1969) studied the flow conditions associated with heavy mineral segregations upon the stoss-sides of ripples, and found large variations in the turbulence intensity and bed shear stress across the bedforms. The turbulence intensity and bed shear stress decreased from the troughs to the crests. Heavy minerals segregated upon the stoss-sides in response to the progressive loss of competence towards the ripple crest. McQuivey & Keefer (1969) suggested that the variation in turbulence intensity across bedforms was a prime factor controlling segregations of heavy from light minerals.

Brady & Jobson (1973) conducted flume experiments to determine the behaviour of heavy minerals within alluvial channels at various stages of discharge. They observed segregations of heavy minerals upon ripples, dunes and upper-flow-regime plane beds at progressively higher flow strengths. The heavy mineral accumulations upon ripples occurred as thin segregations restricted to the stoss-sides. Heavy mineral segregations associated with dunes occurred upon the stoss-sides, and within the topset and foreset laminae. The stoss-side accumulations comprised ripple-like segregations which migrated slowly towards the crest of the dune, where they accumulated as thick, heavy mineral-rich topset laminae. Foreset accumulations of heavy minerals were widely spaced between bundles of light mineral foresets, and resulted from the periodic avalanche of heavy mineral-rich topset deposits down the lee-side of the dune. These accumulations were confined to the upper portions of the foresets, and thinned rapidly down the foreset. Accumulations of heavy minerals within upper-flow-regime plane beds occurred as widespread segregations within the horizontal laminae. These heavy mineral segregations were most richly developed within the laminae deposited first, following a period of scour.

The work of McQuivey & Keefer (1969) and Brady & Jobson (1973) provides an excellent background from which to interpret the concentrations of gold and uranium within the Saaiplaas Quartzite Member. The high concentrations of gold and uranium, associated with conglomerates and scour surfaces, represent heavy mineral segregations developed during the more powerful fluvial discharges. These stream flows transported even the coarsest debris, and reworked large portions of older deposits. The heavy minerals accumulated along with the coarsest clastic debris as lag deposits. The rough surfaces of lag conglomerates would probably have produced large variations in the turbulence intensity of the stream flow, providing ideal conditions for the segregation of heavy minerals

(McQuivey & Keefer, 1969). Heavy mineral particles which collected upon the conglomerate beds were effectively shielded from further entrainment by the high frictional resistance created by the pebbles.

The concentrations of gold and uranium within horizontally laminated, trough cross-bedded and rippled quartz arenites represent segregations associated with the bedload movement of sand across upper-flow-regime plane beds, and lower-flow-regime dunes and ripples, respectively. The heavy minerals were sorted from the light minerals according to differences in entrainment velocity, and congregated in response to either spatial or temporal decreases in the turbulence intensity of the stream flow. In the horizontally laminated quartz arenites, heavy mineral accumulations are greatest above scour surfaces, and were deposited in response to decreasing turbulence intensity of the stream flow during the initial waning phases of floods (Brady & Jobson, 1973).

The segregations of heavy minerals in the upper parts of foreset laminae within trough cross-bedding closely resemble the segregations upon dunes described by Brady & Jobson (1973). Strong erosional discordances between successive trough sets indicate slow sedimentation and rapid rates of bedform migration, during which erosion between the dunes resulted in the severe reworking of the upper parts of the beforms, allowing preservation of only the lower parts of foresets (Allen, 1971). Heavy mineral-rich topset, stoss-side and upper foreset deposits of these ancient dunes are not preserved. The heavy mineral segregations occurring within the trough cross-bedding are therefore portions of foresets which escaped erosion. In contrast, reworking of the upper parts of the upper-flow-regime plane bed deposits would not have affected the rich accumulations of heavy minerals occurring within the lower part of these beds. The greater concentrations of precious minerals within the horizontal lamination, rather than the trough cross-bedding is therefore not due to the selective deposition of heavy minerals upon plane beds rather than dunes, but results from the preservation of only the heavy mineral segregations associated with the lower parts of the bedforms.

Brady & Jobson (1973) observed that ripples develop upon the surface of upper-flow-regime plane beds when the strength of flow declines rapidly. The rippled cabonaceous laminae within the horizontally laminated quartz arenites of the Saaiplaas Quartzite Member are considered to be analogous structures. These changing hydraulic conditions appear to have been favourable for the segregation of gold and uranium upon the stoss-sides of the ripples. Carbon occurring in the lee-sides of the ripple marks probably represents still more sluggish flow conditions, during which organic detritus settled in the lee of the ripples. Thicker carbonaceous laminae within the horizontally laminated quartz arenites are interpreted as thicker and more laterally extensive accumulations of organic detritus. It is possible that algal growth accompanied the accumulation of this organic matter (Hallbauer, 1975).

CONCLUSIONS

The quartz arenites and quartz wackes of the Harmony Formation are the deposits of an extensive alluvial system which prograded northwards into the Witwatersrand Basin. The horizons of quartz arenite channel bodies developed as diverging networks of braided alluvial channels which entrenched the surface of the alluvial fans. Each channel body represents the zone of lateral migration and accretion of a system of sandy braided streams and is the cumulative deposit of many flood episodes. The quartz wackes are the deposits of mud flows and resulted from periods of ephemeral rainfall. Long periods of non-deposition between successive mud flows are marked by channelled and scoured surfaces which developed from the erosional modification of the fan surface. The interbedding of quartz arenites and quartz wackes resulted from alternations in deposition by fluvial and mud-flow processes in response to climatic fluctuations, particularly periods of more frequent and less frequent rainfall.

Gold and uranium were not normal detrital constituents of the sandy braided streams responsible for the deposition of the quartz arenite channel bodies of the Harmony Formation. They were, however, locally introduced from the erosion and reworking of the Steyn Placer during the entrenchment of the channels of the Saaiplaas Quartzite Member. Conditions of slow sedimentation rates associated with a high degree of reworking within the streams led to the re-concentration of gold and uranium as segregations within horizontally laminated, trough cross-bedded and rippled quartz arenites, within conglomerates and upon scour surfaces. The segregations were produced by the sorting of heavy from light minerals according to their difference in entrainment velocity, and probably accumulated in response to spatial and temporal decreases in the turbulence intensity. The low concentration of precious minerals within trough cross-bedded quartz arenites is not due to the sparse deposition of gold and uranium upon the dune bedforms, but probably resulted from the non-preservation of the heavy mineral-rich topset and upper foreset deposits of the dunes.

ACKNOWLEDGMENTS

Part of the data presented in this paper represents some 20 years of observations made during the mining of the Saaiplaas Quartzite Member. The author would therefore like to acknowledge J. P. Pienaar, G. Els, G. Cantello, R. Kidger, D. Dewar and numerous unknown geologists, for the many arduous hours spent in the collection of these data. I would especially like to thank W. E. L. Minter for his advice and encouragment during the preparation of this paper. The permission granted by the Anglo-American Corporation of South Africa to publish this information is gratefully acknowledged.

REFERENCES

ALLEN, J.R.L. (1965) A review of the origin and characteristics of recent alluvial sediments. *Sedimentology*, **5**, 89–91.

ALLEN, J.R.L. (1970). *Physical Processes of Sedimentation.* Allen & Unwin, London. 248 pp.

ALLEN, J.R.L. (1970) Instantaneous sediment deposition rates reduced from climbing ripple cross-lamination. *J. geol. Soc. London*, **127**, 553–561.

ALLEN, J.R.L. (1977) Changeable rivers: some aspects of their mechanics and sedimentation. In: *River Channel Changes* (Ed. By K.J. Gregory), pp. 15–45. Wiley, Chichester.

ANTROBUS, E.S.A. & WHITESIDE, H.C.M. (1964) The geology of certain mines in the East Rand. In: *The Geology of Some Ore Deposits of Southern Africa* (Ed. by S.E. Haughton), pp. 125–160. Geological Society of South Africa.

BLISSENBACH, E. (1954) Geology of alluvial fans in semi-arid regions. *Bull. geol. Soc. Am.* **65**, 175–189.

BRADY, L.L. & JOBSON, H.E. (1973) An experimental study of heavy mineral segregation under alluvial flow conditions. *Prof. Pap. U.S. geol. Surv.* **562-k,** 38 pp.

BULL, W.B. (1977). The alluvial-fan environment. *Prog. Phys. Geogr.* **1**, 222–270.

CANT, D.J. (1978) Development of a facies model for sandy braided river sedimentation: comparison of the South Saskatchewan River and Battery Point Formation. In: *Fluvial Sedimentology* (Ed. by A.D. Miall). *Mem. Can. Soc. Petrol. Geol., Calgary*, **5**, 627–640.

CANT, D.J. & WALKER, R.G. (1976) Development of a braided-fluvial model for the Battery Point Sandstone, Quebec. *Can. J. Earth Sci.* **13**, 102–119.

CLOUD, P.E. (1968) Pre-metozoan evolution and the origin of the Metozoa. In: *Evolution and Environment* (Ed. by E.T. Drake), pp. 1–72. Yale University Press, New Haven.

562 *S. G. Buck*

COETZEE, F. (1965) Distribution and grain size of gold, uraninite and certain other heavy minerals in gold bearing reefs of the Witwatersrand Basin. *Trans. geol. Soc. S. Afr.* **68**, 61–88.

COLLINSON, J.D. (1978) Alluvial sediments. In: *Sedimentary Environments and Facies* (Ed. by H.G. Reading), pp. 15–60. Blackwell Scientific Publications, Oxford.

CONYBEARE, C.E.B. & CROOK, K.A.W. (1968) Manual of sedimentary structures. *Bull. Bur. Miner. Resour. Aust.* **102**, 327 pp.

DE KOK, W.P. (1964) The geology and economic significance of the West Wits Line. In: *The Geology of Some Ore Deposits in Southern Africa* (Ed. by S. H. Haughton), pp. 323–391. Geological Society of South Africa.

FISHER, R.V. (1971) Features of a coarse grained, high concentration fluids and their deposits. *J. sedim. Petrol.* **41**, 916–927.

FLINT, SANDERS & RODGERS (1960a) Symmictite, a name for non-sorted terrigenous sedimentary rocks that contain a wide range of particle sizes. *Bull. geol. Soc. Am.* **71**, 507–510.

FLINT, SANDERS & RODGERS (1960b) Diamictite, a substitute term for symmictite. *Bull. geol. Soc. Am.* **71**, 1809.

GOLE, C.V. & CHITALE, S.V. (1966) Inland delta building activity of the Kosi River. *J. Hydraul. Div. Am. Soc. civ. Engrs* **92**, 111–126.

HALLBAUER, D.K. (1972) Distribution and size of gold particles in reefs. *Rep. Chamb. Mines S. Afr.* **10**, 24–29.

HALLBAUER, D.K. (1975) Plant origin of the Witwatersrand carbon. *Miner. Sci. Engrg* **7**, 111–113.

HAMPTON, M.A. (1972) The role of subaqueous debris flow in generating turbidity currents. *J. sedim. Petrol.* **42**, 775–793.

KOEN, G.M. (1961) The genetic significance of the size distribution of uraninite in the Witwatersrand bankets. *Trans. geol. Soc. S. Afr.* **64**, 23–46.

MCKEE, E.D., CROSBY, E. J. & BERRYHILL, H.L. (1967) Flood deposits, Bijou Creek, Colorado, June, 1965. *J. sedim. Petrol.* **37**, 829–851.

MCKINNEY, J.S. (1964) Geology of the Anglo American group of mines in the Welkom area, Orange Free State Goldfield. In: *The Geology of Some Ore Deposits in Southern Africa* (Ed. by S.H. Haughton), pp. 451–506. Geological Society of South Africa.

MCQUIVEY, R.S. & KEEFER, T.N. (1969) The relation of magnetite over ripples. *Prof. Pap. U.S. geol. Surv.* **650-D**, D244–D247.

MIALL, A. D. (1977) A review of the braided-river depositional environment. *Earth Sci. Rev.* **13**, 1–62.

MIALL, A.D. (1978) Lithofacies types and vertical profile models in braided river deposits: a summary. In: *Fluvial Sedimentology* (Ed. by A.D. Miall). *Mem. Can. Soc. Petrol. Geol., Calgary*, **5**, 597–604.

MIDDLETON, G.V. & HAMPTON, M.A. (1973) Sediment gravity flows: mechanics of flow and deposition. In: *Turbidites and Deep Water Sedimentation* (Ed. by G.V. Middleton and A. Bouma). *Soc. econ. Petrol. Miner. Short Course Notes.*

MINTER, W.E.L. (1970) Gold distribution related to the sedimentology of a Precambrian Witwatersrand conglom-erate, South Africa, as outlined by moving averages analysis. *Econ. Geol.* **65**, 963–969.

MINTER, W.E.L. (1976) Detrital gold, uranium and pyrite concentrations related to the sedimentology of the Precambrian Vaal Reef placer, Witwatersrand, South Africa. *Econ. Geol.* **71**, 157–176.

MINTER, W.E.L. (1978) A sedimentological synthesis of placer gold, uranium and pyrite concentrations in Proterozoic Witwatersrand sediments. In: *Fluvial Sedimentology* (Ed. by A.D. Miall). *Mem. Can. Soc. Petrol. Geol., Calgary*, **5**, 801–829.

MINTER, W.E.L. & TOENS, P.D. (1970) Experimental simulation of gold deposition in gravel beds. *Trans. geol. Soc. S. Afr.* **73**, 89–98.

PICARD, M.D. & HIGH, L.R. (1973) *Sedimentary Structures of Ephemeral Streams. Developments in Sedimentology* **17**, Elsevier, Amsterdam, 215 pp.

PRETORIUS, D.A. (1964) The geology of the South Rand Goldfield. In: *The Geology of Some Ore Deposits in Southern Africa* (Ed. by S.H. Haughton), pp. 219–282. Geological Society of South Africa.

PRETORIUS, D. A. (1976) The nature of the Witwatersrand gold–uranium deposits. In: *Handbook of Strata Bound and Stratiform Ore Deposits* (Ed. by K.H. Wolf), pp. 29–88. Elsevier, Amsterdam.

RUST, B.R. (1978) Depositional models for braided alluvium. In: *Fluvial Sedimentology* (Ed. by A.D. Miall). *Mem. Can. Soc. Petrol. Geol., Calgary*, **5**, 605–626.

SAAGER, R. (1970) Structures in pyrite from the Basal Reef in the Orange Free State Goldfield. *Trans. geol. Soc. S. Afr.* **73**, 29–46.

SCHUMM, S. A. (1977) *The Fluvial System.* Wiley, New York. 338 pp.

SIMMS, J.F.M. (1969) *The stratigraphy and palaeocurrent history of the upper division of the Witwatersrand System on President Steyn Mine and adjacent areas in the Orange Free State Goldfield with specific reference to the auriferous reefs.* Unpublished Ph.D. Thesis, University of Witwater-srand. 181 pp.

SMITH, N.D. (1970) The braided stream depositional environment: comparison of the Platte River with some Silurian clastic rocks, north-central Appalachians. *Bull. geol. Soc. Am.* **81**, 2993–3014.

SMITH, N.D. (1971) Transverse bars and braiding in the Lower Platte River, Nebraska. *Bull. geol. Soc. Am.* **82**, 3407–3420.

SMITH, N.D. & MINTER, W.E.L. (1980) Sedimentological controls of gold and uranium in two Witwatersrand paleoplacers. *Econ. Geol.* **75**, 1–14.

TURNER, B.R. (1978) Sedimentary patterns of uranium mineralization in the Beaufort Group of the Southern Karoo (Gondwana) Basin, South Africa. In: *Fluvial Sedimentology* (Ed. by A.D. Miall). *Mem. Can. Soc. Petrol. Geol., Calgary*, **5**, 831–848.

VOS, R.G. (1975) An alluvial plain and lacustrine model for the Precambrian Witwatersrand deposits of South Africa. *J. sedim. Petrol.* **45**, 480–493.

WALKER, R.G. & CANT, D.J. (1979) Sandy fluvial systems. In: *Facies Models* (Ed. by R.G. Walker). *Geosci. Can.* **1**, 23–32. Geological Association of Canada.

Spec. Publs int. Ass. Sediment. (1983) **6**, 563–575

Gold distribution in relation to depositional processes in the Proterozoic Carbon Leader placer, Witwatersrand, South Africa

M. NAMI

Mining Technology Laboratory, Chamber of Mines of South Africa Research Organisation, P.O. Box 91230, Auckland Park 2006, South Africa

ABSTRACT

A detailed study of 900 m² of the Carbon Leader placer established the relationship between sedimentological features and the distribution pattern of gold.

The Carbon Leader placer at the site investigated displayed many features characteristic of a fluvial environment. Two different subfacies were recognized, one consisting of lenticular-shaped conglomerate/quartzite units considered to have been formed by active channelized flow and the other, a single sheet-like conglomerate, often with underlying thin carbonaceous seams and probably formed by sheet floods. The study of the gold distribution revealed that the stream channels, representing zones of high activity, yielded very erratic gold values while the adjacent plains had consistently high gold values.

The distribution pattern of gold as observed in the area studied is considered to be the result of contrasting hydrodynamic processes which prevailed within the two sub-environments.

INTRODUCTION

Prediction of the gold distribution pattern in a fluvial environment depends largely on the understanding of the processes which are responsible for the concentration of this mineral. Few attempts have been made to identify the factors controlling gold distribution. Mechanisms such as entrapment of heavy minerals in open framework gravel, reworking and winnowing of the sands and gravels and sorting processes in the formation of megaripples and bars are discussed frequently in recent literature concerning the Witwatersrand placer deposits (Minter, 1976, 1978; Pretorius, 1981; Smith & Minter, 1980). These mechanisms could produce small-scale variations in gold distribution but cannot be easily used to predict larger-scale variations.

The controlling factors of gold distribution, particularly in a distal placer deposit, are not clearly understood, partly because of a lack of detailed palaeoenvironmental analysis and partly because of

variation in gold values and changes in sedimentary parameters.

The sedimentological investigation of the Carbon Leader Reef reported here is an attempt, in a selected area, to identify which sedimentological features are related to the gold distribution and to determine the processes which controlled the concentration of gold.

For the purpose of this investigation a site was chosen at the Blyvooruitzicht No. 2 shaft pillar (5-22 drive) where the Carbon Leader Reef exhibits considerable variation in sedimentological features and the gold content fluctuates substantially.

GEOLOGICAL SETTING

The Carbon Leader Reef is one of the major gold-producing conglomerate zones in the Central Rand Group and lies within the lowest portion of the Group in the West Wits Line of Carletonville. The West Wits Line is part of the Witwatersrand basin and stretches between Randfontein and Mooi River (Fig. 1). The Carbon Leader at the site of investigation is approximately 65 m below the Main Reef placer and

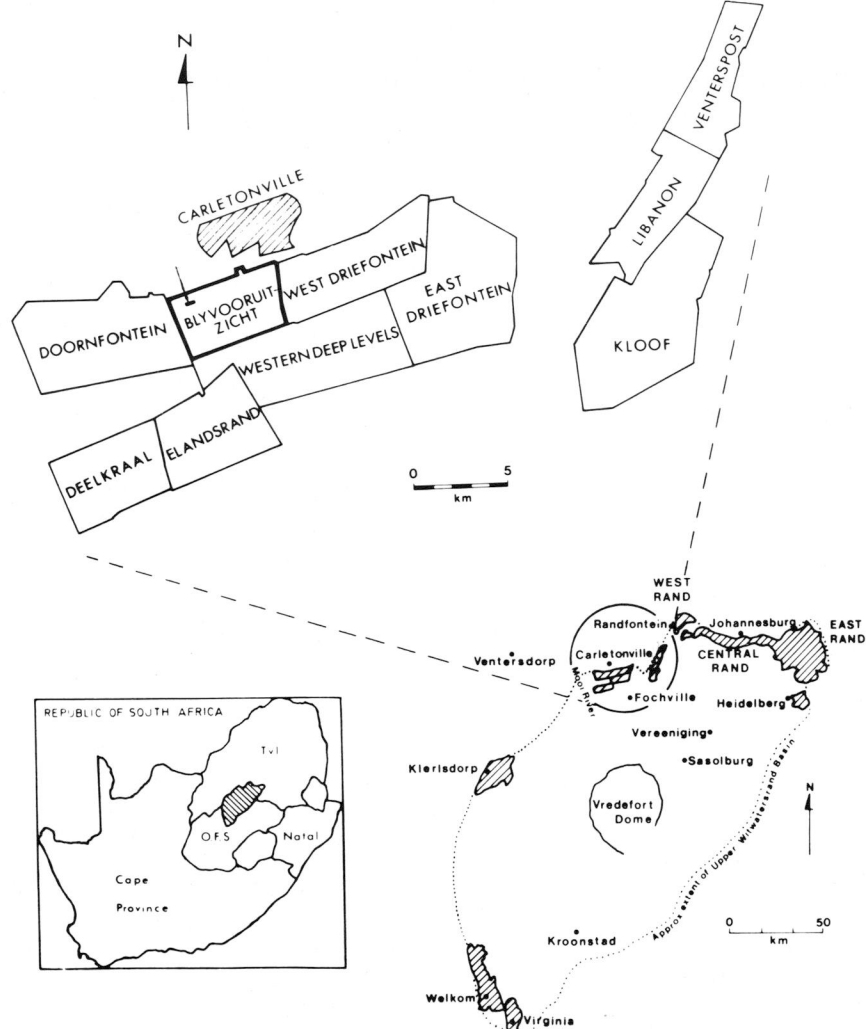

Fig. 1. Location of the Blyvooruitzicht gold mine in the Witwatersrand Basin. An arrow indicates the position of the area studied.

60 m above the Jeppestown Shales of the West Rand Group (Fig. 2). This reef encompasses a sizeable area and is extensively worked in the Blyvooruitzicht, West Driefontein, Doornfontein and Western Deep Levels gold mines. It also occurs in the eastern part of the West Wits Line at East Driefontein, Libanon and Venterspost gold mines, where it has not yet been exploited. This conglomerate horizon dips at an angle of approximately 20° to the south and is truncated in the north by the Black Reef Series of the Transvaal Supergroup. This truncation coincides with the northern boundaries of the Blyvooruitzicht and West Driefontein gold mines. The placer deposit is typically developed in these mines and consists of a thin sheet-like conglomerate with an underlying, friable layer of carbonaceous material. The thickness of the conglomerate is normally about 10 cm and seldom exceeds 30 cm. Towards the east and the west of these mines the thickness of the reef increases, where it consists of several conglomerate bands in the Doornfontein gold mine and numerous layers in the Libanon gold mine.

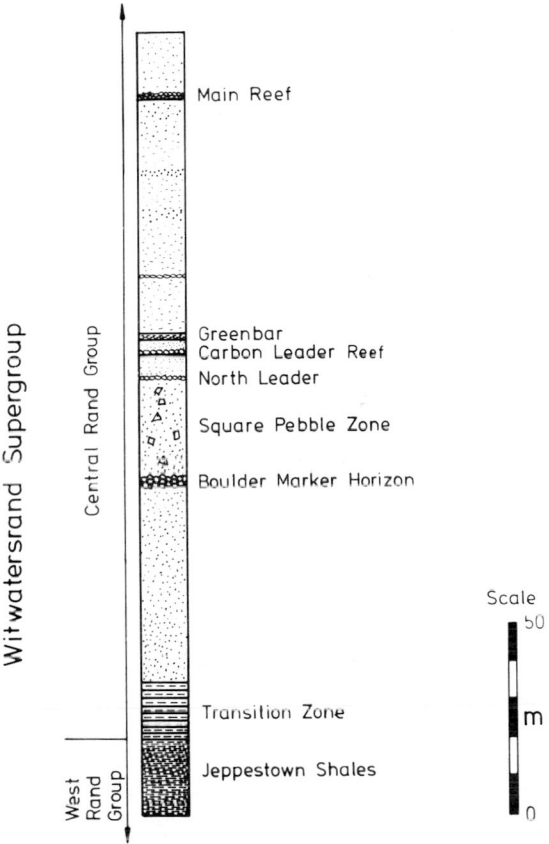

Fig. 2. Stratigraphic divisions of the lower part of the Central Rand Group.

METHODS AND PROCEDURES

In order to obtain quantitative information regarding the detailed gold distribution, the placer deposit was mined in a series of blocks 4 × 4 m parallel to the 5-22 reef drive (Fig. 3).

After ledging the north side of the drive, a 2 m heading was developed from where the mining commenced in both easterly and westerly directions. The blocks were extracted in eight blasts each over the full width (4 m) advancing 0·5 m at a time. The mean stoping width was 0·9 m and approximately 40 tons of rock was mined in each block. At a later stage of the experiment a series of blocks designated north blocks were mined up dip (north raise). Geological mapping was carried out on the exposed faces and a number of sedimentological parameters were measured.

The gold distribution was determined by contiguous chip sampling, and by bulk sampling, to obtain the detailed gold distribution and the overall trend in gold values, respectively.

Contiguous chip samples were taken on the four peripheral faces as well as on the centre face which was exposed by blasting. In addition nine samples were taken in the centre of each face after every blast. These chip samples encompassed the conglomerate layer and were 15 cm wide, 10 cm high and approximately 2·5 cm deep. Where the placer unit was thicker than 10 cm, several samples were taken above each other in order to sample the entire unit.

After each blast the broken rock was collected and transported to a crusher plant on the surface, where it underwent a series of crushing and splitting processes. This resulted in eight representative samples which were assayed to determine the gold content. The collected samples from blasts as well as samples from collecting mud and sweepings determined the gold values of the extracted block.

GENERAL DEPOSITIONAL ENVIRONMENT OF THE CARBON LEADER REEF

A detailed palaeoenvironmental analysis of the whole Carbon Leader Reef has not been attempted here since it would require a much larger-scale investigation, beyond the scope of this study. However, the reef in the project site displays the following features which are consistent with fluvial environments.

(1) Clear evidence of erosion at the base of some conglomerate bands and pronounced channeling.

(2) Cross-stratified quartzite within the pebble layers and a hanging wall with unimodal palaeocurrent directions.

(3) Poorly sorted and sand-supported conglomerates.

(4) Lenticular-shaped and laterally non-persistent nature of some conglomerates.

Two sedimentary sub-facies have been distinguished: sub-facies A, a single sheet-like conglomerate, and sub-facies B, lenticular-shaped conglomerate/quartzite.

Sub-facies A

In the eastern side of the mined-out area and in the central part of the north raise the Carbon Leader Reef exhibits sheet-like deposits of conglomerate with an average thickness of 7 cm. There is rarely evidence of pronounced erosion at the base of the conglomerate

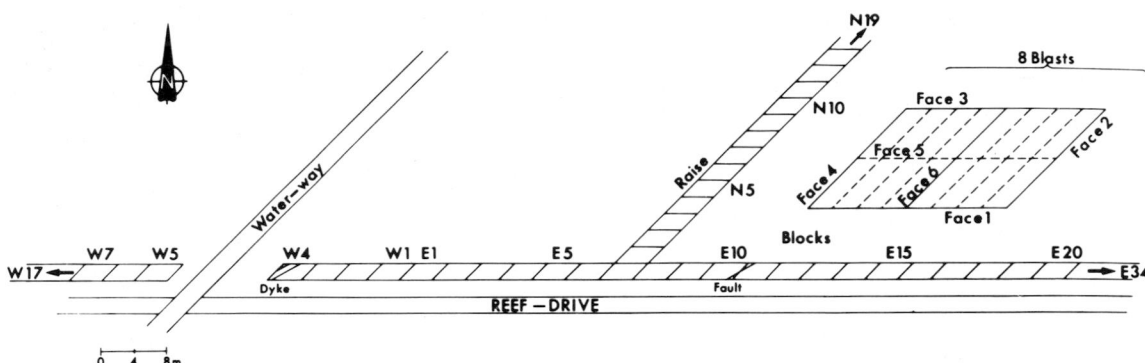

Fig. 3. Schematic perspective representation of the blocks extracted from Blyvooruitzicht 5-22 reef drive.

except in the areas where shallow and small channels occur within these deposits. Texturally the conglomerate is of a sand-supported type, and pebbles are enclosed in a fine- to medium-grained quartzite matrix. Pebble-supported conglomerate occurs infrequently in localities where downcutting exists or the thickness of the reef is reduced substantially. There is a lack of sedimentary structures such as cross-stratification or pebble imbrication in the conglomerate. Internally the conglomerate band in some blocks exhibits two or three consecutive pebble bands each of only one-pebble thickness. These bands are separated by a scoured surface or a thin layer of carbonaceous material. Sub-facies A overlies a light greenish-grey, trough cross-stratified quartzite forming the footwall. Normally a thin sericitic/chloritic shale up to 4 cm thick occurs between the conglomerate and footwall quartzite (Fig. 4A).

In some areas, for example in the lower part of the north raise and in the eastern part of the study site, the deposits underlying the sub-facies consist of isolated trough-shaped dark grey quartzite beds. These beds have a maximum thickness of 20 cm and differ from the footwall quartzite in their internal sedimentary structures and on the basis of lithology (Fig. 4B).

Sub-facies B

Sub-facies B is characteristic of the western part of the study site and upper part of the north raise. This sub-facies consists of several conglomerate bands alternating with quartzite. The thickness varies from 20 to 60 cm. Usually there are only two bands of conglomerate, but three bands may also occur (Fig. 5). The conglomerates are poorly sorted, and both pebble-supported and sand-supported conglomerate can be seen in this sub-facies. The pebble-supported

conglomerates usually occur in the basal and intermediate bands. The lower band frequently displays irregular basal contents, which may have a relief of up to 30 cm. Unlike sub-facies A, imbrication is sometimes apparent in the pebble-supported bands. The conglomerate bands are lenticular in shape, not exceeding 8 m in lateral extent. The basal conglomerate is gradationally overlain by planar or trough cross-bedded quartzite. The quartzite in turn is overlain by another conglomerate band with an erosive base. In one instance there is a marked change from large-scale tabular cross-bedded quartzite to structureless quartzite, which shows horizontal bedding when subjected to X-ray radiography. The thickness of the quartzite is variable and ranges from 10 to 40 cm.

Both sub-facies in the upper part transitionally pass into dark grey, large-scale trough cross-bedded quartzite of the hanging wall. The composition of the conglomerate components is similar in both sub-facies and consists of approximately 91·5% glassy, smokey and milky quartz, 5% blue quartz, 2% black quartz and 1·5% chert. Occasionally quartzite and chloritoid shale pebbles can be seen. There are no distinct variations in the pebble sizes of components of the conglomerate bands. The mean pebble size is 12 mm ($-3·5 \phi$) and the maximum pebble size rarely exceeds 30 mm.

Detailed sedimentological logs of the different sub-facies in the area of investigation are illustrated in Fig. 6.

CARBONACEOUS MATTER

The occurrence of the carbonaceous matter, which is a distinct characteristic of the Carbon Leader Reef, adds another element in distinguishing the two sub-facies. Generally the majority of the carbonaceous

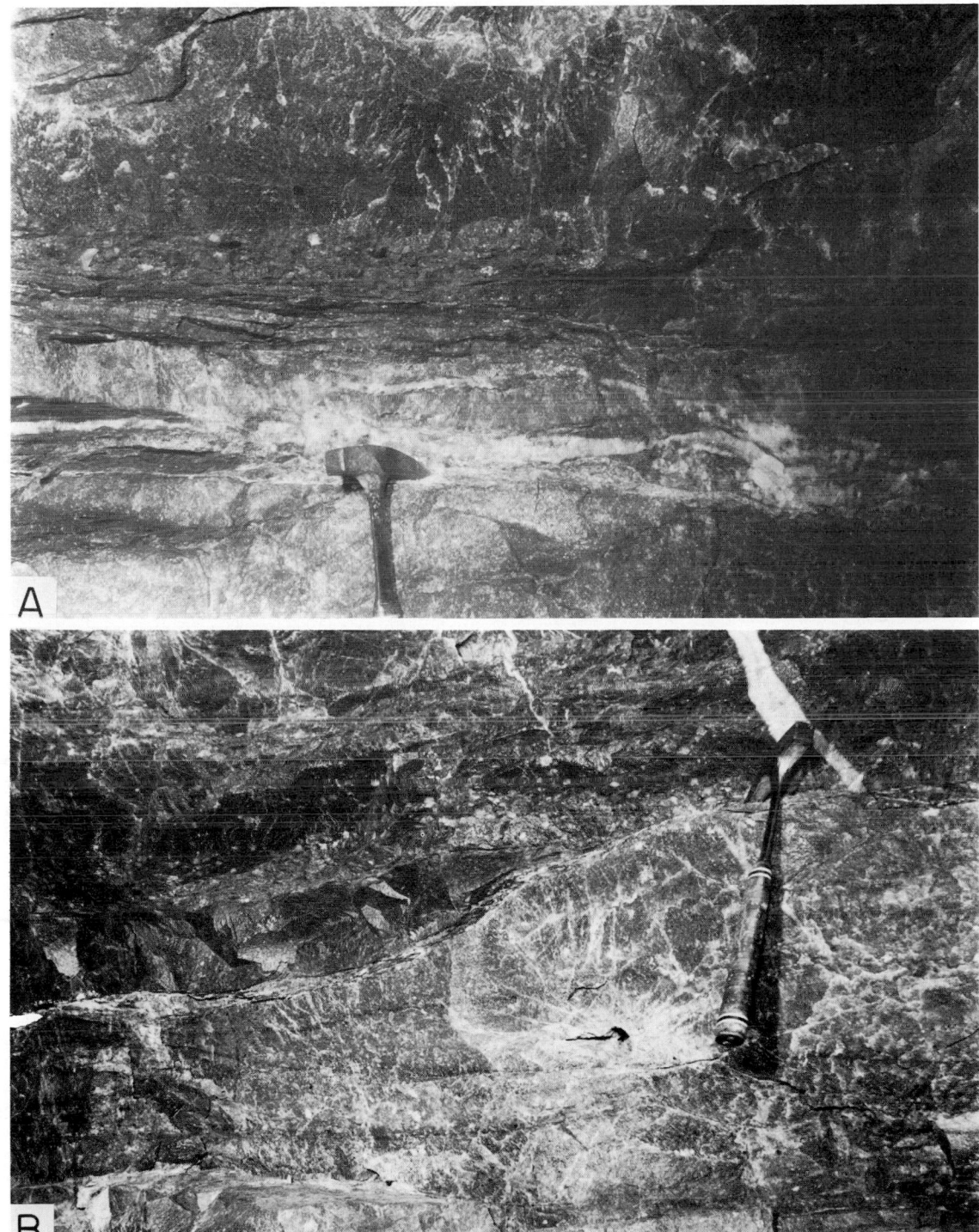

Fig. 4. Examples of sub-facies A in the eastern part of the study area and Central part of north raise. (A) Typical area of sub-facies A, note the presence of sericitic/chloritic shale at the base. (B) Trough-shaped dark grey quartzite in the base of sub-facies A.

Fig. 5. Exposed multi-band conglomerate/quartzite bands of sub-facies B.

matter is associated with sub-facies A. Occasionally small patches of this material can be seen in the western part of the area studied. This material occurs in various forms, namely: (1) at the base of the conglomerate band; (2) on the scoured surfaces and around the pebbles; and (3) as 'fly speck' carbon.

At the base of the conglomerate the carbonaceous matter occurs as single or multiple layers of columnar carbonaceous matter with the filamentous structures perpendicular to the bedding plane (Fig. 7). The thickness of the columnar carbonaceous matter varies considerably and a maximum thickness was recorded in blocks 10, 11 and 18. Often the thicker layers consist of two or more individual layers separated by detrital grains. The distribution pattern of the carbonaceous matter within the study area is illustrated in Fig. 8.

Generally negative correlations exist between the thickness of the carbonaceous matter and thickness of overlying conglomerates. In the blocks where a maximum thickness of carbonaceous matter is evident the overlying pebble bands are either one pebble thick or the conglomerate is very poorly developed.

Several layers of carbonaceous matter which are present on the scoured surfaces occur within the conglomerate band. The thickness of these layers does not exceed 3 mm. The small spherical particles of carbon referred to as 'fly speck' carbon have a diameter of 0·2–1 mm and occur mainly in the columnar carbonaceous seam, but are additionally randomly scattered in the matrix of the overlying conglomerate band.

DEPOSITIONAL MODEL

The two different sub-facies which are developed in this area were probably formed in two different sub-environments as a result of contrasting depositional processes within the general fluvial setting of the Carbon Leader Reef. A number of features exhibited by sub-facies B strongly suggest that it is the result of stream-channel deposition, which took place in the initial stage of Carbon Leader formation. Following partial erosion of the underlying sediments (footwall quartzite), sedimentation occurred by the accumulation of the coarse bedload fraction of the stream as a channel lag. Overlying trough cross-bedded quartzite

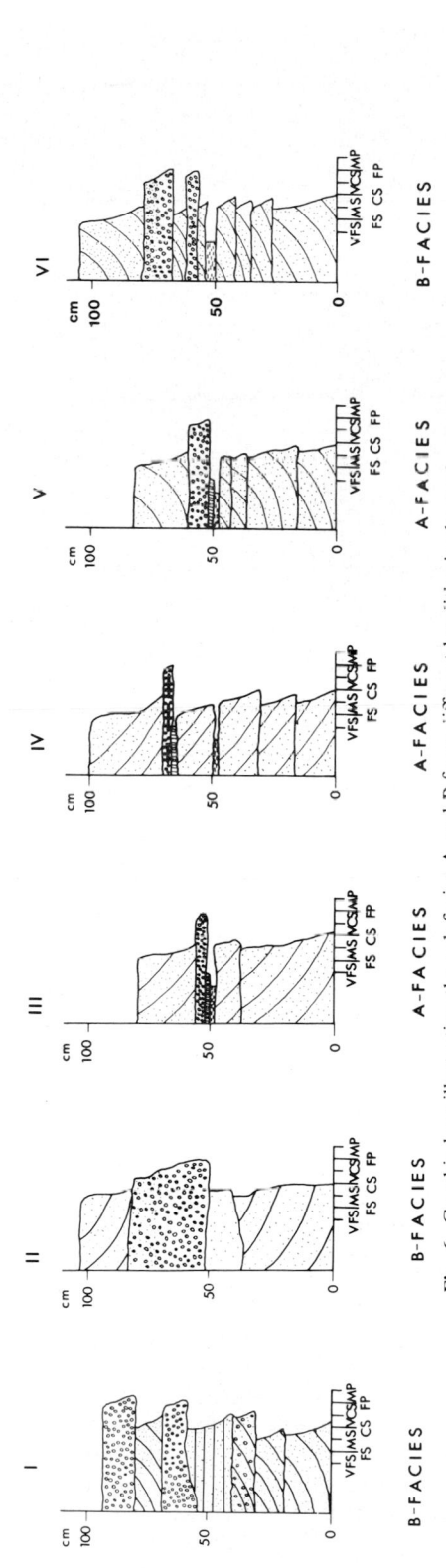

Fig. 6. Graphic logs illustrating the sub-facies A and B from different localities in the study area.

Fig. 7. Columnar carbonaceous matter at base of sub-facies A.

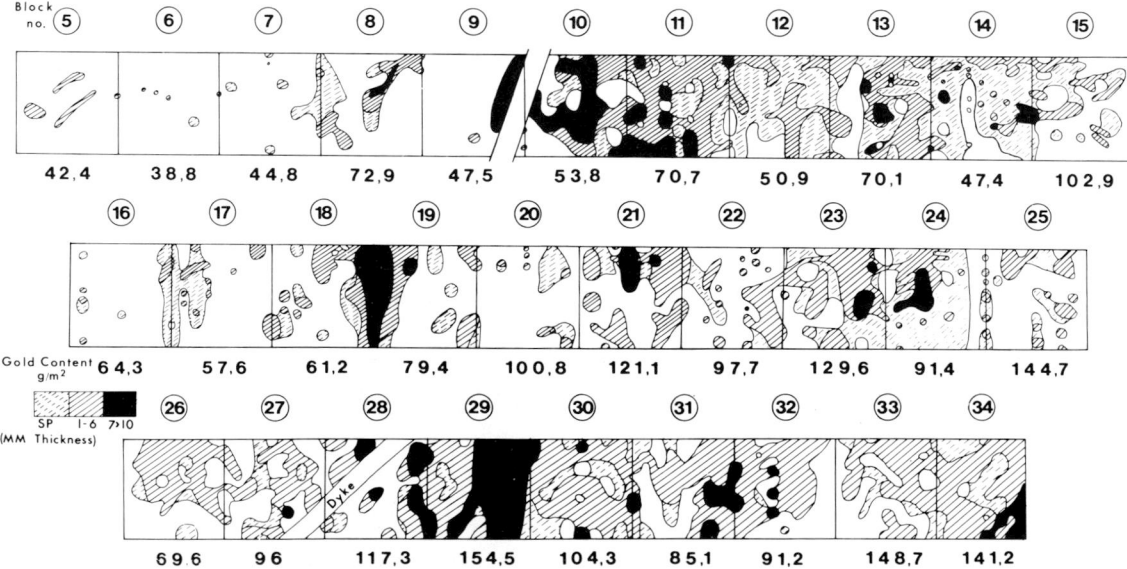

Fig. 8. Contour map of the thickness of carbonaceous matter. Recorded measurements in plan view from 500 geological profiles were classified into three categories of thickness.

was produced by migration of linguoid or lunate dunes on the stream bed (Allen, 1963). Large-scale, tabular-bedded quartzite was formed by accumulation of sand-sized grains in the form of transverse bars or terrace-like features (Allen, 1963, 1964). The horizontal bedded quartzite in the upper part of the tabular cross-beds can be attributed to aggradation of plane beds in the upper flow regime. The presence of horizontally stratified sand beds described by Smith (1971) in the Platte River was also the result of migration of thin sand sheets with downstream foresets. Low-amplitude sand waves which formed in very shallow depth in the lower or transitional flow regime can produce horizontal strata which super-ficially resemble those of the upper flow regime plane beds.

The individual conglomerate bands of sub-facies B with an erosional base indicate a distinct phase of sedimentation. An episode of channel downcutting was followed by aggradation of sediments. Due to continuous fluctuation of flow or multiple stream flows the top of the previously deposited sediments (megaripples or bars) became the new channel floor. Once again the coarse load was deposited as a channel lag. A stream, with a varied flow discharge, which had erosional and depositional phases was responsible for the formation of multi-storey conglomerate/quartzite bands. The stream channels in this area were shallow and wide, in the range of approximately 0·8 m depth and 8 m width.

The dark grey quartzite at the base of sub-facies A also exhibits features of stream channel deposition that are thought to be due to rapid infilling of sand-floored channels.

In contrast, sub-facies A has a planar base, and the lack of evidence of channel downcutting is probably indicative of deposition in the form of sheet-flood rather than channelized flow. During this stage the sediment-laden water followed an unconfined course over a fairly wide area, as shown by the widespread nature of the sub-facies. Sheet-flood sediments, as described by Bull (1972), are a result of widening of the flow into shallow sheets and a concurrent decrease in the depth and velocity of the flow.

The presence of the sericitic/chloritic shale in the base of sub-facies A marks a period of quiet deposition, mainly from suspension load prior to deposition of the sheet-like conglomerate. The occurrence of consecutive pebble bands with scoured surfaces and the lenses of pebble-supported conglom-erate, within sub-facies A illustrates several episodes of activity in a restricted area.

The comprehensive study of carbonaceous matter by Hallbauer, Jahns & Beltmann (1977) showed that this material comprises remnants of a Precambrian mass vegetation with a strong morphological affinity to primitive fungi or filamentous bacteria and algae and was therefore named *Thuchomyces lichenoides*.

The origin of the carbonaceous matter and its position within the depositional setting implies that the low energy and inactive areas had favourable conditions for colonization of these plants. The environment in which carbonaceous matter occurs is similar to that of algal mats in modern fluvial environments, which occur mainly on margins of active channels and in abandoned channels (Button, 1979).

The columnar carbonaceous matter in the base of sub-facies A overlies the thin shale layer and indicates the colonization of areas by these plants prior to deposition of the conglomerate in quiet conditions. The development of carbonaceous matter within the conglomerate and scoured surface provides evidence of an intermittent low-energy period between the flood events leading to the growth of vegetation. The thick carbonaceous matter is probably the result of flourishing primitive plant life in closed depressions or shallow puddles within inter-channel areas, where they were under quiet water for a long period. At a later stage they were rapidly buried, as is evident from the overlying conglomerate.

Relationships between the two sub-facies recognized from Carbon Leader Reef are given as a generalized block diagram in Fig. 9. During deposition of Carbon Leader Reef the different topographic levels of the palaeosurface played a major role. In the first stage the strongest fluvial activity was concentrated in the low-lying areas. In these areas the channels responded to a single rejuvenation event resulting in the formation of multi-storey conglomerate/quartzite. The channels in which the deposition of the dark grey quartzite took place were comparatively smaller, and competency of the flow was probably such that no pebbles were transported. These low-lying areas were either part of the existing relief of the palaeosurface or produced by channel incision. While the fluvial activity was taking place in low-lying areas, the higher levels were receiving sediments during flood events and conditions were generally suitable for growth of primitive algae. In later stages the sheet flood produced sub-facies A, and post-depositional flows gave rise to some degree of erosional and depositional modification.

In this area, although the sediments are composed

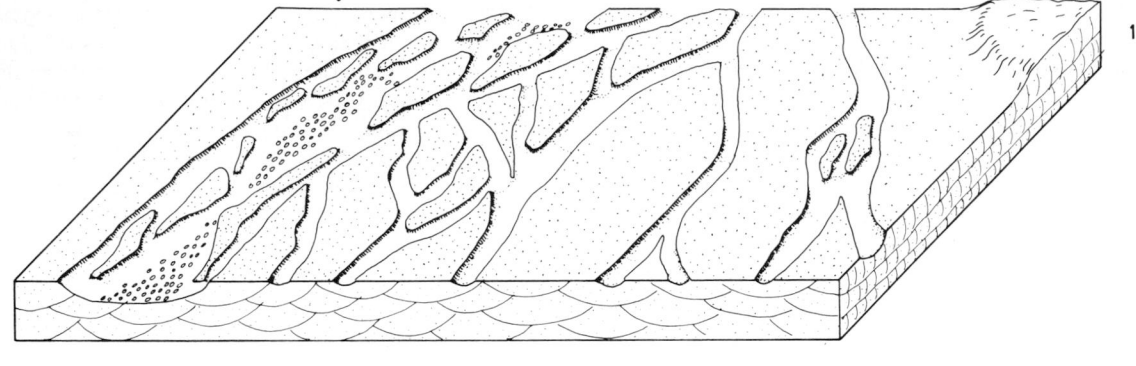

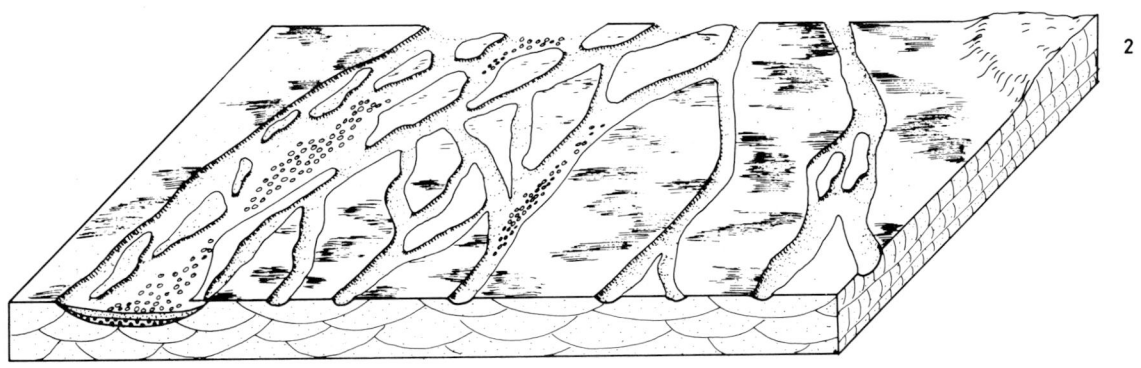

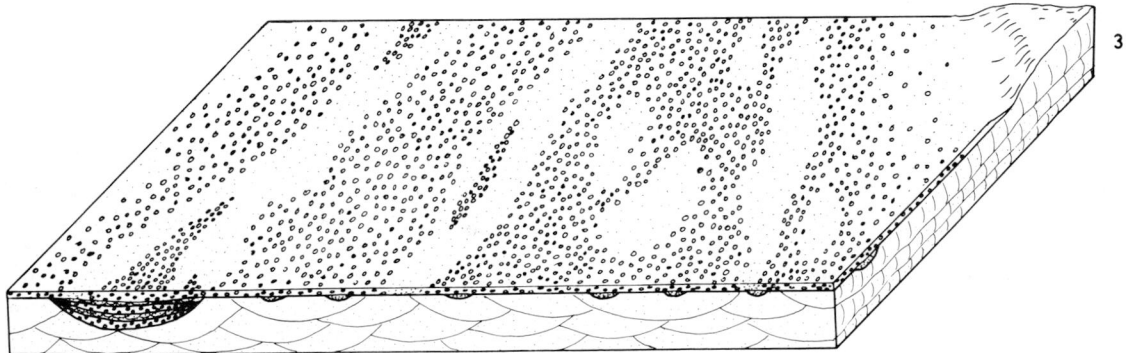

Fig. 9. Generalized block diagram, illustrating the development of Carbon Leader Reef in the study area with trough cross-bedded quartzite as footwall.

entirely of water-laid sediments, differences in hydro-dynamic conditions resulted in marked differences in the sedimentological characteristic of the sub-facies. The transporting medium of sub-facies B could transport sand and pebbles separately and was able to produce downcutting and cross-bedding. In the case of sub-facies A there was a measure of incompatibility between erosional and depositional processes and the transporting medium did not segregate the coarse and fine material. According to Rust (1977) a pebble-supported conglomerate which is not a reworked lag deposit is indicative of a river which, in flood, transported gravel as bedload and sand in suspension. During the falling stage, gravel was deposited first and infiltrated later by sand. In the case of the sand-supported conglomerate, the river has a mixed sand and pebble bedload, and both fractions were deposited simultaneously. In this case the coarser clasts are rarely larger than 15 mm, which is comparable to the sub-facies A of the Carbon Leader Reef.

GOLD DISTRIBUTION

Similar to many of the economic gold-bearing strata of the Witwatersrand basin, the presence of gold is confined to the conglomerate zones. To establish the position and proportion of gold within the conglomerate band, ore microscopy and X-ray radiography were used. Also, for quantitative determination of the gold content, chemical analysis was undertaken.

As Fig. 10 shows, between 65 and 95% of the gold is situated at the base of the conglomerate of sub-facies A, where it is associated mainly with carbonaceous matter. The gold in the carbonaceous matter occurs either in the form of detrital grains in the upper part or as filaments within the fossil plants (Hallbauer, 1978). A small portion of gold is randomly distributed in the matrix of the conglomerate in the form of free particles or intergrown with other minerals such as pyrite and gersdorffite. The average grain size of the gold measured in this area is approximately 70 μm (Fig. 11).

As described previously, the carbonaceous matter also occurs on scoured surfaces within the conglomerate band and usually contains gold particles. Although the gold is associated with carbonaceous matter there is no direct relationship between the gold content and the thickness of the carbonaceous matter. In block 10, where the carbonaceous matter reaches a maximum thickness, the gold

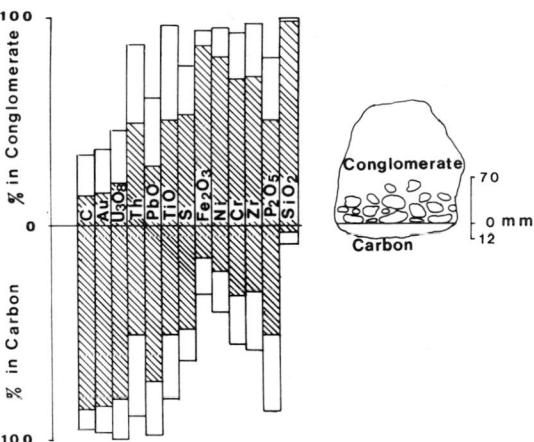

Fig. 10. Concentration of the selected constituents within carbonaceous matter and overlying conglomerate band.

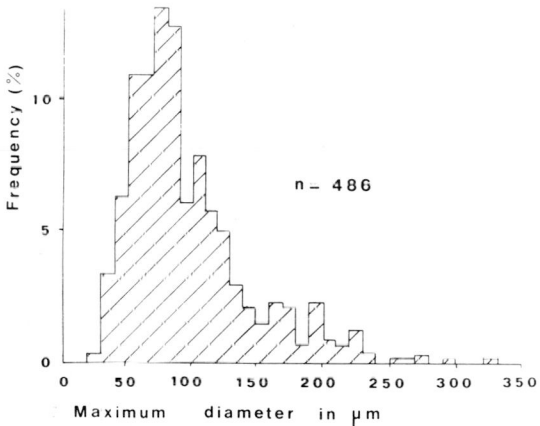

Fig. 11. Grain-size distribution of detrital gold in Carbon. Leader Reef at Blyvooruitzicht 5-22 reef drive.

content is not exceptionally high. Instead, the highest gold content seems to be in the conglomerate layer, which contains several layers of thin carbonaceous matter. In the case of sub-facies B, where the carbonaceous matter is absent, gold is in the matrix of the conglomerate together with other heavy minerals. The highest concentration of gold in this sub-facies is either in the bottom band or well-packed middle conglomerate band.

There is a marked difference in the distribution pattern of gold within the two sub-facies. In sub-facies A the distribution pattern of gold is relatively uniform and the mean gold value is high. In contrast, the distribution pattern of gold in sub-facies B is not uniform and the mean value is low (Fig. 12).

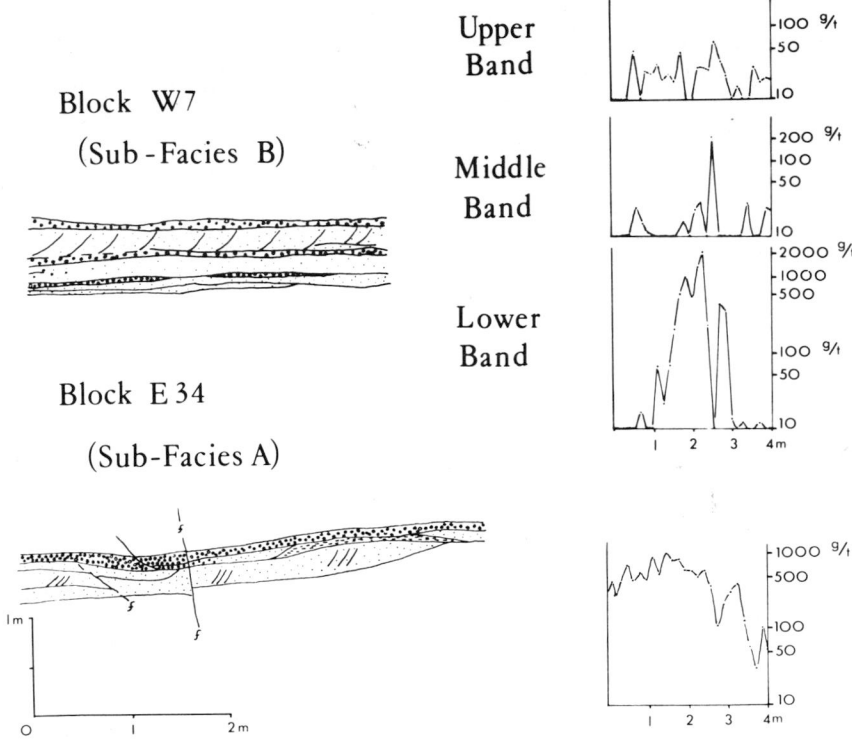

Fig. 12. Geological profiles of sub-facies A and B with gold values in different conglomerate bands as deduced from contiguous chip sampling.

The overall gold distribution pattern in the experimental site is illustrated in Fig. 13. Towards the east (on strike) the gold values increase, and in the north raise the highest values are encountered in the central part. Close inspection of data shows that the gold values fluctuate from block to block.

Faults and dykes have negligible influences on the gold values, except for a slight depletion in their vicinity.

DISCUSSION AND CONCLUSION

The foregoing analysis of the Carbon Leader Reef demonstrates that the observed distribution pattern of gold in the area of this study is the result of separate hydrodynamic conditions which prevailed during deposition of this placer. Occurrence of gold grains in the channel lag deposits of sub-facies B, the high gold content at the base of sub-facies A and the association of gold with carbonaceous matter all imply that a great quantity of gold (65–95%) was concentrated during stream activity. During the initial stage a high rate of flow was concentrated in the channels and gold

particles, being small in size, were probably transported in suspension. Although the conditions were not suitable for concentration of the gold, the presence of open-framework pebbles could have acted as a trap for the gold particles. The resulting sediments in the channels generally have low gold content, but erratic high values will occur, similar to the type of distribution of gold in sub-facies B in the western part and northern part of the north raise.

The areas between the stream channels were periodically covered by unconfined flow, and conditions were suitable for growth of primitive plants. Large amounts of gold were concentrated in these areas; first, due to the presence of mass vegetation which trapped the gold particles, in a similar manner to the corduroy tables in metallurgical practice (Hallbauer, 1978), and secondly, due to the spreading of the flow over a wide area. The concurrent decrease in depth and velocity could result in deposition of gold.

The last episode of flooding which produced the sand-supported single-conglomerate band had no effect on the gold distribution pattern, but was responsible for deposition of additional gold (5–35%),

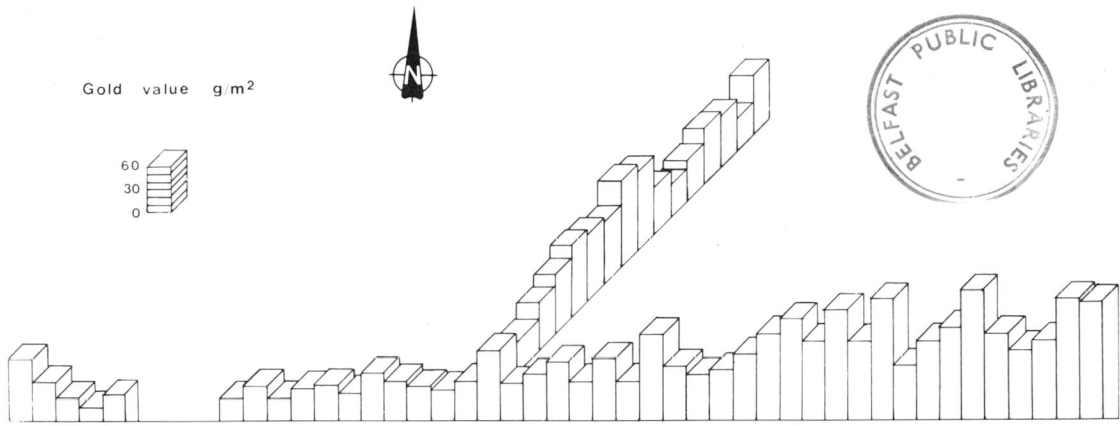

Fig. 13. Gold values of individual blocks illustrating: (1) an increase of values eastwards, and (2) high values in the central part of the north raise.

which occurs in the matrix of this pebble band. Post-depositional flow which was not accompanied by pronounced down-cutting caused reworking and winnowing of lighter material of sub-facies A. In the areas where the downcutting by a later flow was pronounced, complete erosion of the sheet-like conglomerate band took place.

The observed increase in gold values from west to east is the response to the change from channel to inter-channel areas, and the local fluctuations in the eastern part are due to the presence of small channels, where there was no mechanism operative to cause the concentration of gold.

The presence of high gold values in the pebble band containing several layers of carbonaceous matter is attributed to the several flood events which caused the supply of more gold and concentration of this mineral.

At the investigation site, higher concentrations of gold occur in the inter-channel areas than in the channels. Although the inter-channel areas are large compared to the channels, they are not continuous, due to dissection by the channel network. This must be taken into consideration when evaluating the ore reserves and prediction of local pay shoots.

The interpretation of the observed sedimentary processes can be usefully employed in distal placers of similar nature.

ACKNOWLEDGMENTS

The work described in this paper forms part of the geochemical research programme of the Mining Technology Laboratory, Chamber of Mines of South Africa.

I am grateful to the Chamber of Mines for granting permission to present this paper

REFERENCES

ALLEN, J.R.L. (1963) The classification of cross-stratified units, with notes on their origin. *Sedimentology*, **2**, 93–114.

ALLEN, J.R.L. (1964) Studies in fluviatile sedimentation: six cyclothems from the Lower Old Red Sandstone, Anglo-Welsh basin. *Sedimentology*, **3**, 163–198.

BULL, W.B. (1972) Recognition of alluvial-fan deposits in the stratigraphic record. In: *Recognition of Ancient Sedimentary Environments* (Ed. by J. K. Rigby and W. K. Hamblin). *Spec. Publs Soc. econ. Paleont. Miner.*, Tulsa, **16**, 68–83.

BUTTON, A. (1979) Algal concentration in the Sabi River, Rhodesia: deposition model for Witwatersrand carbon. *Econ. Geol.* **74**, 1876–1889.

HALLBAUER, D.K. (1978) Witwatersrand gold deposits. Their genesis in the light of morphological studies. *Gold Bull.* **11**, 18–23.

HALLBAUER, D.K., JAHNS, H.M. & BELTMANN, H.A. (1977) Morphological and anatomical observations on some Precambrian plants from the Witwatersrand, South Africa. *Geol. Rdsch.* **66**, 477–491.

MINTER, W.E.L. (1976) Detrital gold, uraninite and pyrite concentrations related to sedimentology in the Precambrian Vaal Reef placer, Witwatersrand, South Africa. *Econ. Geol.* **71**, 157–175.

MINTER, W.E.L. (1978) A sedimentological synthesis of placer gold, uranium and pyrite concentration in proterozoic Witwatersrand sediments. In: *Fluvial Sedimentology* (Ed. by A.D. Miall). *Mem. Can. Soc. Petrol. Geol.*, Calgary, **5**, 801–824.

PRETORIOUS, D.A. (1981) Gold and uranium in quartz-pebble conglomerate. *Inf. Circ. Econ. Geol. Res. Unit*, **151**, 1–18.

RUST, B.R. (1977) The Cannes de Roche Formation: Carboniferous alluvial fan and floodplain deposits in eastern Gaspe. *Abstr. Geol. Ass. Can.* **2**, 46.

SMITH, N.D. (1971) Transverse bars and braiding in Lower Platte River, Nebraska. *Bull. geol. Soc. Am.* **82**, 3407–3420.

SMITH, N.D. & MINTER, W.E.L. (1980) Sedimentological controls of gold and uranium in two Witwatersrand paleoplacers. *Econ. Geol.* **75**, 1–14.